Related Titles of Interest

Books

CARRUTHERS: Cycloaddition Reactions in Organic Synthesis
CLARIDGE: High-Resolution NMR Techniques in Organic Chemistry
FINET: Ligand Coupling Reactions with Heteroatomic Compounds
GAWLEY & AUBÉ: Principles of Asymmetric Synthesis
HASSNER & STUMER: Organic Syntheses Based on Name Reactions and Unnamed Reactions
LEVY & TANG: The Chemistry of *C*-Glycosides
LI & GRIBBLE: Palladium in Heterocyclic Chemistry
McKILLOP: Advanced Problems in Organic Reaction Mechanisms
OBRECHT & VILLALGORDO: Solid Supported Combinatorial and Parallel Synthesis of Small-Molecular-Weight Compound Libraries
PELLETIER: Alkaloids; Chemical and Biological Perspectives
PERLMUTTER: Conjugate Addition Reactions in Organic Synthesis
SESSLER & WEGHORN: Expanded, Contracted and Isomeric Porphyrins
WONG & WHITESIDES: Enzymes in Synthetic Organic Chemistry

Major Reference Works

BARTON, NAKANISHI & METH-COHN: Comprehensive Natural Products Chemistry
BARTON & OLLIS: Comprehensive Organic Chemistry
KATRITZKY & REES: Comprehensive Heterocyclic Chemistry I CD-Rom
KATRITZKY, REES & SCRIVEN: Comprehensive Heterocyclic Chemistry II
KATRITZKY, METH-COHN & REES: Comprehensive Organic Functional Group Transformations
SAINSBURY: Rodd's Chemistry of Carbon Compounds
TROST & FLEMING: Comprehensive Organic Synthesis

Journals

BIOORGANIC & MEDICINAL CHEMISTRY
BIOORGANIC & MEDICINAL CHEMISTRY LETTERS
CARBOHYDRATE RESEARCH
HETEROCYCLES (distributed by Elsevier)
PHYTOCHEMISTRY
TETRAHEDRON
TETRAHEDRON: ASYMMETRY
TETRAHEDRON LETTERS

Full details of all Elsevier Science publications are available on www.elsevier.com or from your nearest Elsevier Science office.

PROGRESS

IN

HETEROCYCLIC CHEMISTRY

Volume 12

A critical review of the 1999 literature preceded by three chapters on current heterocyclic topics

Editors

GORDON W. GRIBBLE
Department of Chemistry, Darmouth College, Hanover, New Hampshire, USA

and

THOMAS L. GILCHRIST
Department of Chemistry, University of Liverpool, Liverpool, UK

PERGAMON
An Imprint of Elsevier Science

ELSEVIER SCIENCE Ltd
The Boulevard, Langford Lane
Kidlington, Oxford OX5 1GB, UK

First edition 2000

Library of Congress Cataloging in Publication Data
A catalog record from the Library of Congress has been applied for.

British Library Cataloguing in Publication Data
A catalogue record from the British Library has been applied for.

ISBN: 008 0438822 Hardcover
ISBN: 008 0438830 (ISHC members edition)

∞ The paper used in this publication meets the requirements of ANSI/NISO Z39.48-1992 (Permanence of Paper).
Printed in The Netherlands.

Contents

Foreword

This volume of *Progress in Heterocyclic Chemistry* (PHC) is the twelfth annual review of the literature, covering the work published on most of the important heterocyclic ring systems during 1999, with inclusions of earlier material as appropriate. As in PHC-11, there are also three specialized reviews in this year's volume. In the inaugural chapter, Michael Groziak revitalizes the field of boron heterocycles, a relatively obscure class of heterocycles, but with a promising future. Heterocyclic phosphorus ylides are similarly a little known but useful class of compounds and Alan Aitken and Tracy Massil have provided a comprehensive review of them in Chapter 2. In Chapter 3 Jack Li discusses the remarkably versatile palladium chemistry in pyridine alkaloid synthesis.

The subsequent chapters deal with recent advances in the field of heterocyclic chemistry arranged by increasing ring size and with emphasis on synthesis and reactions. The reference format follows the journal code system employed in *Comprehensive Heterocyclic Chemistry*. We thank all authors for providing camera-ready scripts and disks, and we are grateful to Adrian Shell of Elsevier Science for his continuing assistance in producing this volume.

We hope that our readers will find PHC-12 to be a useful and efficient guide to the field of modern heterocyclic chemistry and that this volume will inspire new ideas and directions in this vital field of chemistry. The editors welcome suggestions on how to improve upon PHC and are always seeking topics for future reviews.

Gordon W. Gribble

Tom Gilchrist

Editorial Advisory Board Members
Progress in Heterocyclic Chemistry

2000 - 2001

Information about membership and activities of the International Society of Heterocyclic Chemistry can be found on the World Wide Web; the address of the Society's Home Page is:

http://euch6f.chem.emory.edu/hetsoc.html

Chapter 1

Boron Heterocycles as Platforms for Building New Bioactive Agents

Michael P. Groziak
Pharmaceutical Discovery Division, SRI International, Menlo Park, CA, USA
michael.groziak@sri.com

Chemists working to develop new bioactive compounds try to be alert for new stable heterocycle platforms, but they can easily overlook some of the more, shall we say, exotic ones. When one thinks about the utility of boron in heterocyclic chemistry, the Suzuki cross-coupling reaction typically first comes to mind. In this valuable synthetic reaction <95CRV2457>, a boronic acid group is discarded under basic conditions during a Pd-catalyzed C-C bond formation. There are exceptions, of course, but few chemists appreciate that boron is an element that can be valuable to *retain* in a molecule so that its unique properties can be utilized.

This contribution first surveys some of the attractive properties of boron, briefly describing applications that have been developed mostly with non-aromatic boron-containing compounds. It then examines many of the stable, formally aromatic boron heterocycles that have been reported to date, covering much of the pertinent literature through the end of 1999. With the sum of these two parts, I hope the reader will gain an appreciation of the untapped potential held by boron heterocycles, especially for constructing new bioactive agents.

1.1 WHY BORON?

When selecting atom substitutions for new molecule design, chemists usually look only to the right of carbon in the periodic table. The contrarian looks to the left and finds boron—commonly viewed as a metal, but in fact quite nonmetallic in many respects. In his excellent review of boron analogues of biomolecules, Morin showed why working with boron is so attractive <94T12521>. Here are some of the unique potential applications for any new boron compound:

1.1.1 ^{11}B NMR and MRI

Naturally occurring boron is comprised of the ^{11}B (80.22%) and ^{10}B (19.78%) isotopes. The former is NMR active and fast-relaxing, since it is a quadrupole (angular momentum 3/2 $h/2\pi$).

The determination of the charge, and thereby the valency, of a boron atom in an organic compound is usually straightforward if its ^{11}B NMR chemical shift within the 300+ ppm spectral window is compared to that of a close standard with a firmly established solution structure. But, there is a need for caution: The structure of many boron-containing compounds depends on the nature of the solvent, and so multisolvent (i.e., aprotic *vs.* protic) analyses are often essential for a definitive characterization. Sadly, aqueous solution ^{11}B NMR spectral analyses are seldom reported—even, surprisingly, for compounds clearly prepared for their potential biological value.

In biochemical applications like enzyme inhibition, ^{11}B NMR spectroscopy <B-78MI14, B-83MI49> has proven to be an exceptionally useful tool for detailing the interaction of boron-containing compounds with biomacromolecules <88JA309, 91BMCL9, 93B12651>. Any study of new potential boron-based enzyme inhibitors would likely benefit from using this diagnostic tool. There is a great potential utility for ^{11}B NMR in the more biological and medicinal applications as well. Although likely essential in trace amounts <96MI2441> for proper bone development <90MI61, 99MI335>, boron is not present to any great extent in living tissues, and so there is no background to compete with the detection of the signal from an administered boron-containing compound. The great rapidity of the ^{11}B nuclear relaxation presents some problems in signal acquisition and the spatial resolution may be limited <95MI48>, but clearly ^{11}B MRS (magnetic resonance spectroscopy) and ^{11}B MRI (magnetic resonance imaging) are two of the very exciting potential NMR-based applications for any new boron-based compound. Advances in these fields <88MI231, 90JMR369, 97MI153> have emerged primarily in step with efforts to develop boron neutron capture therapy (BNCT), described next.

1.1.2 ^{10}B Neutron Capture Therapy (BNCT)

The ^{10}B isotope is one of only a handful of nuclides that interact strongly with thermal (slow-moving) neutrons. It has a large capture cross section for them due to a fortuitous resonance between the energy of the thermal neutron "falling" into the lowest unoccupied neutron state in ^{10}B and the energy needed to promote one of the nucleons to an excited state. Once the excited state ^{11}B atom is produced, the powerful nuclear fission reaction $^{10}B(n,\alpha)^{7}Li$ occurs, ejecting a gamma photon together with a 0.87 MeV ^{7}Li particle and a 1.52 MeV ^{4}He particle. These heavy, fast moving particles travel along a mean-free path whose length is close to that of a red blood cell's diameter (5 μm for the ^{7}Li and 9 μm for the ^{4}He), and while so doing can destroy cellular structures like membranes, organelles, and even DNA. There are three separate areas where technological advances are needed to one day make BNCT a routine binary radiation therapy for treating cancer. The first is a high tumor uptake of a boron-containing compound relative to normal tissue. The second is a sufficiently high concentration of boron "target" atoms dispersed within the tumor cell (ideally in the nucleus). It has been estimated that 30 μg of ^{10}B per g of tumor will suffice. The third is the characteristics and quality of the neutron beam. Epithermal (*ca.* 1keV) neutrons are attractive for BNCT, since these readily pass through living tissue without incident as they slow down to become thermal neutrons.

Many review articles highlighting role of chemistry in BNCT are available <93AC(E)950, 94MI119, 94MI849, 97MI41, 98MI174, 98CR1515>. Most of the agents currently under investigation are based on an *o*-carborane ($C_2H_{12}B_{10}$) unit because of its 10 boron atoms. Of course, these 10 atoms are not evenly distributed inside the cell, but there are advantages to the use of carboranes—not the least of which is their virtual lack of reactivity and toxicity. Nucleosides, nucleic acids, amino acids, polyamines, liposomes, and even antibodies equipped with carboranyl units are being developed as BNCT agents. One of the more recent classes of compounds under investigation is the boronated protoporphyrins (BOPP) <99MI761>.

Although attractive, a carborane unit is not required. *p*-Boronophenylalanine (BPA, **1**) has but one boron atom and yet is one of the lead clinical compounds as a BNCT agent to treat glioblastoma multiforme (a form of brain cancer) <99MI1>. BPA, behaving *in vivo* as an analogue of the melanin precursor tyrosine, shows a remarkable selective uptake within these tumor cells. Thus, as long as a boron-containing compound can be delivered selectively and in sufficient quantity to the target group of cells, it has the potential of being a BNCT agent.

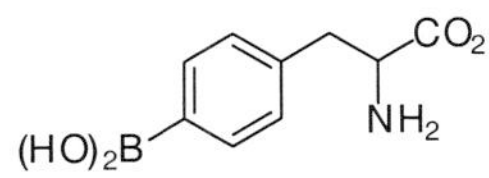

1

1.1.3 Boron Heterocycle-Based Fluorescence

4,4-Difluoro-4-bora-3a,4a-diaza-*s*-indacene (**2**) is the central fluorophore unit of the so-called BODIPY® fluorescent dye compounds <94JA7801>. This boron heterocycle is relatively nonpolar, since with no net ionic charge it is electrically neutral. Useful bioconjugatable dyes with fluorescence emissions spanning the entire visible spectrum were developed by varying the pattern and nature of ring substituents. The extinction coefficients are large (>80,000 $cm^{-1}M^{-1}$) and the quantum yields are close to 1.0—even, importantly, in water. The emission spectra are generally insensitive to solvent polarity and pH and they have a narrow bandwidth. A large two-photon cross section permits multiphoton excitation. New boron-based compounds exhibiting good fluorescence properties like these certainly have the potential to be quite useful as probes in biochemical, biological, or even medical diagnostic applications.

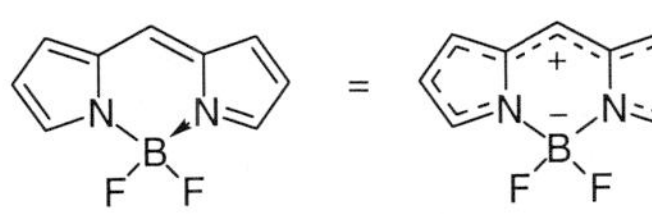

2

1.1.4 Boronic Acid-Based Enzyme Inhibition

Because boronic acids interconvert with ease between the neutral sp^2 (trigonal planar substituted) and the anionic sp^3 (tetrahedral substituted) hybridization states, the B-OH unit has found a unique role as a useful replacement for the C=O one at a site where an acyl group transfer takes place. Boronic acid-based inhibition of proteases and other hydrolytic enzymes capitalizes on the fact that a tetrahedral boronate molecular fragment is an exceptionally close structural mimic of the tetrahedral intermediate of acyl group hydrolysis. Boronic acid-based protease inhibition first emerged in the early 1970s, when phenethylboronic acid (**3**) was found to be a good inhibitor of chymotrypsin <70MI23, 71B2477, 74MI135>.

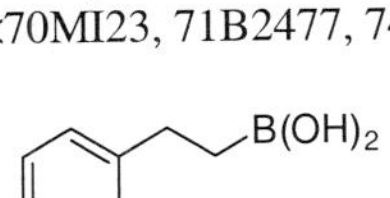

3

Some boronic acid-based enzyme inhibitors undergo strong yet reversible covalent attachment to a nucleophile at the enzyme's active site, while others simply act as competitive inhibitors in their borate conjugate base form. Boronic acid-based inhibition of thrombin has been achieved <93MI109>, and that of β-lactamases has been particularly effective <95TL8399, 96MI688>. When compared to other covalent transition-state analog inhibitors of β-lactamases like phosphonates, silane triols, aldehydes, and α-keto carbonyl compounds, the boronic acids display superior characteristics <97JA1529>. If its structure targets it properly to a hydrolytic enzyme's active site, a new boronic acid-based compound can be a potent enzyme inhibitor.

1.1.5 Bioactive Boron Compounds

It has been known for about two decades that benzo- and hetero-fused 2-alkyl- and arylsulfonylated 2,3,1-diazaborines **4** possess antibacterial properties, particularly against gram negative organisms <84JMC947>. The early indication was that these compounds affected lipopolysaccharide biosynthesis <80AAC549, 81NAT662, 87MI37, 89MI6555, 94MI1937, 94JBC5493, 94MI771, 96EJB689, 97JBC27091>. More recent structural studies have shown that the biomacromolecular target is enoyl acyl carrier protein reductase (ENR), the NAD(P)H-dependent enzyme which catalyzes a latter step of fatty acid biosynthesis <96AX(D)1181, 96SCI2107, 98BP1541, 99MI443, 99JBC30811>. Interestingly, this enzyme is the very same target of the broad-spectrum (bacteria, fungi, viruses) bacteriostatic germicide triclosan <98NAT531, 99JBC11110, 99JMB527, 99JMB859> and the antituberculosis drug isoniazid.

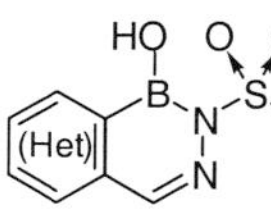
4

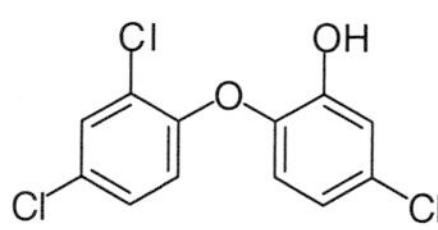
triclosan

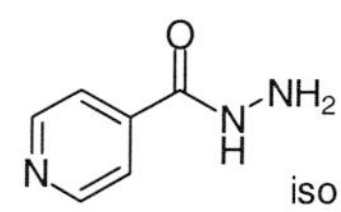
isoniazid

Perhaps because boric acid is a well-known insecticide for cockroaches, boron compounds have been examined as insect chemosterilants <69MI1472, 70JMC128>. Besides this, boron-based compounds have been identified as antivirals <96MI108> and as antituberculosis agents <98BMCL843>. This demonstrates how new boron-based compounds have the potential of exhibiting useful medicinal properties even if there is no predetermined biochemical target or mechanism of action. No boron-based pharmaceutical has yet been developed, but this merely signifies a great opportunity for chemists working with boron compounds <72PHA1>.

Only a few boron-based natural products are known. The ionophoric macrodiolide antibiotics boromycin (**5**) <67HCA1533, 96MI1036>, aplasmomycin (**6**) <76JAN1019, 77JAN714, 80JAN1316>, and tartrolon B (**7**) <94LA283, 95JAN26, 99JA8393> are such potent K^+ carriers that they are highly toxic to both bacteria and to mammalian cells.

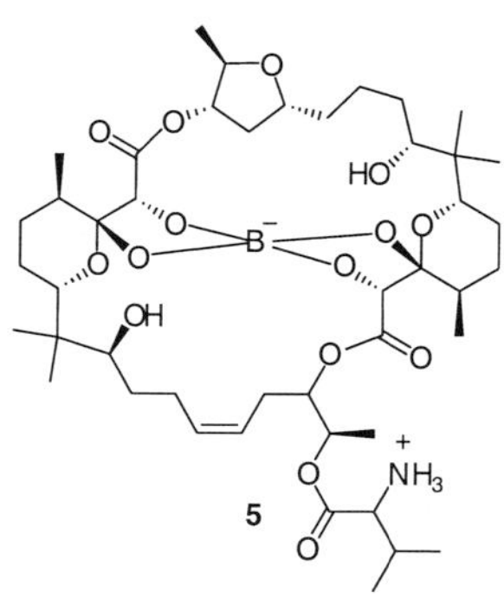
5

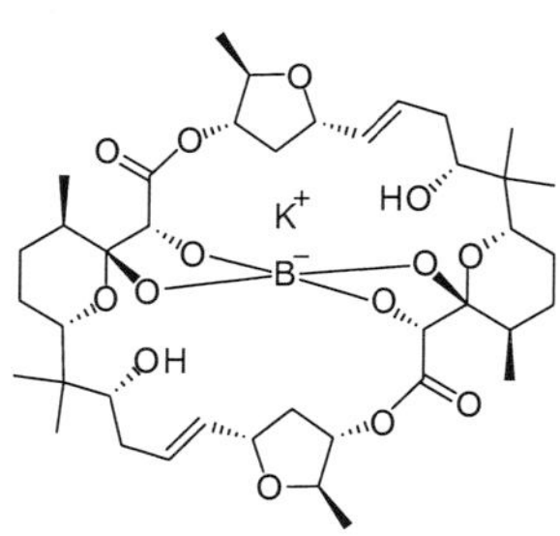
6

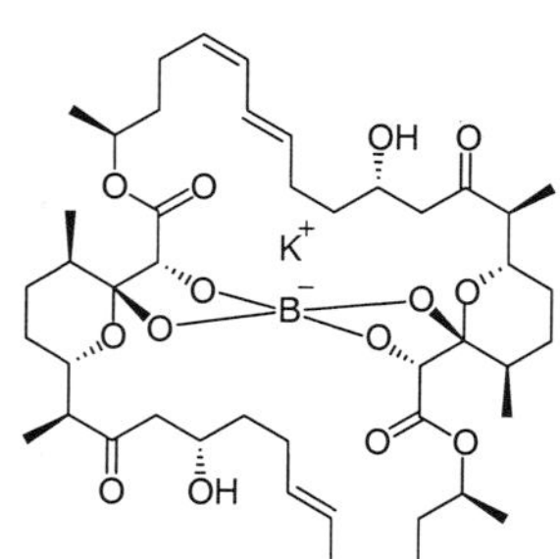
7

1.1.6 Relative Low Toxicity

Most of the boronic acids and other low molecular-weight synthetic boron compounds that have been examined have been found to be relatively nontoxic. The chemistry and biology of simple (mostly inorganic and acyclic organic) boron compounds have been reviewed <92MI229, 98MI2>. Boric acid and borates have been studied in great detail and pose no toxicity threat <98MI1-02>. The published contributions to the International Symposia on the Health Effects of Boron and its Compounds <94MI1, 98MI1-01> are a rich source of health-related information on boric acid and simple organoboron compounds.

There is typically little or no toxicology or metabolism data available for even moderately complicated boron-based compounds. An interesting exception is the collection of tetrahydro-3a,4a,4-diazabora-*s*-indacenes (**8**) <71SRI83> structurally related to the BODIPY® fluorescent dyestuffs. Rather well characterized, these compounds are stable to both water and alcohols at 23 °C and undergo reversible salt formation with HCl and NaOH. Compound **8b**, termed Myborin, was evaluated for its toxicity <75MI434>. The LD_{50} values of 69.5 mg/kg i.p., 180 mg/kg p.o., and 420 mg/kg s.c. in the mouse reveal it to have moderate toxicity.

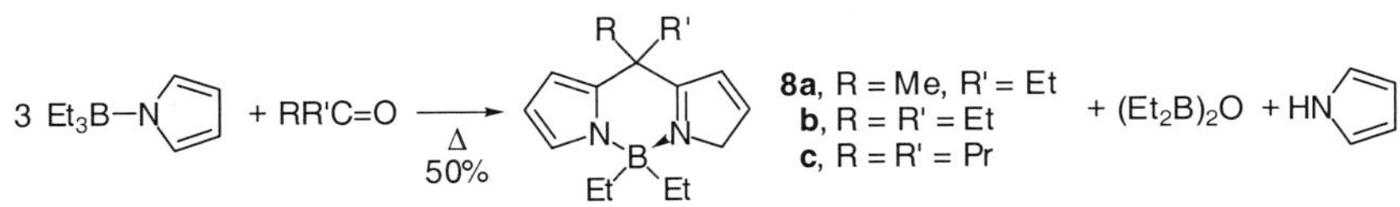

When compared to tin compounds, boronic acids are considerably less toxic. This is particularly striking when one compares the by-products produced by Stille and Suzuki coupling reactions. A Stille coupling generates highly toxic trialkyltin halides which pose a serious waste problem, but a Suzuki coupling generates the comparatively nontoxic boric acid. A look at the MSDS-derived LD_{50} values of two coupling by-products shows the huge difference in toxicity. The LD_{50} of Bu_3SnCl is 60 mg/kg p.o. in the mouse and 129 mg/kg p.o. in the rat. Those of $B(OH)_3$ are 3450 mg/kg in the mouse and 2660 mg/kg in the rat.

1.2 AROMATIC BORON HETEROCYCLES

When a boron atom is connected to the ends of hexatriene, the resulting borepine molecule has circuit of p-orbitals containing a Hückel 4n+2 number of π electrons. Isoelectronic with the tropylium cation, borepine has been shown to exhibit aromatic properties <93OM3225>. Equally fascinating boron heterocycles are produced when the p-electron deficient boron atom is paired with a p-electron excessive one in a ring. In endocyclic and potentially aromatic settings, B-O and B-N single bonds are excellent replacement moieties for C=N and C=C units, respectively. They are isovalent, isoelectronic, and isosteric with these units and maintain enough stability within 4n+2 π-electron circuitry to help establish at least some degree of aromaticity.

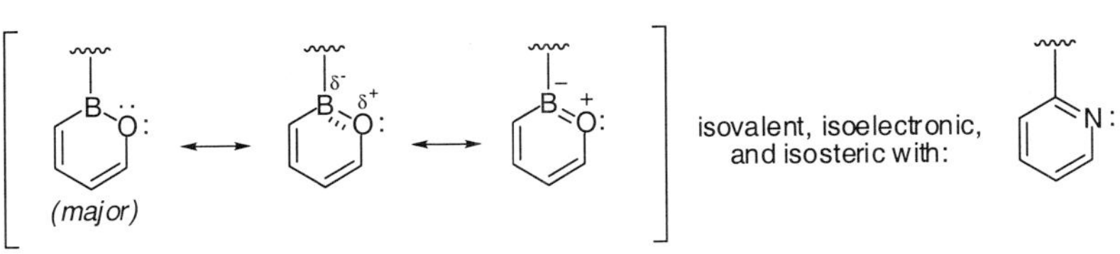

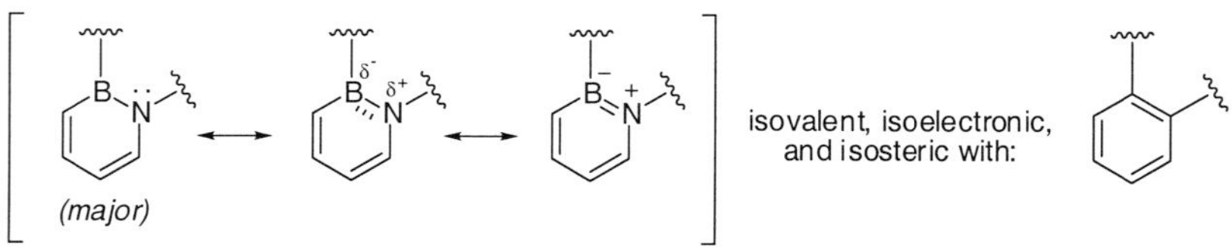

Much of the early literature, reviewed quite well by others <62CRV223, B-64MI227, B-64MI235, B-70MI117, 77HC381, 84CHEC-I(1)629, 96CHEC-II(6)1155>, names these "boroaromatic" compounds using replacement nomenclature (*e.g.*, borazaropyridine instead of diazaborine) and depicts them as zwitterionic species with an endocyclic double-bond from the heteroatom to the boron. However, as the body of ^{11}B NMR chemical shift data has grown <68JA706, 76JOM123, 94JA7597, 97JA7817>, it has become apparent that these species are not major players on the resonance continua. Indeed, except possibly for the borazines (described next), these types of compounds are likely best depicted as nonzwitterionic heteroaromatics with single B-X bonds. Despite the negligible amount of p-electron diffusion from the heteroatom to the boron, though, these compounds display the stability and other attributes expected of them by virtue of their Hückel heteroaromaticity.

1.2.1 Borazines and Boroxins

It is helpful to examine the benzene analogue borazine ($B_3N_3H_6$, **9**) and the *s*-triazine analogue boroxin ($B_3H_3O_3$) so that we can know better what to expect when replacing C=C units with B-N ones or C=N units with B-O ones in more complicated molecules. A direct comparison of the crystal structures of benzene <58PRS1> with **9** <94CB1887> and of 2,4,6-triphenyl-*s*-triazine <84ZSK180> with triphenylboroxin (**10**) <87AX(C)1775> reveals that the B-X replacement bonds are longer by *ca.* 0.05 Å in each case.

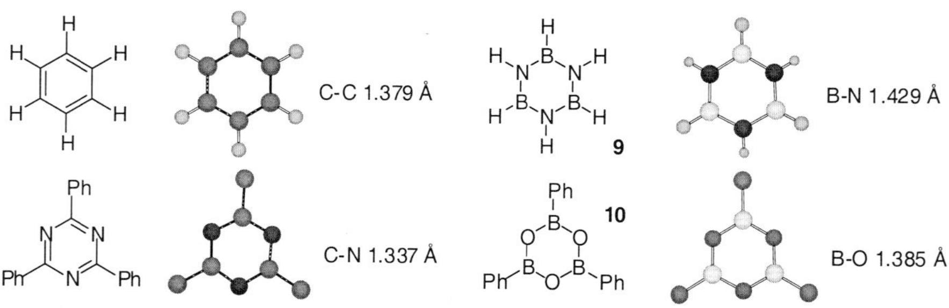

In general, **9** and its derivatives <70JOM323> are known to exhibit less aromatic character than their benzene counterparts <98T14913>, but the electronic excitation and p-electron interaction have very benzene-like features <86JA3602> and the gas phase ion chemistry is remarkably similar to that of benzene <99JA11204>. ^{1}H NMR spectral comparisons of various methylated versions of **9** have been made <73OMR585>, and ^{14}N and ^{11}B NMR spectral analyses of borazines have been conducted as well <76CB3480>. In a study of a series of B-monosubstituted (NMe_2, OMe, OAc, and Cl) borazines, it was concluded that their NH units either do not act as hydrogen bond donors or do so only very weakly <77IC2935>. Highly substituted borazines have been analyzed by X-ray crystallography <95CB1037>.

The electronic structure of benzene, **9**, and **10** have been compared in detail <89JCS(P2)719>. A MNDO semiempirical investigation of **10** concluded that it likely cannot exist in monomeric Ph-B=O form <94JOM31>. B-N for C=C replacement analogs of aromatic hydrocarbons have been the subject of electronic spectral <71CCC1233> and semiempirical <71CCC1248> investigations, and a recent *ab initio* calculation of the various isomers of tandem B-N for C=C replacement analogs of benzene and naphthalene showed that the greatest stability is achieved when the B and N atoms are juxtaposed <97MI65>. *Ab initio* calculations of a collection of 70 known and unknown 6π-electron monocycles containing B and N—including 26 pyridine isosteres—showed that the most stable isomers were those constructed upon the XBHNH unit, where X = N, NH, or O <99JPC(A)2141>.

1.2.2 Relevant Properties of Arylboronic Acids

The properties of phenylboronic acid (**11**) and some of its simple derivatives deserve comment, since boroaromatics are often constructed using these frameworks. In the solid state, **11** self-associates, resembling a carboxylic acid dimer <77CJC3071>. Crystal packing forces can produce some peculiar structures, though, like the one for 2-nitro-4-carboxyphenylboronic acid (**12**) that appears to show an intramolecular association between the NO_2 and $B(OH)_2$ groups <93AX(C)690>. Upon close inspection, however, one finds that little or no concomitant rehybridization of the boron has taken place in response to this apparent interaction.

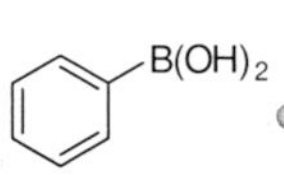

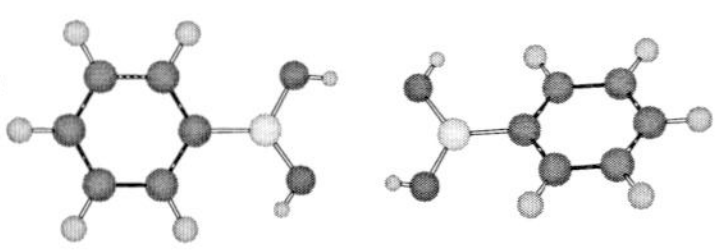

11

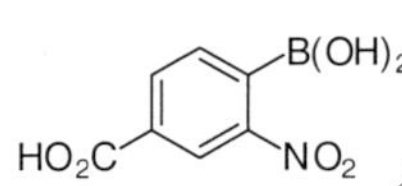

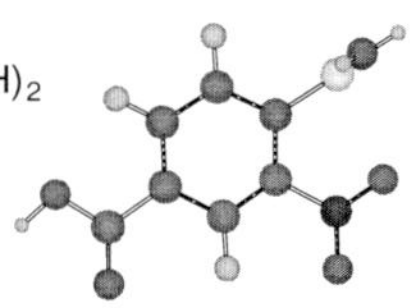

12

By contrast, the X-ray crystal structures of both 2-formylbenzeneboronic acid (**13**) and its O-methyl oxime (**14**) reveal an intramolecular hydrogen bond in which one hydroxyl of the $B(OH)_2$ unit truly acts as a hydrogen bond donor to a heteroatom of the *ortho* side chain <94MI621>. The hydrogen bond distance in the seven-membered ring is 1.562 Å in **13** and 1.614 Å in **14**.

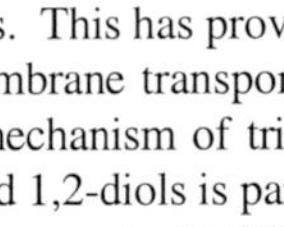

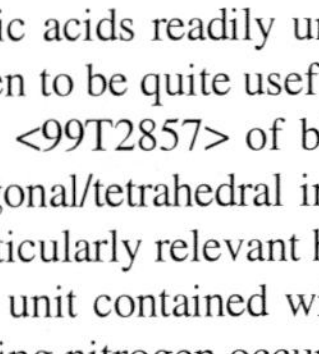

13

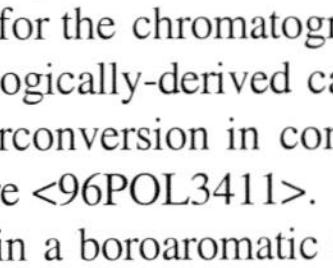

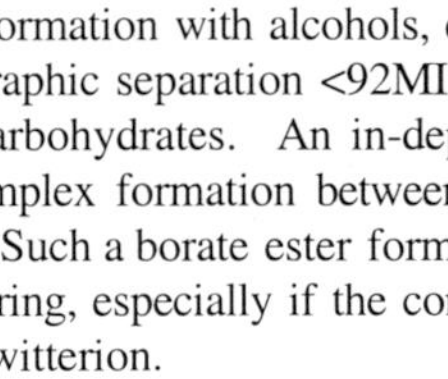

14

In solution, arylboronic acids readily undergo borate ester formation with alcohols, especially 1,2-diols. This has proven to be quite useful for the chromatographic separation <92MI293> and transmembrane transport <99T2857> of biologically-derived carbohydrates. An in-depth study of the mechanism of trigonal/tetrahedral interconversion in complex formation between boronic acids and 1,2-diols is particularly relevant here <96POL3411>. Such a borate ester formation can indeed occur at a B-OH unit contained within a boroaromatic ring, especially if the concomitant protonation of an imine ring nitrogen occurs to afford a stable zwitterion.

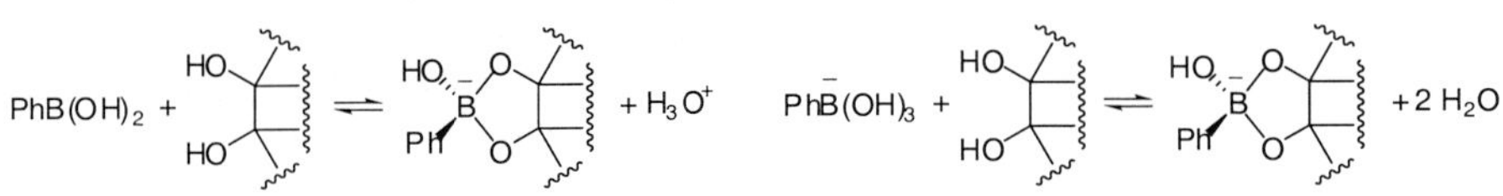

1.2.3 1,3,2-Diheteraboroles and 1,3,2-Diheteraborines

There are many examples of formally aromatic boron heterocycles in which the boron is flanked by two heteroatoms in a ring. Although they exhibit some heteroaromatic stability in nonaqueous environments, the boron atom usually retains enough Lewis acid character to make them unstable in water. In the case of five-membered 1,3,2-diheteraboroles, the crystal and molecular structures of the collection of 5-membered ring 6π electron boron heterocycles **15-19** containing boron, sulfur, and nitrogen show the great variety of heteroatom substitutions possible <80CB3881>. Crystalline **19** was found to exist as a dimer.

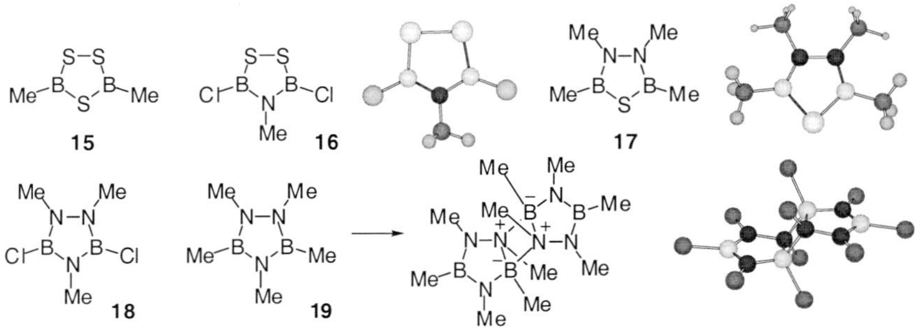

Benzo-fused 1,3,2-diheteraboroles have been prepared from *ortho*-phenylenediamines and *ortho*-aminophenols <59JA2681, 59JA6329, 61JOC4632, 90ZAAC151>, but even in these cases hydrolysis is usually facile <77JOC3545>. Benzo-fused versions of the borane (X-BH-Y) derivatives have been examined extensively by ^{11}B NMR spectroscopy <84SA(A)855>.

As for the six-membered 1,3,2-diheteraborines, one of the earliest examples was reported by Dewar, who found that 1-methyl-4-aminoimidazole-5-carboxamide could be condensed with $PhB(OH)_2$ to give a 2-boradihydropurine (**20**) <59JA6329, 61JA2708>. Unfortunately, this compound hydrolyzes readily in 95% EtOH at 23 °C. The condensation of biuret and $NaBH_4/I_2$ <78IJC(B)85> has been reported to give **21a** and that of N,N'-dialkylureas with dihaloalkylboranes <81JOM17, 82CB3271> gave **21b,c**, all related to **20**.

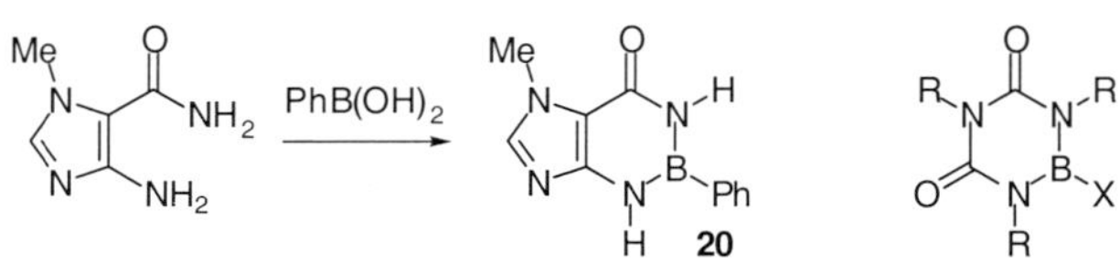

21a, R = X = H;
b, R = Me, X = Cl;
c, R = X = Me

A study of bicyclic 1,3,2-diheteraborin-4-ones derived from *ortho*-amino benzamides revealed a wide range of hydrolytic stability. Fried's group compared the rates of alcoholysis of **22a** and its derivatives **22b-g** <62JA688>, and found that while **22a** is hydrolyzed completely within a few hours at 23 °C, **22b** is stable for at least 144 h! Derivative **22c** hydrolyzes completely in less than 1 h and **22d,e** in less than 2 h, but **22f,g** are stable for at least 120 h. Fried also showed how **22a** could give its water-stable 4-amino-1,3,2-benzodiazaborine counterpart (**23**).

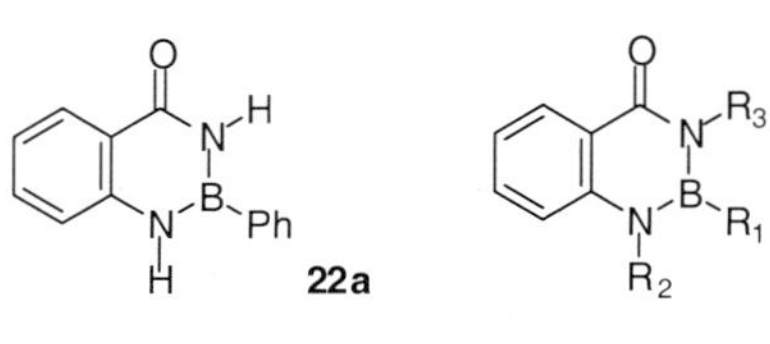

22b, R_1 = mesityl, $R_2 = R_3$ = H;
c, R_1 = Ph, R_2 = H, $R_3 = (CH_2)_3NMe_2$;
d, R_1 = Ph, $R_2 = R_3$ = Me;
e, R_1 = Ph, R_2 = Me, R_3 = H;
f, R_1 = 1-naphthyl, R_2 = H, R_3 = Me;
g, R_1 = 1-naphthyl, R_2 = H, R_3 = Bn

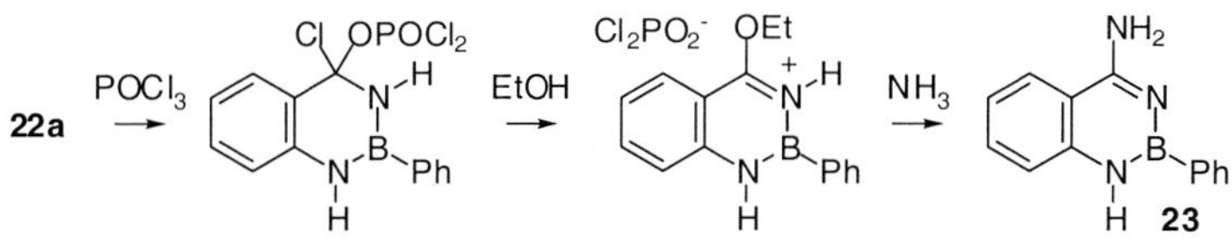

Others have explored the interesting chemistry of the *ortho*-amino-amide/phenylboronic acid adducts <71IJC1167, 77IJC(B)267, 84CZ287>, and one group even managed to replace the C=O unit with an additional boron center <66JCS(A)479>. The preparation of heterocycles **24a-e** by Niedenzu's group shows the variety of heteroatom substitutions possible <73SRI229>.

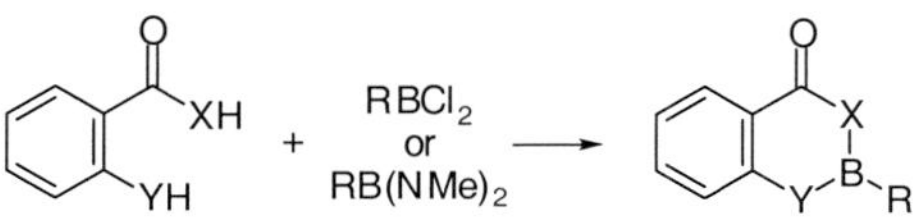

24a, X = NH, Y = NH, R = Ph (20%)
b, X = O, Y = NH, R = Ph (96%)
c, X = O, Y = NH, R = NMe_2 (87%)
d, X = NH, Y = O, R = Ph (85%)
e, X = Y = O, R = Ph (100%)

1.2.4 2,3,1-Diheteraborines and 2,1-Heteraborines

A much greater stability is seen when the boron is anchored to a ring carbon atom and "clamped" into place with a suitable *ortho* side chain. For example, in the benzo-fused 2,3,1-diheteraborines this provides for a great structural robustness in aqueous solution. Within a series of papers on benzeneboronic acids <57AK473, 57AK497, 57AK507, 57AK513>, Torssell provided the starting point for the 2,3,1-benzodiheteraborines by showing how 2-formylbenzeneboronic acid could be prepared by α-dibrominating *o*-tolylboronic acid and then hydrolyzing the *gem*-dibromo product <57AK507>. When Snyder found that a stable intramolecular anhydride—namely, 1-hydroxy-1*H*-2,3,1-benzoxazaborine (**25**)—was obtained

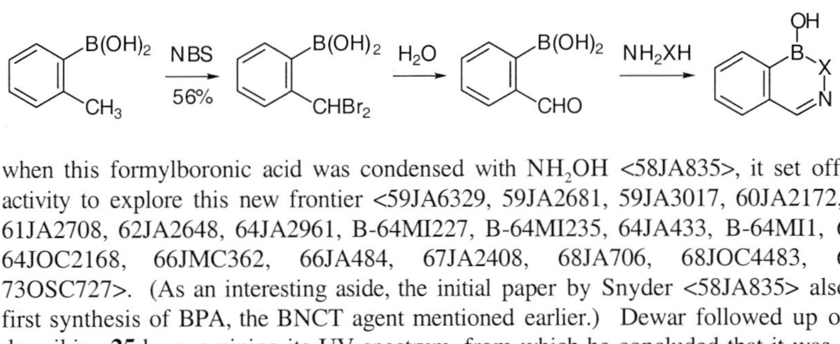

25, X = O
26, X = NH

when this formylboronic acid was condensed with NH_2OH <58JA835>, it set off a flurry of activity to explore this new frontier <59JA6329, 59JA2681, 59JA3017, 60JA2172, 60JA2442, 61JA2708, 62JA2648, 64JA2961, B-64MI227, B-64MI235, 64JA433, B-64MI1, 64JOC3229, 64JOC2168, 66JMC362, 66JA484, 67JA2408, 68JA706, 68JOC4483, 69JOC1660, 73OSC727>. (As an interesting aside, the initial paper by Snyder <58JA835> also relates the first synthesis of BPA, the BNCT agent mentioned earlier.) Dewar followed up on the report describing **25** by examining its UV spectrum, from which he concluded that it was a protic acid like phenol <62JA2648>. Snyder <64JOC2168> and Dewar <64JA433> then both showed that the stable 1,2-dihydro-1-hydroxy-2,3,1-benzodiazaborine (**26**) was obtained in lieu of an open hydrazone.

After the initial flurry, the single largest contributor of knowledge to the field of 2,3,1-diheteraborines was Gronowitz, who focused his studies on the thiophene-fused versions <65ACS1271, 67ACS2151, 68ACS1611, 70AK283, 71APS377, 71APS623, 75ACS457, 75ACS(B)461, 75ACS(B)990, 77CS76, 77ACS(B)765, 77JHC893>. Although his advances—summarized in a review <76JHC(S)76>—are far too numerous to detail here, a key methodological breakthrough deserves mention. In it, he showed how an α-aminoalkoxide species derived from the DMF quench of an organolithium species could be used promptly to direct an *ortho*-metalation, thereby providing speedy access to *ortho*-formylboronic acids like **27** in a four-step, one-pot fashion from 1,2-dihalide precursors <68ACS1353>.

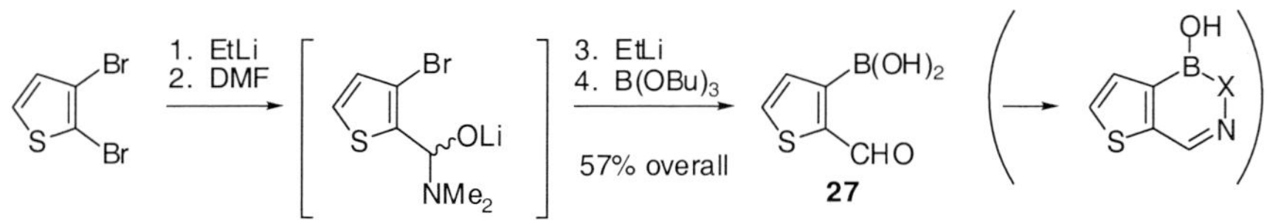

Furan-fused 2,3,1-diheteraborines were investigated by Roques <70MI1898, 70TL4909, 72JOM38, 74BSF2620, 77JCR(S)158>, and 6,7-methylenedioxybenzo-fused <84CZ287> and even selenophene-fused ones have been prepared <72IJS(A)257>. Interestingly, the great utility of ^{15}N and ^{11}B NMR spectroscopy to the solution structure elucidation of these and other boron heterocycles was clearly predicted by Roques in one of his papers <74BSF2620>.

The biocidal 2-arylsulfonylated 2,3,1-benzodiazaborines mentioned earlier are quite stable under most conditions, but Grassberger showed that the benzo- and thieno-fused variants undergo a ring contraction and/or a deboronation when subjected to aq. NaOH at 100 °C <85LA683>. A benzo[*e*]thieno[3,2-*c*]azaborine (**28**) was shown to react with HNO_2, giving the cinnoline precursor to benzo[3,4]cyclobuta[1,2-*b*]thiophene <79TL3571>.

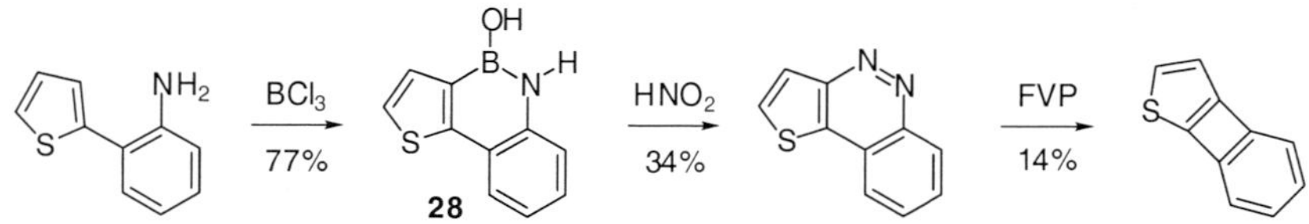

Sharp achieved a methodological breakthrough similar to Gronowitz's by generating a dianion from 2-bromoacetophenone and directly boronating it to obtain the 2-tosylated 2,3,1-benzodiazaborine (**29**) in an excellent overall yield <86TL869>. Under harsher conditions, the condensation of a tosylhydrazone with BCl_3 or BBr_3 will also generate these types of heterocycles <78HCA325>.

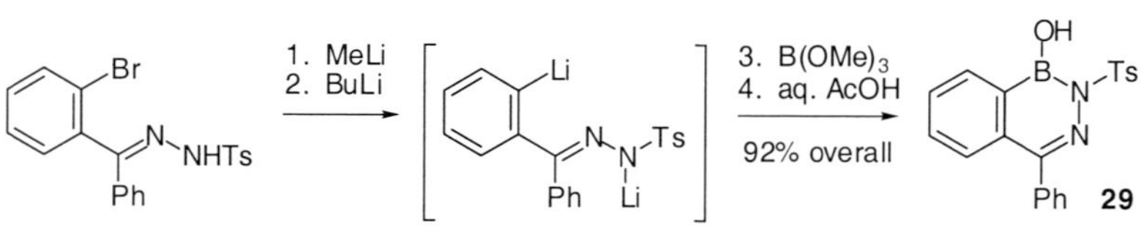

4-Ethyl-3-hydroxy-2-methyl-3,2-borazaropyridine (**31**), one of the very few non-fused (monocyclic) diheteraborines ever reported, was obtained by Gronowitz via desulfurization of 4-hydroxy-5-methyl-4,5-borazarothieno[2,3-*c*]pyridine (**30**) <68ACS1373>. Constitutional isomers **32** and **33** were obtained in a similar fashion <71ACS2435>. The X-ray crystal structure of the precursor to **33**, namely, 7-hydroxy-6-methyl-7,6-borazarothieno[3,2-*c*]pyridine (**34**), was determined <74ACS(B)989>.

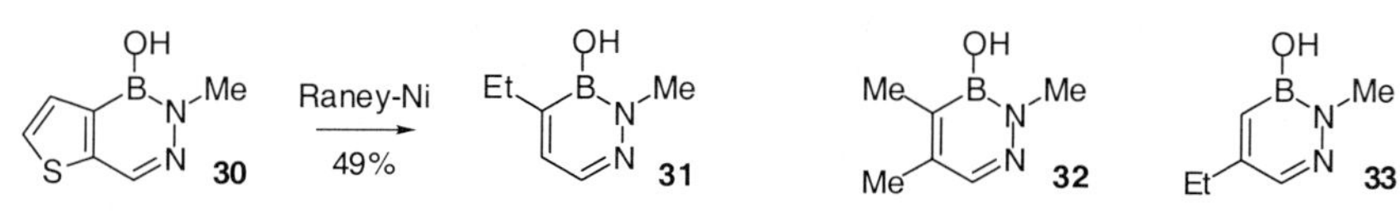

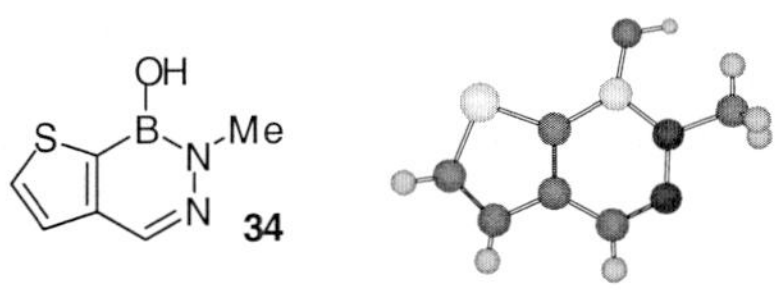

34

Nitrone derivatives of 2-formylphenylboronic acid have been prepared, and these exist in cyclic, trisubstituted boroxin form (**35a-d**) <83JOM247>. Interestingly, treatment with a 1,2-diol converts them into monomers with a rather unique 1,3-zwitterion structure (**36a-d**).

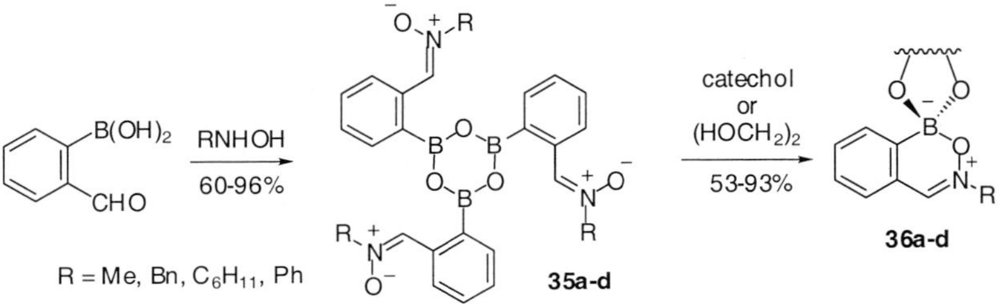

35a-d **36a-d**

Moving from the 2,3,1-diheteraborines to the 2,1-heteraborines, we find that a unique borazaroquinazoline (**37**) was prepared along a multistep route from 4,6-dichloropyrimidine-5-carboxaldehyde by Matteson <78JOC950>. And, in what might have been quite a surprising outcome for a BBr_3-mediated O-demethylation reaction, 4-ethyl-1-hydroxy-(4-hydroxyphenyl)-2-oxa-1-boranaphthalene (**38**) was obtained from a ketone precursor <93JOM139>. Fortunately for us, the investigators determined the X-ray crystal structure of this stable 2,1-benzoxaborine.

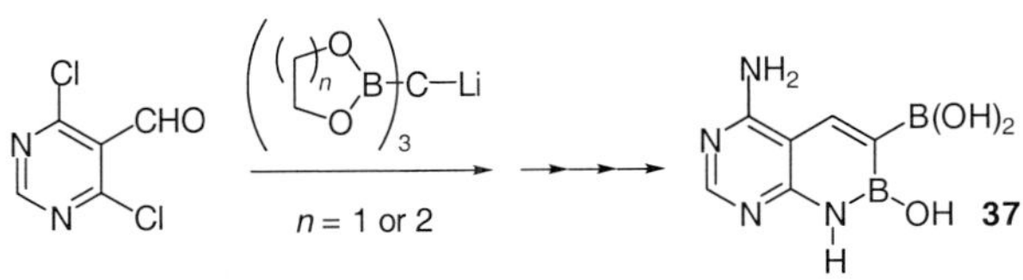

37

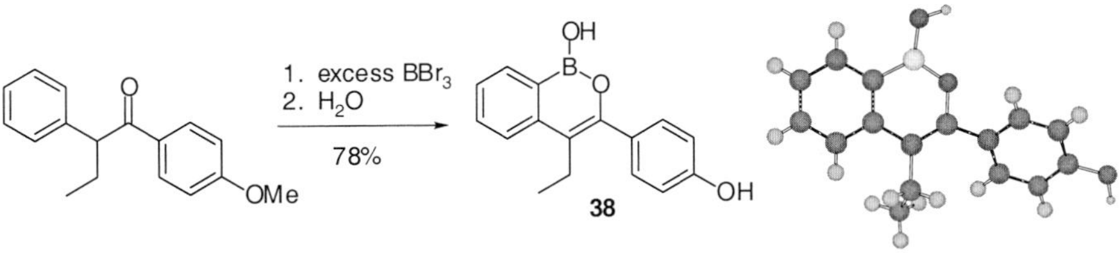

38

Our laboratory conducted the most extensive investigation of the 2,3,1-benzodiazaborines reported to date. We analyzed **25**, 1,2-dihydro-1-hydroxy-2,3,1-benzodiazaborine (**26**), and certain derivatives related to **26** by multisolvent ^{1}H, ^{13}C, ^{11}B, and ^{15}N NMR using isotopically-enriched (^{13}C,^{15}N) compounds <97JA7817>. The X-ray crystal structures of **25** and **26** were obtained first, and that of the 2-methyl derivative **39** was determined soon thereafter <98AX(C)71>. The topography (internal geometry, intramolecular associations) of **39** was found to be most similar to **26**, but some subtle **25**-like characteristics were found. All three boron heterocycles were shown to exist in planar form in protic solution just like they do in the solid state, and were shown to have predominantly Brønsted acidic OH groups. Still, the B-OH group in these heterocycles was found to be Lewis acid-capable under certain circumstances. We demonstrated by VT-NMR that **26** undergoes a triple hydrogen bond solution association with a protected cytidine.

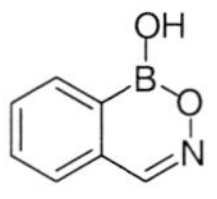
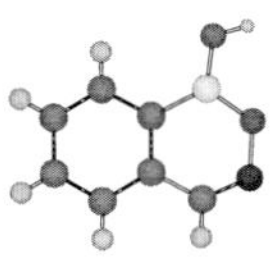

25

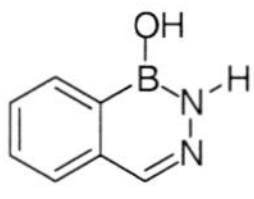
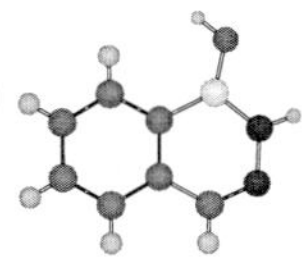

26

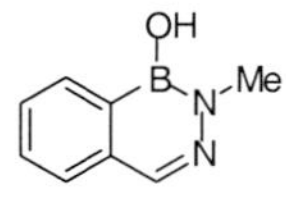
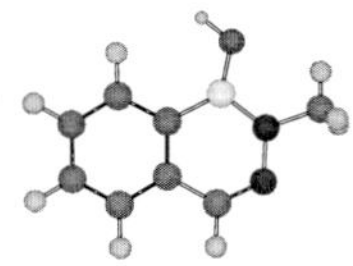

39

When we condensed 2-formylbenzeneboronic acid with 1,1-dimethylhydrazine, we obtained a triphenylboroxin derivative (**41**) instead of the expected monocyclic 1,2-zwitterion (**40**) <96AX(C)2826>. The X-ray crystal structure of **41** revealed it had two intramolecular chelations. By multisolvent ^{1}H, ^{13}C, and ^{11}B NMR spectral analysis, the solution structure in dry CH_3CN is identical to the solid state one, but the one in CH_3OH has only one chelate. The monomeric zwitterion species **40** does appears to exist, but only in water. It can be recovered from aqueous solution unless heated at 100 °C, in which case a seldom-encountered deboronation occurs. The structure of zwitterion **40** in water might well be similar to that of one of the severely ring puckered 2,3,1-benzodiazaborine fragments found in crystalline **41**.

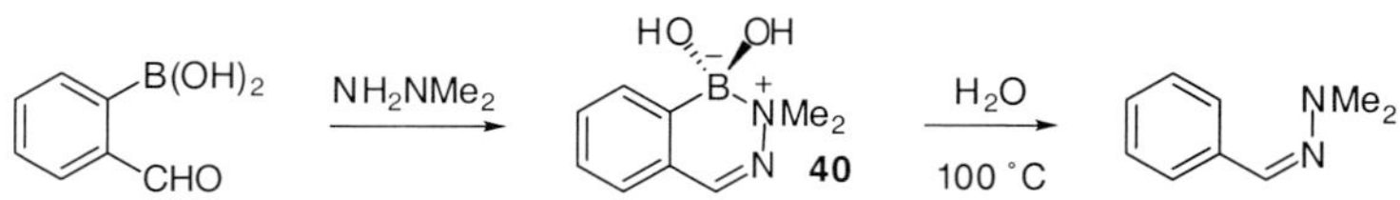

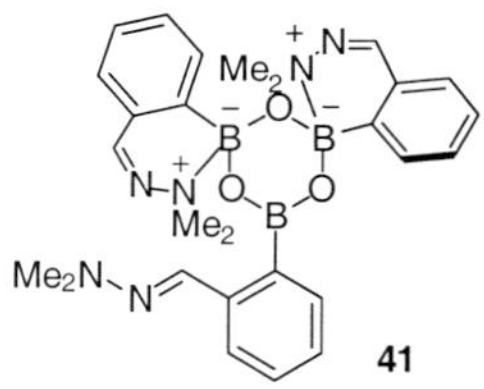

41

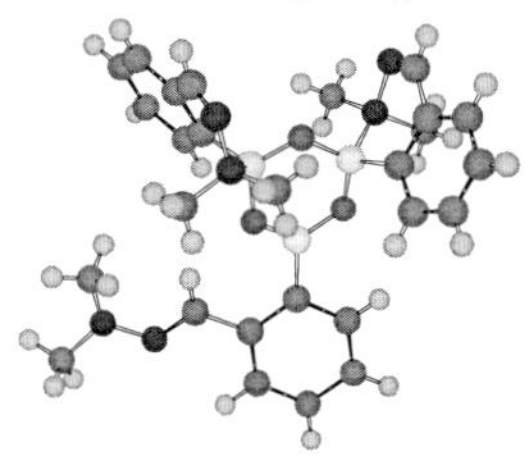
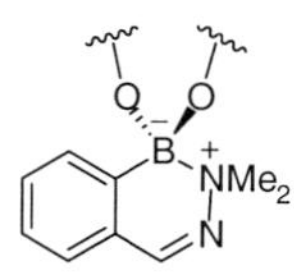

(fragment of **41**)

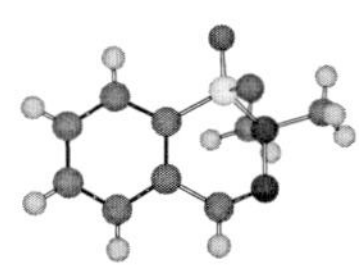

The precise mechanism of action of the aforementioned diazaborine antibacterials is now known, thanks to protein crystallography <96SCI2107>. These heterocycles form a covalent bond with the 2'-hydroxyl group of NAD's ribose unit and assemble a tightly held yet noncovalently bound bisubstrate analog species (**42** or **43**) at the active site of enoyl acyl carrier protein reductase (ENR). The enzyme's ability to generate the tetrahedral borate form of these 2-sulfonylated 2,3,1-diazaborines is noteworthy, since this form is not observed by NMR of the heterocycles alone. Once again, the Brønsted/Lewis acid ambidency of B-hydroxy boron heterocycles becomes evident.

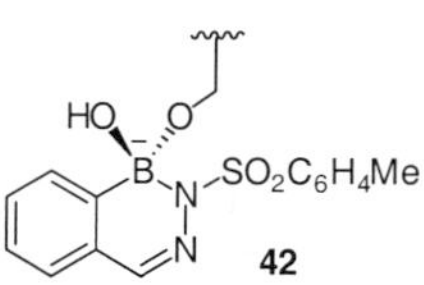

42

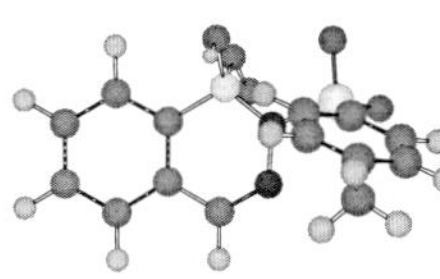
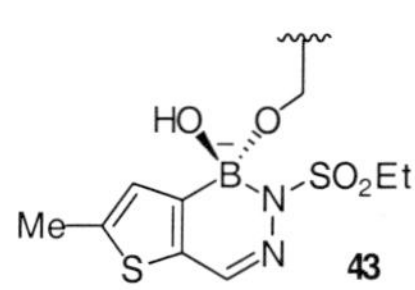

43

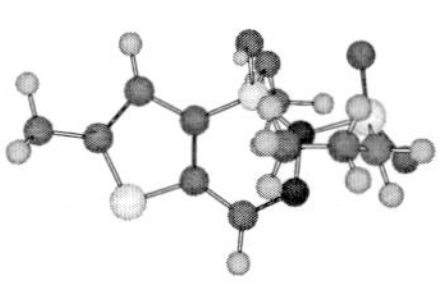

1.2.5 2,4,6,1-Triheteraborines and 2,4,1-Diheteraborines

Inverting the orientation of the C4-N3 imine unit of a 2,3,1-diheteraborine gives a boron heterocycle with a markedly different chemical reactivity. In effect, the weakly basic oxime- or hydrazone-type imine nitrogen in the 2,3,1-diheteraborine is replaced by a much more basic imidate- or amidine-type imine nitrogen in the 2,4,1-diheteraborine. Likely, the Lewis acid tendency of the boron is enhanced by the ready protonation of this basic N4, and the formation of a stable borate-based zwitterion becomes thermodynamically favored.

Monocyclic zwitterionic triheteraborines (**44-47**) were synthesized quite some time ago by treating biguanidine <62JA2529> or guanylurea or its O-alkyl ethers <72JINC3643> with trialkylborates. The borate esters **44a** and **47a** were easily hydrolyzed to the stable corresponding dihydroxy borate anion-based zwitterions **44b** and **47b**, respectively.

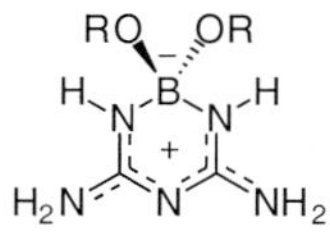

44a, R = Me; **b**, R = H

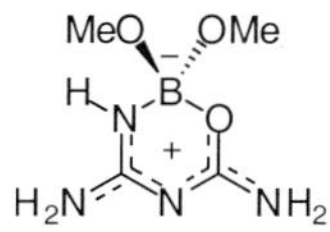

45

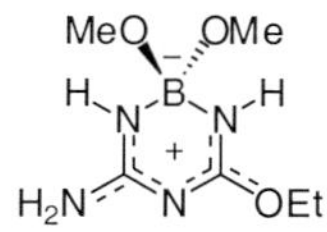

46

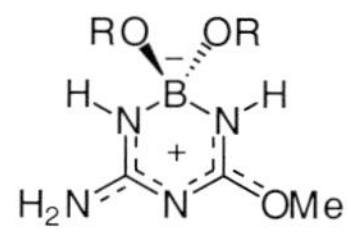

47a, R = Me; **b**, R = H

A tricyclic boron heterocycle (**48**) related to these was synthesized recently along a different route starting from 2-guanidinobenzimidazole <98HAC399>. This time, we are fortunate enough to have an X-ray crystal structure to scrutinize, and can identify features consistent with an extensively delocalized positive charge counterbalancing the borate anion.

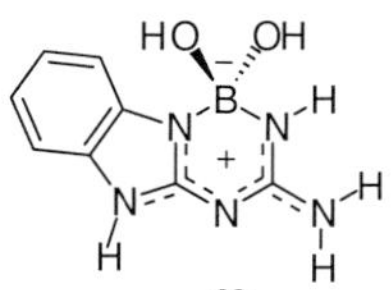

48

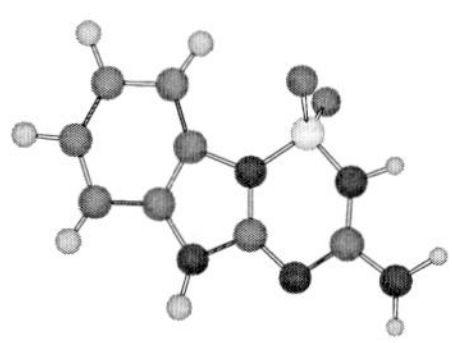

Early on, it was recognized that 2-(acetamido)phenylboronic acid did not exist as an "open" structure in the solid state, but that it was likely the bicyclic 1-hydroxy-*1H*-2,4,1-benzoxazaborine (**49a**) instead <60JA2442>. Later, when 2,5-*bis*(acetamido)phenylboronic acid was prepared, it was seen to be some sort of weakly chelated hydrate form of the 2,4,1-oxazaborine (**49b**) in solution <91MI317>. And within their large body of work on boron heterocycles <79IZV174, 79IZV411, 76JOM123, 85IZV428, 85IZV329, 79IZV80>, Mikhailov's group examined 2-(N^3-phenylacetamidino)phenylboronic acid and formulated it as an internally-chelated hydrate of the 2,4,1-diazaborine (**49c**) <85IZV428>.

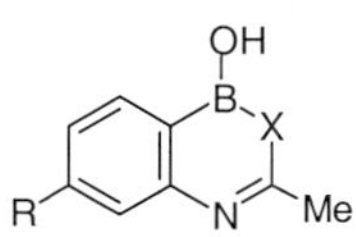

49a, R = H, X = O;
b, R = NHAc, X = O;
c, R = H, X = NPh

Our laboratory conducted the most extensive investigation of the 2,4,1-benzodiazaborines reported to date. We focused attention on 1-hydroxy-1*H*-2,4,1-benzoxazaborine (**50a**), 1,2-dihydro-1-hydroxy-2,4,1-benzodiazaborine (**50b**), and 3-amino-1,2-dihydro-1-hydroxy-2,4,1-benzodiazaborine (**50c**) because their peripheries so closely matched the pyrimidine ring ones of the naturally occurring purines adenine, hypoxanthine, and guanine, respectively <94JA7597>.

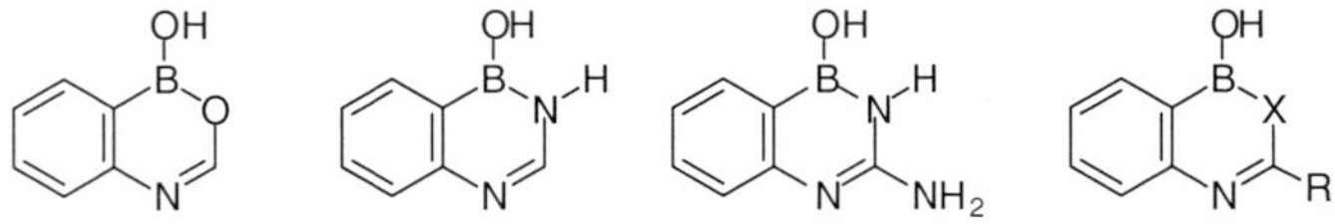

50a 50b 50c

50d, R = H, X = NPr; **e**, R = CCl$_3$, X = NH; **f**, R = CF$_3$, X = O; **g**, R = H, X = NNH$_2$

From multisolvent ^{11}B NMR spectroscopic analyses of **49a** and **50a-g**, we determined that facile 1,4-hydrations and 1,4-alcohol additions were an endemic property of the 2,4,1-oxaza- and diazaborines. Heterocycles **49a**, **50a-e**, and **50g** were all found stable to hydrolysis, but not to a facile hydration that occurs in a 1,4-fashion to give zwitterionic adducts. Only **50f** was found to exist in an "open" form in water, by ^{11}B NMR.

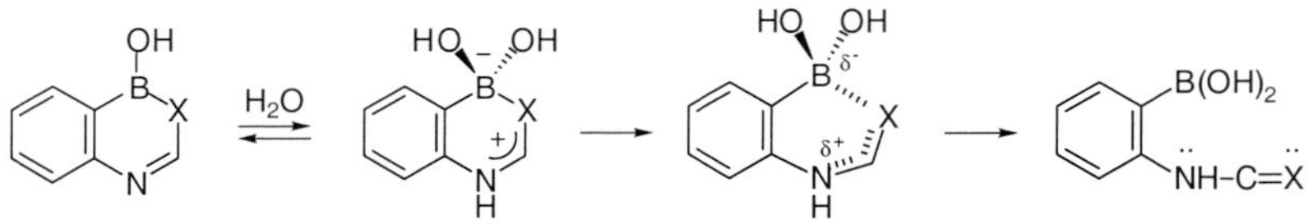

(**49a**, **50a-e**, **50g**) (not observed) (**50f** only)

Similar to their spontaneous 1,4-hydration in water, heterocycles **49a**, **50a-e**, and **50g** all react simply upon dissolution in warm methanol to form *bis*-addition products. The X-ray crystal structure of the *bis*-methanol adduct (**51**) derived from the diazaborine **50b** clearly showed it to be a zwitterion comprised of tetrahedral borate anion and formamidinium cation fragments. Once again, **50f** proved to be the only exception: A weakly chelated dimethylborate ester (not shown) was found to be its structure in MeOH by ^{11}B NMR.

(**49a**, **50a-e**, **50g**) **51**

Moving to the 2,4,1-diheteraborin-3-ones, we find that Martin's group prepared a set of 2,4,1-benzodiazaborin-3-ones and -thiones (**52-54**) as potential antituberculosis agents <98BMCL843>. Compounds **52b** and **53** were determined to be the most active.

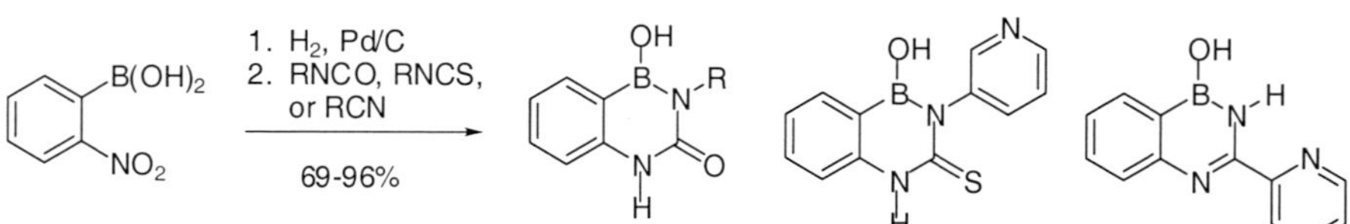

52a, R = nBu; **b**, R = Ph **53** **54**

Smith's group prepared urea-based boron heterocyclic carboxylate binding agents (**55**) <96JOC4510, 97JOC4492> and noted that species **55b** and **57b** present in a 40/60 ratio at equilibrium in MeOH underwent slow exchange, by ^{11}B NMR. Upon condensation with pinacol, the diazaborinones **55a-d** were converted to the oxazaborine zwitterions **56a-d**. Treatment of **55a,b** with KHF$_2$ gave the difluoro zwitterions **58a,b**. Gratefully, X-ray crystal structure determinations of **55a** and **58a** were provided.

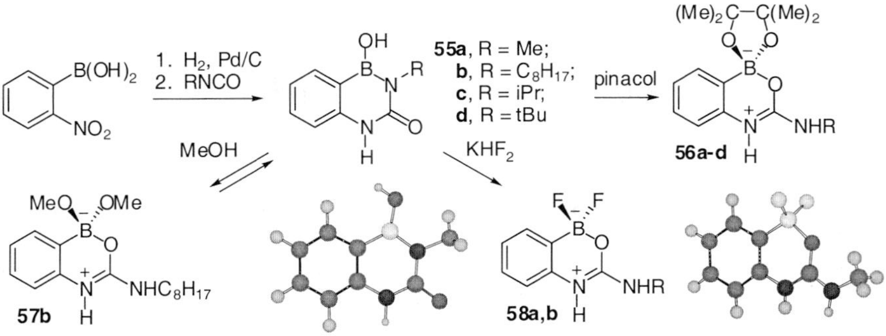

Graham's group accessed 2-benzyl-1,4-dihydro-1-hydroxythieno[2,3-*c*][1,5,2]diazaborin-3(2H)-one (**59**) <98JHC887>, and used ^{11}B NMR to examine its structure in $(CD_3)_2CO$, $CDCl_3$, CD_3CN, and CD_3OD. Curiously, **59** did not appear to form zwitterion **60** in this latter solvent.

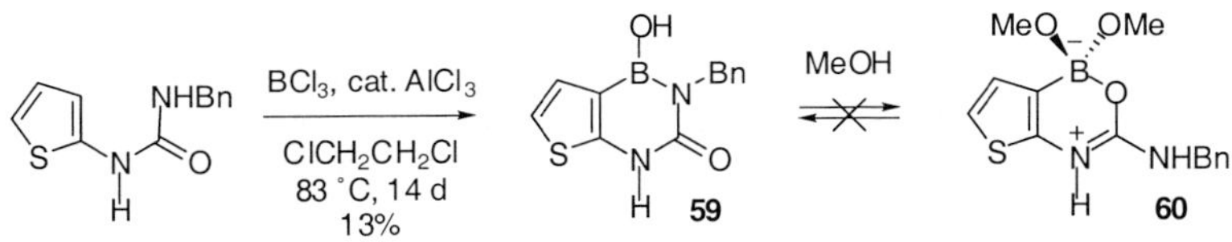

In the most recent publication in the field, Soloway's group reported the preparation of 3-phenyl-1,2-dihydro-2,4,1-benzodiazaborin-3(4*H*)-one (**52b**) <99JOC9566>. They showed by ^{11}B NMR that the addition of MeOH to a $(CD_3)_2SO$ solution of **52b** progressively forms zwitterion **61**. Crystal structure determinations of both **52b** and **61** were obtained.

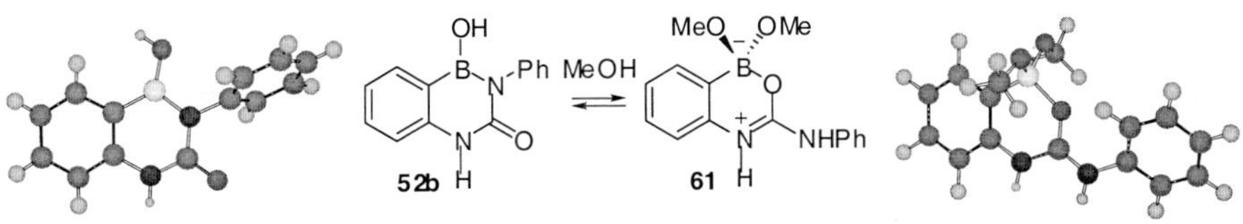

In this same publication, Soloway's group reported what is undoubtedly the first nucleoside built with an endocyclic boron-containing heterocyclic aglycon. Condensation of 2-aminophenylboronic acid with a protected ribose gave, upon treatment with MeNCO, the boron-containing nucleoside **62**. Interestingly, it was not the natural β-anomer but rather the α-anomer that was found to be the more thermodynamically stable. Addition of MeOH to nucleoside **62** reversibly gave a zwitterion in exactly the same way that **52b** had afforded **61**.

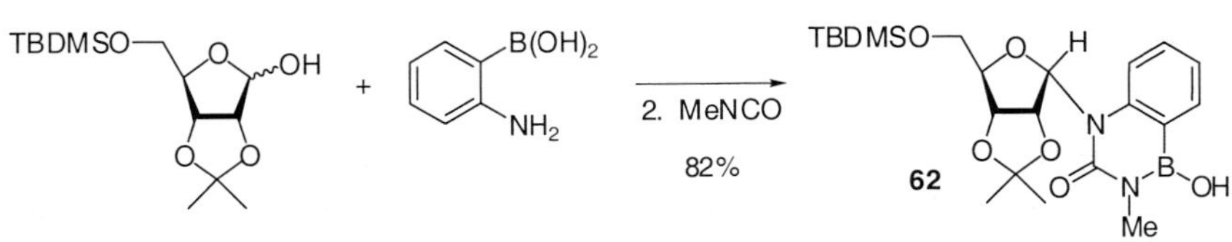

1.3 WHAT DOES THE FUTURE HOLD FOR BORON HETEROCYCLES?

We believe that many new, exciting bioactive agents can now be constructed from the boron heterocycle platforms already well characterized and proven stable, and we have been moving in this direction using a biomimicry-based approach. Contemplating the crystal structures of the 1-hydroxy-2,3,1-benzodiazaborines **26** and **39**, it occurred to us that the hydroxyl group could be forced *syn* coplanar with respect to the B-N bond if it were hydrogen-bonded to an imine nitrogen acceptor atom that we would introduce within a 2-substituent. This hydrogen bond would define a third, "virtual" ring flexibly annelated onto the B–N edge of the bicyclic 2,3,1-benzodiazaborine.

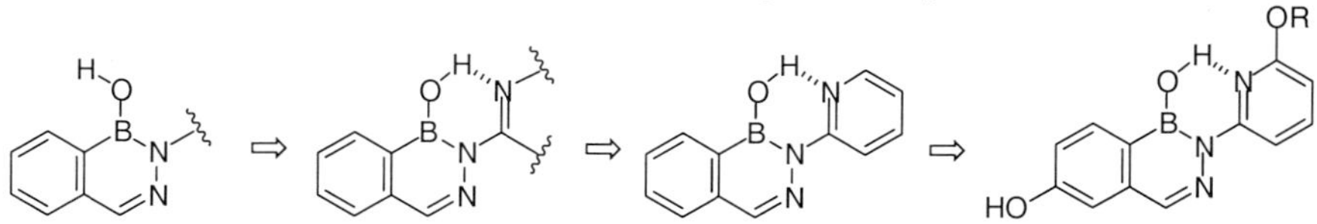

Now, if that imine nitrogen acceptor were further made part of a heterocyclic moiety like a pyridyl one, then the molecule's topography would begin to resemble the familiar steroidal cyclopentanophenanthrene framework. The substitution pattern of an A-ring aromatic estrogen or an AB-ring aromatic equilenin could be readily established by adding oxygen-based substituents to the benzene and pyridine rings. To test the design features of this first-ever class of boron heterocycle estrogen mimics, we prepared a prototype, namely, a boron heterocycle mimic of estradiol or dihydroequilenin O17-methyl ether.

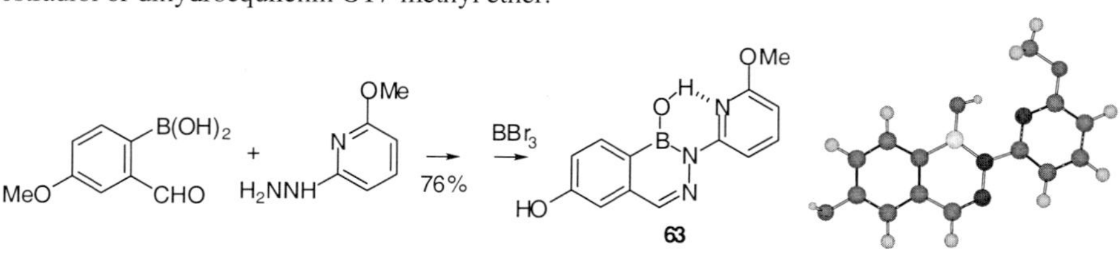

The product of the condensation of 2-formyl-4-methoxybenzeneboronic acid and 2-hydrazino-6-methoxypyridine was regioselectively O-demethylated with BBr_3 to give 1,2-dihydro-1,6-dihydroxy-2-(2-methoxy-6-pyridyl)-2,3,1-benzodiazaborine (**63**). An X-ray crystal structure determination of **63** revealed it has a remarkable estrogen-like topography <99AX(C)1701>. The intentionally engineered intramolecular hydrogen bond between the B-OH donor and the pyridine nitrogen acceptor establishes a transmolecular inter-oxygen distance of 10.6 Å—extremely close to the 10.8 Å one in equilenin. By the downfield location of the B-OH resonance in its ^1H NMR spectrum, the intramolecular hydrogen bond in **63** is maintained in $(CD_3)_2SO$ solution, and it is likely to exist to a large extent in aqueous solution as well. In fact, by the NMR assessment, *both* the B-OH and Ar-OH groups are fairly Brønsted acid labile—the former because of the intramolecular hydrogen bond and the latter because of the electron-withdrawing effect of the boron transmitted through the benzene ring.

Of course, derivatives much more biomimetic than **63** can be readily envisaged and indeed these are being prepared, but since this first prototype possessed at least the biorecognition-critical A-ring hydroxyl group of an estrogen, we decided to see if it had any antiproliferative activity against MCF-7 human breast cancer cells. The IC_{50} value of 5 μM nicely validates our contention that our efforts to construct bioactive boron heterocycles *de novo* along a biomimicry-driven approach will be very fruitful, indeed.

1.4 CONCLUSIONS

A great many endogenous and synthetic bioactive compounds have aromatic or heteroaromatic rings as central cores or key fragments. Many of these, in turn, are suitable candidates for a well-designed B-O for C=N or B-N for C=C unit replacement "therapy" in the pursuit of new boron heterocycle analogues with unusual and useful properties. The preparation and thorough characterization of additional new stable boron heterocycle platforms will certainly facilitate progress in this mostly unexplored and somewhat exotic subfield of heterocyclic chemistry.

1.5 ACKNOWLEDGMENTS

The X-ray crystal structures shown in this article were created with CambridgeSoft's Chem3D Pro™ 5.0 using coordinates obtained from the source, the CSD (Cambridge Structural Database) or the PDB™ (Protein Data Bank), or those determined by Paul D. Robinson at Southern Illinois University (SIU) in collaboration with the author. Most of the author's work on boron heterocycles first at SIU and more recently at SRI International (the nonprofit research institute formerly known as the Stanford Research Institute) was funded by NIH grant GM44819.

1.6 REFERENCES

57AK473 K. Torssell *Ark. Kemi* **1957**, *10*, 473-481.
57AK497 K. Torssell, H. Mayer, B. Zacharias *Ark. Kemi* **1957**, *10*, 497-505.
57AK507 K. Torssell *Ark. Kemi* **1957**, *10*, 507-511.
57AK513 K. Torssell *Ark. Kemi* **1957**, *10*, 513-521.
58JA835 H.R. Snyder, A.J. Reedy, W.J. Lennarz *J. Am. Chem. Soc.* **1958**, *80*, 835-838.
58PRS1 E.G. Cox, D.W.J. Cruickshank, J.A.S. Smith *Proc. R. Soc. London, A* **1958**, *247*, 1.
59JA2681 E. Nyilas, A.H. Soloway *J. Am. Chem. Soc.* **1959**, *81*, 2681-2683.
59JA3017 A.H. Soloway *J. Am. Chem. Soc.* **1959**, *81*, 3017-3019.
59JA6329 S.S. Chissick, M.J.S. Dewar, P.M. Maitlis *J. Am. Chem. Soc.* **1959**, *81*, 6329-6330.
60JA2172 W.J. Lennarz, H.R. Snyder *J. Am. Chem. Soc.* **1960**, *82*, 2172-2175.
60JA2442 A.H. Soloway *J. Am. Chem. Soc.* **1960**, *82*, 2442-2444.
61JA2708 S.S. Chissick, M.J.S. Dewar, P.M. Maitlis *J. Am. Chem. Soc.* **1961**, *83*, 2708-2711.
61JOC4632 R.J. Brotherton, H. Steinberg *J. Org. Chem.* **1961**, *26*, 4632-4634.
62CRV223 P.M. Maitlis *Chem. Rev.* **1962**, *62*, 223-245.
62JA688 H.L. Yale, F.H. Bergeim, F.A. Sowinski, J. Bernstein, J. Fried *J. Am. Chem. Soc.* **1962**, *84*, 688-690.
62JA2529 J.E. Milks, G.W. Kennerly, J.H. Polevy *J. Am. Chem. Soc.* **1962**, *84*, 2529-2534.
62JA2648 M.J.S. Dewar, R.C. Dougherty *J. Am. Chem. Soc.* **1962**, *84*, 2648-2649.
64JA433 M.J.S. Dewar, R.C. Dougherty *J. Am. Chem. Soc.* **1964**, *86*, 433-436.
64JA2961 D.N. Butler, A.H. Soloway *J. Am. Chem. Soc.* **1964**, *86*, 2961.
64JOC2168 P. Tschampel, H.R. Snyder *J. Org. Chem.* **1964**, *29*, 2168-2172.
64JOC3229 R.R. Haynes, H.R. Snyder *J. Org. Chem.* **1964**, *29*, 3229-3233.
B-64MI1 R.L. Letsinger In *Advances in Chemistry Series*; R.F. Gould, Ed.; American Chemical Society: Washington, DC, 1964; Vol. 42, pp. 1-16.
B-64MI227 M.J.S. Dewar In *Advances in Chemistry Series*; R.F. Gould, Ed.; American Chemical Society: Washington, DC, 1964; Vol. 42, pp. 227-250.
B-64MI235 M.J.S. Dewar In *Progress in Boron Chemistry*; H. Steinberg, A.L. McCloskey, Eds.; Pergamon: London, 1964; Vol. 1, pp. 235-263.
65ACS1271 S. Gronowitz, A. Bugge *Acta Chem. Scand.* **1965**, *19*, 1271-1285.
66JA484 D.N. Butler, A.H. Soloway *J. Am. Chem. Soc.* **1966**, *88*, 484-487.
66JCS(A)479 R.K. Bartlett, H.S. Turner, R.J. Warne, M.A. Young, I.J. Lawrenson *J. Chem. Soc. (A)* **1966**, 479-500.
66JMC362 D.N. Butler, A.H. Soloway *J. Med. Chem.* **1966**, *9*, 362-365.
67ACS2151 S. Gronowitz, J. Namtvedt *Acta Chem. Scand.* **1967**, *21*, 2151-2166.

67HCA1533 R. Hütter, W. Keller-Schierlein, F. Knüsel, V. Prelog, G.C. Rodgers, P. Suter, G. Vogel, W. Voser, H. Zähner *Helv. Chim. Acta* **1967**, *50*, 1533-1539.

67JA2408 M.J.S. Dewar, R. Jones *J. Am. Chem. Soc.* **1967**, *89*, 2408-2410.

68ACS1353 U. Michael, S. Gronowitz *Acta Chem. Scand.* **1968**, *22*, 1353-1355.

68ACS1373 J. Namtvedt, S. Gronowitz *Acta Chem. Scand.* **1968**, *22*, 1373-1374.

68ACS1611 J. Namtvedt *Acta Chem. Scand.* **1968**, *22*, 1611-1622.

68JA706 F.A. Davis, M.J.S. Dewar, R. Jones *J. Am. Chem. Soc.* **1968**, *90*, 706-708.

68JOC4483 H.E. Dunn, J.C. Catlin, H.R. Snyder *J. Org. Chem.* **1968**, *33*, 4483-4486.

69JOC1660 J.C. Catlin, H.R. Snyder *J. Org. Chem.* **1969**, *34*, 1660-1663.

69MI1472 A.B. Borkovec, J.A. Settepani, G.C. LaBrecque, R.L. Fye *J. Econ. Entomol.* **1969**, *62*, 1472-1480.

70AK283 S. Gronowitz, U. Michael *Ark. Kemi* **1970**, *32*, 283-294.

70JMC128 J.A. Settepani, J.B. Stokes, A.B. Borkovec *J. Med. Chem.* **1970**, *13*, 128-131.

70JOM323 H. Nöth, M.J. Sprague *J. Organomet. Chem.* **1970**, *23*, 323-327.

70MI23 V.K. Antonov, T.V. Ivanina, I.V. Berezin, K. Martinek *FEBS Lett.* **1970**, *7*, 23-25.

70MI1898 B. Roques, D. Florentin, J.P. Juhasz *C. R. Acad. Sc. Paris, Sér. C* **1970**, *270*, 1898-1900.

70TL4909 M.C. Fournié-Zaluski, B. Roques *Tetrahedron Lett.* **1970**, *56*, 4909-4912.

B-70MI117 D.S. Matteson In *Progress in Boron Chemistry*; Pergamon Press: Oxford, 1970; Vol. 3, pp. 117-176.

71ACS2435 S. Gronowitz, A. Maltesson *Acta Chem. Scand.* **1971**, *25*, 2435-2446.

71APS377 S. Gronowitz, T. Dahlgren, J. Namtvedt, C. Roos, B. Sjoberg, U. Forsgren *Acta Pharm. Suecica* **1971**, *8*, 377-390.

71APS623 S. Gronowitz, T. Dahlgren, J. Namtvedt, C. Roos, G. Rosen, B. Sjoberg, U. Forsgren *Acta Pharm. Suecica* **1971**, *8*, 623-638.

71B2477 K.A. Koehler, G.E. Lienhard *Biochemistry* **1971**, *10*, 2477-2483.

71CCC1233 J. Michl, R. Jones *Collect. Czech. Chem. Commun.* **1971**, *36*, 1233-1247.

71CCC1248 J. Michl *Collect. Czech. Chem. Commun.* **1971**, *36*, 1248-1278.

71IJC1167 V.P. Arya *Indian J. Chem.* **1971**, *9*, 1167.

71SRI83 H. Bellut, C.D. Miller, R. Köster *Synth. React. Inorg. Metal-Org. Chem.* **1971**, *1*, 83-91.

72IJS(A)257 D. Semard, C. Paulmier, J. Morel, P. Pastour *Int. J. Sulfur Chem., Part A* **1972**, *2*, 257-260.

72JINC3643 A.N. Maitra, D. Sen *J. Inorg. Nucl. Chem.* **1972**, *34*, 3643-3649.

72JOM38 B. Roques, D. Florentin *J. Organomet. Chem.* **1972**, *46*, C38-C40.

72PHA1 W. Kliegel *Pharmazie* **1972**, *27*, 1-14.

73OMR585 G. Beaumelou, M. Pasdeloup, J.P. Laurent *Org. Magn. Res.* **1973**, *5*, 585-588.

73OSC727 M.J.S. Dewar, R.B.K. Dewar, Z.L.F. Baibel *Org. Synth., Coll. Vol. V* **1973**, 727-733.

73SRI229 W.L. Cook, K. Niedenzu *Synth. React. Inorg. Metal-Org. Chem.* **1973**, *3*, 229-235.

74ACS(B)989 B. Aurivillius, I. Löefving *Acta Chem. Scand., Ser. B* **1974**, *28*, 989-992.

74BSF2620 D. Florentin, P. Roques Bernard, J.M. Metzger, J.P. Colin *Bull. Soc. Chim. Fr.* **1974**, *11*, 2620-2622.

74MI135 R.N. Lindquist, C. Terry, J.D. Rawn, G.E. Lienhard *Arch. Biochem. Biophys.* **1974**, *160*, 135-144.

75ACS457 S. Gronowitz, A. Maltesson *Acta Chem. Scand.* **1975**, 457-460.

75ACS(B)461 S. Gronowitz, A. Maltesson *Acta Chem. Scand., Ser. B* **1975**, *29*, 461-467.

75ACS(B)990 S. Gronowitz, C. Roos *Acta Chem. Scand., Ser. B* **1975**, *29*, 990-998.

75MI434 H.N. Finkbone, M.W. Williams, C.D. Miller *Proc. Soc. Exp. Biol. Med.* **1975**, *150*, 434-440.

76CB3480 B. Wrackmeyer, H. Noeth *Chem. Ber.* **1976**, *109*, 3480-3485.

76JAN1019 Y. Okami, T. Okazaki, H. Umezawa *J. Antibiot. (Tokyo)* **1976**, *29*, 1019-1025.

76JHC(S)76 S. Gronowitz, C. Glennow *J. Heterocycl. Chem.* **1976**, *3(S)*, 76-82.

76JOM123 B.M. Mikhailov, M.E. Kuimova *J. Organomet. Chem.* **1976**, *116*, 123-133.

77ACS(B)765 S. Gronowitz, A. Maltesson *Acta Chem. Scand., Ser. B* **1977**, *31*, 765-767.

77CJC3071 S.J. Rettig, J. Trotter *Can. J. Chem.* **1977**, *55*, 3071-3075.

77CS76 S. Gronowitz, C. Glennow *Chem. Scr.* **1977**, *11*, 76-82.

77HC381 A.J. Fritsch *The Chemistry of Heterocyclic Compounds* **1977**, *30*, 381-440.

77IC2935 O.T. Beachley, Jr., R.G. Simmons *Inorg. Chem.* **1977**, *16*, 2935-2938.

77IJC(B)267	V.P. Arya *Indian J. Chem.* **1977**, *15B*, 267-268.
77JAN714	H. Nakamura, Y. Iitaka, T. Kitahara, J. Okazaki, Y. Okami *J. Antibiot. (Tokyo)* **1977**, *30*, 714-719.
77JCR(S)158	D. Florentin, M.C. Fournié-Zaluski, B.P. Roques *J. Chem. Res. (S)* **1977**, 158-159.
77JHC893	S. Gronowitz, C. Roos, E. Sandberg, S. Clementi *J. Heterocycl. Chem.* **1977**, *14*, 893-897.
77JOC3545	T. Okyama, T. Takimoto, T. Fueno *J. Org. Chem.* **1977**, *42*, 3545-3549.
78HCA325	B.W. Müller *Helv. Chim. Acta* **1978**, *61*, 325-327.
78IJC(B)85	A. Maitra *Indian J. Chem.* **1978**, *16B*, 85-86.
78JOC950	D.S. Matteson, M.S. Biernbaum, R.A. Bechtold, J.D. Campbell, R.J. Wilcsek *J. Org. Chem.* **1978**, *43*, 950-954.
B-78MI14	H. Nöth, B. Wrackmeyer In *NMR Basic Principles and Progress*, Vol. 14, P. Diehl, E. Fluck, R. Kosfeld, Eds.; Springer-Verlag: Berlin, 1978.
79IZV80	B.M. Zolotarev, V.A. Dorokhov, O.S. Chizhov, L.I. Lavrinovich, B.M. Mikhailov *Izv. Akad. Nauk SSSR, Ser. Khim.* **1979**, 80-84.
79IZV174	V.A. Dorokhov, O.G. Boldyreva, M.N. Bochkareva, B.M. Mikhailov *Izv. Akad. Nauk SSSR, Ser. Khim.* **1979**, *1*, 174-180.
79IZV411	V.A. Dorokhov, M.N. Bochkareva, O.G. Boldyreva, B.V. Rassadin, B.M. Mikhailov *Izv. Akad. Nauk SSSR, Ser. Khim.* **1979**, 411-418.
79TL3571	J.W. Barton, D.J. Lapham *Tetrahedron Lett.* **1979**, *37*, 3571-3572.
80AAC549	P.J. Bailey, G. Cousins, G.A. Snow, A.J. White *Antimicrob. Agents Chemother.* **1980**, *17*, 549-553.
80CB3881	H. Fußstetter, H. Nöth, K. Peters, H.G. von Schnering, J.C. Huffman *Chem. Ber.* **1980**, *113*, 3881-3890.
80JAN1316	T.S. Chen, C.J. Chang, H.G. Floss *J. Antibiot. (Tokyo)* **1980**, *33*, 1316-1322.
81JOM17	W. Maringgele *J. Organomet. Chem.* **1981**, *222*, 17-32.
81NAT662	G. Högenauer, M. Woisetschlaeger *Nature (London)* **1981**, *293*, 662-664.
82CB3271	W. Maringgele *Chem. Ber.* **1982**, *115*, 3271-3289.
83JOM247	W. Kliegel, D. Nanninga *J. Organomet. Chem.* **1983**, *247*, 247-252.
B-83MI49	R.G. Kidd In *NMR of Newly Accessible Nuclei*; P. Laszlo, Ed.; Academic Press: New York, 1983; Vol. 2, pp. 49-77.
84CHEC-I(1)629	I. Ander In *Comp. Heterocycl. Chem., 1st edn.*; A.R. Katritzky, C.W. Reese, Eds.; Pergamon Press: Oxford, 1984; Vol. 1, Chpt. 21, pp. 629-663.
84CZ287	F. Dallacker, E.M. Both-Pollmann, W. Müllners *Chem.-Ztg.* **1984**, *108*, 287-288.
84JMC947	M.A. Grassberger, F. Turnowsky, J. Hildebrandt *J. Med. Chem.* **1984**, *27*, 947-953.
84SA(A)855	H.R. Morales, H. Tlahuext, F. Santiesteban, R. Contreras *Spectrochim. Acta Part A* **1984**, *40*, 855-862.
84ZSK180	S.V. Lindeman, V.E. Shklover, Y.T. Struchkov, L.M. Mitina, V.A. Pankratov *Zh. Strukt. Khim.* **1984**, *25*, 180-181.
85IZV329	L.G. Vorontsova, O.S. Chizhov, O.G. Boldyreva, V.A. Dorokhov, B.M. Mikhailov *Izv. Akad. Nauk SSSR, Ser. Khim.* **1985**, *2*, 329-332.
85IZV428	O.G. Boldyrev, V.A. Dorokhov, B.M. Mikhailov *Izv. Akad. Nauk SSSR, Ser. Khim.* **1985**, *2*, 428-430.
85LA683	M.A. Grassberger *Liebigs Ann. Chem.* **1985**, 683-688.
86JA3602	J.P. Doering, A. Gedanken, A.P. Hitchcock, P. Fischer, J. Moore, J.K. Olthoff, J. Tossell, K. Raghavachari, M.B. Robin *J. Am. Chem. Soc.* **1986**, *108*, 3602-3608.
86TL869	J.T. Sharp, C.E.D. Skinner *Tetrahedron Lett.* **1986**, *27*, 869-872.
87AX(C)1775	C.P. Brock, R.P. Minton, K. Niedenzu *Acta Crystallogr.* **1987**, *C43*, 1775-1779.
87MI37	C. Lam, F. Turnowsky, G. Högenauer, E. Schuetze *J. Antimicrob. Chemother.* **1987**, *20*, 37-45.
88JA309	F. Adebodun, F. Jordan *J. Am. Chem. Soc.* **1988**, *110*, 309-310.
88MI231	G.W. Kabalka, M. Davis, P. Bendel *Magn. Reson. Med.* **1988**, *8*, 231-237.
89JCS(P2)719	D.L. Cooper, S.C. Wright, J. Gerratt, P.A. Hyams, M. Raimondi *J. Chem. Soc., Perkin Trans. 2* **1989**, 719-724.
89MI6555	F. Turnowsky, K. Fuchs, C. Jeschek, G. Högenauer *J. Bacteriol.* **1989**, *171*, 6555-6565.
90JMR369	P. Bendel, M. Davis, E. Berman, G.W. Kabalka *J. Mag. Reson.* **1990**, *88*, 369-375.
90MI61	F.H. Nielsen *Magnesium Trace Elem.* **1990**, *9*, 61-69.
90ZAAC151	I. Uzarewicz, W. Maringgele, A. Meller *Z. Anorg. Allg. Chem.* **1990**, *590*, 151-160.

91BMCL9	J.E. Baldwin, T.D.W. Claridge, A.E. Derome, C.J. Schofield, B.D. Smith *Bioorg. Med. Chem. Lett.* **1991**, *1*, 9-12.
91MI317	S.X. Cai, J.F.W. Keana *Bioconjugate Chem.* **1991**, *2*, 317-322.
92MI229	W.D. Loomis, R.W. Durst *Biofactors* **1992**, *3*, 229-239.
92MI293	R.P. Singhal, S.S. DeSilva *Adv. Chromatogr.* **1992**, *31*, 293-335.
93AC(E)950	M.F. Hawthorne *Angew. Chem., Int. Ed. Engl.* **1993**, *32*, 950-984.
93AX(C)690	S. Soundararajan, E.N. Duesler, J.H. Hageman *Acta Crystallogr.* **1993**, *C49*, 690-693.
93B12651	E. Tsilikounas, C.A. Kettner, W.W. Bachovchin *Biochemistry* **1993**, *32*, 12651-12655.
93JOM139	V.L. Arcus, L. Main, B.K. Nicholson *J. Organomet. Chem.* **1993**, *460*, 139-147.
93MI109	C. Kettner, R.M. Knabb *Adv. Exp. Med. Biol.* **1993**, *340*, 109-118.
93OM3225	A.J. Ashe, III, W. Klein, R. Rousseau *Organometallics* **1993**, *12*, 3225-3231.
94CB1887	R. Boese, A.H. Maulitz, P. Stellberg *Chem. Ber.* **1994**, *127*, 1887-1890.
94JA7597	M.P. Groziak, A.D. Ganguly, P.D. Robinson *J. Am. Chem. Soc.* **1994**, *116*, 7597-7605.
94JA7801	J. Karolin, L.B.-A. Johansson, L. Strandberg, T. Ny *J. Am. Chem. Soc.* **1994**, *116*, 7801-7806.
94JBC5493	H. Bergler, P. Wallner, A. Ebeling, B. Leitinger, S. Fuchsbichler, H. Aschauer, G. Kollenz, G. Högenauer, F. Turnowsky *J. Biol. Chem.* **1994**, *269*, 5493-5496.
94JOM31	A. Sporzynski, H. Szatylowicz *J. Organomet. Chem.* **1994**, *470*, 31-33.
94LA283	D. Schummer, H. Irschik, H. Reichenbach, G. Höfle *Liebigs Ann. Chem.* **1994**, 283-289.
94MI1	International Symposium on the Health Effects of Boron and its Compounds. Irvine, California. September 16-17, 1992. (Proceedings.) *Environ. Health Perspect.* **1994**, *102 (Suppl. 7)*, 1-141.
94MI119	D. Gabel *Radiother. Oncol.* **1994**, *30*, 199-205.
94MI621	W.H. Scouten, X.C. Liu, N. Khangin, D.F. Mullica, E.L. Sappenfield *J. Chem. Crystallogr.* **1994**, *24*, 621-626.
94MI771	M.M. Kater, G.M. Koningstein, H.J.J. Nijkamp, A.R. Stuitje *Plant Mol. Biol.* **1994**, *25*, 771-790.
94MI849	N.M. Goudgaon, G.F. El-Kattan, R.F. Schinazi *Nucleosides Nucleotides* **1994**, *13*, 849-880.
94MI1937	H. Bergler, D. Abraham, H. Aschauer, F. Turnowsky *Microbiology* **1994**, *140*, 1937-1944.
94T12521	C. Morin *Tetrahedron* **1994**, *50*, 12521-12569.
95CB1037	C. Klöfkorn, M. Schmidt, T. Spaniol, T. Wagner, O. Costisor, P. Paetzold *Chem. Ber.* **1995**, *128*, 1037-1043.
95CRV2457	N. Miyaura, A. Suzuki *Chem. Rev.* **1995**, *95*, 2457-2483.
95JAN26	H. Irschik, D. Schummer, K. Gerth, G. Höfle, H. Reichenbach *J. Antibiot. (Tokyo)* **1995**, *48*, 26-30.
95MI48	K.M. Bradshaw, M.P. Schweizer, G.H. Glover, J.R. Hadley, R. Tippets, P.P. Tang, W.L. Davis, M.P. Heilbrun, S. Johnson, T. Ghanem *Magn. Reson. Med.* **1995**, *34*, 48-56.
95TL8399	R. Martin, J.B. Jones *Tetrahedron Lett.* **1995**, *36*, 8399-8402.
96AX(C)2826	P.D. Robinson, M.P. Groziak, L. Yi *Acta Crystallogr.* **1996**, *C52*, 2826-2830.
96AX(D)1181	C. Baldock, J.B. Rafferty, S.E. Sedelnikova, S. Bithell, A.R. Stuitje, A.R. Slabas, D.W. Rice *Acta Crystallogr.* **1996**, *D52*, 1181-1184.
96CHEC-II(6)1155	J.A. Soderquist, A.M. Rane, C.L. Anderson In *Comp. Heterocycl. Chem., 2nd edn.*; A.R. Katritzky, C.W. Reese, E.F.V. Scriven, Eds.; Pergamon Press: Oxford, 1996; Vol. 6, Chpt. 29, pp. 1155-1175.
96EJB689	H. Bergler, S. Fuchsbichler, G. Högenauer, F. Turnowsky *Eur. J. Biochem.* **1996**, *242*, 689-694.
96JOC4510	M.P. Hughes, M. Shang, B.D. Smith *J. Org. Chem.* **1996**, *61*, 4510-4511.
96MI108	X. Chen, K. Sastow, B. Goz, L. Kucera, S.L. Morris-Natschke, K.S. Ishaw *Antiviral Chem. Chemother.* **1996**, *7*, 108-114.
96MI688	N.C.J. Strynadka, R. Martin, S.E. Jensen, M. Gold, J.B. Jones *Nat. Struct. Biol.* **1996**, *3*, 688-695.
96MI1036	J. Kohno, T. Kawahata, T. Otake, M. Morimoto, H. Mori, N. Ueba, M. Nishio, A. Kinumaki, S. Komatsubara, K. Kawashima *Biosci. Biotechnol. Biochem.* **1996**, *60*, 1036-1037.

96MI2441	C.D. Hunt, B.J. Stoecker *J. Nutr.* **1996**, *126*, 2441S-2451S.
96POL3411	R.D. Pizer, C.A. Tihal *Polyhedron* **1996**, *15*, 3411-3416.
96SCI2107	C. Baldock, J.B. Rafferty, S.E. Sedelnikova, P.J. Baker, A.R. Stuitje, A.R. Slabas, T.R. Hawkes, D.W. Rice *Science* **1996**, *274*, 2107-2110.
97JA1529	K. Curley, R.F. Pratt *J. Am. Chem. Soc.* **1997**, *119*, 1529 -1538.
97JA7817	M.P. Groziak, L. Chen, L. Yi, P.D. Robinson *J. Am. Chem. Soc.* **1997**, *119*, 7817-7826.
97JBC27091	F. Wendler, H. Bergler, K. Prutej, H. Jungwirth, G. Zisser, K. Kuchler, G. Högenauer *J. Biol. Chem.* **1997**, *272*, 27091-27098.
97JOC4492	M.P. Hughes, B.D. Smith *J. Org. Chem.* **1997**, *62*, 4492-4499.
97MI41	S. Sjoberg, J. Carlsson, H. Ghaneolhosseini, L. Gedda, T. Hartman, J. Malmquist, C. Naeslund, P. Olsson, W. Tjarks *J. Neurooncol.* **1997**, *33*, 41-52.
97MI65	T. Kar, D.E. Elmore, S. Scheiner *Theochem J. Mol. Struct.* **1997**, *392*, 65-74.
97MI153	G.W. Kabalka, C. Tang, P. Bendel *J. Neurooncol.* **1997**, *33*, 153-161.
98AX(C)71	P.D. Robinson, M.P. Groziak, L. Chen *Acta Crystallogr.* **1998**, *C54*, 71-73.
98BMCL843	M.C. Davis, S.G. Franzblau, A.R. Martin *Bioorg. Med. Chem. Lett.* **1998**, *8*, 843-846.
98BP1541	C. Baldock, G.J. de Boer, J.B. Rafferty, A.R. Stuitje, D.W. Rice *Biochem. Pharmacol.* **1998**, *55*, 1541-1549.
98CR1515	A.H. Soloway, W. Tjarks, B.A. Barnum, F.-G. Rong, R.F. Barth, I.M. Codogni, J.G. Wilson *Chem. Rev.* **1998**, *98*, 1515-1562.
98HAC399	N. Andrade-López, R. Cartas-Rosado, E. García-Baéz, R. Contreras, H. Tlahuext *Heteroat. Chem.* **1998**, *9*, 399-409.
98JHC887	S.M. Graham, L.M. Ohrtman *J. Heterocycl. Chem.* **1998**, *35*, 887-890.
98MI1-01	Proceedings of the 2nd International Symposium on the Health Effects of Boron and its Compounds. Irvine, California. October 22-24, 1997. *Biol. Trace Elem. Res.* **1998**, *66 (Nos. 1-3)*, 1-473.
98MI1-02	P.A. Fail, R.E. Chapin, C.J. Price, J.J. Heindel *Reprod. Toxicol.* **1998**, *12*, 1-18.
98MI2	M. Benderdour, T. Bui-Van, A. Dicko, F. Belleville *J. Trace Elem. Med. Biol.* **1998**, *12*, 2-7.
98MI174	M.F. Hawthorne *Mol. Med. Today* **1998**, *4*, 174-181.
98NAT531	L.M. McMurry, M. Oethinger, S.B. Levy *Nature* **1998**, *394*, 531-532.
98T14913	I.D. Madura, T.M. Krygowski, M.K. Cyranski *Tetrahedron* **1998**, *54*, 14913-14918.
99AX(C)1701	P.D. Robinson, M.P. Groziak *Acta Crystallogr.* **1999**, *C54*, 1701-1704.
99JA8393	M. Berger, J. Mulzer *J. Am. Chem. Soc.* **1999**, *121*, 8393-8394.
99JA11204	B. Chiavarino, M.E. Crestoni, A. Di Marzio, S. Fornarini, M. Rosi *J. Am. Chem. Soc.* **1999**, *121*, 11204-11210.
99JBC11110	R.J. Heath, J.R. Rubin, D.R. Holland, E. Zhang, M.E. Snow, C.O. Rock *J. Biol. Chem.* **1999**, *274*, 11110-11114.
99JBC30811	A. Roujeinikova, S. Sedelnikova, G.-J. de Boer, A.R. Stuitje, A.R. Slabas, J.B. Rafferty, D.W. Rice *J. Biol. Chem.* **1999**, *274*, 30811-30817.
99JMB527	A. Roujeinikova, C.W. Levy, S. Rowsell, S. Sedelnikova, P.J. Baker, C.A. Minshull, A. Mistry, J.G. Colls, R. Camble, A.R. Stuitje, A.R. Slabas, J.B. Rafferty, R.A. Pauptit, R. Viner, D.W. Rice *J. Mol. Biol.* **1999**, *294*, 527-535.
99JMB859	M.J. Stewart, S. Parikh, G. Xiao, P.J. Tonge, C. Kisker *J. Mol. Biol.* **1999**, *290*, 859-865.
99JOC9566	J.-C. Zhuo, A.H. Soloway, J.C. Beeson, W. Ji, B.A. Barnum, F.-G. Rong, W. Tjarks, G.T. Jordan IV, J. Liu, S.G. Shore *J. Org. Chem.* **1999**, *64*, 9655-9574.
99JPC(A)2141	R.J. Doerksen, A.J. Thakkar *J. Phys. Chem. A* **1999**, *103*, 2141-2151.
99MI1	J.A. Coderre, G.M. Morris *Radiat. Res.* **1999**, *151*, 1-18.
99MI335	C.J. Rainey, L.A. Nyquist, R.E. Christensen, P.L. Strong, B.D. Culver, J.R. Coughlin *J. Am. Diet. Assoc.* **1999**, *99*, 335-340.
99MI443	G.J. deBoer, G.J.A. Pielage, H.J.J. Nijkamp, A.R. Slabas, J.B. Rafferty, C. Baldock, D.W. Rice, A.R. Stuitje *Mol. Microbiol.* **1999**, *31*, 443-450.
99MI761	D.E. Callahan, T.M. Forte, S.M. Afzal, D.F. Deen, S.B. Kahl, K.A. Bjornstad, W.F. Bauer, E.A. Blakely *Int. J. Radiat. Oncol. Biol. Phys.* **1999**, *45*, 761-771.
99T2857	S.J. Gardiner, B.D. Smith, P.J. Duggan, M.J. Karpa, G.J. Griffin *Tetrahedron* **1999**, *55*, 2857-2864.

Chapter 2

Heterocyclic Phosphorus Ylides

R. Alan Aitken and Tracy Massil
University of St. Andrews, UK
(e-mail: raa@st-and.ac.uk)

Heterocyclic phosphorus ylides are a rather diverse and little known class of compounds. A variety of such structures are now known and in some cases these are of considerable synthetic value. In this chapter we have attempted to review all heterocyclic compounds containing one or more exocyclic phosphorus ylide functions, i.e. of general structure **1**. It should be noted that in many cases these exist predominantly in the phosphonium ylide (P^+—C^-) form but for simplicity they are represented in the ylene form **1**. Cyclic ylides **2** and **3** in which the phosphorus atom is within the ring are not included.

2.1 THREE-MEMBERED RINGS

A rare example of this type is the metallacycle **4** which is formed by reaction of the carbodiphosphorane $Ph_3P{=}C{=}PPh_3$ with $(Ph_3P)_2Pt{\cdot}CH_2{=}CH_2$ <73JOM(47)391>.

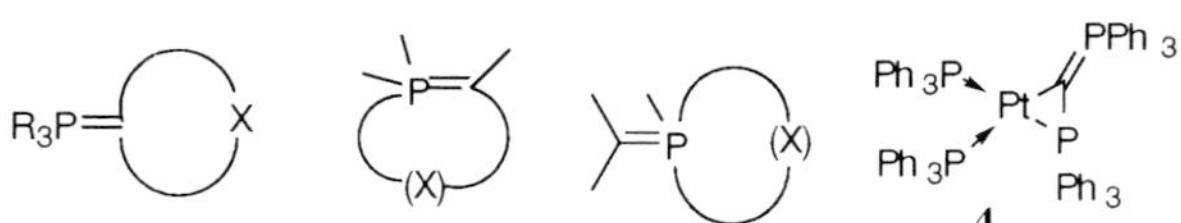

1 **2** **3** **4**

2.2 FOUR-MEMBERED RINGS WITH ONE HETEROATOM

Many of the compounds of this type have been prepared by interaction of cumulene ylides with heterocumulenes and these are covered in a review by Bestmann <77AG(E)349>. The oxetane system **5** results from reaction of $Ph_3P{=}C{=}C{=}S$ with aryl isocyanates, ArNCO <77AG(E)349>. An unusual approach to an oxetane ylide is the reaction of DMAD with CO_2 and two equivalents of triethyl phosphite which gives **6** <81CC607>.

The reaction of $Ph_3P{=}C{=}C{=}O$ with CS_2 results in O/S exchange to give $Ph_3P{=}C{=}C{=}S$ presumably by loss of COS from the thietane ylide **7** <68JA3842>. Treatment of $Ph_3P{=}C{=}C{=}S$ with aryl isothiocyanates, ArNCS gives the isolable thietane ylides **8** <77AG(E)349>. In a similar way, $Ph_3P{=}C{=}C{=}NPh$ reacts with COS, CS_2 <77TL3037> and RNCS <77AG(E)349> to give the products **9**, **10** and **11** respectively. Reaction of

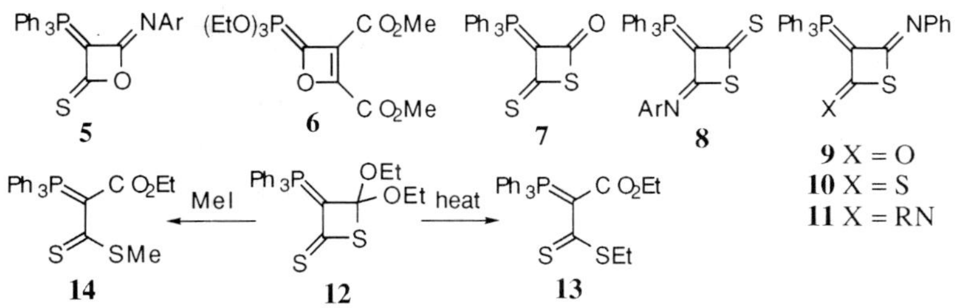

$Ph_3P{=}C{=}C(OEt)_2$ with CS_2 gives the thietane ylide **12** which has a rather interesting reactivity. Upon heating at 100 °C it rearranges to **13** while treatment with methyl iodide results in loss of ethyl iodide to give **14** <79JCR(S)313>.

The reaction of $Ph_3P{=}C{=}C{=}NPh$ with CO_2 takes an unexpected course: addition to give **15** is followed by ring-opening to give **16** which then undergoes ring closure to afford the azetidinedione system **17** <77TL3037>. The same cumulene ylide, $Ph_3P{=}C{=}C{=}NPh$ reacts with isocyanates to give **18** and with the imine ArCH=NAr to give **19** <77AG(E)349>. The formation of **18** with isocyanates is to be contrasted with the formation of **11** with isothiocyanates. Treatment of $Ph_3P{=}C{=}C{=}O$ with $PhSO_2NCO$ also affords access to an azetidinedione ylide **20** <77AG(E)349>.

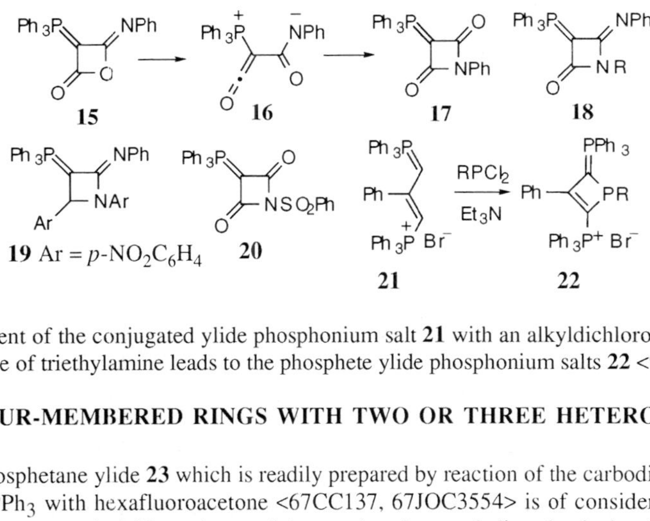

Treatment of the conjugated ylide phosphonium salt **21** with an alkyldichlorophosphine in the presence of triethylamine leads to the phosphete ylide phosphonium salts **22** <96CEJ221>.

2.3 FOUR-MEMBERED RINGS WITH TWO OR THREE HETEROATOMS

The oxaphosphetane ylide **23** which is readily prepared by reaction of the carbodiphosphorane $Ph_3P{=}C{=}PPh_3$ with hexafluoroacetone <67CC137, 67JOC3554> is of considerable interest since it represents an isolable analogue of the oxaphosphetanes believed to be key intermediates in the Wittig reaction. Its X-ray structure has been determined <68JCS(A)568> and as expected it loses Ph_3PO upon heating at 110 °C to give $Ph_3P{=}C{=}C(CF_3)_2$.

The bis(ylide) **25** may be prepared either by treatment of **24** with tri-*N*-morpholinylphosphine <77AG(E)402> or from the dichlorine adduct of **24** and $(Me_2N)_3P$ <78CB2054>. The similar compound **26** is formed both by dimerisation of $Ph_3P{=}C(SiMe_3)PCl_2$ and by reaction of $Ph_3P{=}C(SiMe_3)_2$ with $Ph_3P{=}C(PCl_2)_2$ <95AG(E)1853>. Its X-ray structure has been determined and it reacts with $AlCl_3$ to give **27** and with $AlCl_3$ and Ph_3P to give **28**.

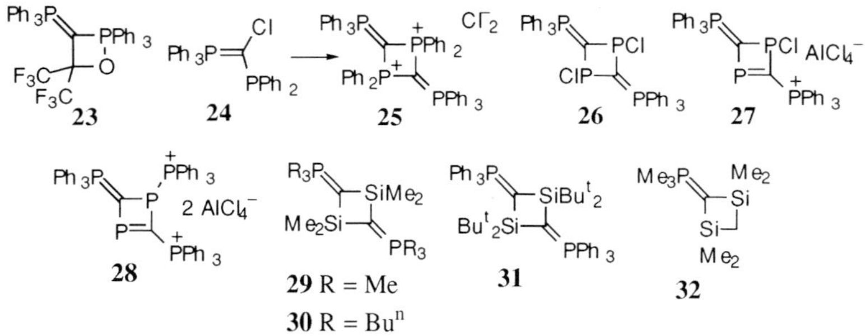

23 24 25 26 27 28 29 R = Me 30 R = Bu^n 31 32

The silicon-containing bis(ylides) **29** and **30** are formed by treatment of Me_2SiCl_2 with an excess of $R_3P{=}CH_2$ <70CB97> and reaction of $Ph_3P{=}CH_2$ with $Bu^t{}_2SiCl_2$ similarly affords **31** <87CB789>. The disilacyclobutane ylide **32** is formed by reaction of dilithiated $Me_3P{=}CH_2$ with $ClMe_2SiCH_2SiMe_2Cl$ <78CB2696>.

A similar approach has allowed preparation of a range of titanium-containing bis(ylides). Thus treatment of $(Me_2N)_2TiCl_2$ with 3 equivalents of $Me_3P{=}CH_2$ gives compound **33** <77ZN(B)858> while reaction of $(Et_2N)_3P{=}CH_2$ and $TiCl_4$ in a 3:2 ratio gives **34** <86AG(E)574>. Reaction of $(Me_2N)_3P{=}CH_2$ and $TiCl_2(NMe_2)_2$ in a 3:2 ratio gives **35** while the same reactants in a 1:1 ratio in the presence of 2,4,6-collidine give **36** <93AG(E)554>. Treatment of tetraneopentylzirconium, $Zr(CH_2Bu^t)_4$, with 3 equivalents of $Me_3P{=}CH_2$ gives the unusual compound **37** <83OM154> while reaction of $Me_3P{=}C(PMe_2)_2$ with $(Me_3P)_4Co^+$ PF_6^- affords the salt **38** <82CB1956>. X-ray structures have been determined for **34**, **35** and **37**.

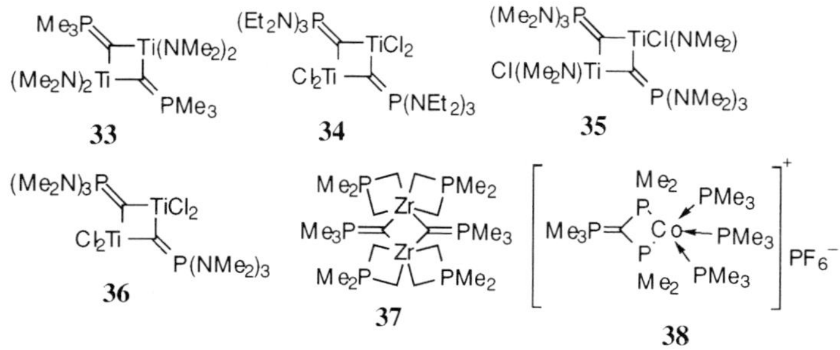

33 34 35 36 37 38

2.4 FIVE-MEMBERED RINGS WITH ONE HETEROATOM

The first ever heterocyclic ylide was reported by Schönberg and Ismail in 1940 from the reaction of maleic anhydride with triphenylphosphine <40JCS1374> although they initially assigned it the dipolar structure **39** and it was not until the advent of IR spectroscopy <61ACS692> and NMR <63HCA2178> that its true structure **40** could be determined with certainty. More recently the structure has been confirmed by X-ray diffraction <93JCS(F)2391>. The formation of **40** is rationalised by a 1,2-proton shift in the initial adduct **41**. The compound is considerably stabilised by being a β-oxo ylide and exists to a significant extent in the phosphonium enolate form **42**. The related compounds **43** <64JOC3721> and **44** <68T2241> have also been described. Wittig reaction of the ylide derived from **40** by ring-opening with ethanol followed by a series of reduction and dehydration steps provides a useful synthesis of 3-substituted furans <77TL2869>. The isomeric ylide **45** has been prepared by treament of $Ph_3P{=}C(CO_2R)COCH_2Cl$ with acetic acid followed by heating the resulting solid complex at its melting point <67HCA1016>.

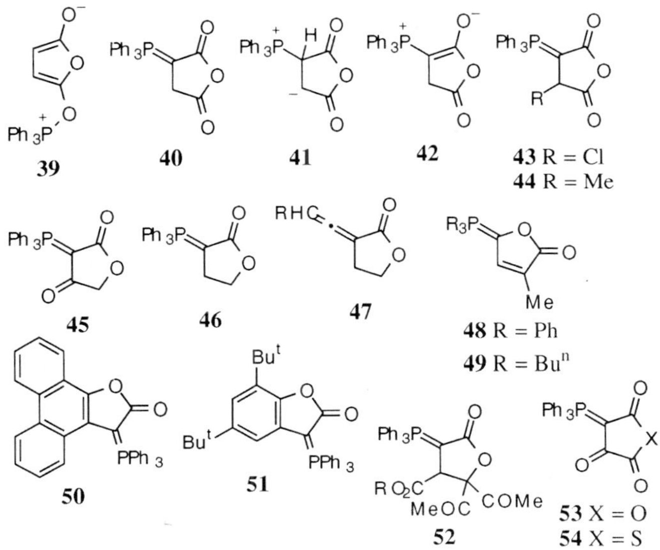

Reaction of α-bromo-γ-butyrolactone with triphenyphosphine followed by base affords the monooxo ylide **46** <63HCA1580> and the Wittig reaction of this has proved to be of some synthetic value. Thus, reaction with ketenes gives the allene products **47** <80HCA1204, 90TL4859> while reaction with an appropriate aldehyde has been used to prepare nucleoside analogues <83JOC1982> and other pharmaceutically active compounds <82EUP57882> containing the α-alkylidene-γ-butyrolactone function.

The unsaturated ylide **48** has been prepared and shown to undergo Wittig reactions <71TL4873> and the Wittig reaction of both it and its tributyl analogue **49** has been used in synthesis of a natural product analogue <74AJC1491>.

Ylides such as **50** and **51** may be prepared from the corresponding *ortho*-quinones by treatment with either $Ph_3P{=}C{=}C(OEt)_2$ <80CB2950> or $Ph_3P{=}CHCO_2Et$ <92JCS(P1)283> followed by Ph_3P. In a recent report, Yavari has described the reaction of dialkyl acetylenedicarboxylates, $RO_2C{-}C{\equiv}C{-}CO_2R$, with 3-chloropentane-2,4-dione and Ph_3P to afford the cyclic ylides **52** <97TL4259>. Finally in this section, the trioxo ylide **53** is readily prepared from dichloromaleic anhydride and Ph_3P either in wet THF or acetic anhydride (with loss respectively of 2 HCl or 2 AcCl) <88S782> and the thio analogue **54** has been similarly prepared from dichlorothiomaleic anhydride and its X-ray structure determined <97MI44>.

Reaction of maleimide and *N*-substituted derivatives with Ph_3P takes the same course as for maleic anhydride to give the heterocyclic ylides **55** <68T2241>. These compounds undergo a Wittig reaction with aldehydes and this has found use in total synthesis of the natural product showdomycin **56** <69CAR(11)574, 70CAR(14)123, 86JOC495>. It is interesting to note that reaction of the "isomaleimides" **57** with Ph_3P is accompanied by a rearrangement to give the same products **55** formed from the maleimides and this approach has allowed preparation of **58** from the dimeric isomaleimide <68T2241>.

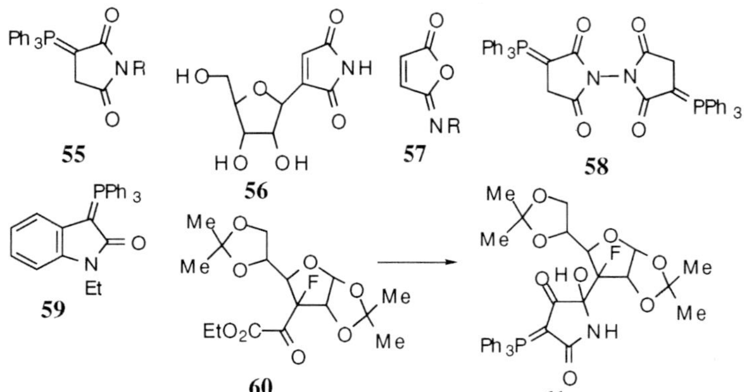

55 56 57 58 59 60 61

A variety of other nitrogen-containing ylides have been prepared by various routes. Thus **59** is formed by treating the corresponding indolone with Ph_3PBr_2 <69JHC265> while reaction of the carbohydrate derived α-ketoester **60** with $Ph_3P{=}CHCONH_2$ gives not only the expected Wittig product but also the cyclic ylide **61** whose X-ray structure was determined <78MI7>.

In our own laboratory we have obtained chiral ylides with the pyrrolidine-2,4-dione (tetramic acid) ring structure **63** by thermally induced cyclisation of the amino acid derived ylides **62**. By using the proline derived ylide the bicyclic system **64** was similarly obtained <000UP1>. The cyano ylide **65** was also found to undergo cyclisation to some extent to give **66** <97TL953>. In the case of the ylide **67** derived from glutamic acid, thermolysis results in cyclisation in both the possible directions to give a mixture of the cyclic ylide **68** and the acyclic ylide **69**. The latter undergoes a further cyclisation upon prolonged storage to give the bicyclic system **70** <000UP1>.

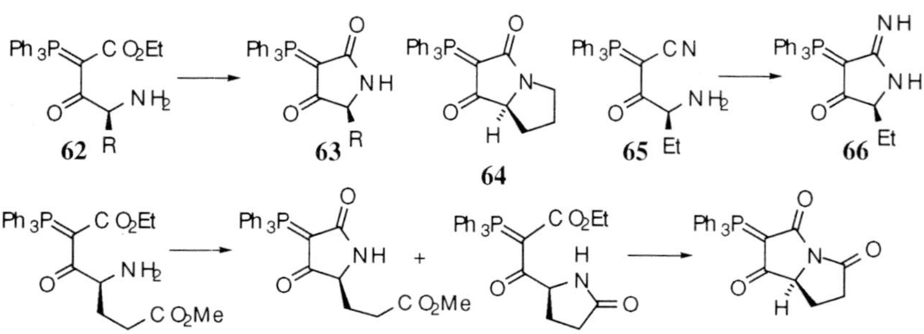

62 63 64 65 66 67 68 69 70

The aminonaphthoquinones **71** react with $Ph_3P{=}CHCO_2R^2$ to afford the cyclic ylides **72** <97T3649>. Reaction of the azo ester **73** with $Ph_3P{=}CHCO_2R$ proceeds with loss of ROH and cyclisation to give **74** whose X-ray structure is reported and a range of similar compounds have been prepared <90T5685>. An unusual example of ylide formation within the coordination sphere of a transition metal is provided by treatment of the isonitrile complex **75** with Et_3N to give **76** whose X-ray structure has been determined <88JCS(D)1803>.

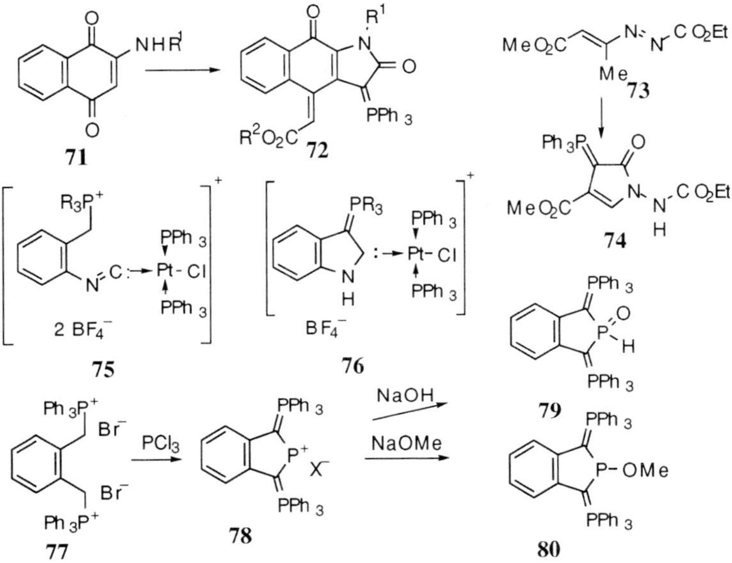

Treatment of the bis(phosphonium salt) **77** with PCl_3 gives the salt **78** which may be trapped with nucleophiles to give stable benzophosphole bis(ylides) such as **79** and **80** <91AG(E)308>.

2.5 FIVE-MEMBERED RINGS WITH TWO HETEROATOMS

The rather unstable dioxolane ylide **81** which may be generated from the corresponding phosphonium salt by treatment with DABCO and used for *in situ* Wittig reaction with aldehydes <84JCS(P1)1531> has proved particularly useful in carbohydrate chemistry <88TL4877>. The 1,3-dipolar cycloaddition of benzonitrile oxide to $Ph_3P{=}C{=}C(OEt)_2$ followed by treatment with aqueous HCl affords the isoxazole ylide **82** <77AG(E)349> while similar addition to $Ph_3P{=}C{=}C{=}NPh$ directly affords the ylide **83** <81TL1679>.

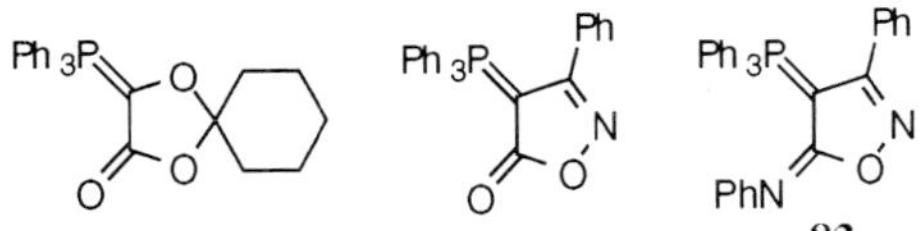

81 **82** **83**

A great deal of use has been made of five-membered ring ylides containing two sulfur atoms, particularly in the synthesis of tetrathiafulvalene (TTF) analogues, but no compound of this type has been isolated and they are always generated and used *in situ*. In 1971 Hartzler described the reaction of the stable tributylphosphine – CS_2 adduct **84** with methyl propiolate to give the ylide **85** which could be trapped by Wittig reaction with an added aldehyde to give the dithioles **86** <71JA4961>. The adducts **87** formed in a similar way from acetylenic acids were found to undergo intramolecular proton transfer to give the phosphonium carboxylates **88**

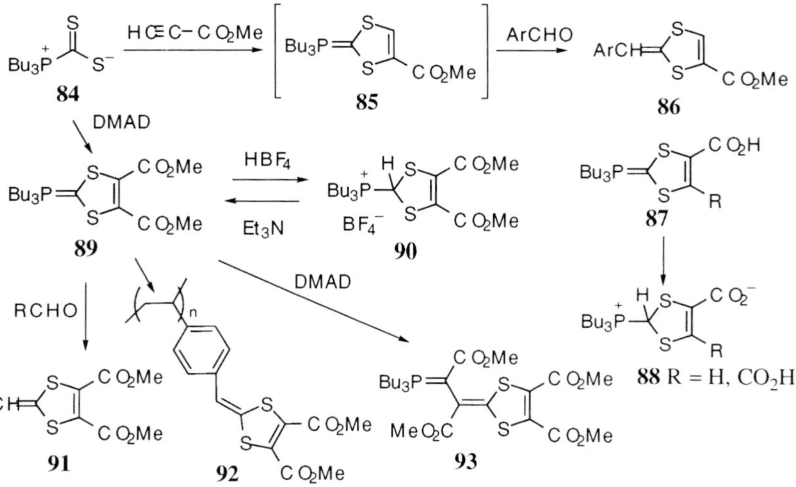

<75CC960>. Reaction of **84** with DMAD gives **89** and a most useful discovery was that this can be trapped by intermolecular protonation to give the stable fluoroborate **90** from which it can be conveniently regenerated by treatment with triethylamine <79JOC930>. This has allowed the widespread use of **89** to construct TTF analogues by reaction with appropriate functionalised aldehydes to give **91** <89CM535, 91TL2897, 92JA5035, 92TL6457, 93TL2131, 93TL4005, 97JOC2616>. Reaction with polymeric aldehydes to give **92** has also been described <80MM240>. Formation of **89** in the absence of any trap leads to reaction with a second molecule of DMAD to give the stable ylide **93** <92T8023>.

The desire to construct a wide range of TTF derivatives has led to the preparation and use of of ylides **94–111** shown below. These are generally prepared by base teatment of the corresponding phosphonium salts which may be obtained from dithiolium salts and the appropriate phosphine or a variety of other methods. Examples of the preparation and use of these ylides may be found as follows: **94** <78BCJ2674, 78JOC369, 91TL2897, 96T3171, 97CC787>, **95** <78JOC369>, **96** <78JOC369, 82TL1813, 91TL2897, 92JA5035, 92TL6457, 93TL7475>, **97** <81S53>, **98** <75JOC2577>, **99** and **106** <92JOC1696>, **100**

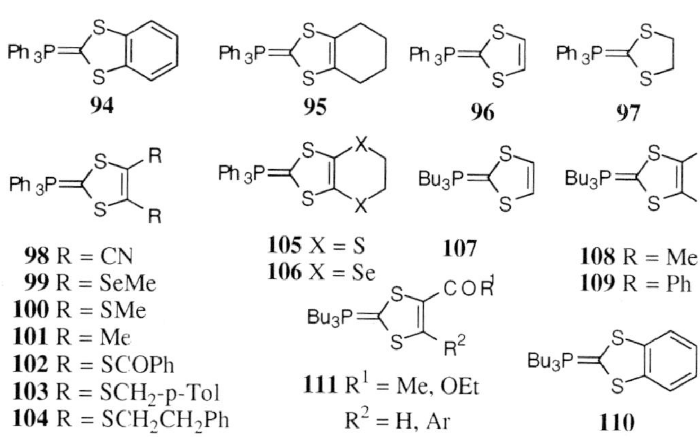

98 R = CN
99 R = SeMe
100 R = SMe
101 R = Me
102 R = SCOPh
103 R = SCH_2-p-Tol
104 R = SCH_2CH_2Ph

105 X = S
106 X = Se

108 R = Me
109 R = Ph

111 R^1 = Me, OEt
R^2 = H, Ar

and **101** <93TL7475>, **102** <94JOC5324>, **103** and **104** <96T3171, 97CC787>, **105** <92JA5035, 92TL6457, 93TL7475, 96T3171, 97CC787>, **107** <93TL4005>, **108** and **109** <89CM535> and **110** <89CM535, 93TL4005>. The various methods available for the preparation of these ylides are illustrated by the examples below.

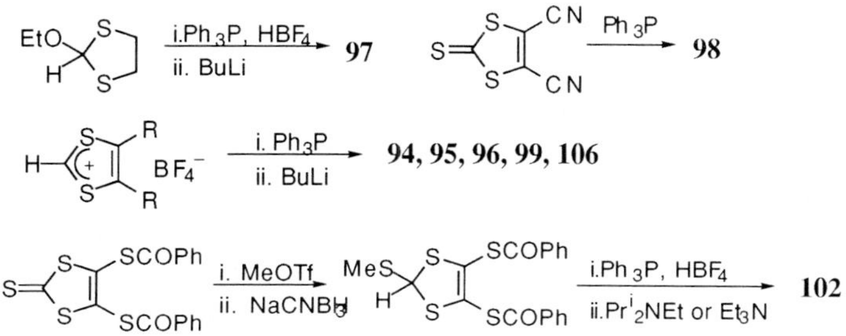

Although most applications involve Wittig reaction with suitable aldehydes, the reaction with nitroso compounds to give 2-imino-1,3-dithiol(an)es has been reported for **97** <81S53> and the unsymmetrical examples **111** <91JOC1816>. Reaction of the ylide with a dithiolium salt to give a TTF is also used in some cases <78JOC369, 82TL1813, 94JOC5324>. The rather unusual reaction of **94** with CS_2 to give the dibenzo-TTF, benzodithiole-2-thione, Ph_3PS and a further heterocyclic compound of uncertain structure has been reported <81BCJ2845>.

Addition of the phosphine – CS_2 adduct **84** to the strained double bond of norbornene initially gives **112** but this then reacts with a second molecule of CS_2 from dissociation of **84** to give the stable solid **113**. However in solution this is in equilibrium with **112** which can be trapped in a Wittig reaction with an added aldehyde. Application of this procedure to a variety of bicyclo[2.2.1]alkenes produces a range of alkylidenedithiolanes **114** <97T2261> and by using the bis(ylide) **115** similarly formed from norbornadiene, together with di- and tri-aldehydes, a range of oligomeric and polymeric dithiolanes were prepared <97T10441>.

The 1,3-diselenole ylide **116** is formed from DMAD, Bu_3P and CSe_2 in the same way as **89** and has also been used in the formation of TTF analogues <89CM535>.

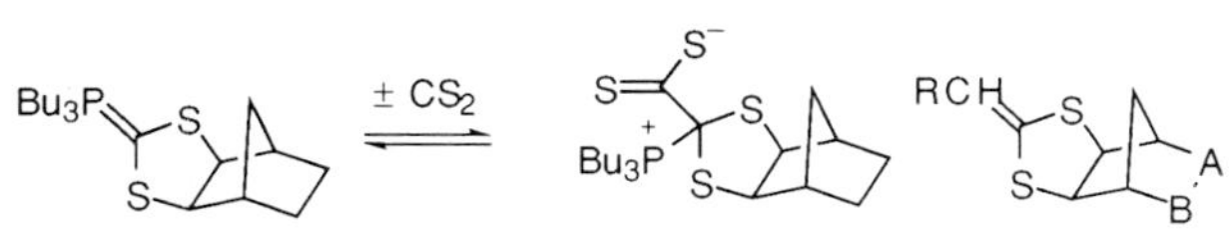

112 **113** **114**

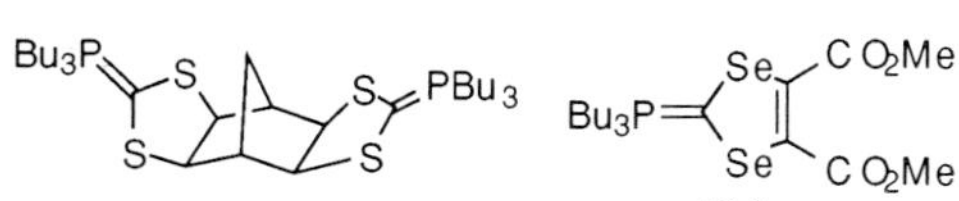

115 **116**

Reaction of the thiazolidinonethiones **117** with Ph_3PBr_2 gives the corresponding ylides **118** <69JHC265>. In a similar way a variety of pyrazolone ylides **120** have been prepared by treatment of the corresponding pyrazolones **119** with Ph_3PBr_2 <69JHC265> and the same type of compounds may alternatively be obtained from the azo compounds **121** by reaction with Ph_3P and then heating in methanol <92T1707>. Hydrolysis of **120** (R^2 = COR^3) to give **120** (R^2 = H) was also reported in the latter case. Further systems containing two nitrogens are

122 which is obtained by cycloaddition of ethyl diazoacetate to $Ph_3P{=}C{=}C{=}NPh$ <81TL1679> and **123** from cycloaddition of $(Pr^i{}_2N)_2P(Cl){=}C{=}N_2$ to tetracyanoethylene <87JA4711>.

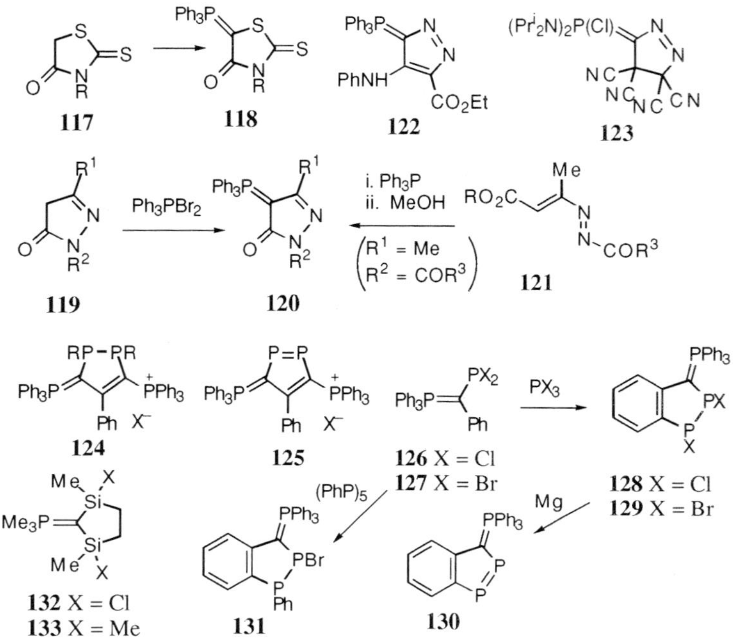

In section 2.2 the reaction of **21** with $RPCl_2$ and Et_3N to give the four-membered ring product **22** was described. If this is conducted in the presence of Ph_3P, the five-membered ring products **124** are instead formed and for R = Cl, subsequent treatment with Bu_3P affords the fully unsaturated diphosphole ylide **125** <96CEJ221>. Reaction of the dihalophosphinyl ylides **126** and **127** with the appropriate phosphorus trihalide gives the ylides **128** and **129** respectively <94AG(E)663>. Treatment of **129** with magnesium gives the benzodiphosphole ylide **130** while reaction of **127** with a cyclopentaphosphine affords **131**. The disilacyclopentane ylide **132** is formed by reaction of $Me_3P{=}C(SiMe_3)_2$ with $Cl_2MeSiCH_2CH_2SiMeCl_2$ and is readily converted into the fully methylated ylide **133** by reaction with MeMgI <78CB2696>.

2.6 FIVE-MEMBERED RINGS WITH THREE OR FOUR HETEROATOMS

Only a few examples of this type have been reported. Cycloaddition of $(Pr^i{}_2N)_2P(Cl){=}C{=}N_2$ to CS_2 gives the thiadiazole system **134** while with phenyl isocyanate the triazole ylide **135** is formed <87JA4711>. Cycloaddition of azides to $Ph_3P{=}C{=}C{=}NPh$

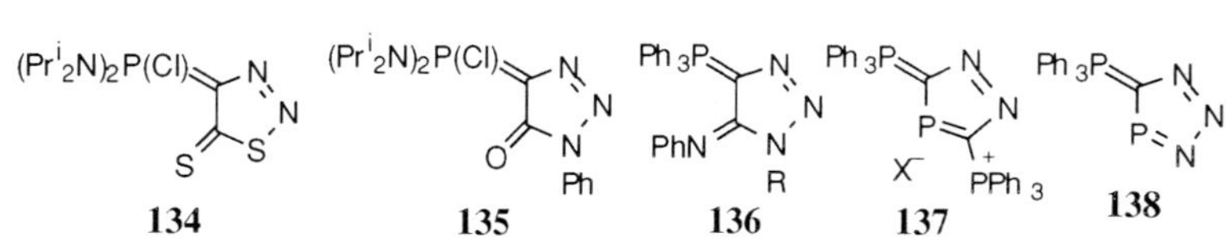

gives the stable triazole ylides **136** <81TL1679>. Treatment of either $Ph_3P{=}(SiMe_3)PCl_2$ with 0.5 equiv. of an azide or the cyclic ylide **27** mentioned earlier in section 2.3 with 1 equiv. of Me_3SiN_3 gives the diazaphosphole ylide salt **137** while the triazaphosphole system **138** results from reaction of either $Ph_3P{=}(SiMe_3)PCl_2$ with 1 equiv. of Me_3SiN_3 or **27** and 2 equiv. of Me_3SiN_3 <97CB89>.

2.7 SIX-MEMBERED RINGS WITH ONE HETEROATOM

The tetrahydropyran ylide **139** is readily formed by treatment of $BrCH_2CH_2OCH_2CH_2Br$ with two equiv. of $Ph_3P{=}CH_2$ <67AG(E)81, 69CB1802> and the bicyclic analogue **140** is similarly prepared from 2,6-bis(bromomethyl)tetrahydropyran. A variety of α,β-unsaturated ketones undergo Diels Alder cycloaddition with the central double bond of $Ph_3P{=}C{=}C{=}NPh$ to give the dihydropyran ylides **141**. More complex enones can also be used and in some cases the reaction is also successful for $Ph_3P{=}C{=}C{=}O$ leading to products such as **142** and **143** <84TL1441>. Similar pyran ylides are formed by using 1-(dimethylaminomethyl)-2-naphthol **144** as an *in situ* source of the *o*-quinone methide to give the products **145** and **146** from $Ph_3P{=}C{=}C{=}O$ and $Ph_3P{=}C{=}C{=}S$ respectively <77AG(E)349> and addition of the former to **147** to give **148** has also been described <68JOC4306>. The interaction of $Ph_3P{=}C{=}C{=}O$ with acid chlorides leads to a range of interesting di- and trioxo-pyran bis(ylides). Thus, reaction with aromatic acid chlorides followed by NaOMe gives **149** while aliphatic acid chlorides, $R^1R^2CH{-}COCl$ give products **150** and methyl chloroformate affords the trioxo compound **151** <80TL257>.

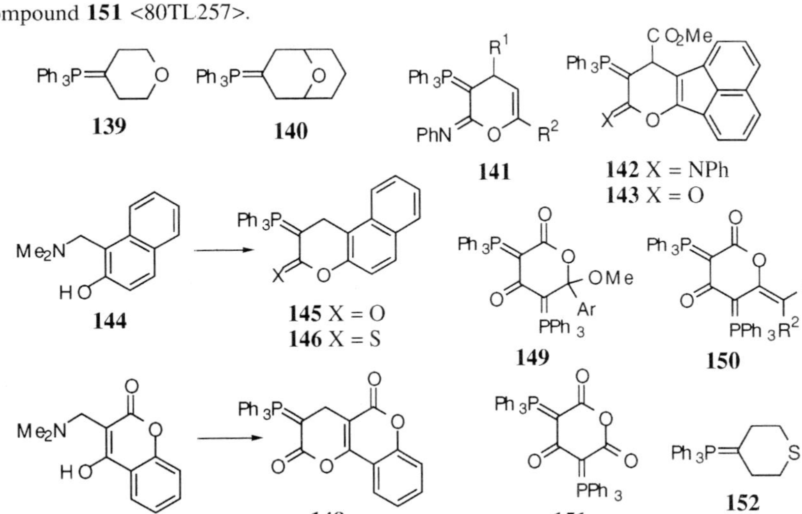

139 **140** **141** **142** X = NPh **143** X = O **144** **145** X = O **146** X = S **149** **150** **147** **148** **151** **152**

The sole example of a six-membered heterocyclic ylide containing one sulfur is the tetrahydrothiapyran compound **152** which is prepared from $BrCH_2CH_2SCH_2CH_2Br$ exactly as for the oxygen analogue **139** <67AG(E)81, 69CB1802>.

There have been relatively few papers describing six-membered heterocyclic ylides with one nitrogen but a good variety of structures have been obtained. The first approach involves conjugate addition of $Ph_3P{=}CHCONH_2$ followed by cyclisation, and treatment of this ylide

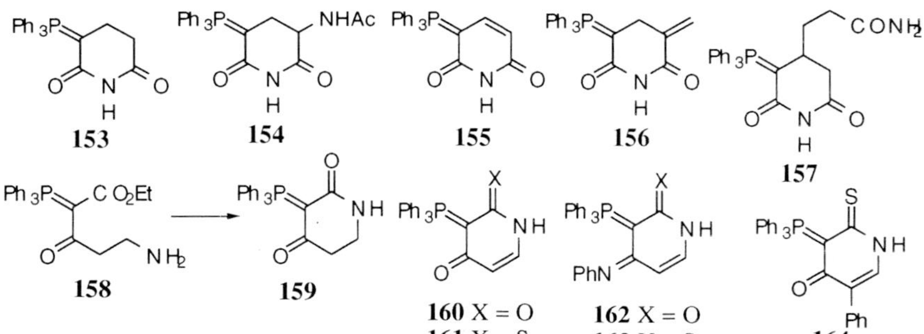

153 154 155 156 157 158 159 160 X = O 161 X = S 162 X = O 163 X = S 164

respectively with methyl acrylate, ethyl α-acetylaminoacrylate and methyl propiolate gives the products **153**, **154** and **155** <88S325>. By using methyl α-(bromomethyl)acrylate the exocylic methylene compound **156** could be obtained and use of the 3-methylenepiperidine-2,6-dione derived from hydrolysis as the electrophile gave the 4-substituted ylide **157** <92TL1513>. In our own laboratory we have obtained the doubly stabilised ylide **159**, isomeric with **153**, by thermal cyclisation of the β-alanine derived ylide **158** <000UP1>.

The other major approach to systems of this type is cycloaddition of cumulene ylides with vinyl isocyanate and vinyl isothiocyanate. Thus, $Ph_3P{=}C{=}C{=}O$ reacts with these two reagents to give **160** and **161** respectively while the corresponding reactions of $Ph_3P{=}C{=}C{=}NPh$ afford **162** and **163** <88T543>. By using styryl isothiocyanate with the ketene ylide, **164** was obtained.

2.8 SIX-MEMBERED RINGS WITH TWO OR THREE HETEROATOMS

Reaction of the 1,3-dithiane-2-ylidene tungsten pentacarbonyl complex **165** with methyldiphenylphosphine affords the tungsten complexed ylide **166** <82JOM(225)253>. As described in section 2.2, reaction of $Ph_3P{=}C{=}C{=}O$ with CS_2 or $PhSO_2NCO$ gives four-membered ring products but this ylide reacts with isothiocyanates in a 1:2 ratio to give the six-membered ring ylides **167** <77AG(E)349>.

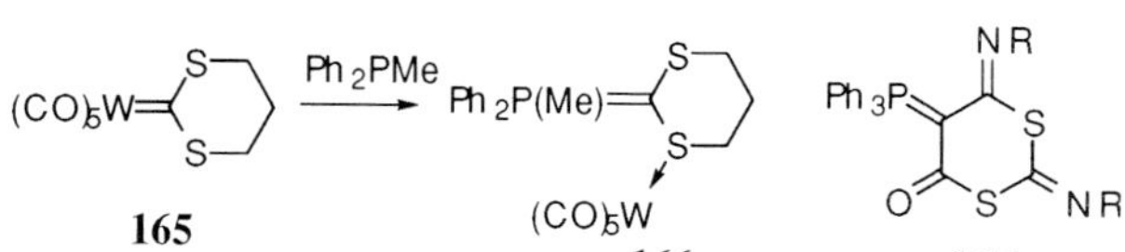

165 166 167

A range of heterocyclic ylides containing two ring nitrogen atoms are accessible by reaction of cumulene ylides with isocyanates and isothiocyanates. Thus, $Ph_3P{=}C{=}C{=}O$ reacts with isocyanates in a 1:2 ratio to give the products **168** <68JA3842, 77AG(E)249>, while $Ph_3P{=}C{=}C(OEt)_2$ reacts with both isocyanates and isothiocyanates to give products **169** and **170** respectively which may be hydrolysed by acid to give **171** and **172** <81CB2661>.

The reaction of $Ph_3P{=}CH_2$ with benzoyl isocyanate takes place in a 1:2 ratio with loss of benzene to afford the ylide **173** <71JOC2029> and treatment of either $Ph_3P{=}CHCO_2Me$ or $Ph_3P{=}CHCO_2Et$ with aryl cyanates, Ar–O–CN, results in formation of the pyrimidine ylides **174** <67CB187>. An alternative means of access to **171** (R = cyclohexyl) is provided by

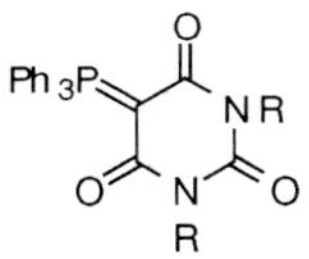

168

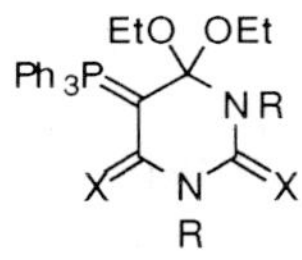

169 X = O
170 X = S

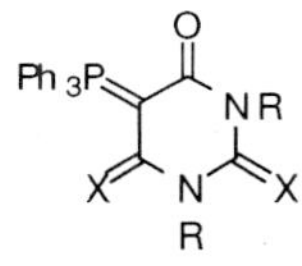

171 X = O
172 X = S

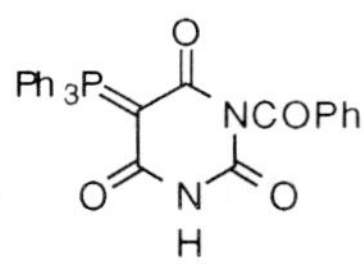

173

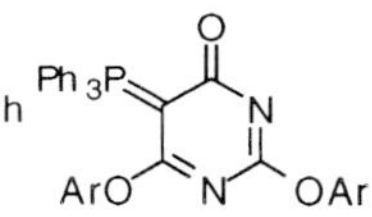

174

reaction of the corresponding barbituric acid with Ph_3PBr_2 <69JHC265>. Reaction of $Ph_3P{=}CHCN$ with acyl isocyanates and isothiocyanates followed by cyclisation and base treatment provides access to **175** and **176** (R = Me, Ph) and **177** has also been obtained <96T8835>. The ylides **175** undergo *N*-alkylation to give **178** whereas both **176** and **177** undergo *S*-alkylation to afford **179**. X-Ray structures of **175**, **176** and **179** (R = Me, X = O) have been determined.

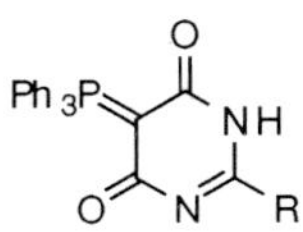

175

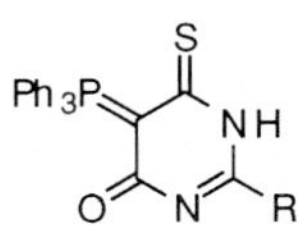

176

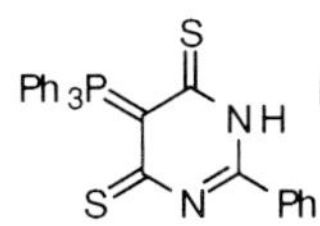

177

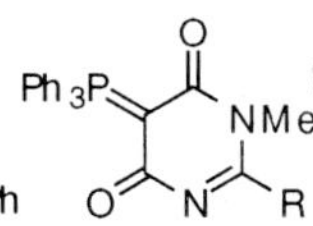

178

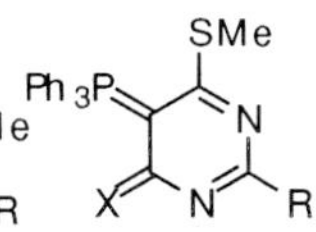

179

The reaction of either $Ph_3P{=}C(SiMe_3)PCl_2$ or $Ph_3P{=}C(SiMe_3)_2$ and $Ph_3P{=}C(PCl_2)_2$ gives not only the four-membered ring product **27** already mentioned in section 2.3 but also the six-membered ring product **180** and the bicyclic structure **181** whose X-ray structure has been determined <95AG(E)1853>. The trisilacyclohexane ylide **182** is formed by treatment of $Me_3P{=}C(SiMe_3)_2$ with $Me_2Si(CH_2SiMe_2Cl)_2$ <78CB2696>.

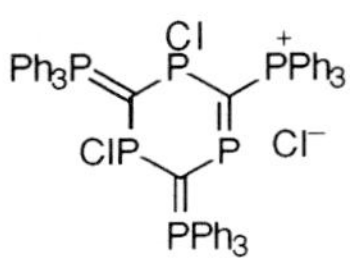

180

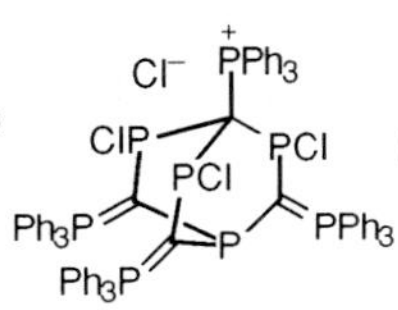

181

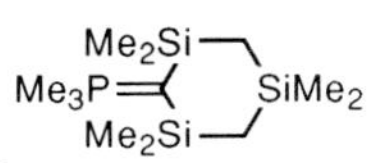

182

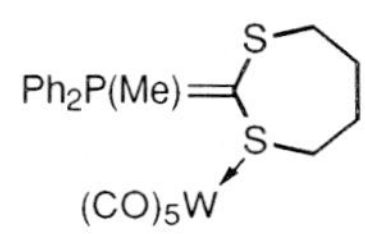

183

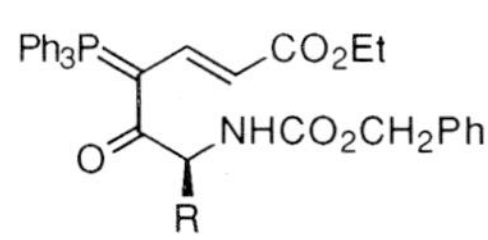

186

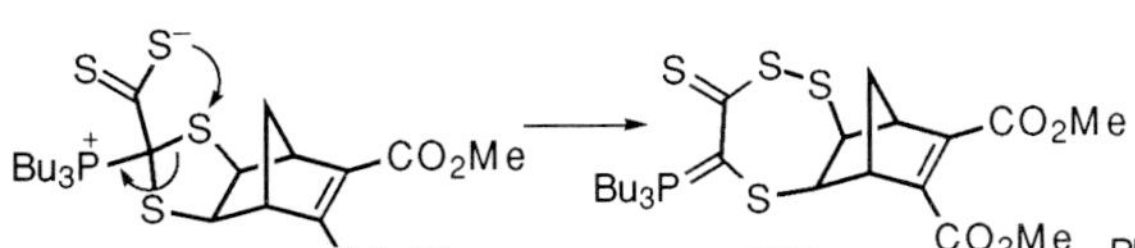

184 **185**

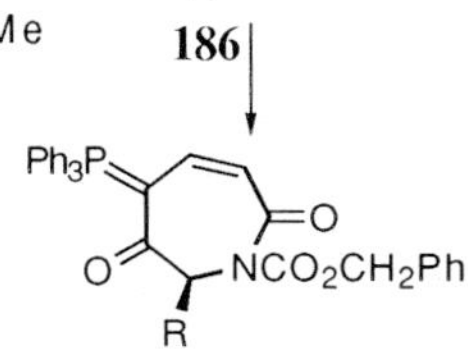

187

2.9 SEVEN-MEMBERED RINGS

There are only very few systems of this type known. The ylide complex **183** is formed in the same way as for the six-membered ring analogue **166** <82JOM(225)253>. The zwitterionic structure **184** which is formed as described in section 2.5 by addition of $Bu_3P{\bullet}CS_2$ **84** to the norbornadiene diester undergoes rearrangement at room temperature to afford the novel 1,2,5-trithiepane stabilised ylide **185** <99TL1061>. Finally we have recently obtained the interesting chiral azepinedione ylides **187** by thermal cyclisation of the amino acid derived ylides **186** <000UP1>. This is remarkable since, in contrast to the cyclisation of **62** or **158** mentioned earlier, it occurs while the nitrogen is still protected as the carbamate and also requires *E–Z* isomerisation of the double bond.

As far as we are aware no larger ring heterocyclic phosphorus ylides have yet been described.

2.10 REFERENCES

40JCS1374 A. Schönberg and A. F. A. Ismail, *J. Chem. Soc.*, **1940**, 1374.
61ACS692 G. Aksnes, *Acta Chem. Scand.*, **1961**, *15*, 692.
63HCA1580 S. Fliszár, R. F. Hudson and G. Salvadori, *Helv. Chim. Acta*, **1963**, *46*, 1580.
63HCA2178 R. F. Hudson and P. A. Chopard, *Helv. Chim. Acta*, **1963**, *46*, 2178.
64JOC3721 C. Osuch, E. Franz and F. B. Zienty, *J. Org. Chem.*, **1964**, *29*, 3721.
67AG(E)81 H. J. Bestmann and E. Kranz, *Angew. Chem., Int. Ed. Engl.*, **1967**, *6*, 81.
67CB187 M. Dieter and H.-J. Niclas, *Chem. Ber.*, **1967**, *100*, 187.
67CC137 G. H. Birum and C. N. Matthews, *J. Chem. Soc., Chem. Commun.*, **1967**, 137.
67HCA1016 P. A. Chopard, *Helv. Chim. Acta*, **1967**, *50*, 1016.
67JOC3554 G. H. Birum and C. N. Matthews, *J. Org. Chem.*, **1967**, *32*, 3554.
68JA3842 G. H. Birum and C. N. Matthews, *J. Am. Chem. Soc.*, **1968**, *90*, 3842.
68JCS(A)568 G. Chioccola and J. J. Daly, *J. Chem. Soc. (A)*, **1968**, 568.
68JOC2993 S. O. Grim and J. H. Ambrus, *J. Org. Chem.*, **1968**, *33*, 2993.
68JOC4306 M. von Strandtmann, M. P. Cohen, C. Puchalski and J. Shavel, Jr., *J. Org. Chem.*, **1968**, *33*, 4306.
68T2241 E. Hedaya and S. Theodoropulos, *Tetrahedron*, **1968**, *24*, 2241.
69CAR(11)574 R. E. Harmon, G. Wellman and S. K. Gupta, *Carbohydr. Res.*, **1969**, *11*, 574.
69CB1802 H. J. Bestmann and E. Kranz, *Chem. Ber.*, **1969**, *102*, 1802.
69JHC265 J. J. Pappas and E. Gancher, *J. Heterocycl. Chem.*, **1969**, *6*, 265.
70CAR(14)123 R. E. Harmon, G. Wellman and S. K. Gupta, *Carbohydr. Res.*, **1970**, *14*, 123.
70CB97 H. Schmidbaur and W. Malisch, *Chem. Ber.*, **1970**, *103*, 97.
71JA4961 H. D. Hartzler, *J. Am. Chem. Soc.*, **1971**, *93*, 4961.
71JOC2029 Y. Ohshiro, Y. Mori, M. Komatsu and T. Agawa, *J. Org. Chem.*, **1971**, *36*, 2029.
71TL4873 J. E. T. Corrie, *Tetrahedron Lett.*, **1971**, 4873.
73JOM(47)391 W. C. Kaska, D. K. Mitchell and R. F. Reichelderfer, *J. Organomet. Chem.*, **1973**, *47*, 391.
74AJC1491 C. F. Ingham and R. A. Massy-Westropp, *Aust. J. Chem.*, **1974**, *27*, 1491.
75CC960 C. U. Pittman and M. Narita, *J. Chem. Soc., Chem. Commun.*, **1975**, 960.
75JOC2577 M. G. Miles, J. S. Wager, J. D. Wilson and A. R. Siedle, *J. Org. Chem.*, **1975**, *40*, 2577.
77AG(E)349 H. J. Bestmann, *Angew. Chem., Int. Ed. Engl.*, **1977**, *16*, 349.
77AG(E)402 R. Appel, F. Knoll and H.-D. Wihler, *Angew. Chem., Int. Ed. Engl.*, **1977**, *16*, 402.
77TL2869 J. E. McMurry and S. F. Donovan, *Tetrahedron Lett.*, **1977**, 2869.
77TL3037 H. J. Bestmann and G. Schmid, *Tetrahedron Lett.*, **1977**, 3037.

77ZN(B)858 H. Schmidbaur, W. Scharf and H.-J. Füller, *Z. Naturforsch., Teil B*, **1977**, *32*, 858.

78BCJ2674 K. Akiba, K. Ishikawa and N. Inamoto, *Bull. Chem. Soc. Jpn.*, **1978**, *51*, 2674.

78CB2054 R. Appel and H.-D. Wihler, *Chem. Ber.*, **1978**, *111*, 2054.

78CB2696 H. Schmidbaur and M. Heimann, *Chem. Ber.*, **1978**, *111*, 2696.

78JOC369 N. C. Gonnella and M. P. Cava, *J. Org. Chem.*, **1978**, *43*, 369.

78MI7 J. C. A. Boeyens, A. J. Brink and A. Jordaan, *S. Afr. J. Chem.*, **1978**, *31*, 7.

79JCR(S)313 H. J. Bestmann and R. W. Saalfrank, *J. Chem. Res. (S)*, **1979**, 313.

79JOC930 M. Sato, N. C. Gonnella and M. P. Cava, *J. Org. Chem.*, **1979**, *44*, 930.

80CB2950 R. W. Saalfrank, E. Ackermann, H. Winkler, W. Paul and R. Böhme, *Chem. Ber.*, **1980**, *113*, 2950.

80HCA1204 R. W. Lang and H.-J. Hansen, *Helv. Chim. Acta*, **1980**, *63*, 1204.

80MM240 J. E. Mulvaney and D. M. Chang, *Macromolecules*, **1980**, *13*, 240.

80TL257 H. J. Bestmann and C. Geismann, *Tetrahedron Lett.*, **1980**, *21*, 257.

81BCJ2845 J. Nakayama, S. Maruyama and M. Hoshino, *Bull. Chem. Soc. Jpn.*, **1981**, *54*, 2845.

81CB2661 H. J. Bestmann and R. W. Saalfrank, *Chem. Ber.*, **1981**, *114*, 2661.

81CC607 D. V. Griffiths and J. C. Tebby, *J. Chem. Soc., Chem. Commun.*, **1981**, 607.

81S53 S. Tanimoto, S. Jo and T. Sugimoto, *Synthesis*, **1981**, 53.

81TL1679 H. J. Bestmann and G. Schmid, *Tetrahedron Lett.*, **1981**, *22*, 1679.

82CB1956 H. H. Karsch, *Chem. Ber.*, **1982**, *115*, 1956.

82EUP57882 I. Katsumi, H. Kondo, K. Yamashita, T. Hidaka, K. Hosoe, Y. Ariki, T. Yamashita and K. Watanabe, *Eur. Pat.* 57 882 (1982) [*Chem. Abstr.*, **1983**, *98*, 34490].

82JOM(225)253 R. A. Pickering and R. J. Angelici, *J. Organomet. Chem.*, **1982**, *225*, 253.

82TL1813 J. M. Fabre, C. Galaine , L. Giral and D. Chasseau, *Tetrahedron Lett.*, **1982**, *23*, 1813.

83JOC1982 J. W. Lyga and J. A. Secrist, *J. Org. Chem.*, **1983**, *48*, 1982.

83OM154 G. W. Rice, G. B. Ansell, M. A. Modrick and S. Zentz, *Organometallics*, **1983**, *2*, 154.

84JCS(P1)1531 R. Ramage, G. J. Griffiths, F. E. Shutt and J. N. A. Sweeney, *J. Chem. Soc., Perkin Trans. 1*, **1984**, 1531.

84TL1441 H. J. Bestmann and G. Schmid, *Tetrahedron Lett.*, **1984**, *25*, 1441.

86AG(E)574 H. Schmidbaur, R. Pichl and G. Müller, *Angew. Chem., Int. Ed. Engl.*, **1986**, *25*, 574.

86JOC495 A. G. M. Barrett, H. B. Broughton, S. V. Attwood and A. A. L. Gunatilaka, *J. Org. Chem.*, **1986**, *51*, 495.

87CB789 H. Schmidbaur, R. Pichl and G. Müller, *Chem. Ber.*, **1987**, *120*, 789.

87JA4711 J.-M. Sotiropoulos, A. Baceiredo and G. Bertrand, *J. Am. Chem. Soc.*, **1987**, *109*, 4711.

88JCS(D)1803 R. A. Michelin, M. Mozzon, G. Facchin, D. Braga and P. Sabatino, *J. Chem. Soc., Dalton Trans.*, **1988**, 1803.

88S325 M. J. Wanner and G. J. Koomen, *Synthesis*, **1988**, 325.

88S782 A. H. Schmidt, W. Goldberger, M. Dümmler and A. Aiméne, *Synthesis*, **1988**, 782

88T543 L. Kniezo, P. Kristian, J. Imrich, F. Ugozzoli and G. D. Andreetti, *Tetrahedron*, **1988**, *44*, 543.

88TL4877 R. Ramage, G. W. Rose and A. M. MacLeod, *Tetrahedron Lett.*, **1988**, *29*, 4877

89CM535 T. Sugimoto, H. Awaji, I. Sugimoto, Y. Misaki, T. Kawase, S. Yoneda and Z. Yoshida, *Chem. Mater.*, **1989**, *1*, 535.

90T5685 O. A. Attanasi, P. Filippone, A. Mei, A. Bongini and E. Foresti, *Tetrahedron*, **1990**, *46*, 5685.

90TL4859 F. Fotiadu, A. Archavlis and G. Buono, *Tetrahedron Lett.*, **1990**, *31*, 4859.

91AG(E)308 A. Schmidpeter and M. Thiele, *Angew. Chem., Int. Ed. Engl.*, **1991**, *30*, 308.

91JOC1816 E. C. Taylor and R. Dötzer, *J. Org. Chem.*, **1991**, *56*, 1816.

91TL2897 M. Salle, A. Belyasmine, A. Gorgues, M. Jubault and N. Soyer, *Tetrahedron Lett.*, **1991**, *32*, 2897.

92JA5035	T. K. Hansen, M. V. Lakshmikantham, M. P. Cava, R. E. Niziurski-Mann, F. Jensen and J. Becher, *J. Am. Chem. Soc.*, **1992**, *114*, 5035.
92JCS(P1)283	D. N. Nicolaides, S. G. Adamopoulos, D. A. Lefkaditis, K. E. Litinas and P. V. Tarantili, *J. Chem. Soc., Perkin Trans. 1*, **1992**, 283.
92JOC1696	M. R. Bryce, M. A. Coffin and W. Clegg, *J. Org. Chem.*, **1992**, *57*, 1696.
92T1707	O. A. Attanasi, P. Filippone and A. Mei, *Tetrahedron*, **1992**, *48*, 1707.
92T8023	R. A. Aitken, S. V. Raut and G. Ferguson, *Tetrahedron*, **1992**, *48*, 8023.
92TL1513	M. J. Wanner and G.-J. Koomen, *Tetrahedron Lett.*, **1992**, *33*, 1513.
92TL6457	A. S. Benahmed-Gasmi, P. Frère, B. Garrigues, A. Gorgues, M. Jubault, R. Carlier and F. Texier, *Tetrahedron Lett.*, **1992**, *33*, 6457.
93AG(E)554	K. A. Hughes, P. G. Dopico, M. Sabat and M. G. Finn, *Angew. Chem., Int. Ed. Engl.*, **1993**, *32*, 554.
93JCS(F)2391	M. Geoffroy, G. Rao, Z. Tancic and G. Bernardinelli, *J. Chem. Soc., Faraday Trans.*, **1993**, *89*, 2391.
93TL2131	A. S. Benahmed-Gasmi, P. Frère, A. Belyasmine, K. M. A. Malik, M. B. Hursthouse, A. J. Moore, M. R. Bryce, M. Jubault and A. Gorgues, *Tetrahedron Lett.*, **1993**, *34*, 2131.
93TL4005	A. Belyasmine, P. Frère, A. Gorgues, M. Jubault, G. Duguay and P. Hudhomme, *Tetrahedron Lett.*, **1993**, *34*, 4005.
93TL7475	M. Sallé, A. J. Moore, M. R. Bryce and M. Jubault, *Tetrahedron Lett.*, **1993**, *34*, 7475.
94AG(E)663	G. Jochem, A. Schmidpeter, M. Thomann and H. Nöth, *Angew. Chem., Int. Ed. Engl.*, **1994**, *33*, 663.
94JOC5324	T. K. Hansen, M. R. Bryce, J. A. K. Howard and D. S. Yufit, *J. Org. Chem.*, **1994**, *59*, 5324.
95AG(E)1853	H.-P. Schrödel, G. Jochem, A. Schmidpeter and H. Nöth, *Angew. Chem., Int. Ed. Engl.*, **1995**, *34*, 1853.
96CEJ221	G. Jochem, A. Schmidpeter and H. Nöth, *Chem. Eur. J.*, **1996**, *2*, 221.
96T3171	R. P. Clausen and J. Becher, *Tetrahedron*, **1996**, *52*, 3171.
96T8835	L. Van Meervelt, O. B. Smolii, N. I. Mishchenko, D. B. Shakhnin, E. A. Romanenko and B. S. Drach, *Tetrahedron*, **1996**, *52*, 8835.
97CB89	H.-P. Schrödel and A. Schmidpeter, *Chem. Ber/Receuil*, **1997**, *130*, 89.
97CC787	L. Yu and D. Zhu, *Chem. Commun.*, **1997**, 787.
97JOC2616	Y. A. Jackson, J. P. Parakka, M. V. Lakshmikantham and M. P. Cava, *J. Org. Chem.*, **1997**, *62*, 2616.
97MI44	V. Bjørnstad, P. Frøyen, H. Hope and J. Skramstad, *Article 044, "Electronic Conference on Heterocyclic Chemistry '96"*, H. S. Rzepa, J. Snyder and C. Leach Eds., Royal Society of Chemistry, 1997.
97T2261	R. A. Aitken, K. Carcas, L. Hill, T. Massil and S. V. Raut, *Tetrahedron*, **1997**, *53*, 2261.
97T3649	L. S. Boulos and M. H. N. Arsanious, *Tetrhedron*, **1997**, *53*, 3649.
97T10441	R. A. Aitken, L. Hill, T. Massil, M. B. Hursthouse and K. M. A. Malik, *Tetrahedron*, **1997**, *53*, 10441.
97TL953	H. H. Wasserman and A. K. Petersen, *Tetrahedron Lett.*, **1997**, *38*, 953.
97TL4259	I. Yavari and R. Baharfar, *Tetrahedron Lett.*, **1997**, *38*, 4259.
99TL1061	R. A. Aitken, L. Hill and N. J. Wilson, *Tetrahedron Lett.*, **1999**, *40*, 1061.
000UP1	R. A. Aitken, G. M. Buchanan, N. Karodia and T. Massil, *Unpublished Results*, **2000**.

Chapter 3

Palladium Chemistry in Pyridine Alkaloid Synthesis

Jie Jack Li
Pfizer Global R&D, 2800 Plymouth Road, Ann Arbor, MI, U.S.A. e-mail: Jack.Li@wl.com

---

3.1 INTRODUCTION

Naturally occurring pyridine alkaloids are important for their biological and pharmacological properties. Palladium chemistry, during the last decade, has rapidly become an indispensable tool for synthetic chemists. In this chapter, applications of palladium chemistry in the total synthesis of naturally occurring pyridine alkaloids are highlighted. Important lessons can be learned while applying known methodology to their total synthesis. The exercise of total synthesis itself, in turn, will spur further interest in developing new methodology.

Pyridine is a π-electron deficient heterocycle. Due to the electronegativity of the nitrogen atom, the α and γ positions bear a partial positive charge, making the C(2), C(4), and C(6) positions prone to nucleophilic attack. A similar trend occurs in the context of palladium chemistry. The α and γ positions of halopyridines are more susceptible to oxidative addition to Pd(0) than are simple all-carbon aryl halides. Even 2-, 4-, and 6-chloropyridine are viable electrophilic substrates for Pd-catalyzed reactions under standard conditions.

3.2 THE NEGISHI COUPLING

Organozinc reagents exhibit greater functional group compatibility than organolithium and Grignard reagents. As a consequence, the Negishi coupling has found wide application in organic synthesis including pyridine alkaloid synthesis.

1-Halo-β-carbolines, readily prepared from tryptamine in three steps, are versatile building blocks. In the total synthesis of nitramarine (**3**), Negishi reaction of 1-bromo-β-carboline (**1**) was the key operation <94T12329>. Thus, 1-bromo-β-carboline (**1**) was sequentially treated with KH and *tert*-butyllithium to generate 1,9-dimetalated β-carboline **2**, which was then transmetalated with $ZnCl_2$ to the corresponding organozinc reagent. The Negishi coupling of the resulting β-carbolinylzinc reagent with 2-chloroquinoline provided a rapid entry to nitramarine (**3**). In another example, Quéguiner *et al.* synthesized eudistomin U (**8**) via a combination of an *ortho*-lithiation and a Negishi reaction <95TL7085>. Organozinc reagent **5** was prepared by regioselective *ortho*-lithiation of 3-fluoropyridine **4** with *n*-butyllithium followed by transmetalation of the resulting lithio species with zinc chloride. Subsequent Negishi coupling of **5** with 3-bromo-1-phenylsufonylindole (**6**) led to hetero triaryl **7**. Ultimately, refluxing **7** with pyridinium chloride followed by a basic workup delivered eudistomin U (**8**).

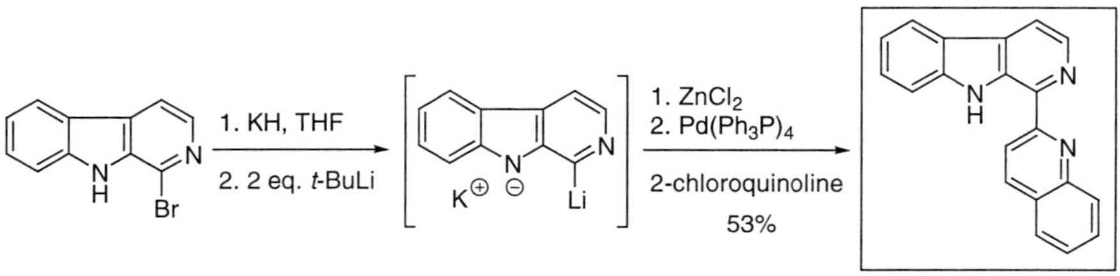

1 2 3

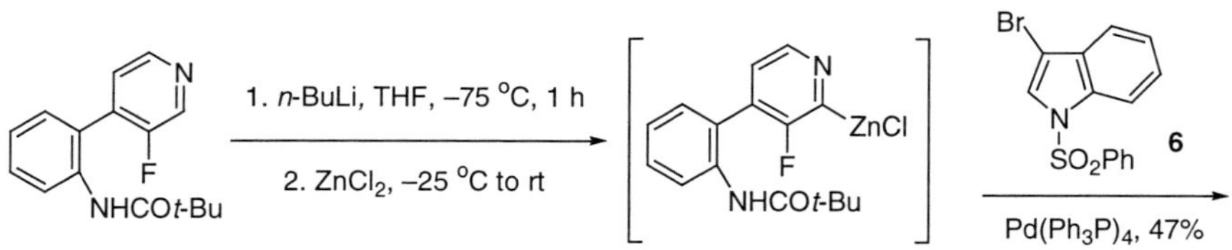

4 5

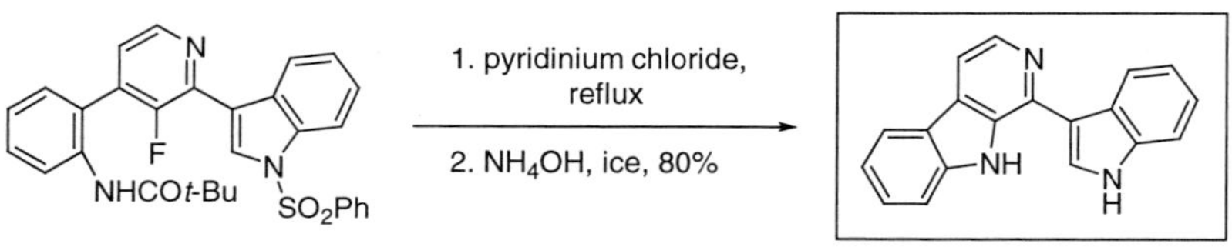

7 8

Quéguiner's preparation of pyridylzinc reagent **5** and its successful coupling is an admirable feat. Pyridyne formation could be a competing pathway at room temperature, although having a 3-fluoro substituent was much more stabilizing than the corresponding chloro- and iodo-analogs. Interchanging the coupling partners, i.e., using pyridyl halides to couple with indolylzinc reagents, may alleviate this problem. In fact, this was the strategy that Bosch's group adapted in their synthesis of pyridylindoles and subsequent total synthesis of (±)-nordasycarpidone (**16**) <93TL5005, 94TL793, 94TL7123, 96TL3071, 96AQ62, 97JOC3158>. They cross-coupled both 2- and 3-indolylzinc derivatives with diversely substituted 2-halopyridines to assemble 2- and 3-(2-pyridyl)indoles, which have become important intermediates in alkaloid total synthesis. Thus, 1-(phenylsulfonyl)-indole (**9**) was converted to 2-indolylzinc reagent **10**, which was then joined with 2-halopyridine **11** to secure 2-(2-pyridyl)indole **12**.

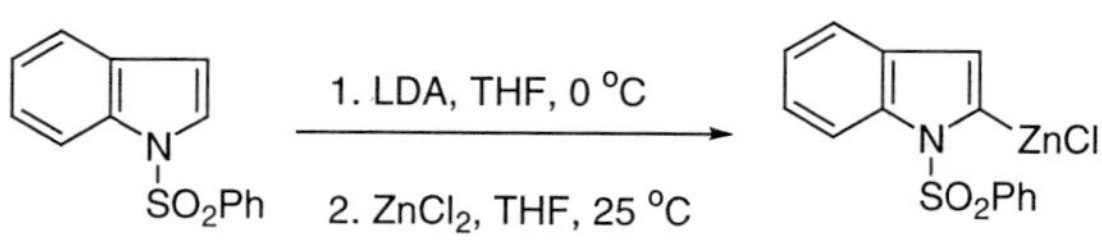

9 10

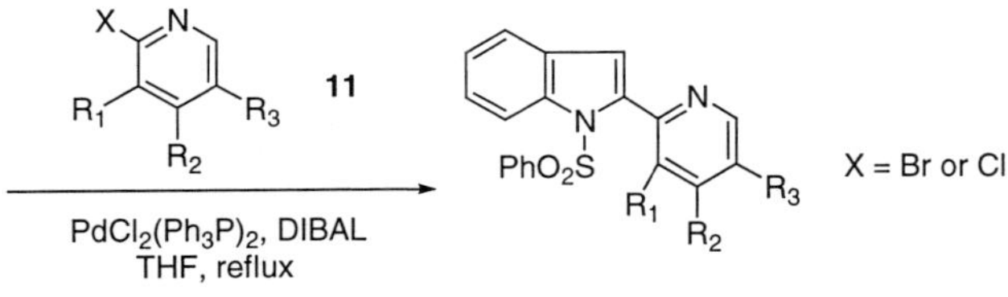

12

Recognizing that 1-(phenylsulfonyl)-3-lithioindole tends to isomerize to the corresponding 2-lithioindole derivative, Bosch *et al.* used a silyl ether protective group to solve the problem. Thereby, they prepared 3-indolylzinc reagent **14** from 3-bromo-1-(*tert*-butyldimethylsilyl)-indole (**13**). Subsequent Negishi coupling of **14** with 2-halopyridine **11** gave 3-(2-pyridyl)indole **15**. The Negishi adduct **15** was then further manipulated to a naturally occurring hexahydropyridine alkaloid, (±)-nordasycarpidone (**16**) <93TL5005, 96AQ62>.

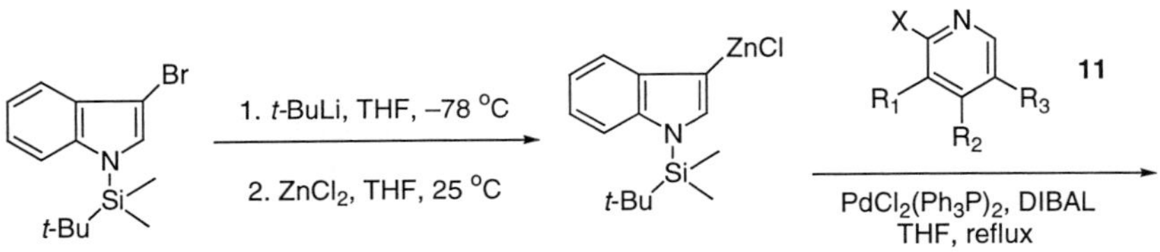

13 **14**

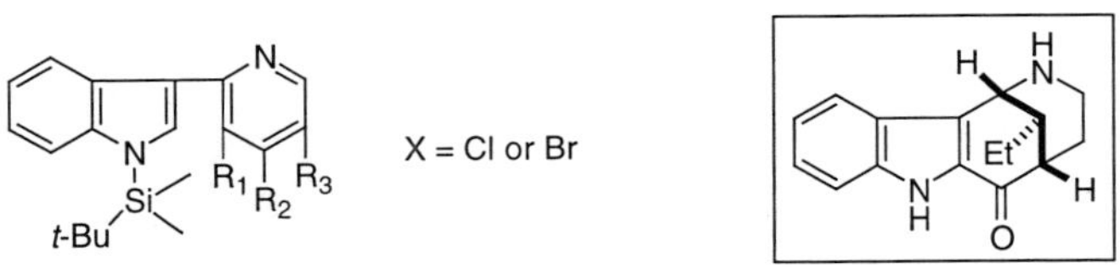

15 **16**

In their formal total synthesis of camptothecin (**21**), Murata and associates employed a Negishi reaction to establish the A, B and D ring linkage <97SL298>. The halogen-metal exchange of 2-chloropyridine **17** was achieved using lithium naphthalenide complex, which was superior to *n*-BuLi because nucleophilic addition to the substrate was avoided. Transmetalation of the resulting lithiopyridine with $ZnCl_2$ generated pyridylzinc reagent **18**, which was then coupled with methyl 2-chloro-3-quinolinecarboxylate (**19**) to provide hetero biaryl **20**, an important intermediate for camptothecin (**21**) synthesis.

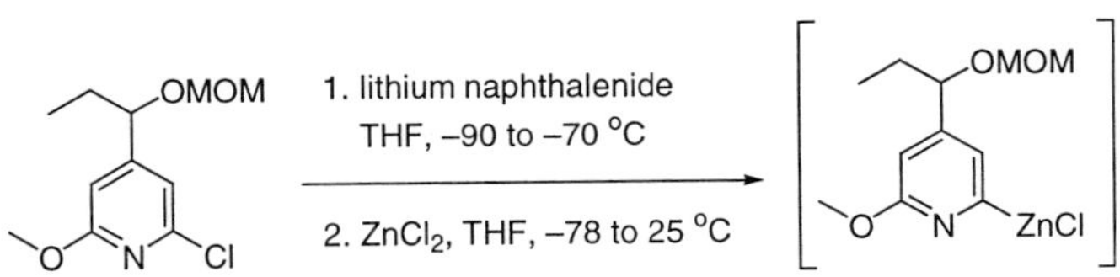

17 **18**

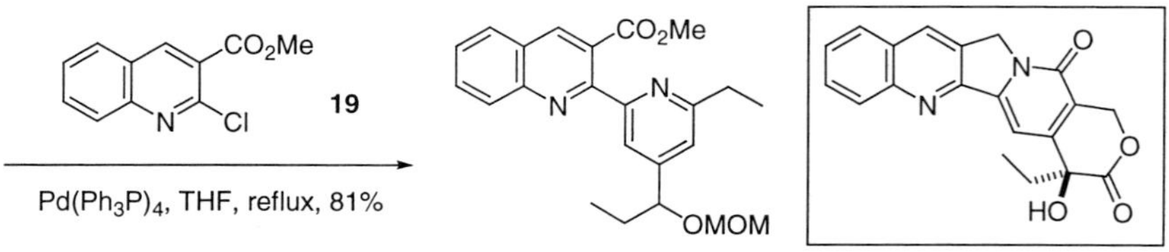

20 **21**

A salient feature of the Negishi coupling is that the reaction also works for *alkylzinc reagents*. This could be exploited in the design of alkylpyridine syntheses.

3.3 THE SUZUKI COUPLING

Miller's synthesis of ellipticine was an example of the total synthesis of a naturally occurring pyridine alkaloid employing a Suzuki reaction <89TL297>. The Suzuki coupling of 6-amino-5,8-dimethyl-7-phenylisoquinoline and phenylboronic acid gave 6-amino-7-bromo-5,8-dimethylisoquinoline, which was transformed into the corresponding azide. Subsequent thermolysis of the resulting azidophenylisoquinoline in dodecane at 180 °C delivered ellipticine via a nitrene intermediate. In another case, Snieckus *et al.* accomplished the total synthesis of onychine (**24**), an azafluorenone alkaloid, in two steps enlisting a Suzuki coupling tactic <88TL2135>. Accordingly, the union of 2-bromonicotine ester **22** with phenylboronic acid afforded heterobiaryl **23**, which upon treatment with polyphosphoric acid cyclized to onychine (**24**).

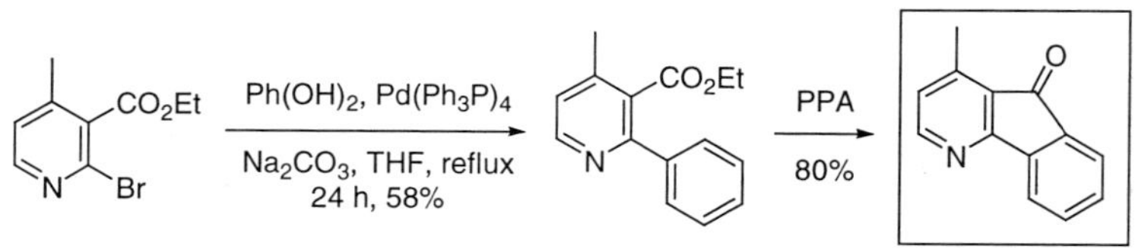

22 **23** **24**

1-Chloro-β-carboline (**25**) has served as a common intermediate in palladium-catalyzed cross-coupling reactions, offering easy access to several pyridine alkaloids. In Bracher's total synthesis of perlolyrine (**27**), a β-carboline alkaloid, the Suzuki reaction of **25** with 5-formylfuranyl-2-boronic acid (**26**) formed the C–C bond between the pyridine and the furan rings <92LA1315>. Reduction of the resulting Suzuki adduct with $NaBH_4$ subsequently

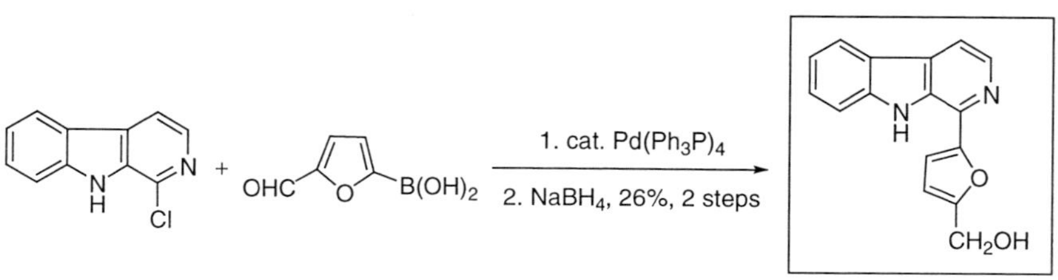

25 **26** **27**

yielded perlolyrine (**27**). In the same manner, the Suzuki reaction of **25** with tri(*m*-propyl)phenylborate (**28**, Ar = *m*-propylphenyl) afforded komaroine (**29**), another β-carboline alkaloid.

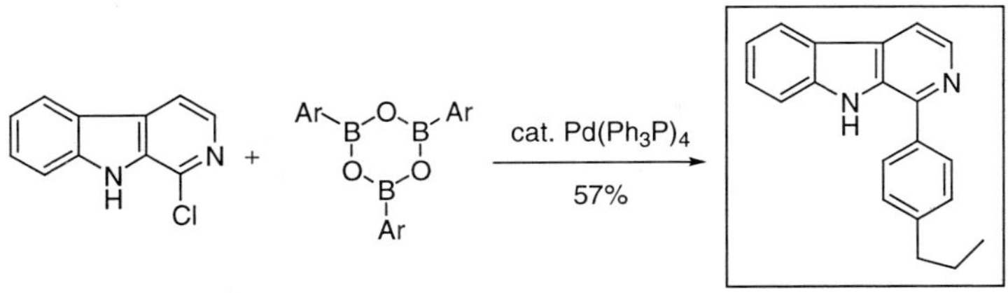

25 **28** **29**

Haminol-A (**35**), a pyridine alkaloid with a pendant polyene side chain, was isolated from a Cephalaspidean mollusc, *Haminoea navicula*. It is believed that **35** and its acetylated analog, haminol-B, are used as alarm pheromones in a defense mechanism against predators. In de Lera's total synthesis of haminols-A and B <98TA3065>, two stereocontrolled Suzuki couplings of 1-alkenylboronic acids and pyridyl/alkenyl halides were employed. Due to the instability of boronic acids derived from ynols, boronate **31** was generated *in situ* by hydroboration of hex-5-yn-1-ol acetate (**30**) with catecholborane. Treatment of the unstable boronate **31** with 3-bromopyridine in the presence of catalytic Pd($Ph_3P)_4$ and aqueous NaOH provided alcohol **32**. As expected, concomitant acetate hydrolysis took place during the Suzuki coupling. Swern oxidation of **32** was followed by a Takai reaction to furnish the *E* alkenyl iodide **33**. The Suzuki coupling of **33** with alkenylboronic acid **34**, derived from hydroboration of the appropriate alkynol with catecholborane and hydrolysis, delivered the unique marine natural product **35**. Subsequent acetylation of **35** readily secured haminol-B.

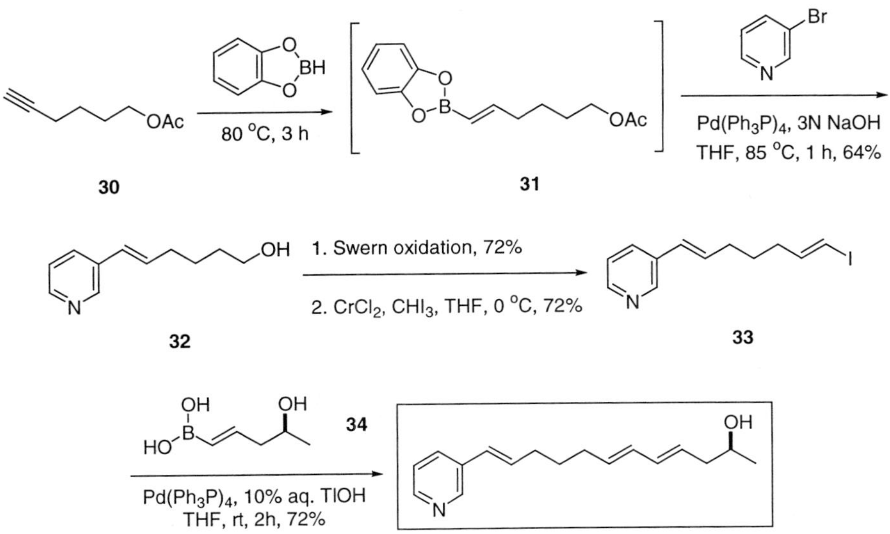

35

Employing a combination of *ortho* metalation and Suzuki coupling, Quéguiner's group has synthesized a number of naturally occurring pyridine alkaloids <95H1055>. One example was the total synthesis of bauerine B, a β-carboline natural product <95SC2901>. Another β-carboline natural product, the antibiotic eudistomin T, and an antimicrobial marine sponge pigment, fascaplysin, <93TL7917> have also been synthesized in the same fashion. The Quéguiner group also prepared a few other hydroxy β-carbolines <92T4123, 95SC3373>, azacarbazoles <93T49>, streptonigrin, lavendamycin analogs <93TL2937, 96JOM25>, and the marine alkaloid amphimedine <95JOC292> utilizing the combination of *ortho* metalation and Suzuki reaction. Furthermore, Rocca and colleagues employed similar methodology in the total synthesis of an indoloquinoline natural product, quindoline (**40**) <98TL6465>. 2-Iodo-3-fluoroquinoline (**37**) was prepared by treating 3-fluoro-4-iodoquinoline (**36**) with LDA followed by quenching with water. The fluoro-directed lithiation was a kinetic process and occured at C(2), but an isomerization ("halogen-dance") occurred to give the more stable 4-lithioquinoline, which was then protonated upon quenching with water to give **37**. The Suzuki reaction of **37** with *N*-pivaloylaminophenylboronic acid (**38**) proceeded using K_2CO_3 as the base to afford the biaryl product **39**, which was then converted to quindoline (**40**).

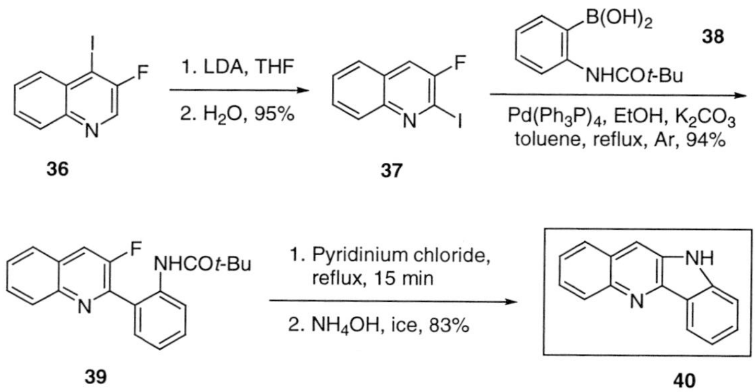

In a model study towards cystodytin alkaloids such as amphimedine (**44**), Quéguiner and colleagues synthesized benzo[c][2,7]naphthyridine **43** <96SC4421>. 4-Iodopyridine **41** was generated from 2-chloro-3-iodopyridine again using the "halogen-dance" (orthomigration) tactic. The adduct from the Suzuki coupling of **41** and boronic acid **38** underwent a spontaneous cyclization, giving rise to benzonaphthyridine **42**. The Stille coupling of **42** with ethoxyvinyltributylstannane and subsequent acidic hydrolysis furnished acetyl benzo[c][2,7]naphthyridine **43**.

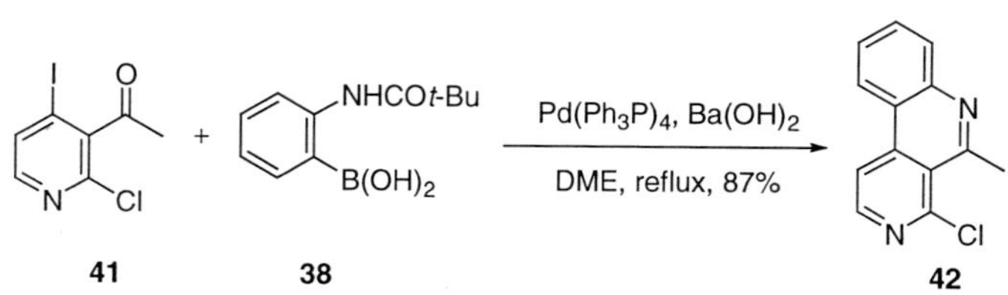

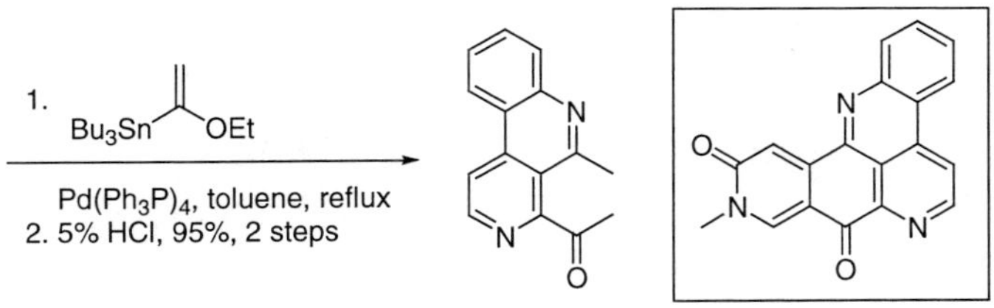

43 **44**

An additional application of the Suzuki reaction is found in the total synthesis of cryptosanguinoline (**48**) by Timári *et al.* <97SL1067>. 3-Bromoquinoline (**45**) was joined with boronic acid **38** to give **46**. Simple functional group transformations of the pivaloylamino group into the corresponding azido group provided **47**, which was then thermolyzed via a nitrene intermediate and subsequently methylated to furnish cryptosanguinoline (**48**).

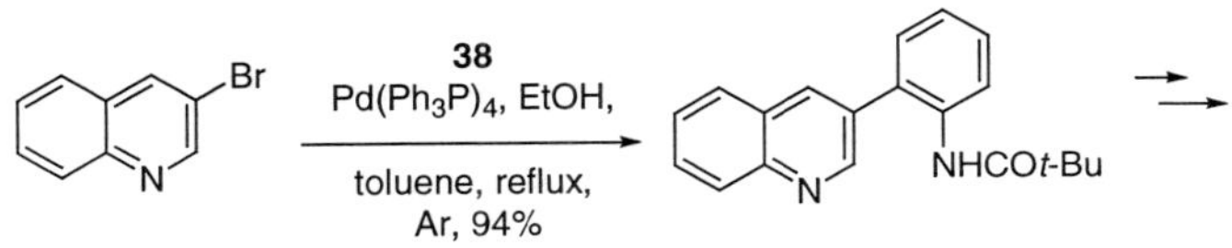

45 **46**

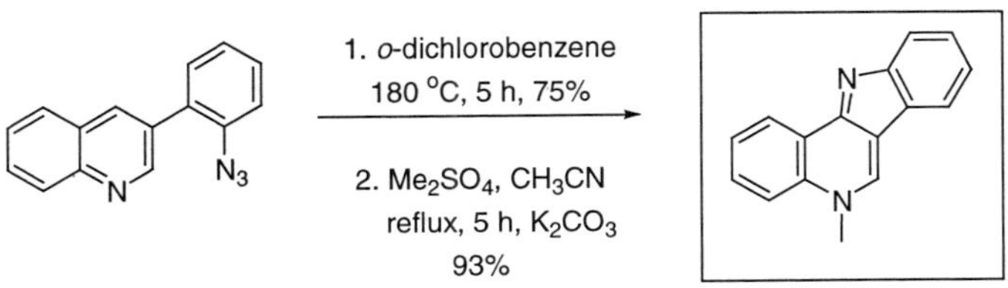

47 **48**

3.4 THE STILLE COUPLING

The Stille coupling is regarded as the most versatile method among all Pd-catalyzed cross-coupling reactions with organometallic reagents. Many naturally occurring pyridine alkaloids have been synthesized using this approach. One simple case involving the coupling of a pyridyl halide and vinyl stannane was utilized by Lavilla *et al.* in their total synthesis of angustine, an indolopyridine alkaloid <95CC1675, 99EJOC373>. In contrast to simple vinylstannane, 1-ethoxy-1-tributylstannylethene provides additional functionality because the resulting vinyl ether is readily hydrolyzed to the corresponding methyl ketone. Pyridine alkaloid **51** from *Rubiaceae* was synthesized using this approach <95MC805>. Thus, the Stille coupling of pyridyl bromide **49** with 1-ethoxy-1-tributylstannylethene gave ketone **50** after acidic hydrolysis. Asymmetric reduction of **50** using baker's yeast provided a quick access to **51**. Similarly, 1-chloro-β-carboline (**25**) was coupled with 1-ethoxy-1-tributylstannylethene to afford ketone **52** <93LA837>. The chlorine α to the nitrogen was sufficiently activated for the Stille coupling. Condensation of **52** with *o*-aminobenzaldehyde then led to nitramarine (**53**), a β-carboline alkaloid. This strategy was also applied to the total synthesis of several other β-carboline alkaloids including pavettine <92LAC1315> and annomontine <93LAC837>.

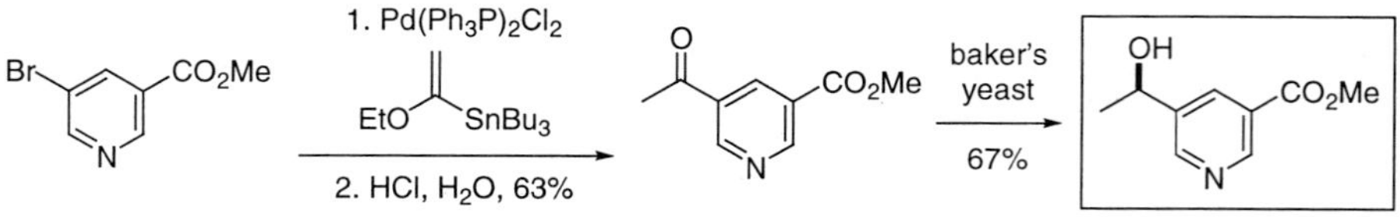

49 **50** **51**

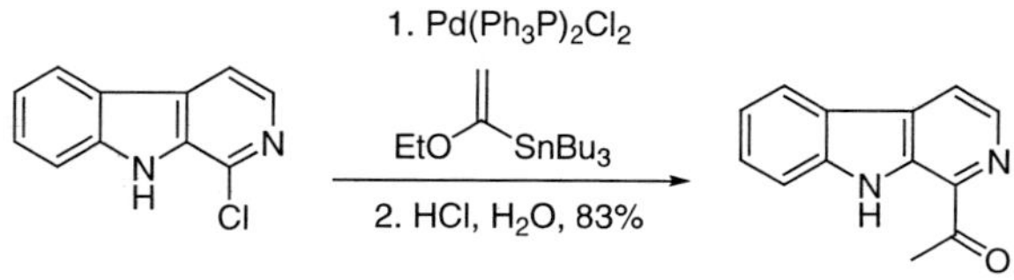

25 **52**

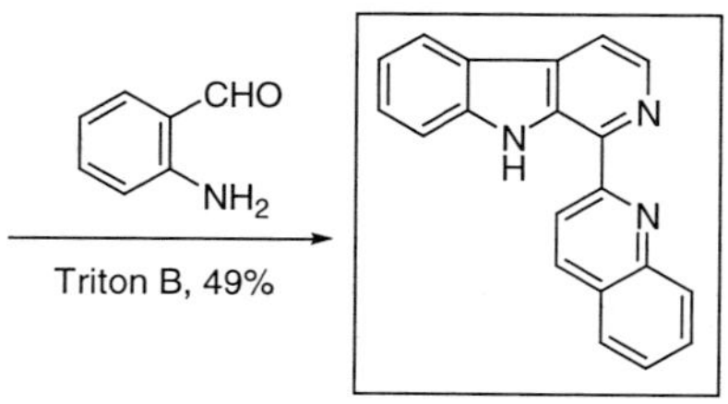

53

In the total synthesis of 1-fluoroellipticine (**56**), 1-ethoxy-1-tributylstannylethene was once again used as a two-carbon building block. The 4-pyridylbromide **54** was assembled by applying a metalation/halogen-dance strategy starting from 2-fluoropyridine <92JOC565>.

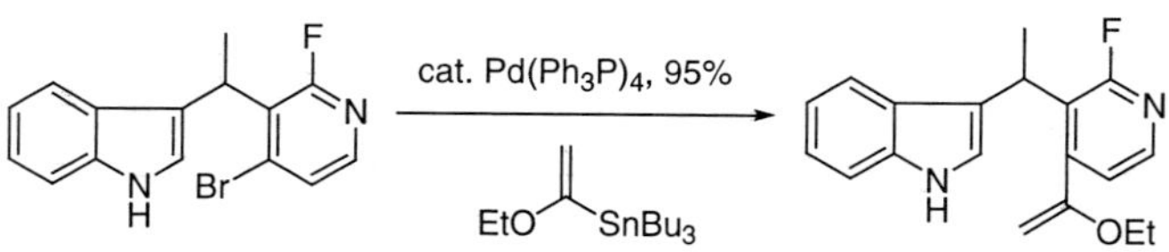

54 **55**

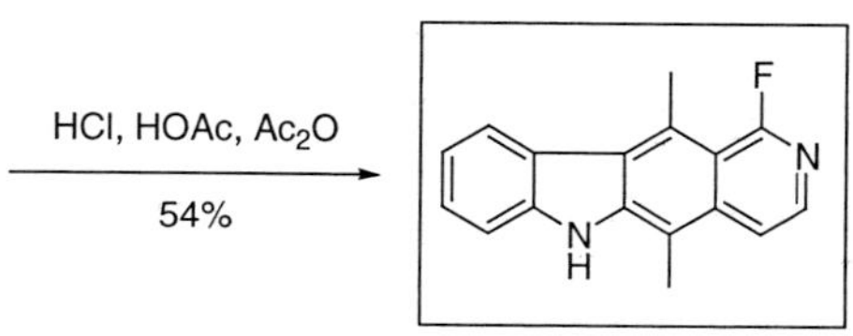

56

Stille coupling of **54** with 1-ethoxy-1-tributylstannylethene constructed 4-(1-ethoxyethenyl)pyridine **55**, which was then annulated to 1-fluoroellipticine (**56**) upon treatment with acid.

An alternative route for installing a methyl ketone is the three component carbonylative-Stille coupling with tetramethyltin. An indolopyridine alkaloid, naucletine (**58**), was prepared using such a sequence from pyridyl bromide **57** <95CC1675>.

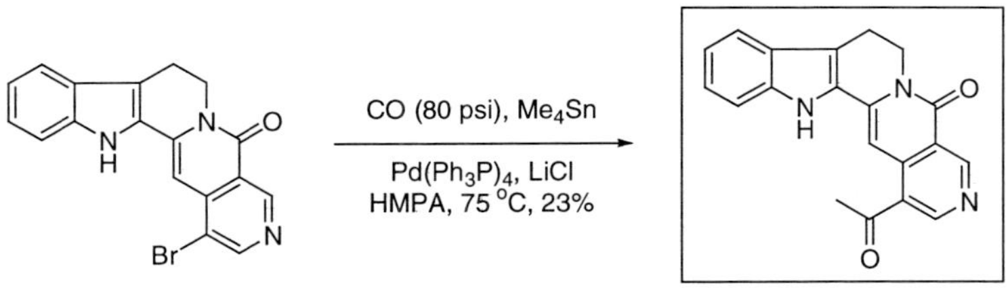

57 **58**

The Stille coupling of α-iodo enones is sluggish under standard conditions. Significant rate enhancement was observed for the Stille reaction of 2-chloro-5-tributylstannyl pyridine with α-iodo enone **59** using triphenylarsine as the soft palladium ligand and CuI as the co-catalyst <99TL557>. Oxygenated functionalities did not affect the efficiency of the reaction provided both Ph_3As and CuI were added. Additional manipulations of **60** resulted in the synthesis of (+)-epibatidine (**61**), the first naturally occurring alkaloid isolated from the skin of an Ecuadorian poisonous frog. (+)-Epibatidine (**61**) exhibits strong non-opioid analgesic activity 200 times more potent than morphine without its addictive effects.

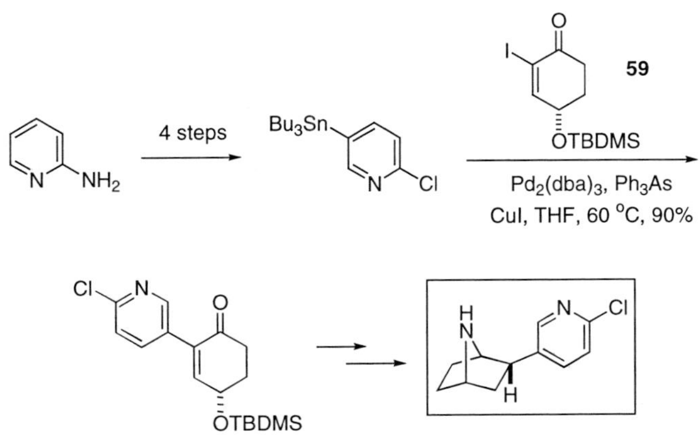

60 **61**

A pyridine-pyridine connection was established by Quéguiner *et al.* in their model study towards a convergent synthesis of the streptonigrin and lavendamycin (**65**) alkaloid skeleton using a Stille coupling strategy <93TL7919, 92T4123>. They coupled chloropyridine **62** with (2-quinolyl)trimethylstannane **63** to form the expected adduct **64**.

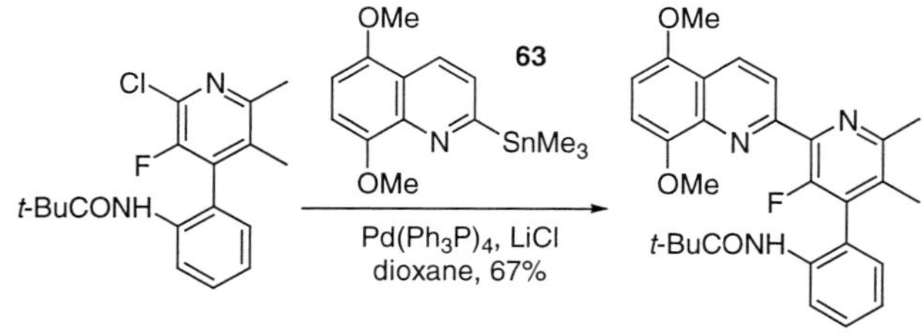

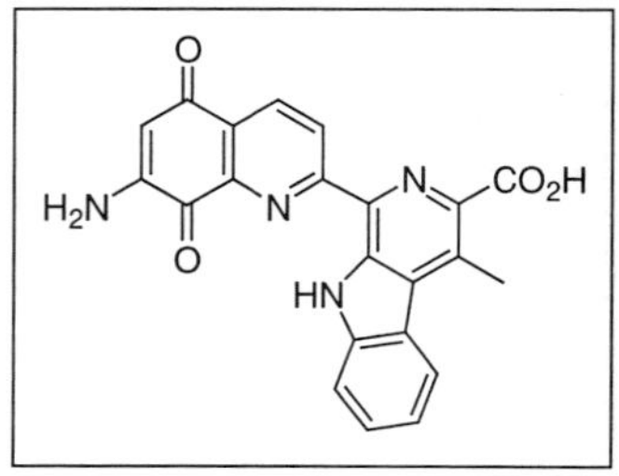

62 64 65

In the total synthesis of amphimedine (**44**) by Echavarren and Stille, a Stille coupling of a quinolinyl triflate was a key operation <88JACS4051>. Quinolinyl triflate **67** was produced from quinolone **66**, and aryl stannane **68** was prepared via *ortho* lithiation of *N-tert*-butoxycarbonylaniline followed by reaction with Me_3SnCl. Subsequent Stille coupling of **67** and **68** in the presence of LiCl gave 4-arylquinoline **69**. In a partial synthesis of amphimedine (**44**), Bracher *et al.* coupled chloroquinoline **70** with 4-trimethylstannylpyridine to furnish **71** <96LA115>. In addition, Achab *et al.* utilized the Stille reaction of stannane **72** and imidazolopyridyl iodide **73** to assemble **74** in a partial synthesis of grossularine-1 <93TL2127>.

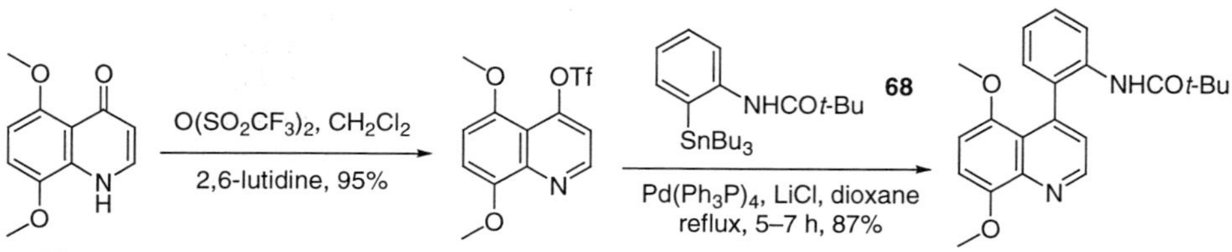

66 67 69

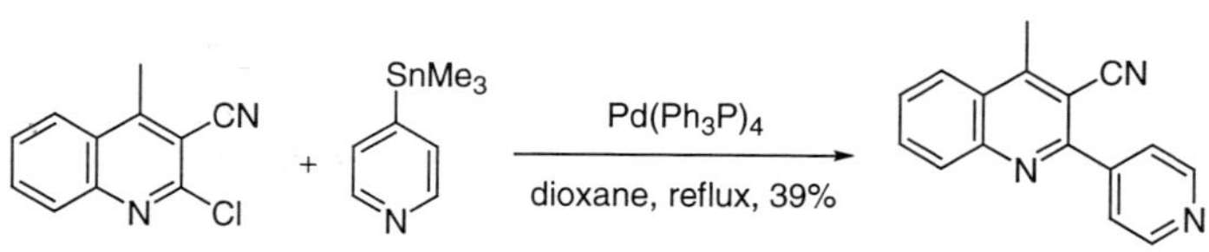

70 71

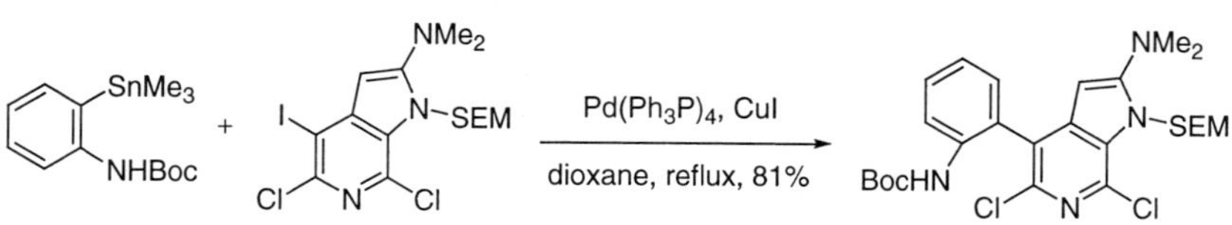

72 73 74

Employing a Stille coupling in which the stannane was generated *in situ*, Kelly's group synthesized dimethyl sulfomycinamate (**77**), the methanolysis product of the antibiotic sulfomycin I <96JOC4623>. Thus, refluxing pyridyl triflate **75** with hexamethyldistannane in the presence of $PdCl_2(Ph_3P)_2$ with slow addition of a dioxane solution of thiazole bromide **76** furnished **77**. However, an effort to prepare the corresponding stannane from **76** failed because dimerization was prevalent when **76** was treated with hexamethyldistannane in the presence of $PdCl_2(Ph_3P)_2$.

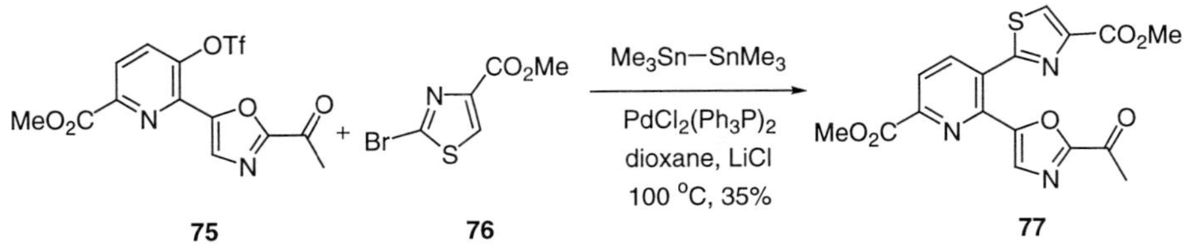

In addition, the Kelly group accomplished an elegant total synthesis of micrococcinic acid (**86**) <91TL4263> using several Stille reactions as the bases for the connections. The first Stille coupling was carried out between bromothiazole **78** and trimethylstannylpyridine **79** with palladium catalysis to give thiazolylpyridine **80**. The ethoxy group in **80** was then modified to provide bromide **81**, which was allowed to react with dithiazole bromide **82** via the Stille coupling with $(Me_3Sn)_2$ and $PdCl_2(Ph_3P)_2$ to furnish the desired cross-coupled **83** in 49% yield. Pyridyl triflate **84**, derived from **83** via conventional functional group transformations, was subjected to another Stille coupling with stannane **85** to secure the pentacycle, which was converted to micrococcinic acid (**86**) via acidic cleavage of the two amides. This total synthesis is a strong testimony for the versatility of Stille reactions even for very complex molecules.

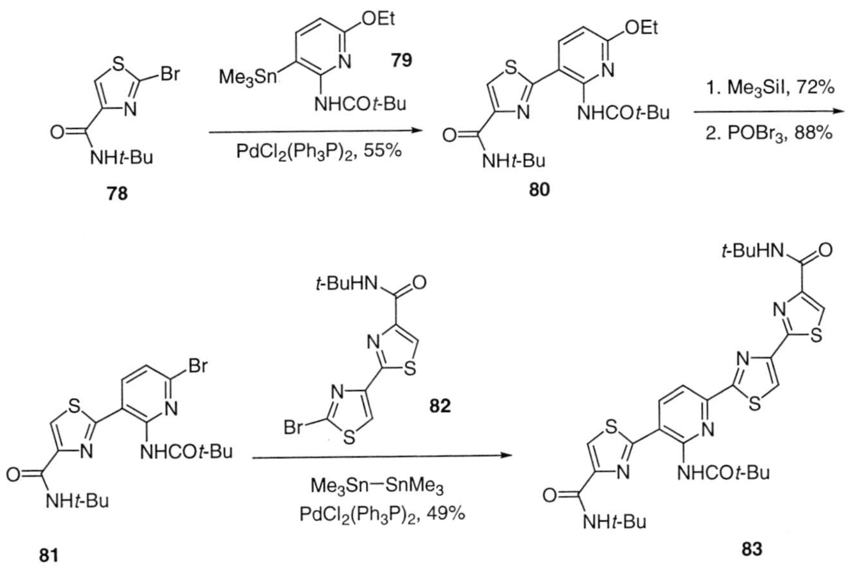

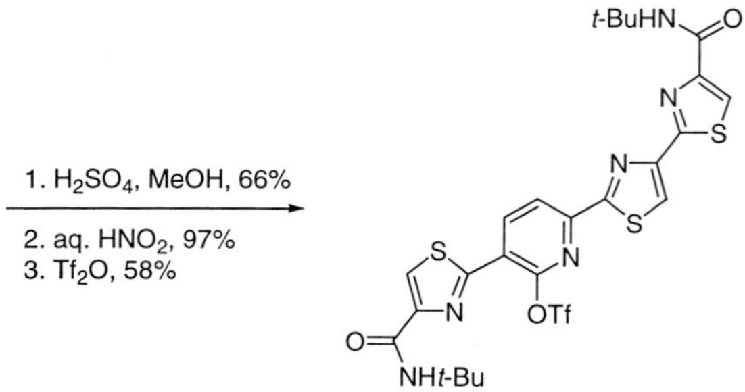

84

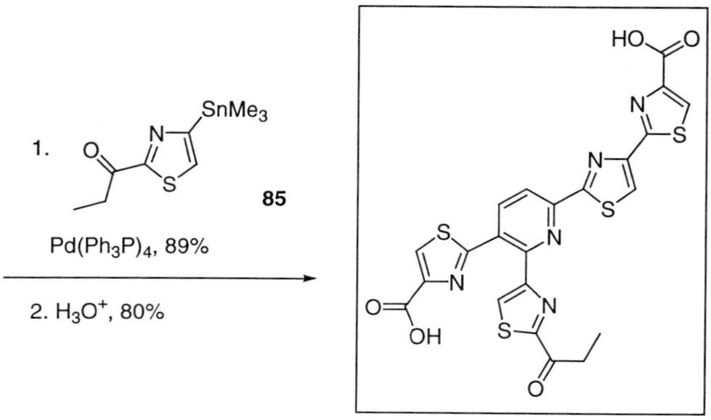

86

The total synthesis of nemertelline (**89**), a neurotoxic tetrapyridine isolated from the hoplonemertine sea worm, is probably the best showcase of Suzuki and Stille couplings in pyridine alkaloid synthesis. Zoltewicz's group <95JOC7491> obtained 3-(tributylstannyl)-2,3'-bipyridine (**88**) from the Suzuki coupling of diethyl (3-pyridyl)-borane with bifunctional 2-bromo-3-(tributylstannyl)pyridine (**87**) *without destannylation under aqueous conditions.* An additional Suzuki coupling between 2,4-dichloropyridine and diethyl (3-pyridyl)-borane took place *preferentially at C(2)* to afford 4-chloro-2,3'-bipyridine (**89**). Finally, bipyridines **88** and **89** were joined via a Stille coupling to deliver nemertelline (**90**).

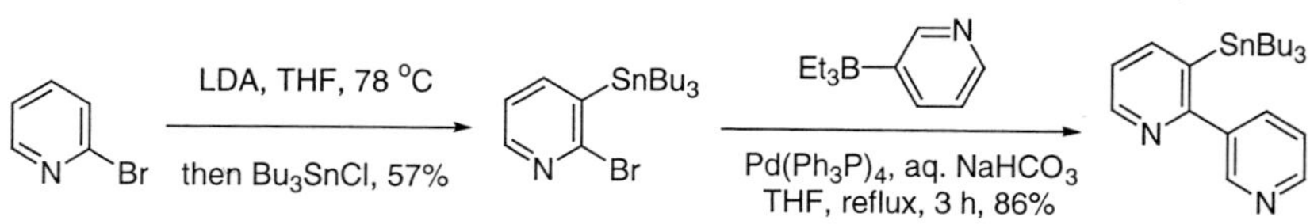

87 **88**

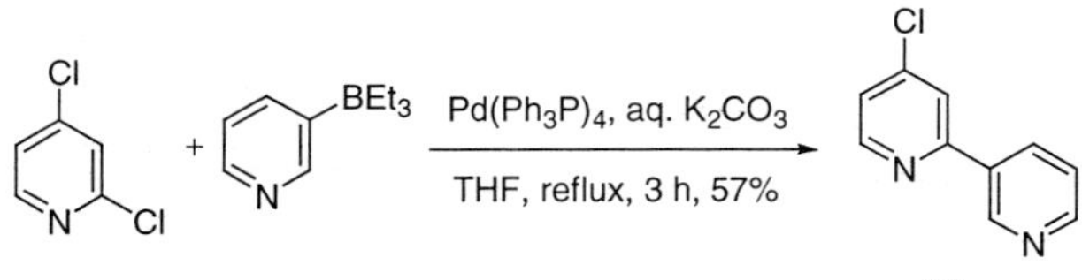

89

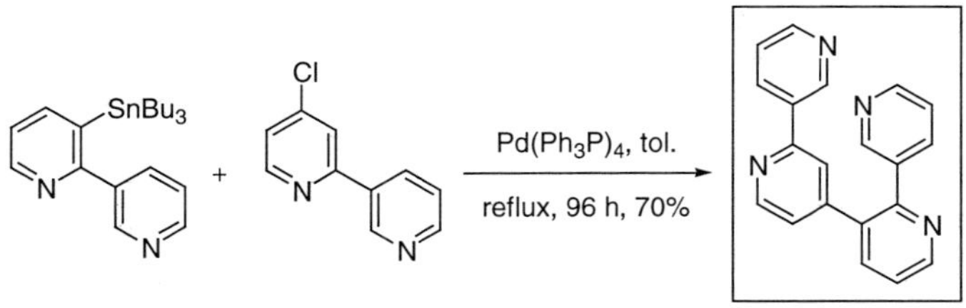

88 **89** **90**

3.5 THE HECK REACTION

In Rao's total synthesis of niphatesines, a key intermediate **91** was elaborated from an intermolecular Heck reaction of 3-bromopyridine with non-8-en-ol <93TL8329>. In another case, Bracher *et al.* synthesized a naturally occurring β-carboline, infractine (**93**), from β-carboline-1-triflate (**92**) in a two step process consisting of a Heck reaction with methyl acrylate followed by a hydrogenation <95PHA182>. Their approach provided an expeditious route to infractine, although the Heck reaction was low yielding.

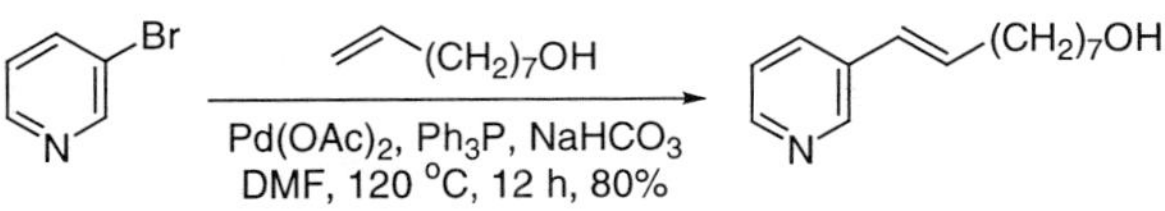

91

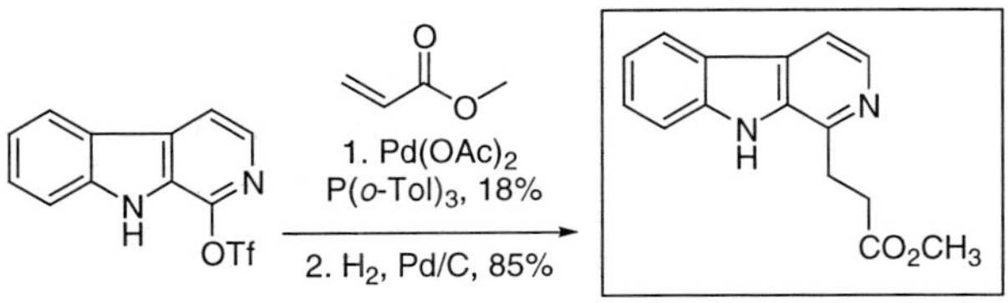

92 **93**

Clayton and Regan reported the total synthesis of (±)-epibatidine (**61**) utilizing a reductive intermolecular Heck heteroarylation (hydroheteroarylation) as the key operation <93TL7493>. The *N*-protected azabicyclo[2,2,1]heptene **94** and 2-chloro-5-iodopyridine were subjected to the reductive Heck reaction conditions to deliver 7-azabicyclo[2.2.1]heptane **95** in moderate yield with the *exo*-stereoisomer as the major product. Subsequent acidic hydrolysis of carbamate **95** using HBr-HOAc secured (±)-epibatidine (**61**) in 74% yield. After Regan's synthesis, numerous analogs of **61** were prepared using the hydroheteroarylation strategy by the groups of Maier

<99T8074>, Malpass <99TL1419, 99T11879>, and Kaufmann <99SL114>. Most noteworthy is the asymmetric synthesis of epibatidine by Kaufmann *et al.* <99SL804>. They synthesized enantiomerically-enriched *N*-protected epibatidine using an asymmetric version of the aforementioned transformation in which the introduction of Noyori's BINAP ligand resulted in 72–81% *ee* and a 53% yield. By using either the (*R*)- or (*S*)-BINAP ligand, both enantiomers were easily accessible.

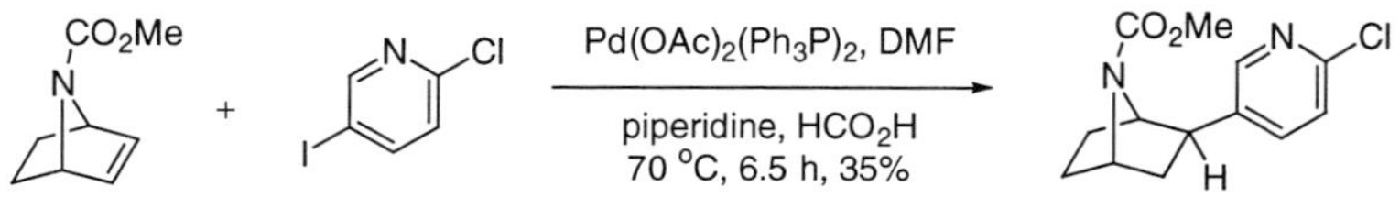

94 **95**

An intramolecular Heck cyclization of substrate **96** was used in Overman's synthesis of pyridinomorphinans <97TL8439>. Under relatively forcing conditions, the desired cyclization of octahydroisoquinoline **96** led to enantiomerically pure **97**, an intermediate for the synthesis of (–)-morphine (**98**) analogs.

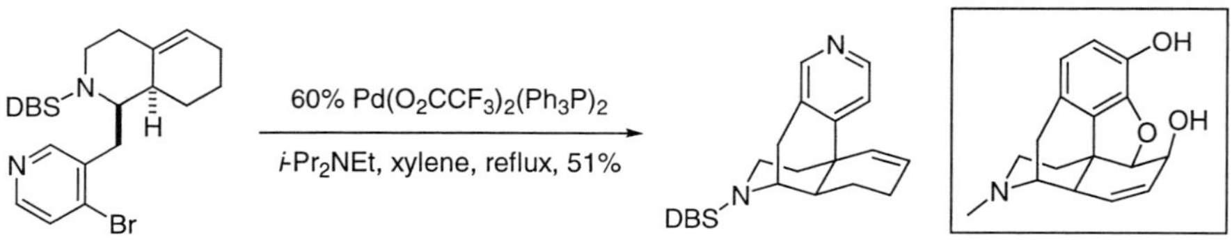

96 **97** **98**

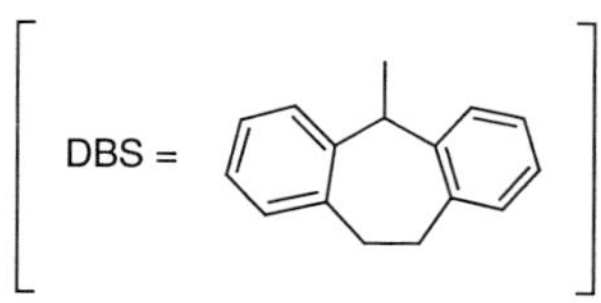

At an early stage of their asymmetric synthesis of camptothecin analog GI147211C, Fang and colleagues exploited an intramolecular Heck approach to prepare the DE ring moiety <94JOC6142>. Thus, crotyl ether **99** was subjected to Jeffery's ligand-free conditions to afford cyclic olefin **100**. In another case, an intramelocular 7-*exo* Heck cyclization was the key step in Kelly's synthesis of maxonine (**103**), isolated from the root of the plant *Simira maxonii* endemic to the Costa Rican tropical forest <93TL6173>. The migratory insertion step of the intramolecular Heck cyclization of substrate **101** took place at the pendant olefin, giving rise to the seven-membered product **102**. Oxidative cleavage of the stilbene double bond in **102** produced maxonine (**103**). Kelly's synthesis also revised the original structural assignment of the natural alkaloid.

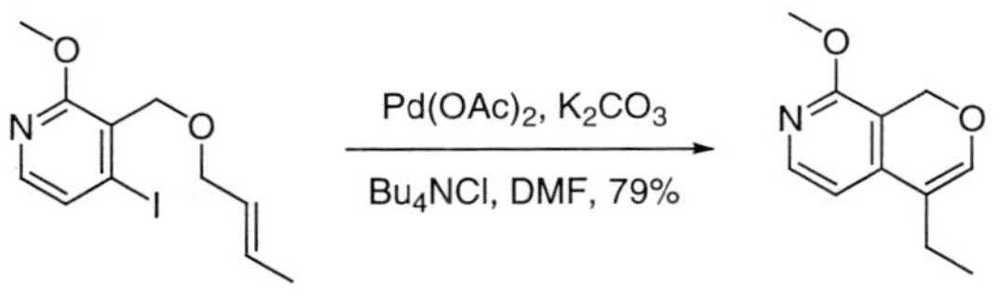

99 **100**

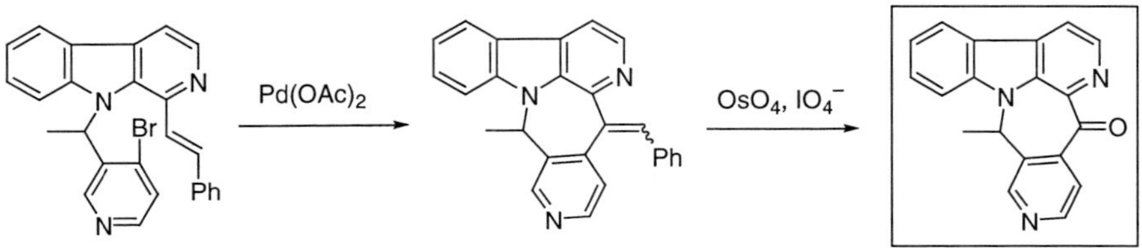

101 **102** **103**

In the synthesis of the *E*-azaeburnane series, an intramolecular "heteroaryl Heck reaction" was the major operation to realize the cyclization <95TL1491>. Under Jeffery's phase-transfer catalysis conditions, the *E*-azaeburnane skeleton **105** was prepared from bromopyridine **104**. The migratory insertion occurs onto the C(2) position of the indole ring.

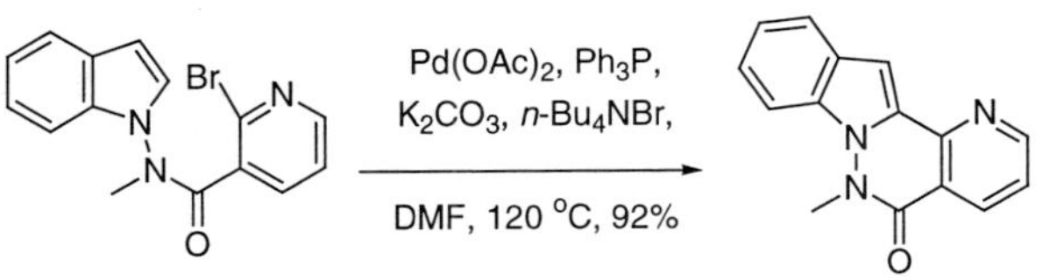

104 **105**

3.6 MISCELLANEOUS

3.6.1 Oxidative cyclization

Oxidative cyclizations are generally facilitated by $Pd(OAc)_2$ in refluxing acetic acid <88AG1113>. The role of acetic acid in such oxidative cyclization processes is to protonate the acetate ligand, making Pd(II) more electrophilic. The initial step in these oxidative cyclization reactions is electrophilic palladation of the aromatic ring. In organic synthesis, oxidative cyclization offers an expeditious route to those target molecules that may not be easily accessible otherwise. In one case, exposure of 6-anilino-5,8-dimethylisoquinoline (**106**) to *two equivalents* of $Pd(OAc)_2$ in TFA/AcOH promoted the oxidative cyclization to afford the desired ellipticine (**107**) <80TL3319>. In another case, quindoline (**109**, an antimalarial agent isolated from a West African plant *Cryptolepis sanguinolenta*) was synthesized in two steps, a remarkably concise synthesis of <97JHC1789>. The precursor, 3-anilinoquinoline (**108**), was prepared by phenylation of 3-aminoquinoline with $Ph_3Bi(OAc)_2$ in the presence of metallic copper. The crucial oxidative cyclization of **108** was then effected by *two equivalents* of $Pd(OAc)_2$ in refluxing trifluoroacetic acid to furnish quindoline (**109**).

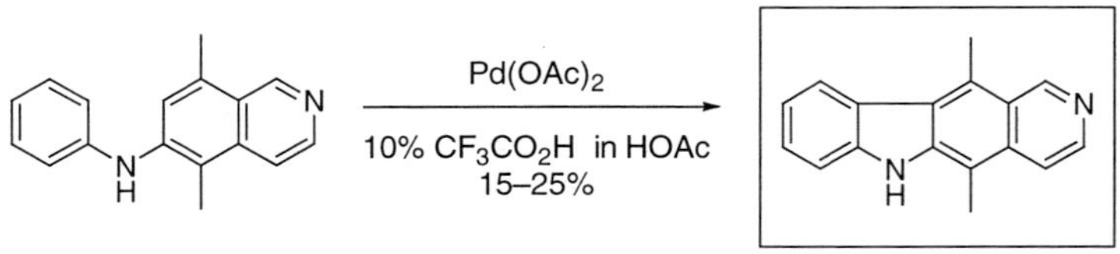

106 107

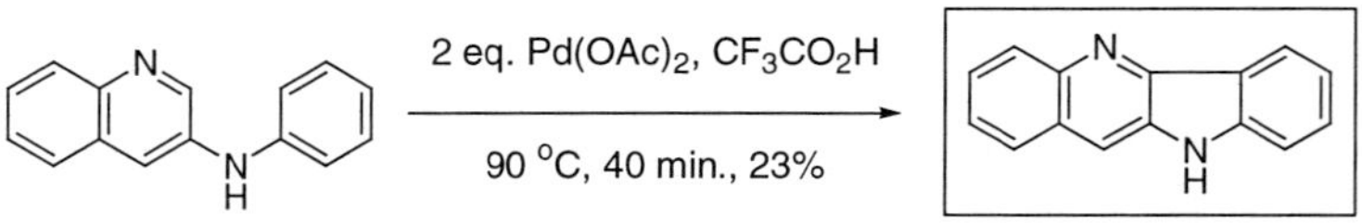

108 109

Judging from literature precedents including the two representative syntheses of pyridine alkaloids shown here, not only are the yields for oxidative cyclizations generally low, but also the reaction consumes at least stoichiometric amounts of expensive $Pd(OAc)_2$. Therefore, this method has limited utility in synthesis. Nonetheless, great progress has been made recently towards *catalytic* palladium oxidative addition <99SL596>, making the method more attractive for preparative purposes.

3.6.2 C—N bond formation

An early example of Pd-catalyzed C—N bond formation involving a pyridine motif was the intramolecular amination of an aryl halide in Boger's synthesis of lavendamycin <84TL3175, 85JOC5782, 85JOC5790>. Boger's initial attempts using copper reagents to construct the aryl-nitrogen bond in **110** failed to give the desired β-carboline **111**. In contrast, treatment of **110** with 1.5 equivalents of $Pd(Ph_3P)_4$ under conditions conducive to oxidative insertion into aryl halide bonds delivered **111** smoothly. The success of the cyclization may be attributed to the reduced nucleophilicity of the aryl amine, the result of methoxycarbonyl delocalization and weakened N—Pd coordination. $Pd(Ph_3P)_4$ was required in stoichiometric amount possibly because of a competitive oxidative addition of liberated HBr with the palladium reagent. In contrast, enlisting Buchwald-Hartwig's *Pd-catalyzed* aryl amination approach, Dodd *et al.* ring-closed amino-bromoindole **112** via an intramolecular C—N bond formation to furnish β-carboline **113**, a key intermediate towards the β-carboline alkaloids grossularine 1 and 2 <98TL2119>.

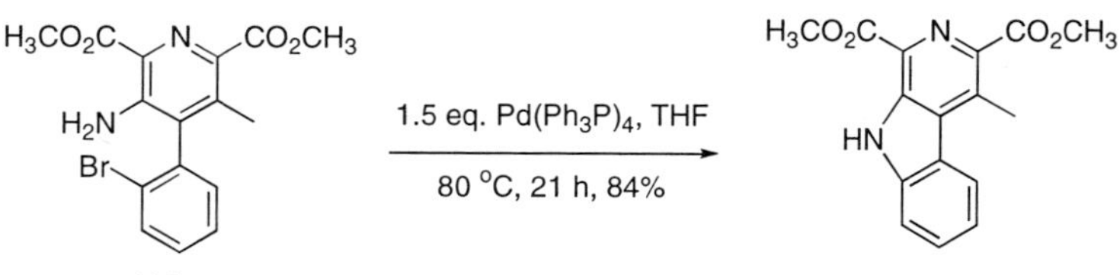

110 111

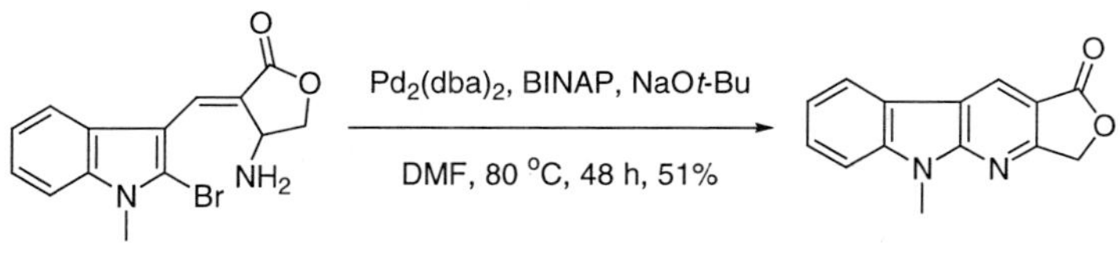

112 **113**

3.6.3 Sonogashira reaction

In a formal total synthesis of matrine (**116**), Yamanaka and coworkers used the Sonogashira reaction of pyridyl halides as the means to form C—C bonds <86CPB2018>. For instance, bromonaphthyridinone **114** was coupled with 3-butyn-1-ol to furnish alkynylnaphthyridinone **115**, an intermediate towards matrine (**116**).

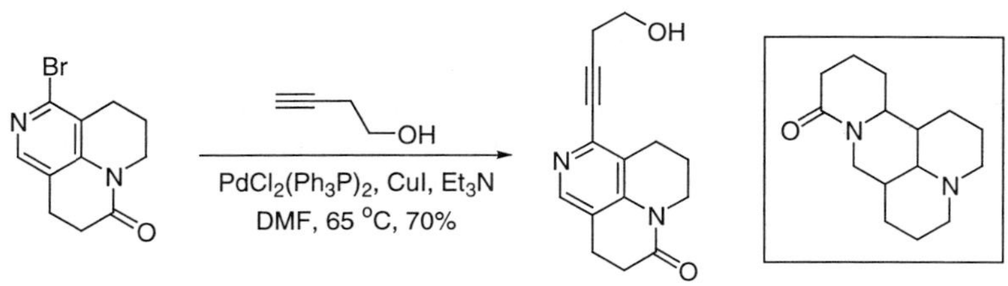

114 **115** **116**

3.6.4 Carbonylation

A Pd-catalyzed alkoxycarbonylation of 2-chloropyridine **117** to ester **118** was reported in the synthesis of camptothecin analogs <97JOC6588>. Another Pd-catalyzed alkoxycarbonylation was the central feature in Hibino's total synthesis of oxopropaline G (**121**), a β-carboline isolated from *Streptomyces* sp G324 <98TL2341>. Triflate **119** was prepared from 2-formyl-3-iodoindole in seven steps. The Pd-catalyzed alkoxycarbonylation of **119** provided the key intermediate, 1-methoxycarbonyl-4-methylcarboline **120**, which was then transformed to oxopropaline G (**121**) in four additional steps.

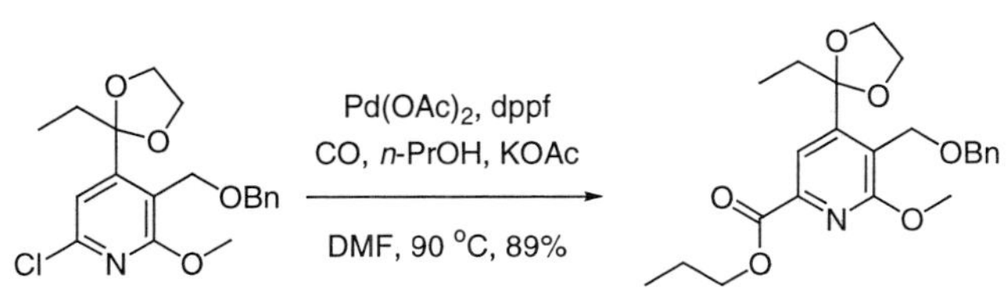

117 **118**

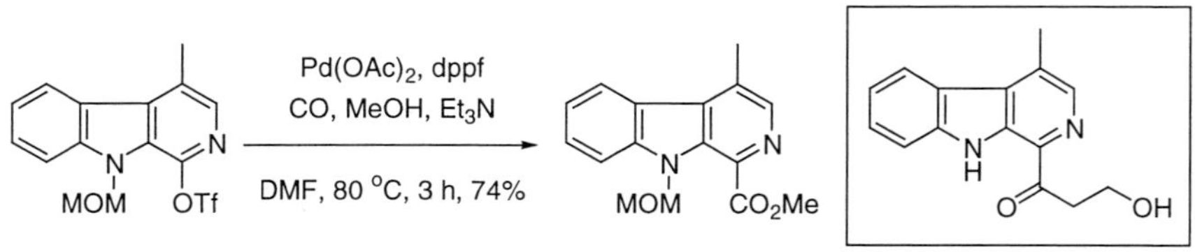

119 120 121

3.6.5 Bi-annulation

The Lycopodium alkaloid huperzine A (**125**) was isolated from Chinese folk medicine *Huperzia serrata*. It has elicited tremendous interest due to its potency and selectivity as a reversible actetylcholinesterase (AChE) inhibitor, and is currently under clinical trial for treating Alzheimer's disease. Extending a palladium-catalyzed bicycloannulation strategy developed by Huang and Lu <88TL5663>, Kozikowski and coworkers accomplished the total synthesis of racemic huperzine A (**125**) <93JCSCC860, 93JOC7660>. With 1,1,3,3,-tetramethylguanidine (TMG) as the base, the dianion generated from β-keto-ester **122** and the bifunctional nucleophile 2-methylene-1,3-propanediol diacetate (**123**) underwent a "Tsuji-Trost-like nucleophilic substitution" *twice* via π-allylpalladium(II) intermediates, giving rise to the methylene-bridged bicyclic intermediate **124** in excellent yield. Furthermore, addition of chiral ligands such as chiral ferrocenyl-phosphines afforded enantiomerically enriched **124** <97TA929, 98T5471, 98T411>.

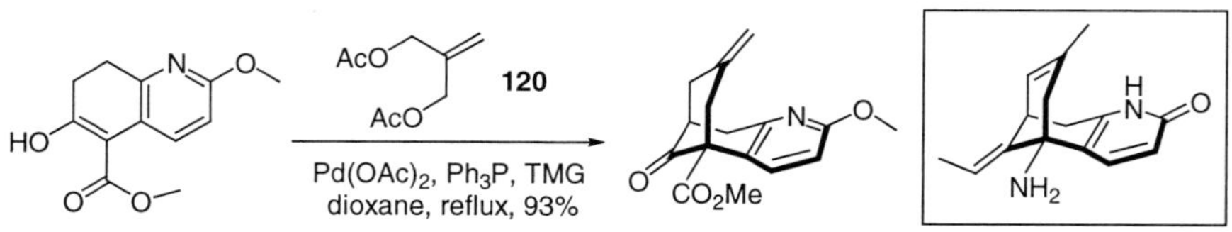

122 123 124 125

3.7 CONCLUDING REMARKS

In conclusion, the Negishi, Suzuki, and Stille couplings are most practical in the synthesis of pyridine alkaloids. The Stille coupling has been most widely applied in aryl and heteroaryl syntheses and it should be the method of choice for small-scale syntheses where toxicity is not a great concern. However, Negishi and Suzuki couplings are better choices for large-scale operations. The palladium oxidative addition method has limited utility since it suffers from both low yields and stoichiometric use of expensive $Pd(OAc)_2$. However, oxidative cyclizations can offer expeditious routes to targets that may not be easily accessible otherwise.

In spite of the abundance of applications of palladium chemistry in the total synthesis of naturally occurring pyridine alkaloids, there are many emerging methods in pyridine synthesis using palladium chemistry yet to be widely employed. For example, although alkyl halides are resistant to undergo oxidative additions, alkylzinc or alkylboron reagents can take part in the Negishi or the Suzuki coupling to install alkyl substituents onto the pyridine rings. Furthermore, many palladium-catalyzed cross-coupling reactions can be conducted regioselectively. With regard to 2,5-dibromopyridine, although halogen/metal exchange takes place at C(5) predominantly, the Kumada, Negishi, Suzuki, and Stille, and the Sonogashira reactions all proceed preferentially at the more activated C(2) position.

In short, more applications of palladium chemistry in the synthesis of pyridine alkaloids are certainly expected as new methods in these areas are being developed, fine-tuned, and optimized.

3.8 ACKNOWLEDGMENTS

I am indebted to Melvin Budzol and the Research Library staff at Parke-Davis Pharmaceutical Research for their support. I am also grateful to W. Howard Roark and Roderick J. Sorenson for proofreading the manuscript.

3.9 REFERENCES

80TL3319 Miller, R. B.; Moock, T. *Tetrahedron Lett.* **1980**, *21*, 3319.

84TL3175 Boger, D. L.; Panek, J. S. *Tetrahedron Lett.* **1984**, *25*, 3175.

85JOC5782 Boger, D. L.; Duff, S. R.; Panek, J. S.; Yasuda, M. *J. Org. Chem.* **1985**, *50*, 5782.

85JOC5790 Boger, D. L.; Duff, S. R.; Panek, J. S.; Yasuda, M. *J. Org. Chem.* **1985**, *50*, 5790.

86CPB2018 Sakamoto, T.; Miura, N.; Kondo, Y.; Yamanaka, H. *Chem. Pharm. Bull.* **1986**, *34*, 2018.

88AG(E)1113 Hegedus, L. S. *Angew. Chem. Int. Ed. Engl.* **1988**, *27*, 1113.

88JACS4051 Echavarren, A. M.; Stille, J. K. *J. Am. Chem. Soc.* **1988**, *61*, 4051.

88TL5663 Huang, Y.; Lu, X. *Tetrahedron Lett.* **1988**, *29*, 5663.

89TL297 Miller, R. B.; Moock, T. *Tetrahedron Lett.* **1989**, *30*, 297.

91TL4263 Kelly, T. R.; Jagoe, C. T.; Gu, Z. *Tetrahedron Lett.* **1991**, *32*, 4263.

92LA1315 Bracher, F.; Hildebrand, D. *Libiegs Ann. Chem.* **1992**, 1315.

92T4123 Godard, A.; Rovera, J. C.; Marsais, F.; Plé, N.; Quéguiner, G. *Tetrahedron* **1992**, *48*, 4123.

93CC860 Kozikowski, A. P.; Campiani, G.; Aagaard, P.; McKinney, M. *J. Chem.Soc., Chem Commun.* **1993**, 860.

93JOC7660 Campiani, G.; Sun, L.-Q.; Kozikowski, A. P.; Aagaard, P.; McKinney, M. *J. Org. Chem.* **1993**, *58*, 7660.

93LA837 Bracher, F.; Hildebrand, D. *Libiegs Ann. Chem.* **1993**, 837.

93T49 Rocca, P.; Marsais, F.; Godard, A.; Quéguiner, G. *Tetrahedron* **1993**, *49*, 49.

93TL2937 Rocca, P.; Marsais, F.; Godard, A.; Quéguiner, G. *Tetrahedron Lett.* **1993**, *34*, 2937.

93TL5005 Amat, M.; Hadida, S.; Bosch, J. *Tetrahedron Lett.* **1993**, *34*, 5005.

93TL6173 Kelly, T. R.; Xu, W.; Sundaresan, J. *Tetrahedron Lett.* **1993**, *34*, 6173.

93TL7917 Rocca, P.; Marsais, F.; Godard, A.; Quéguiner, G. *Tetrahedron Lett.* **1993**, *34*, 7917.

93TL7919 Godard, A.; Rocca, P.; Fourquez, J. M.; Rovera, J. C.; Marsais, F.; Quéguiner, G. *Tetrahedron Lett.* **1993**, *34*, 7919.

93TL7493 Clayton, S. C.; Regan, A. C. *Tetrahedron Lett.* **1993**, *34*, 7493.

93TL8329 Rao, A. V. R.; Reddy, G. R. *Tetrahedron Lett.* **1993**, *34*, 8329.

94JOC6142 Fang, F. G.; Xie, S.; Lowery, M. W. *J. Org. Chem.* **1994**, *59*, 6142.

94T12329 Bracher, F.; Hildebrand, D. *Tetrahedron* **1994**, *50*, 12329.

94TL793 Amat, M.; Hadida, S.; Bosch, J. *Tetrahedron Lett.* **1994**, *35*, 793.

94TL2003 Rocca, P.; Marsais, F.; Godard, A.; Quéguiner, G. *Tetrahedron Lett.* **1994**, *35*, 2003.

94TL7123 Amat, M.; Sathyanarayana, S.; Hadida, S.; Bosch, J. *Tetrahedron Lett.* **1994**, *35*, 7123.

95H1055 Godard, A.; Marsais, F.; Plé, N.; Trecourt, F.; Turck, F.; Quéguiner, G. *Heterocycles* **1995**, *40*, 1055.

95CC1675 Lavilla, R.; Gullón, F.; Bosch, J. *J. Chem. Soc., Chem. Commun.* **1995**, *16*, 1675.

95JHC1171	Rocca, P.; Marsais, F.; Godard, A.; Quéguiner, G. Adams, L.; Alo, B. *J. Heterocycl. Chem.* **1995**, *32,* 1171.
95JOC292	Guillier, F.; Nivoliers, F.; Godard, A.; Marsais, F.; Quéguiner, G. Siddiqui, M. A.; Snieckus, V. *J. Org. Chem.* **1995**, *60*, 292.
95JOC7491	Cruskie, Jr., M. P.; Zoltewicz, J. A.; Abboud, K. *J. Org. Chem.* **1995**, *60*, 7491.
95M805	Bracher, F.; Papke, T. *Monatsh. Chem.* **1995**, *126*, 805.
95PHA182	Bracher, F.; Papke, T. *Pharmazie* **1995**, *50,* 182.
95SC2901	Rocca, P.; Marsais, F.; Godard, A.; Quéguiner, G. *Synth. Commun.* **1995**, *25*, 2901.
95SC3373	Rocca, P.; Marsais, F.; Godard, A.; Quéguiner, G. *Synth. Commun.* **1995**, *25*, 3373.
95TL1491	Melnyk, P.; Legrand, L.; Gasche, J.; Ducrot, P.; Thai, C. *Tetrahedron Lett.* **1995**, *51*, 1491.
95TL7085	Rocca, P.; Marsais, F.; Godard, A.; Quéguiner, G. *Tetrahedron Lett.* **1995**, *36*, 7085.
96JOC4623	Kelly, T. R.; Lang, F. *J. Org. Chem.* **1996**, *61*, 4623.
96JOM25	Godard, A.; Rocca, P.; Pomel, L.; Thomas-dit-Dumont, L.; Rovera, J. C.; Marsais, F.; Quéguiner, G. *J. Organomet. Chem.* **1996**, *517*, 25.
96AQ62	Amat, M.; Hadida, S.; Bosch, J. *An. Quím. Int. Ed.* **1996**, *92*, 62.
96SC4421	Guillier, F.; Nivoliers, F.; Godard, A.; Marsais, F.; Quéguiner, G. *Synth. Commun.* **1996**, *26*, 4421.
96TL3071	Amat, M.; Sathyanarayana, S.; Hadida, S.; Bosch, J. *Tetrahedron Lett.* **1996**, *37*, 3071.
97JHC1789	Fan, P.; Ablordeppey, S. Y. *J. Heterocycl. Chem.* **1997**, *34*, 1789.
97JOC3158	Amat, M.; Hadida, S.; Pshenichnyi, G.; Bosch, J. *J. Org. Chem.* **1997**, *62*, 3158.
97JOC6588	Henegar, K. E.; Ashford, S. W.; Baughman, T. A.; Sih, J. C.; Gu, R.-L. *J. Org. Chem.* **1997**, *62*, 6588.
97SL298	Murata, N.; Sugihara, T.; Kondo, Y.; Sakamoto, T. *Synlett* **1997**, 298.
97SL1067	Timári, G.; Soós, T. Hajós, G. *Synlett* **1997**, 1067.
97TA829	Kaneko, S.; Yoshino, T.; Katoh, T.; Terashima, S. *Tetrahedron: Asymm.* **1997**, *8*, 829.
97TL8439	Hong, C. Y.; Overman, L. E.; Romero, A. *Tetrahedron Lett.* **1997**, *38*, 8439.
98T5471	Kaneko, S.; Yoshino, T.; Katoh, T.; Terashima, S. *Tetrahedron* **1998**, *54*, 5471.
98TA3065	Alvarez, R.; de Lera, A. R. *Tetrahedron: Asymm.* **1998**, *9*, 3065.
98TL411	He, X.-C.; Wang, B.; Bai, D. *Tetrahedron Lett.* **1998**, *39*, 411.
98TL2119	Abouabdellah, A.; Dodd, R. H. *Tetrahedron Lett.* **1998**, *39*, 2119.
98TL2341	Choshi, T.; Matsuya, Y.; Okita, M.; Inada, K.; Sugino, E.; Hibino, S. *Tetrahedron Lett.* **1998**, *39*, 2341.
98TL6465	Arzel, E.; Rocca, P.; Marsais, F.; Godard, A.; Quéguiner, G. *Tetrahedron Lett.* **1998**, *39*, 6465.
99EJOC373	Lavilla, R.; Gullón, F.; Bosch, J. *Eu. J. Org. Chem.* **1999**, 373.
99SL596	Knölker, H.-J.; Reddy, K. R. *Synlett* **1999**, 596; and references cited therein.
99SL114	Namyslo, J.; C.; Kaufman, D. E. *Synlett* **1999**, 114.
99SL804	Namyslo, J.; C.; Kaufman, D. E. *Synlett* **1999**, 804.
99T8047	Kasyan, A.; Wagner, C.; Maier, M. E. *Tetrahedron* **1999**, *54*, 8047.
99T11879	Cox, C. D.; Malpass, J. R. *Tetrahedron* **1999**, *55*, 11879.
99TL557	Barros, M. T.; Maycock, C. D.; Ventura, M. R. *Tetrahedron Lett.* **1999**, *40*, 557.
99TL1419	Malpass, J. R.; Cox, C. D. *Tetrahedron Lett.* **1999**, *40*, 1419.

Chapter 4.1

Three-Membered Ring Systems

Albert Padwa
Emory University, Atlanta, GA, USA
chemap@emory.edu

S. Shaun Murphree
Allegheny College, Meadville, PA, USA
smurphre@alleg.edu

4.1.1 INTRODUCTION

The study of three-membered heterocycles, both as reagents and synthetic targets, has continued to flourish. More than ever, these systems represent an attractive combination of predictable, controllable, and versatile reactivity that has been extensively utilized by the synthetic community. As in earlier volumes, this review seeks not to be comprehensive, but rather to provide an overview of the past year's literature from the synthetic chemist's perspective, focusing on preparatively useful protocols or novel methodology, and highlighting advances in selectivity and ease of use. Organization of the chapter is similar to that seen in previous years.

4.1.2 EPOXIDES

4.1.2.1 Preparation of Epoxides

The transition-metal catalyzed asymmetric epoxidation of unfunctionalized alkenes is now well-established, and chiral (Mn)-salen complexes, such as the Jacobsen (**1**) and the Katsuki (**2**) catalysts, have become the current standards for this approach. In the presence of an oxidant, the manganese(III)-salen catalyst is converted to a manganese(V)-oxo intermediate (**3**), which serves as the active oxygen transfer reagent. Recent models of these systems using density functional theory (Becke3LYP) suggest that this intermediate exists in either a high-spin quintet or triplet state <99OL419>. The known predilection of these catalysts to form *cis*-epoxides has been rationalized by invoking a spin transition from a triplet to a quintet state along the reaction pathway <99JA5083>.

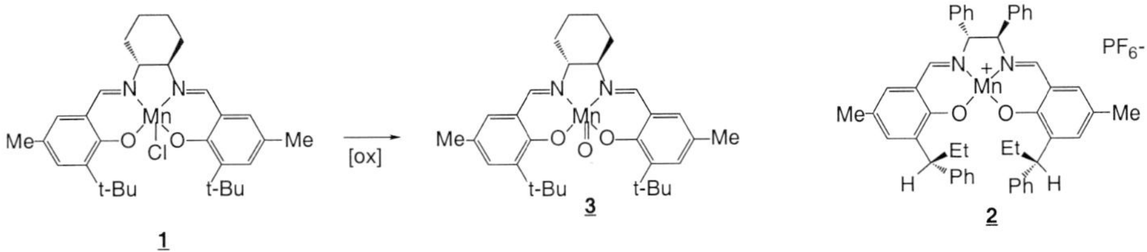

Most of the oxidants used with these catalysts tend to be water soluble. However, the use of tetrabutylammonium monopersulfate (**5**) has been reported to provide relatively smooth oxidation with good *ee's* (*e.g,* **4** → **6**) under monophasic conditions in organic solvents. The enantioselectivity is both substrate and catalyst dependent, with best results being obtained from electron-rich alkenes and Katsuki-type catalysts (*cf.* **2**) <99TL1001>. Che and co-workers <99CC1789> have also reported on the use of an immobilized chromium binaphthyl catalyst, which offers the advantage of simple work-up and catalyst recovery.

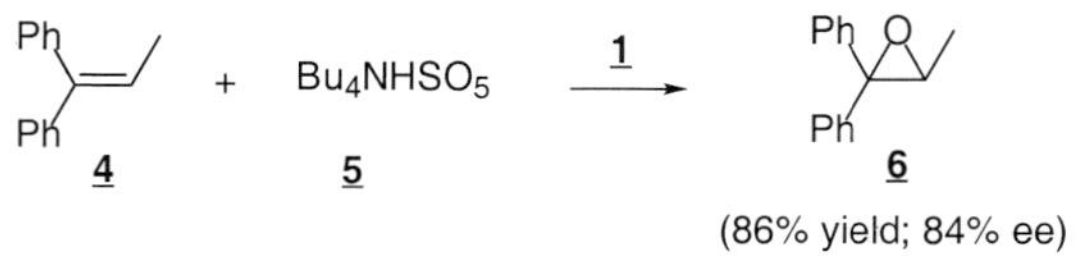

An interesting reversal of chiral induction in chromium(III)-salen complexes using a tartaric derived alicyclic diamine moiety (*i.e.*, **7**) has been observed by Mosset, Saalfrank, and co-workers <99T1063>. Thus, epoxidation of the chromene **8** using catalyst **7** and an oxidant consisting of MCPBA/NMO afforded the *3S,4S* epoxide **9**, whereas the Jacobsen catalyst (**1**) provided the corresponding *3R,4R* enantiomer. A mechanistic rationalization for this curious crossover has not yet been proposed.

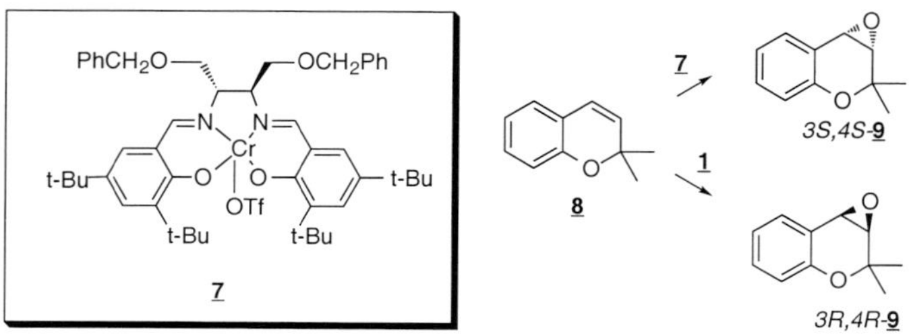

Other studies on the chromium(III)-salen catalysts of type **10** have shown that the 3'-substituent (*i.e.*, R) exhibits relatively little sensitivity with regard to chiral induction, in stark contrast to the analogous Mn-salen complexes, in which the 3'-position must bear a sterically bulky group for acceptable enantiomeric excesses. Thus, the 3'-chloro catalyst with a triphenyl-

Figure 1: Open Side-on Approach for *trans*-Alkenes

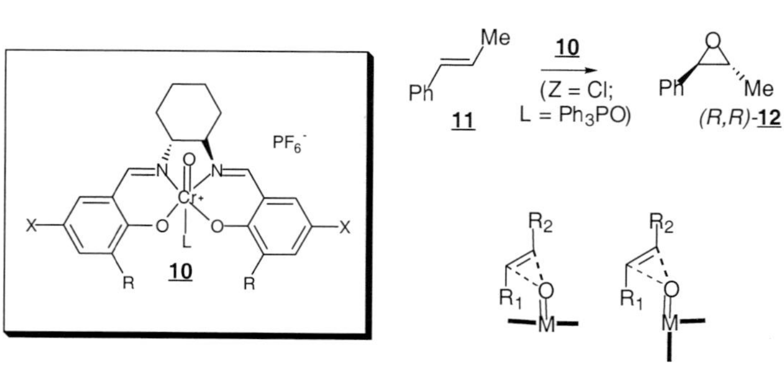

phosphine oxide ligand (L = Ph_3PO) and iodsylbenzene as a stoichiometric oxidant promotes the epoxidation of *E*-β-methylstyrene (**11**) in 71% yield and 79% *ee*. The high level of enantiomeric excess observed when a *trans*-alkene was utilized was rationalized on the basis of providing less steric encumbrance for a side-on approach, as shown in Figure 1 <99TL3613>.

Katsuki has extended his earlier work on asymmetric induction using achiral catalysts such as **13**. In these systems, the stereochemical bias is imbued by a chiral non-racemic axial ligand, such as (+)-3,3'-dimethyl-2,2'-bipyridine *N,N'*-dioxide (**14**), which was purified by crystallization with (*S*)-binaphthol. Epoxidation using these conditions resulted in good *ee's* and fair yields, as exemplified by the preparation of chromene epoxide **16** <99SL783>.

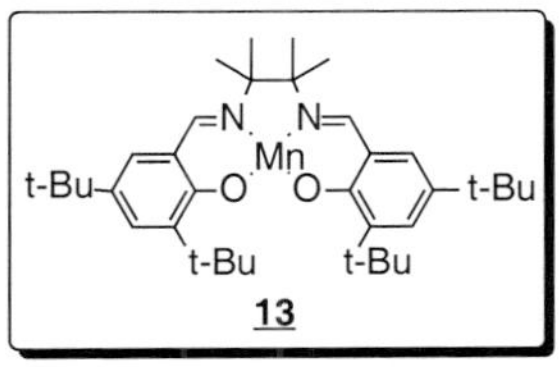

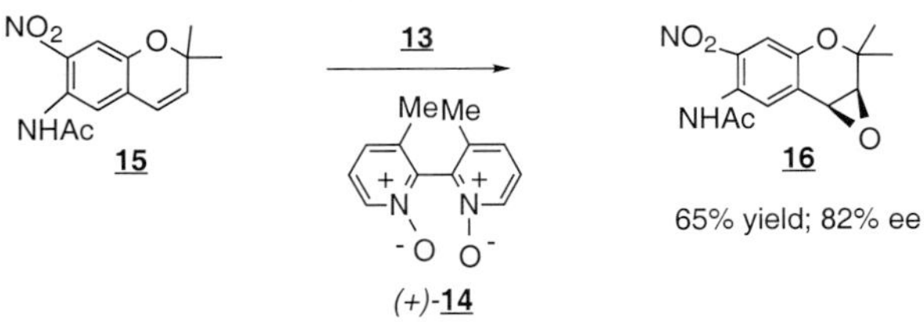

The photocatalyzed asymmetric epoxidation of alkenes using a binap-equipped Ru-salen complex was also reported <99SL1157>.

With a twist on the Sharpless asymmetric epoxidation protocol, Yamamoto and co-workers <99JOC338> have developed a chiral hydroxamic acid (**17**) derived from binaphthol, which serves as a coordinative chiral auxiliary when combined with $VO(acac)_2$ or $VO(i\text{-}PrO)_3$ in the epoxidation of allylic alcohols. In this protocol, triphenylmethyl hydroperoxide (TrOOH) provides markedly increased enantiomeric excess, compared to the more traditional *t*-butyl hydroperoxide. Thus, the epoxidation of *E*-2,3-diphenyl-2-propenol (**18**) with 7.5 mol% $VO(i\text{-}PrO)_3$ and 15 mol% of **17** in toluene (-20 °C; 24 h) provided the *2S,3S* epoxide **19** in 83% *ee*.

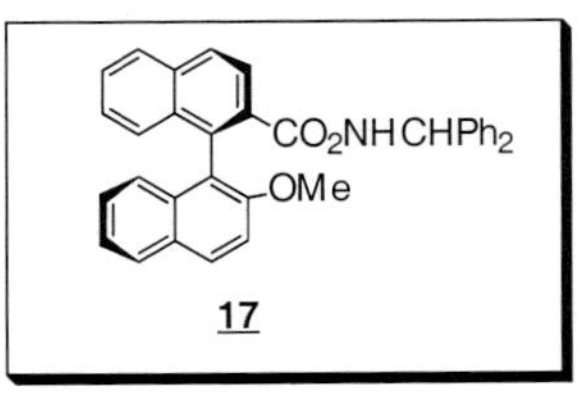

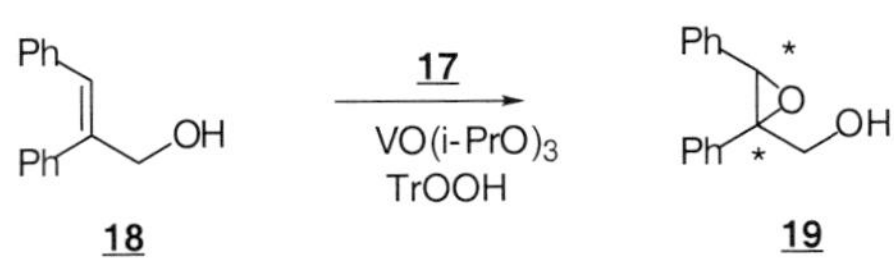

Allylic alcohols can also be epoxidized with methyltrioxorhenium (MTO). However, in contrast to the early transition metal catalysts, metal-alcoholate binding does not appear to be operative, but rather straightforward hydrogen bonding, as demonstrated by the epoxidation of geraniol (**20**)

using MTO with urea/hydrogen peroxide adduct (UHP) as the oxidant. Simple allylic alcohols are efficiently oxidized with preferential formation of the *threo* epoxide (*e.g.*, **23**). This is presumably the result of 1,3-allylic strain in the hydrogen bonded catalyst-substrate complex. Hydrogen peroxide generally gives higher yields <99JOC3699>.

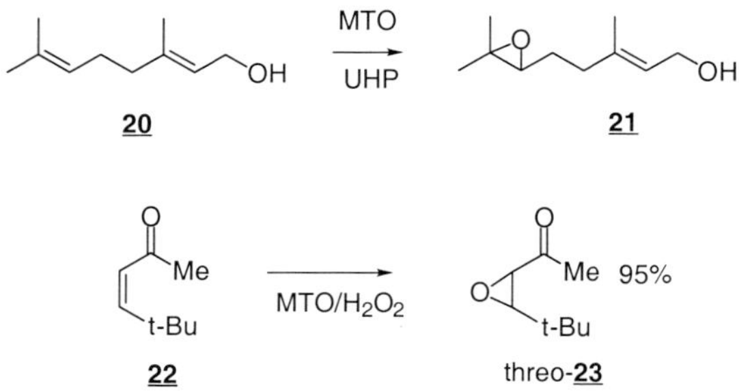

In addition, the use of trifluoroethanol as solvent has been shown to provide very high yields of epoxides, both terminal and internal, as exemplified by the oxidation of vinylcyclohexane (**24**) <99CC821>.

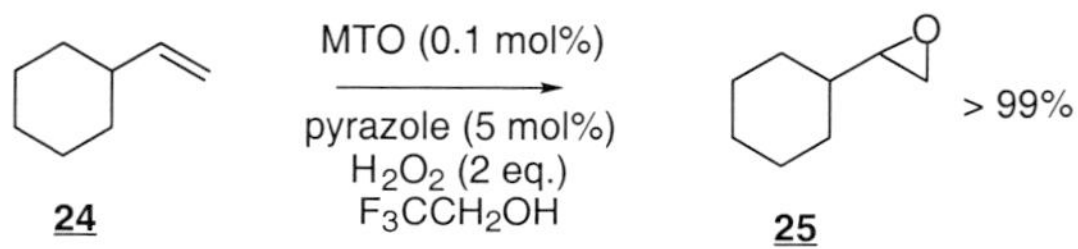

There continues to be an increasing level of activity centered about the use of porphyrin catalysts for the epoxidation of alkenes of various configurations. For example, the sterically encumbered *trans*-dioxoruthenium(VI) porphyrin (**26**) was found to catalyze the epoxidation of a variety of alkenes in yields from fair to excellent (*e.g.*, **27** → **28**). Kinetic studies on a series of *para*-substituted styrenes point to a mechanism which proceeds *via* a rate-limiting benzylic radical formation. The high degree of stereoretention in *cis*-alkenes was attributed to steric crowding which prevents C-C bond rotation of the intermediate radical. This same steric bulk prevents the familiar side-on approach of the alkene substrate, so that a head-on approach is postulated <99JOC7365>.

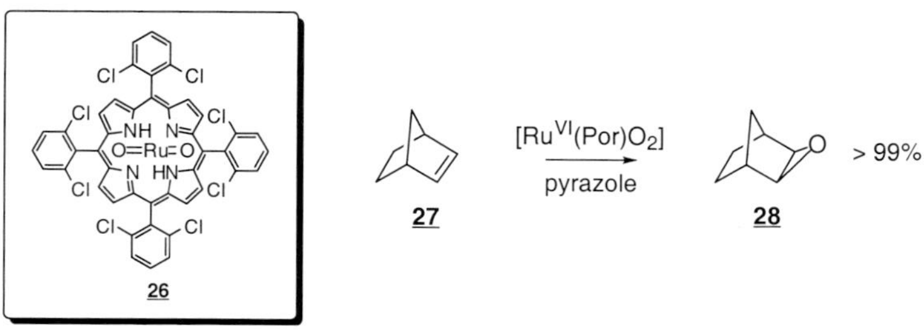

Chiral porphyrins are also effective in the asymmetric epoxidation of alkenes. For example, a C_2-symmetric iron porphyrin (**29**) <99JA460> catalyzes the efficient epoxidation of terminal alkenes, such as **30**, with very good *ee's* and up to 5500 turnovers. Similarly, *trans*-disubstituted alkenes can be epoxidized with fair to good *ee's* using a D_2-symmetric doubly strapped chiral dioxo-ruthenium(VI) porphyrin <99CC409>.

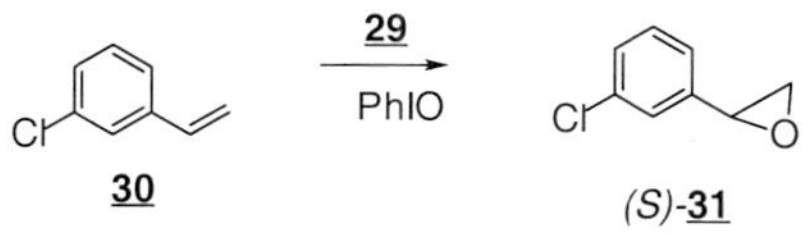

During the past year, chloroperoxidase (CPO) was found to catalyze the smooth asymmetric epoxidation of functionalized *cis*-alkenes, such as the unsaturated ester **32**. The reaction appears to be limited to 2-alkenes (*i.e.*, methyl group on one side of the alkene), although some branching on the longer alkyl chain is tolerated. Allylic alcohols are oxidized to the corresponding unsaturated aldehydes but without epoxide formation <99TL1641>.

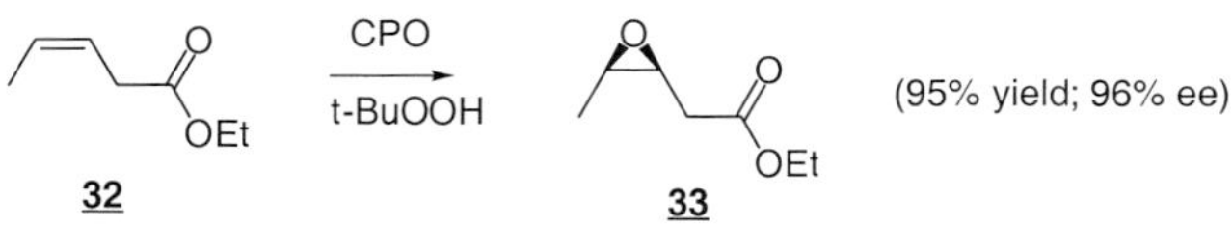

Particularly interesting are the recent contributions toward the development of environmentally benign oxidation protocols, which promise to find broader application in commercial processes. Toward this end, Kaneda and co-workers <99JOC2966> have developed an epoxidation system using hydrotalcite as catalyst and hydrogen peroxide as the oxidant. This protocol, which can show very high selectivity (*e.g.*, **34** → **35**), has the limitation of requiring stoichiometric amounts of isobutyramide. Significant acceleration is observed, however, upon addition of surfactants, such as sodium dodecyl sulfate.

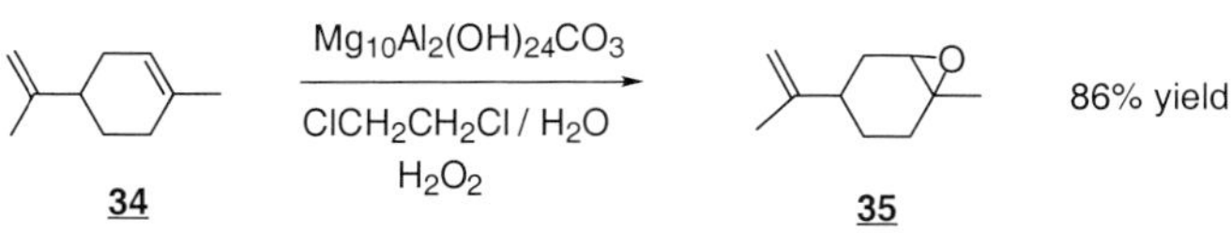

One of the most attractive oxidants for this chemistry is dioxygen, both from an environmental and cost standpoint. In this vein, a metal-free epoxidation protocol was reported, which proceeded by the *in situ* generation of hydrogen peroxide from O_2 through a complex series of steps involving *N*-hydroxyphthalimide (NHPI). Once formed, the H_2O_2 is activated by addition onto the somewhat esoteric solvent, trifluoroacetone, to give 2-hydroperoxy-hexafluoropropan-2-ol (**38**) as the oxygen transfer reagent,. The reaction appears to be general and yields are very good. Regardless of the starting configuration of the alkene, a strong preference for the formation of *trans*-epoxides was observed (*e.g.*, **36** → **37**) <99CC727>.

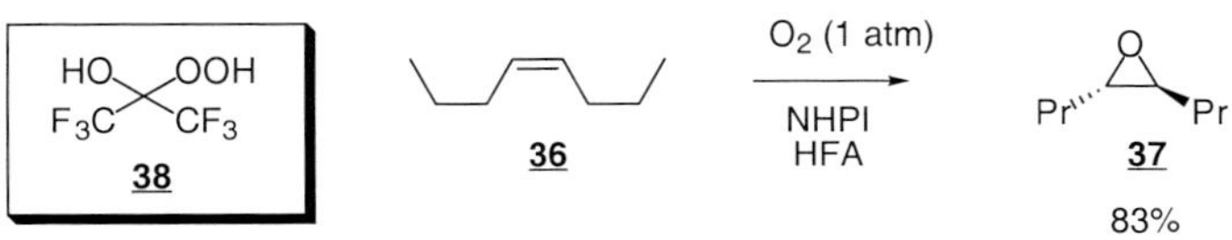

Some interesting solid-phase catalysts were developed last year by the introduction of Co(III) or Mn(III) ions into the precursor gels of molecular sieves before calcining. These novel complexes appear to catalyze the aerobic epoxidation of alkenes, although the methodology requires forcing conditions and the use of a sacrificial aldehyde, such as benzaldehyde <99CC829>.

Electron-deficient alkenes generally require the use of some other epoxidation procedure, owing to their low reactivity under electrophilic addition conditions. Within this category, α,β-unsaturated ketones tend to be the substrates of interest, and basic oxygen transfer reagents are frequently encountered, such as H_2O_2/NaOH, *t*-BuOOH/NaOH, and NaOCl. Much activity has centered around the modification of these traditional conditions to accommodate asymmetric induction. In this regard, variously substituted *Cinchona* alkaloids (*e.g.*, **39** - **41**) have received a fair amount of attention over the past year.

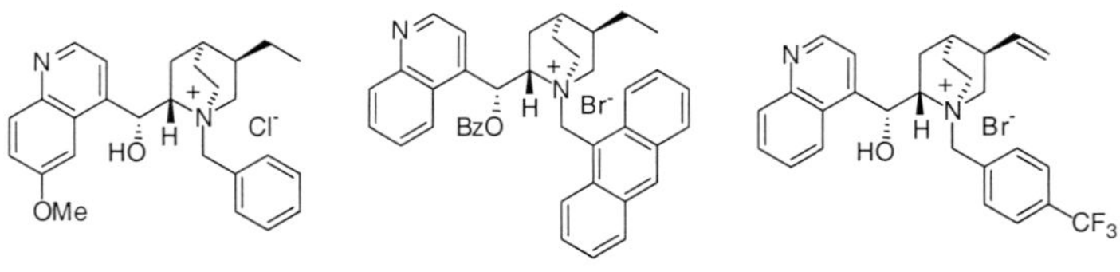

39 40 41

The *N*-anthracenylmethyl derivative **40** was also used under phase transfer conditions (water/toluene) with sodium hypochlorite as the oxidant for the smooth epoxidation of aryl alkyl enones (*e.g.*, **42** → **43**). Enantioselectivities range from 71 – 90% and yields were in the good to excellent range <99T6289>. Studies on the anthracenyl derivative **40** were used to rationalize the the observed rate acceleration and stereochemical outcome. The mechanism proposed involved a three-dimensional arrangement of catalyst, substrate (in this case, benzal-4-fluoroacetophenone), and oxidant leading to the least crowded transition state <99OL1287>.

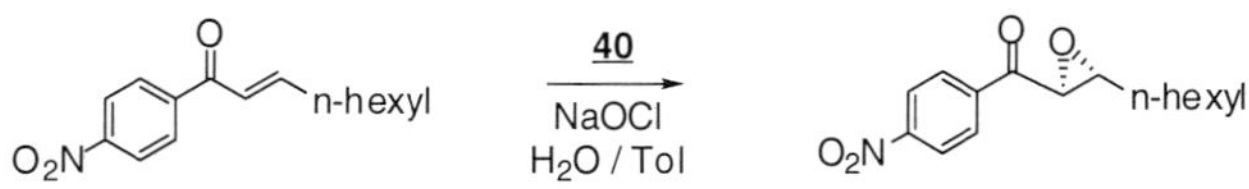

42 43

79% yield; 90% ee

Taylor and co-workers <99SL795> reported some promising results with DBU-like cyclic guanidine derivatives of type **44**, equipped with the appropriate chiral auxiliary. Both yields and *ee's* were modest, but the ready availability of the amine, and the flexibility of substitution on the

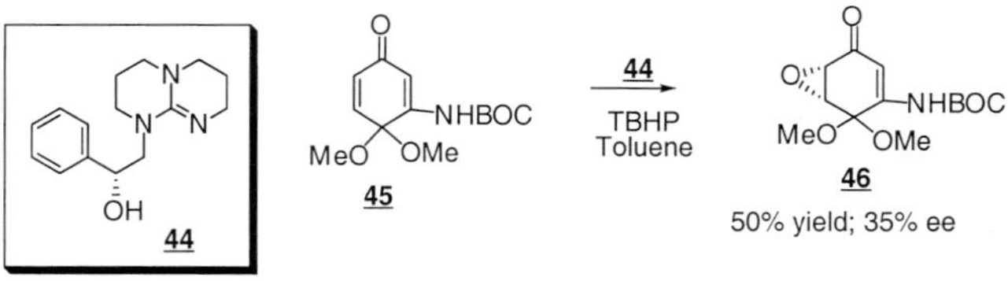

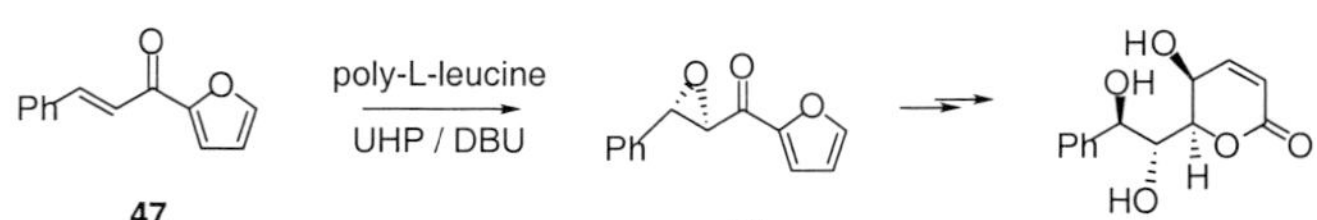

47 48 49

catalyst, should make this protocol amenable to optimization. Other noteworthy catalyst systems include a solid-supported binaphthyl-zinc catalyst, prepared by treating binaphthyl polymers with diethyl zinc <99JOC8149>, and poly-*L*-leucine under biphasic conditions, a system which has been used to prepare naturally-occurring styryl lactones **49** from furyl styryl ketone (**47**) <99JCS(P1)103>. Poly-*L*-leucine has also been adsorbed onto silica to provide a very active solid-phase catalyst <99JCS(P1)1397>.

While the oxygenation of alkenes represents the lion's share of methodology development, new and improved protocols for the [C=O + C] approach to epoxides have also been reported. For example, the phase-transfer-catalyzed asymmetric Darzens reaction using the chiral *Cinchona* derivative **41** afforded spirocyclic epoxide **52** in good to excellent yields and fair to good *ee's* <99T6375>.

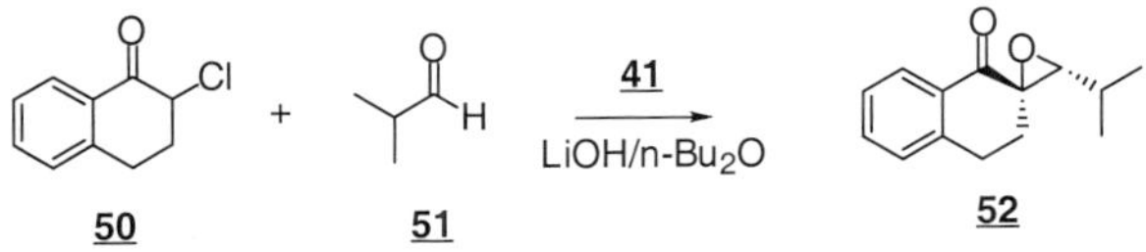

The carbon fragment used in this approach can also be provided by sulfur ylides. In this arena, Metzner and co-workers <99JCS(P1)731> developed a novel asymmetric variant employing (+)-(*2R,5R*)-2,5-dimethylthiolane (**53**) as the chiral auxiliary to prepare *trans*-(*S,S*)-stilbene oxide (**56**). Chiral epoxides have also been prepared from aldehydes using sulfur ylides derived from the products of Baker's yeast reductions <99SL1328>.

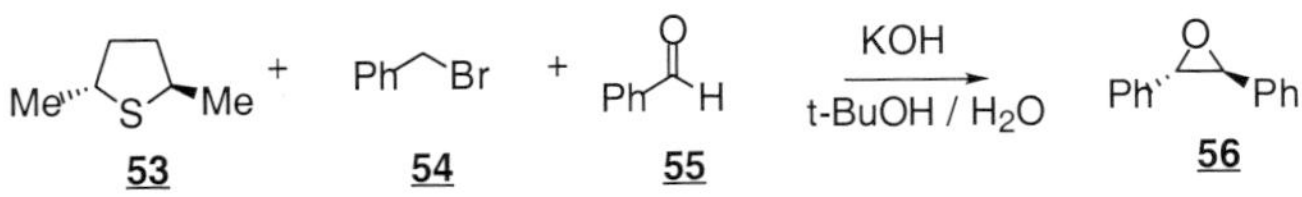

4.1.2.2 Reactions of Epoxides

Epoxides easily undergo nucleophilic ring-opening reactions, with hydrolysis and alcoholysis being perhaps the first examples to spring to mind. In this regard, the titanium salts $TiCl_3(OTf)$ and $TiO(TFA)_2$ have been reported to function as efficient catalysts for the stereospecific acid-catalyzed alcoholysis of asymmetric monosubstituted epoxides. Thus, treatment of *R*-(+)-styrene oxide (**57**) with methanol and $TiO(TFA)_2$ resulted in the rapid formation of the corresponding alkoxyalkanol (**58**) with high yield (96%) and enantiomeric excess (95%) <99SC1017>.

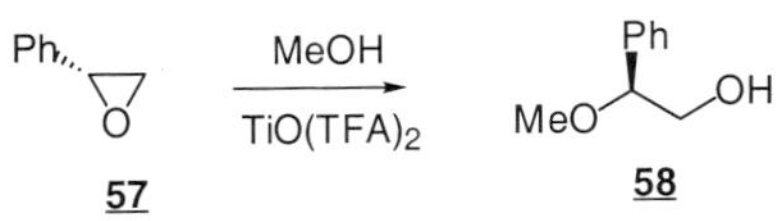

In the realm of hydrolytic reactions, Jacobsen has applied his work with chiral salen complexes to advantage for the kinetic resolution of racemic epoxides. For example, the cobalt salen catalyst **59** gave the chiral bromohydrin **61** in excellent *ee* (>99%) and good yield (74%) from the racemic bromo-epoxide **60**. The higher than 50% yield, unusual for a kinetic resolution, is attributed to a bromide-induced dynamic equilibrium with the dibromo alcohol **62**, which allows for conversion of unused substrate into the active enantiomer <99JA6086>. Even the recalcitrant 2,2-disubstituted epoxides (*e.g.*, **64**) succumbed to smooth kinetic resolution upon treatment with trimethylsilyl azide and the chromium salen catalyst **63**. Interestingly, these substrates proved to be unreactive to the corresponding cobalt salen complexes <99TL7303>.

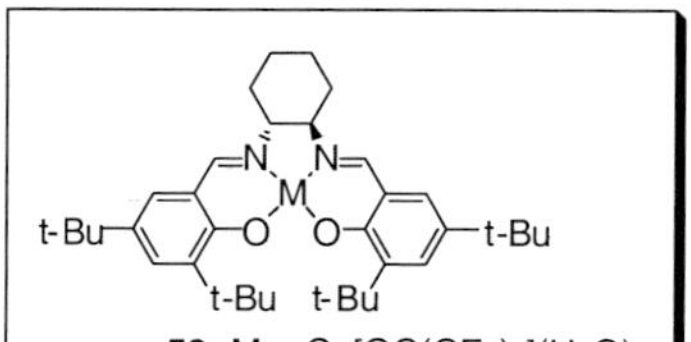

59, M = $Co[OC(CF_3)_3](H_2O)$

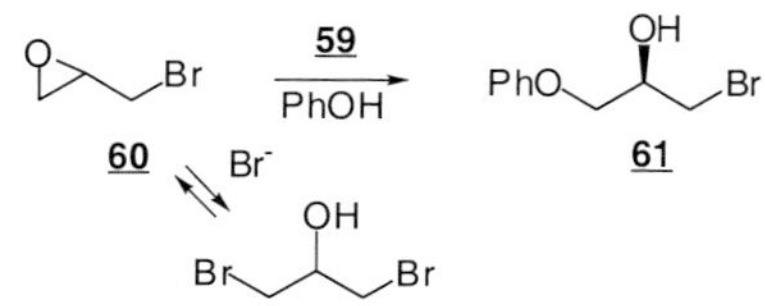

60 **61** **62**

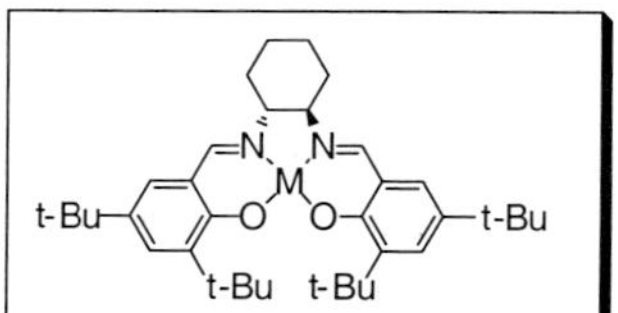

63, M = $Cr(N_3)$

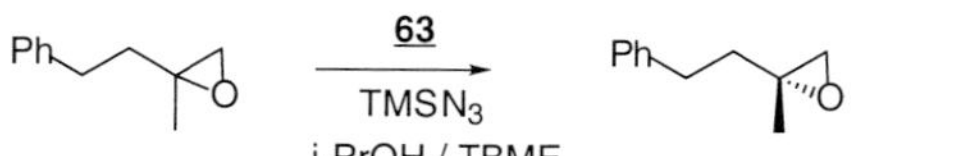

(±) **64**

44% yield; 97% ee

Over the past year, racemic 1,2-dialkyl epoxides were resolved enzymatically using soluble epoxide hydrolase (sEH), although the outcome of the reaction is characteristically substrate-dependent. In an example of the best enantioselection exhibited, epoxide **65** afforded the (*3R,4R*)-diol **66** upon treatment with sEH at pH 7.4. The course of these reactions is different from those in which the same substrates were treated with microsomal epoxide hydrolase <99T11589>.

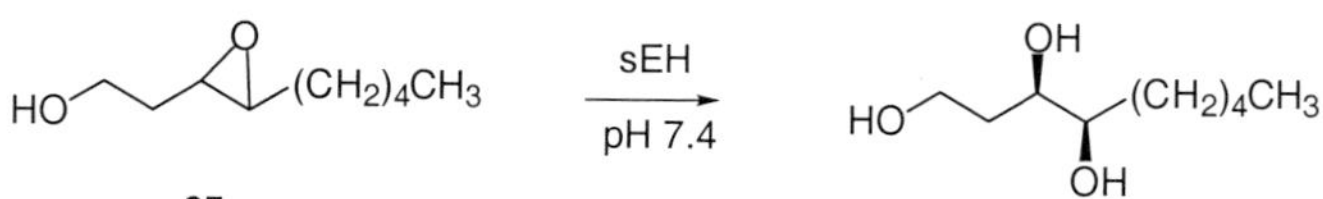

65 **66**

More structurally complex epoxides can be ring-opened intramolecularly in a synthetically useful fashion. Thus, in their approach to methyl-substituted *trans*-fused tetrahydropyran subunits found in marine natural products, Mori and co-workers <99TL8019> treated the polyfunctional arylsulfonyl epoxide **67** with Lewis acid to induce a 6-*endo* cyclization onto the epoxide moiety, with concomitant ejection of arylsulfinate, to provide the bicyclic ether **68**. This system was found to be highly sensitive to the nature of the Lewis acid catalyst used.

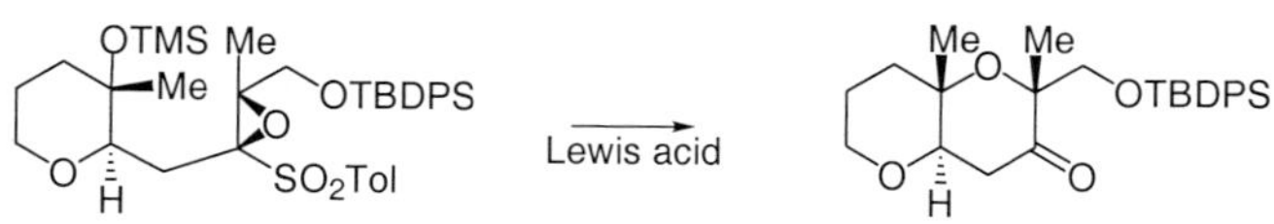

67 **68**

Along these lines, Jacobsen and co-workers <99AG(E)2012> published an interesting enantioselective cyclization of meso epoxy alcohols which were catalyzed by the cobalt(III)salen complex **69**. Thus, epoxy alcohol **70** was converted to the chiral bicyclic hydroxy ether **71** in 96% yield and 98% *ee*.

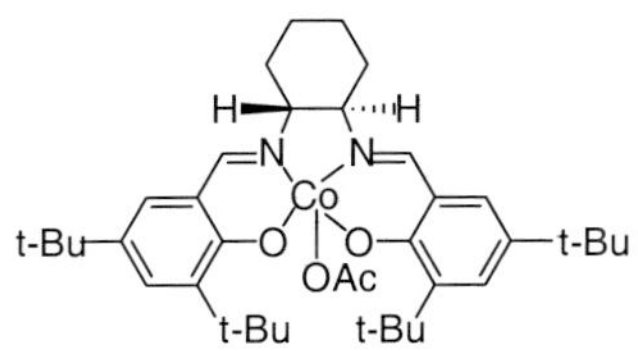

69

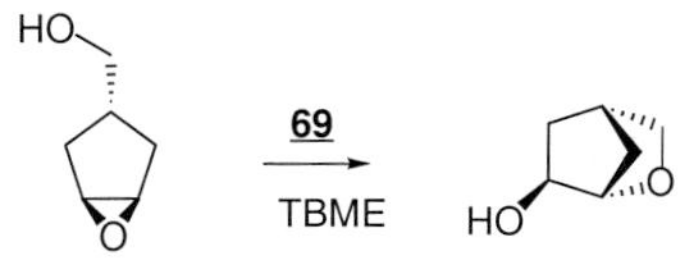

70 71

Aside from water and alcohols, a wide variety of nucleophiles can also induce synthetically useful epoxide ring-opening reactions. In this regard, the azide anion is often encountered in this role. For example, diphenyl phosphorazidate (**73**) was found to cleave epoxides in the presence of 4-dimethylaminopyridine (DMAP) and lithium perchlorate to give *O*-diphenylphosphoryl vicinal azidohydrins **74**, which are precursors for β-amino alcohols and aziridines. The reaction proceeds with high regio- and stereoselectivity, where the less substituted epoxide carbon generally undergoes nucleophilic attack. Epoxyketone **75** gave α-azido enone **76** <99TL7105>.

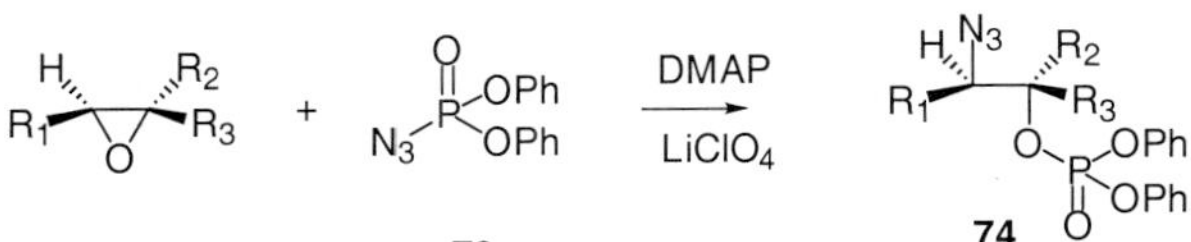

72 73 74

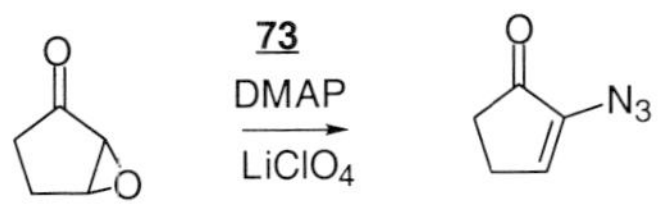

75 76

1,2-Azidoalcohols (**79**, **81**) can be accessed directly through the cerium-catalyzed addition of sodium azide onto mono-substituted epoxides. When the substituent is a simple alkyl or aryl group, nucleophilic attack at the more substituted epoxide carbon was observed (*i.e.*, **78** → **79**). However, when a phenoxy group was incorporated into the side chain (*e.g.*, **80**), a crossover to attack on the unsubstituted methylene carbon was encountered <99SC561>.

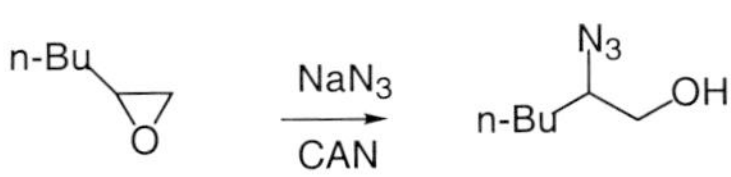

78 79

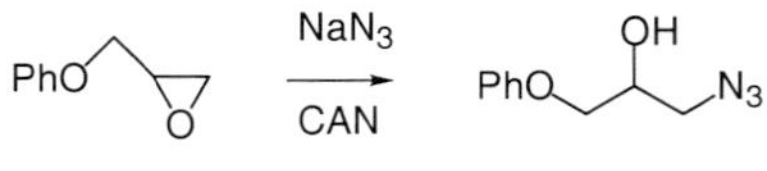

80 81

The site of attack can also be directed by functionality on the substrate itself, as in the phenylboronate-mediated C-2 selective azide ring-opening reaction of *trans*-2,3-epoxy alcohols (**82**) by sodium azide. In this reaction, the nucleophile is delivered intramolecularly from the azidoboronate intermediate **83**. Yields are generally good to excellent <99TL4589>.

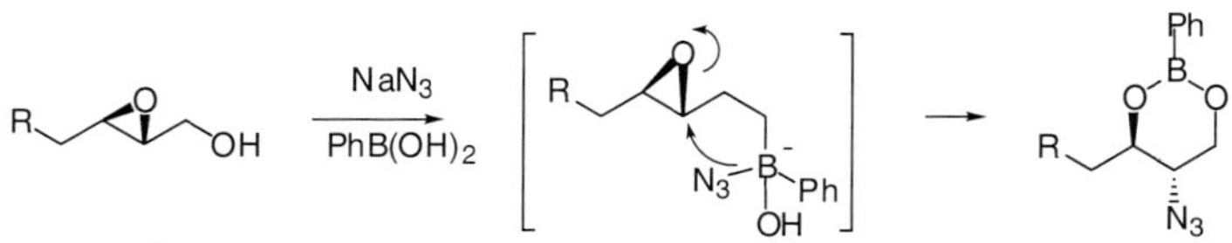

82 83 84

Wrapping up the nitrogen-based nucleophiles, aromatic amines were also noted to cleave epoxides in the presence of stannic or cupric triflate to form β-amino alcohols such as **86** directly. This protocol appears to be general for aromatic amines, even strongly electron-deficient ones. Aliphatic amines are completely unreactive under the same experimental conditions <99JOC287>.

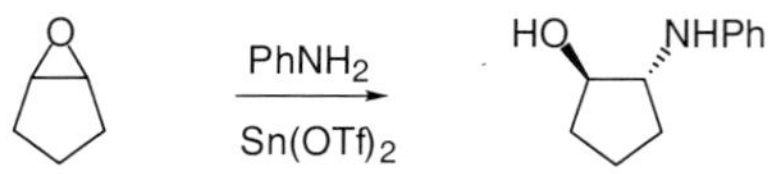

85 86

The epoxide moiety has often been utilized for carbon-carbon bond forming reactions. One of the simplest examples involves the ring-opening of epoxides with cyanide anion. Benedetti and co-workers <99TL1041> reported that diethylaluminum cyanide is a particularly useful reagent for the synthesis of 1-cyano-2,3-diols (**88**) by the addition of cyanide onto the functionalized epoxide **87**. The reaction is both regio- and stereoselective, with nucleophilic attack occurring at C-3 — being directed by the coordination of Lewis acid to the oxygen atom — with inversion of configuration.

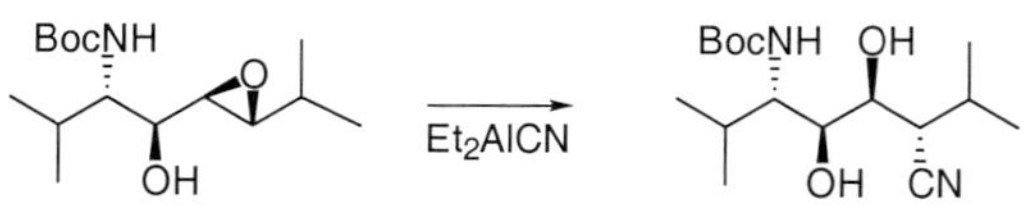

87 88

Terminal epoxides undergo ring-opening with sodium cyanide using a cerium(IV)triflate catalyst to provide β-hydroxy nitriles (**90**) <99SC2249>, or with trimethylsilyl cyanide using Mn-salen immobilized on MCM-41 mesoporous material, to give the trimethylsiloxy derivatives **92** in high yield <99SC1121>. In both cases, attack takes place at the less substituted epoxide carbon.

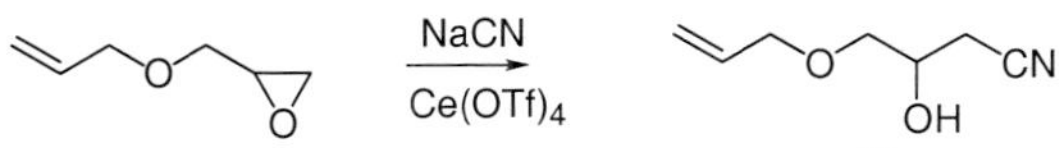

89 90

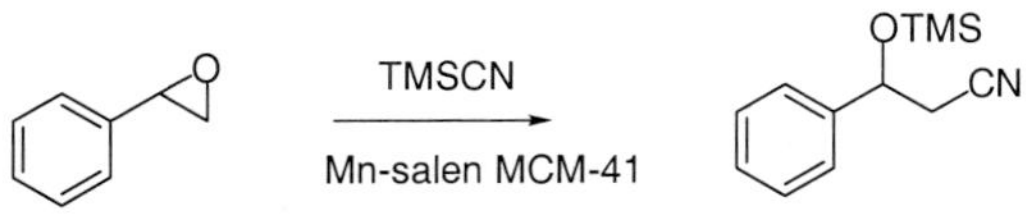

91 92

Other reagents have also been used to effect carbon-carbon bond formation. For example, chiral monosubstituted epoxides (**93**) can be regioselectively carbomethoxylated under relatively mild conditions with CO/H_2 in the presence of the salen complex **69**. The reaction proceeds with retention of chirality about the secondary epoxide carbon, and represents a new route to chiral β-hydroxy esters **94** <99JOC2164>.

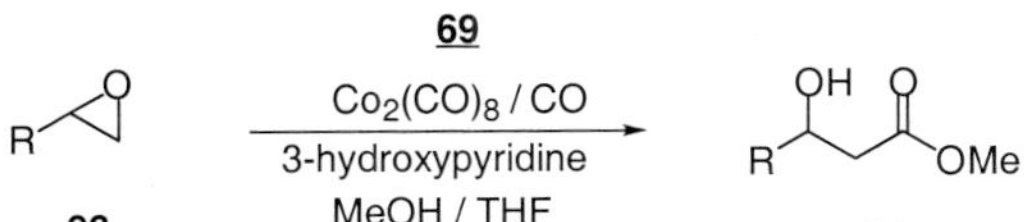

93 94

Trimethylaluminum was found to catalyze the addition of alkynyllithiums onto heterosubstituted epoxides **95** to give the alkynyl alcohols **96** <99JA3328>. In the presence of water, trimethylaluminum will induce smooth methylation of epoxides to give the corresponding alcohols (**98**, **99**) in good to excellent yields, although the regioselectivity is somewhat ambiguous <99TL5369>.

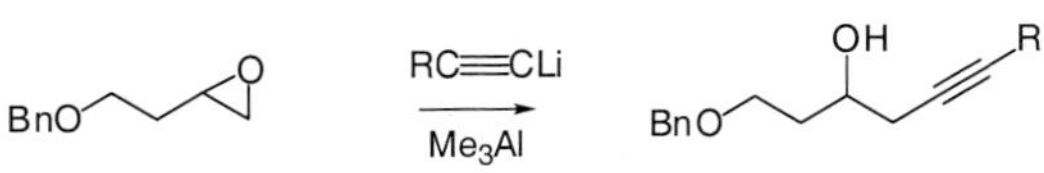

95 96

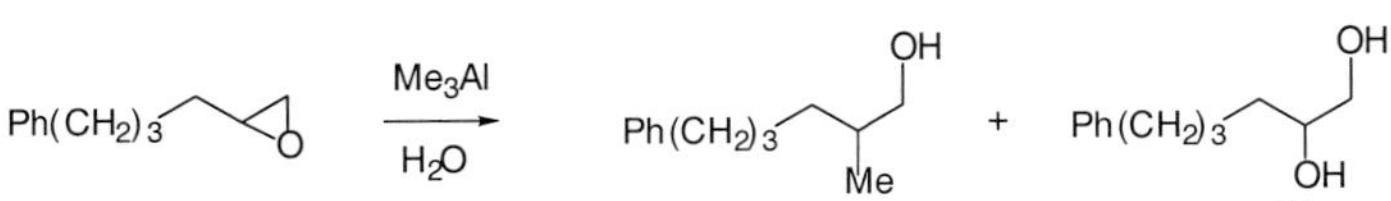

97 98 99

Epoxides can also serve as effective carbocyclization promotors, either through a polyene cyclization, as in the biomimetic epoxy-olefin cyclization of **100** in the presence of boron trifluoride etherate <99CC325>, or by a Friedel-Crafts approach, as exemplified by the cyclialkylation of arylalkyl epoxides **102** under the influence of solid acid catalysts <99EJOC837>.

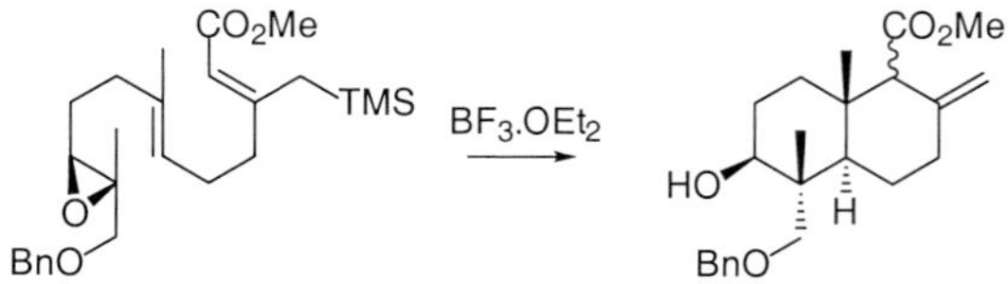

100 **101**

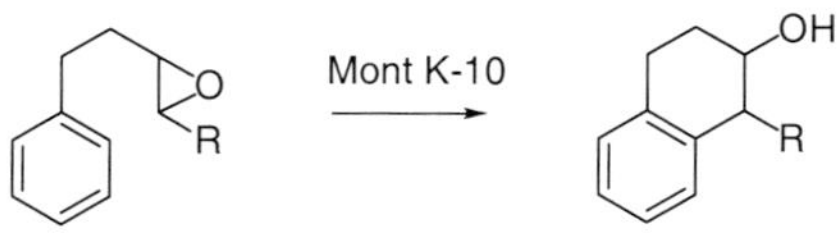

102 **103**

Epoxides can also undergo interesting and synthetically useful rearrangements to carbonyl groups in the presence of Lewis or Bronsted acids. The course of these rearrangements is highly dependent upon the nature of the substrate. For example, the simple monoalkyl-substituted epoxide **104** undergoes regioselective rearrangement in the presence of iron(III)tetraphenylporphyrin to give the corresponding aldehyde (**105**) *via* a 1,2-hydride shift <99TL7243>. On the other hand, rearrangement of the allyl epoxide **106** proceeds *via* a 1,2-alkyl shift to give the corresponding multifunctional ketone **107** <99TL3129>.

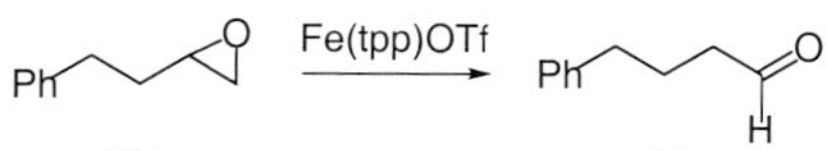

104 **105**

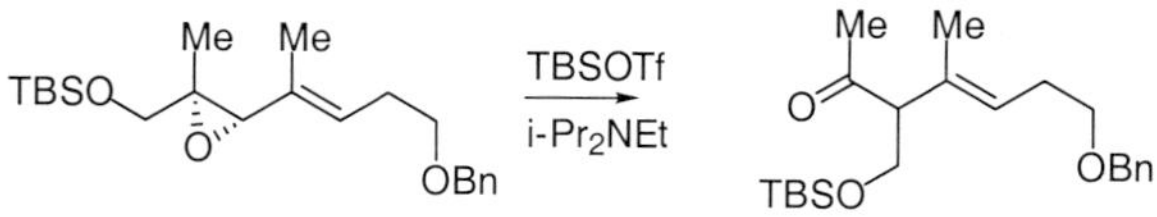

106 **107**

4.1.3 AZIRIDINES

4.1.3.1 Preparation of Aziridines

In contrast to the epoxides, preparative routes to the aziridines are fairly evenly split between the [C=N + C] and the [C=C + N] routes. Among contributions in the former category, aziridine carboxylate derivatives **110** can be prepared through the lanthanide-catalyzed reaction of imines with diazo compounds, such as ethyl diazoacetate (EDA). In this protocol, *N*-benzyl aryl aldimines and imines derived from aromatic amines and hindered aliphatic aldehydes are appropriate substrates <99T12929>. An intramolecular variant of this reaction (*e.g.*, **111** → **112**) has also been reported <99OL667>.

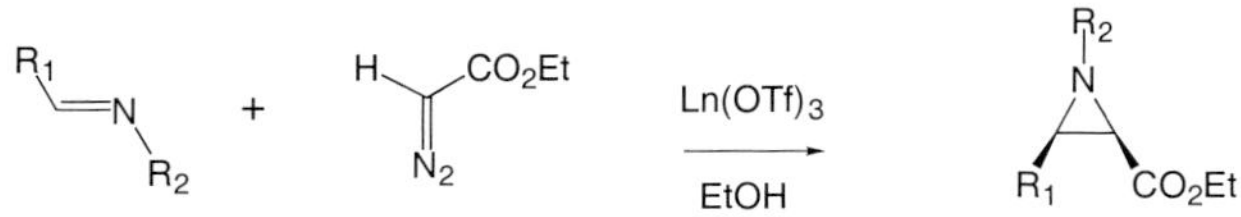

108 109 110

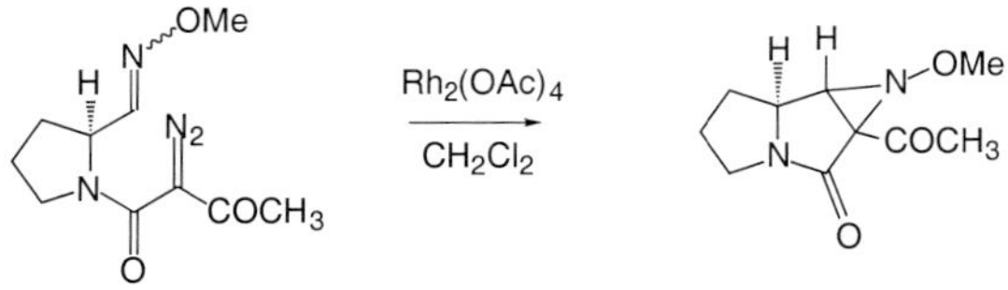

112

111

As with epoxide synthesis, formation of optically pure aziridines is of ever-increasing interest. In this regard, the asymmetric aziridination of α-imino esters **113** can be promoted by copper(I) catalysts equipped with chiral BINAP or *bis*-oxazoline ligands. In this case, the asymmetric induction is believed to occur through a pre-coordination of the imino ester with the catalyst <99JCS(P1)2293>. Simple imines, such as **117**, undergo aziridination under the influence of the chiral boron Lewis acid derived from S-VAPOL (**116**) to provide scalemic aziridines in excellent *ee's* in almost all cases. Yields are fair to good <99JA5099>.

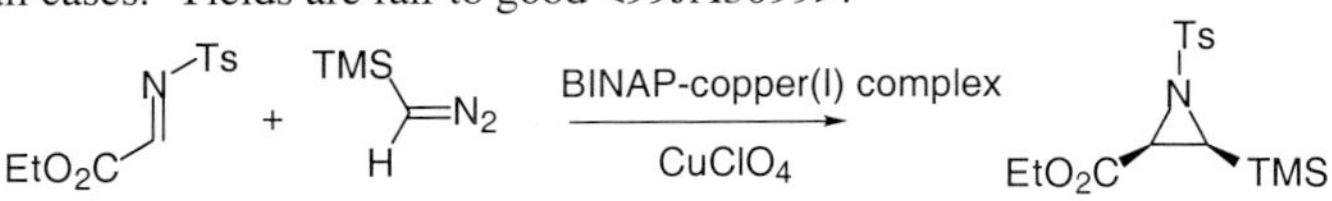

113 114 115

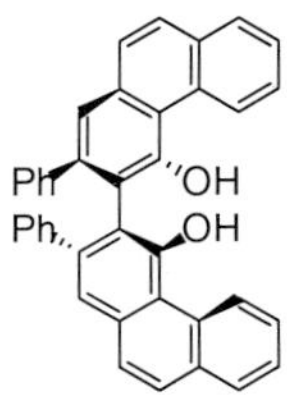

116, S-VAPOL

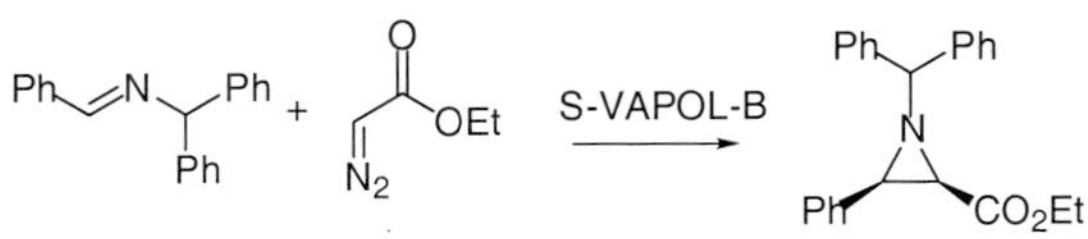

117 109 118

Alkenes can be aziridinated using a variety of nitrogen sources. Among the recently reported systems are Chloramine T (*N*-chloro-*N*-sodio-*p*-toluenesulfonamide) with pyridinium hydrobromide perbromide catalyst (*e.g.*, **119** → **120**) <99OL705>, the *N*-chloramine salt of *t*-butyl-sulfonamide (**121**), which serves as both nitrogen source and terminal oxidant, in the presence of phenyltrimethylammonium tribromide (PTAB) <99OL783>, and *N*-[2-(trimethylsilyl)ethanesulfony]iminophenyliodinane (**124**) <99JOC5304>. The last example is particularly interesting, inasmuch as it represents the first such *N*-alkylsulfonyl derivative used for such purposes. The trimethylsilylethanesulfonyl (SES) group has the advantage of being easily removed under conditions which are amenable to substrates with sensitive functionality.

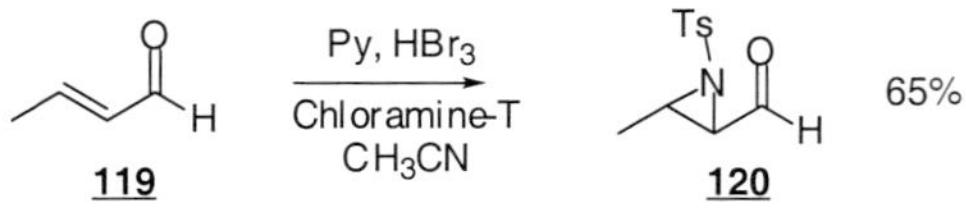

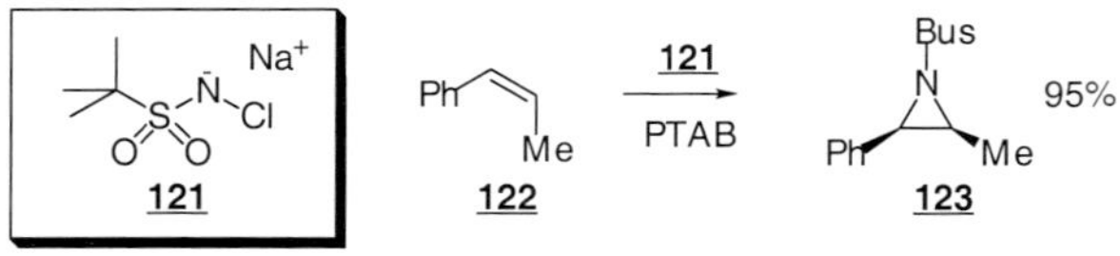

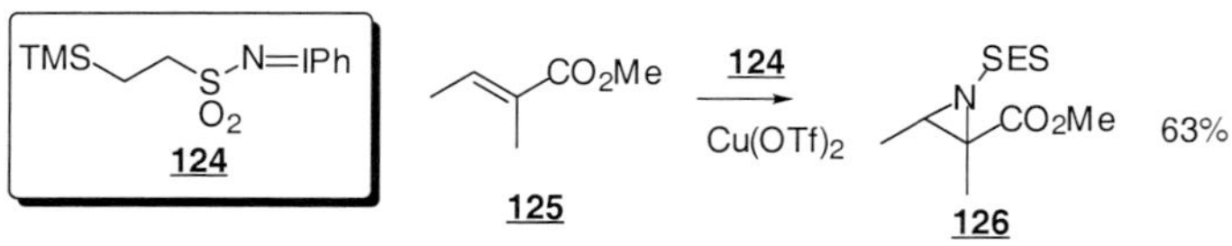

In the realm of heterogeneous catalysis, a copper-exchanged zeolite (CuHY) modified with *bis*-oxazoline was found to exhibit modest asymmetric induction in the aziridination of alkenes using [*N*-(*p*-tolylsulfonyl)imino]phenyliodinane (PHI=NTs) as the nitrene donor <99JCS(P2)-1043>. Oligopeptides and amino acids containing an aziridine 2-carboxylate group have been prepared using a solid phase version of the Gabriel-Cromwell reaction (*i.e.*, **127** → **129**) <99TL6503>.

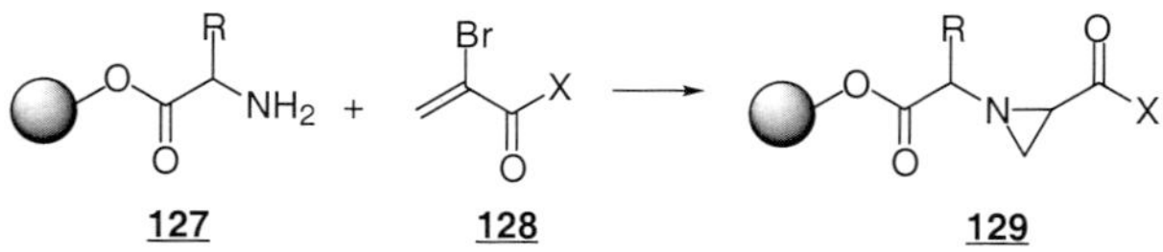

The Gabriel-Cromwell approach proceeds through the intramolecular displacement of the halide in the cyclization step, and this end game can be approached from more than one starting point. Thus, Davis and co-workers <99JOC7559> reported on a one-step aza-Darzens reaction of sulfinimines **130** with lithium α-bromoenolates **131** to give the corresponding aziridines (**132**) in fair to good yield and good to excellent diastereomeric excess. The *cis/trans*-isomer ratio is dependent upon the nature of the bromoenolate, with the anion of α-bromoacetate itself giving rise to predominantly the *cis*-isomer (**132**), and substituted analogs producing mainly the *trans*-isomer. This selectivity was rationalized on the basis of a chair-like transition state.

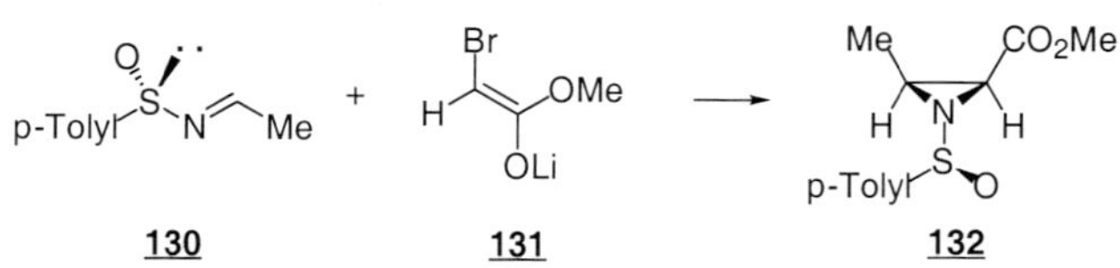

An interesting anionic aziridination of α,β-unsaturated amides was reported this past year <99TL5207> utilizing lithiated 3,3-pentamethylenediaziridine (**134**) as the nitrogen atom donor. Formation of *cis*-aziridines was generally observed, regardless of the stereochemistry of the starting material, a phenomenon which is in keeping with a stepwise mechanism of conjugate addition and subsequent ring-closure.

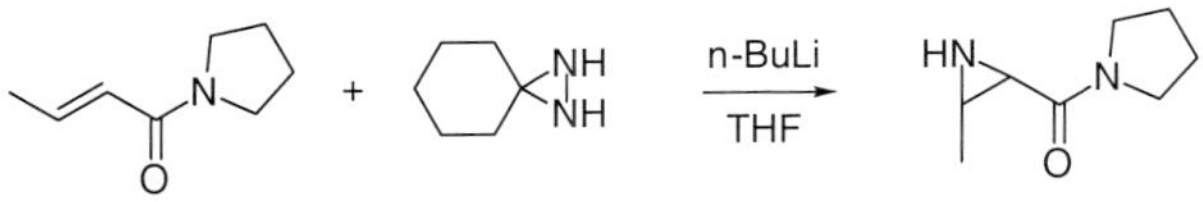

133 134 135

Sterically congested *cis*-aziridines such as **137** were prepared from the derivatized amino allyl alcohol precursor **136** through a palladium-catalyzed cyclization reaction <99TL1331>. This methodology has also been extended to the cyclization of amino allenes <99JOC2992>.

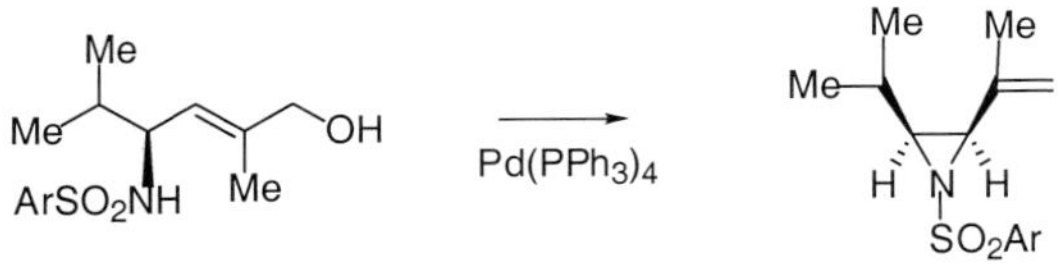

136 137

4.1.3.2 Reactions of Aziridines

Aziridines undergo a variety of synthetically useful transformations, not least of which are simple ring-opening reactions with nucleophiles. For example, the bicyclic aziridine **138** was found to undergo smooth ring cleavage by aniline in the presence of $Sn(OTf)_2$ to give the corresponding 1,2-diamino compound **139** <99JOC2537>. The chiral trifluoromethyl aziridine of type **140** can be ring-opened even with relatively weak nucleophiles (in this example, water) to give opticlly active amines **141** in good yields with excellent retention of configuration <99JOC7323>.

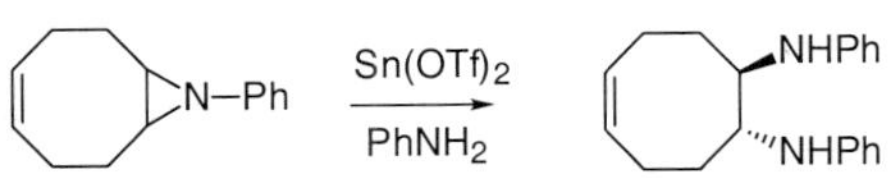

138 139

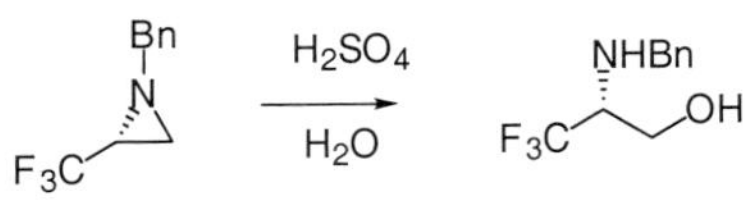

140 141

98% yield; > 99% ee

The ring opening of an aziridine can also occur in an intramolecular fashion, as observed in the formal [3+2] aziridine-allylsilane cycloaddition reaction of **142**. These substrates were used for the preparation of both 5-5 and 6-5 fused ring systems <99T8025>.

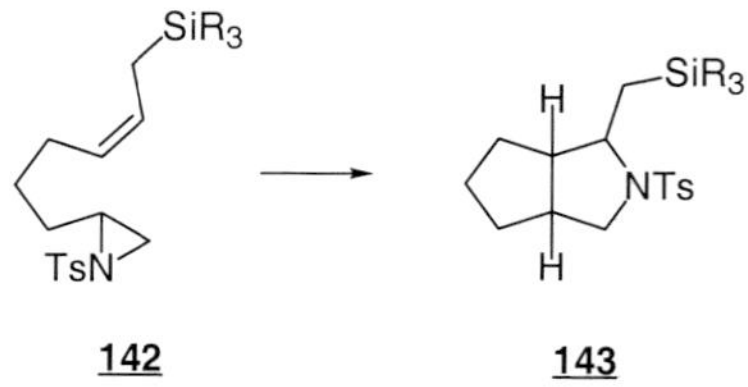

Certain aziridines have been shown to engage in some interesting ring expansion reactions. For example, phenylaziridine **144** behaved as a 1,3-dipole toward dihydropyran (**145**) in the presence of boron trifluoride etherate to give the bicyclic species **146**, which can be subsequently converted to substituted pyrrolidines <99TL5315>. The silylated hydroxymethyl aziridine **147** undergoes carbonylative ring expansion promoted by dicobalt octacarbonyl to provide the functionalized β-lactam **148**, a process which proceeds with inversion of configuration <99JOC518>.

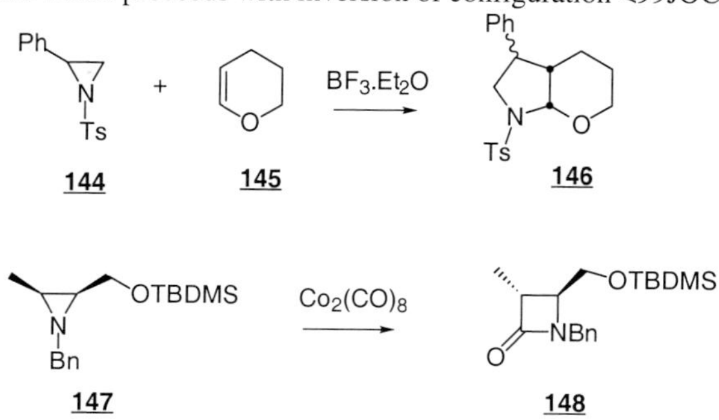

Aziridinylcarbinyl radicals (*e.g.*, **150**) are interesting reactive intermediates and were shown to undergo β-cleavage to form aminoalkenes (*e.g.*, **151**), which are the products of C-N bond cleavage. The selectivity of the ring-opening was rationalized on the basis of more effective overlap of the singly occupied *p*-orbital on the radical center with the C-N bond <99TL4873>.

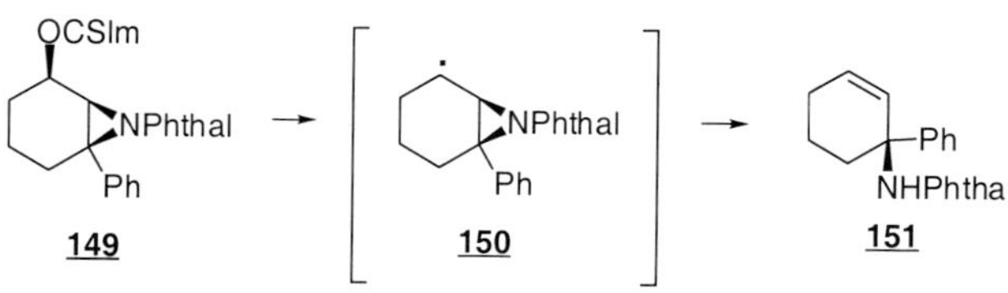

Finally, an interesting deamination reaction of aziridines was reported, in which treatment of *N*-unsubstituted aziridines (**152**) with dinitrogen tetroxide (2 equiv) in the presence of Et_3N results in clean deamination to provide the corresponding alkenes (**154**) with remarkably high yields (>90%). The reaction is believed to proceed *via* the *N*-nitroso intermediate **153**, so that the driving force for the reaction is liberation of N_2O <99SC1241>.

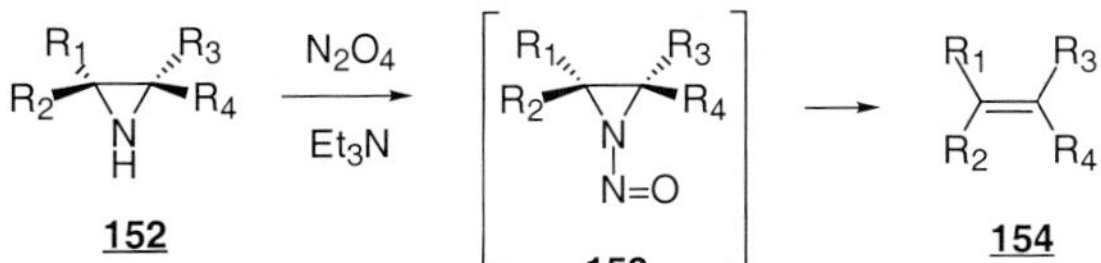

152 153 154

4.1.4 AZIRINES

Reaction of 1-azirine-3-methylacrylates (**155**) with imidazoles and pyrazoles under mild conditions results in the formation of 2-aza-1,3-dienes (**156**), which are useful as dienes in hetero Diels-Alder reactions with electron-deficient dienophiles <99JOC49>. When the related methyl 2-aryl-2*H*-azirine-3-carboxylate (**157**) was used as the substrate, reaction with an amine induced a ring opening by addition of the amino group onto the C=N bond followed by cleavage to provide enediamine **158** <99JCS(P1)1305>.

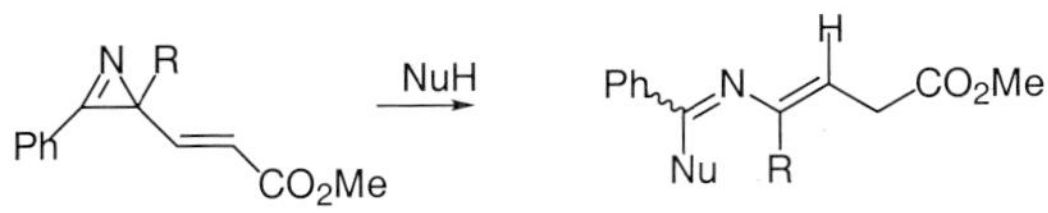

155 156

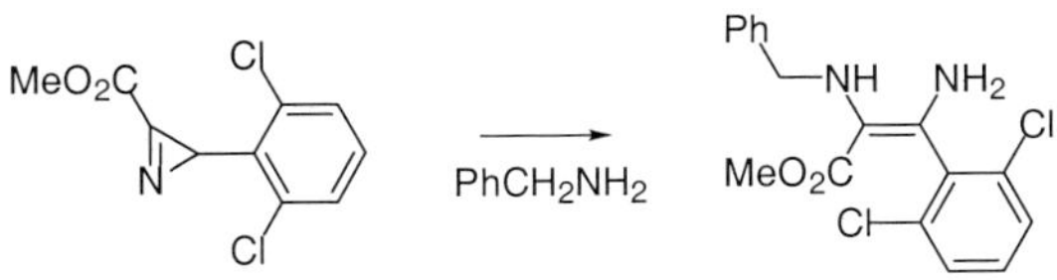

157 158

4.1.5 DIOXIRANES

In recent years, dioxiranes have become workhorses for a variety of selective transformations in organic synthesis, from epoxidation of alkenes to the conversion of alcohols into the corresponding ketones <99CJC308>. Dioxirane-mediated epoxidation continues to be the method of choice for complex substrates with acid-sensitive functionality. Thus, the dimethyl-dioxirane (DMD)-mediated epoxidation of the silylated stilbene lactam **159** has been reported as a key step in the synthesis of protoberberines <99JOC877>.

159 160

The development of chiral ketone precursors for asymmetric induction in dioxirane-promoted epoxidations has been the subject of intense and fruitful study in recent years, and Denmark and Wu have recently published a very useful review on the topic <99SL847>. One such chiral ketone (*i.e.*, **161**) has been used to advantage for the kinetic resolution of the racemic cyclic olefin **162** <99JA7718>.

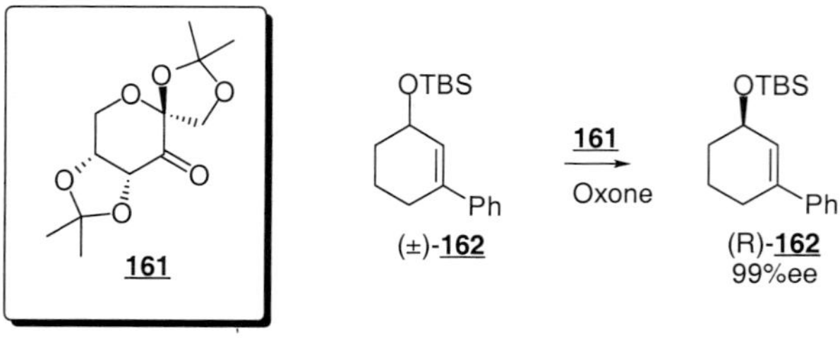

A sometimes nagging aspect of dioxirane-based oxidations is the degradation of catalyst. In this regard, Carnell and co-workers <99TL8029> have reported on the use of *N,N*-dialkyl-alloxan **163** as a particularly robust dioxirane precursor, which can be recovered in high yield with no evidence of catalyst decomposition. Attempts thus far to parlay this catalyst into an asymmetric induction paradigm (*e.g.*, *via* **164**) have been unsuccessful.

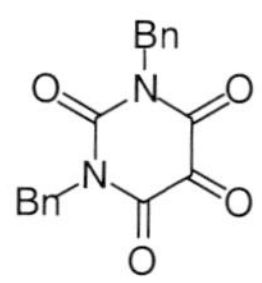

163

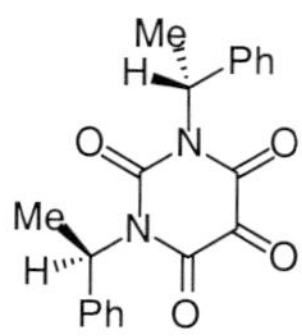

164

4.1.6 OXAZIRIDINES

Chiral oxaziridines have become very handy oxygen-transfer reagents for the asymmetric epoxidation of unfunctionalized olefins. In this regard, the synthesis of an optically pure oxaziridium salt (**165**) from (+)-norephedrine has recently been described, as well as its use in the epoxidation of alkenes <99T141>. The transfer of oxygen from *N*-sulfonyl oxaziridines has been investigated using the endocyclic restriction test, which points toward a transition state with more advanced N-O than C-O bond cleavage (*i.e.*, **166**) <99OL1415>.

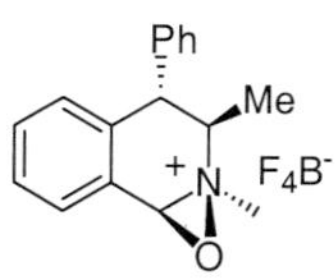

165

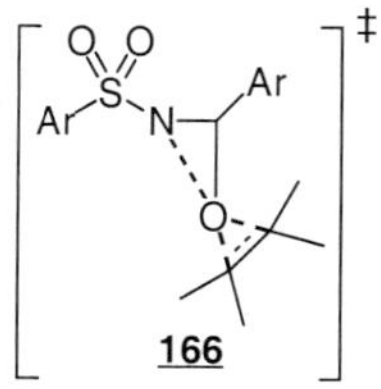

166

An interesting intramolecular variant of this epoxidation procedure is represented in the reaction of the unsaturated oxaziridine **167**, which undergoes highly stereoselective oxygen transfer through a spiro transition state to provide the epoxyaldehyde **168** <99TL4453>.

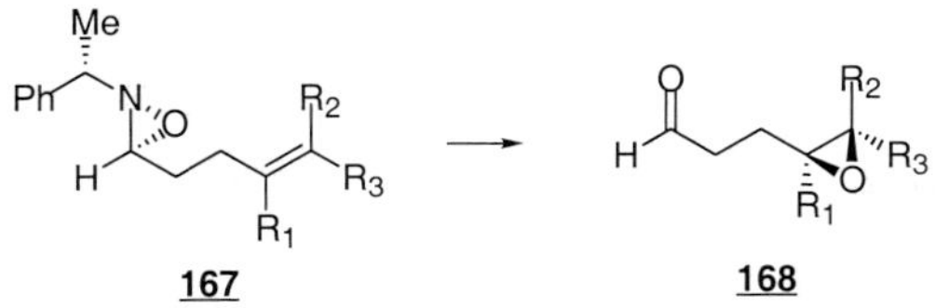

4.1.7 REFERENCES

99AG(E)2012 M. H. Wu, K. B. Hansen, E. N. Jacobsen, *Angew. Chem., Int. Ed. Engl.* **1999**, *38*, 2012.

99CC325 V. K. Aggarwal, P. A. Bethel, R. Giles, *J. Chem. Soc., Chem. Commun.* **1999**, 325.

99CC409 R. Zhang, W.-Y. Yu, T.-S. Lai, C.-M. Che, *J. Chem. Soc., Chem. Commun.* **1999**, 409.

99CC727 T. Iwahama, S. Sakaguchi, Y. Ishii, *J. Chem. Soc., Chem. Commun.* **1999**, 727.

99CC821 M. C. A. van Vliet, I. W. C. E. Arends, R. A. Sheldon, *J. Chem. Soc., Chem. Commun.* **1999**, 821.

99CC829 R. Raja, G. Sankar, J. Thomas, *J. Chem. Soc., Chem. Commun.* **1999**, 829.

99CC1789 X.-G. Zhou, X.-Q. Yu, J. S. Huang, S.-G. Li, L. S. Li, C. M. Che, *J. Chem. Soc., Chem. Commun.* **1999**, 1789.

99CJC308 A. L. Baumstark, F. Kovac, P. C. Vasquez, *Can. J. Chem.* **1999**, 308.

99EJOC837 J. A. Elings, R. S. Downing, R. A. Sheldon, *Eur. J. Org. Chem.* **1999**, 837.

99JA460 J. P. Collman, Z. Wang, A. Straumanis, M. Quelquejeu, *J. Am. Chem. Soc.* **1999**, *121*, 460.

99JA3328 T. Ooi, N. Kagoshima, H. Ichikawa, K. Maruoka, *J. Am. Chem. Soc.* **1999**, *121*, 3328.

99JA5083 C. Linde, B. Akermark, P.-O. Norrby, M. Svensson, *J. Am. Chem. Soc.* **1999**, *121*, 5083.

99JA5099 J. C. Antilla, W. D. Wulff, *J. Am. Chem. Soc.* **1999**, *121*, 5099.

99JA6086 J. M. Ready, E. N. Jacobsen, *J. Am. Chem. Soc.* **1999**, *121*, 6086.

99JA7718 M. Frohn, X. Zhou, J.-R. Zhang, Y. Tang, Y. Shi, *J. Am. Chem. Soc.* **1999**, *121*, 7718.

99JCS(P1)103 W.-P. Chen, S. M. Roberts, *J. Chem. Soc., Perkin Trans. 1* **1999**, 103.

99JCS(P1)731 K. Julienne, P. Metzner, V. Henryon, *J. Chem. Soc., Perkin Trans. 1*, **1999**, 731.

99JCS(P1)1305 M. J. Alves, T. L. Gilchrist, J. H. Sousa, *J. Chem. Soc., Perkin Trans 1* **1999**, 1305.

99JCS(P1)1397 T. Geller, S. M. Roberts, *J. Chem. Soc., Perkin Trans. 1* **1999**, 1397.

99JCS(P1)2293 K. Juhl, R. G. Hazell, K. A. Jorgensen, *J. Chem. Soc., Perkin Trans. 1* **1999**, 2293.

99JCS(P2)1043 C. Langham, S. Taylor, D. Bethell, P. McMorn, P. C. Bulman Page, D. J. Willock, C. Sly, F. E. Hancock, F. King, G. J. Hutchings, *J. Chem. Soc., Perkin Trans. 2* **1999**, 1043.

99JOC49 M. T. Barroso, A. Kascheres, *J. Org. Chem.* **1999**, *64*, 49.

99JOC287 G. Sekar, V. K. Singh, *J. Org. Chem.* **1999**, *64*, 287.

99JOC338 N. Murase, Y. Hoshino, M. Oishi, H. Yamaoto, *J. Org. Chem.* **1999**, *64*, 338.

99JOC518 P. Davoli, I. Moretti, F. Prati, H. Alper, *J. Org. Chem.* **1999**, *64*, 518.

99JOC877 G. Rodríguez, L. Castedo, D. Dominguez, C. Saá, W. Adam, C. R. Saha-Möller, *J. Org. Chem.* **1999**, *64*, 877.

99JOC2164 K. Hinterding, E. N. Jacobsen, *J. Org. Chem.* **1999**, *64*, 2164.

99JOC2537 G. Sekar, V. K. Singh, *J. Org. Chem.* **1999**, *64*, 2537.

99JOC2966 K. Yamaguchi, K. Ebitani, K. Kaneda, *J. Org. Chem.* **1999**, *64*, 2966.

99JOC2992 H. Ohno, A. Toda, Y. Miwa, T. Taga, E. Osawa, Y. Yamaoka, N. Fujii, T. Ibuka, *J. Org. Chem.* **1999**, *64*, 2992.

99JOC3699 W. Adam, C. M. Mitchell, C. R. Saha-Möller, *J. Org. Chem.* **1999**, *64*, 3699.

99JOC5304 P. Dauban, R. H. Dodd, *J. Org. Chem.* **1999**, *64*, 5304.

99JOC7323 T. Katagiri, M. Takahashi, Y. Fujiwara, H. Ihara, K. Uneyama, *J. Org. Chem.* **1999**, *64*, 7323.

99JOC7365	C.-J. Liu, W.-Y. Yu, C.-M. Che, C.-H. Yeung, *J. Org. Chem.* **1999**, *64*, 7365.
99JOC7559	F. A. Davis, H. Liu, P. Zhou, T. Fang, G. V. Reddy, Y. Zhang, *J. Org. Chem.* **1999**, *64*, 7559.
99JOC8149	H. B. Yu, X. F. Zheng, Z. M. Lin, Q. S. Hu, W. S. Huang, L. Pu, *J. Org. Chem.* **1999**, *64*, 8149.
99OL419	T. Strassner, K. N. Houk, *Org. Lett.* **1999**, *1*, 419.
99OL667	D. L. Wright, M. C. McMills, *Org. Lett.* **1999**, *1*, 667.
99OL705	S. L. Ali, M. D. Nikalje, A. Sudalai, *Org. Lett.* **1999**, *1*, 705.
99OL783	A. V. Gontcharov, H. Liu, K. B. Sharpless, *Org. Lett.* **1999**, *1*, 783.
99OL1287	E. J. Corey, F. Y. Zhang, *Org. Lett.* **1999**, *1*, 1287.
99OL1415	D. R. Anderson, K. W. Woods, P. Beak, *Org. Lett.* **1999**, *1*, 1415.
99SC561	N. Iranpoor, F. Kazemi, *Synth. Commun.* **1999**, *29*, 561.
99SC1017	N. Iranpoor, B. Zeynizadeh, *Synth. Commun.* **1999**, *29*, 1017.
99SC1121	M. L. Kantam, B. M. Choudary, B. Bharathi, *Synth. Commun.* **1999**, *29*, 1121.
99SC1241	K. Lee, Y. H. Kim, *Synth. Commun.* **1999**, *29*, 1241.
99SC2249	N. Iranpoor, M. Shekarriz, *Synth. Commun.* **1999**, *29*, 2249.
99SL783	K. Miura, T. Katsuki, *Synlett* **1999**, 783.
99SL795	T. Genski, G. Macdonald, X. Wei, N. Lewis, R. J. K. Taylor, *Synlett* **1999**, 795.
99SL847	S. E. Denmark, Z. Wu, *Synlett* **1999**, 847.
99SL1157	T. Takeda, R. Irie, Y. Shinoda, T. Katsuki, *Synlett* **1999**, 1157.
99SL1328	R. Hayakawa, M. Shimizu, *Synlett* **1999**, 1328.
99T141	L. Bohé, M. Lusinchi, X. Lusinchi, *Tetrahedron* **1999**, *55*, 141.
99T1063	A. Scheurer, P. Mosset, M. Spiegel, R. W. Saalfrank, *Tetrahedron* **1999**, *55*, 1063.
99T6289	B. Lygo, P. G. Wainwright, *Tetrahedron* **1999**, *55*, 6289.
99T6375	S. Arai, Y. Shirai, T. Ishida, T. Shioiri, *Tetrahedron* **1999**, *55*, 6375.
99T8025	S. C. Bergmeier, S. L. Fundy, P. P. Seth, *Tetrahedron* **1999**, *55*, 8025.
99T11589	C. Chiappe, C. D. Palese, *Tetrahedron* **1999**, *55*, 11589.
99T12929	W. Xie, J. Fang, J. Li, P. G. Wang, *Tetrahedron* **1999**, *55*, 12929.
99TL1001	P. Pietikäinen, *Tetrahedron Lett.* **1999**, *40*, 1001.
99TL1041	F. Benedetti, F. Berti, S. Norbedo, *Tetrahedron Lett.* **1999**, *40*, 1041.
99TL1331	H. Ohno, A. Toda, N. Fujii, Y. Miwa, T. Tage, Y. Yamaoka, E. Osawa, T. Ibuka, *Tetrahedron Lett.* **1999**, *40*, 1331.
99TL1641	S. Hu, L. P. Hager, *Tetrahedron Lett.* **1999**, *40*, 1641.
99TL3129	M. E. Jung, R. Marquez, *Tetrahedron Lett.* **1999**, *40*, 3129.
99TL3613	K. M. Ryan, C. Bousquet, D. G. Gilheany, *Tetrahedron Lett.* **1999**, *40*, 3613.
99TL4453	A. Armstrong, A. G. Draffan, *Tetrahedron Lett.* **1999**, *40*, 4453.
99TL4589	H. Hayakawa, N. Okada, M. Miyazawa, M. Miyashita, *Tetrahedron Lett.* **1999**, *40*, 4589.
99TL4873	B. A. Marples, R. C. Toon, *Tetrahedron Lett.* **1999**, *40*, 4873.
99TL5207	K. Hori, H. Sugihara, Y. N. Ito, T. Katsuki, *Tetrahedron Lett.* **1999**, *40*, 5207.
99TL5315	I. Ungureanu, C. Bologa, S. Chayer, A. Mann, *Tetrahedron Lett.* **1999**, *40*, 5315.
99TL5369	N. Abe, H. Hanawa, K. Maruoka, *Tetrahedron Lett.* **1999**, *40*, 5369.
99TL6503	S. N. Filigheddu, S. Masala, M. Taddei, *Tetrahedron Lett.* **1999**, *40*, 6503.
99TL7105	M. Mizuno, T. Shioiri, *Tetrahedron Lett.* **1999**, *40*, 7105.
99TL7243	K. Suda, K. Baba, S.-I. Nakajima, T. Takanami, *Tetrahedron Lett.* **1999**, *40*, 7243.
99TL7303	H. Lebel, E. N. Jacobsen, *Tetrahedron Lett.* **1999**, *40*, 7303.
99TL8019	Y. Mori, H. Furuta, T. Takase, S. Mitsuoka, H. Furukawa, *Tetrahedron Lett.* **1999**, *40*, 8019.
99TL8029	A. J. Carnell, R. A. W. Johnstone, C. C. Parsy, W. R. Sanderson, *Tetrahedron Lett.* **1999**, *40*, 8029.

Chapter 4.2

Four-Membered Ring Systems

L. K. Mehta and J. Parrick
Brunel University, Uxbridge, UB8 3PH, UK
E-mail: Lina.Mehta@brunel.ac.uk and John.Parrick@brunel.ac.uk

---

4.2.1 INTRODUCTION

Much care has been taken in the selection of work to be included but, necessarily, the choice is subjective and the review is in no way comprehensive. β-Lactam chemistry dominates the field in terms of the number of publications. In contrast, studies of heterocycles containing two different heteroatoms appear to be neglected. In general, reviews are mentioned in the appropriate section but mention of a survey of heterocycles bearing fluorine or trifluoromethyl substituents <98MI355> is more appropriate here.

4.2.2 AZETINES, AZETIDINES AND 3-AZETIDINONES

The Diels-Alder reaction of cyclopentadiene or isobenzofuran derivatives with *N*-acetyl-2-azetine gives cycloadducts (e.g., **1**) in high yield by *endo* addition <99TL443>.

Efforts have been made to find stereoselective routes which provide disubstituted azetidines. Palladium catalysed cyclization of an enantiomer of allene-substituted amines and amino acids gives the azetidine ester **2** and a tetrahydropyridine in variable yield and ratio, depending on the substituents and conditions <99OL717>. The (2*R*,3*S*)- and (2*S*,3*R*)-isomers of the substituted azetidine-2-carboxylic acids **3** (R = CO_2H) are obtained in several steps from the corresponding **3** (R = $CH_2OSiMe_2Bu^t$) which, in turn, is produced in high yield by photochemical intramolecular cyclization <98HCA1803>.

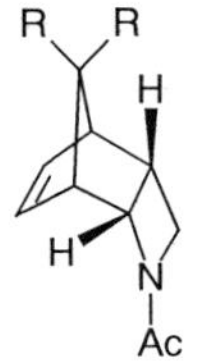

1

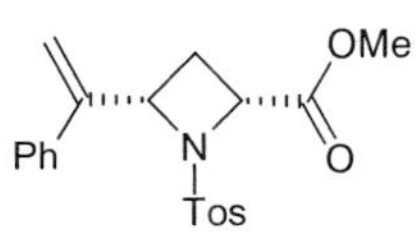

2

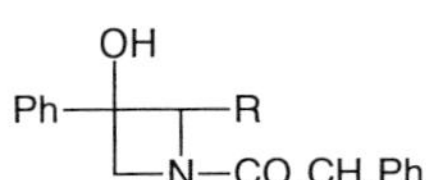

3

The chiral ligand **4** is used in the asymmetric addition of diethyl zinc to aldehydes to give *sec*-alcohols in high yield having *S*-absolute configuration <99TA1673>.

A convenient, one-pot, two-step synthesis of 1-azabicyclo[1.1.0]butane (**5**, R = H) from *N*-chlorosuccinimide is reported and its application to the synthesis of 1,3,3-trinitroazetidine (TNAZ) is discussed <98SC3949>. Another novel and efficient synthesis of 1-azabicyclo[1.1.0]butane (**5**, R = H) and its derivatives is from 2,3-dibromopropylamine. The bicyclic **5** (R = H) is also useful in the synthesis of the pendant group of a 1β-methylcarbapenem antibiotic <99TL3761>. The reaction of **5** (R = Et and Ph) with tosyl chloride and tosyl azide are described <98T15127, 99H131>.

Intramolecular N-H bond insertion of α-diazocarbonyls $R^1NHCH_2COCHN_2$ is an efficient route to derivatives of 3-azetidinones **6** (R^1 = Tos, Cbz and Boc, R^2 = O) and the reaction is catalysed by $Cu(acac)_2$ <99JCS(P1)2277>.

N-Acylazetidin-3-ones **6** (R^1 = acyl, R^2 = O) have been synthesised from **5** (R = CH_2Br) by ring opening with acyl halide, followed by zinc promoted dehalogenation and ozonolysis of the 3-methylene azetidine **6** (R^1 = acyl, R^2 = CH_2) <99SC885>. Readily available epichlorohydrin was used in the synthesis of **6** (R^1 = $CHPh_2$, R^2 = O), which was converted in three high yielding steps to 3-azetidinylidene acetic acid (**6**, R^1 = H, R^2 = $CHCO_2H$) <99CL605>. A spiro annulated heterocyclic oxirane **7** was obtained in two steps through samarium iodide reductive coupling of **6** (R^1 = Tos, R^2 = O) <98H149>.

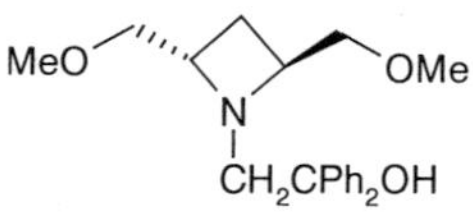

4

5

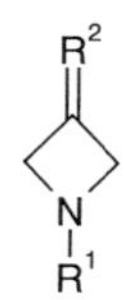

6

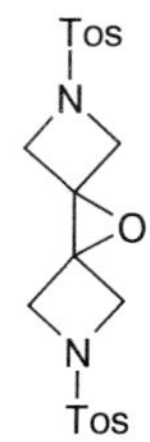

7

4.2.3 THIETES, THIETANES, OXETANES AND 2-OXETANONES

Reviews of thietanes <99MI138> and saturated oxygen heterocycles including oxetanes are available <98JCS(P1)4175>.

Flash vacuum pyrolysis has been used in the synthesis of 2*H*-naphtho[1,2-*b*]thiete **8** <98JHC1505> and benzobisthietes **9** and **10** <98TL9639>, and their Diels-Alder cycloaddition reactions have been studied.

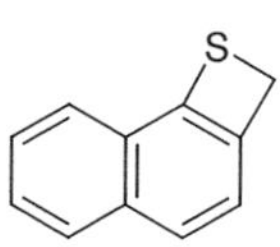

8

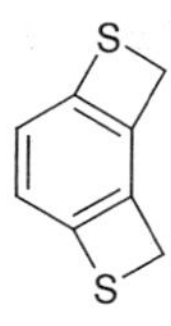

9

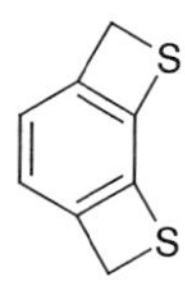

10

Solid state photocyclization of **11** gave optically active pyrrolidinylthietane **12** in high yield <98CC2315>.

Photocycloaddition of an alkene to the thione group of **13** gave the thietane **14** which was stable at room temperature, but on refluxing **14** in toluene an iminothietane **15** and/or a 2-substituted benzoxazole were obtained by rearrangement processes <99JCS(P1)1151>.

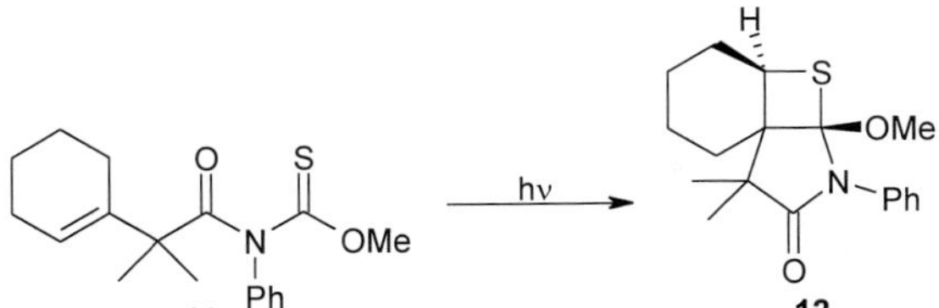

11 **12**

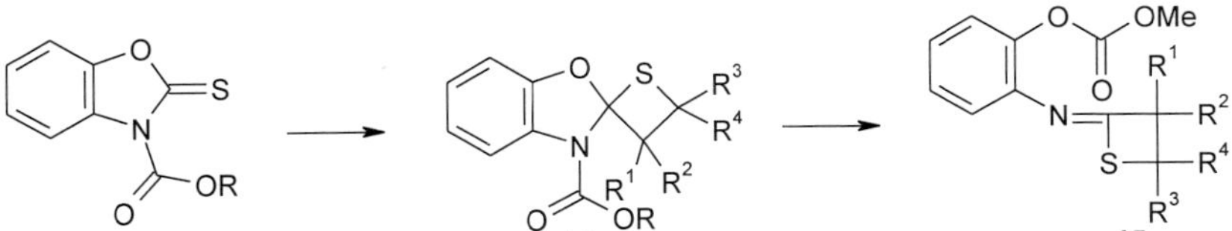

13 **14** **15**

Photocycloaddition of aromatic carbonyl compounds with β,β-dimethylketene silylacetals gave 2-alkoxy-4-aryloxetanes <98JCS(P1)3253>. A similar photoaddition of derivatives of stilbene to chloranil gave the isomeric spiro oxetanes **16** in very high total yields <99JOC2250>.

An intramolecular substitution of trimethylamine from **17** gave a bicyclic oxetane **18** in a diastereoselective process <98MI2185>. A [2+2]cycloaddition of 2,2,4,5-tetrasubstituted 2,3-dihydrofuran to aryl aldehydes gave the bicyclic oxetane **19** <98JCS(P1)3261>. 2,2-Disubstituted-3-bromooxetane was obtained by a 4-*endo*-trig cyclization process of 3,3 -disubstituted allyl alcohol in the presence of bis(collidine)bromine hexafluorophosphate <99JOC81>.

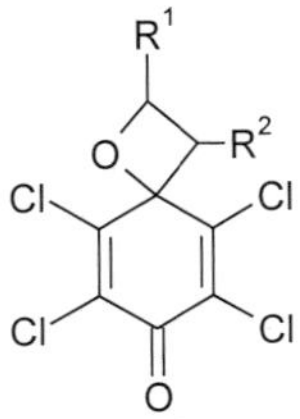

16

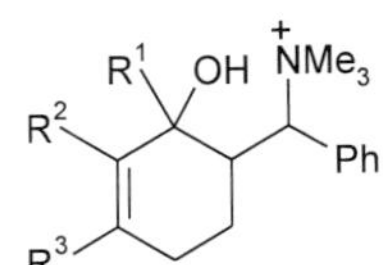

17

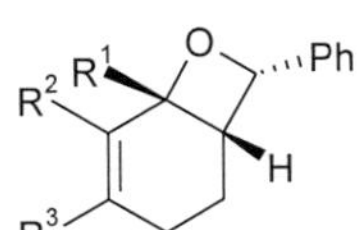

18

2,2-Disubstituted-3-phenyloxetanes were obtained by regioselective ring opening of the 1,5-dioxaspiro[3,2]hexanes **20** <99OL825>. Pinacol-type rearrangement reactions of 2-phenyl-3-silyloxyoxetanes in the presence of Lewis acid catalysts have been discussed <99JOC8041>. Palladium catalysed ring expansion of 2-vinyloxetanes **21** with carbodiimides (e.g., PhC:N:CPh) gave 4-vinyl-1,3-oxazin-2-imines (e.g., **22**) <99JOC4152>. 2-Methyleneoxetanes undergo ring-opening reaction to give substituted 4-hydroxy-1-butynes <99JOC7074>.

The synthesis of optically active 2-oxetanones (β-lactones) has been reviewed <99T6403>. Cross-aldol reactions catalysed by aluminium tris(hexafluoroantimonate) of acyl halides and aldehydes gave 4-substituted 2-oxetanones. Possibilities for the development of chemo- and regio-specific catalysed cross-aldol reactions are discussed <99TL6535>. In another investigation 4-substituted β-lactones were most conveniently prepared either by a catalysed [2+2]cycloaddition of aldehydes and ketenes or by the tandem Mukaiyama-aldol lactonization <98BMC1255>. A key intermediate **23** for the synthesis of the enzyme inhibitors tetrahydrolipstatin and tetrahydroesterastin has been obtained in its diastereomerically pure forms by a tandem aldol lactonization <99JOC 5301>.

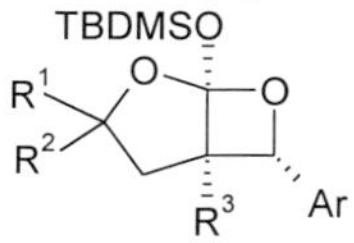

19

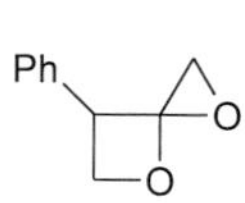

20

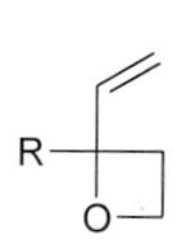

21

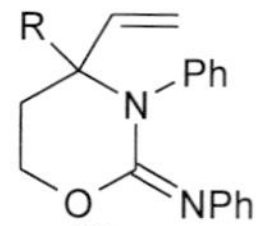

22

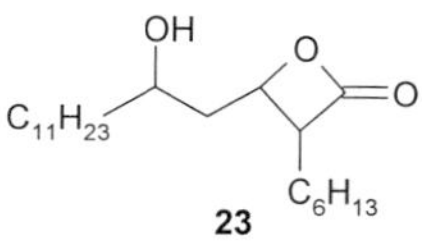

23

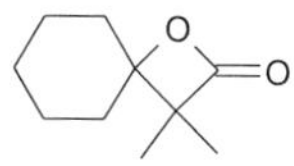
24

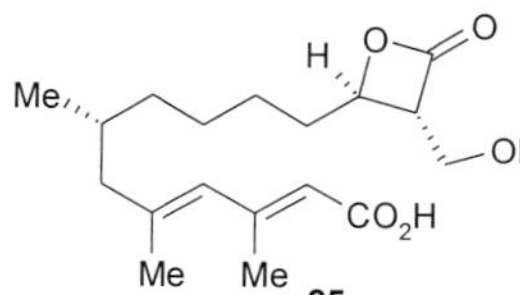

25

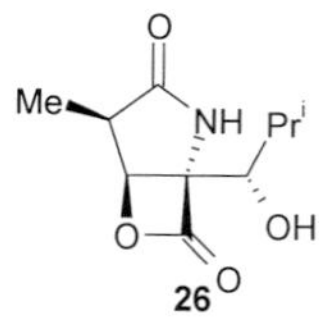

26

The spirolactone **24** was obtained in high yield from cyclohexanone and an enolate of phenyl 2-methylpropanoate <98OS116>. 3,3-Disubstituted propenoic acid in the presence of bis(collidine)bromine(I) hexafluorophosphate gave 3-bromo-4,4-disubstituted β-lactones <99JOC81>. The reactions of chromium aryl(alkoxy)carbenes with propargylic alcohols to give functionalised β-lactones under thermal and ultrasonic conditions have been investigated and compared <99TL3481, 99TL3485>. Syntheses of the cholesterol biosynthesis inhibitor 1233A **25** have been reported <98S1655, 99JCS(P1)1917>. The latter report describes the reaction of ceric ammonium nitrate with a (π-allyl)tricarbonyliron lactone complex to form the β-lactone. The total synthesis of omuralide **26** and some analogues have been reported <99CPB1>. The absolute stereochemistry of 16-methyloxazolomycin **27** produced by *streptomyces sp* has been determined <99JAN193>.

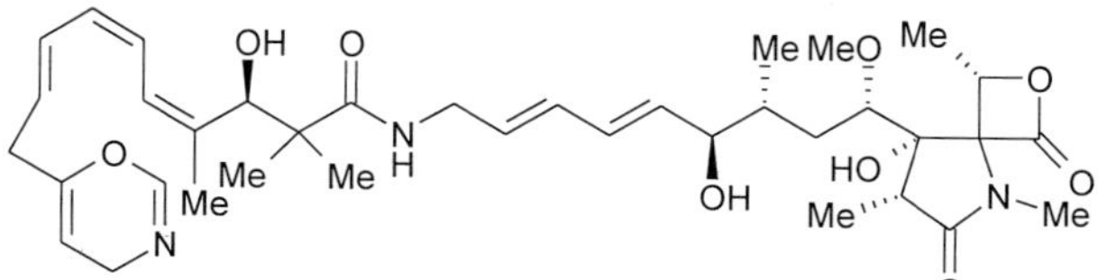

27

2-Methyleneoxetanes are readily obtained from the corresponding 2-oxetanones (yields 20-86%) by reaction with Cp_2TiMe_2 even when a substituent includes olefinic or carbonyl functions <99JOC7074>. A one-pot conversion of β-lactones into β-lactams in a two-step process in good to excellent yields is reported <99JOC7657>.

4.2.4 DIAZETIDINES, DIOXETANES, DITHIETES AND DITHIETANES

Aryl diisocyanates have been dimerised to give 1,3-disubstituted 1,3-diazetidin-2,4-diones **28** <99JPR616>. Tetracycles **29** have been prepared <99JOC1121>.

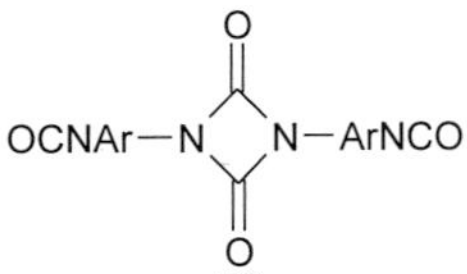

28

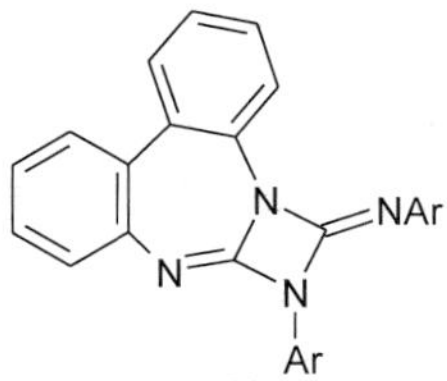

29

A [2+2]cycloaddition reaction of singlet oxygen generated by irradiation of O_2 in the presence of 5,10,15,20-tetrakis(pentafluorophenyl)porphine (TPFPP) to a chiral allylic alcohol gave *erythro* and *threo* (mainly) forms of the dioxetane **30** (R = adamantyl) by a highly diastereoselective process. The observed *threo*-selectivity (*threo*:*erythreo*, 89-95:11-5) is thought to be due to the hydrogen bonding in the exciplex between singlet oxygen and the OH group <99JA1834>.

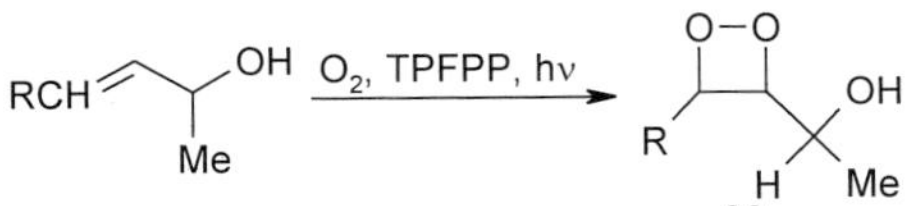

30

Dioxetanes bearing an electron donating group, e.g., **31**, show charge transfer induced decomposition which produces light efficiently <99T4287>.

Tetrabutylammonium fluoride caused chemically initiated electron exchange luminescence through deprotection of **32** to give the phenolate and light emission (λ_{em} max. ≈ 550 nm) <99TL2443>. The dioxetane **33** decomposes at low temperature but gives the normal fragmentation to a ketoester at high temperature <98CC2319>.

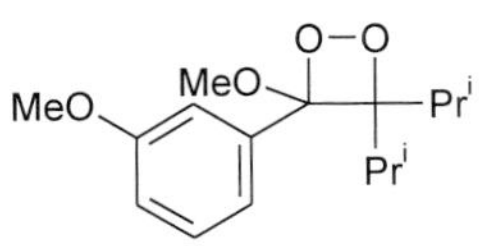

31

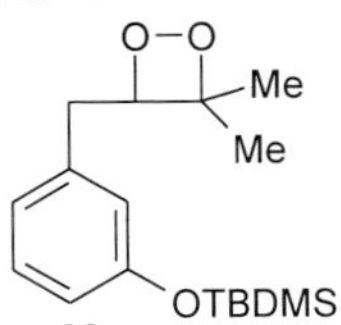

32

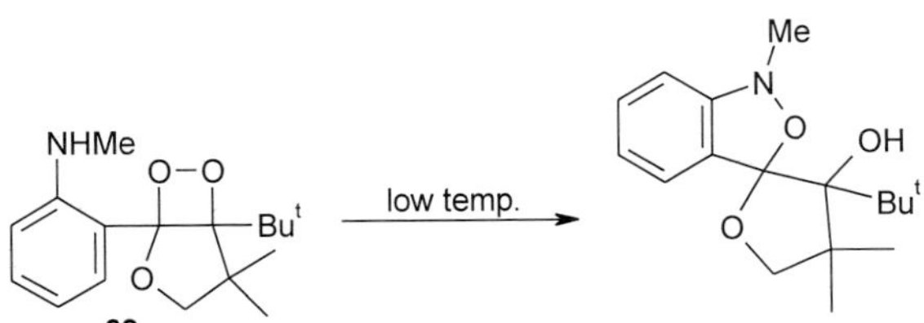

33

The 1,2-dithiete **34** has been prepared <98JOC8192> and X-ray crystal structure determination shows the ring to be planar as had previously been shown for sterically hindered 1,2-dithietes <98BCJ1181>. Prolonged reaction of (alkylthio)chloroacetylenes (RSC ⋮ CCl) with Na_2S in DMSO yields 1,3-dithietanes **35** <99SUL57>.

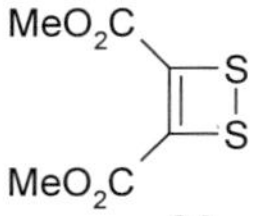

34

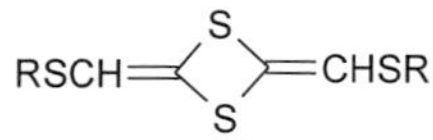

35

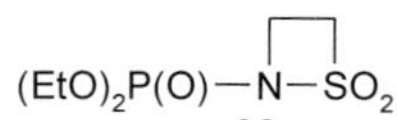

36

4.2.5 THIAZETIDINES AND THIAZETIDINONES

A review of combinatorial synthesis including β-sultams has appeared <98AJC875>.

A two step synthesis of 2-phosphoro-1,2-thiazetidine 1,1-dioxide **36** from phosphorochloridites has been reported <99HAC61>. The action of ammonia and primary amines ($RCH_2(CH_2)_nNH_2$) on 4,4-dimethyl-1,2-thiazetidin-3-one 1,1-dioxide **37** gives ring-opened products, but ring-enlarged products were obtained for **37**, (R = $CH_2)_nCHR^1NHCO_2Bu^t$, n = 2-4) <99HCA354>.

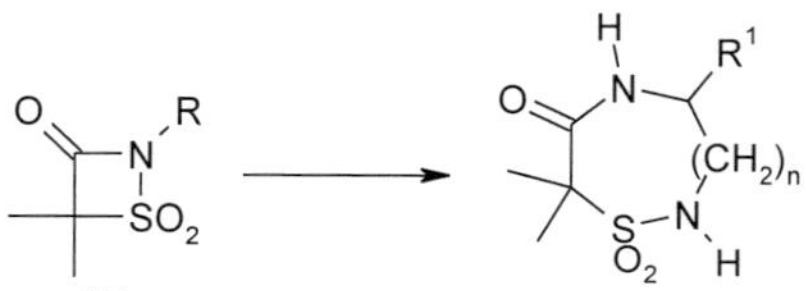

37

4.2.6 SILICON AND PHOSPHORUS HETEROCYCLES

The molecular structure of sila-heterocycles is reviewed <98MI649>. The Raman and infrared spectra of 1-chloro-1-methylsilabutane have been studied in detail <99MI399>. The first allenic coompounds with both silicon and phosphorus doubly bonded (ArP:C:Si(Ph)Tip, Tip = 2,4,6-triisopropylphenyl) have been prepared and dimerised to give **38** and **39** in the ratio of 3:2 <99MI774>.

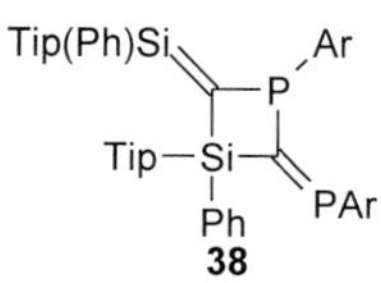

38

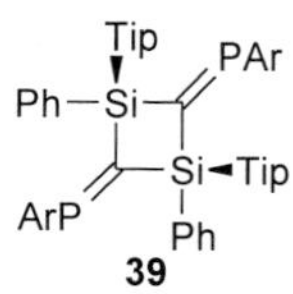

39

Silyl substituted 1*H*-phosphirene **40** undergoes photochemical ring expansion to the 1,2-dihydro-1,2-phosphasilete **41** with cleavage of a silicon-silicon bond <99MI1581>.

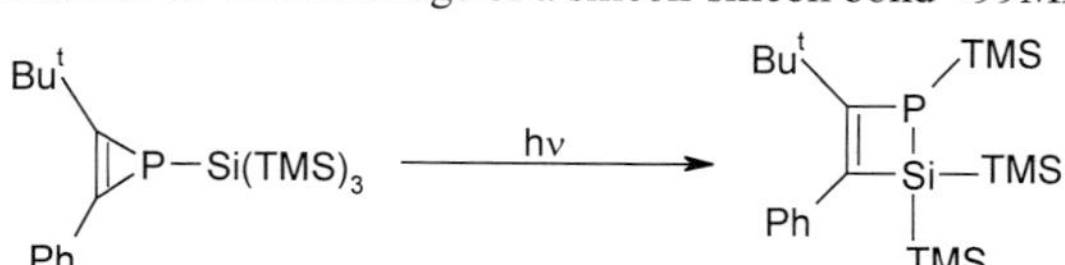

40 **41**

Synthetic approaches to chiral phosphetanes and the chemistry of cyclobutanes containing at least one highly coordinate main group element have been reviewed <98MI755, 99MI1>. The potential of phosphetanes as chiral ligands in organometallic catalysis is of current interest <98S1539>. Addition of a phenyl phosphinidene complex ($PhPW(CO)_5$) to 2,4-hexadiyne gives *cis*- and *trans*-1,2-dihydro-1,2-diphosphate **42** in the ratio of 2:1, respectively <99OM796>. 2-Phosphino-2*H*-phosphirene **43** undergoes thermal or photochemical rearrangement to give diphosphete **44** which can be rearranged to the 1,2-dihydrodiphosphete **45** <99MI274>. The first experimental data for the pseudorotation of unstabilised and unconstrained 1,2-oxaphosphetanes **46** (Wittig intermediates) has been obtained <98JA10653>.

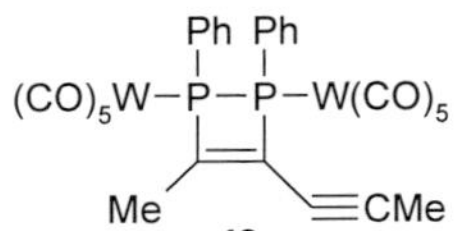

42

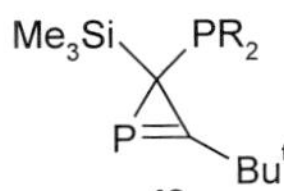

43

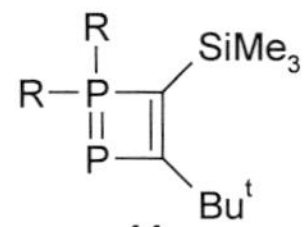

44

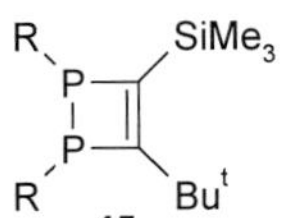

45

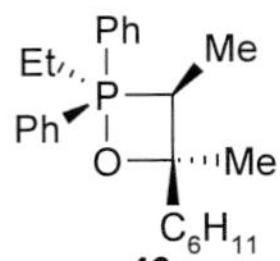

46

4.2.7 MONOCYCLIC 2-AZETIDINONES (β-LACTAMS) AND 2,3-AZETIDIN-DIONES

Reviews including aspects of β-lactam chemistry are ketene-imine cycloaddition reactions <98CHE1222>, radical cyclization processes <98MI169>, combinatorial synthesis <98AJC875>, electrophilic cyclization of unsaturated amides <98T13681> and theoretical studies on the synthesis of β-lactams <98MI245>.

Tert-butyl magnesium chloride has been used in the ring closure of β-amino esters to derivatives of 3-(2-hydroxyethyl)-β-lactams in three investigations <98OS106, 99JOC3790, 99TL6995>. A one-pot procedure was used to obtain α-phenylseleno-β-aminoesters and then 3-(phenylseleno)-3-alkyl-β-lactams, which are easily converted into 3-alkylidene-β-lactams <98T15657>. Intermolecular cyclization has been used to obtain 3-phenylthio-β-lactams **47** from α-chloro-α-phenylthioacetyl chlorides <98IJC(B)1114>. A salen-copper(II) complex derived from (1*R*,2*R*)-(-)-1,2-diaminocyclohexane was used in a novel asymmetric synthesis of 3-phenylsulfonyl-4-phenylazetidin-2-ones **48** <99TL585>.

Azetidinones on a solid support **49** have been prepared in high yield by Staudinger reaction of a supported imine with an acid chloride in the presence of a base. The liberated β-lactams were of high purity <99TL1249>. Cycloaddition of a ketene intermediate, derived from an azo compound, to an imine having an oxidatively cleavable chiral auxillary *N*-substituent was used to obtain β-lactams **50**. The *trans*:*cis* ratio which varied between 69:31 and 93:7, depended on the nature of the substituents R^1 and R^2 <99S650>.

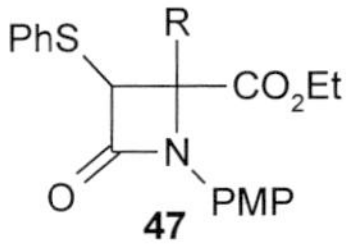
47

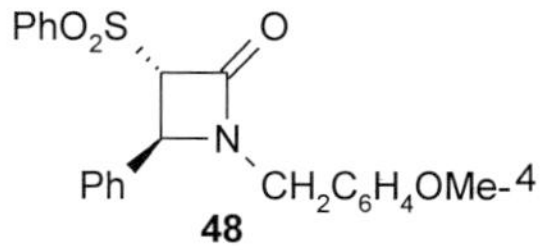
48

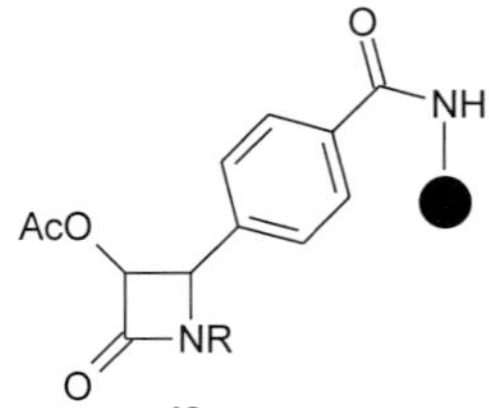
49

A systematic investigation of chiral ligand mediated addition of imines to lithium ester enolates to give β-lactams has been carried out to study the effects of the variation of the alkoxy group in the latter reagent. A maximum of 93% ee was obtained <98TL9055>. The Seebach synthetic principle of self-regeneration of stereocentres has been used in the synthesis of 3-alkyl-3-hydroxy-β-lactams from imines and (2*S*)-chiral enolates of 1,3-dioxolan-4-ones <99JOC4643>. Tridentate chiral amines <99CC715, 99T11219> and chiral bisoxazolines <99CPB720> have been used to increase the enantioselectivity of lithium ester enolate and imine cycloadditions. Lithium ynolates ($Bu\text{-}\equiv\text{-}O^-Li^+$) undergo cycloaddition with *N*-sulfonylimines to give 3,4-disubstituted-β-lactams **51** <98H113>.

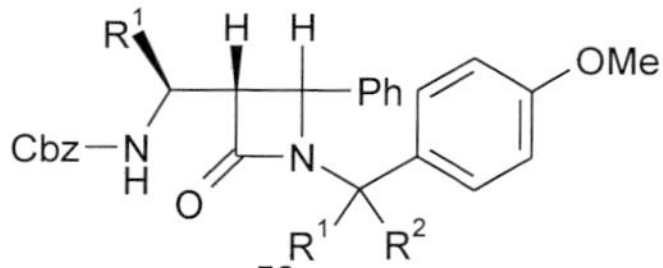
50

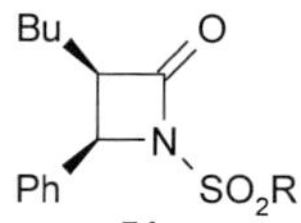
51

Microwave irradiation of a mixture of an imine and trichloroacetic anhydride in the presence of diiron nonacarbonyl gives high yields of 3,3-dichloro-β-lactam **52** <98JCR(S)724>. Carbonylative cycloaddition of benzyl halide (or allyl halide) with imines in the presence of a palladium catalyst ($PdCl_2(PPh_3)_2$) yields β-lactams <99SC2695>. Functionalised π-allyltricarbonyl-iron lactam complexes **53** derived from aziridines, have been used in the synthesis of β-lactams **54** in good yields <98CC1995>. Carbonylative ring expansion of silylated hydroxymethylaziridines is catalysed by dicobalt octacarbonyl and proceeds with inversion of configuration to give β-lactams <99JOC518>.

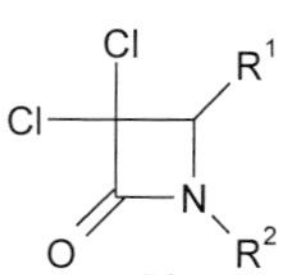
52

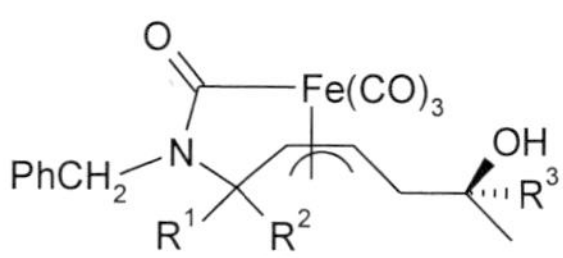
53

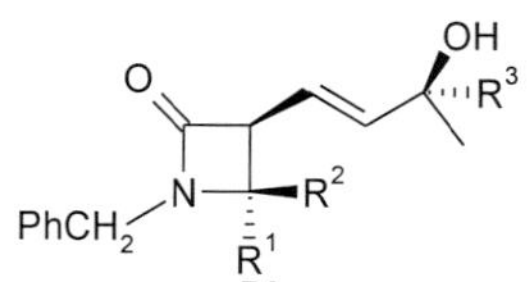
54

Theoretical studies of the ester enolate-imine reaction <98MI1826>, the effects of solvation on barriers of reaction <99JPC8628>, interactions of β-lactams in aqueous solution <99MI1401> and their ammonolysis and aminolysis <99JOC3281, 99MI1045> are available.

N-Phenylazetidin-2-ones are prepared by reaction of a β-lactam with bromobenzenes in the presence of palladium(II) acetate and 1,1-bis(diphenylphosphino)ferrocene <99TL2035>.

The first aza-Wittig reaction of the β-lactam carbonyl group gives the tricyclic **56** from **55** <98SL1288>.

55 **56**

Indium mediated allylation of 4-acetoxy-2-azetidinones gave the products **57** in high yield <99SL447> and similar reactions with azetidin-2,3-diones gave 3-substituted 3-hydroxy-β-lactams **58** <98H97>.

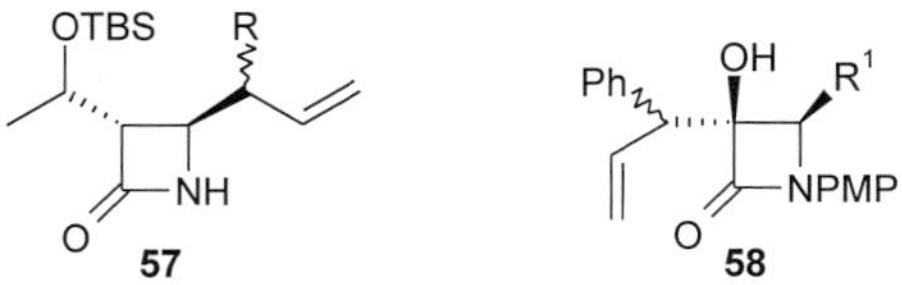

57 **58**

Monobactams have been investigated as β-lactamase inhibitors <98CHE1308, 98CHE1319>. The ketene-imine route to β-lactams was used to obtain 1,3,4-trisubstituted derivatives with high *trans* selectivity. The enolate from 4-hydroxy-γ-lactone reacted with the imine ($Ar^1CH:NAr^2$) to give **59**, which cyclized in the presence of lithium chloride at low temperature to yield **60**. The compounds were assayed for cholesterol absorption inhibition and **61** ($R^1 = R^2 = OH$, $R^3 = F$) was found to be a potent inhibitor of 3-hydroxy-3-methylglutaryl-coenzyme A <98BMC1429, 99JOC3714>.

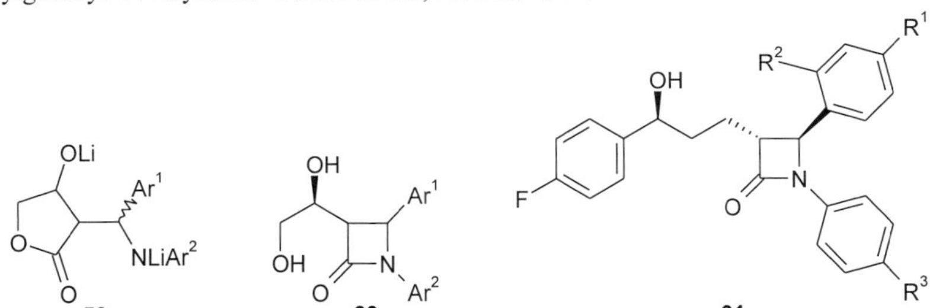

59 **60** **61**

4.2.8 BI-, TRI- AND TETRA-CYCLIC β-LACTAMS

Reviews of the synthetic strategies to β-lactams and their medicinal properties <99MI195>, advances in carbapenem chemistry <98CHE1249> and electrophilic cyclization of unsaturated amides <98T13681> are available.

The 2-formyl-1β-methylcarbapenem **62** has been obtained in five steps from a readily available β-lactam in 23-26% overall yield <98MI1294>. Suzuki-Miyaura cross-coupling of arylboronic acids and vinyl triflates is a convenient route to 2-arylcarbapenems on a small scale but may present problems on a large scale. Vinyl phosphates, mesylates or tosylates are convenient alternatives to triflates <99SL471>. Radical cyclizations of readily available enyne-2-azetidinones (e.g., **63**) with a tin hydride, R_3SnH, provides a route to the stereoselective synthesis of fused bicyclic lactams (e.g., **64**). With suitably disposed substituents, cyclization may occur across the C_3-C_4 positions. In the absence of a double bond, other reactions may occur including C_3-C_4 bond cleavage of the β-lactam <99JOC5377>.

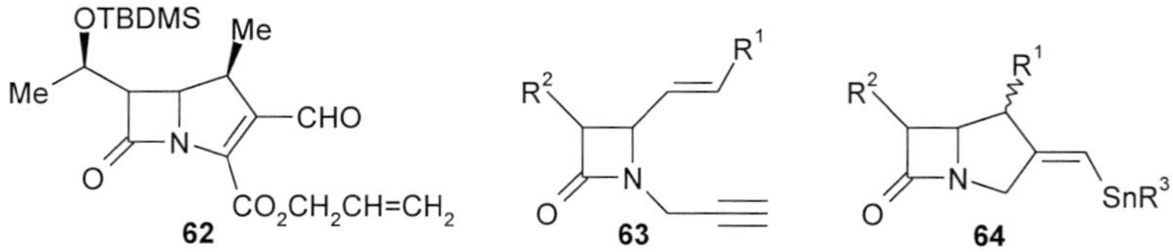

62 **63** **64**

Oxapenems **67** may be obtained by thermolysis of 1-oxa-4-azabicyclo[3.2.0]hept-2,5-diones **65** with aldehydes or ketones. The mechanism is thought to involve 1,3-dipolar addition of the aldehyde or ketone to the intermediate **66** <99CC249>.

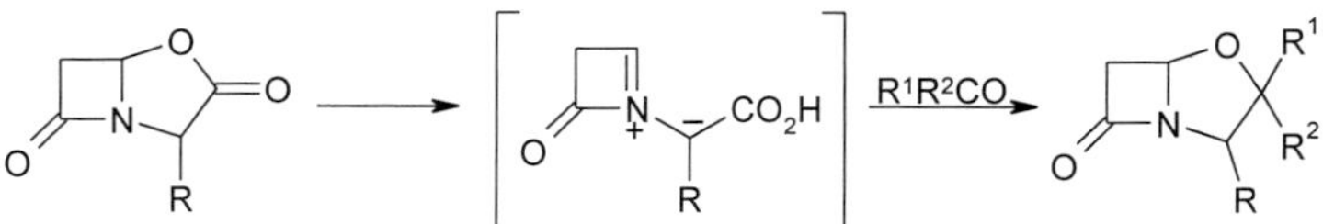

65 **66** **67**

A comparison of microwave with conventional heating for the synthesis of cephalosporins has shown that the former method gives the higher yields and in a shorter time <99CL487>. The iodotrimethylsilane-triphenylphosphine reagent has been shown to be a useful method for the removal of *p*-methoxybenzyl and diphenylmethyl ester protecting groups to yield the acids of the cephalosporin series <99SC3533>.

Carbacephem **69** is obtained in excellent yield by ring-closure of the 1,4-diallyl-β-lactam **68** by a metathesis reaction in the presence of a ruthenium catalyst <99JCS(P1)1695>. Annulated carbacephams **71** is obtained by cyclization of the piperidine **70** in the presence of a 2-chloro-1-methylpyridinium salt <98JOC8170>.

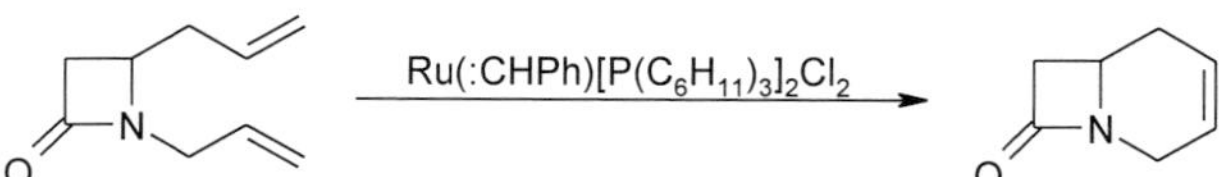

68 **69**

70 **71**

A new acetal resin valuable for solid-phase synthesis has been used for the preparation of 1-oxacephams <99AG(E)1121, 99TL5909>.

A route to **72**, a known precursor for thienamycin, relies on stereoselectivity provided by the dimethyl(phenyl)silyl group <99JCS(P1)2663>. High pressure promotes the tandem cycloaddition of β-nitrostyrene to enol ethers ($R^1CH{:}CHOR^2$) to give **73** which in the presence of base rearranges to **74** <99CC855>.

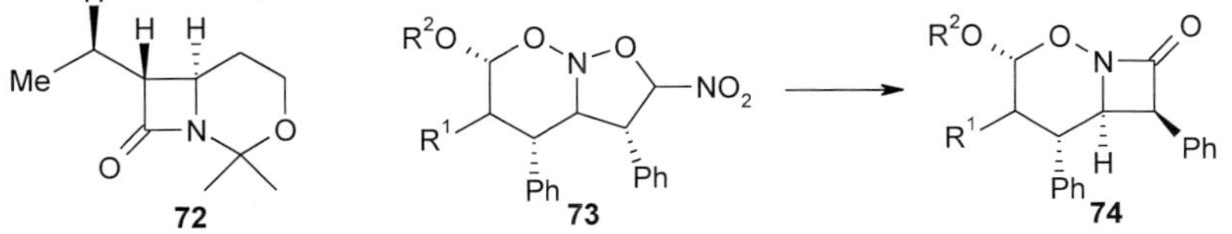

Bicyclic β-lactams not containing a bridgehead nitrogen are mechanism-based inhibitors of class C β-lactamases <98JMC3961>. Calculations giving thermodynamic stabilities and LUMO energies have been made. These studies suggest that the *N*-fused compounds are more electrophilic than the *C*-fused isomers e.g., **75**. Some compounds were synthesised from disubstituted β-lactams <98JOC8898>.

The tricyclic **76** was obtained from an intramolecular Wittig reaction <98H2287>. The tricyclic **78** was obtained by reaction of **77** with benzyl bromoacetate in the presence of base followed by intramolecular Michael reaction <99T3427>.

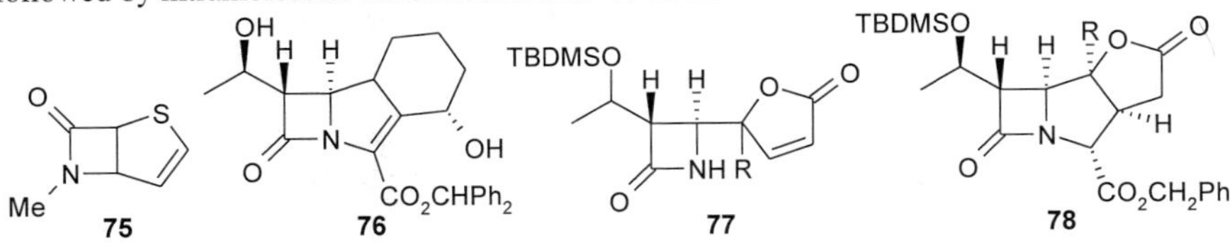

Tricyclic β-lactams not having a bridgehead nitrogen atom have been obtained by intramolecular Friedel-Crafts reactions <99T5567> and from the intramolecular Diels-Alder reactions of 1,3-dienes generated from a mesylate **79** <99TL1015>. Other tricyclic β-lactams e.g., **80** have been obtained by intramolecular nitrone-alkene cycloaddition <99TL5391>.

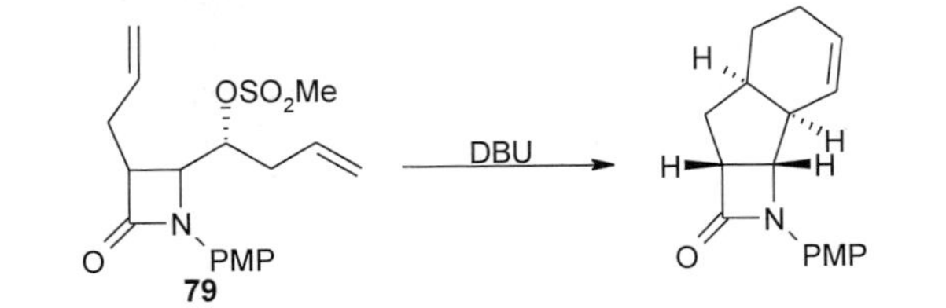

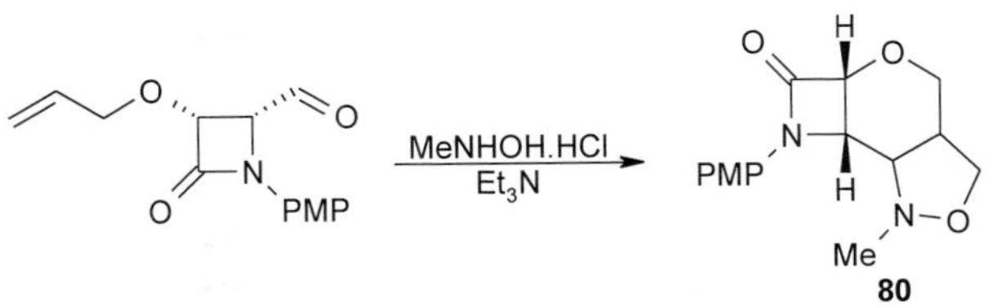

A tetracyclic β-lactam **82** was formed in an attempted deprotection of **81** using lead tetraacetate <99T8457>. Other tetracyclic β-lactams (e.g. **83**) have been obtained by ketene addition to the C_9-N_{10} bond of phenanthridine <98TL7431, 99TL2005>.

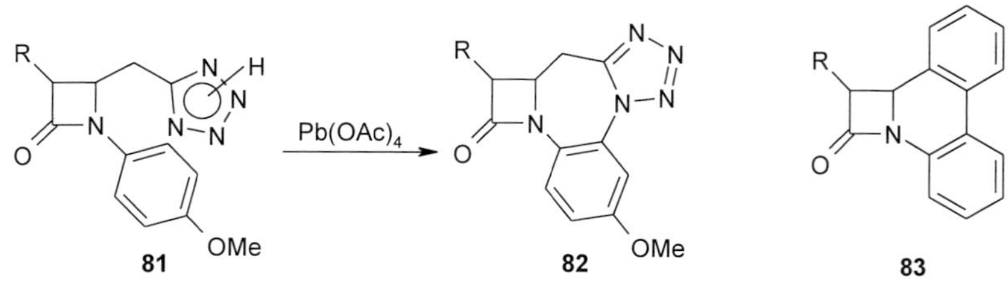

4.2.9 REFERENCES

98AJC875 W. A. Loughlin, *Aust. J. Chem.* **1998**, *51*, 875.

98BCJ1181 J. Nakayama, N. Masui, Y. Sugihara, A. Ishii, *Bull. Chem. Soc. Jpn.* **1998**, *71*, 1181.

98BMC1255 D. Romo, P. H. M. Harrison, S. I. Jenkins, R. W. Riddoch, K. Park, H. W. Yang, C. Zhao, G. D. Wright, *Bioorg. Med. Chem.* **1998**, *6*, 1255.

98BMC1429 W. D. Vaccaro, R. Sher, H. R. Davis, Jr., *Bioorg. Med. Chem.* **1998**, *6*, 1429.

98CC1995 S. V. Ley, B. Middleton, *Chem. Commun.* **1998**, 1995.

98CC2315 M. Sakamoto, M. Takahashi, T. Arai, M. Shimizu, K. Yamaguchi, T. Mino, S. Watanabe, T. Fujita, *Chem. Commun.* **1998**, 2315.

98CC2319 M. Matsumoto, H. Murakami, N. Watanabe, *Chem. Commun.* **1998**, 2319.

98CHE1222 C. Palomo, J. M. Aizpurua, *Chem. Heterocycl. Compd.* **1998**, *34*, 1222.

98CHE1249 A. Sasaki, M. Sunagawa, *Chem. Heterocycl. Compd.* **1998**, *34*, 1249.

98CHE1308 O. A. Phillips, D. P. Czajkowski, K. Alchison, R. G. Micetich, S. N. Maiti, C. Kunugita, A. Hyodo, *Chem. Heterocycl. Compd.* **1998**, *34*, 1308.

98CHE1319 O. A. Phillips, E. L. Setti, A. V. N. Reddy, R. G. Micetich, C. Kunigita, A. Hyodo, S. M. Maiti, *Chem. Heterocycl. Compd.* **1998**, *34*, 1319.

98H97 M. Jayaraman, M. T. Batista, M. S. Manhas, A. K. Bose, *Heterocycles* **1998**, *49*, 97.

98H113 M. Shindo, S. Oya, Y. Sato, K. Shishido, *Heterocycles* **1998**, *49*, 113.

98H149 A. P. Marchand, A. Devasgayaraj, S. G. Bolt, *Heterocycles* **1998**, *49*, 149.

98H2287 Y. Iso, Y. Nishitani, *Heterocycles* **1998**, *48*, 2287.

98HCA1803 P. Wessig, J. Schwarz, *Helv. Chim. Acta* **1998**, *81*, 1803.

98IJC(B)1114 S. S. Bari, A. K. Sharma, M. K. Sethi, *Indian J. Chem., Sect. B, Org. Chem. Incl. Med. Chem.* **1998**, *37B*, 1114.

98JA10653 F. Bangerter, M. Karpf, L. A. Meier, P. Rys, P. Skrabal, *J. Am. Chem. Soc.* **1998**, *120*, 10653.

98JCS(P1)3253 M. Abe, Y. Shirodai, M. Nojima, *J. Chem. Soc., Perkin Trans. 1* **1998**, 3253.

98JCS(P1)3261 M. Abe, M. Skeda, M. Nojima, *J. Chem. Soc., Perkin Trans. 1* **1998**, 3261.

98JCS(P1)4175 M. C. Elliott, *J. Chem. Soc., Perkin Trans. 1* **1998**, 4175.

98JCR(S)724 M. S. Khajavi, F. Sefidkon, S. S. S. Hosseini, *J. Chem. Res., Synop.* **1998**, 724.

98JHC1505 N. Rumpf, D. Groschl, H. Meier, D. C. Oniciu, A. R. Katritzky, *J. Heterocycl. Chem.* **1998**, *35*, 1505.

98JMC3961 I. Heinze-Krauss, P. Angehrn, R. L. Charnas, K. Gubernator, E.-M. Gutknecht, C. Hubschwerlen, M. Kania, C. Oefner, M. G. P. Page, S. Sogabe, J.-L. Specklin, F. Winkler, *J. Med. Chem.* **1998**, *41*, 3961.

98JOC8170 J. J. Folmer, C. Acero, D. L. Thai, H. Rapaport, *J. Org. Chem.* **1998**, *63*, 8170.

98JOC8192 T. Shimizu, H. Murakami, Y. Kobayashi, K. Iwata, N. Kamigata, *J. Org. Chem.* **1998**, *63*, 8192.

98JOC8898 X.-F. Ren, M. I. Konaklieva, H. Shi, S. Dickey, D. V. Lim, *J. Org. Chem.* **1998**, *64*, 8898

98MI169 M. Ikeda, T. Sato, H. Ishibashi, *Rev. Heteroat. Chem.* **1998**, *18*, 169.

98MI245 R. Lopez, E. Del Rio, N. Diaz, D. Suarez, M. I. Menendez, T. L. Sordo, *Recent Res. Dev. Phys. Chem.* **1998**, *2*, 245.

98MI355 G. G. Furin, *Targets Heterocycl. Syst.* **1998**, *2*, 355.

98MI649 E. Lukevics, O. Pudova, *Main Group Met. Chem.* **1998**, *21*, 649.

98MI755 A. Marinetti, V. Kruger, F.-X. Buzin, *Coord. Chem. Rev.* **1998**, *178-180*, 755.

98MI1294 J. S. Lee, D. K. Pyun, W. K. Lee, C. H. Lee, *Bull. Korean Chem. Soc.* **1998**, *19*, 1294.

98MI1826 E. Del Rio, R. Lopez, M. I. Menendez, T. L. Sordo, M. F. Ruiz-Lopez, *J. Comput. Chem.* **1998**, *19*, 1826.

98MI2185 D. Moelm, U. Floerke, N. Risch, *Eur. J. Org. Chem.* **1998**, 2185.

98OS106 K. Manabe, K. Koga, *Org. Synth.* **1998**, *75*, 106.

98OS116 R. Pal, P. Belica, S. Wolff, *Org. Synth.* **1998**, *75*, 116.

98S1539 A. Marinetti, V. Kruger, B. Couetoux, *Synthesis* **1998**, 1539.

98S1655 B. W. Dymock, P. J. Kocienski, J.-M. Pons, *Synthesis* **1998**, 1655.

98SC3949 R. Bartnik, D. Cal, A. P. Marchand, S. Alihodzic, A. Devasagayaraj, *Synth. Commun.* **1998**, *28*, 3949.

98SL1288 M. Alajarin, P. Molina, A. Vidal, F. Tovar, *Synlett* **1998**, 1288.

98T13681 S. Robin, G. Rousseau, *Tetrahedron* **1998**, *54*, 13681.

98T15127 A. P. Marchand, S. Alihodzic, *Tetrahedron* **1998**, *54*, 15127.

98T15657 G. C. Torchiarolo, F. D'Onofrio, R. Margarita, L. Parlanti, G. Piancatelli, M. Bella, *Tetrahedron* **1998**, *54*, 15657.

98TL7431 A. Afonso, S. B. Rosenblum, M. S. Puar, A. T. McPhail, *Tetrahedron Lett.* **1998**, *39*, 7431.

98TL9055 T. Kambara, M. A. Hussein, H. Fujieda, A. Iida, K. Tomioka, *Tetrahedron Lett.* **1998**, *39*, 9055.

98TL9639 H. Meier, N. Rumpf, *Tetrahedron Lett.* **1998**, *39*, 9639.

99AG(E)1121 B. Furman, R. Thurmer, Z. Kaluza, R. Lysek, W. Voelter, M. Chmielewski, *Angew. Chem., Int. Ed.* **1999**, *38*, 1121.

99CC249 M. D. Andrews, G. A. Brown, J. P. H. Charmant, T. M. Peakman, A. Rebello, K. E. Walsh, T. Gallagher, N. J. Hales, *Chem. Commun.* **1999**, 249.

99CC715 K. Tomioka, H. Ahmed, K. T. Mostafa, H. Fujieda, S. Hayashi, Y. Nomura, M. Kanai, K. Koga, *Chem. Commun.* **1999**, 715.

99CC855 G. J. Kuster, F. Kalmoua, H. W. Scheeren, R. de Gelder, *Chem. Commun.* **1999**, 855.

99CL487 M. Kidwai, K. R. Bhushan, P. Misra, *Chem. Lett.* **1999**, 487.

99CL605 M. Vuilhorgne, A. Commercon, S. Mignani, *Chem. Lett.* **1999**, 605.

99CPB1 E. J. Corey, W.-D. Li, *Chem. Pharm. Bull.* **1999**, *47*, 1.

99CPB720 T. Kambara, K. Tomioka, *Chem. Pharm. Bull.* **1999**, *47*, 720.

99H131 A. P. Marchand, S. Alihodzic, *Heterocycles* **1999**, *50*, 131.

99HAC61 G. Mielniczak, A. Lopusinski, *Heteroat. Chem.* **1999**, *10*, 61.

99HCA354 T. R. Todorova, A. Linden, H. Heimgartner, *Helv. Chim. Acta* **1999**, *82*, 354.

99JA1834 W. Adam, C. R. Saha-Moeller, S. B. Schambony, *J. Am. Chem. Soc.* **1999**, *121*, 1834.

99JAN193 G. Ryu, S.-K. Kim, *J. Antibiot.* **1999**, *52*, 193.

99JCS(P1)1151 T. Nishio, *J. Chem. Soc., Perkin Trans. 1* **1999**, 1151.

99JCS(P1)1695 C. A. Tarling, A. B. Holmes, R. E. Markwell, N. D. Pearson, *J. Chem. Soc., Perkin Trans. 1* **1999**, 1695.

99JCS(P1)1917 R. W. Bates, E. Fernandez-Megia, S. V. Ley, K. Ruck-Braun, D. M. G. Tilbrook, *J. Chem. Soc., Perkin Trans. 1* **1999**, 1917.

99JCS(P1)2277 J. Wang, Y. Hou, P. Wu, *J. Chem. Soc., Perkin Trans. 1* **1999**, 2277.

99JCS(P1)2663 I. Fleming, J. D. Kilburn, *J. Chem. Soc., Perkin Trans. 1* **1999**, 2663.

99JOC81 F. Homsi, G. Rousseau, *J. Org. Chem.* **1999**, *64*, 81.

99JOC518 P. Davoli, I. Moretti, F. Prati, H. Alper, *J. Org. Chem.* **1999**, *64*, 518.

99JOC1121 M. Alajarin, P. Molina, P. Sanchez-Andrada, M. C. Foces-Foces, *J. Org. Chem.* **1999**, *64*, 1121.

99JOC2250 D. Sun, S. M. Hubig, J. K. Kochi, *J. Org. Chem.* **1999**, *64*, 2250.

99JOC3281 N. Díaz, D. Suárez, T. L. Sordo, *J. Org. Chem.* **1999**, *64*, 3281.

99JOC3714 G. Wu, Y. Wong, X. Chen, Z. Ding, *J. Org. Chem.* **1999**, *64*, 3714.

99JOC3790 H. Ohtake, Y. Imada, S.-I. Murahashi, *J. Org. Chem.* **1999**, *64*, 3790.

99JOC4152 C. Larksarp, H. Alper, *J. Org. Chem.* **1999**, *64*, 4152.

99JOC4643 G. Barbaro, A. Battaglia, A. Guerrini, C. Bertucci, *J. Org. Chem.* **1999**, *64*, 4643.

99JOC5301 C. Wedler, B. Costisella, H. Schick, *J. Org. Chem.* **1999**, *64*, 5301.

99JOC5377 B. Alcaide, I. M. Rodriguez-Campos, J. Rodriguez-Lopez, A. Rodriguez-Vicente, *J. Org. Chem.* **1999**, *64*, 5377.

99JOC7657 H. W. Yang, D. Romo, *J. Org. Chem.* **1999**, *64*, 7657.

99JOC7074 L. M. Dollinger, A. J. Ndakala, M. Hashemzadeh, G. Wang, Y. Wang, I. Martinez, J. T. Arcari, D. J. Galluzzo, A. R. Howell, A. L. Rheingold, J. S. Figuero, *J. Org. Chem.* **1999**, *64*, 7074.

99JOC8041 T. Bach, F. Eilers, *J. Org. Chem.* **1999**, *64*, 8041.

.99JPC8628 I. Massova, P. A. Kollman, *J. Phys. Chem. B* **1999**, *103*, 8628.

99JPR616 N. Risch, U. Westerwelle, J. Kiene, R. Keuper, *J. Prakt. Chem.* **1999**, *341*, 616.

99MI1 T. Kawashima, P. Okazaki, *Adv. Strained Interesting Org. Mol.* **1999**, *7*, 1.

99MI138 N. I. Korotkikh, *Ross. Khim. Zh.* **1999**, *43*, 138.

99MI195 M. Kidwai, P. Sapra, K. R. Bhusan, *Curr. Med. Chem.* **1999**, *6*, 195.

99MI274 M. Sanchez, R. Reau, C. J. Marsden, M. Regitz, G. Bertrand, *Chem.--Eur. J.* **1999**, *5*, 274.

99MI399 T. K. Gounev, G. A. Guirgis, T. A. Mohamed, P. Zhen, J. R. Durig, *J. Raman Spectrosc.* **1999**, *30*, 399.

99MI774 L. Rigon, H. Ranaivonjatovo, J. Escudie, A. Dubourg, J.-P. Declercq, *Chem.--Eur. J.* **1999**, *5*, 774.

99MI1045 N. Díaz, D. Suárez, T. L. Sordo, *Chem.--Eur. J.* **1999**, *5*, 1045.

99MI1401 J. Pitarch, J.-L. Pascual-Ahair, E. Silla, I. Tunon, M. F. Ruiz-Lopez, *J. Comput. Chem.* **1999**, *20*, 1401.

99MI1581 S. Haber, M. Schmitz, U. Bergstrasser, J. Hoffmann, M. Regitz, *Chem.--Eur. J.* **1999**, *5*, 1581.

99OL717 F. P. J. T. Rutjes, K. C. M. F. Tjen, L. B. Wolf, W. F. J. Karstens, H. E. Schoemaker, H. Hiemstra, *Org. Lett.* **1999**, *1*, 717.

99OL825 A. R. Howell, A. J. Ndakala, *Org. Lett.* **1999**, *1*, 825.

99OM796 B. Wang, K. A. Nguyen, G. N. Srinivas, C. L. Watkins, S. Menzer, A. L. Spek, K. Lammertsama, *Organometallics* **1999**, *18*, 796.

99S650 J. Podlech, S. Steurer, *Synthesis* **1999**, 650.

99SC885 A. P. Marchand, A. Devasagayaraj, *Synth. Commun.* **1999**, *29*, 885.

99SC2695 C. S. Cho, L. H. Jiang, S. C. Shim, *Synth. Commun.* **1999**, *29*, 2695.

99SC3533 K. H. Cha, T. W. Kang, D. O. Cho, H.-W. Lee, J. Shin, K. Y. Jin, K.-W. Kim, J.-W. Kim, C. Hong, *Synth. Commun.* **1999**, *29*, 3533.

99SL447 S.-K. Kang, T.-G. Baik, X.-H. Jiao, K.-J. Lee, C. H. Lee, *Synlett* **1999**, 447.

99SL471 M. A. Huffman, N. Yasuda, *Synlett* **1999**, 471.

99SUL57 S. G. D'Yachkova, A. V. Afonin, N. A. Kalinina, E. A. Beskrylaya, A. G. Mal'Kina, E. I. Kositsina, B. A. Trofimov, *Sulfur Lett.* **1999**, *52*, 57.

99T3427 S. Hanessian, B. Reddy, *Tetrahedron* **1999**, *55*, 3427.

99T4287 N. Watanabe, H. Suganuma, H. Kobayashi, H. Mutoh, Y. Katao, M. Matsumoto, *Tetrahedron* **1999**, *55*, 4287.

99T5567 F. Bertha, J. Fetter, M. Kajtár-Peredy, K. Lempert, *Tetrahedron* **1999**, *55*, 5567.

99T6403 H. W. Yang, D. Romo, *Tetrahedron* **1999**, *55*, 6403.

99T8457 L. T. Giang, J. Fetter, M. Kajtár-Peredy, K. Lempert, F. Bertha, G. M. Keserü, G. Czira, T. Czuppon, *Tetrahedron* **1999**, *55*, 8457.

99T11219 M. A. Hussein, A. Iida, K. Tomioka, *Tetrahedron* **1999**, *40*, 11219.

99TA1673 M. Shi, J.-K. Jiang, *Tetrahedron: Asymmetry* **1999**, *10*, 1673.

99TL443 P. R. Dave, R. Duddu, R. Surapaneni, R. Gilardi, *Tetrahedron Lett.* **1999**, *40*, 443.

99TL585 F. Zhou, M. R. Detty, R. J. Lachicotte, *Tetrahedron Lett.* **1999**, *40*, 585.

99TL1015 B. Alcaide, P. Almendros, *Tetrahedron Lett.* **1999**, *40*, 1015.

99TL1249 R. Singh, J. M. Nuss, *Tetrahedron Lett.* **1999**, *40*, 1249.

99TL2005 B. Alcaide, A. Rodriguez-Vicente, *Tetrahedron Lett.* **1999**, *40*, 2005.

99TL2035 W. C. Shakespeare, *Tetrahedron Lett.* **1999**, *40*, 2035.

99TL2443 A. L. P. Nery, S. Ropke, L. H. Catalani, W. J. Baader, *Tetrahedron Lett.* **1999**, *40*, 2443.

99TL3481 J. J. Caldwell, J. P. A. Harrity, N. M. Heron, W. J. Kerr, S. McKendry, D. Middlemiss, *Tetrahedron Lett.* **1999**, *40*, 3481.

99TL3485 J. J. Caldwell, W. J. Kerr, S. McKendry, *Tetrahedron Lett.* **1999**, *40*, 3485.

99TL3761 K. Hayashi, C. Sato, S. Hiki, T. Kumagai, S. Tamai, T. Abe, Y. Nagao, *Tetrahedron Lett.* **1999**, *40*, 3761.

99TL5391 B. Alcaide, J. M. Alonso, M. F. Aly, E. Saez, M. P. Martinez-Alcazar, F. Hernandez-Cano, *Tetrahedron Lett.* **1999**, *40*, 5391.

99TL5909 B. Furman, R. Thurmer, Z. Kaluza, W. Voelter, M. Chmielewski, *Tetrahedron Lett.* **1999**, *40*, 5909.

99TL6535 S. G. Nelson, Z. Wan, T. J. Peelen, K. L. Spencer, *Tetrahedron Lett.* **1999**, *40*, 6535.

99TL6995 C. De Risi, G. Pöllini, A. C. Veronese, V. Bertolasi, *Tetrahedron Lett.* **1999**, *40*, 6995.

Chapter 5.1

Five-Membered Ring Systems: Thiophenes & Se, Te Analogs

Erin T. Pelkey
Stanford University, Stanford, CA, USA
pelk@saurus.stanford.edu

5.1.1 INTRODUCTION

Reports detailing the chemistry and syntheses of thiophenes, benzothiophenes, and related ring systems that have appeared during the past year are the primary focus of this review. The synthesis of heterocycles including thiophenes has been reviewed <99JCS(P1)2849, 99JH1469>. As always, the author apologizes in advance for all errors and omissions.

5.1.2 THIOPHENE RING SYNTHESIS

The first total synthesis of the potent cytotoxic marine natural product makaluvamine F (**5**) involved the preparation of 2,3-dihydrobenzothiophene **2** <99CC143>. Debenzylation and subsequent acid-catalyzed cyclization of thioether **1** gave **2** which was converted in four steps to 2-azido-2,3-dihydrobenzothiophene **3** using a combination of PhI=O and Me_3SiN_3 for installation of the azide. Reduction of the azide followed by coupling the resultant amine with pyrroloiminoquinone **4** then gave makaluvamine F (**5**).

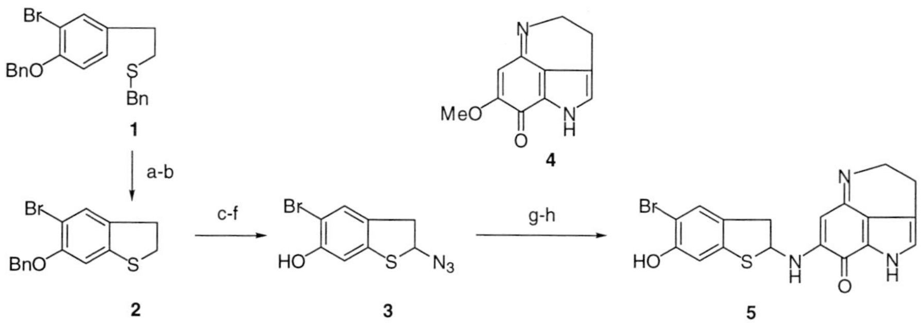

Reagents: (a) $PhI(OCOCF_3)_2$, BF_3-Et_2O; (b) aq $MeNH_2$; (c) BF_3-Et_2O, EtSH; (d) Ac_2O, NaOAc; (e) PhI=O, Me_3SiN_3, MeCN; (f) NaOH, MeOH; (g) H_2, 10% Pd/C, EtOH, TFA; (h) **4**, MeOH

An unexpected aryl-group rearrangement was observed during the attempted preparation of 3-arylbenzothiophenes <99TL2909>. For example, the cyclization of thioether **6** in the presence of PPA gave 2-arylbenzothiophene **7** rather than the expected product **8**. The rearrangement was eventually avoided by performing the cyclization of **6** in neat BF_3-Et_2O which gave the desired product **8**.

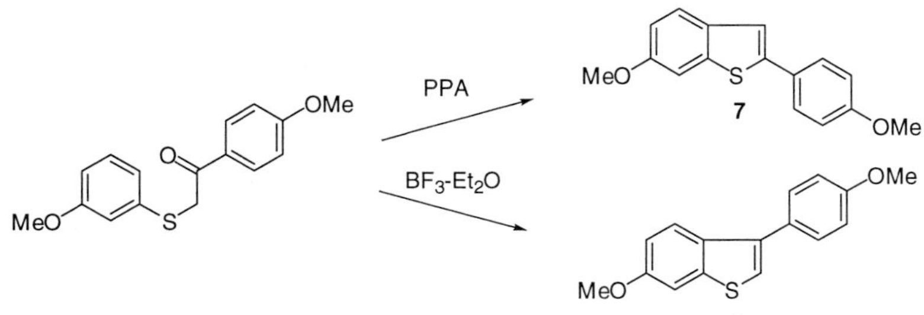

6 **8**

A novel route to 2,3-dihydrothiophenes involved a titanocene-promoted carbene formation and subsequenct intramolecular cyclization onto a thiol ester <99SL1029>. Treatment of thioacetal **9** with the low-valent titanium complex **10** gave 2,3-dihydrothiophene **12** by intramolecular olefination of the thiol ester of titanium-carbene intermediate **11**. Another metal-mediated cyclization onto the thiophene ring system involved the palladium-catalyzed cyclization of 1,6-diynes <99T485>. For example, treatment of thioether 1,6-diyne **13** with PdI_2 in the presence of CO and O_2 in methanol followed by treatment with base gave **14**.

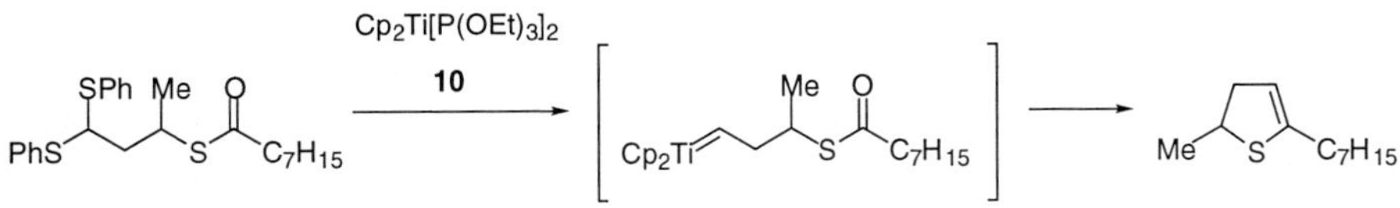

9 **11** **12**

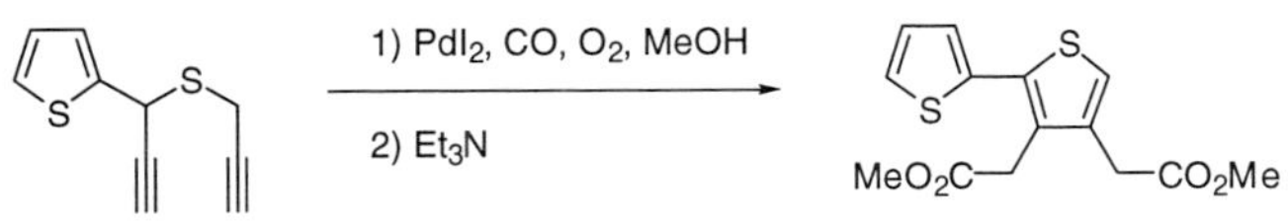

13 **14**

The preparation of norbornadiene-fused thiophene (**17**) involved a double-Wittig reaction of 1,2-dione **15** with the bis-ylide derived from phosphonium salt **16** <99BCSJ1597>. The effect of the fused heteroaromatic ring of **17** (neighboring group participation) on electrophilic substitution of the norbornadiene ring was examined.

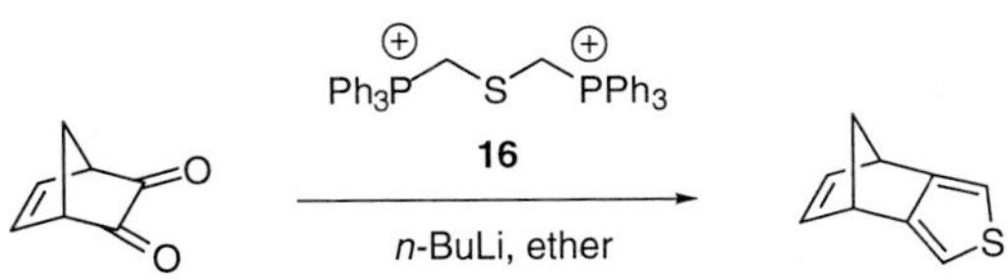

15 **17**

A couple of interesting rearrangements and transformations leading to the thiophene ring system have been reported. An extension of a previously reported rearrangement, oxidation of thioether **18** with *m*-CPBA gave the *S*-oxide **19** which underwent successive [2,3] and [3,3] sigmatropic rearrangements leading to dihydrothiophene **20** <99T1449>. Heating dithiete **21** in refluxing *p*-xylene led to thiophene **23** by the cycloaddition of **21** with ring-opened dithione intermediate **22** followed by extrusion of sulfur <99JOC8489>.

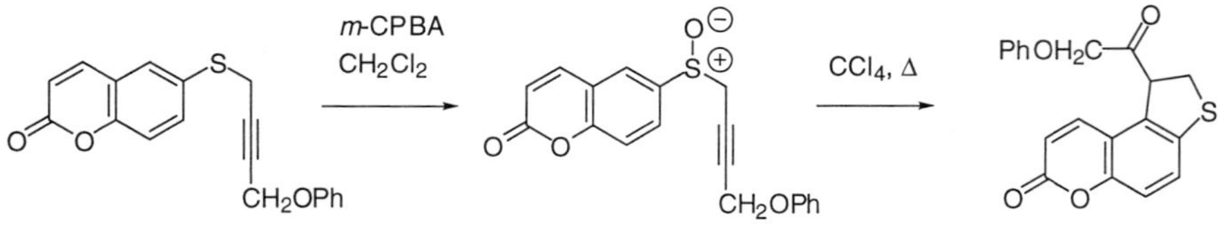

18 **19** **20**

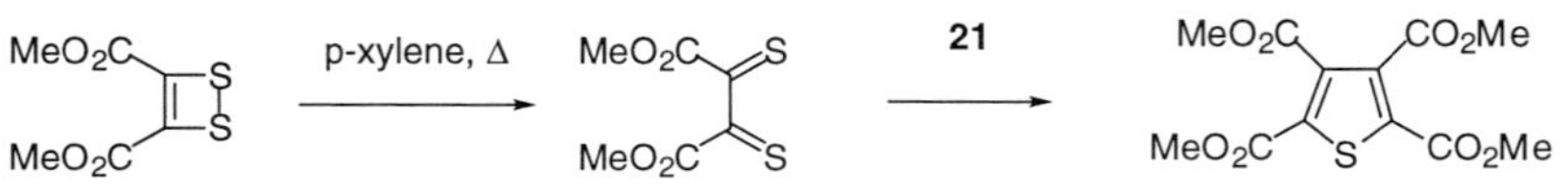

21 **22** **23**

Many reports detailing the condensation of activated α-thiol-substituted compounds onto adjacent carbonyls for the synthesis of complex thiophenes appeared during the past year including the novel thiophene-fused morphine analogue **26**. Condensation of diketone **24** with α-thioglycolic acid gave intermediate **25** which underwent an intramolecular cyclization giving the product **26** <99OL513>. This type of reaction sequence was used to synthesize benzothiophene-2-carboxylates <99CPB1221>.

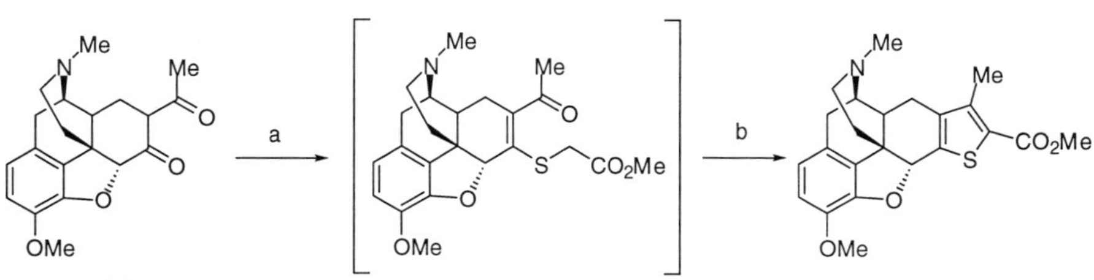

24 **25** **26**

Reagents: (a) $HSCH_2CO_2H$, HCl, MeOH; (b) NaOMe, MeOH

Similarly, the intramolecular cyclization of α-thioglycolates or α-thioketones onto adjacent nitriles remains a popular method for the synthesis of β-amino-substituted thiophenes. For example, treatment of acrylonitrile **27** with methyl α-thioglycolate gave β-aminothiophene **28** <99JH659>. Additional β-aminothiophenes have been synthesized using this type of reaction sequence including thiophene **29** <99JH423> and α-thiothiophene **30** <99JMC1849>. The preparation of α-aminothiophenes is still often accomplished using the classic Gewald synthesis <99JH333>. For example, treatment of β-ketoester **31** with α-cyanoacetate **32** and elemental sulfur gave α-aminothiophene **33** which was utilized as a molecular scaffold for library generation <99TL5471>. A fused thiophene quinone **34** was also synthesized using this type of sequence <99JH15>. A similar reaction sequence involved the condensation of ethyl acetoacetate, sulfur, and 3-aminocrotonitrile which gave a 2-cyanothiophene <99JCR(S)536>.

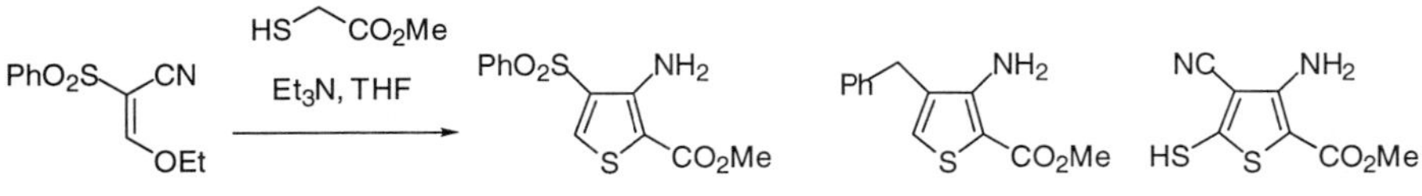

27 28 29 30

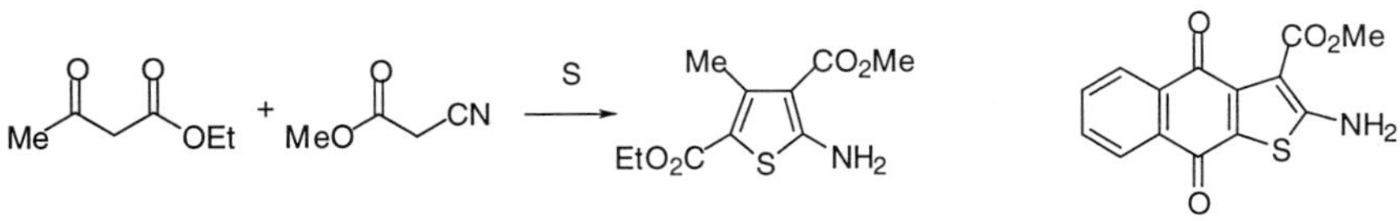

31 32 33 34

5.1.3 THIOPHENE RING SUBSTITUTION

The unsubstituted α-positions of the thiophene ring system continue to be elaborated using standard electrophilic aromatic substitution reactions including bromination <99CM2533, 99JH241, 99JMAT2109, 99SC1607, 99T485>, nitration <99JCS(P1)3691, 99MC573>, and Friedel-Crafts acylation <99H819, 99JMC3199, 99JOC7890, 99SL1621>. Friedel-Crafts acylation of the unsubstitutued β-position of thiophene **35** with acetic anhydride proceeded readily to give β-acetylthiophene **36** <99CL1071>. Vinylogous aldol additions with α-silyloxythiophenes (e.g., **37**) were reported <99SL1333>. Bis-bromomethylation of thiophene with paraformaldehyde and HBr gave **38** (albeit in low yield) which was utilized to prepare macrocyclic thiophenophane **40** for the study of novel host-guest chemistry <99T4709>.

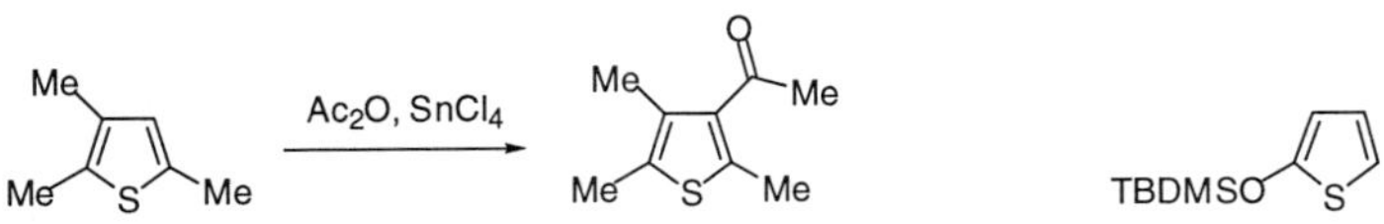

35 36 37

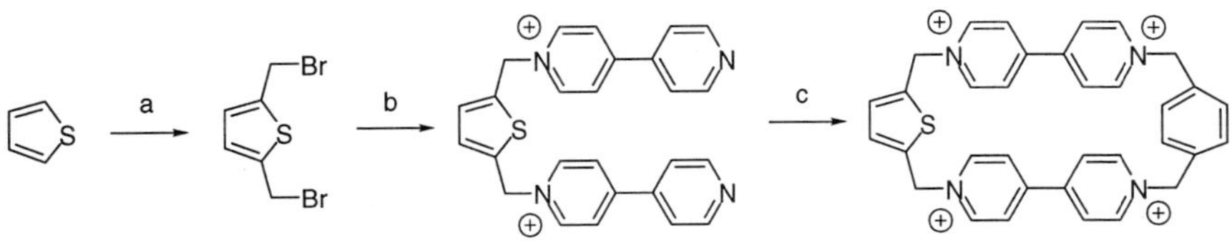

38 39 40

Reagents: (a) $(CH_2O)_n$, HBr, HOAc; (b) 4,4'-bipyridine, CH_3CN; (c) 1,3-bis(bromomethyl)benzene, aq. CH_3NO_2

Fluorine-substituted thiophenes have been prepared by electrophilic fluorination. Fluorination of 3-acetamidothiophene (**41**) with Selectfluor™ was directed by the acetamido group to give α-fluorothiophene **42** <99JH1247>. Similarly, fluorination of 2-acetamidothiophene (**43**) gave β-fluorothiophene **44**. In the latter case, the activating effect of the acetamido group exceeded the intrinsic reactivity of the thiophene ring. Additional syntheses of fluorothiophene derivatives have appeared <99IJC648, 99JF(93)73, 99JF(99)73>.

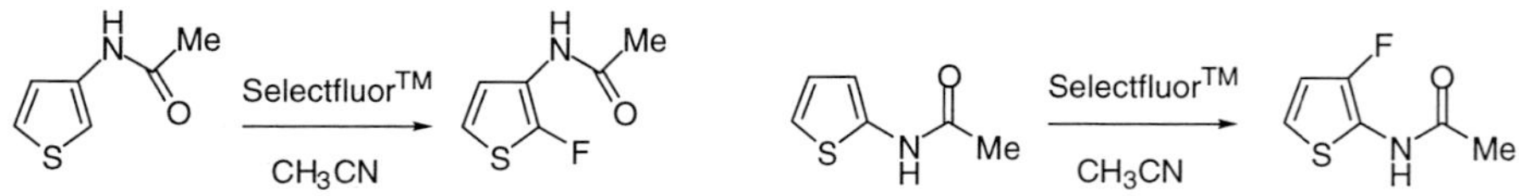

41 **42** **43** **44**

A novel route to fused thiophene-1-oxides involved the zirconocene-mediated cyclization of 1,6-diynes. For example, treating 1,6-diyne **45** with a zirconocene followed by sulfur dioxide gave thiophene-1-oxide **46** <99JA9744>. Oxidation of **46** with *m*-CPBA gave thiophene-1,1-dioxide **47**. The titanocene-mediated cyclization tolerated the presence of aryl bromides thus allowing for further elaboration of the thiophene-1-oxide and thiophene-1,1-dioxide products using cross-coupling reactions for the synthesis of oligomers and polymers. Several additional reports involving the synthesis and reactions of thiophene-1-oxides and thiophene-1,1-dioxides have appeared during the past year <99BCSJ1919, 99CM2533, 99JF(93)73, 99JH249, 99JH1077, 99O3187>.

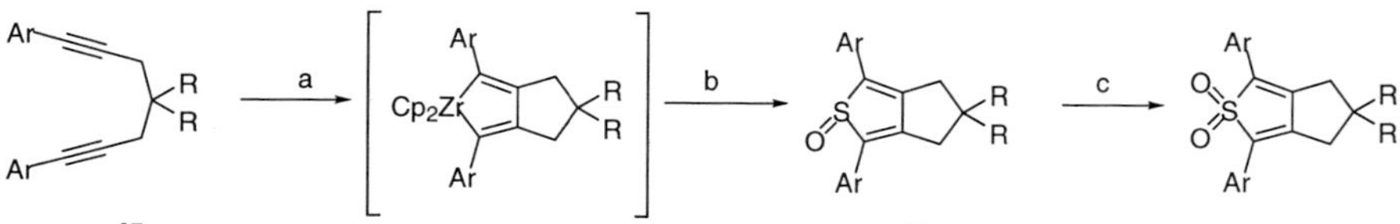

45 **46** **47**

R = $CH_2OC_6H_{13}$. Reagents: (a) "Cp_2Zr", THF; (b) SO_2; (c) *m*-CPBA, CH_2Cl_2

Synthetic routes to the nitrogen analogues of thiophene-1-oxides and thiophene-1,1-dioxides have been developed. Treatment of thiophene-1-oxide **48** with TsN=IPh gave sulfoximide **49** <99TL3785>, while a similar reaction involving thiophene **50** provided a mixture of thiophene-1-imine **51** and thiophene-1,1-diimine **52** amongst several products <99TL5549>.

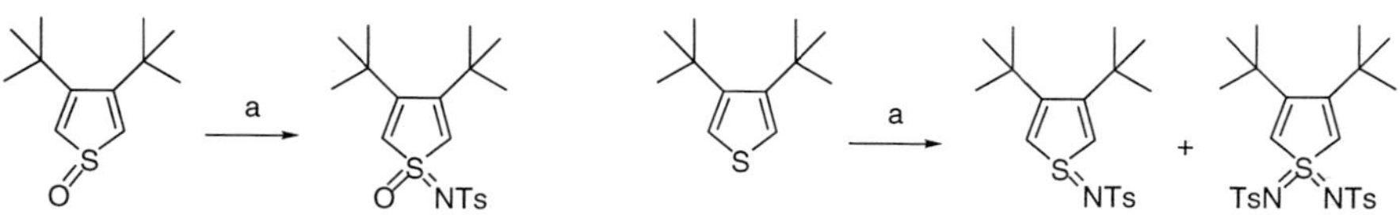

48 **49** **50** **51** **52**

Reagents: (a) TsN=IPh, MeCN, $Cu(MeCN)_4PF_6$

Nucleophilic substitution of electron-deficient thiophenes continues to be an important method for preparing functionalized thiophenes. The nucleophilic addition of alkoxide **54** to α-chlorothiophene **53** gave intermediate thiophene **55** which was further elaborated into thiophene **56**, an analogue of the TXA_1-synthase inhibitor, dazoxiben <99MC645>. A similar addition-elimination reaction of an α-sulfonylthiophene by a phenolate was utilized to prepare α-phenoxythiophenes <99JMC1849>. Nucleophilic addition-elimination of weakly activated thiophenes by amine nucleophiles in aqueous media was reported <99T6511>. Finally, the Michael-addition of diethylamine onto nitro-activated α-ethenylthiophene **57** gave push-pull thiophene **59** <99T13973>.

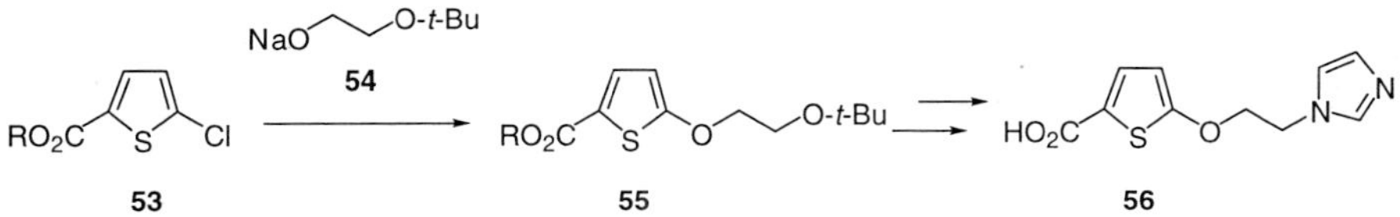

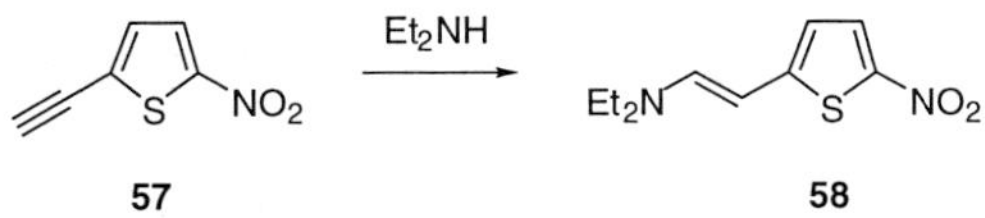

The α-lithiation of thiophene by organolithium bases is one of the most utilized methods for the preparation α-substituted thiophenes and several applications appeared during the past year <99BMCL759, 99BMCL1733, 99BMCL2837, 99JH65, 99JH249, 99JOC7890, 99SL1582, 99SL1621>. For example, α-lithiation of 2,2'-bithiophene (**59**) followed by quenching with 2-chloropyrimidine gave dihydropyrimidine **60** <BMCL2643>. Treatment of thiophene-3-carboxaldehyde (**61**) with LiNMP followed by *sec*-BuLi gave the α-lithiated thiophene **62** which was quenched with benzaldehyde to give thiophene **63** <99TL8457>. Lithiation of the adjacent α-position was blocked by the bulky amine substituent. Addition of PhMgBr to thiophene **63** gave **64**, the required building block for a porphyrin synthesis.

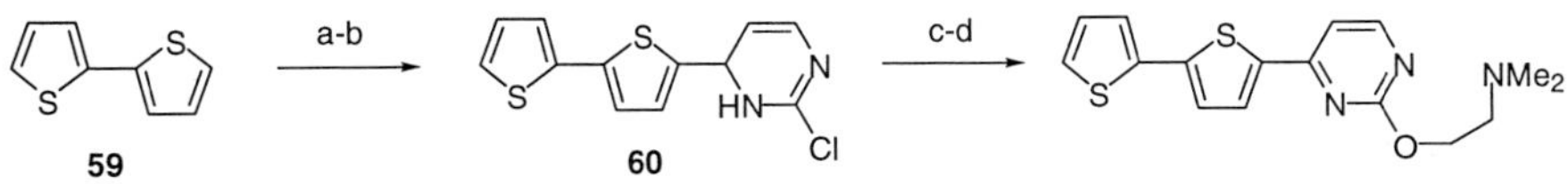

Reagents: (a) *n*-BuLi; (b) 2-chloropyrimidine; (c) $KMnO_4$, acetone; (d) NaH, THF, $HOCH_2CH_2NMe_2$

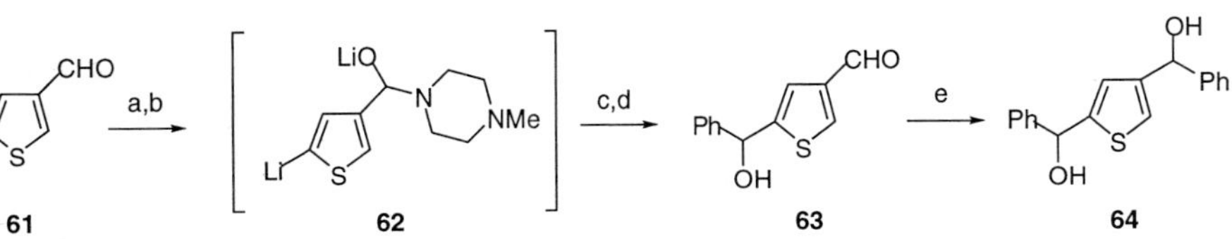

Reagents: (a) LiNMP; (b) *sec*-BuLi, TMEDA; (c) PHCHO; (d) H_2O; (e) PhMgBr

The utilization of halogen-metal exchange for the regioselective functionalization of thiophenes appeared during the past year <99BMC231, 99CM1957, 99JH241, 99O2760>. A low yield was obtained in the preparation of thiophene-3-carboxylic acid **67** by halogen-metal exchange of 3,4-dibromothiophene (**65**) (and subsequent trapping with carbon dioxide) which prompted further investigation into this reaction including the characterization of all the by-products <99BMC297>. The optimized conditions for obtaining **67** involved a two-step procedure. Halogen-metal exchange at –116 °C in ether followed by quenching with methyl chloroformate gave ester **66** as the major product which was hydrolyzed under basic conditions to give the desired thiophene **67**.

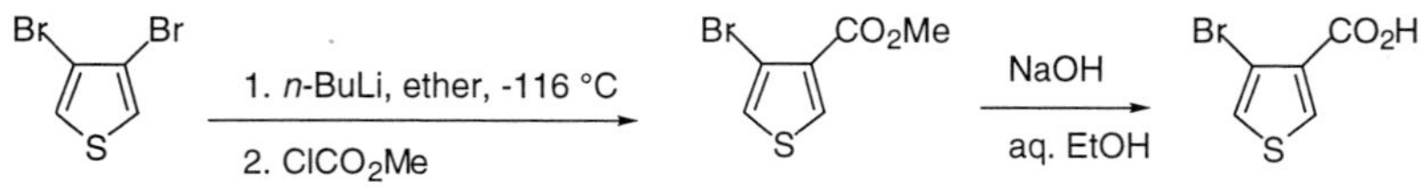

65 **66** **67**

Organometallic cross-coupling reactions of metallated thiophenes continue to be a powerful tool for the preparation of highly functionalized thiophenes <99BMC231, 99BMCL759, 99CM458, 99SL1621, 99T485>. Solid-phase organometallic cross coupling of thiophene-2-boronic acid derivatives has been utilized to prepare novel phosphodiesterase-4 (PDE-4) inhibitors <99T11669>. Specifically, treatment of aryl bromide **68** attached to Wang resin with thiophene-2-boronic acid **69** and palladium(0) gave **70**.

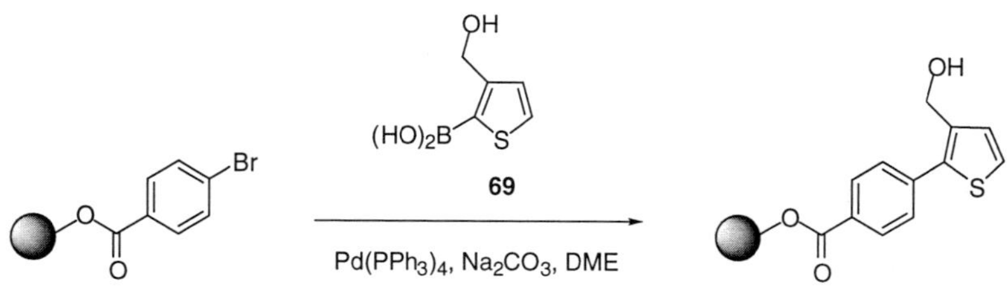

68 **70**

Another related method for preparing functionalized thiophenes that is regularly utilized is the organometallic cross-coupling of halogenated thiophenes <99CM867, 99CM2533, 99CM3050, 99JMAT2109, 99JMAT2155, 99T11669, 99TL857>. A mild method for the palladium-catalyzed cross-coupling of α-bromothiophenes with aryl boronic acids involved the use of aqueous media <99OL965>. For example, treatment of α-bromothiophene **71** with aryl boronic acid **72** in the presence of palladium(II)acetate in water gave α-arylthiophene **73**. This reaction was also extended to prepare 2,3-diarylthiophenes from 2,3-dibromothiophenes. An interesting rearrangement was discovered during the attempted synthesis of a 3-arylbenzothiophene **77** using a Suzuki palladium-catalyzed cross-coupling reaction <99TL2909>. Specifically, treatment of 3-iodobenzothiophene **74** with aryl boronic acid **75** unexpectedly resulted in the formation of 2-arylbenzothiophene **76** rather than the expected 3-arylbenzothiophene **77**. The mechanism of the unprecedented rearrangement is unclear.

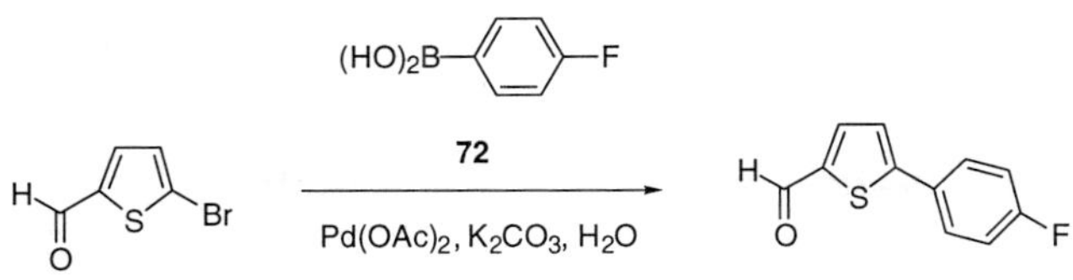

71 73

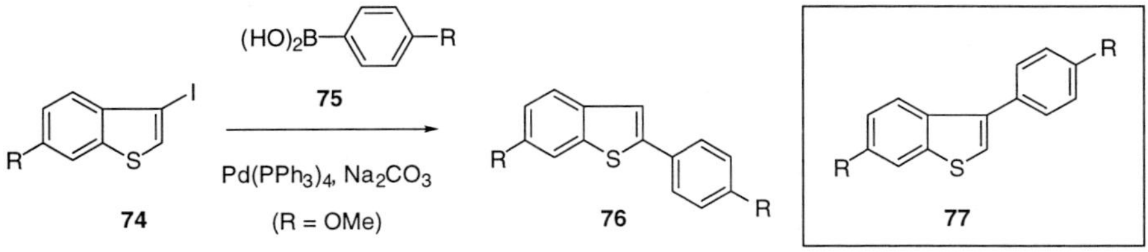

The thiophene analogue **80** of the immunosuppressive drug, prodigiosin, was synthesized using a photochemical coupling reaction <99SC35, 99T2013>. Specifically, irradiation of thiophene in the presence of iodopyrrole **78** gave 2-arylthiophene **79** which was converted to **80** in two steps. Interestingly, irradiation of thiophene in the presence of the corresponding mono-iodinated pyrrole gave no arylation product.

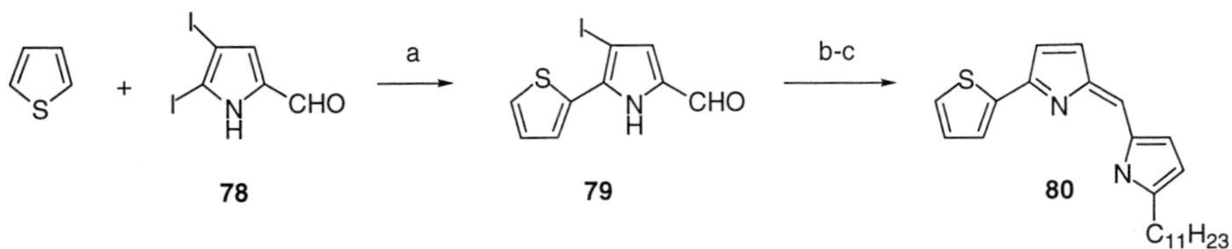

Reagents: (a) CH_3CN, hv; (b) $NiCl_2$, PPh_3, NaI, Zn, DMF, H_2O; (c) 2-undecylpyrrole, HCl

The preparation of vinyl-substituted thiophene **83** involved the Wittig reaction between thiophene phosphonium salt **82** and thiophene-2,5-dicarboxaldehyde <99T11745>. Phosphonium salt **82** was prepared from thiophene **81** in three steps including a Mannich reaction, quaternization of the resulting amine, and displacement with triphenylphosphine. A facile synthesis of 3-vinylthiophene from commercially available 3-(β-hydroxyethyl)thiophene was reported <99JH1105>. Other side-chain related reactions of thiophenes that have been reported include side-chain oxidations <99BMC1025, 99CL1071, 99TL1635> and benzotriazole-mediated functionalization <99JH927>.

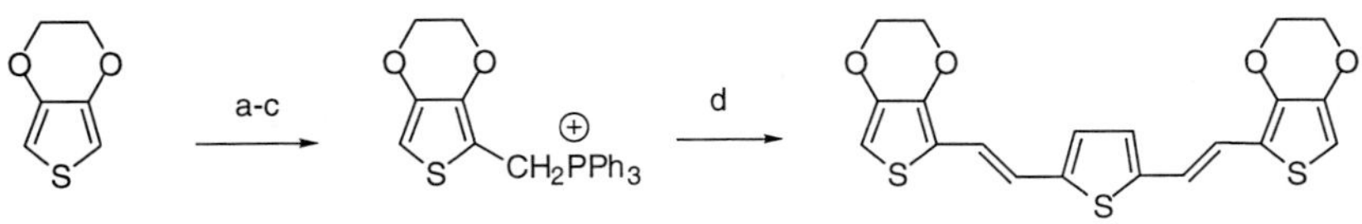

81 82 83

Reagents: (a) CH_2O, Me_2NH, H^+; (b) MeI; (c) PPh_3, DMF; (d) thiophene-2,5-dicarboxaldehyde

The synthesis and chemistry of metal complexes of thiophenes have been reported including the electrophilic additions to osmium-thiophene complexes <99O2988> and nucleophilic additions to ruthenium-thiophene complexes <99JOMC242>. The selectivity for the insertion of ruthenium into 3-substituted thiophenes was studied <99CC1793>. For example, treatment of 3-acetylthiophene (**84**) with Ru(cod)(cot) led to a regioselective 1,2-insertion of ruthenium giving thiaruthenacycle **85**.

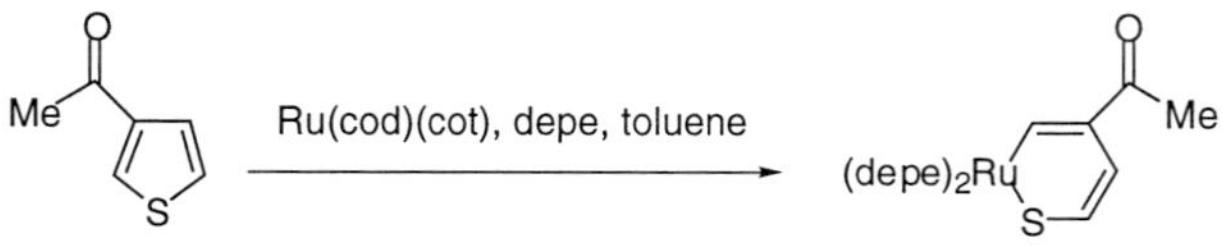

84 **85**

5.1.4 RING ANNELATION ON THIOPHENE

The electron-rich thiophene ring system can be annelated by intramolecular Friedel-Crafts acylation reactions to give fused thiophenes <99IJC648, 99JMC2774>. The synthesis of a thiophene isostere of ninhydrin involved an intramolecular Friedel-Crafts acylation <99SL1450>. Specifically, treatment of thiophene **86** with thionyl chloride followed by aluminum chloride gave annelated thiophene **87**. The synthesis of isothianinhydrin **88** was then accomplished in six steps from **87**.

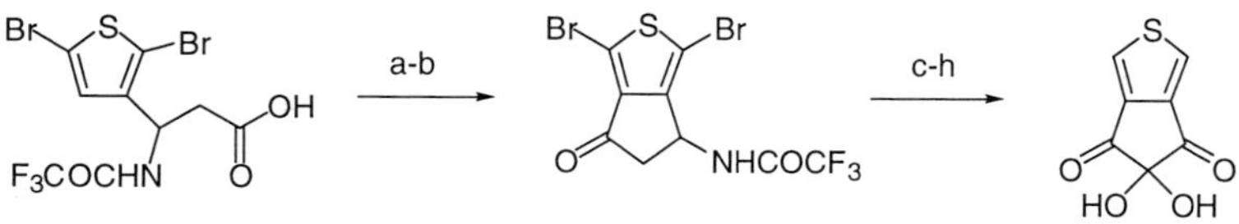

86 **87** **88**

Reagents: (a) $SOCl_2$; (b) $AlCl_3$; (c) $(EtO)_2P(O)H$, Et_3N, THF; (d) HCl, aq. EtOH; (e) $NaNO_2$, aq. AcOH; (f) CrO_3; (g) Br_2, AcOH; (h) DMSO, toluene

Novel [4+2] cycloadditions of 1-phenyl-1-benzothiophenium triflate salt **89** with dienophiles (cyclopentadiene and 1,3-diphenylbenzofuran) have been reported <99OL257>. Heating **89** and cyclopentadiene in CH_2Cl_2 gave a single product, *endo* cycloadduct **90**. The stereochemistry of **90** was confirmed by single-crystal X-ray analysis. The cycloaddition of 3-methyl-1-phenyl-benzothiophenium triflate salt with cyclopentadiene proceeded only in low yield and required much more rigorous reaction conditions (CH_3CN, sealed tube, Δ).

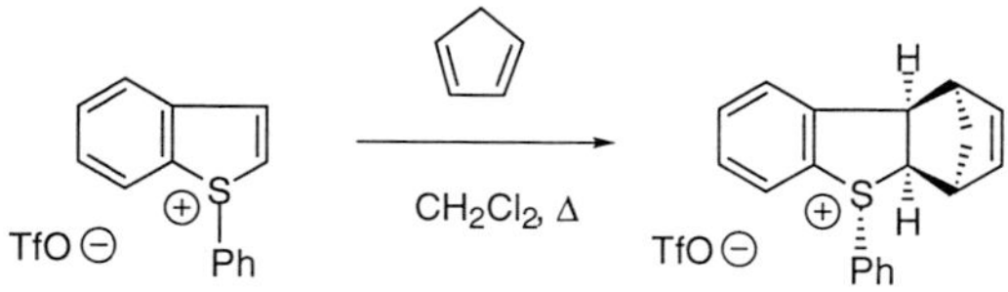

89 **90**

Flash-vacuum pyrolysis has been utilized to synthesize complex thiophene-containing polycyclic hydrocarbons from alkynyl-substituted thiophenes and chlorovinyl-substituted thiophenes <99TL2789>. For example, the "bowl-shaped" heteroaromatic thiophene **92** was prepared by flash-vacuum pyrolysis of benzotrithiophene **91** <99CC1859>.

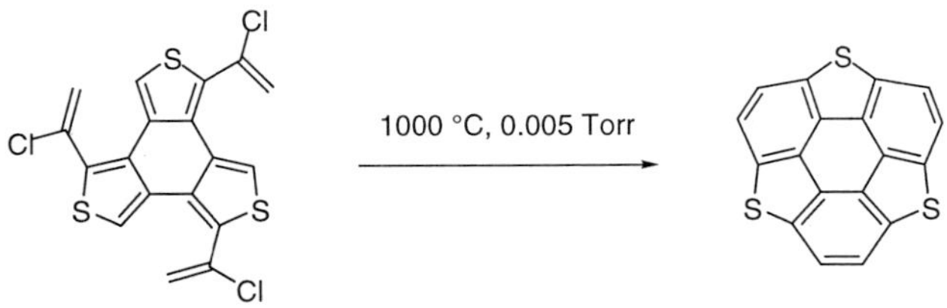

91 **92**

Photocyclizations are another method for preparing complex thiophene-containing polycyclic hydrocarbons <99JCS(P1)2391>. For example, irradiation of napthalene **93** in the presence of iodine gave naphtho[2,3-*g*]thiopheno[3,2-*e*]benzothiophene **94** <99S1303>. An interesting solvent dependency has been reported with the photochemical rearrangement of 2-styrylthiophenes <99OL1039>. Specifically, irradiation of thiophene **95** in methylene chloride gave benzothiophene **96** by photochemical cyclization, 1,9-hydrogen shift, lateral ring opening, and enol-ketone tautomerization. On the other hand, when the same reaction was performed in dehydrated benzene the rearrangement occurred without formation of the ketone giving instead benzothiophene **97**.

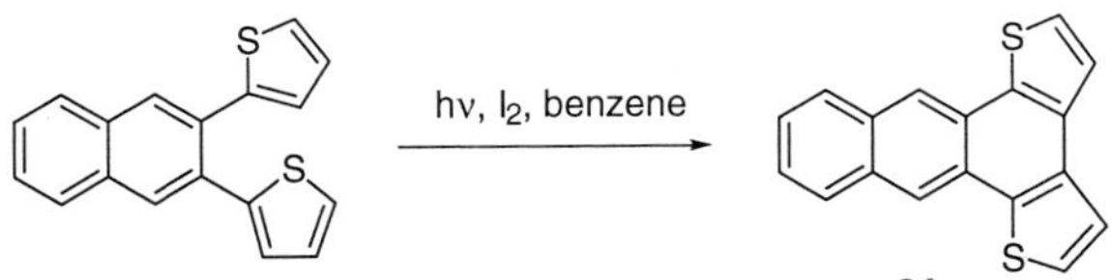

93 **94**

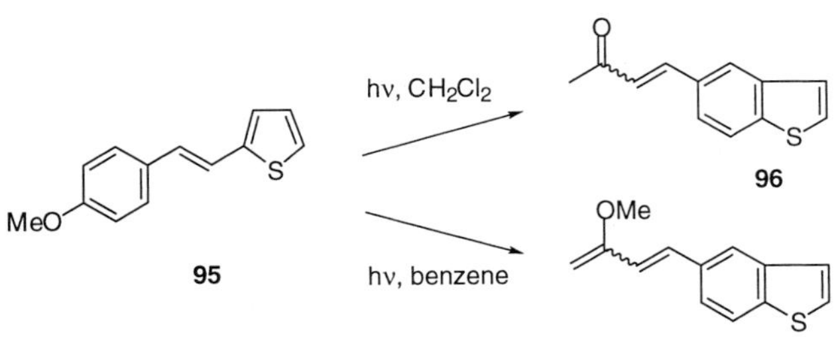

95 **96** **97**

An interesting thermal electrocyclization leading to dibenzothiophenes has been reported <99CC541>. Heating benzothiophene **98** in triethylene glycol gave dibenzothiophene **100** by electrocyclization to intermediate **99** followed by loss of carbamic acid. Compound **98** was prepared in five steps from 2-acetyl-3-hydroxybenzothiophene.

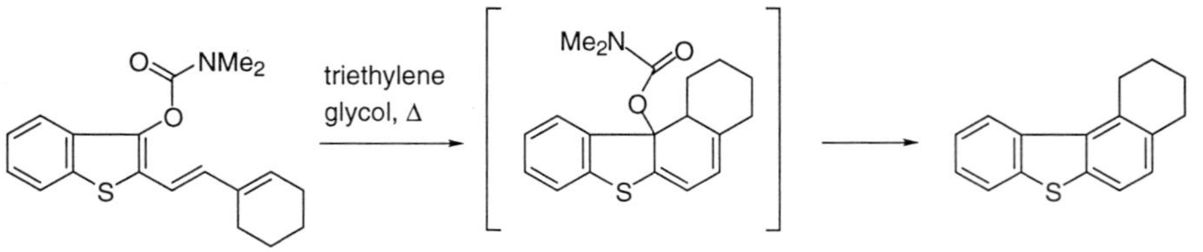

98 99 100

Synthetic routes to a variety of fused thiophene derivatives have appeared during the last year including thieno[2,3-*d*][1,2,3]thiadiazole **101** <99JH761, 99MC573>, thieno[2,3-*e*]-4,1,2-oxathiazepine **102** <99JH65>, benzo[*e*]thieno[2,3-*b*]-4-thiazepine **103** <99JH659>, pyrido[4',3':4,5]thieno[2,3-*d*]pyrimidine **104** <99JH1119>, benzophen[2,3-*d*]-1',2',3'-selenadiazole **105** <99IJC308>, thieno[2,3-*b*]imidazole **106** , thieno[2,3-*d*]pyridazine **107** <99JOC394>, thieno[2,3-*d*]pyrimidine **108** <99JH423>, naphtho[2,3-*b*]thiophene-4,9-diones <99CL503>, furo[2,3-*e*]benzothiophenes <99JCS(P1)3705>, thieno[2,3-*h*]pyrimido[2,1-*f*][1,6]naphthyridines <99JH461>, thieno[2',3':6,7][1,3]diazocino[2,1-*a*]isoindolediones <99JH735>, benzothieno[2,3-*d*]pyrimidine <99T13757>, and di[1]benzothieno[3,2-*b*:2,3-*e*]pyran <99JCR(S)542>.

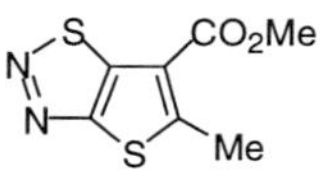

101

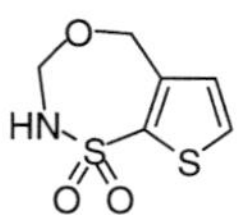

102

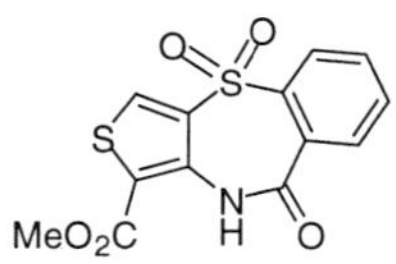

103

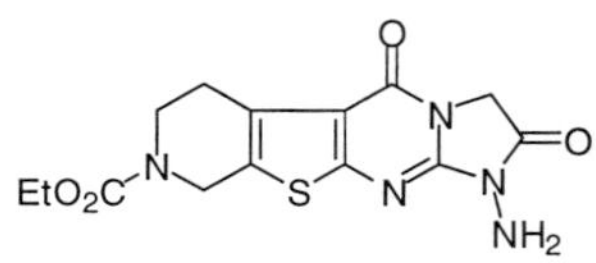

104

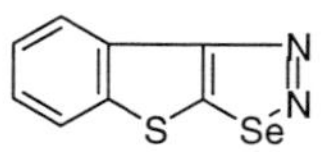

105

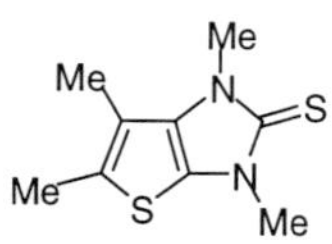

106

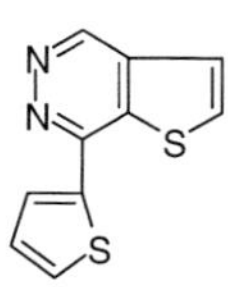

107

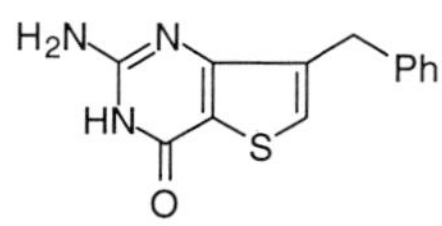

108

5.1.5 THIOPHENE INTERMEDIATES IN SYNTHESIS

The thiophene ring system can be utilized as a synthetic scaffold for the preparation of other complex molecules as the sulfur can be removed during the synthesis by reduction (desulfurization) or extrusion (loss of SO_2). The synthesis and chemistry of thiophene-1,1-dioxides including their cycloaddition chemistry leading to aromatic compounds has been reviewed <99BCSJ1919, 99TCC131>. Another application for preparing substituted butadienes and related materials by the nucleophilic ring-opening addition to 3,4-dinitrothiophene has appeared <99EJOC431>. The ring opening of benzothiophen-3-yllithium and related derivatives has been exploited to prepare functionalized alkynes. For example, lithiation of thieno[3,2-*b*]thiophene **109** with *n*-butyllithium followed by warming and quenching with ammonium chloride gave α-alkynylthiophene **110** <99JCS(P1)1273>.

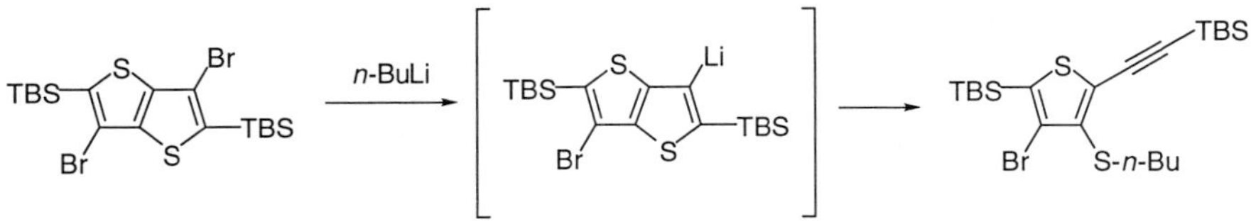

109 **110**

The enantiomeric synthesis of *trans*-3,4-disubstituted tetrahydrothiophenes using a sulfur ylide cycloaddition has been reported <99OL1667>. The sulfur ylide derived from the action of cesium fluoride on sulfide **111** underwent an asymmetric cycloaddition with chiral α,β-unsaturated camphorsultam amide **112** giving tetrahydrothiophene **113** (80% de). The configuration was confirmed by cleavage of the chiral auxiliary followed by reductive desulfurization with Raney-Ni which gave known carboxylic acid **114**.

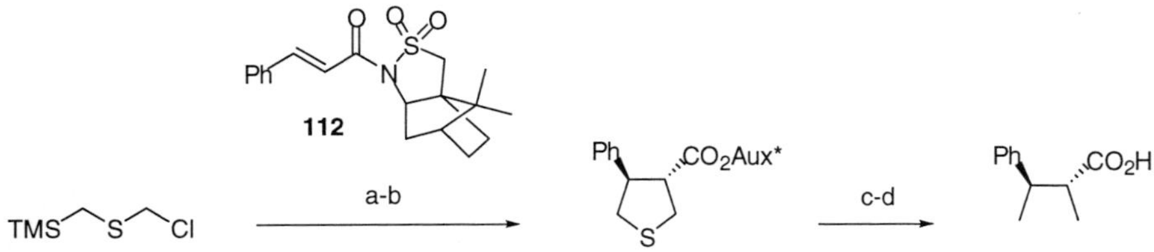

111 **113** **114**

Reagents: (a) CsF, CH_3CN; (b) **112**; (c) LiOH, aq. THF; (d) Raney Ni, EtOH

5.1.6 BIOLOGICALLY IMPORTANT THIOPHENE DERIVATIVES

A large number of biologically active thiophene-containing compounds have been synthesized and evaluated. The benzothiophene moiety has served as a scaffold for a variety of classes of compounds displaying a range of biological activity including estrogen receptor modulators (e.g., raloxifene **115**) <99TL5155>, thrombin inhibitors (e.g., **116**) <99BMCL363, 99BMCL759, 99BMCL775>, multi-drug resistance modulators (e.g., **117**) <99BMCL3381>, tubulin polymerization inhibitors (e.g., **118**) <99BMCL1081>, and HIV-1 protease inhibitors (e.g., **119**) <99BMCL2019>.

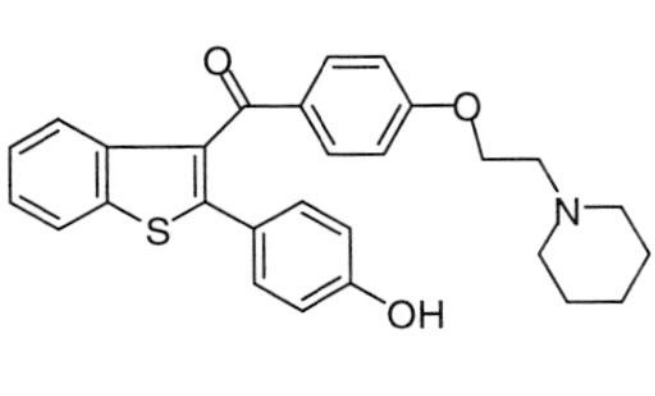

115

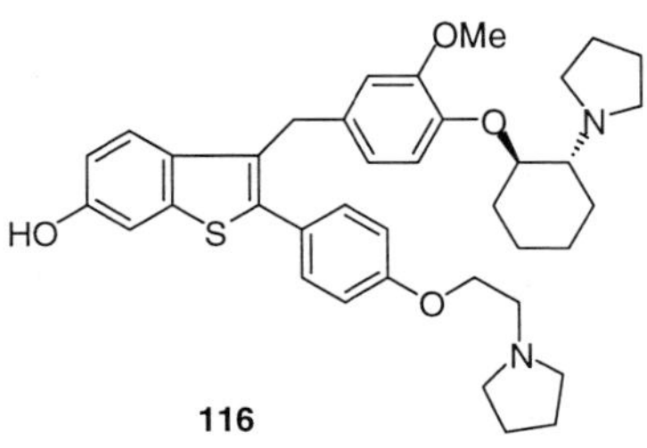

116

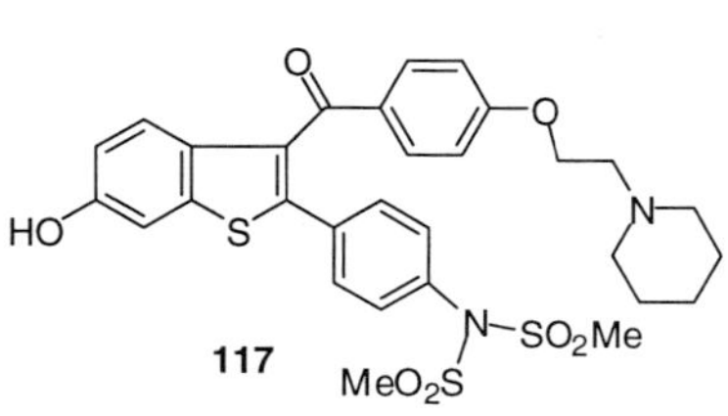

117

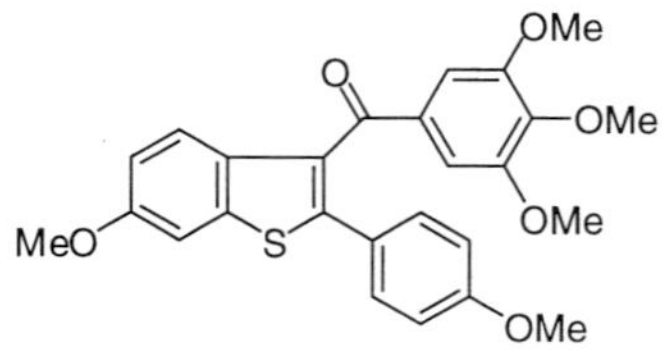

118

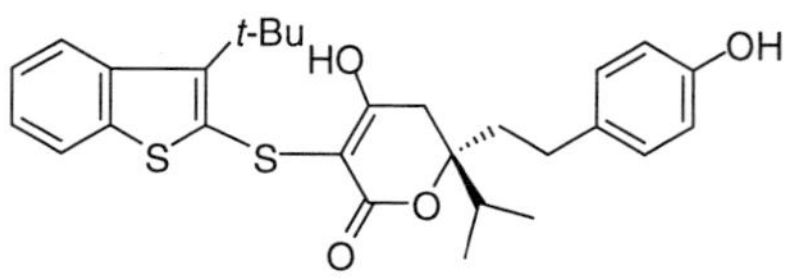

119

Additional common fused thiophene structural motifs with biological activity, include thienopyridines, thienopyrimidines, and thiophene-quinones. These ring systems display a wide range of biologically active including antiprotozoal (e.g., **120**) <99CPB1221>, anti-cancer (e.g., **121** and **122**) <99BMC1025, 99BMCL797>, orally bioavailable thrombin inhibitors (e.g., **123**) <99BMCL2837>, and poly(ADP-ribose)polymerase inhibitors (e.g., **124**) <99BMC297>. Other biologically active fused thiophenes include antipsychotics (e.g., **125** and **126**) <99JMC1106, 99JMC2774>, human cytomegalovirus protease inhibitors (e.g., **127**) <99BMCL449>, vasopressin receptor antagonists (e.g., **128**) <99BMCL1733>, and those with antihyperglycemic activity (e.g., benzo[*b*]naphtho[2,3-*d*]thiophenes) <99JMC3199>.

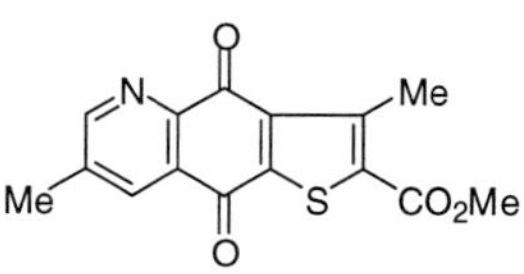

120

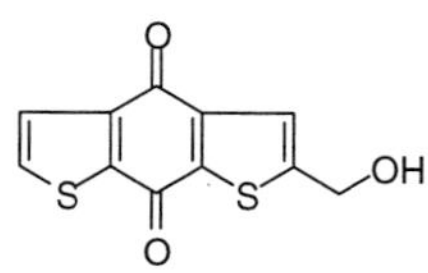

121

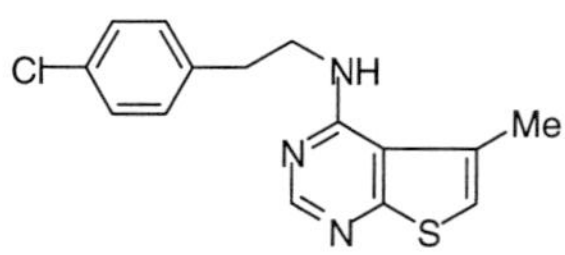

122

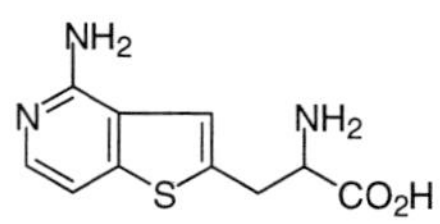

123

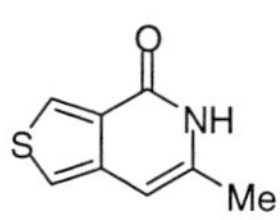

124

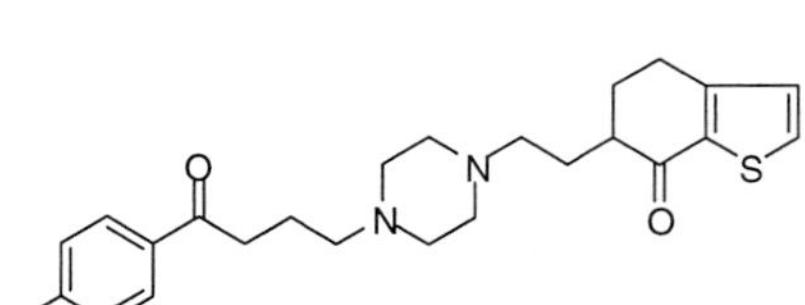

125

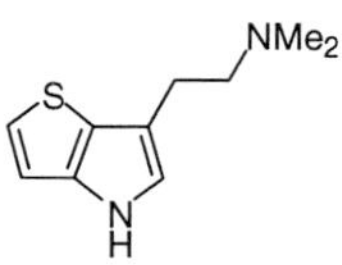

126

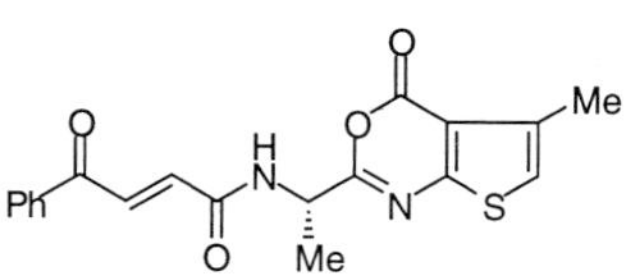

127

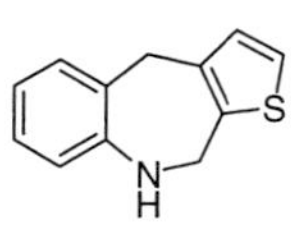

128

A number of non-fused thiophene derivatives also show biological activity including thiophenes which are HIV-1 strain MDR inhibitors (e.g., **129**) <99BMCL3411>, protein kinase C inhibitors (e.g., **130**) <99BMCL2279>, antidepressants (e.g., **131**) <99BMC1349>, and a GABA-AT inactivator (4-amino-4,5-dihydrothiophene-2-carboxylic acid) <99JA7751>.

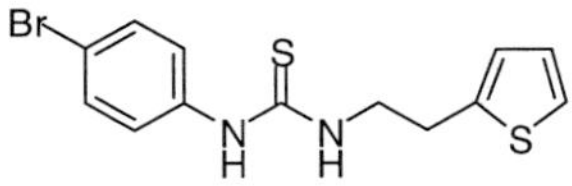

129

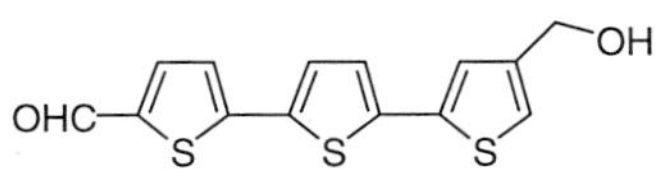

130

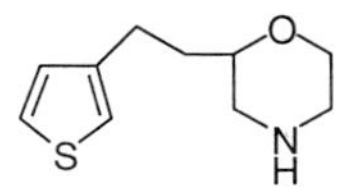

131

5.1.7 NOVEL THIOPHENE DERIVATIVES

The structure of a novel benzothiophene glycoside, echinothiophene **132**, isolated from the roots of *Echinops grijissii* was elucidated by a combination of spectroscopic methods (MS and NMR) <99OL197>. Interestingly, **132** undergoes a facile epimerization to **134**, perhaps through enol intermediate **133**.

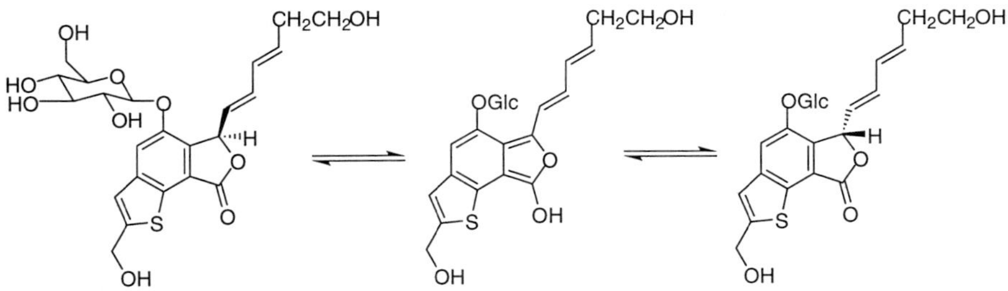

132 **133** **134**

In order to tune the electronic properties of porphyrin ring systems, thiophene-substituted analogues have been designed, synthesized, and evaluated including porphyrins **135** <99JOC7890>, **136** <99TL8457>, **137** <99TL1921>, and thiophene-containing core-expanded porphyrins <99TL1921>.

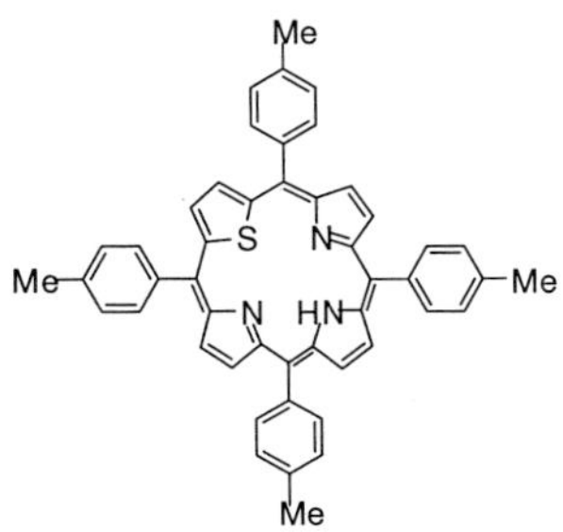

135

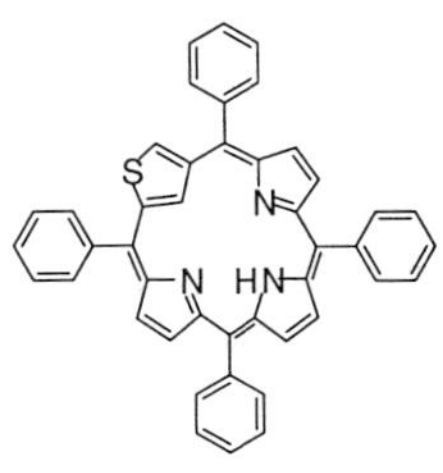

136

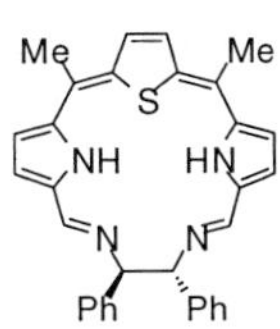

137

In addition to porphyrins, a variety of novel macrocyclic thiophene-containing ring systems have been prepared including cyclotriyne **138** <99CM3050>, homooxacalix[*n*]thiophenes **139** <99TL3749>, silacalix[*n*]phosphinines **140** <99CE2109>, 1,1'-binaphthyl thiophene **141** <99TL9065>, thiophenophane **142** <99H1059>, and thiophene-substituted crown ethers <99JMAT2139>. A theoretical approach to studying the rotationally dynamics of interlocking thiophene-containing [2]catenanes has been reported <99JA2364>.

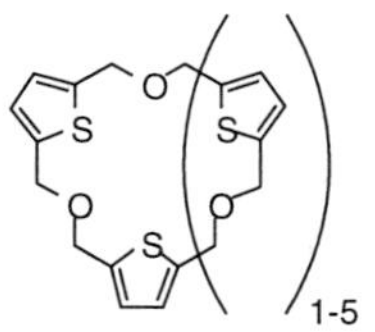

138

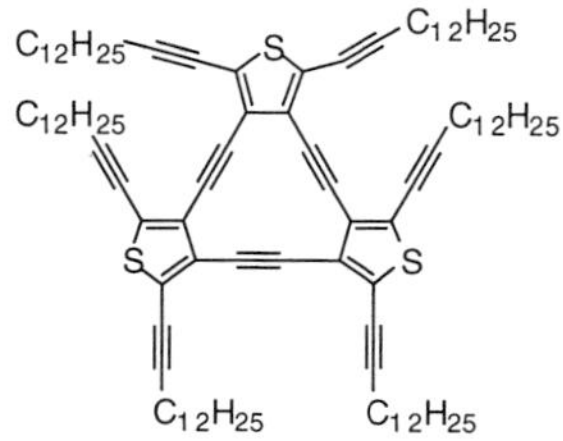

139

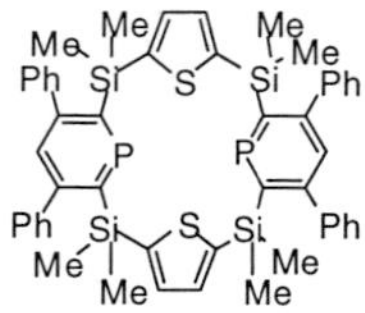

140

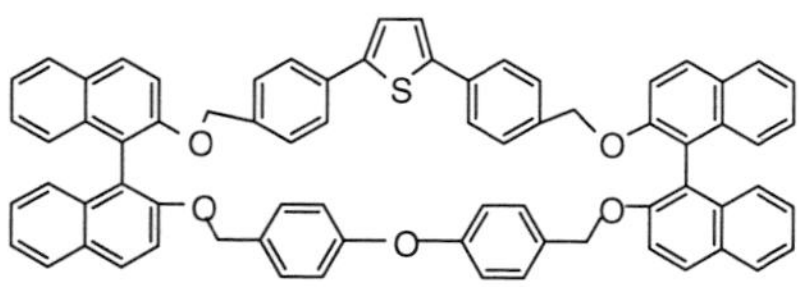

141

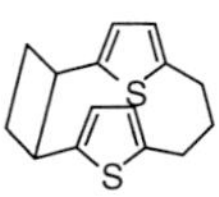

142

The unique electronic properties of the thiophene ring system are often utilized to manipulate the electronic and optical properties of various materials. Some examples of compounds with special electronic properties include androstene **143** <99CE96>, thiophene-linked ruthenium complexes <99CC869>, and tetrathiafulvene **144** <99JOC4267>. Some examples of compounds with special optical properties, often which are thiophenes with "push-pull" substitution, include triarylamine **145** <99JOC4289>, arylamine **146** <99JMAT1449>, bridged dithiophenes <99CM1541, 99T14985>, sulfonyl-substitituted iminothiophenes <BMCL2643>, thiophene-europium complexes <99JMAT3023>, ferrocene complexes (e.g., **147**) <99CM2995, 99JOMC301>, and C_{60}-thiophene conjugates <99CC429, 99CL443, 99JOC4884>. Interestingly, a dimeric C_{60}-fullerene has been prepared with thiophene bridging units <99CC465>. The photochemical properties of perfluorocyclopentene-linked benzothiophenes and thiophenes have been studied further <99BCSJ1139, 99CC1487, 99CL1071, 99JA8450, 99TL1345>.

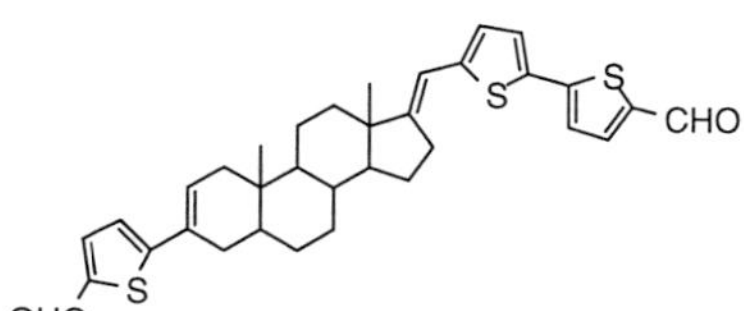

143

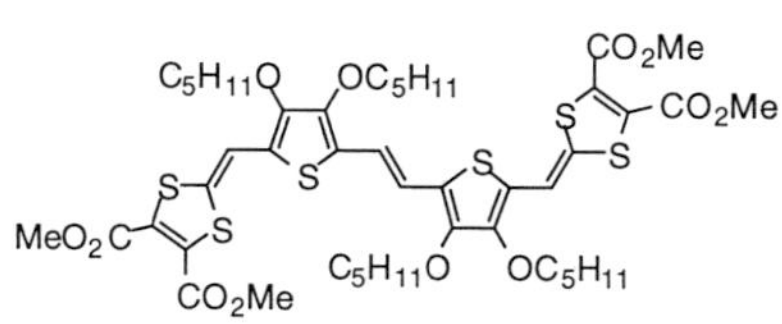

144

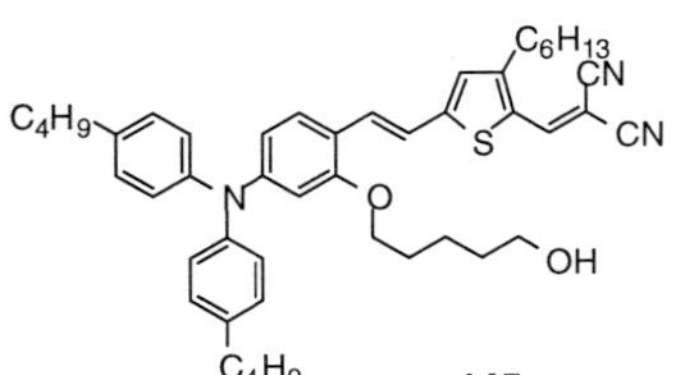

145

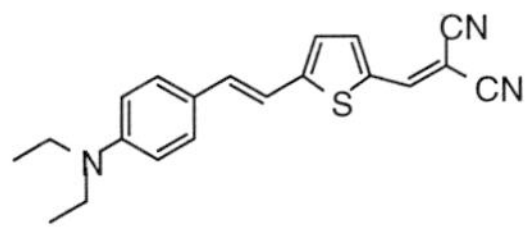

146

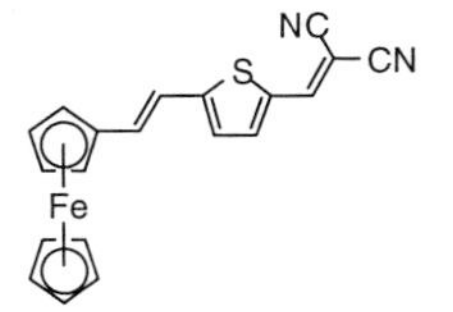

147

Finally, a synthesis of thiophene-2-phosphate (**150**), an interesting organic-inorganic hybrid molecule, has been reported <99JMAT2559>. Treatment of 2-bromothiophene (**148**) with triethylphosphite and $NiCl_2$ gave **149** which was converted into **150** by a two step procedure.

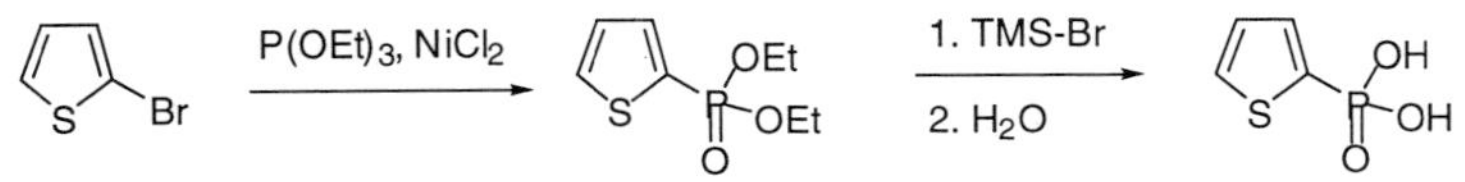

148 **149** **150**

5.1.8 THIOPHENE OLIGOMERS AND POLYMERS

The thiophene ring is a common building block for novel oligomeric and polymeric materials. The preparation of monodisperse thiophene-containing oligomers has been reviewed <99AC(E)1350>. The synthesis and/or evaluation of thiophene-containing oligomers that have appeared during the past year include terthiophene-substituted ruthenium complexes <99AC(E)2565, 99JMAT865, 99JMAT2123, 99O1930>, novel fused terthiophenes (e.g., **151**) <99JOC6418>, tetrathiophenes <99JH241>, ethylenedioxy-tetrathiophenes <99T11745>, sexithiophenes <99JA8920>, and dendrimer-substituted oligothiophenes (e.g **152**) <99CM3420>. Some mixed heterocyclic thiophene-containing oligomers that have been prepared and/or evaluated include mixed pyrrole-thiophenes <99TL8887>, thiazole-thiophenes <99CM458>, pyridine-thiophenes <99CC2121, 99CM3342>, and carbazole-thiophenes (e.g., **153**) <99JMAT2189>. A solid-phase synthetic approach was utilized to synthesize a set of impressive linear homooligomers comprised of thiophenes, alkynes, and aromatic rings <99JOC8898>. Interesting symmetrical oligomers have been prepared including C3-symmetric pyrimdine-dithiophene (e.g., **154**) <99CC2083> and C6-symmetric benzene-terthiophene (e.g., **155**) <99JCR(S)596>.

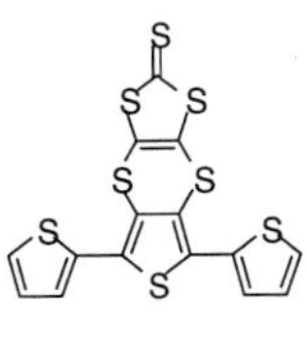

151

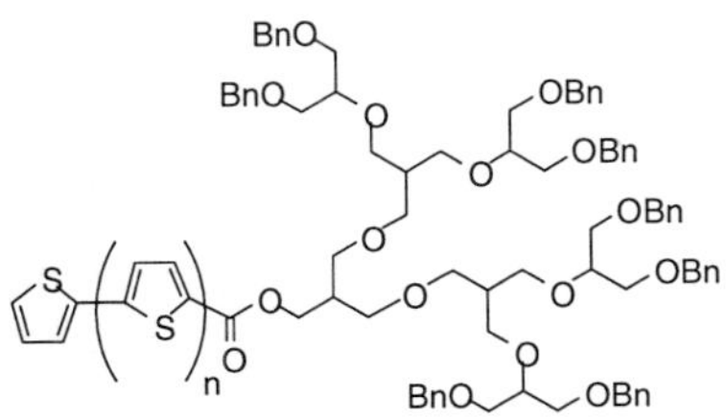

152

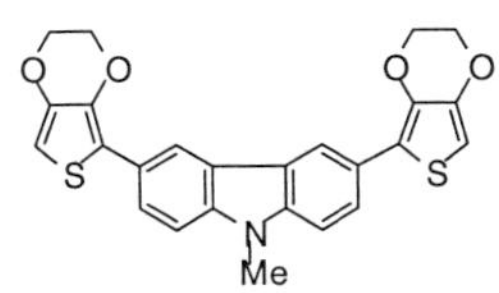

153

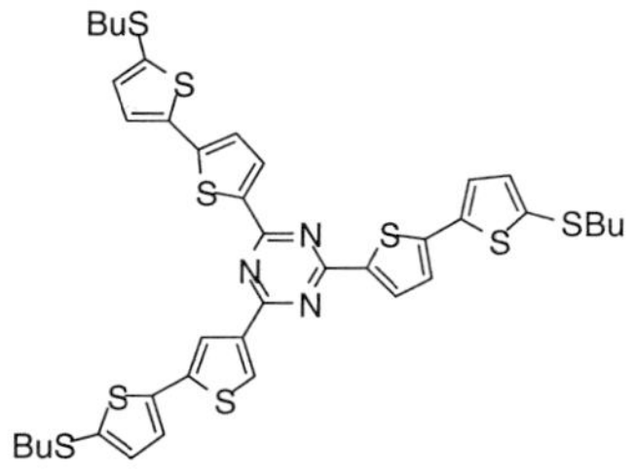

154

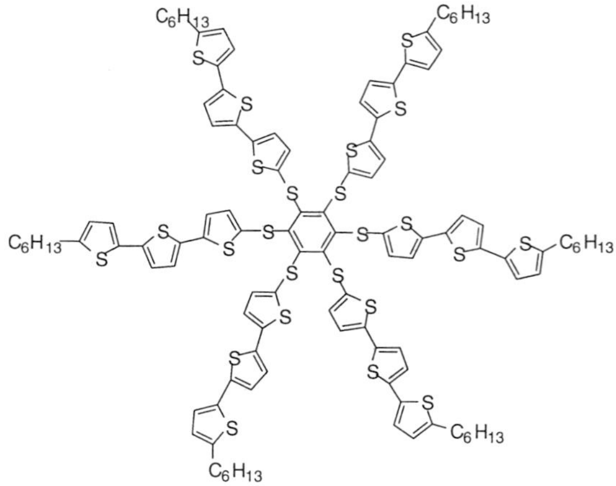

155

The synthesis of novel thiophene-containing polymers has been reviewed during the previous year <99BCSJ621, 99JMAT1875>. A method for capping polyethylene with thiophene has appeared <99JA6082>. A novel method for preparing polythiophenes utilizes Heck-type couplings of α-iodothiophenes <99TL5873>. Types of polythiophenes that have been prepared during the past year include cross-linked polymers <99JMAT2109>, polymetallaacetylene polymers <99JOMC210>, dithieno[3,4-*b*:3',4'-*d*]thiophene polymers <99JMAT1719>, fluorine-substituted phenylenes <99CM1075>, regioregular polythiophenes with chiral side chains <99CC867>, thiophene-1,1-dioxide polythiophenes <99CM2533>, azobenzene-substituted polythiophenes <99JMAT2215>, crown ether bearing polythiophenes <99JMAT2133>, and polythiophene-3-carboxylates <99JMAT2155>.

5.1.9 SELENOPHENES AND TELLUROPHENES

A small number of reports on the chemistry of selenophenes and tellurophenes appeared during the past year. The preparation of selenophene analogues of tryptophan (e.g., **156**) involved the use of tryptophan synthase <99BMCL637>. Nucleophilic substitution reactions of 5-bromoselenophene-2-carboxaldehyde in aqueous media were reported <99T6511>. Electrophilic additions to osmium-selenophene complexes have been studied <99O1559>. Antioxidants, selenophene **158** and tellurophene **159**, have been prepared from a common starting material **157** <99JOC6764>. Finally, the structures of isotellurophenes and isoselenophenes have been elucidated by single crystal X-ray methods <99JOMC15>.

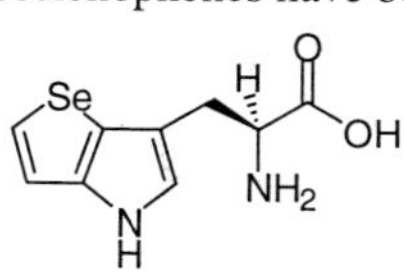

156

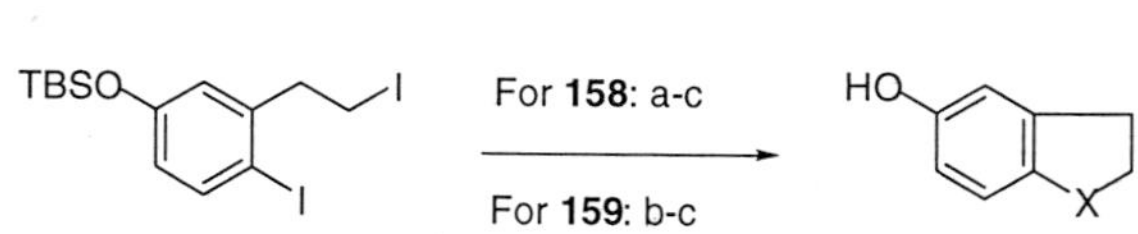

157

Reagents: (a) $(BnSe)_2$, $NaBH_4$; (b) $(BuTe)_2$, $NaBH_4$; (c) TBAF

158 X = Se
159 X = Te

5.1.10 REFERENCES

<99AC(E)1350> Martin, R. E.; Diederich, F. *Angew. Chem., Int. Ed. Engl.* **1999**, *38*, 1350.

<99AC(E)2565> Weinberger, D. A.; Higgins, T. B.; Mirkin, C. A.; Liable-Sands, L. M.; Rheingold, A. L. *Angew. Chem., Int. Ed. Engl.* **1999**, *38*, 2565.

<99BMC231> Yu, J.; Castagnoli, N. *Bioorg. Med. Chem.* **1999**, *7*, 231.

<99BMC297> Shinkwin, A. E.; Whish, W. J. D.; Threadgill, M. D. *Bioorg. Med. Chem.* **1999**, *7*, 297.

<99BMC1025> Chao, Y.-H.; Kuo, S.-C.; Ku, K.; Chiu, I.-P.; Wu, C.-H.; Mauger, A.; Wang, H.-K.; Lee, K.-H. *Bioorg. Med. Chem.* **1999**, *7*, 1025.

<99BMC1349> Corral, C.; Lissavetzky, J.; Manzanares, I.; Darias, V.; Expósito-Orta, M. A.; Conde, J. A. M.; Sánchez-Mateo, C. C. *Bioorg. Med. Chem.* **1999**, *7*, 1349.

<99BMCL363> Bastian, J. A.; Chirgadze, N.; Denney, M. L.; Gifford-Moore, D. S.; Sall, D. J.; Smith, G. F.; Wikel, J. H. *Bioorg. Med. Chem. Lett.* **1999**, *9*, 363.

<99BMCL449> Pinto, I. L.; Jarvest, R. L.; Clarke, B.; Dabrowski, C. E.; Fenwick, A.; Gorczyca, M. M.; Jennings, L. J.; Lavery, P.; Sternberg, E. J.; Tew, D. G.; West, A. *Bioorg. Med. Chem. Lett.* **1999**, *9*, 449.

<99BMCL637> Welch, M.; Phillips, R. S. *Bioorg. Med. Chem. Lett.* **1999**, *9*, 637.

<99BMCL759> Takeuchi, K.; Kohn, T. J.; Sall, D. J.; Denney, M. L.; McCowan, J. R.; Smith, G. F.; Gifford-Moore, D. S. *Bioorg. Med. Chem. Lett.* **1999**, *9*, 759.

<99BMCL775> Zhang, M.; Bailey, D. L.; Bastian, J. A.; Briggs, S. L.; Chirgadze, N. Y.; Clawson, D. K.; Denney, M. L.; Gifford-Moore, D. S.; Harper, R. W.; Johnson, L. M.; Klimkowski, V. J.; Kohn, T. J.; Lin, H.-S.; McCowan, J. R.; Richett, M. E.; Sall, D. J.; Smith, A. J.; Smith, G. F.; Snyder, D. W.; Takeuchi, K.; Utterback, B. G.; Yan, S.-C. B. *Bioorg. Med. Chem. Lett.* **1999**, *9*, 775.

<99BMCL797> Katada, J.; Iijima, K.; Muramatsu, M.; Takami, M.; Yasuda, E.; Hayashi, M.; Hattori, M.; Hayashi, Y. *Bioorg. Med. Chem. Lett.* **1999**, *9*, 797.

<99BMCL1081> Pinney, K. G.; Bounds, A. D.; Dingeman, K. M.; Mocharla, V. P.; Pettit, G. R.; Bai, R.; Hamel, E. *Bioorg. Med. Chem. Lett.* **1999**, *9*, 1081.

<99BMCL1733> Aranapakam, V.; Albright, J. D.; Grosu, G. T.; Chan, P. S.; Coupet, J.; Saunders, T.; Ru, X.; Mazandarani, H. *Bioorg. Med. Chem. Lett.* **1999**, *9*, 1733.

<99BMCL2019> Ellsworth, E. L.; Domagala, J.; Prasad, J. V. N. V.; Hagen, S.; Ferguson, D.; Holler, T.; Hupe, D.; Graham, N.; Nouhan, C.; Tummino, P. J.; Zeikus, G.; Lunney, E. A. *Bioorg. Med. Chem. Lett.* **1999**, *9*, 2019.

<99BMCL2279> Xu, W.-C.; Zhou, Q.; Ashendel, C. L.; Chang, C.-t.; Chang, C.-j. *Bioorg. Med. Chem. Lett.* **1999**, *9*, 2279.

<99BMCL2643> Chou, Y.-M.; Lai, M.-C.; Hwang, T.-M.; Ong, C. W. *Bioorg. Med. Chem. Lett.* **1999**, *9*, 2643.

<99BMCL2837> Rewinkel, J. B. M.; Lucas, H.; Smit, M. J.; Noach, A. B. J.; van Dinther, T. G.; Rood, A. M. M.; Jenneboer, A. J. S. M.; van Boeckel, C. A. A. *Bioorg. Med. Chem. Lett.* **1999**, *9*, 2837.

<99BMCL3381> Norman, B. H.; Dantzig, A. H.; Kroin, J. S.; Law, K. L.; Tabas, L. B.; Shepard, R. L.; Palkowitz, A. D.; Hauser, K. L.; Winter, M. A.; Sluka, J. P.; Starling, J. J. *Bioorg. Med. Chem. Lett.* **1999**, *9*, 3381.

<99BMCL3411> Uckun, F. M.; Pendergrass, S.; Maher, D.; Zhu, D.; Tuel-Ahlgren, L.; Mao, C.; Venkatachalam, T. K. *Bioorg. Med. Chem. Lett.* **1999**, *9*, 3411.

<99BCSJ621> Yamamoto, T. *Bull. Chem. Soc. Jpn.* **1999**, *72*, 621.

<99BCSJ1139> Irie, S.; Yamaguchi, T.; Nakazumi, H.; Kobatake, S.; Irie, M. *Bull. Chem. Soc. Jpn.* **1999**, *72*, 1139.

<99BCSJ1597> Kobayashi, T.; Tasuzuki, T.; Saitoh, M. *Bull. Chem. Soc. Jpn.* **1999**, *72*, 1597.

<99BCSJ1919> Nagasawa, H.; Sugihara, Y.; Ishii, A.; Nakayama, J. *Bull. Chem. Soc. Jpn.* **1999**, *72*, 1919.

<99CC143> Kita, Y.; Egi, M.; Tohma, H. *Chem. Commun.* **1999**, 143.

<99CC429> Pasimeni, L.; Maniero, A. L.; Ruzzi, M.; Prato, M.; Da Ros, T.; Barbarella, G.; Zambianchi, M. *Chem. Commun.* **1999**, 429.

<99CC465> Giesa, S.; Gross, J. H.; Hull, W. E.; Lebedkin, S.; Gromov, A.; Gleiter, R.; Krätschmer, W. *Chem. Commun.* **1999**, 465.

<99JOC4289>	Thayumanavan, S.; Mendez, J.; Marder, S. R. *J. Org. Chem.* **1999**, *64*, 4289.
<99JOC4884>	Liu, S.-G.; Shu, L.; Rivera, J.; Liu, H.; Raimundo, J.-M.; Roncali, J.; Gorgues, A.; Echegoyen, L. *J. Org. Chem.* **1999**, *64*, 4884.
<99JOC6418>	Skabara, P. J.; Serebryakov, I. M.; Roberts, D. M.; Perepichka, I. F.; Coles, S. J.; Hursthouse, M. B. *J. Org. Chem.* **1999**, *64*, 6418.
<99JOC6764>	Engman, L.; Laws, M. J.; Malmström, J. *J. Org. Chem.* **1999**, *64*, 6764.
<99JOC7890>	Cho, W.-S.; Kim, H.-J.; Littler, B. J.; Miller, M. A.; C.-H., L.; Lindsey, J. S. *J. Org. Chem.* **1999**, *1999*, 7890.
<99JOC8489>	Shimizu, T.; Murakami, H.; Kamigata, N. *J. Org. Chem.* **1999**, *64*, 8489.
<99JOC8898>	Huang, S.; Tour, J. M. *J. Org. Chem.* **1999**, *64*, 8898.
<99JOMC15>	Majeed, Z.; McWhinnie, W. R.; Paxton, K.; Hamor, T. A. *J. Organometal. Chem.* **1999**, *577*, 15.
<99JOMC210>	Antonelli, E.; Rosi, P.; Lo Sterzo, C.; Viola, E. *J. Organometal. Chem.* **1999**, *578*, 210.
<99JOMC242>	Birri, A.; Steed, J. W.; Tocher, D. A. *J. Organometal. Chem.* **1999**, *575*, 242.
<99JOMC301>	Thomas, K. R. J.; Lin, J. T.; Wen, Y. S. *J. Organometal. Chem.* **1999**, *575*, 301.
<99MC573>	Stannety, P.; Mihovilovic, M. D. *Monatsch. Chem.* **1999**, *130*, 573.
<99MC645>	Binder, D.; Pyerin, M.; Pusterer, F. *Monatsh. Chem.* **1999**, *130*, 645.
<99OL197>	Koike, K.; Jia, Z.; Nikaido, T.; Liu, Y.; Zhao, Y.; Guo, D. *Org. Lett.* **1999**, *1*, 197.
<99OL257>	Kitamura, T.; Zhang, B.-X.; Fujiwara, Y.; Shiro, M. *Org. Lett.* **1999**, *1*, 257.
<99OL513>	Ronzoni, S.; Cerri, A.; Dondio, G.; Fronza, G.; Petrillo, P.; Raveglia, L. F.; Gatti, P. A. *Org. Lett.* **1999**, *1*, 513.
<99OL965>	Bussolari, J. C.; Rehborn, D. C. *Org. Lett.* **1999**, *1*, 965.
<99OL1039>	Wu, J.-Y.; Ho, J.-H.; Shih, S.-M.; Hsieh, T.-L.; Ho, T.-I. *Org. Lett.* **1999**, *1*, 1039.
<99OL1667>	Karlsson, S.; Högberg, H.-E. *Org. Lett.* **1999**, *1*, 1667.
<99O1559>	Spera, M. L.; Harman, W. D. *Organometallics* **1999**, *18*, 1559.
<99O1930>	Zhu, Y.; Millet, D. B.; Wol, M. O.; Rettig, S. J. *Organometallics* **1999**, *18*, 1930.
<99O2760>	Riedmiller, F.; Jockisch, A.; H., S. *Organometallics* **1999**, *18*, 2760.
<99O2988>	Spera, M. L.; Harman, W. D. *Organometallics* **1999**, *18*, 2988.
<99O3187>	Lukevics, E.; Arsenyan, P.; Belyakov, S.; Popelis, J.; Pudova, O. *Organometallics* **1999**, *18*, 3187.
<99SL1029>	Rahim, M. A.; Fujiwara, T.; Takeda, T. *Synlett* **1999**, 1029.
<99SL1333>	Rassu, G.; Battistini, L.; Casiraghi, G. *Synlett* **1999**, 1333.
<99SL1450>	Khanh, L. P.; Dallemagne, P.; Rault, S. *Synlett* **1999**, 1450.
<99SL1582>	Inoue, A.; Shinokubo, H.; Oshima, K. *Synlett* **1999**, 1582.
<99SL1621>	Kotha, S.; Chakraborty, K.; Brahmachary, E. *Synlett* **1999**, 1621.
<99SC35>	D'Auria, M.; De Luca, E.; Mauriello, G.; Racioppi, R. *Synth. Commun.* **1999**, *29*, 35.
<99SC1607>	Hoffmann, K. J.; Carlsen, H. J. *Synth. Commun.* **1999**, *29*, 1607.
<99S1303>	Brooks, P.; Donati, D.; Pelter, A.; Poticelli, F. *Synthesis* **1999**, 1303.
<99T485>	Fazio, A.; Gabriele, B.; Salerno, G.; Destri, S. *Tetrahedron* **1999**, *55*, 485.
<99T1449>	Majumdar, K. C.; Biswas, P. *Tetrahedron* **1999**, *55*, 1449.
<99T2013>	D'Auria, M.; Ferri, T.; Mauriello, G.; Pesce, A.; Racioppi, R. *Tetrahedron* **1999**, *55*, 2013.
<99T4709>	Scheytza, H.; Reissig, H.-U.; Rademacher, O. *Tetrahedron* **1999**, *55*, 4709.
<99T6511>	Prim, D.; Kirsch, G. *Tetrahedron* **1999**, *55*, 6511.
<99T11669>	Han, Y.; Giroux, A.; Lépine, C.; Laliberté, F.; Huang, Z.; Perrier, H.; Bayly, C. I.; Young, R. N. *Tetrahedron* **1999**, *55*, 11669.
<99T11745>	Mohanakrishnan, A. K.; Hucke, A.; Lyon, M. A.; Lakshmikantham, M. V.; Cava, M. P. *Tetrahedron* **1999**, *55*, 11745.
<99T13757>	Bilokin, Y. V.; Vasylyev, M. V.; Branytska, O. V.; Kovalenko, S. M.; Chernykh, V. P. *Tetrahedron* **1999**, *55*, 13757.
<99T13973>	Wu, I.-Y.; Lin, J. T.; Li, C.-S.; Wang, W. C.; Huang, T. H.; Wen, Y. S.; Chow, T.; Tsai, C. *Tetrahedron* **1999**, *55*, 13973.
<99T14985>	Pasini, M.; Destri, S.; Botta, C.; Porzio, W. *Tetrahedron* **1999**, *55*, 14985.
<99TL857>	Hassan, J.; Lavenot, L.; Gozzi, C.; Lemaire, M. *Tetrahedron Lett.* **1999**, *40*, 857.
<99TL1345>	Takeshita, M.; Irie, M. *Tetrahedron Lett.* **1999**, *40*, 1345.
<99TL1635>	Kernag, C. A.; Bobbitt, J. M.; McGrath, D. V. *Tetrahedron Lett.* **1999**, *40*, 1635.
<99TL1921>	Lee, C.-H.; Oh, K.-T. *Tetrahedron Lett.* **1999**, *40*, 1921.

<99TL2789> Imamura, K.; Hirayama, D.; Yoshimura, H.; Takimiya, K.; Aso, Y.; Otsubo, T. *Tetrahedron Lett.* **1999**, *40*, 2789.

<99TL2909> Kim, S.; Yang, J.; DiNinno, F. *Tetrahedron Lett.* **1999**, *40*, 2909.

<99TL3749> Komatsu, N.; Taniguchi, A.; Suzuki, H. *Tetrahedron Lett.* **1999**, *40*, 3749.

<99TL3785> Nakayama, J.; Sano, T.; Sugihara, Y.; Ishii, A. *Tetrahedron Lett.* **1999**, *40*, 3785.

<99TL5155> Bradley, D. A.; Godfrey, A. G.; Schmid, C. R. *Tetrahedron Lett.* **1999**, *40*, 5155.

<99TL5471> McKibben, B. P.; Cartwright, C. H.; Castelhano, A. L. *Tetrahedron Lett.* **1999**, *40*, 5471.

<99TL5549> Otani, T.; Sugihara, Y.; Ishii, A.; Nakayama, J. *Tetrahedron Lett.* **1999**, *40*, 5549.

<99TL5873> Sévignon, M.; Papillon, J.; Schulz, E.; Lemaire, M. *Tetrahedron Lett.* **1999**, *40*, 5873.

<99TL8457> Sprutta, N.; Latos-Grazynski, L. *Tetrahedron Lett.* **1999**, *40*, 8457.

<99TL8887> Ogura, K.; Yanai, H.; Miokawa, M.; Akazome, M. *Tetrahedron Lett.* **1999**, *40*, 8887.

<99TL9065> Ogura, K.; Matsui, H.; Matsumoto, S.; Akazome, M.; Harada, N. *Tetrahedron Lett.* **1999**, *40*, 9065.

<99TCC131> Nakayama, J.; Sugihara, Y. *Topics Curr. Chem.* **1999**, *205*, 131.

Chapter 5.2

Five Membered Ring Systems: Pyrroles and Benzo Derivatives

Daniel M. Ketcha
Wright State University, Dayton, OH, USA
daniel.ketcha@wright.edu

5.2.1 INTRODUCTION

A survey of computational methods was used to calculate the homolytic bond dissociation energies of the C-H and N-H bonds in monocyclic aromatic molecules including pyrrole and indole which are representative of the functionalities present in coal <99JA491>. On a more conceptual level, an interesting commentary appeared on the treatment of aromaticity and the acid-base character of pyridine and pyrrole in contemporary organic chemistry textbooks. This article pointed out that while most texts state that the electron pair on the ring nitrogen in pyrrole is unavailable for protonation, discussions still tend to dwell on the "relative basicity" rather than the weak acid character of this heterocycle, thereby presenting a confusing picture to students <99JCE1151>. In terms of review articles relevant to the subject matter, Gilchrist published a review on the synthesis of aromatic heterocycles including pyrroles, indoles and carbazoles covering the period of March 1997 to February 1999 <99JCS(P1)2849>, while Trofimov has written reviews on the synthesis of arylpyrroles <98RJO1691> and the synthesis of heterocycles including pyrroles *via* the reaction of alkynylamines and isothiocyanates <99JHC1469>.

5.2.2 SYNTHESIS OF PYRROLES

In what might fairly be called the Brandsma/Trofimov pyrrole synthesis (although the general process has been applied to a number of heterocyclic systems), a one-pot synthesis of 1,2,3,5-tetrasubstituted pyrroles was reported starting from 2-alkynylamines and isothiocyanates <99EJO2663>. In this process, 2-alkynylamines **1** were metallated with nBuLi to produce the unsaturated carbanions **2**, which undergo regiospecific reaction with alkyl isothiocyanates followed by S-methylation leading to the azatrienes **3**. Addition of catalytic amounts of copper(I) bromide then induces cyclization to the tetrasubstituted pyrroles **4**.

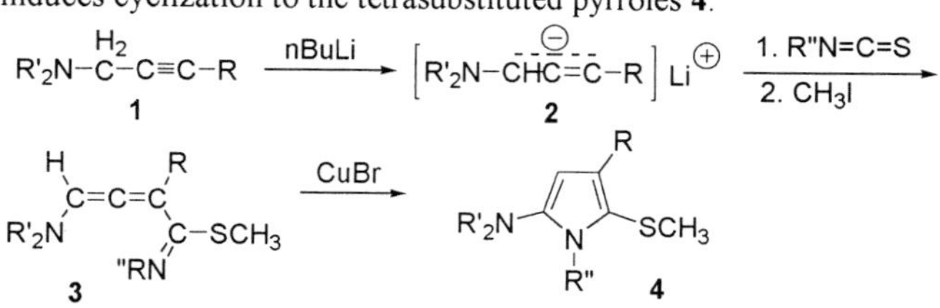

In another one-pot process, Nakamura has developed a novel synthesis of pyrroles through the [3+2] coupling of a ketone hydrazone **5** and a vinylstannane <99OL1505>. In this procedure, zincated hydrazone **6** is converted to the *gem*-Zn/Sn dimetallic intermediate **7** which upon exposure to oxygen directly affords the 1-(dimethylamino)pyrroles **8**.

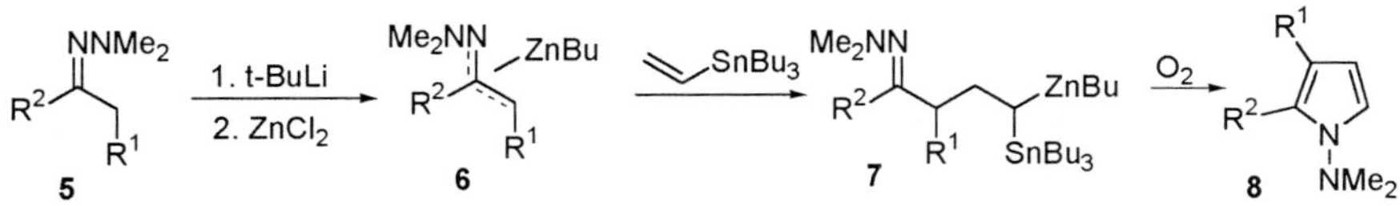

Narasaka has utilized a Heck-type intramolecular aminopalladation of the olefinic moiety of γ,δ-unsaturated ketone *O*-pentafluorobenzoyloximes **9** to afford 2,5-disubstituted pyrroles **10** (R^1 = CH_3, CO_2Et) <99CL45>. The use of the *O*-pentafluorobenzoyl group was found necessary to preclude competing Beckmann rearrangements.

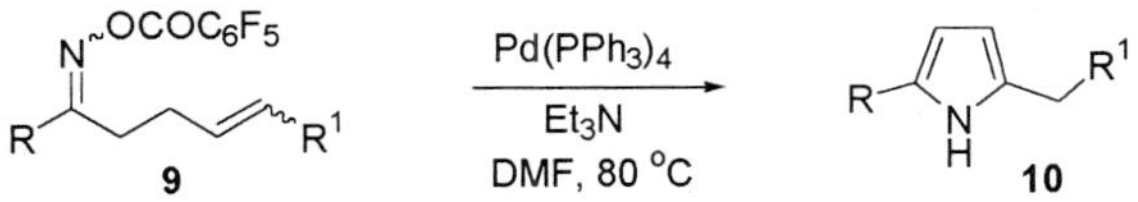

The application of microwave irradiation (100-200 watts, neat, 0.5-2 min) has been shown effective in promoting the reaction of hexane-2,5-dione with primary amines to afford *N*-alkylpyrroles *via* the Paal-Knorr route <99TL3957>. This cyclization was employed as the key steps in two separate syntheses of the tricyclic ketopyrrole right hand half of roseophilin <99OL649, 99CC1455>. In the latter example, it was found that heating dione **11** with benzylamine in ethanolic acetic acid followed by addition of dilute HCl resulted in ring closure accompanied by an unusual *in situ* oxidation of the hydroxy group to yield the ketopyrrole **12**.

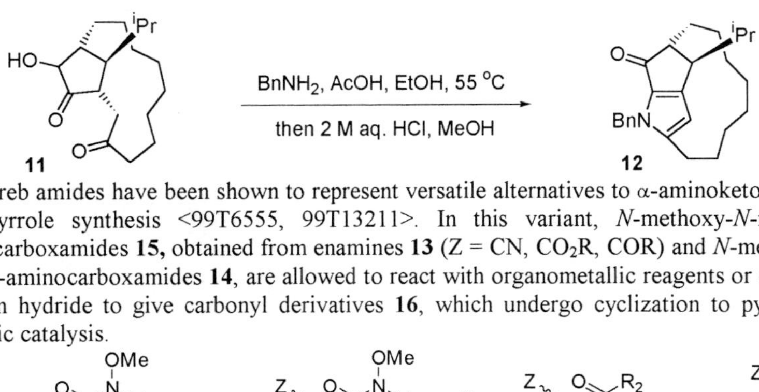

Weinreb amides have been shown to represent versatile alternatives to α-aminoketones in the Knorr pyrrole synthesis <99T6555, 99T13211>. In this variant, *N*-methoxy-*N*-methyl-α-enaminocarboxamides **15,** obtained from enamines **13** (Z = CN, CO_2R, COR) and *N*-methoxy-*N*-methyl-α-aminocarboxamides **14**, are allowed to react with organometallic reagents or diisobutyl aluminum hydride to give carbonyl derivatives **16**, which undergo cyclization to pyrroles **17** upon basic catalysis.

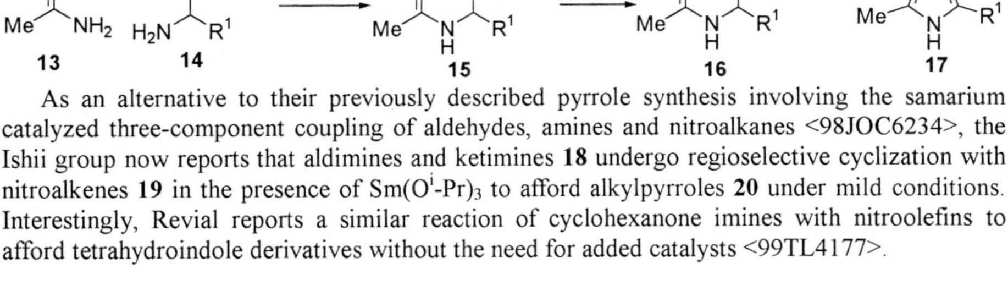

As an alternative to their previously described pyrrole synthesis involving the samarium catalyzed three-component coupling of aldehydes, amines and nitroalkanes <98JOC6234>, the Ishii group now reports that aldimines and ketimines **18** undergo regioselective cyclization with nitroalkenes **19** in the presence of $Sm(O^i\text{-}Pr)_3$ to afford alkylpyrroles **20** under mild conditions. Interestingly, Revial reports a similar reaction of cyclohexanone imines with nitroolefins to afford tetrahydroindole derivatives without the need for added catalysts <99TL4177>.

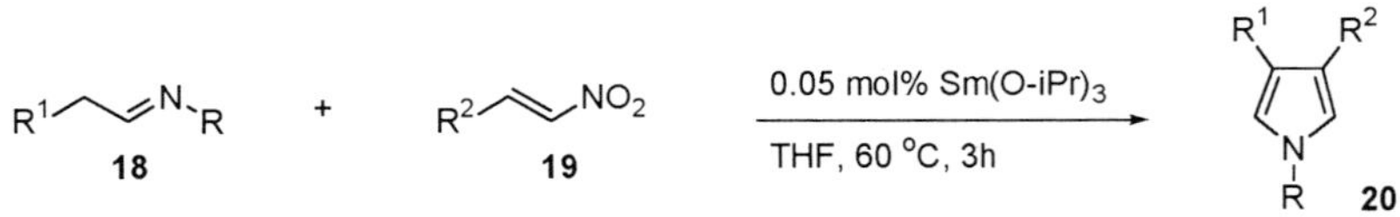

Although vinyl sulfones are less effective than the corresponding nitro compounds as anion stabilizers in the Barton-Zard synthesis, Ono finds that when another electron-withdrawing group is present as in the case of α-trifluoromethyl, α-cyano and α-ethoxycarbonyl alkenyl sulfones **21** (R^1 = CF_3, CN, CO_2Et), condensation with ethyl isocyanoacetate leads to formation of pyrrole carboxylates **22** with electron-withdrawing groups at the C-4 position <99S471>. Ono also employed the Barton-Zard procedure with β-nitroacetates for the synthesis of the lipid peroxidation inhibitor pyrrolostatin **23**, which consists of a pyrrole-2-carboxylic acid with a geranyl group at the C-4 position <99JOC6518>.

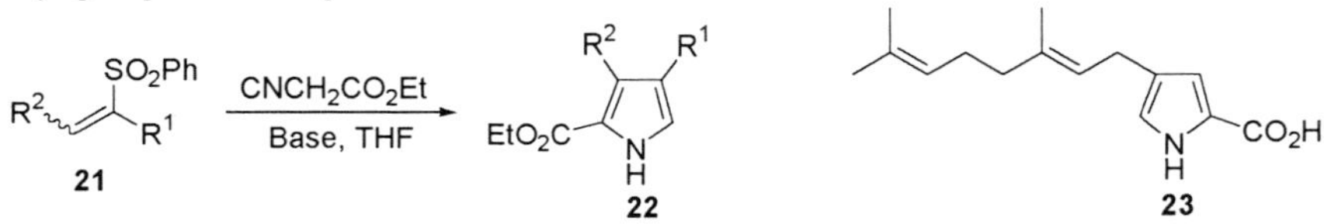

In one of a series of publications featuring multi-step reaction sequences using polymer supported reagents and/or solid sequestering agents termed "clean synthesis", Ley demonstrated the use of this technology for the preparation of an array of 1,2,3,4-tetrasubstituted pyrrole derivatives *via* the Barton-Zard reaction <99JCS(P1)107>. In this sequence, oxidation of the benzylic alcohols **24** using polymer supported permanganate in dichloromethane furnished the benzaldehydes **25** which underwent a Henry reaction with nitroalkanes in the presence of the polymer-supported base Amberlite IRA-420 (OH). Elimination of the nitroaldol adducts **26** was accomplished by treatment with 50% trifluoroacetic anhydride followed by triethylamine induced elimination of the corresponding trifluoroacetates to afford the nitrostyrenes **27**. Workup of the elimination steps involved treatment with aminomethyl polystyrene and acidic Amberlyst A-15. The pyrrole ring **28** was then constructed by 1,3-dipolar cycloaddition of *tert*-butyl isocyanoacetate in the presence of the polymer supported guanidine base 1,5,7-triazabicyclo[4.4.0]dec-5-ene (TBD). Further *N*-substitution could later be achieved by reaction of the pyrroles with alkyl halides in the presence of a polymer supported phosphazene base.

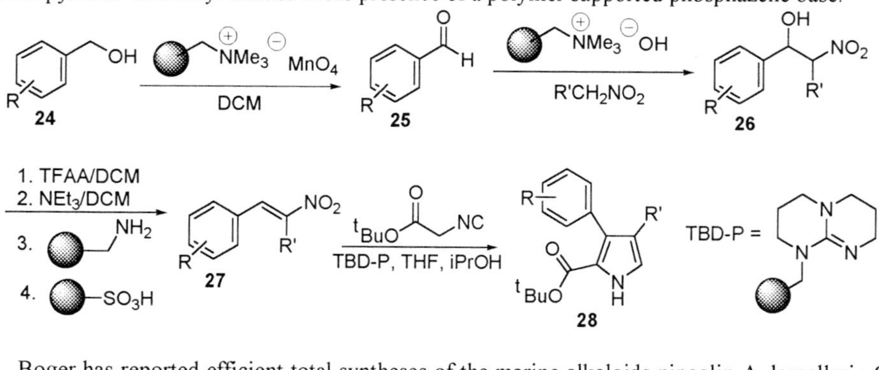

Boger has reported efficient total syntheses of the marine alkaloids ningalin A, lamellarin O, lukianol A, and storniamide A each of which possess a common 3,4-diaryl-substituted pyrrole nucleus bearing 2- or 2,5-carboxylates <99JA54>. A key step in each of these syntheses utilized a zinc mediated reductive ring contraction of 1,2-diazines such as **29** to pyrrole **30**, a precursor in the synthesis of lukianol A. The requisite 1,2-diazines were, in turn, prepared by a Diels-Alder reaction between alkynes and tetrazines.

Alternatively, Gupton reported an approach to these alkaloids utilizing vinylogous iminium salts as building blocks for pyrroles. Using this methodology, vinylogous amides were converted to β-chloroenal **31** which then were condensed with amino acid esters under acidic, neutral or basic conditions to afford precursors such as **32** <99T14515>.

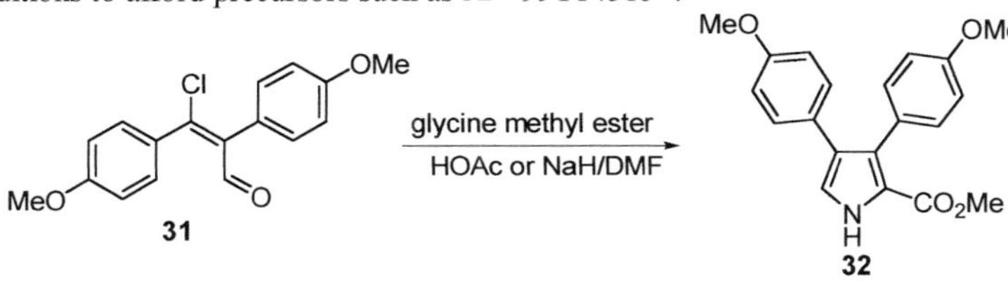

5.2.3 REACTIONS OF PYRROLES

Electrochemically generated tetraethylammonium peroxydicarbonate (TEAPC) has been shown to be capable of deprotonating the pyrrole nitrogen as a means of preparing *N*-alkylated pyrroles under mild conditions wherein competing *C*-alkylation was not observed<99EJO955>. Interestingly, attempted electrochemical reduction of pyrrole in the presence of 2,4,7-trinitrofluoren-9-one led to the isolation of a Meisenheimer complex under expectedly non-basic conditions <99JOC4572>.

In an extension of atom-transfer radical reactions to heterocyclic systems, Byers has introduced a novel methodology for the addition of electron-deficient radicals to unprotected pyrroles and indoles in a stannane-free, non-oxidative process <99TL2677>. For example, photochemical reaction of pyrrole (**33**) with ethyl iodoacetate (**34**) in presence of thiosulfate as an iodine reductant, phase transfer catalyst and propylene oxide led to high yields of the 2-alkylated pyrrole **35** <99TL2677>.

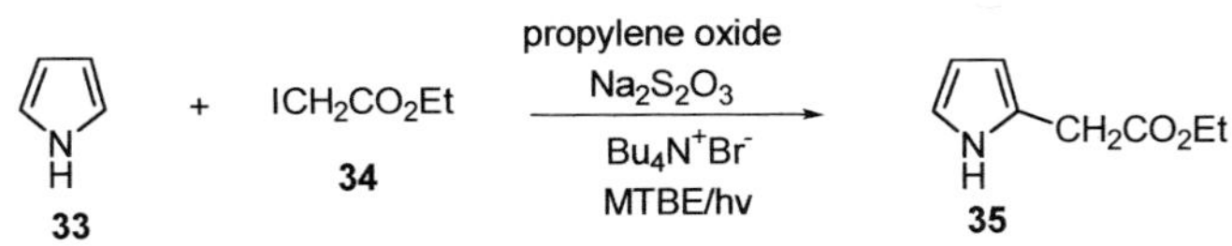

Whereas pyrroles normally undergo Friedel-Crafts acylation predominantly at the C-2 position, this regioselectivity can be perturbed by the presence of *N*-protecting groups. In this regard, it is interesting note to that in the course of preparing the pyrrolo[1,2-*c*]pyrimidine system of the marine alkaloid variolin, a variety of *N*-carbamoylpyrroles were prepared and found to afford mixtures of C-2 and C-3 acylated products in the presence of aluminum chloride

<99JCS(P1)249>. In an interesting alternative to direct acylation of this heterocycle, Pierini and Rossi examined the photostimulated $S_{RN}1$ reaction of the enolate of 2-acetyl-1-methylpyrrole **36** with haloarenes (e.g., iodobenzene, **37**) as well as neopentyl iodides to afford α-substituted acetyl pyrroles **38** <99JOC6487>.

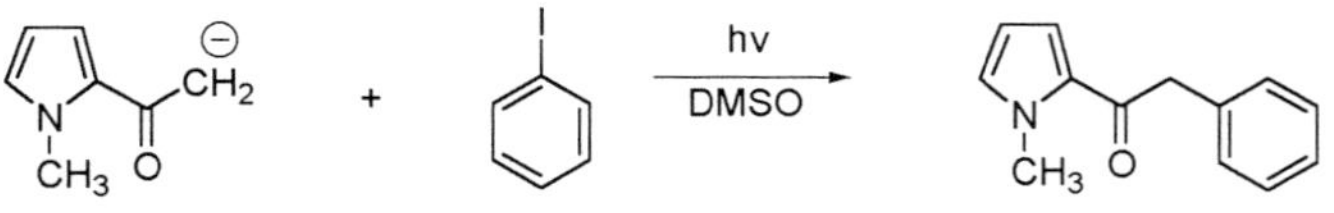

36 **37** **38**

Another method of functionalizing the pyrrole ring appearing with increased frequency involves the coupling of boronic acid derivatives. For instance, Burgess employed the Suzuki couplings of arylboronic acids to *N-tert*-butoxycarbonyl-2-bromopyrrole to prepare 2-arylpyrrole precursors of new "BODIPY" dyes that fluoresce at relatively long wavelengths <99JOC7813, 99CC1889>. In the alternate mode of reaction, Furstner achieved the successful cross-coupling of the pyrrole boronic acid **39** with the electron-rich pyrrolyl triflate **40** to produce the diene **41** in anticipation of a ring-closing metathesis reaction to form the macrocyclic ring of the cyclic tripyrrole pigment, nonylprodigiosin <99JOC8275>.

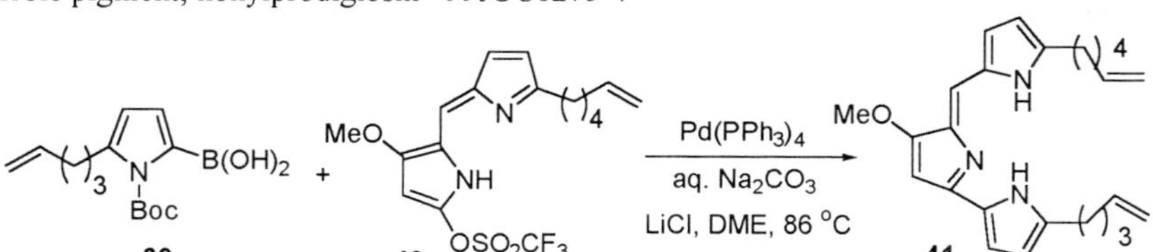

39 **40** **41**

In order to overcome shortcomings in methods amenable to the preparation of 2-substituted-3-arylpyrroles required for the total synthesis of the alkaloid (-)-rhazinilam and analogs which mimic the properties of taxol, Ghosez developed a short synthesis of 2-formyl-3-iodo-1-tosylpyrrole <99TL4555>. Thus, cyclization of the alkyne **42** (derived from from cinnamaldehyde) with concentrated HI led to the styryl pyrrole **43** which underwent oxidative cleavage of the double bond with permanganate to afford the aforementioned aldehyde **44**. This substrate was shown to undergo facile Suzuki coupling with a variety of aryl boronic acids to give the corresponding biaryl compounds **45** in high yields.

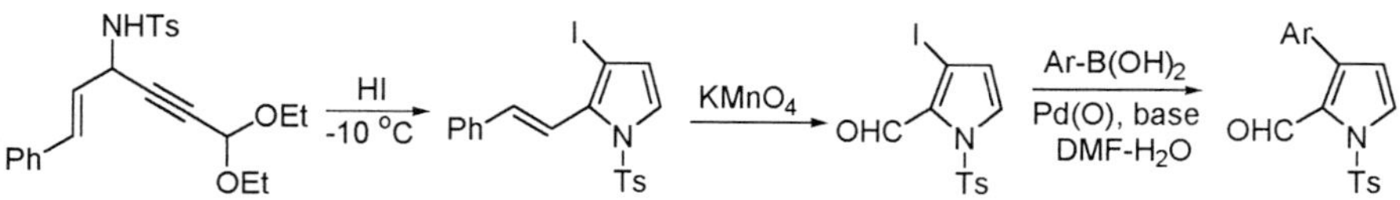

42 **43** **44** **45**

Banwell has developed a new approach to the core associated with several members of the lamellarin class of marine natural products. This approach utilized some interesting pyrrole arylation reactions including a Negishi cross coupling of the iodopyrrole **46** followed by a double-barrelled Heck cyclization of the resultant arylpyrrole **47** yielding the core structure **48** <99AJC755>.

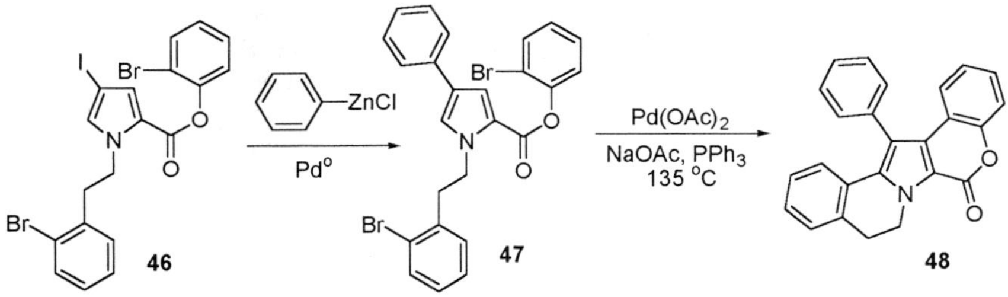

The stereoselective Birch reduction of electron-deficient pyrroles bearing chiral auxiliaries at the C-2 position has been demonstrated by two groups <99TL435, 99T12309>. In the former, Donohue reports that treatment of the (-)-8-phenylmenthol derived ester **49** with lithium metal in liquid ammonia and THF at –78 °C in the presence of the additive $(MeOCH_2CH_2)_2NH$ followed by quenching with a range of electrophiles afforded the dehydroproline derivatives **50** in up to 20:1 ds. To circumvent the relative inaccessibility of the (+)-antipode of this auxiliary, utilization of (+)-*trans*-2-(α-cumyl)cyclohexanol (TCC) resulted in the opposite facial selectivity. Donohue has also found that Birch reduction of pyrroles bearing electron withdrawing groups at the 3 and 4-positions occurs with a double reductive alkylation to yield *cis*-3,4-disubstituted pyrrolidines <99CC141>.

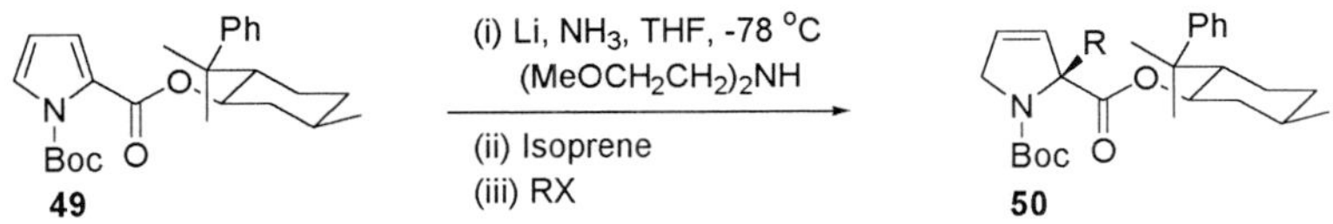

A preliminary investigation into the oxidation of the benzylic methylene groups on 1,2-dialkyl-3-nitropyrroles **51** and 3-alkylaminosulfonylpyrroles was conducted using a wide variety of oxidants. At present, it appears that atmospheric oxygen and NBS/hv represent the most efficient methods for the introduction of either a hydroxyl or bromo substituent at this site yielding the derivatives **52** (X = OH, Br) <99S447>.

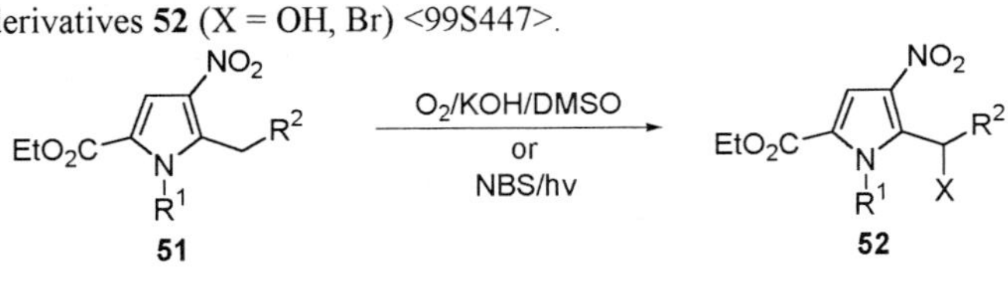

5.2.4 SYNTHESIS OF INDOLES

Kita has introduced a novel one-pot preparation of 5-methoxylated indoline **55** and indole **56** derivatives by intermolecular addition followed by cyclization between *N*-tosylaniline derivatives **53** and activated olefins **54** using phenyliodine(III) bis(trifluoroacetate) (PIFA) <99H511785>. In the reaction of **53** with phenyl vinyl sulfides, indoles were produced directly by the spontaneous elimination of thiophenol.

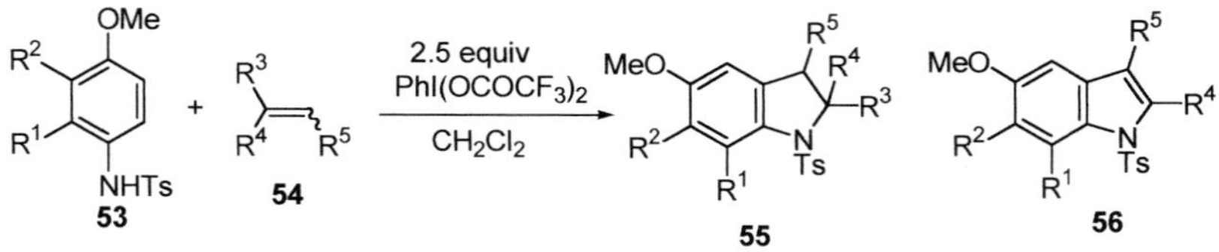

In terms of novel conditions for the Fischer indole synthesis, montmorillonite K10 impregnated with $ZnCl_2$ ("clayzic") was found to be effective for the preparation of 2-(2-pyridyl)indole derivatives in solvent-free medium under microwave irradiation <99SC1349>. Naito has reported that *N*-trifluoroacetyl enehydrazines (prepared by acylation of the corresponding hydrazones with trifluoroacetic anhydride) undergo thermal cyclization to indoles at or below temperatures of 90 °C <99TL3601>. Moreover, these authors have also examined the thermal cyclization of *o*-substituted *N*-trifluoroacetyl enehydrazines **57** and succeeded in the isolation and structure determination of dienylimine intermediates **58** in this indolization process along with C-7 substituted indolines and indoles (e.g., **59**) <99CC2429>.

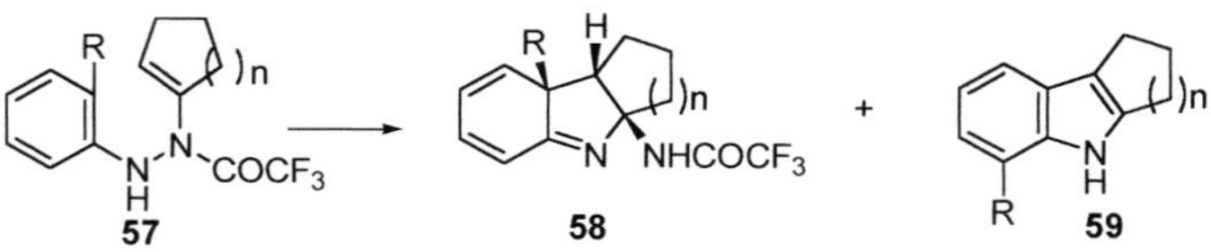

In order to circumvent the restrictions imposed on the Fischer indole synthesis by the limited availability of arylhydrazines, Buchwald reported a variant of the reaction involving the Pd-cross-coupling reaction of benzophenone hydrazones **60** with aryl bromides to afford *N*-arylhydrazones **61** <98JA6621>. The *N*-arylhydrazones are then converted to indole products **62** via an *in situ* hydrolysis/Fischer cyclization protocol in the presence of *p*-TsOH-H_2O. A full account of this process has now been disclosed, wherein notable improvements include facile processes for the preparation of *N*-alkyl or *N*-arylindoles by appropriate prior functionalization of the N-arylhydrazone precursors <99JA10251>. Buchwald also reports further improvements on the palladium catalyzed intramolecular cyclization of aryl bromides with pendant secondary amide groups to afford indolines or oxindoles <99OL35>.

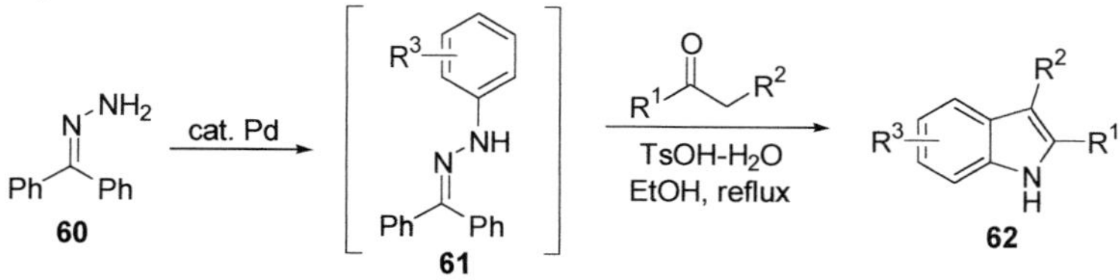

Undoubtedly, 2-haloaniline derivatives still maintain most favored precursor status for the preparation of the indole nucleus. Larock now reports the full details of his examination of the asymmetric addition of *N*-tosyl-2-iodoaniline (**63**) to allenes **64** (e.g. 1,2-undecadadiene) in the presence of palladium catalysts and chiral bisoxazoline ligand to afford chiral indolines **65** in up to 88% ee <99JOC7312>. Cook has utilized the palladium-catalyzed heteroannulation of iodoanilines with alkynes derivatized with the Schollkopf chiral auxiliary as a reliable route to optically active ring-A substituted tryptophans <99TL657>.

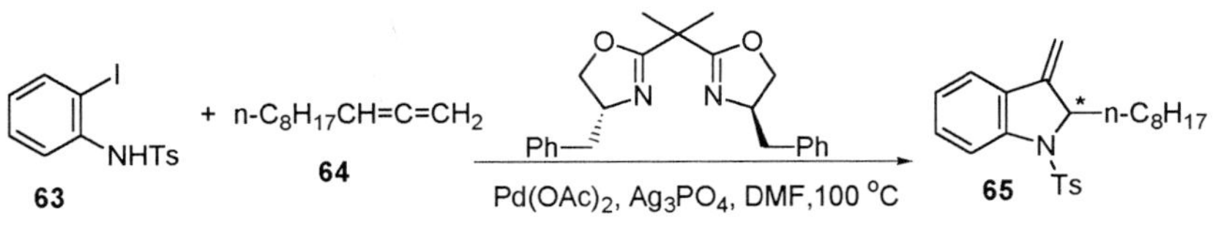

The related cyclization of 2-ethynylanilines **67** also represents one of the useful methods for the synthesis of 2-substituted indoles since the precursors are easily prepared from 2-haloanilines **66** by Pd-catalyzed cross-coupling with terminal alkynes. Although cyclizations of such alkynes are normally effected using Cu(I) or Pd(II) species, Sakamoto showed that in the absence of such metals, base catalysis (e.g., NaOEt) alone can accomplish the same goal. This author now reports that tetrabutylammonium fluoride (TBAF) is capable of inducing cyclization to the indoles **68** without affecting functionalities such as bromo, cyano, ethoxycarbonyl, and ethynyl <99JCS(P1)529>.

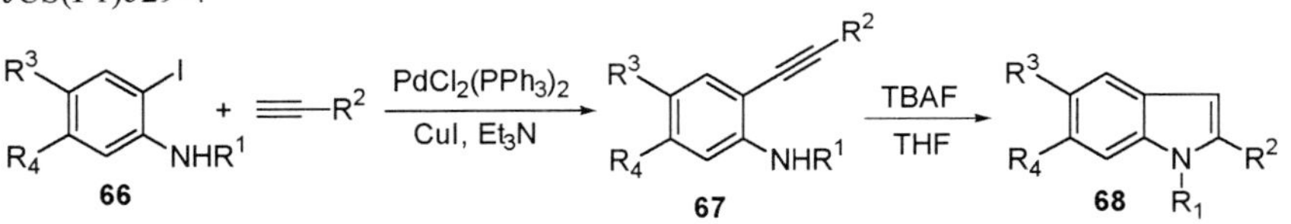

Barluenga has extended the intramolecular carbolithiation reaction of *N*-allyl-*N*-(2-lithioallyl)amines to the synthesis of 4-substituted indoles using a benzyne-cyclization methodology <99TL1049>. Thus, treatment of *N*-(2-bromoallyl)-*N*-methyl-2-fluoroaniline (**69**) with 3.3 equiv of *t*-butyllithium effects halogen-metal exchange followed by abstraction of the proton adjacent to the fluorine to yield intermediate **70**. Subsequent elimination of LiF produces a benzyne intermediate **71** which is trapped intramolecularly by the 2-lithioallyl moiety, affording a 3-methyleneindoline lithiated at C-4 which isomerizes to the corresponding 4-lithio indole derivative **72**. Quenching the 4-lithio species with electrophiles then leads to a variety of C-4 substituted indoles.

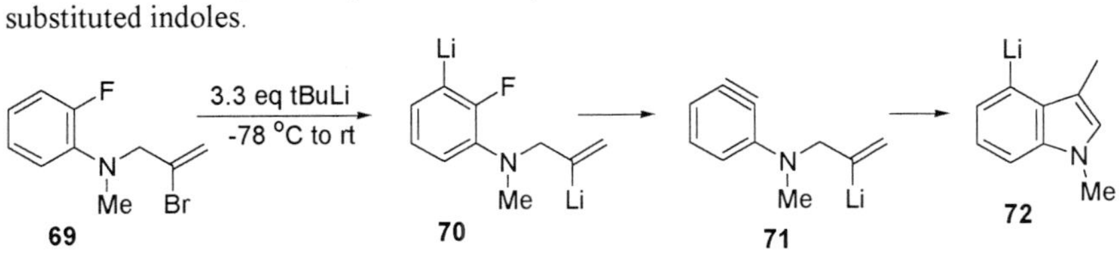

Grigg has reported a new battery of sequential cyclization-anion capture procedures for the preparation of indolines and oxindoles. Most notably, in what may represent the first report of a palladium catalyzed carbonylation using a resin-bound capture reagent, the Wang derived hydroxylamine **73** was subjected to a termolecular reaction involving carbon monoxide and the aryl iodides **74** to yield the resin bound indolines **75** which could be cleaved from the resin with TFA to afford the free hydroxamic acids <99TL7709>. Additionally, the palladium catalyzed cyclization-carbonylation-anion capture of aryl iodides with unsaturated secondary amines led to amides which could undergo sequential olefin metathesis to afford fused and spirocyclic rings <99TL3021>. In another variant, cyclization was coupled to a Wilkinson's catalyst [2+2+2] alkyne cycloaddition to afford *N*-tethered aromatic systems<99TL8277>

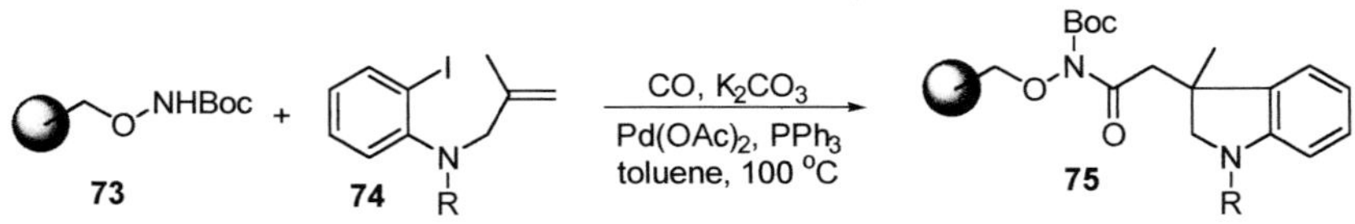

Curran has employed microwave flash-heating for the radical mediated cyclization of **76** to the indoline **77** with $HSn(CH_2CH_2C_{10}F_{21})_3$ in benzotrifluoride (BTF) wherein the organic and fluorous phases were found to coalesce upon heating giving a homogeneous solution <99JOC4549, 99JA6607>. Sample purification was achieved by filtration through standard silica gel since the fluorous tin compounds were insoluble in the organic eluting solvent and simply adhered to the top of the column.

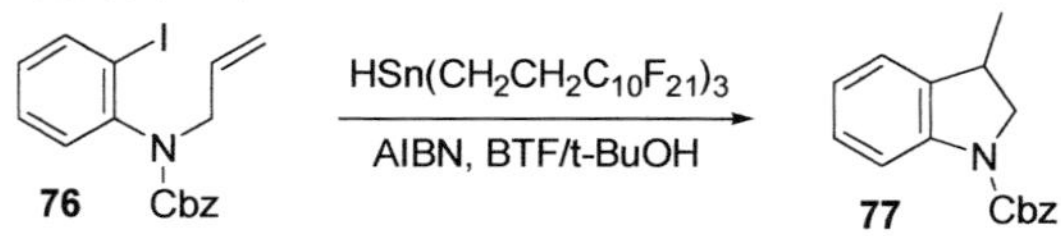

In the alternate mode of reaction, Zard now reports an expedient and versatile synthesis of indolines by the radical addition of a xanthate to an *N*-allylanilide **78** (R = Ms, Ac) yielding an adduct **79** which upon further exposure to peroxide generates a radical capable of cyclization onto the aromatic ring to produce the indolines **80** <99TL2533>.

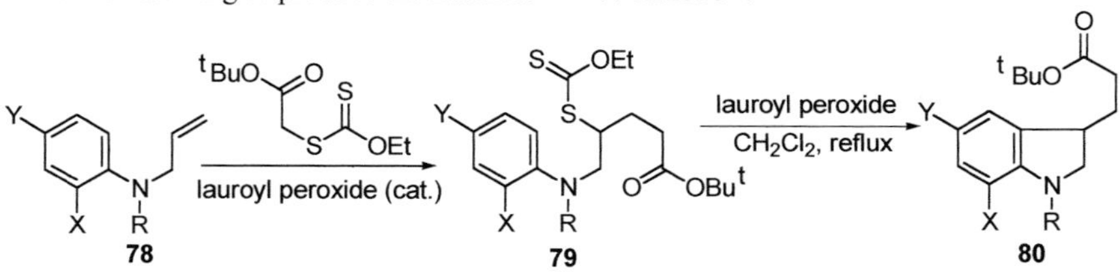

As originally described, the Fukuyama indole synthesis entailed the addition of tri-*n*-butyl tin radical to 2-alkenylphenylisonitriles leading to the formation of 2-stannylindoles. Recently, the synthetic potential of this indolization process was well demonstrated in total syntheses of vincadifformine, tabersonine <99TL1519> and catharanthine <99OL973>. Moreover, since tin radicals are also known to add to thioamides, Fukuyama now reports that 2-alkenylthioanilides **81** (prepared from 2-iodoanilines) react similarly yielding an imidoyl radical species such as **82** which can then undergo radical cyclization to afford 2,3-disubstituted indoles **83** in high yields <99JA3791>. This process allows for the introduction of both acid and base sensitive functionalities (esters, THP ethers) at the 2- and 3-positions and is also well suited for indoles bearing substituents on the carbocyclic ring.

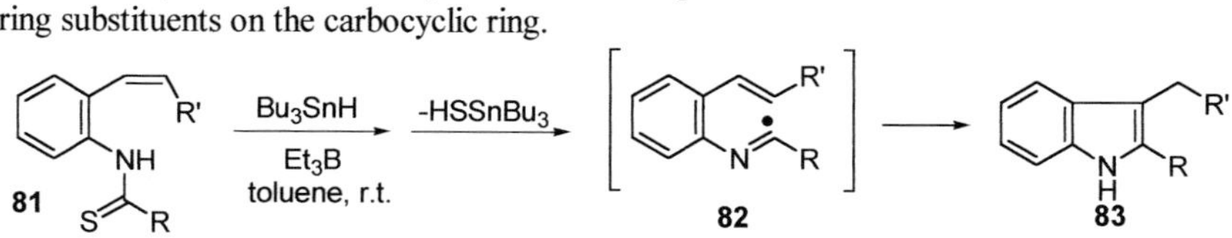

Selenium has now been added to the growing list of transition metals shown capable of catalyzing the reductive *N*-heterocyclization of 2-nitrostyrenes to indoles with carbon monoxide <99TL5717>. Utilizing the palladium-phosphine catalyzed methodology, Soderberg achieved the synthesis of several 4-substituted 2-methylindole natural products isolated from *Trichloma*

species <99JOC9731>. For instance, Stille coupling of the aryl bromide **84** with (tri-*n*-butyl-1-propen-1-yl)stannane yielded the nitrostyrene **85** as a mixture of isomers. Treatment of the isomeric mixture with carbon monoxide (60 psi) in the presence of palladium diacetate at 70 °C gave the expected alkaloid **86**. This author has also extended this approach towards a novel route to fused indoles *via* two consecutive palladium catalyzed reactions, namely an intramolecular Heck reaction followed by *N*-heterocyclization <99TL3657>.

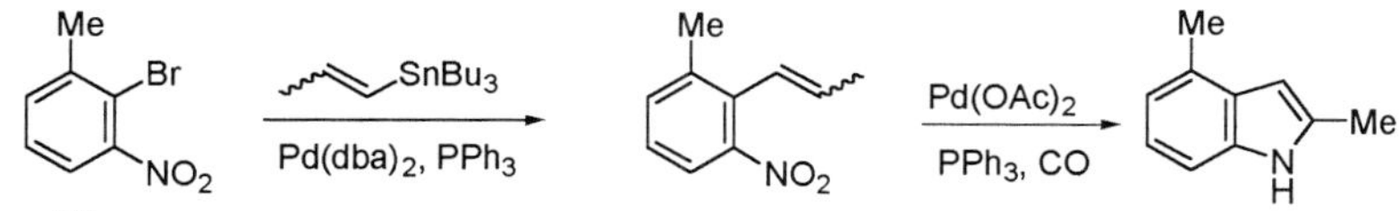

84 **85** **86**

The reductive cyclization of an *o*,β-dinitrostyrene with Fe/HOAc was employed by Corey to synthesize 6,7-dimethoxyindole enroute to the indole alkaloid, aspidophytine <99JA6771>. Prota has designed an imaginative route to 5,6-dihydroxyindole which features a Zn(II) assisted cyclization and avoids the need for hydroxyl protection <99S793>. In this approach, regioselective nitration of nitrostyrene **87** was accomplished with tetranitromethane in the presence of chelating Zn(II) ions at the distal 6-position yielding **88**. Reductive cyclization of this sensitive catechol was then effected by treatment with sodium dithionite in the presence of Zn(II) ions to yield the indole **89**.

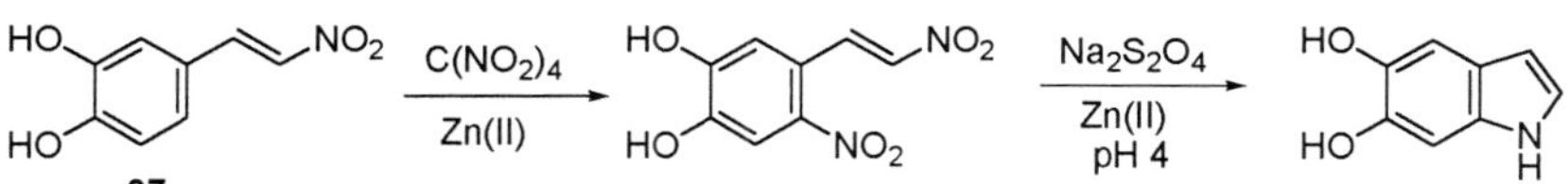

87 **88** **89**

Murphy has utilized radical <99JOC7856, 99JS(P1)995> as well as tetrathiafulvalene (TTF) mediated "radical-polar crossover" reactions <99TL161> in the construction of the tetracyclic ring system of the *Aspidosperma* alkaloids. Moreover, a polymer-supported TTF reagent **91** was prepared and its reactivity in radical polar crossover reactions validated by preparing an intermediate in the solution-phase synthesis of aspidospermidine. Thus, electron transfer from the TTF derivative to diazonium salt **90** (R = $CH_2CH_2NHCOCF_3$) is followed by loss of N_2, whereupon the aryl radical cyclizes and the resulting alkyl radical couples to $TTF^{+\cdot}$ to form an intermediate sulfonium salt which undergoes hydrolysis to the alcohol **92**.

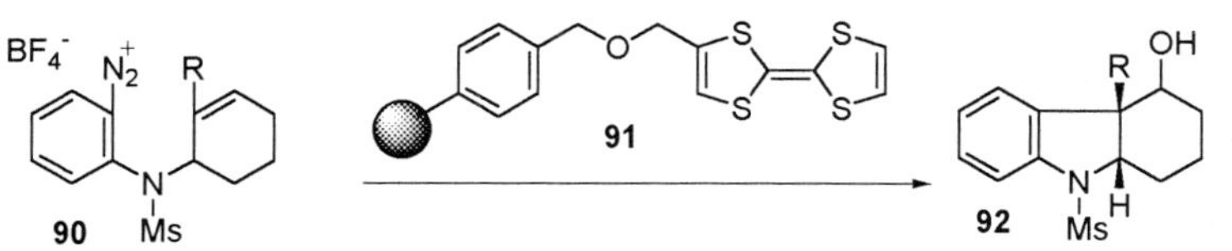

90 **92**

Sulikowski reported an enantioselective synthesis of a 1,2-aziridinomitosene which made clever use of the Jacobsen asymmetric ring opening of a *meso* epoxide with $TMSN_3$ to produce a chiral pyrrolidine precursor later elaborated as an oxazolidinone and subsequently cross-coupled to 2-bromoiodobenzene using Buchwald-Hartwig conditions *enroute* to the functionalized diazoester **93**. Chemoselective carbon-hydrogen insertion reaction of the metal carbene derived from **93** using a copper(I) complex led to the mitsosene **94** <99JOC4224>.

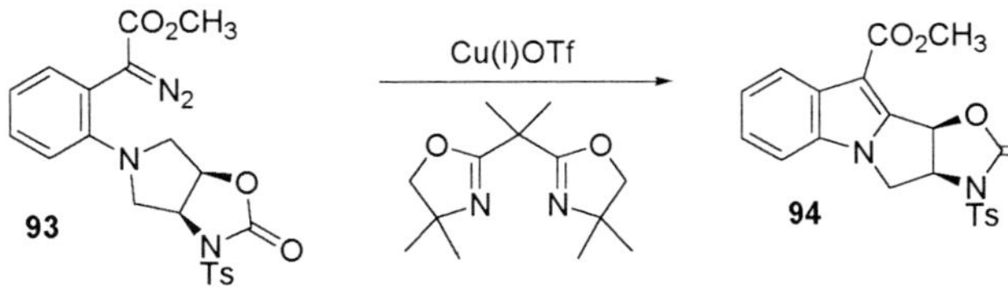

5.2.5 REACTIONS OF INDOLES

In terms of *N*-substitution, Hartwig reported improved conditions for the Pd(0) catalyzed *N*-arylation of indoles and pyrrole <99JOC5575>. It was found that when commercially available $P(t\text{-}Bu)_3$ was employed as ligand and cesium carbonate as base, the reaction between indoles **95** and unhindered aryl bromides **96** or chlorides occurred under milder conditions than the $Pd(OAc)_2$/DPPF system previously reported yielding the *N*-arylated products **97**. Alternatively, it has been found that pyrrole- and indole-2-carboxylic acid esters can be selectively *N*-arylated with phenylboronic acids in the presence of cupric acetate and either triethylamine or pyridine <99T12757>.

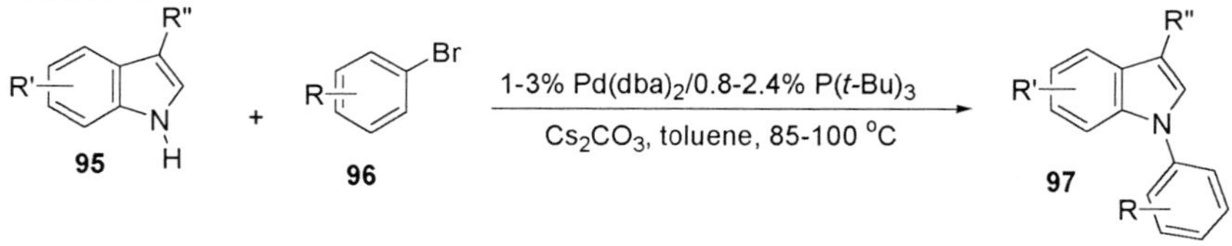

Macor has developed what appears to be a quite general method for the acylation (protection) of the indole nitrogen using 1,1'-carbonyldiimidazole (CDI) in the presence of DMAP. Reaction of substituted indoles **98** with CDI presumably occurs to form the imidazolyl amide of indole **99** which upon treatment *in situ* with either amines, alcohols or thiols affords the desired indole-1-carboxamides, -1-carboxylates, or 1-thiocarboxylates derivatives **100**, respectively.

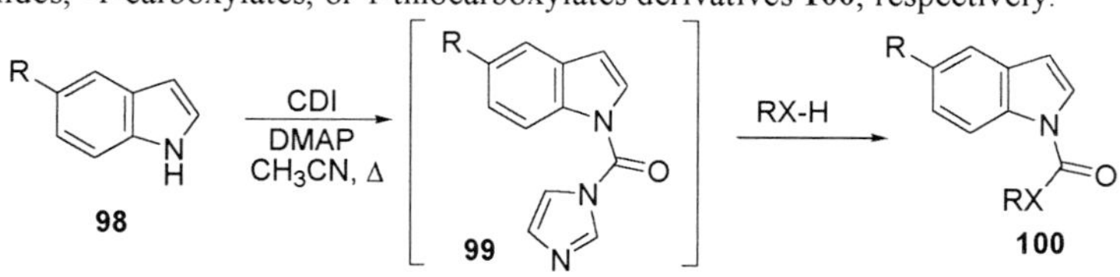

In terms of radical methods for the selective C-2 functionalization of indoles, in addition to the Byers stannane-free atom-transfer radical process described earlier <99TL2677>, Jones reports the intermolecular trapping of indol-2-yl radicals with electron deficient alkenes <99CC1761>. Although the intramolecular cyclizations of radicals derived from 2-bromoindoles have been previously demonstrated, it was found that reaction of 2-iodoindoles **101** under catalytic tin hydride conditions (tributyltin chloride and $NaCNBH_3$) using AIBN as initiator effectively generates indol-2-yl radicals which undergo addition to acrylonitrile or acrylate esters **102** (Z = CN, CO_2R) to afford 2-substituted indoles **103** in moderate yields. In an alternate use of 2-haloindoles in a non-radical process, Passarella found that modified Sonogashira conditions (PPh_3-CuI in the presence of K_2CO_3) proved optimal in promoting the coupling of *N*-SEM-2-iodoindole with methyl propynoate to yield methyl indol-2-ylpropiolate.

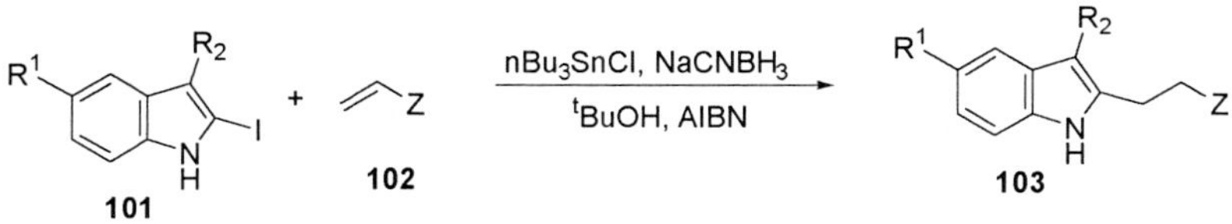

101 102 103

A novel tandem carbonylation/cyclization radical process has been developed for the intramolecular acylation of 1-(2-iodoethyl)indoles and pyrroles <99TL7153>. In this process, an acyl radical is formed when CO is trapped by an alkyl radical formed from the AIBN-induced radical reaction of 1-(2-iodoethyl)indoles **104** with Bu_3SnH. Intramolecular addition of the acyl radical to the C-2 position of the heteroaromatic system presumably affords a benzylic radical which undergoes *in situ* oxidative rearomatization to the bicycloketones **105**.

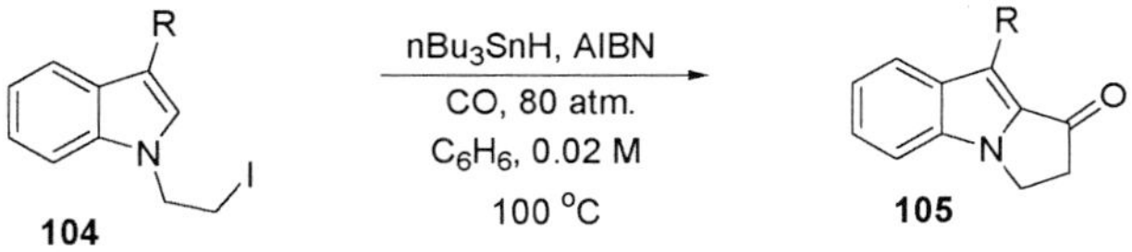

104 105

In continuation of studies on the synthesis and reactions (e.g., Barton-Zard) of 2-and 3-nitroindoles, Gribble now reports that a variety of N-protected indoles undergo regioselective nitration with acetyl nitrate generated *in situ* at low temperatures thereby providing a general and reliable route to the corresponding 3-nitroindoles <99S1117>. Additionally, this author also finds that 2-nitro-1-(phenylsulfonyl)indole (**106**) undergoes formal S_N2' displacement of phenylsulfinate with nucleophiles such as enolates (e.g., of diethyl malonate) as well as lithium dimethyl cuprate to give the corresponding 3-substituted-2-nitroindoles **107** <99TL7615>.

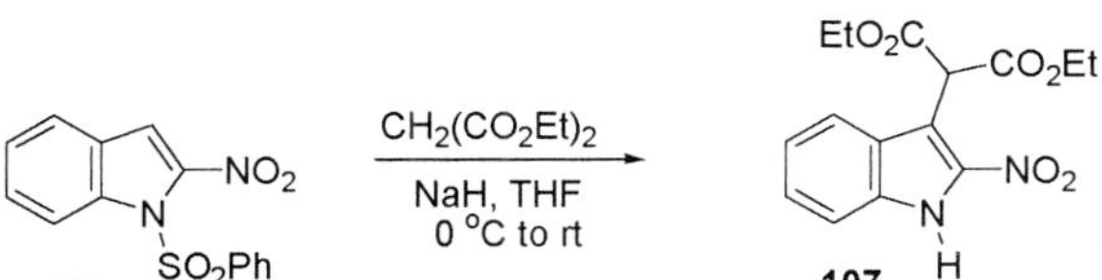

106 107

While lanthanide triflates have been demonstrated to promote the reaction of indoles with imines <99SL498>, Johannsen has developed a new synthesis of optically active β-indolyl *N*-tosyl α-amino acids **110** *via* the enantioselective addition of *N*-tosylimnio esters of ethyl glyoxylate **109** to indoles **108** bearing both electron-donor and electron-acceptor substituents at C-5 using 1-5 mol% of a chiral copper(I)-Tol-BINAP catalyst <99CC2233>.

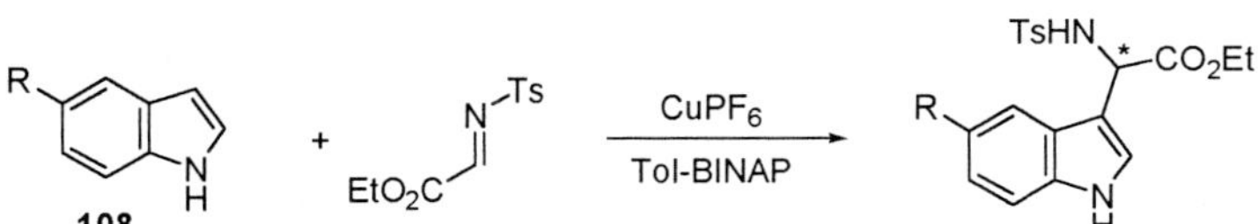

108 109 110

In terms of Michael additions at the indole C-3-position, montmorillonite K10 has been shown to be effective in promoting the conjugate addition of indoles to α,β-unsaturated carbonyl compounds <99MC929>. In what is also a Michael process, microwave irradiation has been shown to be effective in promoting the trimolecular condensation of substituted indoles **111** with paraformaldehyde **112** and Meldrum's acid **113** to afford the masked propanoic acid precursors **114** <99S254>. Such Meldrum's acid derivatives were similarly prepared by the conventional approach using aqueous formaldehyde and subsequently oxidized with NBS in *tert*-butyl alcohol/water to afford the corresponding oxindole derivatives which are potent aldose reductase inhibitors <99JOC1369>.

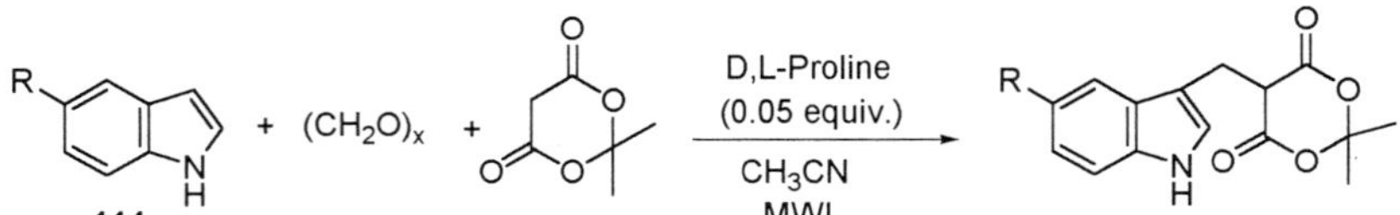

111 **112** **113** **114**

The previously described lanthanide triflate promoted ring opening of indoles **115** with 1,1-cyclopropanediesters **116** has been extended to provide an annulation reaction of indoles, wherein nucleophilic ring opening by the indole C-3 followed by attack of the intermediate malonic enolate **117** on the iminium moiety to afford the annulated products **118** <99TL5671>.

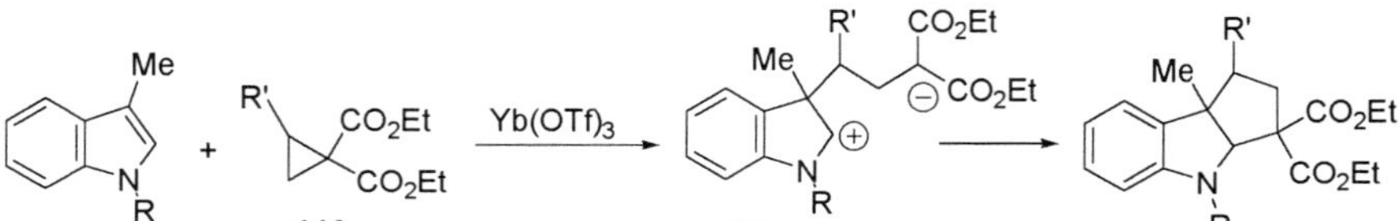

115 **116** **117** **118**

Nakagawa investigated the diastereoselective Pictet-Spengler (P-S) reaction of chiral tryptamines bearing an α-naphthylethyl auxiliary group with various aldehydes which afforded the corresponding tetrahydro-β-carbolines at a diastereoselectivity of up to 97:3 <99H(50)1033>. The asymmetric Pictet-Spengler reactions of D-tryptophan derivatives have been employed by Cook as key steps in total syntheses of (+)-ajmaline, alkaloid G and norsuaveoline <99JA6998>. In terms of solid-phase P-S reactions, L-tryptophan immobilized on polystyrene Wang resin was sequentially treated with aldehydes and Fmoc-amino acid chlorides to generate transient *N*-acyliminium species **119** which undergo P-S condensation affording resin-bound tetrahydro-β-carbolines which upon removal of the Fmoc protecting group undergo cyclizative cleavage from the resin yielding the diketopiperazines **120** <99OL1647>.

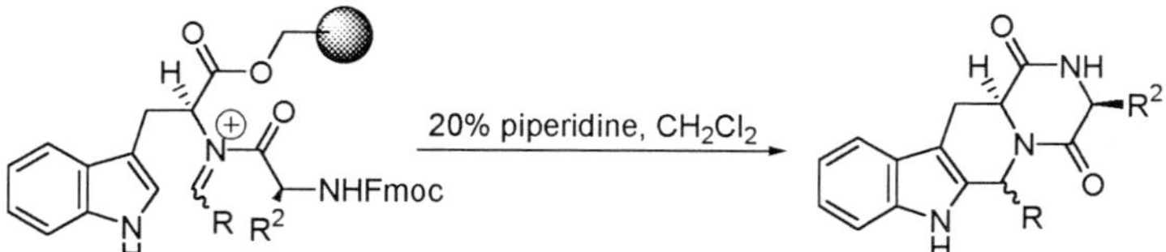

119 **120**

The photoinduced electron transfer (PET) catalyzed radical cation Diels-Alder reaction of electron-rich indoles has previously been limited to exocyclic dienes necessarily affording ring-annellated derivatives. This reaction has now been extended to the use of exocyclic 1,3-dienes containing an N-O or N-N bond as intentional cleaving points which after cleavage yield highly functionalized tetrahydrocarbazole derivatives; products of formal Diels-Alder reaction between indole and open chain 1,3-dienes <99CC433>. *N*-Tosyl-3-nitroindole (**121**) undergoes Diels-Alder reactions with 1-(*N*-acyl-*N*-alkylamino)-1,3-butadienes **122** (e.g., R = *N*-acetyl-*N*-propyl amino) with thermal extrusion of nitrous acid to afford advanced intermediates **123** for the synthesis of *Aspidosperma* alkaloids <99TL3343>. The observed regiocontrol observed in this reaction being completely dominated by the acylamino substituent in the diene and the powerful electron-withdrawing nitro group in the dienophile.

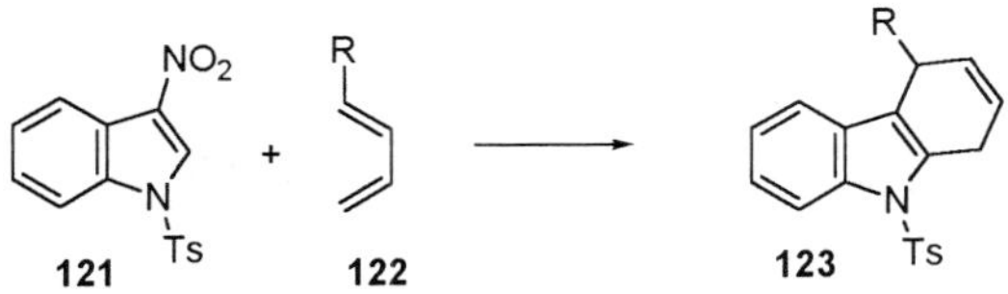

The use of indole-2,3-quinodimethanes or their synthetic equivalents continues to represent a facile approach to fused ring derivatives. Levy utilized the thermolysis of gramines **124** (R^1 = CO_2Et, Ph, H, etc.) to generate indole-2,3-quinodimethanes **125** which could be trapped with dienophiles such as maleimides to yield carbazoles **126** <99TL7549,TL997463>. The versatility of this reaction was extended to gramine derivatives with substituents on the 3-methylene group, allowing in turn for intramolecular Diels-Alder reaction. An intramolecular Diels-Alder of furo[3,4-*b*]indoles was employed by Gribble to synthesize benzocarbazoles <99SC729>, while pyrano[3,4-*b*]indoles were utilized in syntheses of carbazole-fused pyridazines <99H(51)2703>.

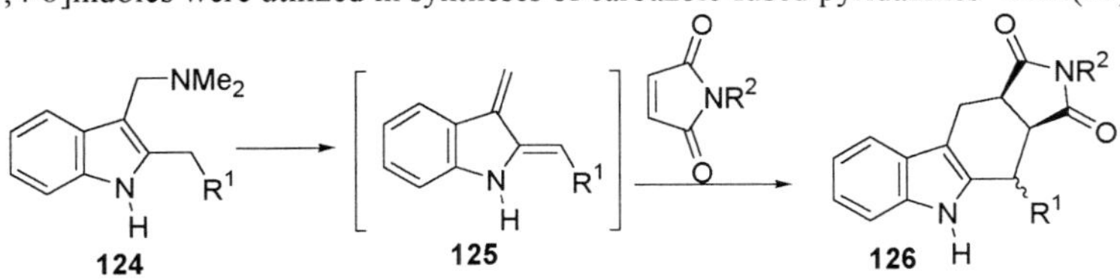

Snieckus achieved the first example of the generation of an indole-4,5-quinodimethane by a sequence involving directed *ortho*-metalation-carboxamidation of the carbamate **127** followed by cross-coupling with $TMSCH_2MgCl$ under Ni catalyzed conditions leading to the benzylsilane **128** <99TL2453>. Dibal reduction followed by quaternization of the resulting amine with MeI and switching the N-protecting group to an N-Boc derivative produced the quinodimethane precursor **129** which was found to undergo cycloaddition with a number of dienophiles.

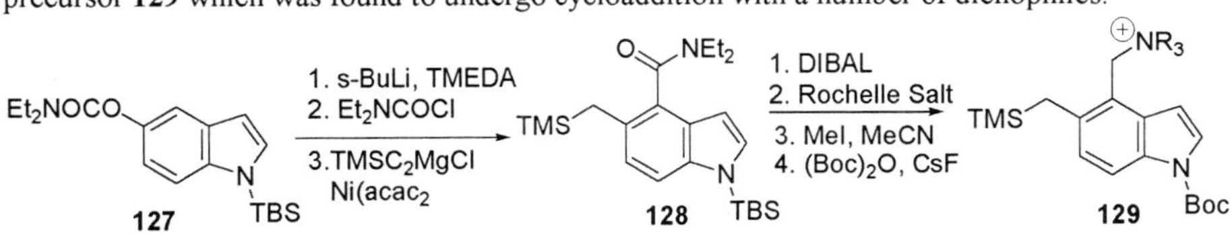

Rapoport reports an efficacious route to the tricyclic core of the ergot nucleus by initial deprotonation of bromotryptophan **130** with excess *n*-BuLi to remove the three acidic protons followed by halogen-metal exchange with *t*-BuLi for formation of the 4-lithio species, which in turn led to the cyclization to the ketone **131** <99JOC225>. In another example of halogen-metal exchange upon an *N*-unsubstituted 4-bromoindole, Doll introduced a boronic acid function at this position in anticipation of a Suzuki coupling towards the 8-azaergoline ring system <99JOC1372>.

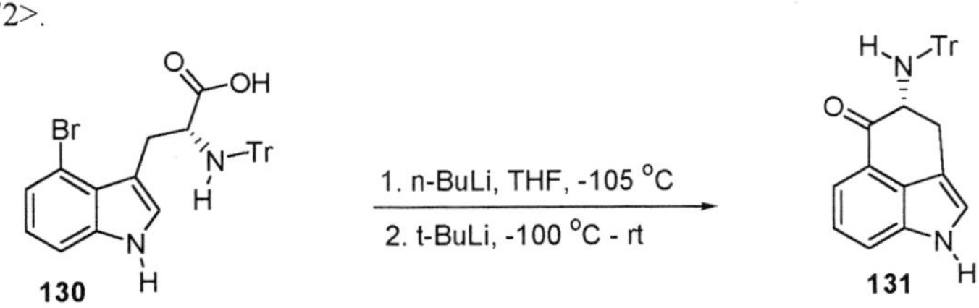

Iwao has done a masterful job in effecting the functionalization of the carbocyclic ring of indoles through directed lithiation of remote directing groups. The method devised for the selective C-4 functionalization of indoles ring *via* directed lithiation of 1-(triisopropylsilyl) gramine was utilized as key steps in syntheses of (-)-*cis*- and (-)-*trans*-clavicipitic acid methyl esters <99T10989>, as well as in the first total synthesis of veiutamine, a new type of pyrroloiminoquinone marine alkaloid <99TL1713>. Moreover, while the latter paper describes an interesting aryne mediated intramolecular cyclization of a 4-chlorotryptamine to a 1,3,4,5-tetrahydropyrrolo[4,3,2-*de*]quinoline, this author also investigated the directed lithiation of the *N*-Boc derivative **132** under standard conditions whereupon the resulting 6-lithio species **133** was regioselectively generated and allowed to react with a range of electrophiles providing the corresponding 6-substituted compounds **134**. This selectivity is apparently due to the powerful *ortho* directing effects of the *N*-Boc group and to the steric shielding of the C-2 and C-8 protons with the bulky 1-TIPS group. Interestingly, Iwao now finds that a bulky 2,2-diethylbutanoyl (DEB) group promotes the unusual C-7 lithiation of indoles, the greatest regioselectivity being observed in the case of 3-substituted derivatives <99T9151>.

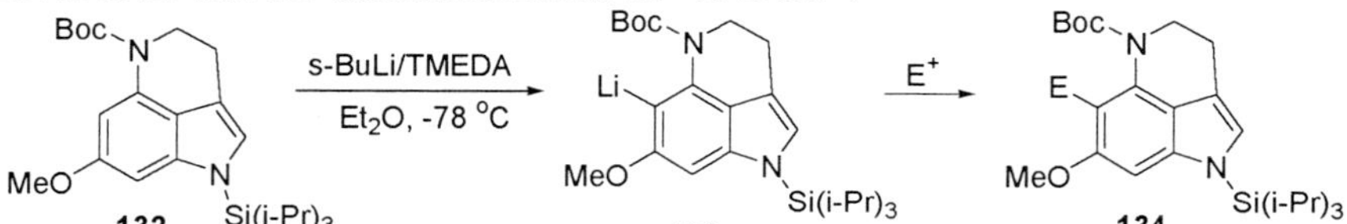

132 **133** **134**

As regards other fused ring indole derivatives, Jimenez investigated the preparation and reaction of several dialkyl- and diarylvinylsulfonium salts with the sodium salt of indole-2-carboxaldehyde <99T10659> and now finds that the reaction of aldehydes **135** with diisopropylvinylsulfonium triflate (**136**) in the presence of sodium hydride followed by the addition of sodium azide leads to an improvement over dimethylvinylsulfonium iodide for construction of the C ring of the mitomycin skeleton **137** *enroute* to a synthesis of a fully functionalized 7-methoxyaziridinomitosene <99JOC2520>.

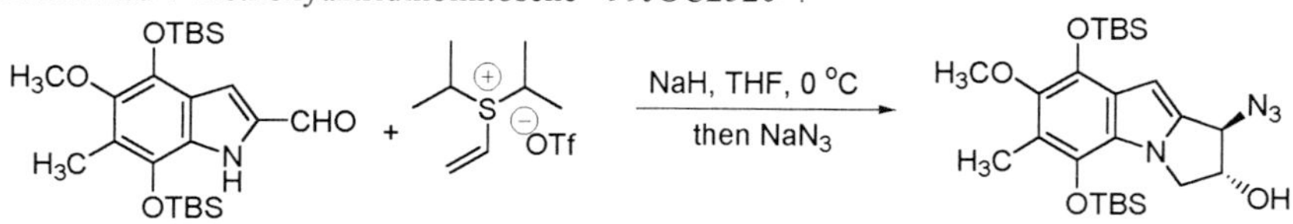

135 **136** **137**

The intramolecular cyclization of tryptophans is an important step in the syntheses of a number of biologically relevant alkaloids, and it has been well established by Crich and Hino that the protonically induced closure of tryptophan methyl esters results in preferential formation of the thermodynamically more stable 2-*endo* carbomethoxy hexahydropyrroloindole isomer. Significantly, in chemistry relevant to the synthesis of the multidrug resistant reversal agent 5-*N*-acetylardeemin, Danishefsky reports that selenocyclization of L-tryptophan methyl ester protected as its bis(Boc)derivative **138** with *N*-phenylselenophthalimide (N-PSP) in the presence of anhydrous pyridinium *p*-toluenesulfonate (PPTS), provides *exo*-isomer **139** as the major kinetic product (18:1) over the *endo*-isomer **140** <99JA11953>. This unusual reversal in stereoselectivity was further confirmed by Crich, who has utilized this finding to develop a simple protocol for the alkylation of tryptophan, leading to α-substituted tryptophans with clean inversion of configuration <99JOC7218>.

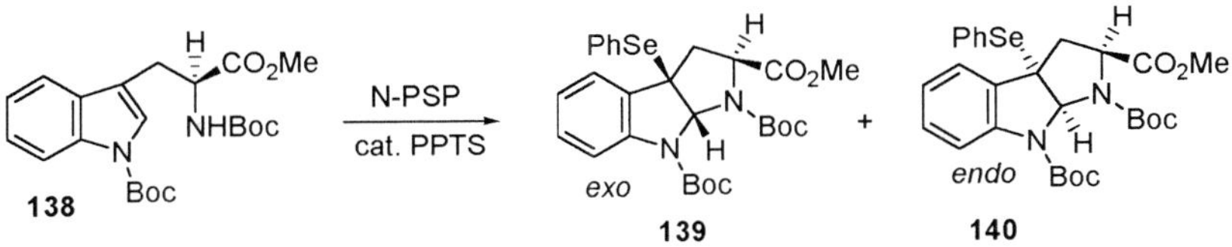

Additionally, *enroute* to achieving the total syntheses of gypsetin, deoxybrevianamide E, brevianamide E, and tryptostatin B, Danishesfsky devised a novel method for the synthesis of 2,3-disubstituted indoles involving initial treatment of 3-substituted indoles (e.g., N,N-dibenzyltryptophan methyl ester **141**) with *tert*-butylhypochlorite to generate chloroindolenine intermediates, e.g., **142**. Such intermediates have been shown capable of reaction at the C-2 position with a variety of nucleophiles such as prenyl 9-BBN to afford 2,3-disubstituted indoles such as **143** <99JA11964>. This chloroindolenine method was generalized to produce a range of 2-substituted tryptophan congeners not easily obtained by other means.

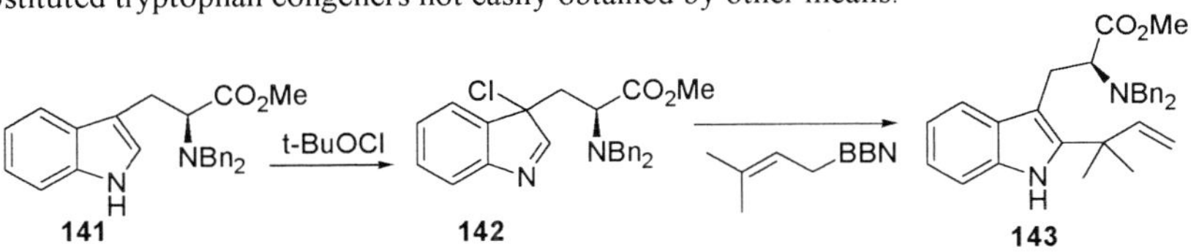

In terms of reduction processes, Van Vranken has developed a general two-step procedure for the reduction of indoles (e.g., yohimbinol **144**) to the corresponding 4,5,6,7-tetrahydroindoles **145** utilizing a regioselective Birch reduction followed by catalytic hydrogenation <99TL8039>, thus allowing access to chiral complex pyrroles from the chiral pool of alkaloids.

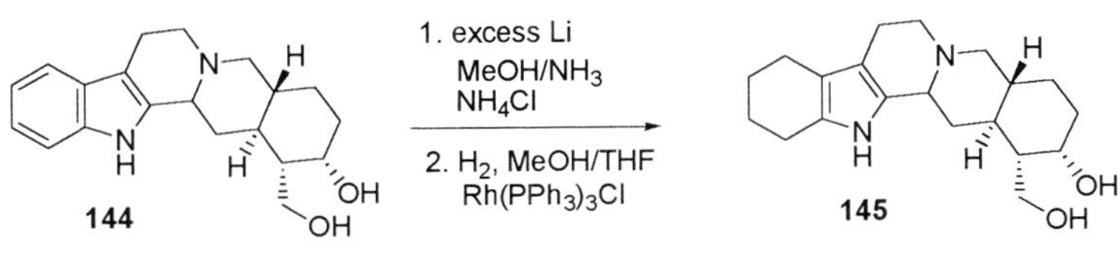

5.2.5 PYRROLE AND INDOLE ALKALOIDS

In terms of pyrrole derived natural products, Gribble describes the synthesis and identification of two halogenated bipyrroles (major component **146**) found present in seabird eggs <99CC2195> and has written a review on the diversity of naturally occurring organobromine compounds <99CSR335>. The identification of such halogenated species leads to speculation that some of the anthropogenic chlorinated compounds in seabirds, fish and marine mammals were in reality naturally occurring compounds. In examples of other halogenated alkaloids, Weinreb <99JOC9574> has achieved the total synthesis of the brominated cytotoxic marine sponge alkaloid agelastatin A (**147**), while Furstner <99JOC8275> details a second generation synthesis of roseophilin (**148**) and a has written an account on catalysis-based natural product synthesis as relates to such objectives <99SL1523>. Other synthetic approaches to **148** include routes utilizing the Paal-Knorr synthesis <99OL649, 99CC1455>, and Knight's endeavor relying on the iodocyclization of a homopropargylic sulfonamide <99TL6117>. Additionally, Tanaka reported the first total synthesis of the chlorinated pyrrole antibiotic

99JCS(P1)995	O. Callaghan, C. Lampard, A.R. Kennedy, J.A. Murphy, *J. Chem. Soc., Perkin Trans. 1* **1999**, 995.
99JCS(P1)2669	D. Passarella, G. Lesma, M. Deleo, M. Martinelli, A. Silvani, *J. Chem. Soc., Perkin Trans. 1* **1999**, 2669.
99JCS(P1)2849	T.L. Gilchrist, *J. Chem. Soc., Perkin Trans. 1* **1999**, 2849.
99JHC1469	B.A. Trofimov, *J. Heterocyclic Chem.* **1999**, *36*, 1469.
99JOC16	K.M. Aubart, C.H. Heathcock, *J. Org. Chem.* **1999**, *64*, 16.
99JOC225	C.R. Hurt, R. Lin, H. Rapoport, *J. Org. Chem.* **1999**, *64*, 225.
99JOC1369	W.G. Rajeswaran, R.B. Labroo, L.A. Cohen, M.M. King, *J. Org. Chem.* **1999**, *64*, 1369.
99JOC1372	M.K.-H. Doll, *J. Org. Chem.* **1999**, *64*, 1372.
99JOC2361	A. Furstner, T. Gastner, H. Weintritt, *J. Org. Chem.* **1999**, *64*, 2361.
99JOC2465	M.M. Faul, L.L. Winneroski, C.A. Krumrich, *J. Org. Chem.* **1999**, *64*, 2465.
99JOC2520	W. Dong, L.S. Jimenez, *J. Org. Chem.* **1999**, *64*, 2520.
99JOC4224	S. Lee, W.-M. Lee, G. Sulikowski, *J. Org. Chem.* **1999**, *64*, 4224.
99JOC4539	K. Olafsson, S.-Y. Kim, M. Larhed, D.P. Curran, A. Hallberg, *J. Org. Chem.* **1999**, *121*, 4539.
99JOC4572	K. Hoshino, N. Ozawa, H. Kokado, H. Seki, T. Tokunaga, T. Ishikawa, *J. Org. Chem.* **1999**, *64*, 4572.
99JOC4697	S. Mahboobi, E. Eibler, M. Koller, S. Kumar KC, A. Popp, D. Schollmeyer, *J. Org. Chem.* **1999**, *64*, 4697.
99JOC5575	J.F. Hartwig, M. Kawatsura, S.I. Hauck, K.H. Shaughnessy, L.M. Alcazar-Roman, *J. Org. Chem.* **1999**, *64*, 5575.
99JOC6487	M.T. Baumgartner, A.B. Pierini, R.A. Rossi, *J. Org. Chem.* **1999**, *64*, 6487.
99JOC6518	Y. Fumoto, T. Eguchi, H. Uno, N. Ono, *J. Org. Chem.* **1999**, *64*, 6518.
99JOC7218	D. Crich, X. Huang, *J. Org. Chem.* **1999**, *64*, 7218.
99JOC7312	J.M. Zenner, R.C. Larock, *J. Org. Chem.* **1999**, *64*, 7312.
99JOC7586	P.A. Grieco, M.D. Kaufman, *J. Org. Chem.* **1999**, *64*, 7586.
99JOC7813	A. Burghart, H. Kim, M.B. Welch, L.H. Thoresen, J. Reibenspies, K. Burgess, F. Bergstrom, L.B.-A. Johansson, *J. Org. Chem.* **1999**, *64*, 7813.
99JOC7856	M. Kizil, B. Patro, O. Callaghan, J.A. Murphy, M.B. Hursthouse, D. Hibbs, *J. Org. Chem.* **1999**, *64*, 7856.
99JOC8275	A. Furstner, J. Grabowski, C.W. Lehmann, *J. Org. Chem.* **1999**, *64*, 8275.
99JOC9605	M.-L. Bennesar, J.-M. Jimenez, B. Vidal, B.A. Sufi, J. Bosch, *J. Org. Chem.* **1999**, *64*, 9605.
99JOC9731	B.C. Soderberg, A.C. Chisnell, S.N. O'Neil, J.A. Shriver, *J. Org. Chem.* **1999**, *64*, 9731.
99M929	J.H. Poupaert, J. Bukuru, A. Gozzo, *Monatsh. Chem.* **1999**, *130*, 929.
99OL35	B.H. Yanh, S.L. Buchwald, *Org. Lett.* **1999**, *1*, 35.
99OL79	S.F. Martin, K.X. Chen, C.T. Eary, *Org. Lett.* **1999**, *1*, 79.
99OL649	P.E. Harrington, M.A. Tius, *Org. Lett.* **1999**, *1*, 649.
99OL1505	M. Nakamura, K. Hara, G. Sakata, E. Nakamura, *Org. Lett.* **1999**, *1*, 1505.
99OL1647	H, Wang, A. Ganesan, *Org. Lett.* **1999**, *1*, 1647.
99OL973	M.T. Reding, T. Fukuyama, *Org. Lett.* **1999**, *1*, 973.
99S254	C. Nemes, J.-Y. Laronze, *Synthesis* **1999**, 254.
99S275	S. Eils, E. Winterfeldt, *Synthesis* **1999**, 275.
99S447	C. Moranta, M.D. Pujol, A.M. Molins-Pujol, J. Bonal, *Synthesis* **1999**, 447.
99S793	L. Novellino, M. d'Ischia, G. Prota, *Synthesis* **1999**, 793.
99S885	Y. Kita, M. Egi, T. Takeda, H. Tohma, *Synthesis* **1999**, 885.
99S1117	E.T. Pelkey, G.W. Gribble, *Synthesis* **1999**, 1117.
99S1523	A. Furstner, *Synthesis* **1999**, 1523.
99SC729	G.W. Gribble, R.A. Silva, M.G. Saulnier, *Synth. Commun.* **1999**, *29*, 729.
99SC1349	T. Lipinska, E. Guibe-Jampel, A. Petit, A. Loupy, *Synth. Commun.* **1999**, *29*, 1349.
99SL498	W. Xie, K.M. Broomfield, Y. Jin, N.Y. Dolney, P.G. Wang, *Synlett.* **1999**, 498.
99SL561	M.J. Sung, C.-W. Lee, J.K. Cha, *Synlett.* **1999**, 561.
99SL1651	P.M. Fresneda, P. Molina, M.A. Saez, *Synlett.* **1999**, 1651.
99T2363	J. Bergman, E. Desarbre, E. Koch, *Tetrahedron* **1999**, *55*, 2363.
99T2371	T. Janosik, J.. Bergman, *Tetrahedron* **1999**, *55*, 2371.
99T6555	A. Alberola, A.G. Ortega, M.L. Sadaba, C. Sanudo, *Tetrahedron* **1999**, *55*, 6555.
99T9151	T. Fukuda, R. Maeda, M. Iwao, *Tetrahedron* **1999**, *55*, 9151.
99T10659	Y. Wang, W. Zhang, V.J. Colandrea, L.S. Jimenez, *Tetrahedron* **1999**, *55*, 10659.
99T10989	H. Shinohara, T. Fukuda, M. Iwao, *Tetrahedron* **1999**, *55*, 10989.
99T11095	A. Urrutia, J. G. Rodriguez, *Tetrahedron* **1999**, *55*, 11095.
99T12309	A. Schafer, B. Schafer, *Tetrahedron* **1999**, *55*, 12309.
99T12757	W.W.K.R. Mederski, M. Lefort, M. Germann, D. Kux, *Tetrahedron* **1999**, *55*, 12757.

99T13211 A. Alberola, R. Alvaro, A.G. Ortega, M.L. Sadaba, M.C. Sanudo, *Tetrahedron* **1999**, *55*, 13211.
99T13957 H. Shiraishi, T. Nishitani, T. Nishihara, S. Sakaguchi, Y. Ishii, *Tetrahedron* **1999**, *55*, 13957.
99T14515 J. T. Gupton, K.E. Krumpe, B.S. Burnham, T.M. Webb, J.S. Shuford, J.A. Sikorski, *Tetrahedron* **1999**, *55*, 14515.
99TL161 O. Callaghan, C. Lampard, A.R. Kennedy, J.A. Murphy, *Tetrahedron Lett.* **1999**, *40*, 161.
99TL435 T.J. Donohue, P.M. Guyo, M. Helliwell, *Tetrahedron Lett.* **1999**, *40*, 435.
99TL657 C. Ma, X. Liu, S. Yu, S. Zhao, J.M. Cook, *Tetrahedron Lett.* **1999**, *40*, 657.
99TL1049 J. Barluenga, F.J. Fananas, R. Sanz, Y. Fernandez, *Tetrahedron Lett.* **1999**, *40*, 1049.
99TL1519 S. Kobayashi, G. Peng, T. Fukuyama, *Tetrahedron Lett.* **1999**, *40*, 1519.
99TL1713 Y. Moro-oka, T. Fukuda, M. Iwao, *Tetrahedron Lett.* **1999**, *40*, 1713.
99TL1929 K. Tatsuta, M. Takahashi, N. Tanaka, *Tetrahedron Lett.* **1999**, *40*, 1929.
99TL2453 A.C. Kinsman, V. Snieckus, *Tetrahedron Lett.* **1999**, *40*, 2453.
99TL2533 T.-M. Ly, B. Quiclet-Sire, B. Sortais, S.Z. Zard, *Tetrahedron Lett.* **1999**, *40*, 2533.
99TL2677 J.H. Byers, J.E. Campbell, F.H. Knapp, J.G. Thissell, *Tetrahedron Lett.* **1999**, *40*, 2677.
99TL2733 J.E. Macor, A. Cuff, L. Cornelius, *Tetrahedron Lett.* **1999**, *40*, 2733.
99TL3021 P. Evans, R. Grigg, M.I. Ramzan, V. Sridharan, M. York, *Tetrahedron Lett.* **1999**, *40*, 3021.
99TL3343 B. Biolatto, M. Kneeteman, P. Mancini, *Tetrahedron Lett.* **1999**, *40*, 3343.
99TL3601 O. Miyata, Y. Kimura, K. Muroya, H. Hiramatsu, T. Naito, *Tetrahedron Lett.* **1999**, *40*, 3601.
99TL3657 B.C. Soderberg, S.R. Rector, S.N. O'Neil, *Tetrahedron Lett.* **1999**, *40*, 3657.
99TL3957 T.N. Danks, *Tetrahedron Lett.* **1999**, *40*, 3957.
99TL4177 S. Lim, I. Jabin, G. Revial, *Tetrahedron Lett.* **1999**, *40*, 4177.
99TL4555 C. Franc, F. Denonne, C. Cuisinier, L. Ghosez, *Tetrahedron Lett.* **1999**, *40*, 4555.
99TL5671 M.A. Kerr, R.G. Keddy, *Tetrahedron Lett.* **1999**, *40*, 5671.
99TL5717 Y. Nishiyama, R. Maema, K. Ohno, M. Hirose, N. Sonoda, *Tetrahedron Lett.* **1999**, *40*, 5717.
99TL6117 M.A. Fagan, D.W. Knight, *Tetrahedron Lett.* **1999**, *40*, 6117.
99TL6211 G. McCort, O. Duclos, C. Cadilhac, E. Guilpain, *Tetrahedron Lett.* **1999**, *40*, 6211.
99TL6909 P. Magnus, T.E. Mansley, *Tetrahedron Lett.* **1999**, *40*, 6909.
99TL7153 L.D. Miranda, R. Cruz-Almanza, M. Pavon, E. Alva, J.M. Muchowski, *Tetrahedron Lett.* **1999**, *40*, 7153.
99TL7459 K. Diker, M. Doe de Maindreville, J. Levy, *Tetrahedron Lett.* **1999**, *40*, 7459.
99TL7463 K. Diker, M. Doe de Maindreville, D. Royer, F. Le Provost, J. Levy, *Tetrahedron Lett.* **1999**, *40*, 7463.
99TL7587 H.H. Wasserman, A.K. Petersen, M. Xia, J. Wang, *Tetrahedron Lett.* **1999**, *40*, 7587.
99TL7615 E.T. Pelkey, T.C. Barden, G.W. Gribble, *Tetrahedron Lett.* **1999**, *40*, 7615.
99TL7709 R. Grigg, J.P. Major, F.M. Martin, M. Whittaker, *Tetrahedron Lett.* **1999**, *40*, 7709.
99TL7857 B. Patro, M. Merrett, J.A. Murphy, D.C. Sherrington, M.G.J.T. Morrison, *Tetrahedron Lett.* **1999**, *40*, 7857.
99TL8039 C.C. McComas, D.L. Van Vranken, *Tetrahedron Lett.* **1999**, *40*, 8039.
99TL8141 J. Leonard, A.B. Hague, G. Harms, M.F. Jones, *Tetrahedron Lett.* **1999**, *40*, 8141.
99TL8255 P.D. Bailey, K.M. Morgan, G. Rosair, R.L. Thomas, *Tetrahedron Lett.* **1999**, *40*, 8255.
99TL8277 R. Grigg, V. Sridharan, J. Zhang, *Tetrahedron Lett.* **1999**, *40*, 8277.

Chapter 5.3

Five-Membered Ring Systems

Furans and Benzofurans

Stefan Greve and Willy Friedrichsen
Institute of Organic Chemistry, University of Kiel, 24098 Kiel, Germany
(email: friedrichsen@oc.uni-kiel.de)

5.3.1 INTRODUCTION

The chemistry of furans and benzofurans was a field of lively research in the last year. There are a number of reasons for this activity. The furan ring system – both in its native as well as in its reduced form – occurs in a great number of natural products and a wide variety of these compounds has been isolated from natural sources. This subject is treated regularly with care and accuracy in *"Heterocycles"* and will not be repeated here. Therefore only a few examples will be given in this chapter. Several new furan derivatives were isolated from natural sources: (-)-wistarin **1** from the marine sponge *Ircinia sp.* <99TA3869>,

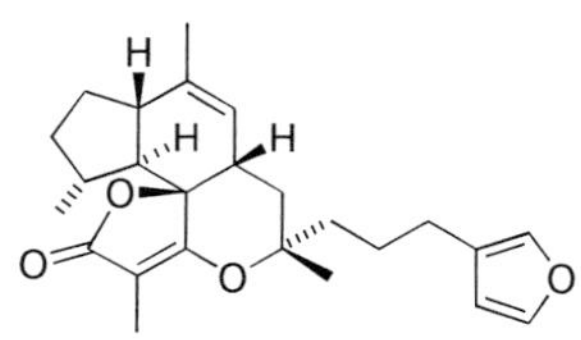

1

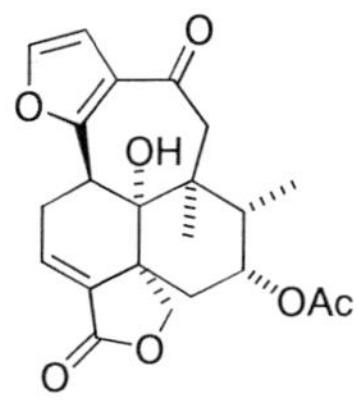

2

a new tyrosine kinase inhibitor from *Ulocladium botrytis* and new metabolites from the marine fungi *Asteromyces cruciatus* and *Varicosporina ramulosa* <99EJO2949>, a new diterpenoid named salvimexicanolide **2** from the aerial parts of *Salvia mexicana var. major* (besides the known flavonoid naringenin) <99H(51)1647>, a new coumarin (mashrin, **3**) and a new dimeric acridone alkaloid (citbismine-F, **4**) from the roots of marsh grapefruit (*Citrus paradisi macf.*) <99H(51)851>.

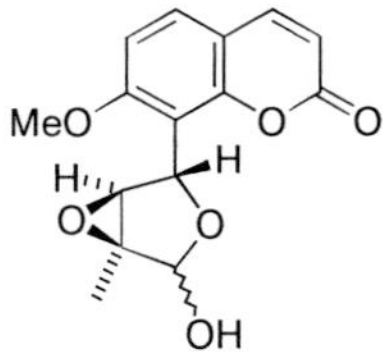

3

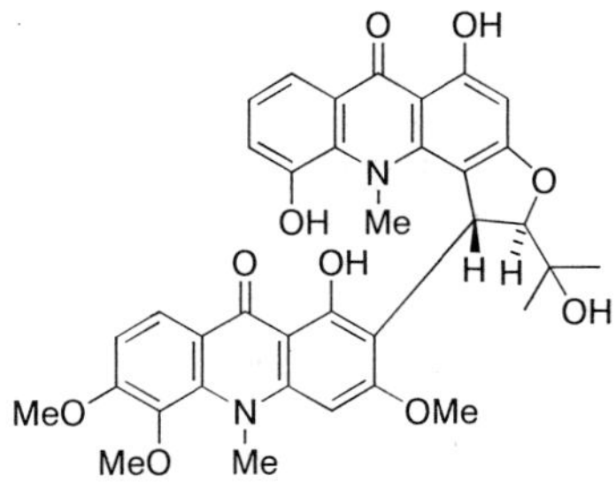

4

Morus alba callus and cell suspension cultures specifically produce chalomoracin **5** and kuwanon J **6** from cinnamoylpolyketide intermediates. Administration of [2-^{13}C]cinnamic acid *N*-acetylcysteamine thioester to the *M. alba* cell cultures revealed the cinnamoyl CoA intermediate to be a significant precursor <99H(50)989>. For further biosynthetic investigations see <99H(51)231>.

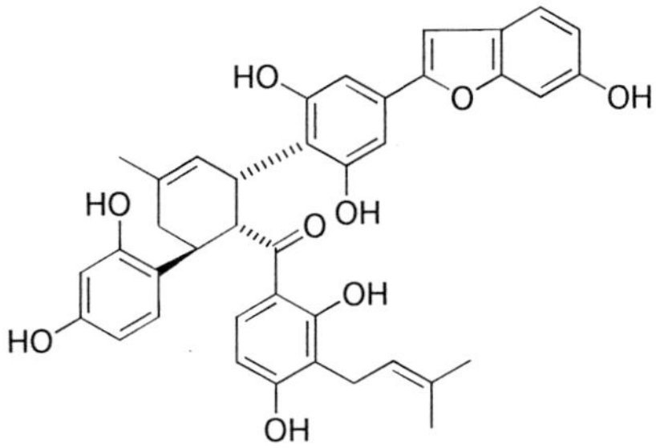

5

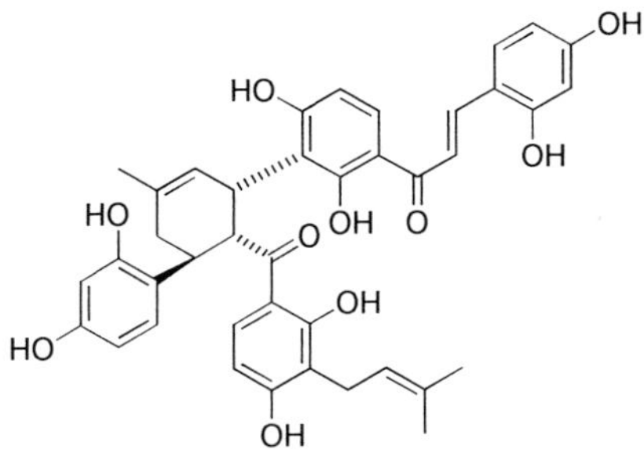

6

Twenty two new limonoids have been isolated from a Meliaceae plant *Melia toosendan*. Their antifeeding activity was tested against the larvae of *Spodoptera* insects <99H(50)595>. Noelaquinone **7**, a new hexacyclic triazine quinone, has been isolated from the Indonesian sponge *Xestospongia sp*. <98H(49)355>.

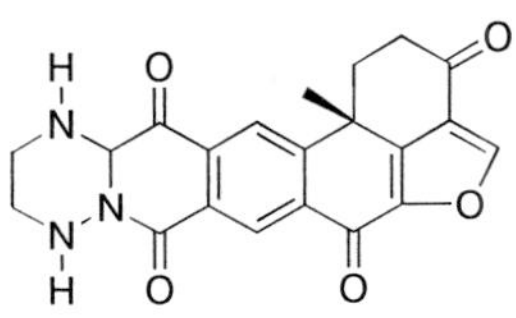

7

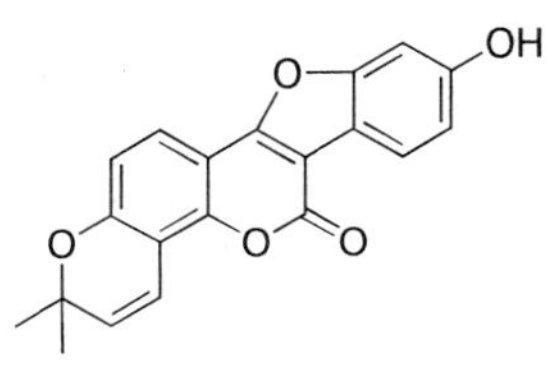

8

The isolation and characterization of four new coumarins (racemosone, furanoracemosone, mammea A/BC, isoracemosol) from the leaves of *Mesua racemosa* has been reported <99H(51)2183>. A survey of oxygenated 2,11-cyclized cembranoids of marine origin (cladiellins, asbestinins, sacrodictyins) has been compiled in tabular form <98H(49)531>. Several total syntheses of naturally occurring furan derivatives (prehispanolone, sphydrofuran, secosyrins, syringolides) have been reviewed <99EJO1757>. The total syntheses of plicadin (**8**, probably a coumestan from *Psoralea plicata*) <99AG1528>, of (-)-mucocin <99AG1341>, of solanin <99H(50)981> and an effective route to the *trans*-tetrahydrofuran units in Annonaceous acetogenins and nonnatural ion channels have been described <99TA2551>. Corey *et al.* reported a highly stereoselective synthesis of Amprenavir (**9**, a HIV protease inhibitor) and its C(2) diastereomer <99AG2057>.

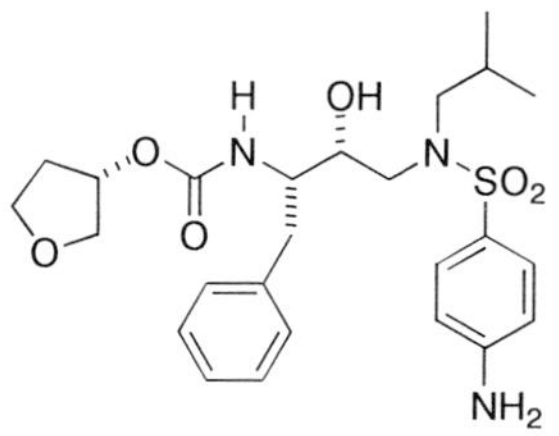

9

A review concerning the synthesis of optically active α-furfuryl amine derivatives and their application to asymmetric synthesis appeared <99T11959>.

5.3.2 REACTIONS

5.3.2.1 Furans

Diels-Alder reactions with furans constitute a major class of cycloaddition reactions, which have been used for the preparation of a variety of new compounds. Some examples will be given in this section. A one-pot stereoselective synthesis of tricyclic γ-lactones was achieved *via* Diels-Alder reactions of 2-methoxyfuran with masked *o*-benzoquinones, which are in turn available from the corresponding 2-methoxyphenols <99CC713>.

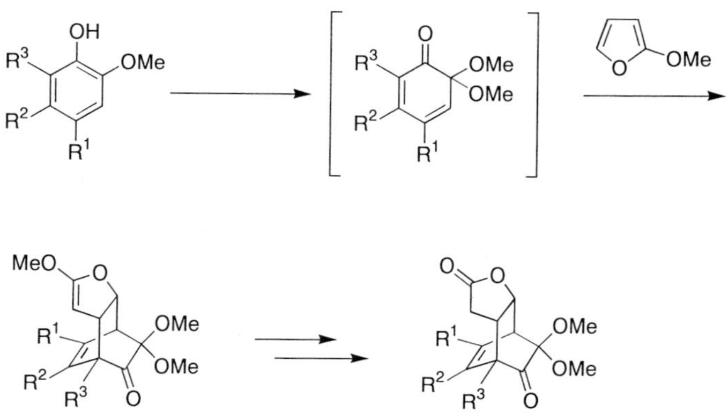

An enantioselective synthesis of (+)-**11** has been achieved in four steps starting with furan and dienophile (+)-**10** *via* an asymmetric Diels-Alder reaction <99TA2237>.

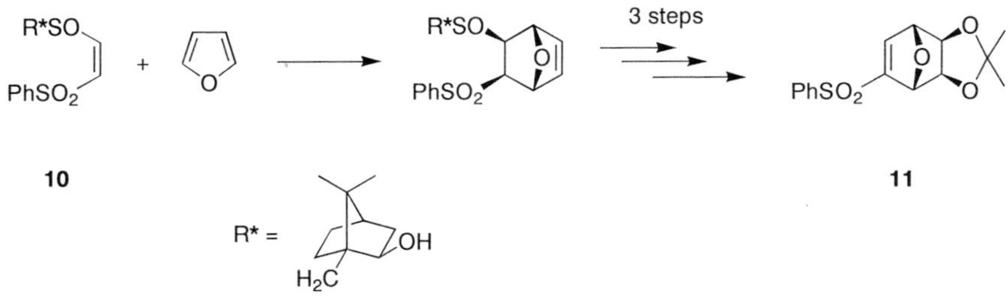

10 **11**

Normal *vs*. tandem 'pincer' reactions of furan derivatives with phenylsulfonylacetylene have been studied <99H(50)653>. It has been found that furans react with methyl 3-nitro-acrylate to concurrently give Diels-Alder and Michael adducts, and the former turn into the latter <99H(51)243>. Diels-Alder reactions of bis-protected 2,5-furan-dimethanol provide a new entry into the core of zaragozic acids <99TL2777>. For further studies on the synthesis of zaragozic acids see <99SL1407>. Synthetic approaches towards phorbols via the ultra-high-pressure (19 kbar) mediated intramolecular Diels-Alder reaction of furans and an investigation of furan substitution on this reaction have been reported <99JCS(P1)913>. It is well known that 2-phenylfuran derivatives are notoriously poor dienes in Diels-Alder reactions. A design of a system which is capable of accelerating and facilitating such a disfavoured cycloaddition between a furan and a maleimide was recently presented. Thus after heating diene **12** and maleimide **13** in $CDCl_3$ (50 °C, 15 h), a significant quantity of the *exo*-cycloadduct could be detected. The equilibrium constant was determined as 15.3 M^{-1}.

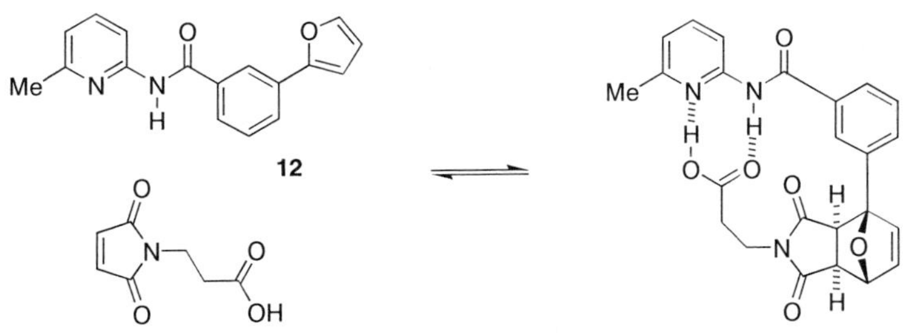

12

13 **14**

Molecular mechanic calculations suggest that the origin of the strong preference for the formation of **14** as its *exo* stereoisomer arises from the ability of the cycloadduct to form intra-molecular hydrogen bonds <99OL1087>. Firestone and co-worker report on an interesting case of reduced stereospecificity in a pericyclic reaction. The aqueous Diels-Alder reaction between 2-methylfuran and maleic acid is 99.9% stereospecific. Addition of heavy atom salts to the *retrodiene* reaction reduces the degree of stereospecificity significantly implicating a diradical mechanism <99T14269>. An efficient solid phase synthesis of highly substituted rigid tricyclic nitrogen heterocycles through acylation of resinbound furylamine with fumaric acid derivatives has been studied. The product is formed *via* an initial *N*-acylation followed by intramolecular Diels-Alder reaction <99SL1609>, see also <99T7433>.

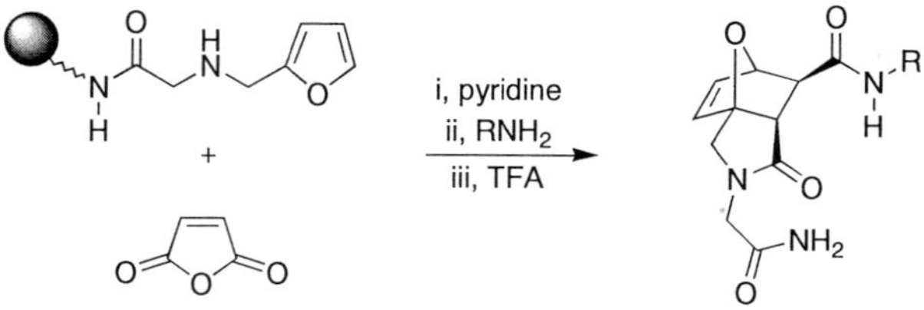

[4+3]-Cycloadditions of dienes with oxyallyl offer the opportunity to prepare seven-membered ring systems. This reaction has also proved to be of importance in the furan series. A few examples may illustrate the value of this methodology. A tandem Pummerer rearrangement and intramolecular [4+3]-cycloaddition with a furan derivative has been reported <99TL545>. For a similar reaction see <99T13999>.

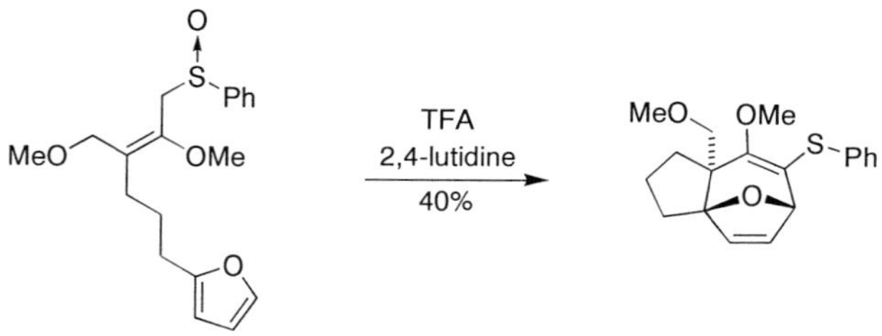

A series of α,α-dimethoxysilyl enol ethers has been prepared and shown to undergo diastereoselective [4+3]-cycloaddition with furan and cyclopentadiene in the presence of catalytic amounts of trimethylsilyl triflate <99SL213>. Furo[3,4-*d*]oxazoles and furo[3,4-*d*]thiazoles react with 1,3-dimethyloxyallyl to give [4+3]-cycloadducts. The ring opening reaction of these compounds with H_2SO_4/Et_2O yields annulated hydroxytropones <99H(51)1225>.

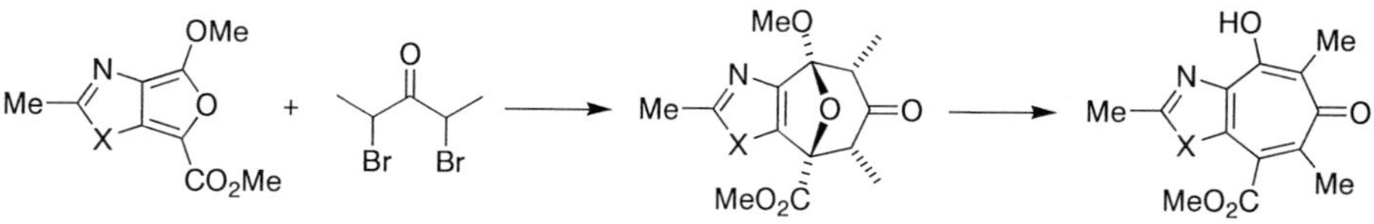

X = O, S

[4+3]-Cycloadducts have also been used as starting materials for the synthesis of the C29-C37 segment of a spongistatin <99TA1539> and the synthesis of the cyathin core <99OL1535>.

2,2-Dichlorocyclopentanone and 2,5-dibromocyclopentanone react with various dienes in the presence of base to afford cycloadducts derived from [4+3]-cycloaddition of the corresponding cyclopentenyl oxyallyl. Treating these compounds with carbon nucleophiles results in formation of cyclobutane containing products. This transformation makes dihalogenated cyclopentanones synthetic equivalents of a variety of substituted cyclobutenes and allows easy access to Diels-Alder adducts of various compounds of this type <99S1534>.

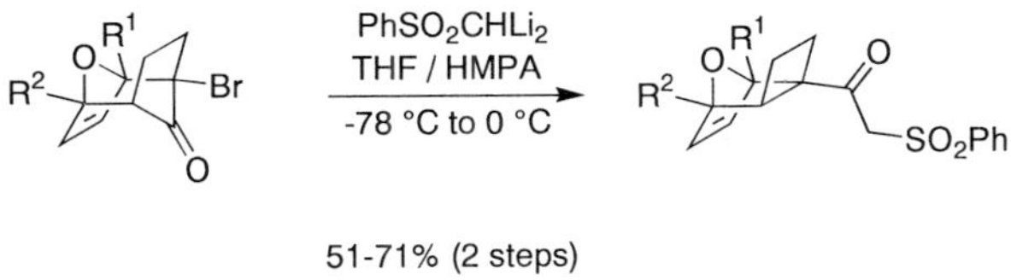

As has been reported previously <97T9313> furfural reacts with DEAD to give an unexpected product believed to be **15**. The mechanism for the formation of the analogous dimethyl dicarboxylate has now been investigated computationally <99JCS(P2)73>.

15

Photochemical cycloadditions of furans have also been reported occasionally. The photochemical reaction of furan and 3-cyano-2-methoxypyridines (in benzene solution) results in the formation of 1:1 adducts **16** and **17** (accompanied by a transpositional pyridine and a pyridine dimer) <99JCS(P1)171>.

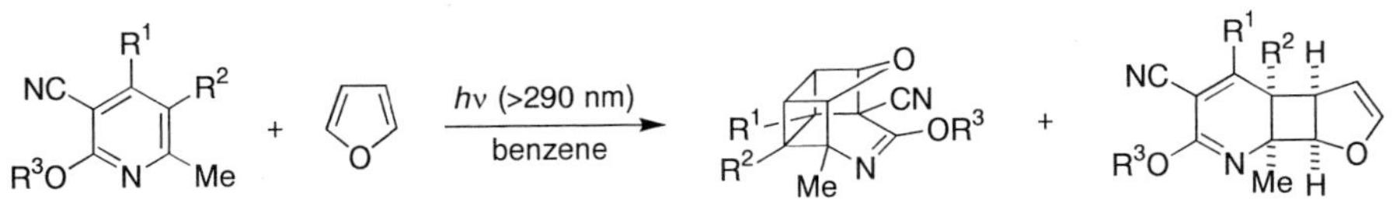

16 **17**

Stereoselective intramolecular [2+2]-cycloaddition reactions were observed during irradiation (λ = 313 nm) of some neoclerodane and labdane diterpene ketones <99EJO2011>.
Besides cycloaddition reactions the furan skeleton has been used in a number of cases as building block for other molecules. Some examples will be given. Reaction of lithiated butenolide with 2,3-*O*-alkylidene-D-glycerinaldehyde followed by successive elimination, hydrolysis, benzoylation and acetylation enabled the total synthesis of (*S*)-(+)-melodorinol and (*S*)-(+)-acetylmelodorinol <99H(51)795>. Reactions of a 3(2*H*)-furanone lithium enolate with 4-halocrotonates have also been investigated <99SL1391>. Hydroxymethyl-substituted indolizidines, representative members of a ring-B-expanded alexine-australine subclass, are readily accessible by starting with a furan-based silyloxydiene and a hydroxymethyl hemiaminal <99EJO1395>.

A vinylogous Mannich reaction of trisisopropylsilyloxyfuran and aldimines in the presence of catalytic amounts of $Ti(O\textit{i}Pr)_4$/*S*-BINOL complex gave substituted aminoalkyl butenolides in good yield. Enantioselectivities up to 54% ee were observed <99TL8949>. For a similar reaction of a O-TIPS substituted furan with an iminium ion (generated *in situ*) see <99T8905>. A mechanistic study of the reactions between 2-trimethylsilyloxyfuran and (*S*,*S*)-2-(arylsulfinyl)-1,4-benzoquinones has been reported <99TA4357>. Several coupling reactions with suitably substituted furan derivatives have been described. 5-Acceptor-substituted 2,3-dibromofurans underwent a regioselective Pd(0)-catalyzed coupling reaction at the C-2 carbon atom. With alkynes the corresponding 2-alkynylfurans were accessible. Alkyl-, aryl- and alkenylzinc reagents gave the corresponding 2-substituted furans. 2-Allylfurans were obtained by a regioselective Stille coupling. Subsequent Pd(0)-catalyzed reactions to introduce a methyl group in 3-position by a methyldebromination were successfully performed with MeZnCl and $PdCl_2(PPh_3)_2$ as the catalyst in THF and with $SnMe_4$ and $PdCl_2[P(o\text{-tol})_3]_2$ as the catalyst in DMA for the 2-allyl-3-bromofuran <99EJO2045>. 2-Alkyl-substituted furans (and thiophenes) were efficiently synthesized from organostannyl derivatives as starting materials using Cu(I) as catalyst. The effect of DMF and NMP as solvents were discussed, as well as other solvent and temperature effects <99SL1942>. A convenient amination of weakly activated furans (thiophenes, selenophenes) in aqueous media was reported. Thus treatment of 5-bromofurfural with various amines (morpholine, piperidine, dimethylamine) yields the corresponding 5-amino-substituted furfurals <99T6511>. The synthesis of *cis*- and *trans*-crobarbatic acid was reported. The five-step sequence proceeds in high yield and with control of both relative and absolute stereochemistry. The key step in the synthesis was a Birch reductive alkylation of a chiral furoic acid which sets the absolute stereochemistry of the products <99TA1315>. Oxidative ring opening reactions of furans are well known. This methodology has been used in an asymmetric total synthesis of (+)-desoxyprosophylline and prosophylline starting from chiral 1-(α-furyl)-2-phenylmethoxy-*N*-tosylethylamine <99TA2311>. For a similar oxidative ring opening see <99TA3649>. α-(Methoxycarbonyl)furfuryl amide was prepared by Lewis acid catalyzed allylic substitution of the corresponding α-(methoxycarbonyl)furfuryl carbinol acetate with various nitriles as nucleophiles. The amides were subjected to oxidative cleavage of the furan ring to afford *N*-protected polyhydroxy amino acids <99SL1399>. An enantio- and diastereo-controlled route to both enantiomers of eight possible hexoses has been explored starting from furfural, by employing the Sharpless asymmetric dihydroxylation as a key step <99S341>. An enantio- and diastereoselective synthesis of (+)-asperlin was achieved by employing the Sharpless asymmetric kinetic resolution of an unsymmetrical carbinol <99H(50)433>.

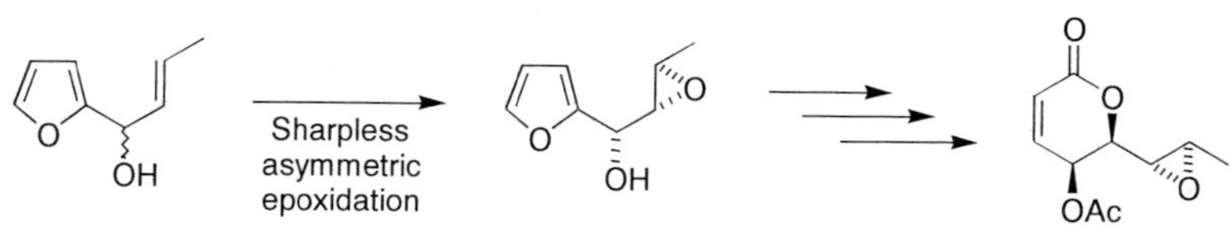

A chiral sugar building block having a dioxabicyclooctane framework has been prepared in enantiomerically pure form, along with analogues carrying a different protecting group, in an enantiocomplementary way by employing lipase-mediated kinetic resolution as the key step <99S1325>.

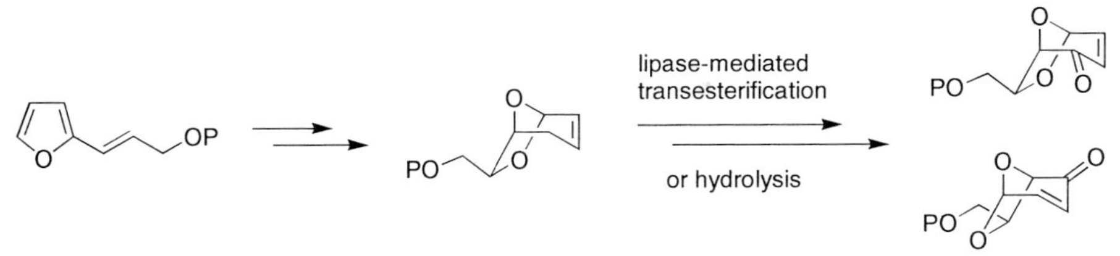

P = 2-naphthylmethyl; benzyl; *tert.*-butyldimethylsilyl

Furfural has been used as a convenient starting material for the synthesis of D- and L-mannose, gulose, and talose *via* diastereo- and enantioselective dihydroxylation reaction of the corresponding vinylfuran <99JOC2982>, see also <99JOC2984>. Non-racemic 6-(furan-3-yl)-5,6-dihydropyran-2-ones are accessible in high yields and enantioselectivities by a five-step sequence involving in the key-step asymmetric aldol-condensations of masked acetoacetic esters to 3-formylfuran <99TA3659>. An asymmetric total synthesis of prostaglandin E_1 has been achieved in a two-component coupling process. The chiral hydroxycyclopentenone was readily available from furan with 96% ee <99EJO2655>.

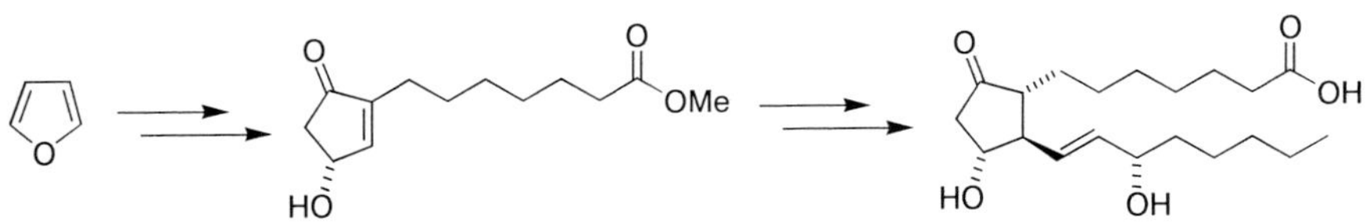

96% ee PGE$_1$

Simple one-pot reactions have been developed for the synthesis of 4-bromopyrimidines and condensed derivatives. Under the catalytic influence of dry hydrogen bromide, *N*-(cyanovinyl)-amidines cyclized to 4-bromopyrimidines. 2-Aminonitrile compounds with halogenoacetonitriles gave condensed furopyrimidines in good yields <99H(51)2723>.

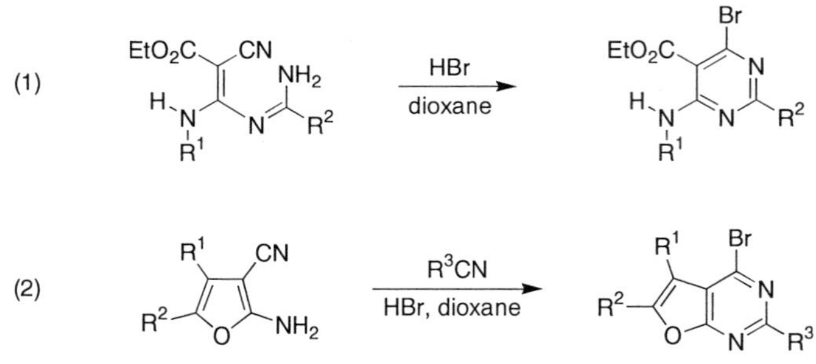

(1) R^1, R^2 = H, Ph, Me

(2) R^1, R^2 = H, Ph, Me, $-(CH_2)_4-$

$R^3 = CH_2Cl, CHCl_2$

Some total syntheses targeting structurally diverse and biologically active pyrrole alkaloids, furanoterpenes, macrolide antibiotics and carbohydrates have been summarized <99SL1523>.

5.3.2.2 Di- and Tetrahydrofurans

Dihydrofurans (and other cyclic ketene trimethylacetals) add electron deficient alkynes to give the corresponding [2+2]-cycloadducts. Ring opening with $BF_3{\cdot}OEt_2$ yields tetrahydrofuran derivatives <99TL839>.

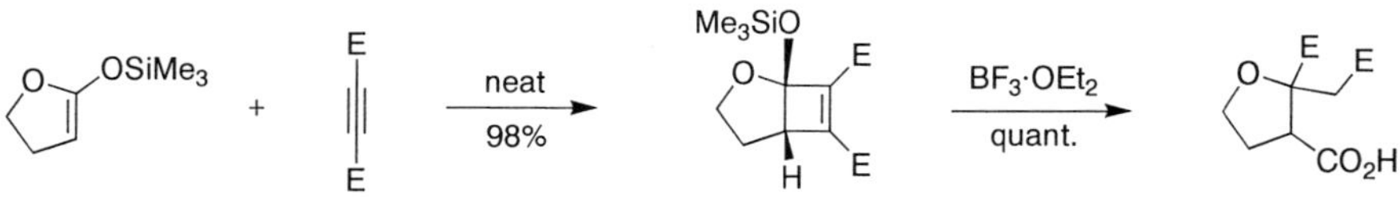

$E = CO_2Me$

3,4-Dialkoxypyrroles can be obtained in four steps from commercially available 2,5-dimethoxy-2,5-dihydrofuran <99S94>. For corresponding reactions of tetrahydrofurans see <99S74>. Dihydrofurans bearing a 3-hydroxymethyl group may be efficiently combined with aldehyde dimethylacetals to provide functionalized 2,6-disubstituted 3,7-dioxabicyclo [3.3.0]octanes in a one-pot procedure with high stereocontrol <99SL474>.

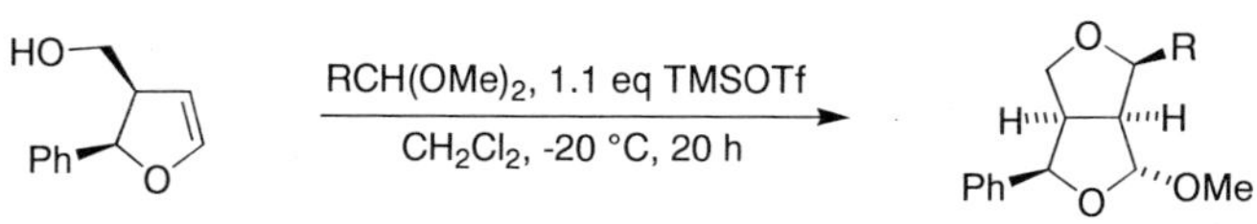

The synthesis of butylene crown ethers was accomplished by acid catalyzed ring opening of tetrahydrofuran. Trifluoromethanesulfonic acid is an effective and convenient catalyst for initiating the polymerisation reaction (also suitable for other cyclic ethers) <99S1193>.

Rearrangements of tetrahydrofurans (and tetrahydropyrans) having a C1'-mesyloxy group on the C2-side chain with zinc acetate in aqueous acetic acid were investigated. Rearrangement-ring expansion and/or rearrangement-ring opening reactions took place depending on the stereostructure of the substrates <99H(50)919>. The preparation of *N*- and *C*-protected derivatives of 2,5-disubstituted *trans*- and *cis*-THF amino acids in enantiomerically pure form from L-alanine was reported <99EJO2977>.

5.3.2.3 Benzofurans and Related Compounds

New β-substituted benzo- and naphthofuryl derivatives of *o*-divinylbenzene were synthesized and irradiated in order to form annulated bicyclo[3.2.1]octadienes. The mechanism of the intramolecular [2+2]-cycloaddition was explained <99H(51)1355>.

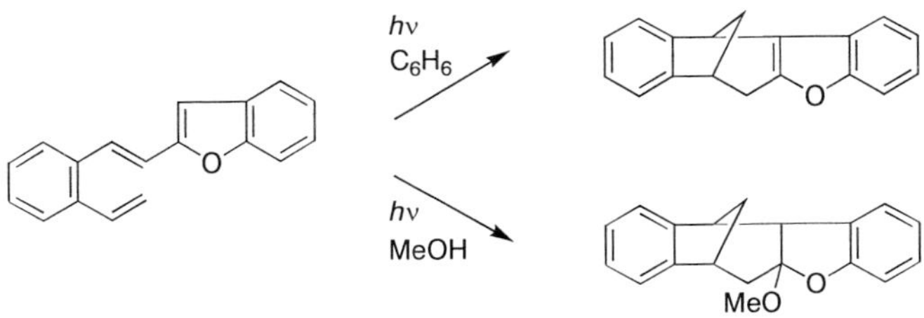

A much improved synthesis of a dioxino[2,3-*e*]indolemethanol was described. This procedure now allows the preparation of multigram quantities of the enantiomerically pure compound. Application of this methodology to different 5-hydroxy heterocycles enables access to furo[3,2-*f*][1,4]benzodioxin, thieno[3,2-*f*][1,4]benzodioxin and 1,4-dioxino[2,3-*e*]indazole ring systems <99S1181>.

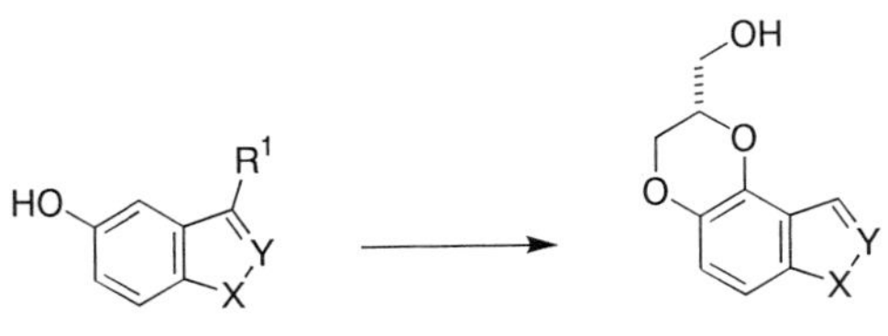

X = O, S, NH Y = C-CO_2Et, CH, N R^1 = H, Cl

Diels-Alder reactions of benzo[*b*]furan-4,5-diones and benzo[*b*]furan-4,7-diones have been reported <99H(50)1137>. The activation of a C-O bond in a $Mn(CO)_3^+$ complexed benzo[*b*]furan was examined <99AG2343>. The enzyme-catalyzed dealkylation of (+)-marmesin to the phototoxic psoralene (and acetone) has been investigated <99AG413>. A high yielding synthesis of bidentate bisoxazoline DBFOX/Ph [(*R*,*R*)-4,6-dibenzofurandiyl-2,2'-bis(4-phenyloxazoline)] was developed. DBFOX/Ph was subsequently tested in enantioselective conjugate radical additions onto 3-(3-phenyl-2-propenoyl)-2-oxazolidinone <99TA2417>.

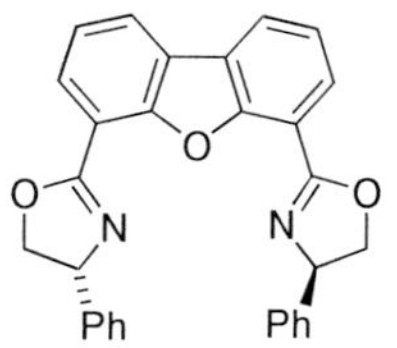

R,R-DBFOX/Ph

5.3.3 SYNTHESIS

5.3.3.1 Furans, Di- and Tetrahydrofurans

A simple synthesis of 3-substituted and 2,3-disubstituted 4-chlorofurans was accomplished. It involves a CuCl/bipy-catalyzed regioselective cyclization of 1-acetoxy-2,2,2-trichloroethyl allyl ether followed successively by dechloroacetoxylation with Zn dust and tandem dehydrohalogenation-aromatization with *t*BuOK/18-crown-6 <99CC2267>.

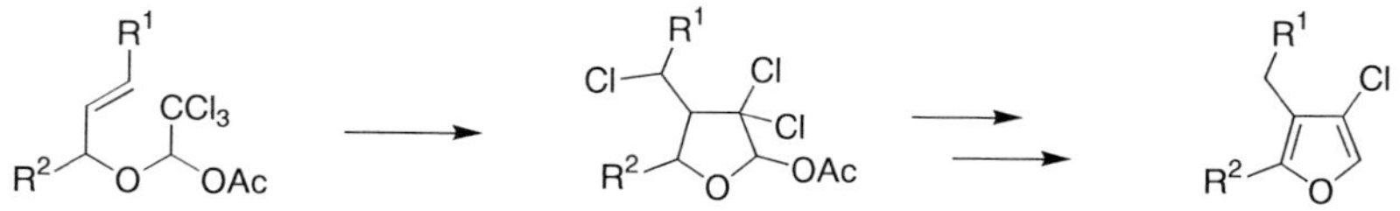

A facile three step process, under mild conditions, for the synthesis of 2,4-diarylfurans with the same or different substituents on the two aryl rings, starting from benzaldehydes and acetophenones, employing Haller-Bauer type cleavage of 2-aroyl-3,5-diarylfurans was described <99S61>.

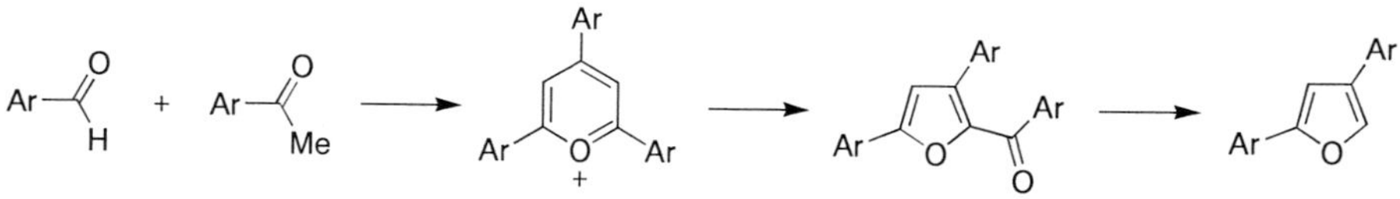

Treatment of D-glucal with a catalytic amount of $Sm(OTf)_3$ or $RuCl_2(PPh_3)_3$ in the presence of one equivalent of water afforded an optically active furan diol <99CC965>.

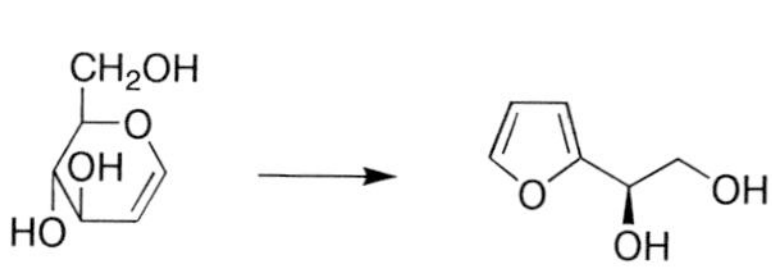

Conjugate addition of alkynylborates to α,β-unsaturated ketones, followed by acid-catalyzed cyclization of the resulting γ-alkynyl ketones afforded trisubstituted furans in moderate to high yields <99T14233>.

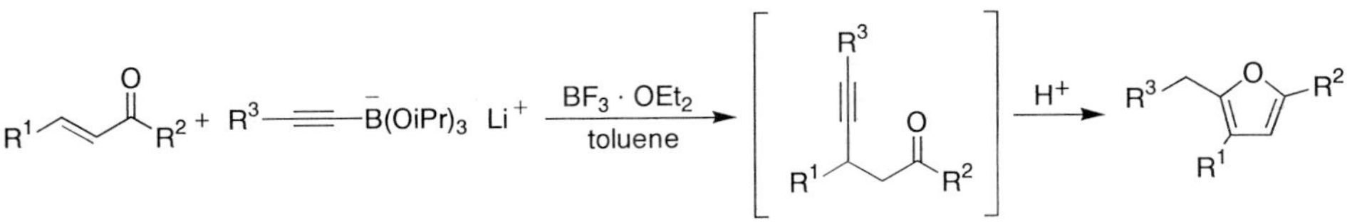

The preparation of 5-arylfurfurals (and arylthiophene-2-carboxaldehydes) *via* Pd-catalyzed C-C bond formation in aqueous media was carried out <99OL965>. Full details have been reported for the Pd-catalyzed cycloisomerization of (*Z*)-2-en-4-yn-1-ols (a facile synthesis of a variety of substituted furans) <99JOC7687>

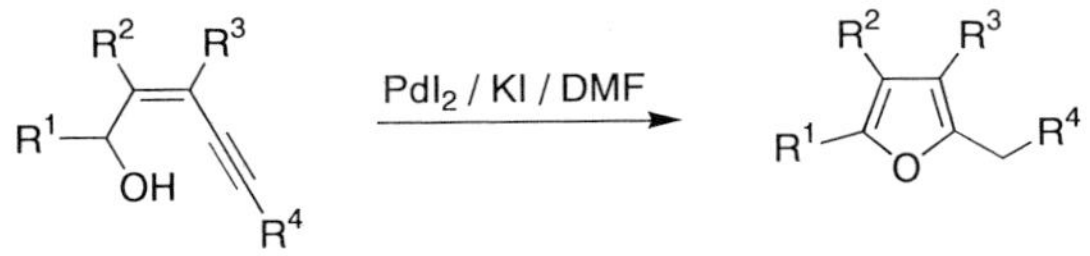

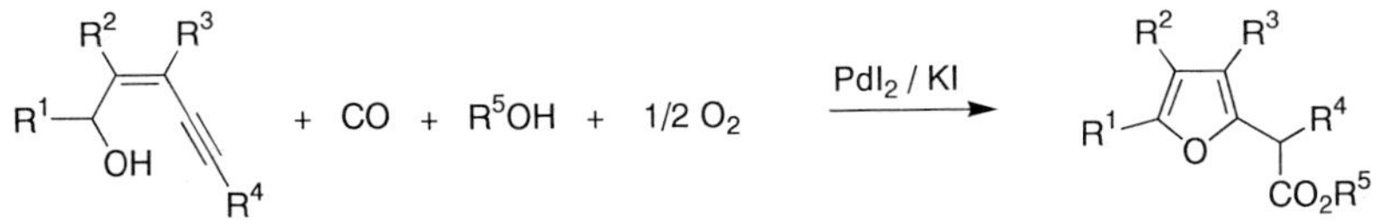

as well as for the synthesis of furan-2-acetic esters *via* a carbonylation reaction <99JOC7693>. The addition of vinyl and aryl Grignard reagents to propargyl alcohols with subsequent reaction of the intermediate Mg chelate with nitriles gave access to furans <00TL17>.

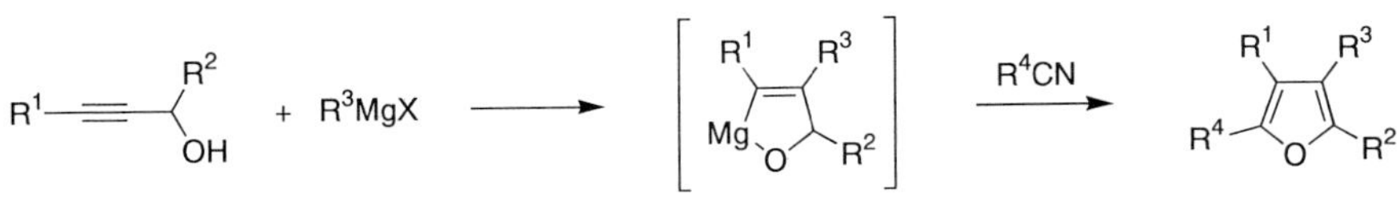

A convenient synthesis of 3,3'-difurans was presented. Treating alkynones with $Pd(PPh_3)_4$ (THF, Et_3N) at room temperature gave 2,5-disubstituted furans, but under similar conditions with $PdCl_2(PPh_3)_2$ by a tandem dimerization and cyclization 3,3'-difurans were obtained predominantly <99TL4841>.

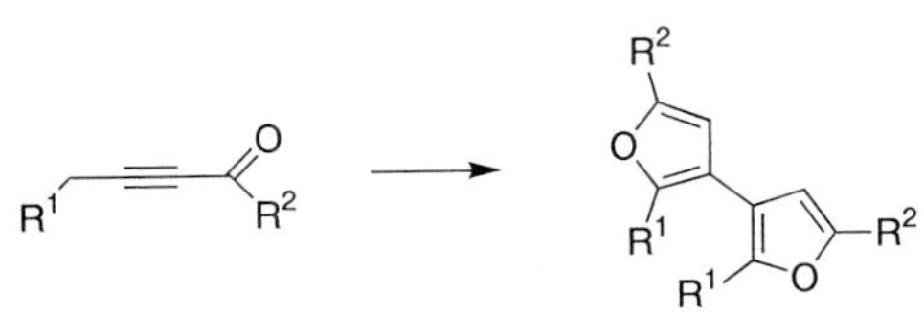

A Pd-catalyzed tandem cyclization and dimerisation of (*Z*)-3-iodo-3-alken-1-ones opens the way to 3,3'-difuran derivatives, too <99JOC1738>.

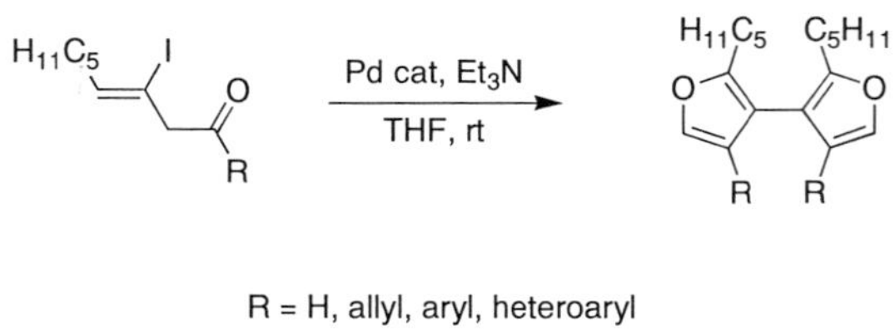

Treatment of 2,3-disubstituted 1,4-dioxenes with a catalytic amount of camphorsulfonic acid in dichloromethane at room temperature afforded substituted furans <99TL2521>.

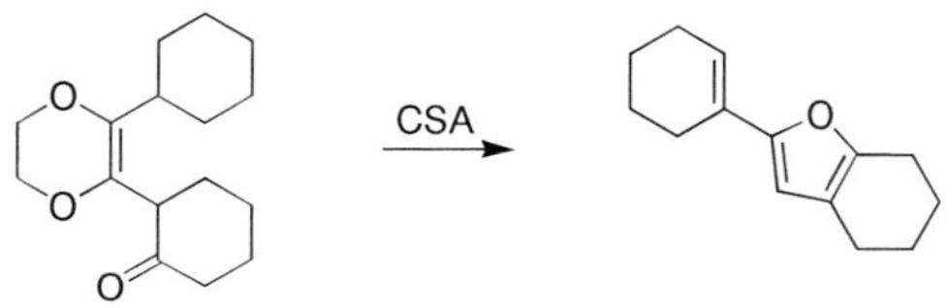

Furans having a double bond at position 2 can be obtained by a phosphine-initiated cyclization of enyne ketones in the presence of an aldehyde. The reaction may involve the 1,6-addition of the phosphine toward the enyne, ring closure and Wittig reaction of the ylide with the aldehyde <99TL3753>.

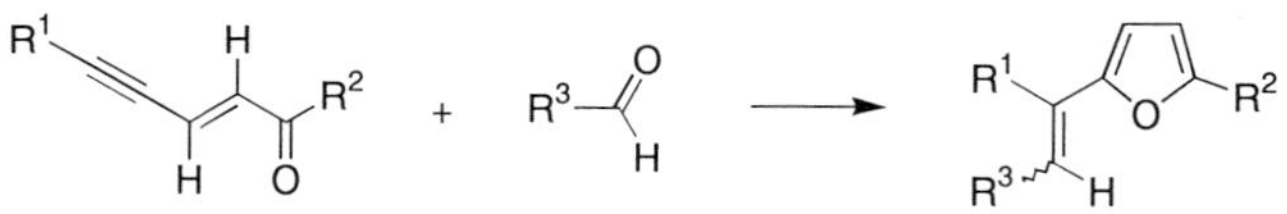

The diastereoselective synthesis of enantiopure 5-[2-(alkoxyalkyl)-1-(hydroperoxypropyl)]-3-alkoxycarbonyl-2-alkylfurans has been reported <99TA2023>.

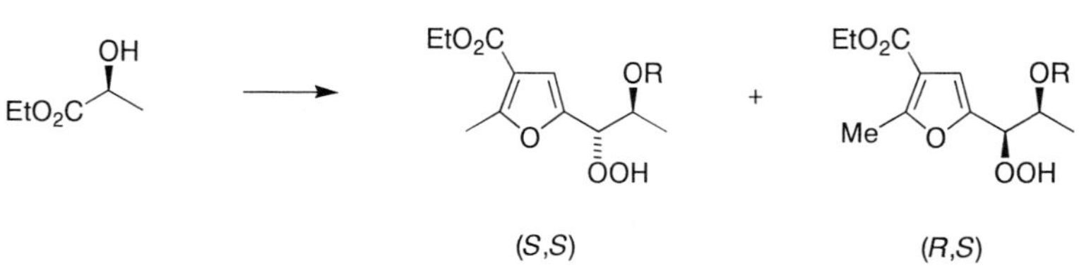

Highly enantioenriched 4-alken-1-yn-3-ol moieties present in many bioactive acetylenic metabolites from sponges have been efficiently obtained by reduction of the parent 1-trimethylsilyl-4-alken-1-yn-3-one **18** with Alpine-borane or with $BH_3{\cdot}SMe_2$ in the presence of chiral oxazaborolidines, followed by desilylation of the resulting alcohol. This strategy has been applied to the first stereoselective synthesis of petrofuran **19** <99SL429>.

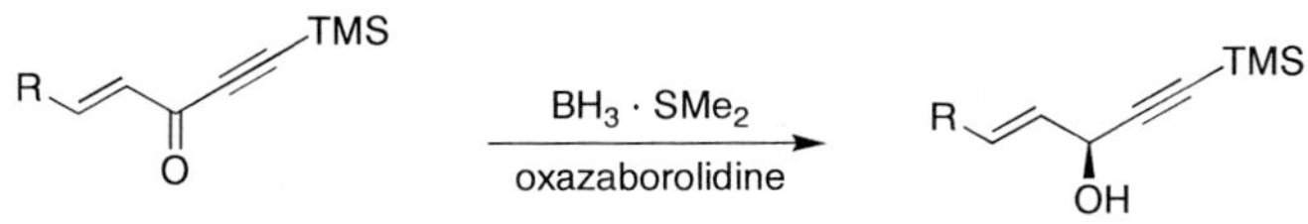

18

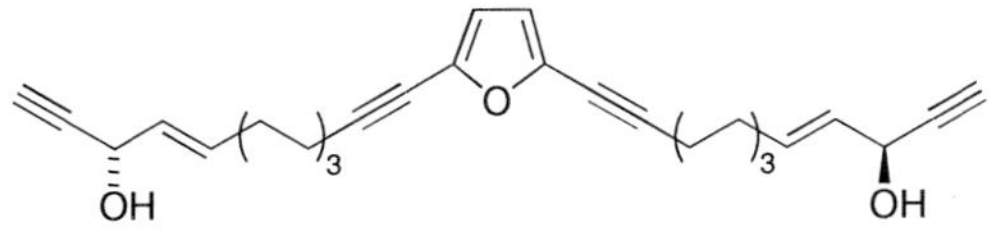

19

The reaction of alkynyl ketones with strong potassium base (KO*t*Bu, KHMDS, KH) yields substituted furans in moderate to good yields <99T2847>.

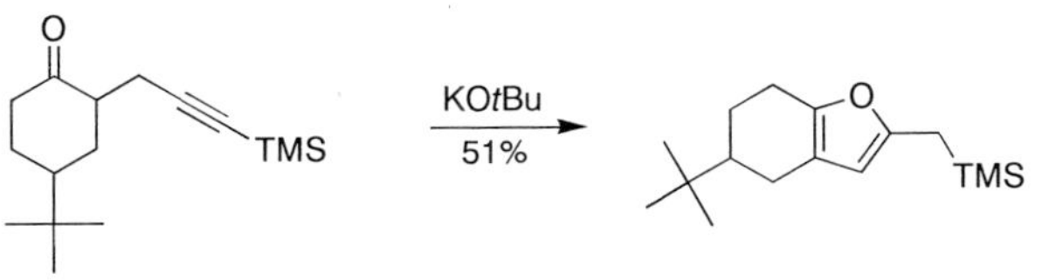

A convenient synthesis of 2-furylsugars starting from readily available sugar phosphonates and TBDPS-protected glycolaldehyde has been reported <99SL313>.

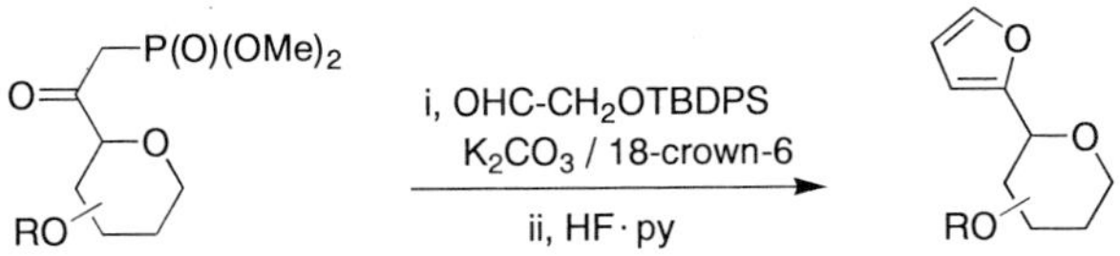

A flexible entry into 2,4-disubstituted furan derivatives through condensation of the sulfur ylide derived from **20** with aldehydes, Pd-catalyzed opening of the vinyl oxirane thus formed, and a final oxidative cyclization of the furan ring was reported. Its utility was exemplified by the first total synthesis of the marine natural product ircinin-4 (**21**) <99SL29>.

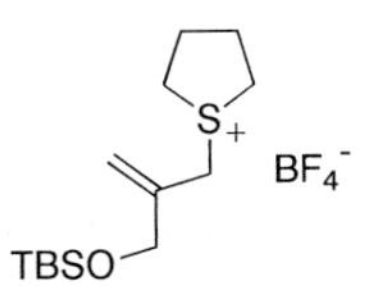

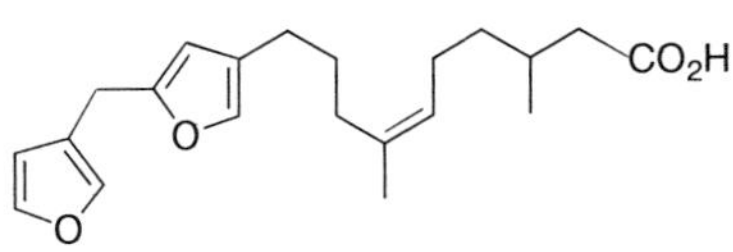

ircinin-4

20 **21**

An efficient synthesis of 5-arylfurfurals (and 5-aryl-2-vinylfurans) by Pd-catalyzed cross coupling strategies was reported. Thus, treating protected furfural with n-BuLi/Bu_3SnCl and then with an aryl halide under the catalytic action of $Pd_2(dba)_3{\cdot}CHCl_3$ (5%) yields 5-arylfurfurals. Various functional groups on the arene ring are tolerated (OR, CH_2OH, CO_2Me) <99TL4769>.

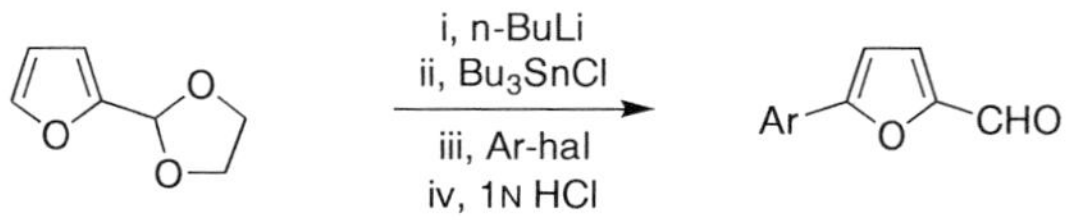

5-Azidopyrazoles of type **22** extrude nitrogen upon heating (toluene, 95 °C) to give the corresponding nitrenes, which immediately rearrange to produce a mixture of furans and pyrazoles <99JOC2814>.

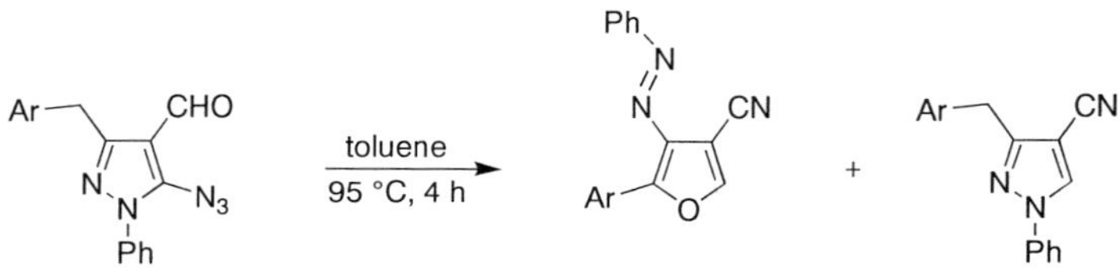

22

Again there is a number of interesting syntheses of di- and tetrahydrofuran derivatives. Due to space limitations only a few examples can be mentioned. The transformation of enantiomerically enriched butyn-1-ol monoesters into 2,2,5-trisubstituted 3-acyloxy-2,5-dihydrofurans with complete stereospecificity was achieved by a Ag(I)-mediated rearrangement *via* allenic intermediates. The starting materials were prepared by an enantioselective reduction of the corresponding ketone, followed by acylation <99BCJ279>.

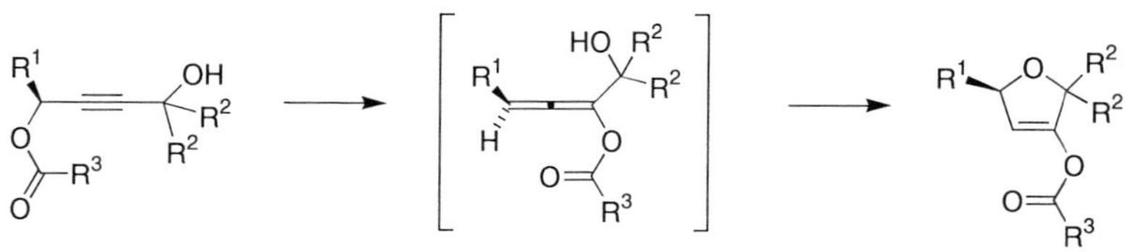

Various types of diethylphosphono-substituted α-allenic alcohols have been transformed to the corresponding 4-(diethylphosphono)-2,5-dihydrofurans by treatment with a catalytic amount of silver(I) nitrate under a nitrogen atmosphere <99S463>.

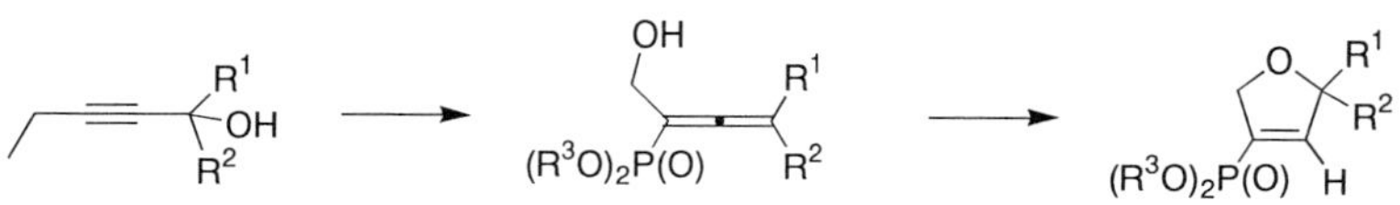

Reaction of oxazirconacyclopentenes (*e.g.*, **23**) with propynoates provides a new pathway for the formation of 2,5-dihydrofuran derivatives <99TL2375>.

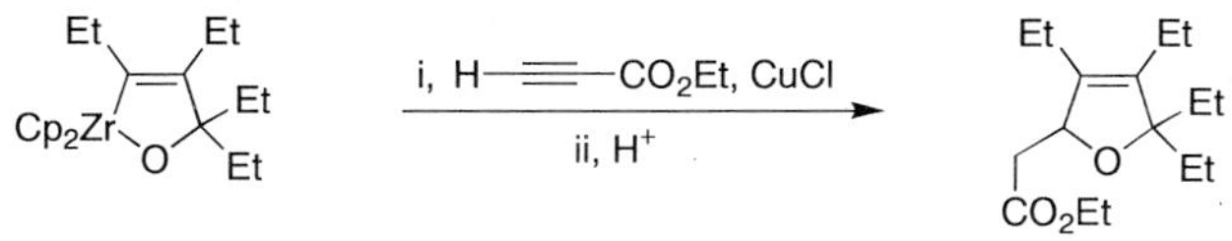

23

The Ru-catalyzed cycloisomerization of heptadiyne **24** in the presence of a functionalized terminal olefin offers an access to dihydrofurans **25** <99CC237>.

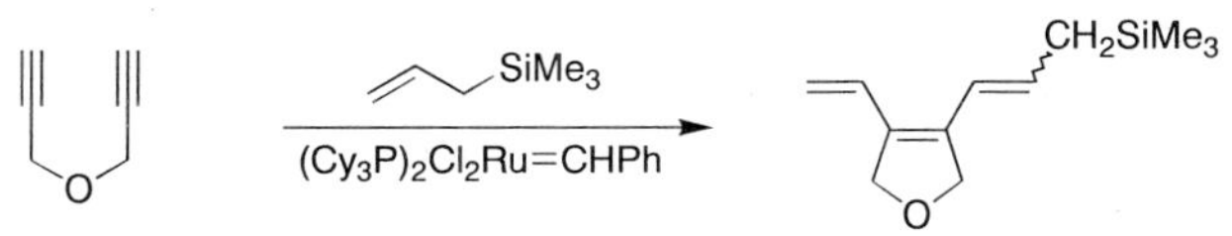

24 **25**

Regio- and enantioselective Heck reactions of 2,3-dihydrofuran with aryl and alkenyl triflates in the presence of the chiral ligand (*R*)-BITIANP provides 2-substituted 2,3-dihydrofurans with complete regioselectivity, high enantioselectivity (86-96% ee) and good yields (76-93%) <99CC1811>. A catalytic oxyselenylation-deselenylation reaction of alkenes offers a stereoselective one-pot conversion of alkenes into 2,5-dihydrofurans <99EJO797>.

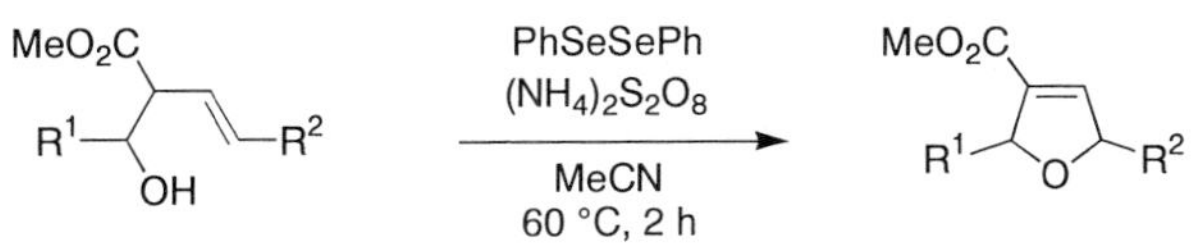

A Tellurium and iodine promoted cyclofunctionalization of alkenyl substituted β-ketoesters was reported >99SL567>.

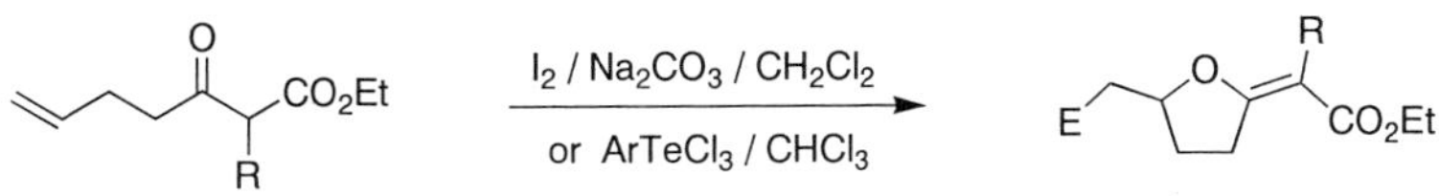

R = H, Me
E = ArTeCl$_3$, I

3-Substituted furan-2(5*H*)-ones are available regioselectively by reductive carbonylation of alk-1-ynes <99TL989>.

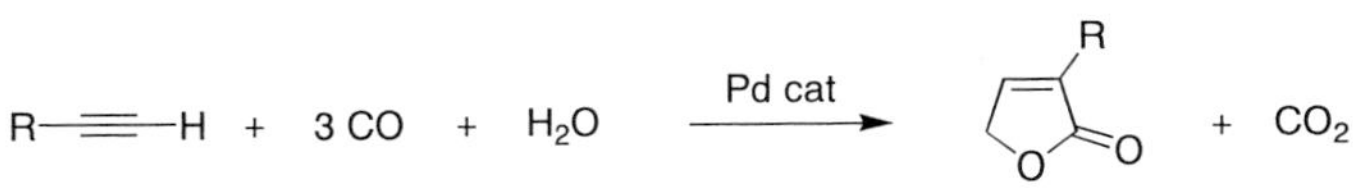

R = alkyl, aryl

For radical cyclizations (on solid support) for the synthesis of γ-butyrolactones see <99TL3411>. For γ-lactone formation in the addition of benzenesulfonylbromide to diene and enyne esters see <99JOC2066>. On irradiation a cyclic β-dicarbonyl iodonium ylide reacts with substituted 1,3-dienes in acetonitrile to give substituted dihydrofurans in moderate to high yields, presumably *via* a stepwise mechanism <99SL1925>. An asymmetric synthesis of tetrasubstituted tetrahydrofuran, 2-epigoniothalesdiol, employing stereoselective hydrogenation was reported <99SL1969>.

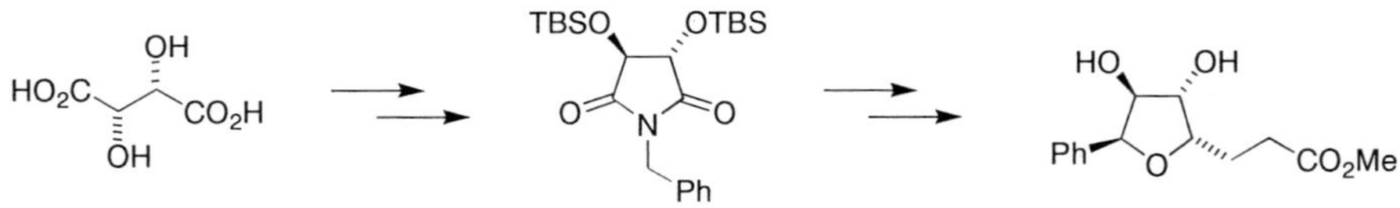

An extremely facile coupling reaction between arylmagnesium compounds and THF by means of an iodoalkane-EtMgBr system provides 2-aryltetrahydrofurans <99SL1582>. Treatment of 4-phenylsulfanyl-1,5,7-triols with TsCl in pyridine gives substituted tetrahydrofurans by [1,4]-SPh participation *via* a five-membered sulfonium salt and then capture by a tethered nucleophile. Acid catalyzed rearrangement gives other THFs by [1,2]-SPh migration and cyclization <99SL1215>.

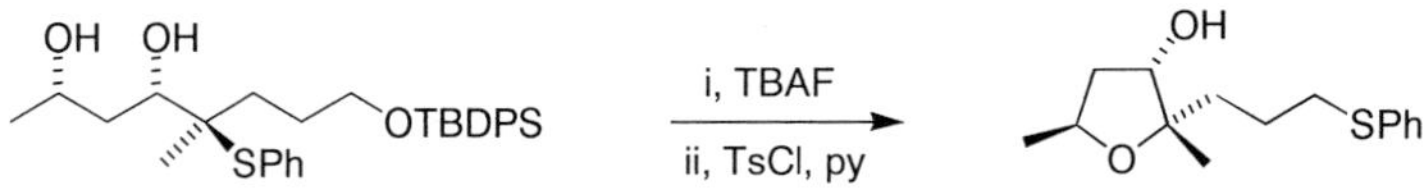

Regioselective ring-opening and cross-coupling metathesis of 2-substituted 7-oxanorbornenes (for a review of 7-oxanorbornanes and related compounds see <99T13521>) offers a new stereoselective entry into trisubstituted tetrahydrofurans <99JOC9739>.

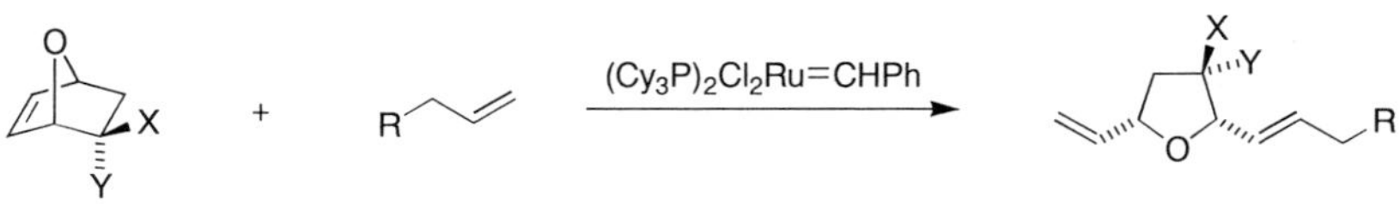

X, Y = O; X = H, Y = OH, OAc, OCOR', etc.

R = OAc, OBn

Sulfonyl-substituted homoallylic alcohols undergo 5-*endo-trig* cyclization reactions on treatment with base, with cyclization stereoselectively depending on double bond geometry, to give substituted tetrahydrofurans <99T13471>.

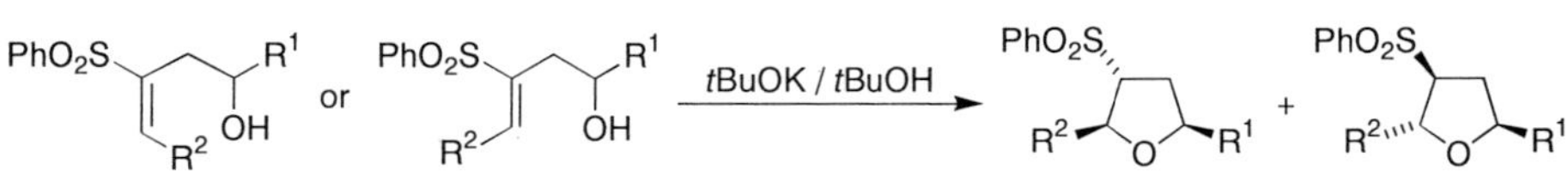

Radical cylization of a β-alkoxymethacrylate leads to the stereoselective preparation of the benzylether of (+)-methyl nonactate, demonstrating "2,5-*cis*" selectivity in the radical cyclization step and "*threo*" selectivity in the hydrogen abstraction step <99OL1127>.

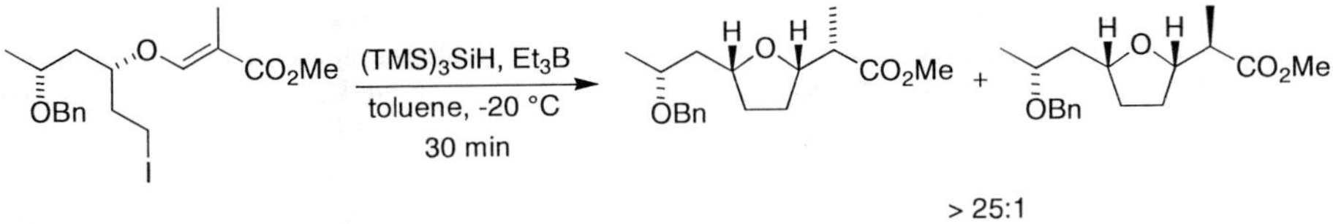

A double intramolecular $S_{N'}$ *O*-cyclization for the stereoselective synthesis of the bistetrahydrofuran core of Acetogenins has been achieved <99JOC2259>.

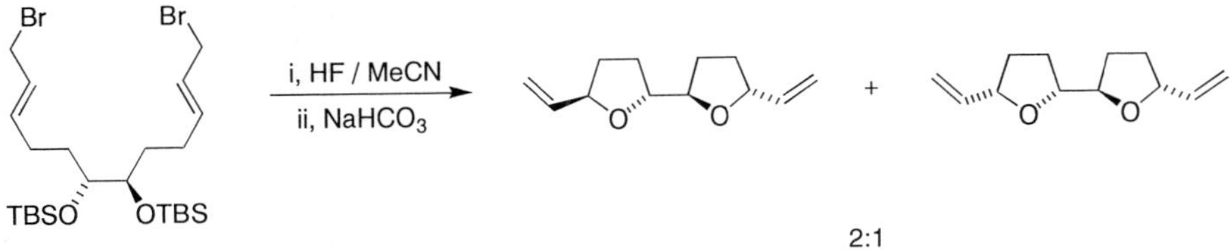

Titanium tetrachloride promoted coupling of ethyl glyoxylate and dihydrofuran with subsequent reaction of the intermediate with a nucleophilic trapping agent provides a new route to functionalized tetrahydrofurans <99TL1083>.

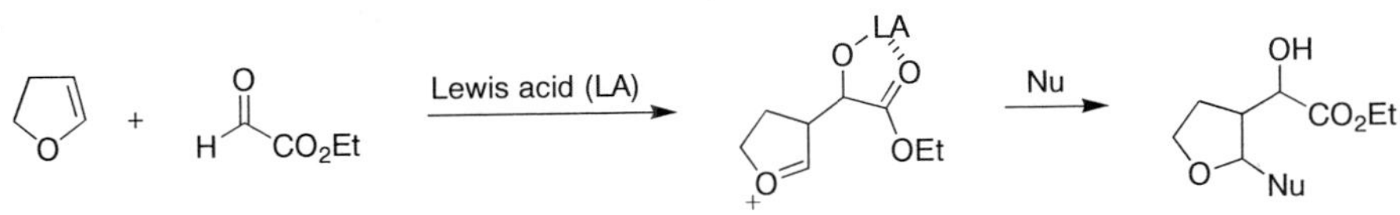

The annulation reaction of bis(trimethylsilyloxy)enol ether **26** with 1,4-pentanedione in the presence of a catalytic quantity of trimethylsilyltriflate at –50 °C in dry dichloromethane furnishes oxabicyclo[3.2.1]heptanone **27** in 92% yield. Compound **27** is used as starting material for (±)-davanone, a major component of the South Indian plant *Artemisia pallens* <99T617>.

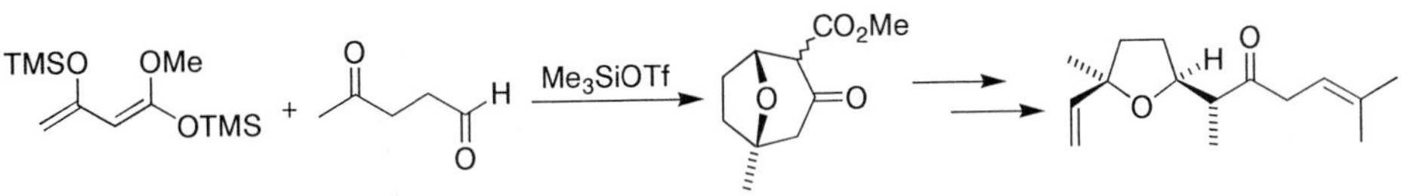

26 **27**

Several further syntheses of tetrahydrofurans have been reported, *e.g.*, a stereoselective synthesis of (±)-botryodiplodin (application of the Ueno-Stork reaction) <99TL3375>, see also <99TL3371>, an enantioselective synthesis of (-)-kumausallen <99AG3370>, a synthesis of *meso*- and *d,l*-2,2'-methylenebis(tetrahydro-2-furanmethanol) (as potential building block for the construction of ionophores housing spirotetrahydrofuranyl motifs) <99H(50)27>, the chemical conversion of a labdane-type diterpenoid isolated from the liverwort *Porella perrottetiana* into (-)-Ambrox <98H(49)315>, an enantioselective construction of the tetrahydrofuran (and tetrahydropyran) fragments of the antitumor agent mucocin <99TL1253>, the total synthesis of two acetogenins, squamocin A and D by a multiple Williamson reaction <99TL5979>, the preparation of the "Eastern" part (C8-C18 fragment) of pamamycin-607 <99EJO2303>, synthetic studies towards the B,C,D,E fragment of antibiotic CP44,161 <99SL295>, an enantioselective formal total synthesis of phytotuberin <99SL1395>, bis(tetrahydrofurans) from carbohydrates *via* iodoetherification and iodine substitution <99S330>, and a stereoselective synthesis of tetrahydrofurans (and tetrahydropyrans) by acid-catalyzed cyclization of hydroxy selenides and hydroxy sulfides <99T14097>. The oxidative cyclization of a monoprotected diol to a tetrahydrofuran is a main step in the synthesis of gigantetrocin A <99TA667>.

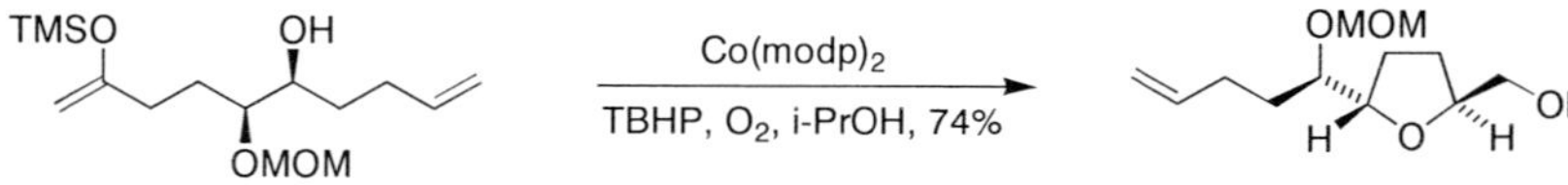

The synthesis of (6*S*,7*S*,9*R*,10*R*)-6,9-epoxy-nonadec-18-ene-7,10-diol **28**, a marine epoxy lipid isolated from the brown algae *Notheia anomala*, by an oxiranyl anion strategy, was reported. One main step is the construction of the furan ring <99TL731>.

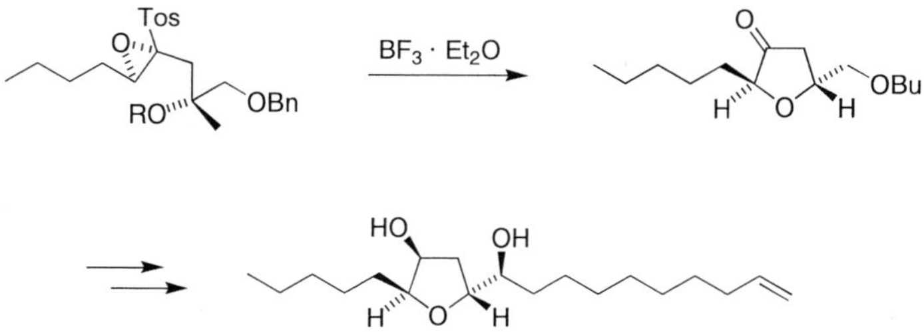

28

The PET generated electrophilic species $[PhSeSePh]^{+\bullet}$ was found to effect stereoselective oxyselenylation of 1,n-diolefins leading to a novel strategy for the synthesis of α,α'-*trans*-dialkyl cyclic ethers in good yields <99SL1257>.

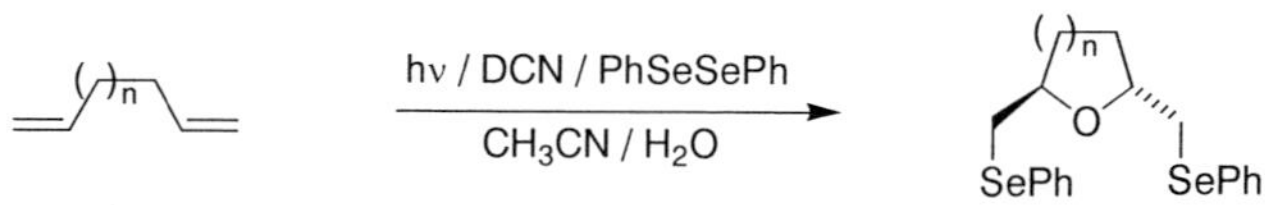

n = 1,2,3

It was reported that Pd(0)-catalyzed coupling reactions of allenic alcohols, amines and acids with hypervalent iodonium salts afforded cyclized heterocyclic tetrahydrofurans, tetrahydropyrans, pyrrolidines, piperidines, or lactones under mild conditions <99SL324>. Intramolecular 1,5-hydrogen atom transfer radical cyclization reaction of pyrrolidine derivatives was examined. Reaction of 3,4-diallyloxy-*N*-(*O*-bromobenzyl)pyrrolidine gave hexahydrofuro[3,2-*b*]pyrroles together with dimeric compounds <98H(49)305>. Heteroatom radical and heteroatom-mediated addition-cyclizations of several substrates including various types of multiple bonds have been reviewed from a synthetic point of view <99H(50)505>. It is well known that β-alkoxyacrylates are excellent precursors for the stereoselective preparation of *cis*-2,5-disubstituted tetrahydrofurans (and pyranes) *via* radical cyclizations. In an extension of this methodology a double radical cyclization strategy for the preparation of fused derivatives was reported <99JCS(P1)3395>. Annulated tetrahydrofurans (fused to isoxazolidines) have been prepared by intramolecular silyl nitronate olefin cycloaddition starting from nitroolefins in a one-pot protocol (Michael reaction, silylation, cycloaddition, and desilylation) <99EJO2689>. Factors controlling the cyclization of 1,4-diols with PhS migration to give THPs rather than THFs have been reassessed <99SL1211>.

5.3.3.2 Benzofurans and Related Compounds

Benzofuran derivatives are an important class of heterocyclic compounds that are known to possess interesting biological properties. Compounds of this type find wide application in many fields of chemistry and agriculture. Therefore the development of simple, fast and flexible syntheses remains a challenge for organic chemists. Ley and co-workers reported an efficient route to 3-phenylbenzofurans. Bromination of acetophenone by polymer supported pyridinium bromide perbromide, substitution of the bromide by phenols using 1,5,7-triaza-bicyclo[4.4.0]dec-5-ene and cyclodehydration of the resulting α-phenoxyacetophenone using Amberlyst 15 afforded pure products without need for any chromatographic purification step <99JCS(P1)2421>.

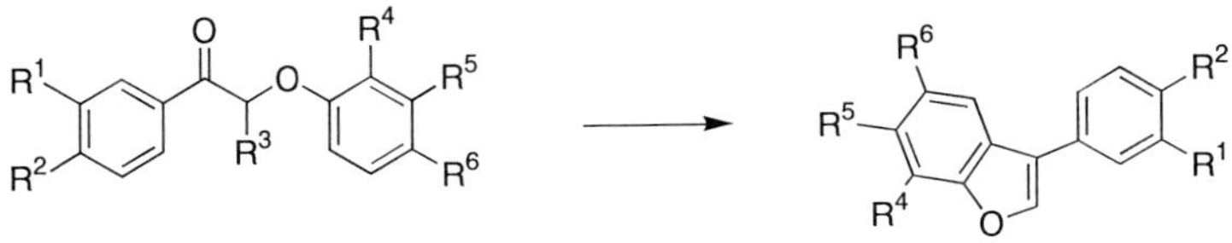

Carbonylative cyclization of *o*-hydroxytolans yields 3-acylbenzo[*b*]furans <99SL1079>.

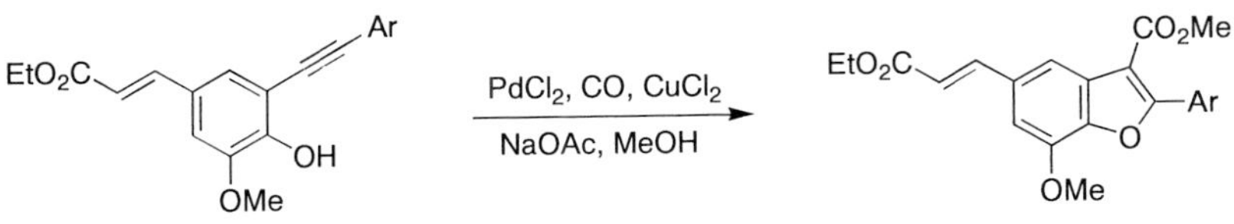

A new approach to 2,3-disubstituted benzo[*b*]furans from *o*-alkynylphenols *via* 5-*endo-dig*-iodocyclization reaction was reported. The iodinated compounds are starting materials for a variety of substituted benzo[*b*]furans <99SL1432>.

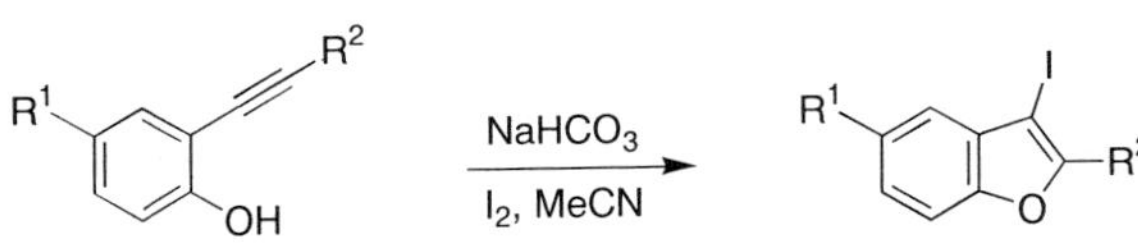

A highly efficient synthesis of 1-alkylidene-1,3-dihydroisobenzofurans and 3-alkylidene isobenzofuran-1(3*H*)-ones (phthalides) through Pd-Cu-catalysis using acetylenic carbinols as synthons was reported <99SL456>.

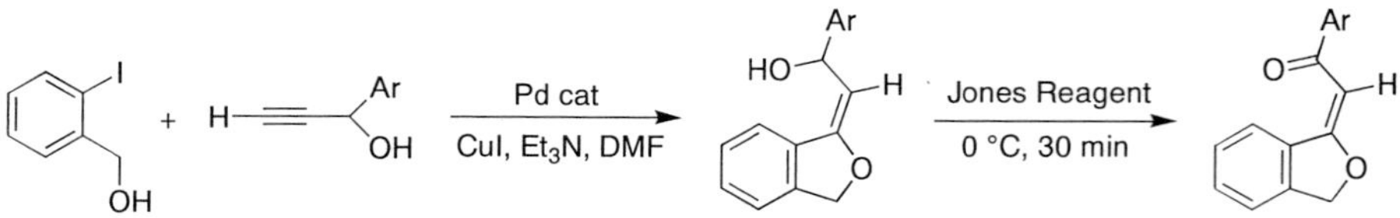

Benzo[*b*]furans have also been prepared by a photocyclization of styrylfurans <99AG2753>. The first direct transformation of 2,2'-dihydroxychalcones into coumestans was reported. Thus, treatment of **29** with $Tl(NO_3)_3$ in methanol and subsequently with methanol/HCl in the presence of oxygen yields **30** <99T861>.

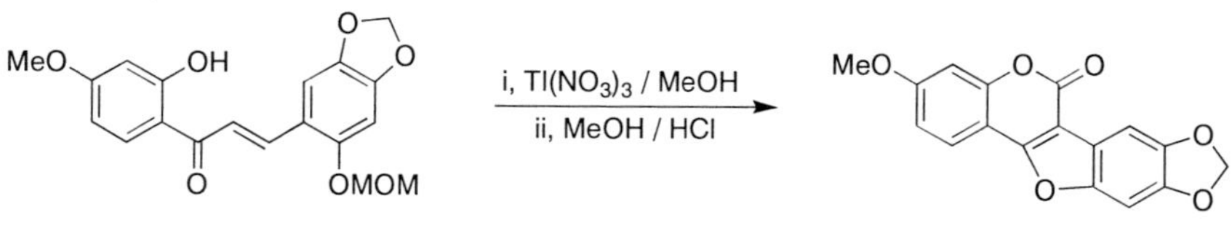

29 **30**

Rh-mediated dipolar cycloadditions of diazoquinolinediones **31** with alkenes and alkynes have been investigated. Because of the unsymmetrical nature of the diazo compounds, both linear and angular furoquinoxaline products are possible. In most cases a mixture of regioisomers was obtained. This methodology has been used for the synthesis of naturally occurring alkaloids, *e.g.*, isodictamnine **32** (R = H) <99JOC3642>.

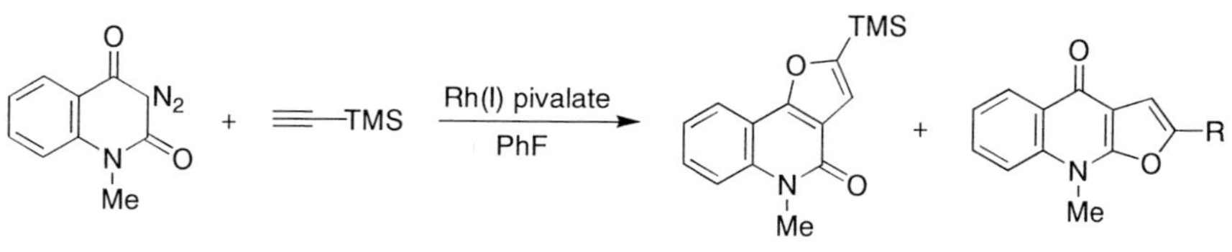

31 **32**

77:23 (R = TMS)

Unusual ring contraction reactions of 3*H*-pyrano[2,3-*c*]quinolin-5(6*H*)-ones to furo[2,3-*c*] quinolin-4(5*H*)-ones have been observed <99H(51)471>, see also <99H(51)2399>. It has been found that the intramolecular nucleophilic addition of 2-allylphenol can be catalyzed by $RuCl_3 \cdot nH_2O/AgOTf–PPh_3–Cu(OTf)_2$ to afford 2,3-dihydro-2-methyl-benzo[*b*]furans in good yield <98CL1083>. Dihydrobenzofurans are also accessible by an anodic [3+2] cycloaddition of phenols with unactivated alkenes <99JOC7654>. A one-step synthesis of ethyl 1,2-dihydronaphtho[2,1-*b*]furan-2-carboxylates from substituted 2-naphthols and ethyl 2,3-dibromopropanoate was described <99S1241>.

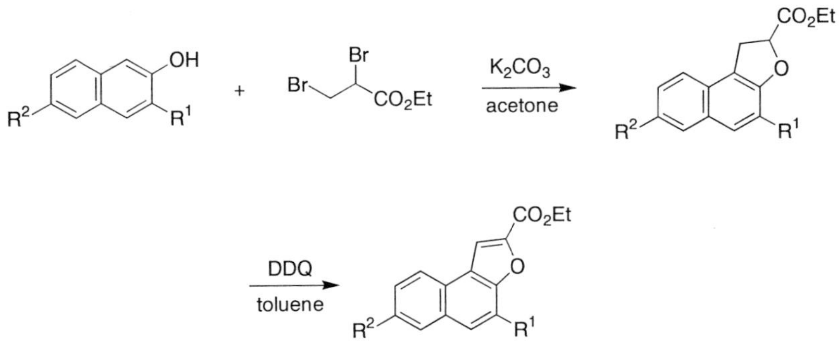

CAN-mediated formation of furo[3,2-*c*]-, furo[2,3-*b*]pyranones and furo[3,2-*c*]quinolinones was reported <99H(51)2881>. The electrochemical reduction of a series of 2-haloaryl ethers containing allyl and propargyl groups under CO_2 atmosphere allows the synthesis of benzofuranacetic acid derivatives. This novel intramolecular cyclization-carboxylation reaction was carried out in single compartment cells and was catalyzed by [Ni(cyclam)Br_2] <99EJO1885>.The reaction of enolates of 1,3-dicarbonyl compounds with 3,4-dibromo-2-butanone afforded hydrofuran derivatives in the presence of DBU in THF. This reaction was applied to the one-pot synthesis of antitumor naphthofuran natural products <99H(51)497>. 3-Aryl-3*H*-benzofuran-2-ones have recently gained an industrial importance as an extremely powerful heat stabilizer family for polymers. A versatile new synthesis of these compounds was reported starting with an acid-catalyzed cyclocondensation of phenols with glyoxylic acid to give the 3-hydroxy-3*H*-benzofuran-2-ones. Treating these compounds with aromatic or heteroaromatic hydrocarbons under Friedel-Crafts conditions yielded the corresponding 3-aryl derivatives <99SL863>.

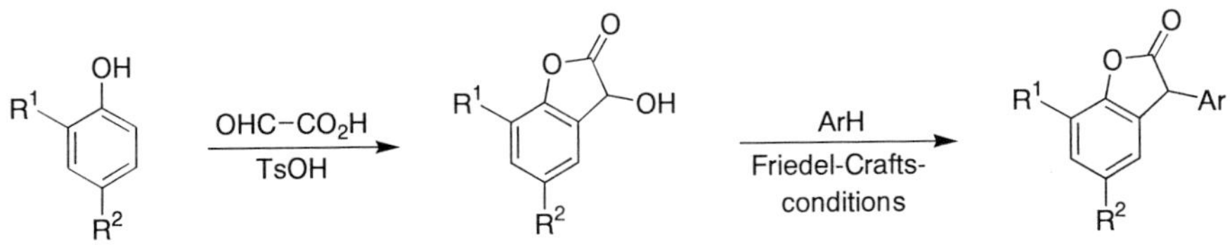

For further work in the field of benzofurans see <99TA1521> (synthesis of TAK-218 using (*R*)-2-methylglycidyl tosylate as a chiral building block), <99SL495> (heterocyclic rubicene analogues) and <99S751> (synthesis of benzofuro[2,3-*b*]benzofuran derivatives under Hoesch reaction conditions).

5.3.4 MISCELLANEOUS

The synthesis of furan-(thiophene-, benzo[*b*]thiophene-)bridged macrocycles of 4,4'-bipyridine was reported <99T4709>.

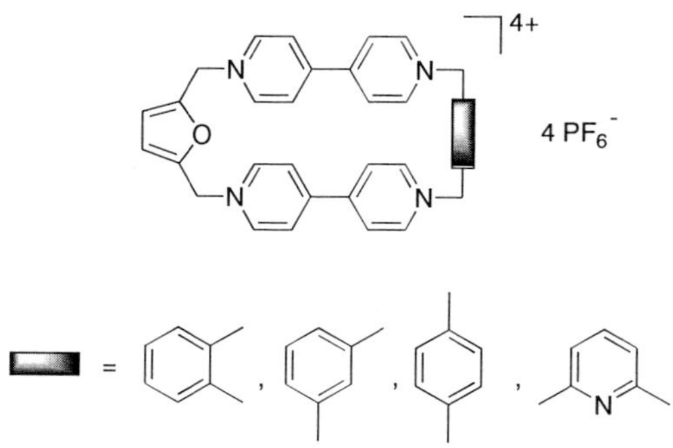

The preparation of bis(thienyl)furans of type **33** and **34** has been descibed. These compounds have been proposed as suitable precursors for electroactive molecular species and conducting polymers <99JOC6418>.

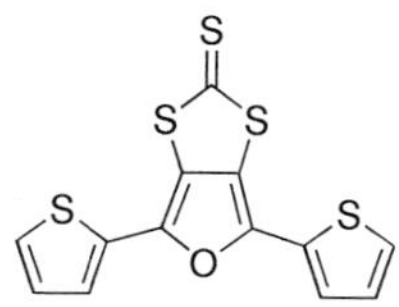

33

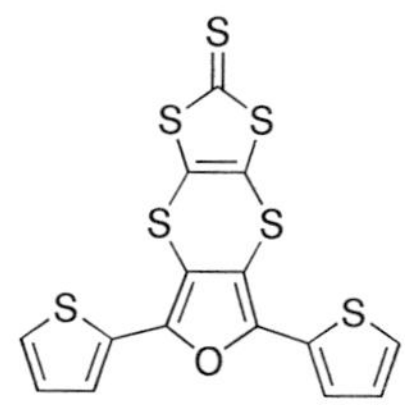

34

Syn and *anti* conformational isomers of 3-furylchlorocarbene have been characterized in low-temperature matrices by IR, UV/VIS, chemical trapping and calculational modeling (B3LYP/6-31G**). Displaying significantly different electronic spectra, the two isomers could be photochemically interconverted <99OL1091>. It has been observed that photochemically generated 2-furylcarbenes suffer from a ring opening reaction to give alkynones. The effect of substituents upon this reaction has been studied computationally by G2 (MP2, SVP) <99JOC9170>. A synthesis of new (phthalocyanine)Ni complexes was reported by the Hanack group. *In situ* generated reactive isobenzofurans are intermediates in these syntheses <99EJI693>. The first example of a Wittig rearrangement of furfuryl ethers and its application to the preparation of 3-(2-furyl)-3-hydroxy-2-methylpropionates has been reported <99CC2263>. Triethylborane-induced radical reduction of organic halides with tri-2-furanylgermane provides the corresponding reduced compounds in good yields <99SL1415>. PM3 calculations have been carried out to study the origin of stereoselectivity in the intramolecular Michael reaction for a highly efficient construction of the B and A rings of halichondrin B <98H(49)89>. A DFT study of the geometry and vibrational spectrum of benzo[*b*]furan (and other heterocyclic analogues) has been published <99SA(A)2437>. A theoretical study on the molecular mechanism of the domino cycloadditions between dimethyl acetylenedicarboxylate and naphthaleno- and anthracenofuranophanes has been published <99JOC3026>. It has been investigated whether density functional theory contributes to an understanding of excited states of unsaturated organic compounds. Besides furan other compounds have been studied (tetrazine, cyclopentadiene, pyrrole, thiophene, acetone, and a dipeptide) <99MP859>.

Acknowledgment: Help and assistance of Timm Graening is gratefully acknowledged.

5.3.5 REFERENCES

97T9313	E. Zaballos-Garcia, M. E. Gonzalez-Rosende, J. M. Jorda-Gregori, J. Sepúlveda-Arques, W. B. Jennings, D. O'Leary, S. Twomey, *Tetrahedron* **1997**, *53*, 9313.
98CL1083	K. Hori, H. Kitagawa, A. Miyoshi, T. Ohta, I. Furukawa, *Chem. Lett.* **1998**, 1083.
98H(49)89	O. Yonemitsu, T. Yamazaki, J.-i. Uenishi, *Heterocycles* **1998**, *49*, 89.
98H(49)305	M. Ishizaki, H. Takano, O. Hoshino, *Heterocycles* **1998**, *49*, 305.
98H(49)315	T. Hashimoto, K. Shiki, M. Tanaka, S. Takaoka, Y. Asakawa, *Heterocycles* **1998**, *49*, 315.
98H(49)355	Y. Zhu, W. Y. Yoshida, M. Kelly-Borges, P. J. Scheuer, *Heterocycles* **1998**, *49*, 355.
98H(49)531	P. Bernardelli, L. A. Paquette, *Heterocycles* **1998**, *49*, 531.
99AG413	V. Stanjek, M. Miksch, P. Lueer, U. Matern, W. Boland, *Angew. Chem.* **1999**, *111*, 413.
99AG1341	S. Bäurle, S. Hoppen, U. Koert, *Angew. Chem.* **1999**, *111*, 1341.
99AG1528	B. A. Chauder, A. V. Kalinin, N. J. Taylor, V. Snieckus, *Angew. Chem.* **1999**, *111*, 1528.
99AG2057	E. J. Corey, F.-Y. Zhang, *Angew. Chem.* **1999**, *111*, 2057.
99AG2343	X. Zhang, E. J. Watson, C. A. Dullaghan, S. M. Gorun, D. A. Sweigart, *Angew. Chem.* **1999**, *111*, 2343.
99AG2753	T.-I. Ho, J.-Y. Wu, S.-L. Wang, *Angew. Chem.* **1999**, *111*, 2753.
99AG3370	P. A. Evans, V. S. Murthy, J. D. Roseman, A. L. Rheingold, *Angew. Chem.* **1999**, *111*, 3370.
99BCJ279	H. Saimoto, M. Yasui, S.-i. Ohrai, H. Oikawa, K. Yokoyama, Y. Shigemasa, *Bull. Chem. Soc. Jpn.* **1999**, *72*, 279.
99CC237	R. Stragies, M.Schuster, S. Blechert, *J. Chem. Soc., Chem. Commun.* **1999**, 237.
99CC713	P. D. Rao, C.-H. Chen, C.-C. Liao, *J. Chem. Soc., Chem. Commun.* **1999**, 713.
99CC965	M. Hayashi, H. Kawabata, K. Yamada, *J. Chem. Soc., Chem. Commun.* **1999**, 965.
99CC1811	L. F. Tietze, K. Thede, F. Sannicolò, *J. Chem. Soc., Chem. Commun.* **1999**, 1811.
99CC2263	M. Tsubuki, T. Kamata, H. Okita, M. Arai, A. Shigihara, T. Honda, *J. Chem. Soc., Chem. Commun.* **1999**, 2263.
99CC2267	R. N. Ram, I. Charles, *J. Chem. Soc., Chem. Commun.* **1999**, 2267.
99EJI693	B. Hauschel, R. Jung, M. Hanack, *Eur. J. Inorg. Chem.* **1999**, 693.
99EJO797	M. Tiecco, L. Testaferri, C. Santi, *Eur. J. Org. Chem.* **1999**, 797.
99EJO1395	G. Rassu, P. Carta, L. Pinna, L. Battistini, F. Zanardi, D. Acquotti, G. Casiraghi, *Eur. J. Org. Chem.* **1999**, 1395.
99EJO1757	H. N. C. Wong, *Eur. J. Org. Chem.* **1999**, 1757.
99EJO1885	S. Olivero, E. Duñach, *Eur. J. Org. Chem.* **1999**, 1885.
99EJO2011	G. Fontana, G. Savona, N. Vivona, B. Rodríguez, *Eur. J. Org. Chem.* **1999**, 2011.
99EJO2045	T. Bach, L. Krüger, *Eur. J. Org. Chem.* **1999**, 2045.
99EJO2303	G. Mandville, R. Bloch, *Eur. J. Org. Chem.* **1999**, 2303.
99EJO2655	A. Rodríguez, M. Nomen, B. W. Spur, J.-J. Godfroid, *Eur. J. Org. Chem.* **1999**, 2655.
99EJO2689	Q. Cheng, T. Oritani, T. Horiguchi, Q. Shi, *Eur. J. Org. Chem.* **1999**, 2689.
99EJO2949	U. Höller, G. M. König, A. D. Wright, *Eur. J. Org. Chem.* **1999**, 2949.
99EJO2977	A. Schrey, F. Osterkamp, A. Straudi, C. Rickert, H. Wagner, U. Koert, B. Herrschaft, K. Harms, *Eur. J. Org. Chem.* **1999**, 2977.
99H(50)27	L. A. Paquette, B. A. Skarns, P. A. Mooney, J. Tae, *Heterocycles* **1999**, *50*, 27.
99H(50)433	K. Kanai, N. Sano, T. Honda, *Heterocycles* **1999**, *50*, 433.
99H(50)505	T. Naito, *Heterocycles* **1999**, *50*, 505.
99H(50)595	M. Nakatani, *Heterocycles* **1999**, *50*, 595.
99H(50)653	O. Arjona, F. Iradier, R. Medel, J. Plumet, *Heterocycles* **1999**, *50*, 653.
99H(50)919	K. Nagasawa, N. Hori, H. Koshino, T. Nakata, *Heterocycles* **1999**, *50*, 919.
99H(50)981	W. Kuriyama, K. Ishigami, T. Kitahara, *Heterocycles* **1999**, *50*, 981.
99H(50)989	Y. Hano, M. Shimazaki, T. Nomura, S. Ueda, *Heterocycles* **1999**, *50*, 989.
99H(50)1137	P. Nebois, H. Fillion, *Heterocycles* **1999**, *50*, 1137.
99H(51)231	Y. Hano, T. Nomura, S. Ueda, *Heterocycles* **1999**, *51*, 231.

99H(51)243 K. Itoh, K. Kitoh, A. Sera, *Heterocycles* **1999**, *51*, 243.

99H(51)471 K. C. Majumdar, A. K. Kundu, P. Biswas, *Heterocycles* **1999**, *51*, 471.

99H(51)497 H. Hagiwara, K. Sato, T. Suzuki, M. Ando, *Heterocycles* **1999**, *51*, 497.

99H(51)795 M. Pohmakotr, P. Tuchinda, P. Premkaisorn, A. Limpongpan, V. Reutrakul, *Heterocycles* **1999**, *51*, 795.

99H(51)851 Y. Takemura, M. Ju-ichi, M. Omura, C. Ito, H. Furukawa, *Heterocycles* **1999**, *51*, 851.

99H(51)1225 S. Reck, C. Näther, W. Friedrichsen, *Heterocycles* **1999**, *51*, 1225.

99H(51)1355 M. Šindler-Kulyk, I. Škorić , S. Tomšić, Ž. Marinić, D. Mrvoš-Sermek, *Heterocycles* **1999**, *51*, 1355.

99H(51)1647 B. Esquivel, N. Ramírez-Dávalos, G. Espinosa-Peréz, *Heterocycles* **1999**, *51*, 1647.

99H(51)2183 C. Morel, C. Dartiguelongue, T. Youhana, J.-M. Oger, D. Séraphin, O. Duval, P. Richomme, J. Bruneton, *Heterocycles* **1999**, *51*, 2183.

99H(51)2399 K. C. Majumdar, A. K. Kundu, P. Biswas, *Heterocycles* **1999**, *51*, 2399.

99H(51)2723 M. T. Chhabria, C. J. Shishoo, *Heterocycles* **1999**, *51*, 2723.

99H(51)2881 K. Kobayashi, K. Sakashita, H. Akamatsu, K. Tanaka, M. Uchida, T. Uneda, T. Kitamura, O. Morikawa, H. Konishi, *Heterocycles* **1999**, *51*, 2881.

99JCS(P1)171 M. Sakamoto, A. Kinbara, T. Yagi, M. Takahashi, K. Yamaguchi, T. Mino, S. Watanabe, T. Fujita, *J. Chem. Soc., Perkin Trans. 1* **1999**, 171.

99JCS(P1)913 A. C. Brickwood, M. G. B. Drew, L. M. Harwood, T. Ishikawa, P. Marais, V. Morisson, *J. Chem. Soc., Perkin Trans. 1* **1999**, 913.

99JCS(P1)2421 J. Habermann, S. V. Ley, R. Smits, *J. Chem. Soc., Perkin Trans. 1* **1999**, 2421.

99JCS(P1)3395 E. Lee, H. Y. Song, H. J. Kim, *J. Chem. Soc., Perkin Trans. 1* **1999**, 3395.

99JCS(P2)73 M. E. Gonzalez-Rosende, J. Sepúlveda-Arques, E. Zaballos-Garcia, L. R. Domingo, R. J. Zaragozá, W. B. Jennings, S. E. Lawrence, D. O'Leary, *J. Chem. Soc., Perkin Trans. 2* **1999**, 73.

99JOC1738 F.-T. Luo, A. C. Bajji, A. Jeevanandam, *J. Org. Chem.* **1999**, *64*, 1738.

99JOC2066 C. Wang, G. A. Russell, *J. Org. Chem.* **1999**, *64*, 2066.

99JOC2259 P. Li, J. Yang, K. Zhao, *J. Org. Chem.* **1999**, *64*, 2259.

99JOC2814 N. Svenstrup, K. B. Simonsen, N. Thorup, J. Brodersen, W. Dehaen, J. Becher, *J. Org. Chem.* **1999**, *64*, 2814.

99JOC2982 J. M. Harris, M. D. Keranen, G. A. O'Doherty, *J. Org. Chem.* **1999**, *64*, 2982.

99JOC2984 M. L. Bushey, M. H. Haukaas, G. A. O'Doherty, *J. Org. Chem.* **1999**, *64*, 2984.

99JOC3026 L. R. Domingo, M. T. Picher, J. Andrés, M. Oliva, *J. Org. Chem.* **1999**, *64*, 3026.

99JOC3642 M. C. Pirrung, F. Blume, *J. Org, Chem.* **1999**, *64*, 3642.

99JOC6418 P. J. Skabara, I. M. Serebryakov, D. M. Roberts, I. F. Perepichka, S. J. Coles, M. B. Hursthouse, *J. Org. Chem.* **1999**, *64*, 6418.

99JOC7654 K. Chiba, M. Fukuda, S. Kim, Y. Kitano, M. Tada, *J. Org. Chem.* **1999**, *64*, 7654.

99JOC7687 B. Gabriele, G. Salerno, E. Lauria, *J. Org. Chem.* **1999**, *64*, 7687.

99JOC7693 B. Gabriele, G. Salerno, F. De Pascali, M. Costa, G. P. Chiusoli, *J. Org. Chem.* **1999**, *64*, 7693.

99JOC9170 Y. Sun, M. W. Wong, *J. Org. Chem.* **1999**, *64*, 9170.

99JOC9739 O. Arjona, A. G. Csákÿ, M. C. Murcia, J. Plumet, *J. Org. Chem.* **1999**, *64*, 9739.

99MP859 D. J. Tozer, R. D. Amos, N. C. Handy, B. O. Roos, L. Serrano-Andrés, *Mol. Phys.* **1999**, *97*, 859.

99OL965 J. C. Bussolari, D. C. Rehborn, *Org. Lett.* **1999**, *1*, 965.

99OL1087 R. Bennes, D. Philp, N. Spencer, B. M. Kariuki, K. D. M. Harris, *Org. Lett.* **1999**, *1*, 1087.

99OL1091 T. Khasanova, R. S. Sheridan, *Org. Lett.* **1999**, *1*, 1091.

99OL1127 E. Lee, S. J. Choi, *Org. Lett.* **1999**, *1*, 1127.

99OL1535 D. L. Wright, C. R. Whitehead, E. H. Sessions, I. Ghiviriga, D. A. Frey, *Org. Lett.* **1999**, *1*, 1535.

99S61 I. Francesconi, A. Patel, D. W. Boykin, *Synthesis* **1999**, 61.

99S74 G. Verardo, A. Dolce, N. Toniutti, *Synthesis* **1999**, 74.

99S94 A. Merz, T. Meyer, *Synthesis* **1999**, 94.

99S330 P. Bertrand, H. El Sukkari, J.-P. Gesson, B. Renoux, *Synlett* **1999**, 330.

99S341 M. Takeuchi, T. Taniguchi, K. Ogasawara, *Synthesis* **1999**, 341.

99S463 V. K. Brel, *Synthesis* **1999**, 463.

99S751 R. Kawęcki, A. P. Mazurek, L. Kozerski, J. K. Maurin, *Synthesis* **1999**, 751.

99S1181 M. Andrew, A. M. Birch, P. A. Bradley, *Synthesis* **1999**, 1181.

99S1193 M. J. Poss, M. A. Nordhaus, K. W. Stagliano, *Synthesis* **1999**, 1193.

99S1241 A. Arrault, F. Touzeau, G. Guillaumet, J.-Y. Mérour, *Synthesis* **1999**, 1241.

99S1325 T. Taniguchi, M. Takeuchi, K. Kadota, A. S. ElAzab, K. Ogasawara, *Synthesis* **1999**, 1325.

99S1534	M. Harmata, L. Shao, *Synthesis* **1999**, 1534.
99SA(A)2437	A. A. El-Azhary, *Spectrochim. Acta* **1999**, *55A*, 2437.
99SL29	A. Fürstner, T. Gastner, *Synlett* **1999**, 29.
99SL213	S. Pierau, H. M. R. Hoffmann, *Synlett* **1999**, 213.
99SL295	P. A. Allen, M. A. Brimble, H. Prabaharan, *Synlett* **1999**, 295.
99SL313	S. Janosz, M. Mach, S. Skóra, *Synlett* **1999**, 313.
99SL324	S.-K. Kang, T.-G. Baik, A. N. Kulak, *Synlett* **1999**, 324.
99SL429	J. Garcia, M. López, J. Romeu, *Synlett* **1999**, 429.
99SL456	M. W. Khan, N. G. Kundu, *Synlett* **1999**, 456.
99SL474	D. J. Aldous, W. M. Dutton, P. G. Steel, *Synlett* **1999**, 474.
99SL495	M. Smet, J. Van Dijk, W. Dehaen, Synlett **1999**, 495.
99SL567	H. M. C. Ferraz, M.K. Sano, A. C. Scalfo, *Synlett* **1999**, 567.
99SL863	P. Nesvadba, L. Bugnon, P. Dubs, S. Evans, *Synlett* **1999**, 863.
99SL1079	H. Lütjens, P. J. Scammels, *Synlett* **1999**, 1079.
99SL1211	J. Eames, N. Kuhnert, F. H. Sansbury, S. Warren, *Synlett* **1999**, 1211.
99SL1215	J. Eames, N. Kuhnert, S. Warren, *Synlett* **1999**, 1215.
99SL1257	G. Pandey, R. Sochanchingwung, S. K. Tiwari, *Synlett* **1999**, 1257.
99SL1391	D. S. Caine, M. A. Paige, *Synlett* **1999**, 1391.
99SL1395	G. A. Kraus, X. Wang, *Synlett* **1999**, 1395.
99SL1399	X.-L. Sun, T. Kai, H. Takayanagi, K. Furuhata, *Synlett* **1999**, 1399.
99SL1407	P. J. Parsons, T. Montagnon, M. Giles, P. Hitchcock, *Synlett* **1999**, 1407.
99SL1415	T. Nakamura, H. Yorimitsu, H. Shinokubo, K. Oshima, *Synlett* **1999**, 1415.
99SL1432	A. Arcadi, S. Cacchi, G. Fabrizi, F. Marinelli, L. Moro, *Synlett* **1999**, 1432.
99SL1523	A. Fürstner, *Synlett* **1999**, 1523.
99SL1582	A. Inoue, H. Shinokubo, K. Oshima, *Synlett* **1999**, 1582.
99SL1609	K. Paulvannan, T. Chen, J. W. Jacobs, *Synlett* **1999**, 1609.
99SL1925	I. Alexiou, E. P. Gogonas, L. P. Hadjiarapoglou, *Synlett* **1999**, 1925.
99SL1942	N. S. Nudelman, C. Carro, *Synlett* **1999**, 1942.
99SL1969	H. Yoda, T. Shimojo, K. Takabe, *Synlett* **1999**, 1969.
99T617	G. A. Molander, J. Haas, *Tetrahedron* **1999**, *55*, 617.
99T861	B. I. Kamara, E. V. Brandt, D. Ferreira, *Tetrahedron* **1999**, *55*, 861.
99T2847	D. I. MaGee, J. D. Leach, S. Setiadyi, *Tetrahedron* **1999**, *55*, 2847.
99T4709	H. Scheytza, H.-U. Reissig, O. Rademacher, *Tetrahedron* **1999**, *55*, 4709.
99T6511	D. Prim, G. Kirsch, *Tetrahedron* **1999**, *55*, 6511.
99T7433	K. Paulvannan, J. W. Jacobs, *Tetrahedron* **1999**, *55*, 7433.
99T8905	S. F. Martin, S. K. Bur, *Tetrahedron* **1999**, *55*, 8905.
99T11959	W.-S. Zhou, Z.-H. Lu, Y.-M. Xu, L.-X. Liao, Z.-M. Wang, *Tetrahedron* **1999**, *55*, 11959.
99T13471	D. Craig, N. J. Ikin, N. Mathews, A. M. Smith, *Tetrahedron* **1999**, *55*, 13471.
99T13521	P. Vogel, J. Cossy, J. Plumet, O. Arjona, *Tetrahedron* **1999**, *55*, 13521.
99T13999	L. Ou, Y. Hu, G. Song, D. Bai, *Tetrahedron* **1999**, *55*, 13999.
99T14097	M. Gruttadauria, P. Lo Meo, R. Noto, *Tetrahedron* **1999**, *55*, 14097.
99T14233	C. D. Brown, J. M. Chong, L. Shen, *Tetrahedron* **1999**, *55*, 14233.
99T14269	L. A. Telan, R. A. Firestone, *Tetrahedron* **1999**, *55*, 14269.
99TA667	Z.-M. Wang, S.-K. Tian, M. Shi, *Tetrahedron: Asymmetry* **1999**, *10*, 667.
99TA1315	T. J. Donohoe, C. A. Stevenson, M. Helliwell, R. Irshad, T. Ladduwahetty, *Tetrahedron: Asymmetry* **1999**, *10*, 1315.
99TA1521	K. Fukatsu, N. Fujii, S. Ohkawa, *Tetrahedron: Asymmetry* **1999**, *10*, 1521.
99TA1539	R. Dunkel, J. Treu, H. M. R. Hoffmann, *Tetrahedron: Asymmetry* **1999**, *10*, 1539.
99TA2023	A. Lattanzi, F. Sagulo, A. Scettri, *Tetrahedron: Asymmetry* **1999**, *10*, 2023.
99TA2237	O. Arjona, F. Iradier, R. Medel, J. Plumet, *Tetrahedron: Asymmetry* **1999**, *10*, 2237.
99TA2311	C. Yang, L. Liao, Y. Xu, H. Zhang, P. Xia, W. Zhou, *Tetrahedron: Asymmetry* **1999**, *10*, 2311.
99TA2417	U. Iserloh, D. P. Curran, S. Kanemasa, *Tetrahedron: Asymmetry* **1999**, *10*, 2417.
99TA2551	S.-K. Tian, Z.-M. Wang, J.-K. Jiang, M. Shi, *Tetrahedron: Asymmetry* **1999**, *10*, 2551.
99TA3649	L.-X. Liao, Z.-M. Wang, H.-X. Zhang, W.-S. Zhou, *Tetrahedron: Asymmetry* **1999**, *10*, 3649.
99TA3659	M. De Rosa, R. Dell'Anglio, A. Soriente, A. Scettri, *Tetrahedron: Asymmetry* **1999**, *10*, 3659.
99TA3869	A. Fontana, I. Fakhr, E. Mollo, G. Cimino, *Tetrahedron: Asymmetry* **1999**, *10*, 3869.
99TA4357	M. C. Carreño, J. L. Ruano, A. Urbano, C. Z. Remor, Y. Arroyo, *Tetrahedron: Asymmetry* **1999**, *10*, 4357.
99TL545	Y. Hu, L. Ou, D. Bai, *Tetrahedron Lett.* **1999**, *40*, 545.
99TL731	Y. Mori, T. Sawada, H. Furukawa, *Tetrahedron Lett.* **1999**, *40*, 731.

99TL839 M. Miesch, F. Wendling, M. Franck-Neumann, *Tetrahedron Lett.* **1999**, *40*, 839.
99TL989 B. Gabriele, G. Salerno, M. Costa, G. P. Chiusoli, *Tetrahedron Lett.* **1999**, *40*, 989.
99TL1083 A. K. Ghosh, R. Kawahama, *Tetrahedron Lett.* **1999**, *40*, 1083.
99TL1253 P. A. Evans, V. S. Murthy, *Tetrahedron Lett.* **1999**, *40*, 1253.
99TL2375 C. Xi, M. Kotora, T. Takahashi, *Tetrahedron Lett.* **1999**, *40*, 2375.
99TL2521 I. Hanna, *Tetrahedron Lett.* **1999**, *40*, 2521.
99TL2777 H. Koshimizu, T. Baba, T. Yoshimitsu, H. Nagaoka, *Tetrahedron Lett.* **1999**, *40*, 2777.
99TL3371 R. Nouguier, S. Gastaldi, D. Stien, M. Bertrand, P. Renaud, *Tetrahedron Lett.* **1999**, *40*, 3371.
99TL3375 F. Villar, O. Andrey, P. Renaud, *Tetrahedron Lett.* **1999**, *40*, 3375.
99TL3411 Y. Watanabe, S. Ishikawa, G. Takao, T. Toru, *Tetrahedron Lett.* **1999**, *40*, 3411.
99TL3753 H. Kuroda, E. Hanaki, M. Kawakami, *Tetrahedron Lett.* **1999**, *40*, 3753.
99TL4769 D. Balachari, L. Quinn, G. A. O'Doherty, *Tetrahedron Lett.* **1999**, *40*, 4769.
99TL4841 A. Jeevanandam, K. Narkunan, C. Cartwright, Y.-C. Ling, *Tetrahedron Lett.* **1999**, *40*, 4841.
99TL5979 U. Emde, U. Koert, *Tetrahedron Lett.* **1999**, *40*, 5979.
99TL8949 S. F. Martin, O. D. Lopez, *Tetrahedron Lett.* **1999**, *40*, 8949.
00TL17 P. Forgione, P. D. Wilson, A. G. Fallis, *Tetrahedron Lett.* **2000**, *41*, 17.

Chapter 5.4

Five Membered Ring Systems: With More than One N Atom

Larry Yet
Albany Molecular Research, Inc., Albany, NY
larryy@albmolecular.com

5.4.1 INTRODUCTION

Major advancements in the chemistry of pyrazoles, imidazoles, triazoles, tetrazoles, and related fused heterocyclic derivatives appeared in 1999. Solid-phase combinatorial chemistry of benzimidazoles and triazoles has been particularly active. Synthetic routes to all areas continue to be pursued vigorously with improvements and applications. In medicinal chemistry, synthesis and structure-activity relationship (SAR) studies utilizing these core structures have been exploited heavily. The physical organic chemistry of pyrazoles and imidazoles continue.

5.4.2 PYRAZOLES AND RING-FUSED DERIVATIVES

A review on the synthesis of pyrazoles and condensed pyrazoles covering the period of 1989-1998 was written <99JHC321>.

Reactions of 3,5-dichloropyrazoles and 3,3'-dichloro-5,5'-bipyrazoles with zero-valent nickel complex to afford novel bis(pyrazolyl)nickel(II) complexes were studied <99BCSJ1629>. The crystal structure determination of 3,5-dimethyl-4-aminopyrazole and of 3(5)-aminopyrazolium picrate salt was studied <99JHC595>. The first report of the synthesis, structure and properties of a series of ruthenium(II)-bearing η^5-pyrazolato ligands was disclosed <99JACS4536>. A synthesis of novel 4-acylpyrazol-5-one-substituted crown ethers, used as metal-chelating extracting reagents, was described <99JCS(P1)693>. Thermolysis of 5-azido-4-formylpyrazoles led through a pyrazole to furan rearrangement products <99JOC2814>. The kinetics of methanolysis of acetyl pyrazole in the presence of zinc(II) perchlorate and cobalt(II) perchlorate were studied <99CJC1005>. The quaternization and dequaternization of pyrazoles in solvent-free conditions with conventional heating versus microwave irradiation were examined <99JHC889>. NMR and X-ray structure investigations of 4-acyl-5-methyl-2-phenylpyrazolones were reported <99H(50)799>. The synthesis and mesogenic properties of Schiff bases derived from aminopyrazoles were investigated <99H(51)751>. Photolysis of 2-(Δ^1-pyrazolinyl)-Δ^3(1,3,4)oxadiazolines afforded *gem*-dimethylcyclopropane ketone products <99TL9027>.

4-Nitro-1*H*-pyrazole-1-[*N,N'*-bis(*tert*-butoxycarbonyl)]carboxamidine (**1**) has been developed as a new reagent for the rapid and efficient solid and solution phase synthesis of bis(carbamate)-protected guanidines from primary and secondary amines <99TL53>.

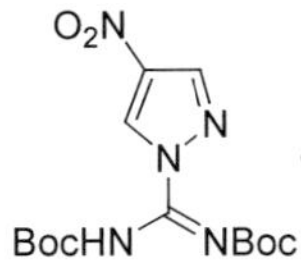

1

Synthetic approaches to pyrazoles continued in 1999. A large number of methodology developments were based on the condensation of a 1,3-difunctional species with hydrazine derivatives. A one-pot procedure, utilizing β-alkoxyvinyl trichloromethyl ketones with methyl- and phenylhydrazines, provided 1-methyl- and 1-phenylpyrazole-3-(5)-ethyl esters <99JHC217>. 3-Halo-4-methoxycoumarins were transformed into 4-halo-5-(2-hydroxyphenyl)-3-oxo-2,3-dihydropyrazoles with hydrazines <99JHC767>. 4-Alkylamino-2-chloroquinoline-3-carbonitriles reacted with hydrazine hydrate to give 3-amino-4-hydrazino-1*H*-pyrazolo[3,4-*b*]quinoline <99JCS(P1)2183>. Phenylhydrazine and hydrazine were reacted with 3-acetyl-4-hydroxy-6-methyl-2*H*-pyran-2-one to afford 4-aceto-acetyl-3-methylpyrazolin-5-ones, which were employed in the synthesis of bipyrazoles and pyrazoloisoxazoles <99JHC1291>. Reactions of hydrazines with *N*-acetoacetyl derivatives of (4*S*)-4-benzyloxazolidin-2-one (Evans oxazolidinone) and (2*R*)-bornane-10,2-sultam (Oppolzer sultam) in very acidic media gave pyrazoles retaining the C-3(5) chiral moiety <99S157>. Phosphazenes, derived from hydrazines and acetylenic esters, were employed in the regioselective synthesis of 5-pyrazolones and pyrazoles <99T14451>. Trisubstituted pyrazole carboxylic acids were prepared by polymer-bound arylidene- or alkylidene-β-oxo esters with phenylhydrazines <99S1961>. A series of stable 5-hydroxy-1*H*-pyrazolines were prepared from semicarbazide additions to haloacetylated enol ethers <99T345>. A novel synthesis of thieno[2,3-*c*]pyrazole from ethyl acetoacetate was reported <99HC303, 99SC2355>. A new heterocyclization of 4-trifluoroacetyl-2,3-dihydropyrroles (**2**) with hydrazines provided entry to trifluoromethylated pyrazoles **3** bearing a β-aminoethyl side chain <99TL2541>. Tandem Michael addition/cyclization of the anion of 2-phenylazo-1,3-dicarbonyl substrates to 1,2-diaza-1,3-butadienes gave 1-phenylpyrazoles <99TL3891>. Reactions of polymer-bound 1,2-diaza-1,3-butadienes with triphenylphosphine provided convenient entry to 4-triphenylphosphoranylidene-4,5-dihydropyrazol-5-ones <99TL9277>. A versatile and efficient synthesis of 3-substituted-1*H*-indazoles **5** from aryl mesylates **4** and hydrazines through the intermediate hydrazone was disclosed <99S588>. The preparation of 2-(1-phenyl-5-phenyl or 5-substituted phenyl-1*H*-pyrazole-3-yl)phenols **7** from trilithiated 2'-hydroxyacetophenone phenylhydrazone (**6**) and aromatic esters was reported <99SC495>. Dilithiated 2-tetralone phenylhydrazone and aromatic esters were employed in the synthesis of 4,5-dihydro-2*H*-benz[*e*]indazoles <99JHC1231>. β-Tosylhydrazono phosphonates **8**, novel and bifunctional reagents, were employed in the concise approach to give polysubstituted pyrazoles **9** in the presence of aromatic and aliphatic aldehydes under basic conditions <99SL299>.

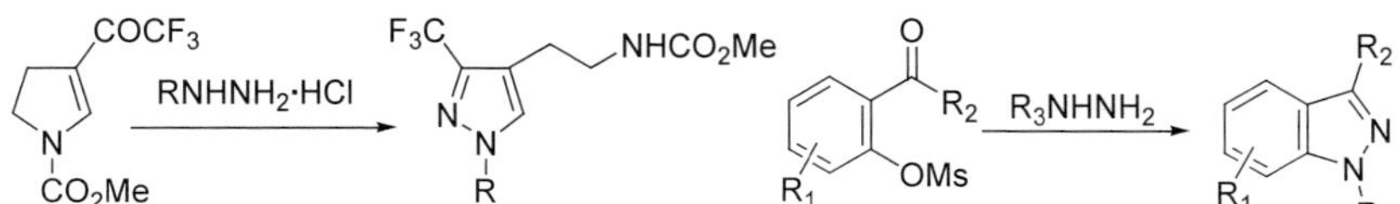

2 **3** **4** **5**

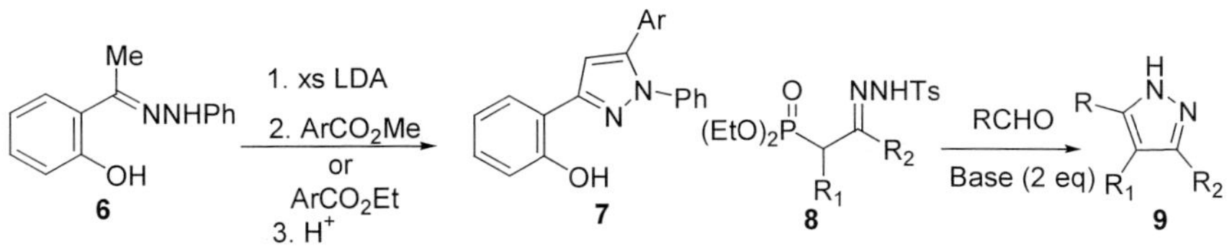

Cycloaddition approaches to pyrazoles and cycloadditions with pyrazoles continued to be active areas of research in 1999. Reactions of aroyl-substituted heterocyclic ketene aminals with nitrile imines to give fully substituted pyrazoles via 1,3-dipolar cycloaddition were reported <99TL7399>. Synthesis of enantiomerically pure 3,3a-dihydropyrazolo[1,5-*a*][1,4]benzodiazepine-6(4H)-ones from stereoselective intramolecular cycloadditions of homochiral nitrile imines were investigated <99TA2203>. The first example of asymmetric induction in intramolecular nitrile imine cycloadditions was shown in the synthesis of enantiopure 3-substituted 6-oxo-2,3,3a,5-tetrahydro-4-carbomethoxy-furo[3,4-*c*]pyrazoles <99TA487>. The first example of a *(1R,2S,5R)*-(-)-menthyl chiral auxiliary in the intramolecular nitrilimine cycloaddition in syntheses of enantiopure pyrazolo[1,5-*a*][4,1]benzoxazepines and pyrazolo[1,5-*a*][4,1]benzodiazepines was developed <99TA3873>. Asymmetric induction by the (*S*)-1-phenylethyl group in intramolecular nitrile imine cycloadditions gave enantiopure 3,3a-dihydropyrazolo[1,5-*a*][1,4]benzodiazepine-4(*6H*)-ones <99TA4447>. 1,3-Dipolar cycloaddition of 2-diazopropane with diarylideneacetones afforded diastereomeric bis-Δ^2-pyrazolines <99T449>. In addition, 1,3-dipolar cycloaddition of diazomethane and ethyl diazoacetate to α-(diethoxyphosphoryl)vinyl *p*-tolyl sulfoxide furnished 3-phosphorylpyrazoles <99T14791>. The effect of lithiation on the reactivity of diazo derivatives with sulfonylalkynes to the synthesis of three isomeric trisubstituted pyrazoles was studied <99TL883>. Azomethine imines, generated from α-silylnitrosamines via 1,4-silatropic shift, underwent 1,3-dipolar cycloadditions with dipolarophiles to give pyrazole derivatives <99TL8849>. A facile synthesis of pyrazolo[4',5'][60]fullerenes from pyrazolyl hydrazones with [60]fullerene under microwave irradiation with evidence of intramolecular charge transfer interaction was shown <99TL1587>. Novel C_{60}-fused isoxazolines have been synthesized from 1,3-dipolar cycloadditions of pyrazole nitrile oxides to C_{60} under thermal or microwave irradiation <99T4889>.

The synthesis of 4-substituted 1-(benzyloxyl)pyrazoles **11** via iodine-magnesium exchange of 1-(benzyloxy)-4-iodopyrazole (**10**) has been reported <99JOC4196>. Regioselective introduction of electrophiles to the 4-position of 4-bromo-1-[(*tert*-butyldiphenylsilyl)oxy]pyrazole (**12**) via bromine-lithium exchange to give 4-substituted 5-(*tert*-butyldiphenylsilyl)-1-hydroxypyrazole (**13**) with silyl group migration was also studied <99JOC5366>. The solid-state Michael addition of indole to 4-arylidene-3-methyl-5-pyrazolone was investigated <99JHC697>. *rel*-(4*R*,5*R*)-4-Benzoylamino-5-phenyl-3-pyrazolidinones were reductively ring-cleaved with Raney-nickel alloy in methanolic potassium hydroxide to furnish *rel*-(4*R*,5*R*)-*N*-benzoyl-3-alkylamino-3-phenylalanine amides <99JHC607>. In contrast, *rel*-(2*R*,3*R*,5*S*)-5-aryl-2-benzoylamino-6,7-bis(methoxycarbonyl)-2,3-dihydro-1-oxo-3-phenyl-1*H*,5*H*-pyrazolo[1,2-*a*]pyrazoles were oxidatively ring-cleaved with methanolic bromine to give *rel*-(2*R*,3*R*)-3-phenyl-3-[5-aryl-3,4-bis(methoxycarbonyl)pyrazolyl-1]alanine esters <99JHC799>. The regioselective benzylation of an indazo-substituted pyrazole under the influence of inorganic solid supported bases was studied <99JCR(S)274>. Vinylpyrazoles, under microwave irradiation, reacted with ethyl *N*-trichloroethylidene carbamate to give Michael addition product of the exocyclic double bond to the imine <99T9623>. Suzuki cross-coupling of 3-iodoindazoles with aryl boronic acids provided a general and flexible route to 3-arylindazoles <99T6917>.

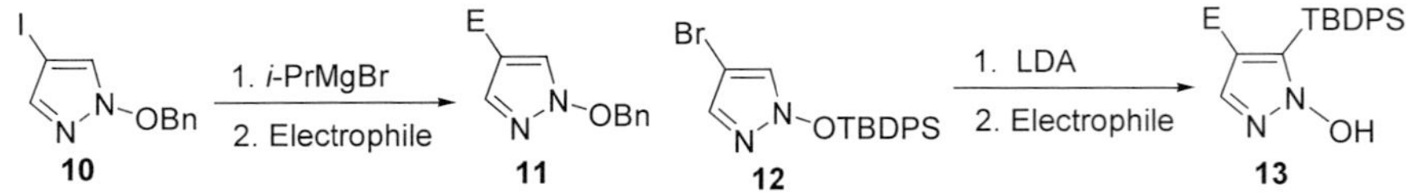

The synthesis of mono- and disubstituted 1*H*-imidazo[1,2-*b*]pyrazoles **14** from different precursors was reported <99SC311>. A facile route to pyrazolo[3,4-*b*]pyridines **15** was developed <99SC655>. Fused [1,2,3]triazolium salts were heated in trifluoroacetic acid to yield new pyrazole-fused heterocycles <99JOC5499>. Indazolo[2,3-*a*]quinoline (**16**) was prepared from thermolysis of 2-[2-(1-azidophenyl)]quinoline <99SC3961>. New 4-aryl-6-methyl-8-phenyl-2,3-dihydropyrazolo[3,4-*b*]diazepines and 4-aryl-8-methyl-6-phenyl-2,3-dihydropyrazolo[4,3-*b*]diazepines were obtained from the reaction of 4,5-diamino-3-methyl-1-phenylpyrazole with substituted 3-dimethylaminopropiophenones <99JHC635>. Synthesis and characterization of 4,5-dihydro-1*H*-pyrazolo[3,4-*b*][1,4]azaphosphinines **17** has been described <99HC391>. 3-Amino-4-(4'-arylthiazol-2'-yl)-5-phenylaminopyrazole reacted with different reagents to afford pyrazolo[1,5-*a*]pyrimidines and pyrazolo [5,1-*c*]triazines <99HC508>. The synthesis of 5-cyanopyrazolo[3,4-*b*]pyridines was achieved from the reaction of 5-amino-3-methyl-1-phenylpyrazole with arylidene derivatives of malonodinitrile and ethyl cyanoacetate <99JHC1311>. Reaction of lithium 5-lithiomethyl-3-methylpyrazole-1-carboxylate with α-oxoketene dithioacetals provided a general method for preparing substituted and annelated pyrazolo[1,5-*a*]pyridines **18** <99T7645>. The synthesis, reactivity, and ^{13}C NMR analysis of 1*H*-pyrazolo[3,4-*c*]pyridine derivatives were investigated <99H(51)1661>.

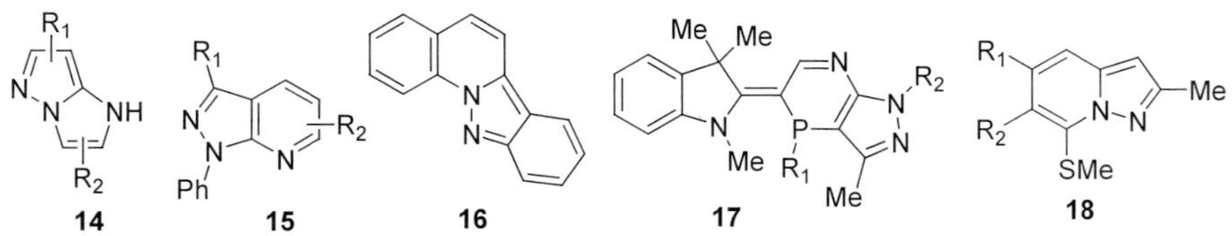

Structure-activity relationships of pyrazole-containing structures have been investigated in medicinal chemistry for active biological profiles. A series of 6-aryl-substituted-5*H*-pyrazolo[1,5-*d*]-1,2,4-triazine-4,7-diones **19** have been found to be active as inhibitors of serine protease enzymes <99S453>. Monocyclic pyrazolidinones **20** showed moderate antibacterial activity <99BML2205>. An efficient synthesis of a new DNA intercalating agent, KW-2170 (**21**), a 7,10-dihydroxy-6*H*-pyrazolo[5,5,1-*de*]acridin-6-one derivative, was desribed <99S947>. Biarylpyrazoles **22** were synthesized to investigate the antagonism studies of the brain cannabinoid receptor CB1 <99JMC769>. An enantioselective reduction of pyrazole phenyl ketone **23** to (*R*)-alcohol **24** led to the synthesis of (*R*)-cizolirtine (**25**), an analgesic <99SL765>. Novel benzoyl nitrogen mustard derivatives of pyrazole analogues of distamycin A were prepared for *in vitro* and *in vivo* studies of L1210 leukemia cell lines <99BML251>. 3-Heteroaryl-2-phenylpyrazolo[1,5-*a*]pyridines were prepared as novel adenosine A_1 receptor antagonists with potent diuretic activity <99JMC779>. Several new pyrazolo[1,5-*a*]pyrimidines were found to have *in vitro* anthelmintic activity against *Nippostrongylus brasiliensis* <99JHC11> and to be ligands for the benzodiazepine receptor <99BMC2705>. 3-Heteroaryl-8-chloropyrazolo[5,1-*c*][1,2,4]benzotriazine 5-oxides were synthesized and tested for their binding activities at the central benzodiazepine receptor

<99JMC2218>. (*Z*)- and (*E*)-2-(5-Arylpyrazol-3-yl)-3-(pyrrol-2-yl)acrylonitriles were evaluated as novel antioxidants <99BMC1425>. A facile and general syntheses of 3- and/or 5-substituted 7*H*-pyrazolo[4,3-*e*]-1,2,4-triazolo[4,3-*c*]pyrimidines led to a new class of potential xanthine oxidase inhibitors <99CC1461>. An efficient synthesis of FR166124, a water-soluble adenosine A_1 receptor antagonist containing a pyridazinone pyrazolo[1,5-*a*]pyridine core was reported <99T10351>. Novel optically active pyrazoles containing bicarboxylic α-amino acids were prepared as building blocks for peptidomimetics with a *cis* amide bond <99TL8701>. Unsymmetrical cyclic ureas bearing novel biaryl indazoles as P2/P2' substituents were evaluated as HIV-1 protease inhibitors <99BML3217>.

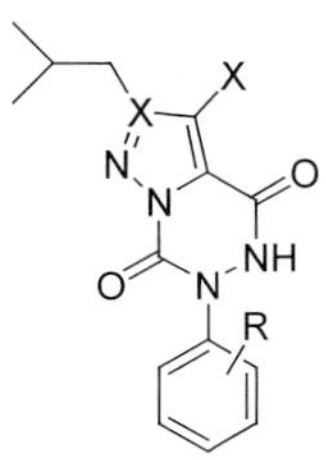

19

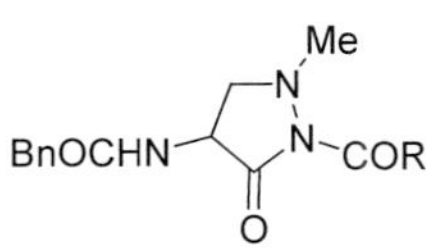

20

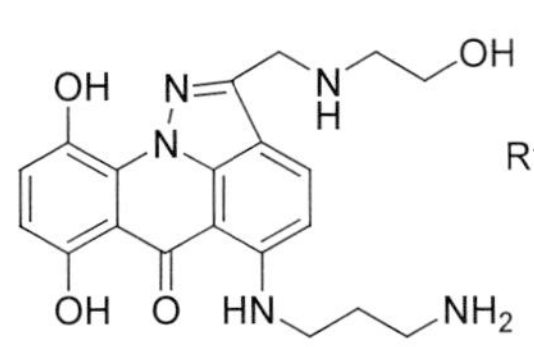

21

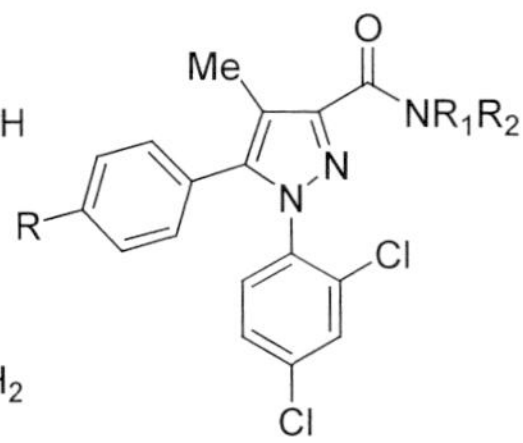

22

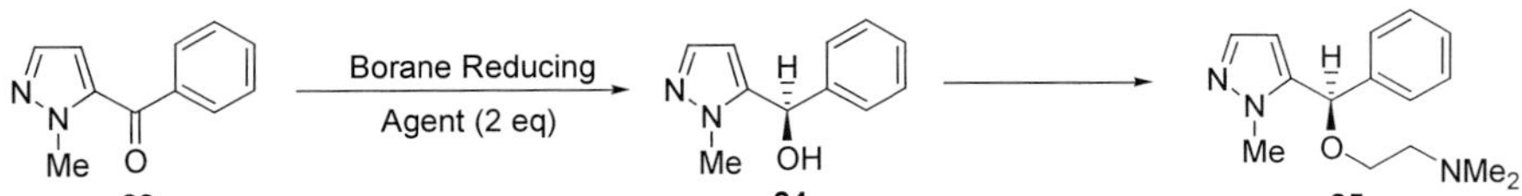

23 **24** **25**

5.4.3 IMIDAZOLES AND RING-FUSED DERIVATIVES

The solvent effects on the decarboxylation reaction of neutral *N*-carboxy-2-imidazolidinone in aqueous solution have been investigated by a combined quantum mechanical and molecular mechanical (QM/MM) Monte Carlo simulation method <99JOC4492>. Spectral characteristics of 2-(4'-*N,N*-dimethylaminophenyl)pyrido[3,4-*d*]imidazole have been studied in two different solvents using absorption, fluorescence, and fluorescence excitation spectral analyses <99JOC6566>. Substituent effects on fluorescent properties of imidazo[1,2-*a*]pyridine-based compounds were studied with absorption and fluorescence spectral detection <99BCSJ1327>. A series of gold(I)-carbene complexes with benzimidazol-2-ylidene were synthesized and studied spectroscopically <99O1216>. An anion template-directed synthesis of dicationic $[1_4]$imidazoliophanes was reported <99OL1035>. Diaza-18-crown-6 with two benzimidazole ligands attached was prepared <99JHC771>. The observation of the 1-methyl-2-imidazolylnitrenium ion was seen in the photochemistry of 2-azido-1-methylimidazole in aqueous solutions <99JACS1459>. Complexes derived from nickelocene and carbenes, 1,3-bis(2,6-dimethyl-4-bromo)imadazol-2-ylidene and tetramethylimidazol-2-ylidene, were studied <99JACS2329>. An unprecedented example of the *in situ* direct observation of a light-induced radical pair in a crystal of 2,2'-di(orthochlorophenyl)-4,4',5,5'-tetraphenylbiimidazole by X-ray diffraction was reported <99JACS8106>. 2-Aryl-1*H*-imidazo[4,5-*b*]porphyrins, useful precursors for multi-porphyrin arrays, were synthesized <99JCS(P1)2429>. Dianions derived from chiral imidazolines underwent selective one-electron oxidation reactions in the presence of TEMPO to form metallated radical species that were trapped stereoselectively <99TL4035>. The synthesis and properties of *N,N*-bis(4,5-dicyano-1-methyl-2-imidazolyl)cyanimide and

N,N,N',N',N'',N''-hexakis(4,5-dicyano-1-methyl-2-imidazolyl)melamine were explored as high nitrogen-containing compounds for thermally stable materials of high temperature applications and advanced materials <99T353>. The synthesis and characterization of derivatives of 4,5-dicyanoimidazole dimers was explored <99T2811>. A carbene analogue with low-valent gallium as a heteroatom in a quasi-aromatic imidazolate anion has been reported <99JACS9758>. Imidazolylidenes, imidazolinylidenes and imidazolidines were each synthesized from glyoxal <99T14523>. An efficient preparation of novel ferrocenylimidazole derivatives from thermal heterocyclization of β-ferrocenylvinylcarbodiimides or β-ferrocenylvinylureas has been reported <99T1417>.

Imidazole-containing reagents have found useful applications in a variety of organic transformations. Imidazolyl-derived (arylsulfonylimino)iodobenzene **26** has been utilized as a nitrene intermediate in an aziridination reaction with styrene under copper catalysis <99TL5459>. 2-Chloro-1,3-dimethylimidazolinium chloride (**27**) has been developed as a powerful dehydrating equivalent to DCC <99JOC6984,6989>. It has found applications in the preparation of esters, amides, and anhydrides, as well as for the construction of heterocycles through dehydration reactions. Carbamoyl imidazolium salts **28** were exploited as carbamoylation reagents for the formation of carbamates and thiocarbamates with alcohols and thiols, respectively <99TL2669>. A second generation of new ruthenium-based olefin metathesis catalysts coordinated with imidazolin-2-ylidene ligands **29** and 1,3-dimesityl-4,5-dihydroimidazol-2-ylidene ligands **30** have been developed <99JACS2674, 99O2370, 99OL953, 99TL2247, 99TL4787>. These catalysts have proven to be more air- and water-tolerant than the earlier ruthenium-based catalysts and their ring-closing metathesis activity has greatly exceeded the previous analogs. Ruthenium catalysts **30** allow the formation of tetrasubstituted alkenes that were previously out of reach with earlier catalysts. The bulky imidazolium salt **31** was found to greatly accelerate the amination of aryl chlorides with palladium catalysts as well as to provide efficient cross-coupling of aryl chlorides with aryl Grignard reagents <99OL1307, 99JACS9889>. Benzimidazolium bromochromate was developed as a new reagent for the bromination of aromatic compounds and for the oxidation of alcohols <99SC763>. New thioacylating reagents based on thioacylfluoro-*N*-benzimidazolone were developed <99JMC2046>. Photoreactions of epoxy ketones, aromatic ketones, haloketones, and aromatic halides with 1,3-dimethyl-2-phenylbenzimidazoline (DMPBI, **32**) were studied <99T12957>.

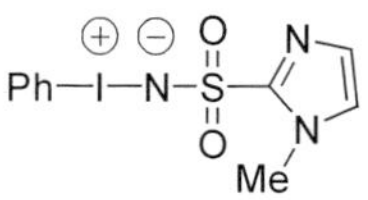

26

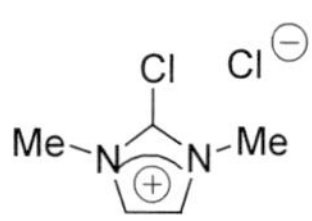

27

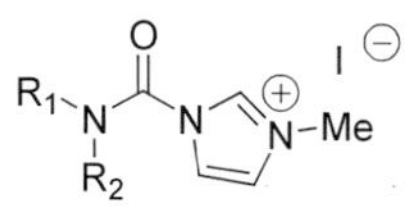

28

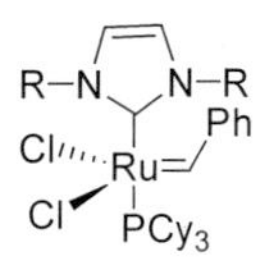

29

R = CHMePh

R = C_6H_2-2,4,6-Me_3

R = C_6H_2-2,6-Me_2-4-OMe

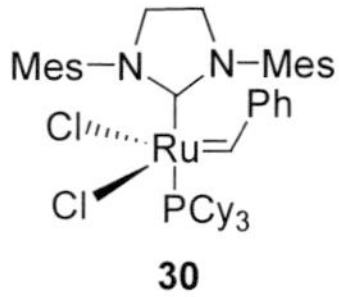

30

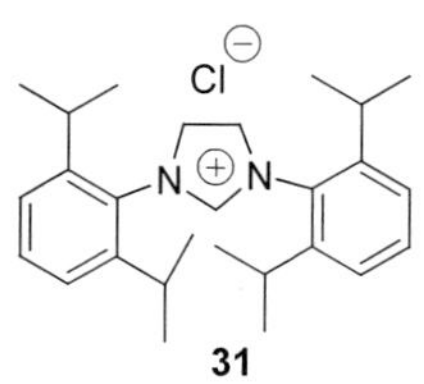

31

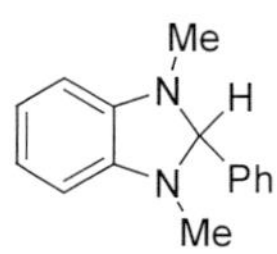

32

General synthetic investigations into the core imidazole structure were continued in 1999. Trifluoromethylaryl imines **33**, under basic conditions, provided a convenient synthesis of 2-arylbenzimidazoles **34** <99TL4119>. A facile synthesis of 2-aryloxyl-4*H*-imidazolin-4-ones **36** was achieved from carbodiimides **35** with phenols in the presence of catalytic potassium carbonate <99SC1171>. Highly regioselective nucleophilic/cycloaddition reactions of *N*-arylamino 1,3-diazabuta-1,3-dienes with α-nitrosostyrenes were employed in the synthesis of functionalized imidazoles <99JCS(P1)615>. 2-Substituted benzimidazoles were prepared from direct condensation of β-alkoxyvinyl trichloromethyl ketones with *o*-phenylenediamine <99JHC45>. A catalytic enantioselective access to optically active 2-imidazolines from *N*-sulfonylimines and ethyl isocyanoacetate in the presence of a chiral ferrocenylphosphine-Me_2SAuCl complex was developed <99JOC1331>. 2-Imidazolidinones were prepared from 1,2-amino alcohols through the regiocontrolled cyclization reaction with *N*-(2-hydroxyethyl)ureas <99JOC2941>. Intramolecular amination of diazenes **37** with zirconium catalyst afforded *N*-aminobenzimidazole-2-ones **38** <99JOC2558>. 4-Amino-2-oxazolines **39** were directly converted to 2-imidazolidinones **40** in the presence of *p*-toluenesulfonyl isocyanates <99TL8163>. 2-Imidazolidinone was prepared from ethylenediamine with iodine as the oxidant, $W(CO)_6$ as the catalyst, and carbon monoxide as the carbonyl source at room temperature <99OL961>. The 2-nitrophenylhydrazones of D-arabino-2-hexulopyranosonic acid, D-arabinose, D-galactose, and D-galacturonic acid were used as precursors to form chiral functionalized benzimidazoles by reductive cyclization methods <99JHC589>. Stereoselective syntheses of imidazolidine, imidazoline, and imidazole *C*- and *N*-pseudonucleosides were published <99TA3011>. Reactions of 3(,5)-(di)chloro-2*H*-1,4-(benz)oxazin-2-ones with α-aminoketones provided new bi- and tricyclic imidazo-fused intermediates via intramolecular cyclization reactions <99T3987>. An efficient synthesis of N^1- or N^3-substituted thieno[2,3-*d*]imidazol-2-ones from thiaisatoic anhydrides was reported <99T6167>. Microwave-assisted *N*-alkylation of 2-halopyridines provided entry to 2-aminoimidazo[1,2-*a*]pyridine derivatives after reaction of the alkylated substrates with cyanamide under basic conditions <99T2317>. 1,3-Dipolar cycloaddition of 3,4-dihydro-6,7-dimethoxyisoquinoline-*N*-methoxycarbonyl methylide to Schiff bases resulted in the synthesis of imidazo[2,1-*a*]isoquinolines as single racemates <99T7279>. A remarkable three-step one solution-phase preparation of novel imidazolines utilized a UDC (Ugi/de-Boc/cyclize) strategy <99TL7925>. Stable axially chiral atropisomeric carbohydrate imidazolidines were prepared from reactions of 2-aminosugars with *o,o'*-disubstituted aryl isocyanates and isothiocyanates <99TA4071>. A novel synthesis of 2-butyl-5-chloro-3*H*-imidazole-4-carbaldehyde was developed as a key intermediate for the synthesis of the Angiotensin II antagonist Losartan <99JOC8084>. Electrochemical mercury cathodic reduction of phenacyl azides, in aprotic DMF-lithium perchlorate medium, afforded 2-aroyl-4-arylimidazoles <99OL1521>. 4-Dimethylamino 2-aza-1,3-dienes reacted with hydrazines or amines to give imidazole-4-carboxylates <99TL8097>.

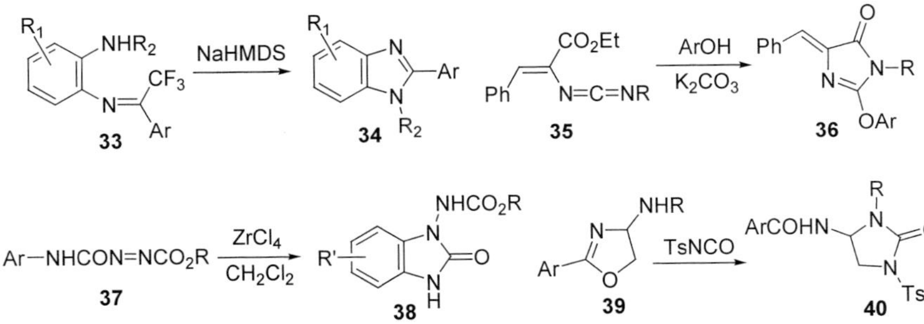

Chemical manipulations of the imidazole ring are still under intense investigation. Reaction of aminobenzimidazoles with 4-hydroxy-6-methyl-2-pyrone and 4-hydroxycoumarin to give 6- and 7-(4-hydroxy-6-methyl-2-oxo-1-pyridyl)benzimidazoles and 1-(6- and 7-benzimidazolyl)-4-hydroxy-2-oxoquinolines, respectively, were reported <99SC2435>. Halogenation followed by nitration of several benzimidazoles was investigated <99SC387>. *N*-Alkylation of 2-methyl-2-imidazoline with organic halides in the presence of phase transfer catalysis and absence of solvent afforded 1-alkyl-2-methyl-2-imidazolines <99SC3025>. New 5-nitroimidazoles bearing a tetrasubstituted ethylenic double bond in the 2-position were synthesized in high yields by the reaction of secondary nitronate anions with a new reductive alkylating agent, 1-methyl-2-trichloromethyl-5-nitroimidazole, via electron transfer processes <99SL801>. Azaazulenes were aminated at the *N*-1 position with *O*-mesitylsulfonylhydroxylamine to give azaazulenium salts <99JCS(P1)1339>. Imidazole and benzimidazole were added regioselectively in the cesium hydroxide catalyzed alkynylation of phenylacetylene to give exclusively *cis*-substituted olefins <99TL6193>. Solid supported imidazoles or benzimidazoles were *N*-arylated under microwave irradiation in the presence of copper(II) acetate with arylboronic acids <99TL1623>. Vincinal diamination of *N*-alkenylbenzimidazoles with palladium catalysts has been investigated <99TL1739>. 4-Iodo-1-tritylimidazole underwent magnesium-iodine exchange with a Grignard reagent to give the 4-magnesioimidazole derivative selectively, which in turn reacted with esters to form a variety of bi-, tri-, and pentadentate ligands for biomimetic studies <99OL949>. An improved synthesis of N^4,N^4-disubstituted-2-cyano-N^1,N^1-dimethyl-1*H*-imidazole-1,4-disulfonamides, which are fungicidal agents, was described <99OPRD104>. A series of ether bis-imidazoles and their derivatives were conveniently and efficiently synthesized <99SC1217>. 2,3-Dihydroimidazo[2,1-*b*]oxazoles were prepared from an intramolecular nucleophilic *ipso*-substitution of 2-alkylsulfonylimidazoles <99S1613>. The synthesis of hitherto unknown 2-(2,4,6-triarylphenyl)substituted 4,5-dihydro-1*H*-imidazolium perchlorates from 2-methyl imidazolium perchlorates and 2,4,6-triarylpyrylium perchlorate salts was published <99JC813>. Substituted aromatic heterocycles from reactions of 2,3-dihydrospiro[1*H*-4-and 5-azabenzimidazole-2,1'cyclohexane] with various nucleophiles were examined <99JHC117>. Ethyl 2-(1*H*-benzimidazol-2-yl)-3-dimethylamino-acrylate reactions with various amines stereoselectively afforded (2*E*)-3-amino-2-(1*H*-benzimidazol-2-yl)acrylates <99HC446>. Copper(I) triflate•benzene catalyzed *N*-arylation of imidazoles, with cesium carbonate as the base and 1,10-phenanthroline and *trans,trans*-dibenzylidineacetone as additives in warm xylene, was successfully applied to a range of aryl halides <99TL2657>. The Ullmann coupling reaction between 2-trifluoroacetamido-3-iodo-6-benzoylimidazo[1,2-*a*]pyridine and dialkyldisulfides, mediated by copper bronze, led to efficient incorporation of alkylthio groups at the C-3 position <99T541>. Tributyltin hydride radical cyclizations onto imidazoles and benzimidazoles via oxidative radical cyclization mechanisms <99T8111> and aromatic radical *ipso* substitutions <99T4109> have been proposed. Reactions of an imidazo[4,5-*c*]isoxazole-6-carboxylate with electron deficient acetylenes afforded 2-pyrrol-2-ylimidazoles <99TL8157>.

2-(Alkylamino)benzimidazoles are well-known biologically active aminoheterocycles, particularly in the antihistamine area. The first fluoride-catalyzed general amination process of chloroimidazole **41** to 2-(alkylamino)benzimidazoles **42** was developed as part of a program towards the antihistamine norastemizole <99TL6875>. Similarly, chloroimidazole **41** was converted to 2-(alkylamino)benzimidazoles **42** under high pressure $\mathrm{S_NAr}$ reactions <99TL2439>. Reaction of thioureas **43** in the presence of mercuric(II) oxide and catalytic sulfur generated 2-(alkylamino)benzimidazoles **44** <99TL1103>. A necklace-coded polymer-supported combinatorial synthesis of 2-arylaminobenzimidazoles from resin-supported *o*-phenyldiamine with arylisothiocyanates was investigated <99JCC368>. Synthesis of marine alkaloids isonaamine A, dorimidazole A, and preclathridine A via iminophosphoranemediated preparation of 2-amino-1,4-disubstituted imidazoles from α-azido esters was described <99JOC2540>. *m*-Nitrophenylguanidines, carrying a vicarious leaving group, cyclized under basic conditions to yield nitro-2-(alkylamino)benzimidazoles <99JCS(P1)1153>. A solid-phase synthesis of 2-aminoimidazolones from resin-bound *S*-methyl isothioureas has been developed <99OL1351>.

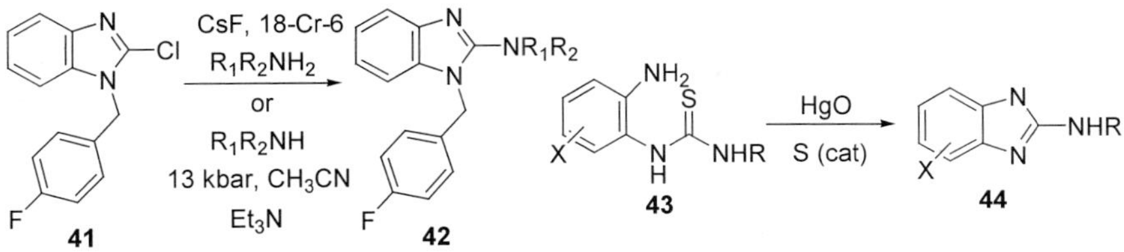

2-(Alkylthio)benzimidazoles are important substrates in the synthesis of pharmaceutical agents. An efficient liquid-phase synthesis of 2-(alkylthio)-5-(carbamoyl)benzimidazoles from *S*-alkylation of a PEG-bound benzimidazole-2-thione was developed <99TL7247, 99SL810>. 2-Mercaptobenzimidazoles were prepared from 1,2-diaminobenzene and methyl isothiocyanate in refluxing ethanol <99SC289>. Alkylations of 2-mercaptobenzimidazole derivatives have been achieved efficiently in the presence of phase transfer reagent, Dq-Br, 2-benzilidine-*N,N,N,N',N',N'*-hexaethyl propane-1,3-diammonium dibromide <99SC4087>.

Solid-phase and liquid-phase combinatorial synthesis of benzimidazole libraries continue. A new "traceless" solid phase synthesis of a benzimidazole library with trifluoroacetic acid cleavage of solid-supported aniline **45** in the presence of trimethylorthoformate to give substituted benzimidazole **46** was disclosed <TL2665>. Addition of aldehydes to resin **47** followed by acidic cleavage produced substituted benzimidazoles **48** <99TL6185>. Similar studies using a solid-phase traceless synthesis of benzimidazoles with three combinatorial steps was also reported <99TL7633>. The parallel synthesis of polyethylene glycol-supported biaryl benzimidazoles and imidazopyridines were prepared from polymer-bound aldehydes with *o*-arylenediamines <99SL307>. An efficient liquid phase synthesis of substituted benzimidazolones **50** prepared from cyclization of resin-bound *o*-arylenediamine **49** and triphosgene followed by methanolic cleavage was reported <99TL6443>. Microwave-accelerated condensation of aldehydes, amines, and isocyanides on montmorillonite K10 clay led to a rapid and solvent-free method for synthesis of imidazo[1,2-*a*]pyridines, imidazo[1,2-*a*]pyrazines and imidazo[1,2-*a*]pyrimidines <99TL7665>. An efficient method for the solid-phase synthesis of 1,3,4-trisubstituted-2-imidazolidones (or imidazolidinethiones) from reduced *N*-acylated dipeptides was reported <99JCC195>.

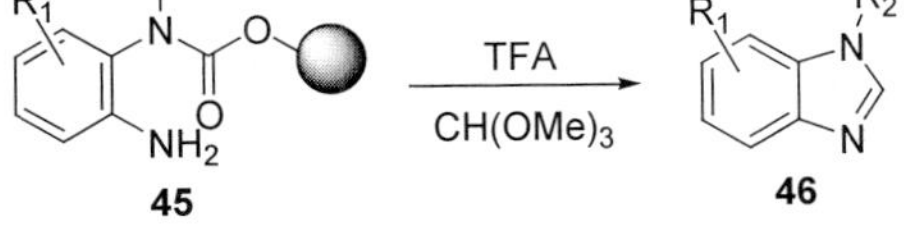

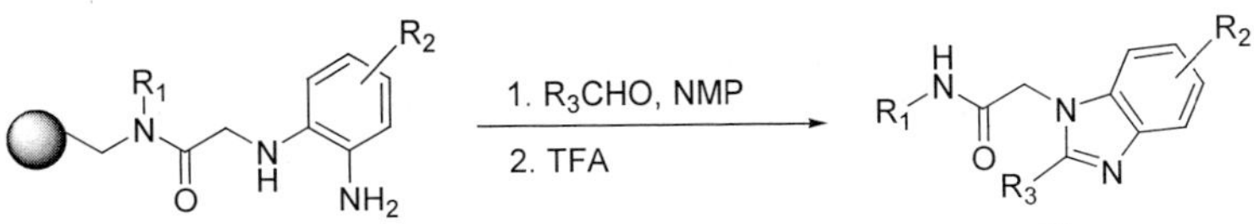

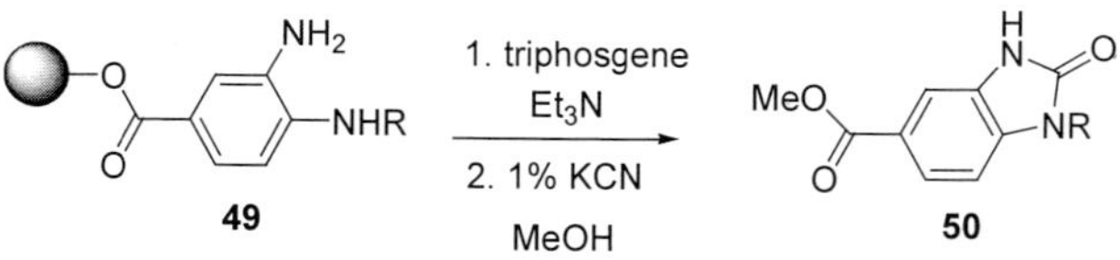

Solid phase attachment of histidine-containing peptides by anchoring the imidazole ring to trityl resins has been developed for combinatorial library preparation of diketopiperazines <99TL809>. Histidine, histamine, and urocanic acid are all imidazole-containing molecules that have been attached to a trityl-type resin to allow their application to combinatorial chemistry <99TL2825>.

Fused heterocyclic rings with imidazoles or benzimidazoles have been synthesized. Thus, a facile synthesis of 6-substituted benzimidazo[1,2-*c*]quinazolines under microwave irradiation was reported <99SC2617>. A new synthesis of substituted imidazo[4,5-*b*]pyridinones by reductive cyclization of 4-nitro-1*H*-imidazol-5-yl di- and tri-carbonyl compounds provided a unique access to these class of compounds <99JCS(P1)629>. The first example of the imidazo[4,5-*c*]isoxazole ring system was prepared from the thermolysis of 2-(4-nitro-1*H*-imidazol-5yl)acetate and malonate derivatives <99JCS(P1)817>. Also, a convenient approach to the synthesis of the imidazo[5,1-*b*]oxazole ring system was disclosed <99H(50)1081>. Imidazo[4,5-*c*]pyridines were prepared from the acid catalyzed reaction of 3-amino-4-acylaminopyridines <99JHC1143>. The synthesis of β-L-xylofuranosyl benzimidazoles was observed by an unexpected intramolecular displacement of 2-chlorobenzimidazole with 1,2-di-*O*-acetyl-3,5-di-*O*-benzolyl-L-xylofuranose <99JOC4169>.

Structure-activity relationships of imidazole-containing structures have dominated investigations in medicinal chemistry in 1999 for active biological entities. Palladium(0)-catalyzed amination of bromobenzimidazoles with piperazines provided entry to new (benzimidazolyl)piperazines for affinity studies of the 5-HT_{1A} receptor ligands <99BML2339>. 5-Substituted 1-phenylbenzimidazoles were synthesized as selective inhibitors of the platelet-derived growth factor receptor <99JMC2373>. Imidazole-substituted biphenyls were found to be a new class of highly potent *in vivo* active inhibitors of P450 17 as potential therapeutics for treatment of prostate cancer <99BMC913>. 2-Amino-3-substituted-6-[(*E*)-1-phenyl-2-(*N*-methylcarbamoyl)vinyl]imidazol[1,2-*a*]pyridines were discovered as a novel class of inhibitors of human rhinovirus <99JMC50>. 2-Nitroimidazoles, substituted with aspirin or salicylic acid as leaving groups linked through the (imidazol-5-ly)methyl position, were prepared as studies of bioreductively-activated prodrugs for targeting hypoxic tissues <99BML1267>. Structure-activity relationship of a series of diaminoalkyl substituted benzimidazoles as neuropeptide Y1 receptor antagonists and as serine protease thrombin inhibitors were prepared from combinatorial libraries <99BML647, 99T11619, 99T11641>. 3-Alkyl-(5,5'-diphenyl)imidazolidinediones were synthesized and evaluated as new cannabinoid receptor ligands <99BML2233>. Novel 2'-deoxy-4'-thioimidazole nucleosides were synthesized and evaluated for *in vitro* antiviral activity <99BMC481>. New imidazo[1,2-*a*]pyrazine derivatives with bronchodilatory and cyclic nucleotide phosphodiesterase inhibitory activities were prepared either by direct cyclization from pyrazines or by electrophilic substitutions <99BMC1059>. A series of 4-(1-imidazolyl)-2,2-diphenylbutyramides were discovered to be potent and subtype-selective antimuscarinic agents <99BMC1151>. Several imidazolium salt derivatives were also prepared for evaluation as subtype-selective antimuscarinic agents <99BML3003>. Imidazothiadiazine dioxides were prepared, representing new lead compounds for antiviral drug research

<99BMC1617>. Imidazole-containing amino acids, larger homologues and analogues of L-histidine, were prepared and found to be selective inhibitors of nitric oxide synthases <99BMC1941>. Guanidine-substituted imidazoles were also prepared and evaluated as inhibitors of the three isoforms of nitric oxide synthases <99BML2953>. Novel chiral *N*-alkylcarbamate derivatives of 3-(1*H*-imidazol-4-yl)propanol were prepared as selective H_3-receptor antagonists <99JMC593, 4269>. Acetylene-based and acetylene-cyclopropane based imidazole-containing compounds have also been employed as selective H_3-receptor antagonists <99JMC903>. 2-(*R* and *S*)-Amino-3-(1*H*-imidazol-4(5)-yl)propyl ether derivatives were prepared as a series of new chiral histamine H_3 receptor ligands <99JMC1193>. The synthesis and histamine H_3 receptor activity of 4-(*n*-alkyl)-1*H*-imidazoles and 4-(ω-phenylalkyl)-1*H*-imidazoles were investigated <99BMC3003>. Various approaches to the synthesis of 2-(1*H*-imidazol-4-yl)cyclopropylamines were investigated in the characterization of the binding site of the histamine H_3 receptor <99JMC1115>. Imifuramine and its stereoisomers, novel imidazole *C*-nucleoside derivatives, were found to exhibit histamine H_3-agonistic activity <99TL2561>. Also, 4(5)-[5-(aminomethyl)-2,5-dihydrofuran-2yl]imidazoles were found to be histamine H_3-agonists <99JOC8608>. A family of peptidomimetic inhibitors of protein geranylgeranyltransferase I-containing imidazole derivatives at the *N*-terminus of the dipeptide was reported <99JMC1333>. Similarly, inhibitors of farnesyl protein transferase based upon a pseudotripeptide template comprising of an imidazole group substituted with a hydrophobic component has been described <99JMC3356>. Piperazine imidazo[1,5-*a*]quinoxaline ureas were developed as high-affinity $GABA_A$ ligands for the γ-aminobutyric acid A/benzodiazepine receptor complex <99JMC1123>. 1-(3-Cyanobenzylpiperidin-4-yl)-5-methyl-4-phenyl-1,3-dihydroimidazol-2-one was found to be a selective high-affinity antagonist for the human dopamine D_4 receptor with excellent selectivity over ion channels <99JMC2706>. The design of highly active analogues of pyrrolo[1,2-*a*]benzimidazole anti-tumor agents were reported <99JMC3324>. 2-(2-Phenylcyclopropyl)imidazolines were examined as I_1 and I_2 imidazoline receptors <99JMC2737>. A series of novel potent and selective diaryl tetrasubstituted imidazole inhibitors of p38 mitogen-activated protein kinase were prepared <99JMC2180>. Anti-inflammatory and analgesic activity evaluation of some mercapto pyrimidobenzimidazole derivatives were determined <99S878>. Both enantiomers of 1-[2-benzofuranyl(4-chlorophenyl)methyl]-1-*H*-imidazole were synthesized in the study of potent aromatase inhibitors, which show promise as chemotherapeutic agents for the treatment of estrogen-dependent tumors <99TL7289>. Several *N*-sulfonylated benzimidazoles were synthesized as potential antiviral agents <99BML2525>. The first potent, specific, and cell-penetrable AMP deaminase inhibitors were discovered through an investigation of 3-substituted 3,6,7,8-tetrahydroimidazo[4,5-*d*][1,3]diazepin-8-ol analogues <99JACS308>. A cellulose-derived imidazole was synthesized as part of studies of inhibitors of glycosidase activity <99JACS2621>. A rational design of sequence-specific DNA alkylating agents, based on duocarmycin A and pyrrole-imidazole hairpin polyamides, were synthesized <99JACS4961>. Novel 2-phenylimidazo[1,2-*a*]pyridine derivatives were found to be potent and selective ligands for peripheral benzodiazepine receptors <99JMC3934>. Process research in the improvement of RWJ-51204, a lead compound in treating anxiety-related disorders, with a pyrido[1,2-*a*]benzimidazole core was reported <99OPRD260>. 4,5-Diamide imidazoles were evaluated for HIV-1 protease inhibitor activity <99OL249>. 4-Benzyl-1-[4-(1*H*-imidazol-4-yl)but-3-ynyl]piperidine has been identified as a potent antagonist of the NR1A/2B subtype of the NMDA receptor <99BML2815>. Imidazole sugars were prepared and tested in inhibitory assays with glycosidases <99EJOC893>. A highly chemoselective route to the imidazolyl trichloroacetimidate fragment was applied to the convergent, practical synthesis of the immunosuppressant L-733,725 <99JOC1859>. New benzimidazole-4-carboxylic acid derivatives were found to possess affinity for serotoninergic 5-HT_3 receptors <99BMC2271,

99JMC5020>. Imidazole-containing diarylether and diarylsulfone inhibitors of farnesyl-protein transferase has been described <99BML3301>.

5.4.4 1,2,3-TRIAZOLES AND RING-FUSED DERIVATIVES

Zinc chloride-doped natural phosphate was shown to have catalytic behavior in the 1,3-dipolar cycloadditions of nucleoside acetylenes with azides to form triazolonucleosides <99SC1057>. A soluble polymer-supported 1,3-dipolar cycloaddition of carbohydrate-derived 1,2,3-triazoles has been reported <99H(51)1807>. 2-Styrylchromones and sodium azide were employed in the synthesis of 4(5)-aryl-5(4)-(2-chromonyl)-1,2,3-triazoles <99H(51)481>. Lead(IV) acetate oxidation of mixed bis-aroyl hydrazones of biacetyl led to 1-(α-aroyloxyarylideneamino)-3,5-dimethyl-1,2,3-triazoles <99H(51)599>. Reaction of 1-amino-3-methylbenzimidazolium chloride with lead(IV) acetate afforded 1-methyl-1*H*-benzotriazole <99BML961>. Hydrogenation reactions of some [1,2,3]triazolo[1,5-*a*]pyridines, [1,2,3]triazolo[1,5-*a*]quinolines, and [1,2,3]triazolo[1,5-*a*]isoquinolines were studied <99T12881>.

Substituents located in the *N*-1 position of 1,2,3-benzotriazoles have been useful as reagents in a variety of reactions. 1-Cyanobenzotriazole (**51**) has been found to participate in electrophilic cyanations of sp^2 and sp carbanions <99JOC313>. 1*H*-Benzotriazole-1-yl methanesulfonate (**52**) was explored as a regioselective *N*-mesylating reagent <99TL117>. Reagent **52** mesylated molecules containing both primary and secondary amines on the primary amino position and mesylation occurred on the amino group in molecules containing both amino and hydroxy groups. 1*H*-Benzotriazole-1-yl alkyl carbonates (**53**) were convenient and inexpensive coupling agents in preparation of active esters for the synthesis of amides <99SC2547>. A general and efficient route to thionoesters via thionoacyl nitrobenzotriazoles **54** was reported <99JOC1065>.

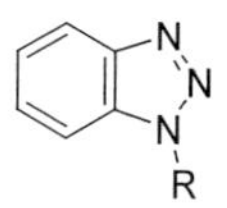

51 R = CN
52 R = OMs
53 R = OCO_2R

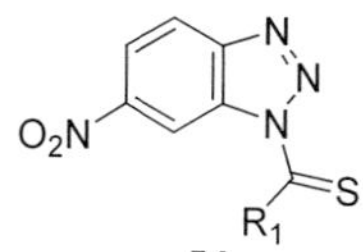

54

Benzotriazole-based methodologies continued to be dominant in 1999. Benzotriazole-thionyl chloride combination proved to be an efficient system for transformation of aldoximes to nitriles <99SC1741>. A new method of removal of the benzotriazole moiety with lithium naphthalide was developed <99T4271>. Vilsmeier-type reagent **55** with β-enaminonitriles provided a regioselective route to the preparation of nicotinonitriles <99JOC6076>. *N*-[α-Amino(hetero)arylmethyl]benzotriazoles **56** reacted with sodium phenolates to afford amino(hetero)arylmethylated phenols <99JOC6071>. A one-pot synthesis of *N*-imidoylbenzotriazoles **57**, useful substitutes for imidoyl chlorides via benzotriazole-tosylate mediated Beckmann rearrangement of oximes under basic conditions was reported <99OL577>. Nucleophilic substitutions of benzotriazolylalkyl chlorides with Grignard reagents provided direct routes to benzotriazoloalkyl(hetero)aromatic compounds **58** <99S1437>. A convenient and stereospecific (*E*)-stilbene synthesis from tosylhydrazones of benzaldehydes and *N*-benzylbenzotriazoles via Shapiro-type transformation was described <99JOC3332>. A new method of synthesis of propargylic amines and ethers via benzotriazole derivatives using sodium dialkynyldiethylaluminates was exploited <99JOC488>. A stereospecific synthesis of 2-azapodolphyllotoxin analogues based on the benzotriazole methodology was investigated <99TA255>. 1-(Chloromethyl)benzotriazole reacted with sodium dialkyl phosphites to give dialkyl-(1-benzotriazolmethyl)phosphonates **59**, potential Horner-Emmons reagents <99SC803>. Treatment of 10-(benzotriazol-1-yl)alkylphenothiazines with electron-rich alkenes in the presence of zinc bromide gave tetra- and pentacyclic phenothiazines <99JHC473>. *N*-Aryl-1*H*-benzotriazolyl-1-methanamines underwent Lewis-acid assisted reactions with allyltrimethylsilanes to give 4-(trimethylsilyl)methyl-1,2,3,4-tetrahydroquinolines <99JHC371>. 1-(Benzotriazol-1-yl)alkyl esters were obtained from the treatment of various aldehydes with the corresponding *N*-acylbenzotriazoles in the presence of a catalytic amount of potassium carbonate <99JHC777>. Benzotriazole methodology was used to prepare novel analogs of indolizines and pyrrolo[2,1-*a*]isoquinolines <99JOC7618> and polycyclic fused phenanthridines <99JHC927>. Ethyl (2*S*,4*S*)-*N*-(benzotriazol-1-yl)methyl-4-phenyloxazolidine-2-carboxylate was converted to chiral *N*-substituted oxazolidines via regiospecific substitutions of the benzotriazolyl residue <99JCR(S)162>. Various (hetero)aromatic amides were efficiently synthesized by direct carbamoylation of benzotriazole-1-carboxamides with organometallic reagents <99JCR(S)230>. A general route to 1-(1-alkenyl)benzotriazoles from 1-(1-silyl-alkylsubstituted)benzotriazoles via the Peterson olefination process <99T11903> and its application in the synthesis of hindered spiro-oxindoles by photolysis of 1-(1-alkenyl)benzotriazoles has been developed <99T11927>. Benzotriazole methodology was applied to the synthesis of ethyl 6-substituted-2,3,3a,4,5,6-hexahydro-1*H*-indolo[3,2,1-*de*][1,5]naphthyridine-2-carboxylates from tryptophan <99T3489>. *N*-(arylprop-2-en-1-yl)benzotriazoles were prepared by the Heck reaction <99H(50)767>. A direct synthesis of α-(benzotriazol-1-yl)alkyl ethers has been described <99H(51)1877>. Triazolo[4,5-*f*]quinolines were derived from 5-amino-(1*H* and 2-methyl-2*H*)benzotriazoles with β-diketones and 3-buten-2-one <99H(51)2171>. 1-[(Dialklylamino)methyl]benzotriazoles reacted with allyl- and propargyltrimethylsilanes in the presence of aluminum chloride to give 4-(trimethylsilyl)aminoalkanes and 4-(trimethylsilyl)aminoalk-2-enes <99O4270>. Amidoalkylations of enamines, silyl enol ethers, and vinyl ethers with *N*-(1-benzotriazol-1-ylalkyl)amides provided novel syntheses of β-amidoalkyl ketones and aldehydes <99JOC7622>. Palladium(0)-catalyzed reactions of allylic benzotriazoles with enamines allowed a novel entry into the stereoselective synthesis of (4*E*)-γ,δ-unsaturated ketones <99JOC7625>. α-Benzotriazolyl ketones and α,β-unsaturated ketones were employed in the novel and efficient synthesis of the 2,4,6-trisubstituted pyridine ring system <99S2114>.

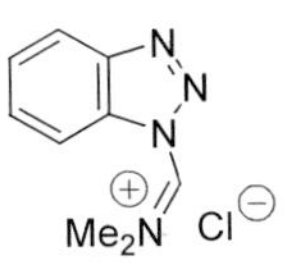

55

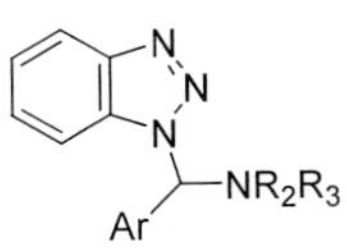

56

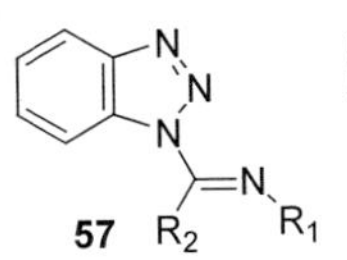

57

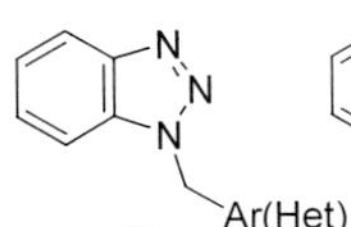

58

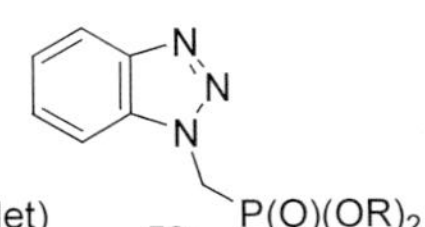

59

Solid-supported benzotriazole methodologies have also appeared. A polymer-supported benzotriazole **60** was employed as a novel traceless linker in the reaction of amines and aldehydes to form Mannich-type adducts **61**, which were cleaved with Grignard reagents to provide a small library of homologated secondary and tertiary amine products **62** <99JOC4972>. Similarly, a benzyl or amide polymer-bound 1*H*-benzotriazole support was used to prepare tertiary amines **62** with analogous reagents <99JCC173, 99JCC317>.

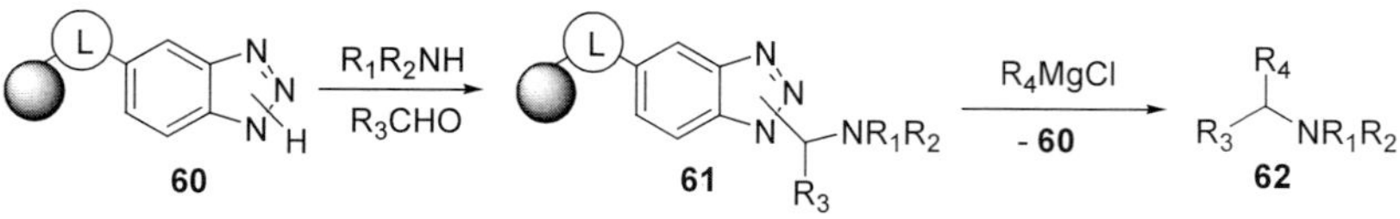

The synthesis, characterization, and *in vitro* anti-tumor activity of two novel podophyllotoxins, 4β-(5-methyl-1,2,3-triazol-1-ly)podophyllotoxin, and 4β-(5-phenyl-1,2,3-triazol-1-ly)podophyllotoxin were described <99SC2053>. Triazolopyrimidines were synthesized and examined for the corticotropin-releasing factor receptor binding affinity <99JMC833>. The synthesis and biological activity of 1-, 2- or 3-substituted benzothieno[2,3-*d*]triazole derivatives structurally related to trazodone, one of the most-used antidepressants, was investigated <99JHC549>. Triazolo[4,5-*d*]pyridazine nucleosides were evaluated for HIV-1 activity <99BMC2373>.

5.4.5 1,2,4-TRIAZOLES AND RING-FUSED DERIVATIVES

1-Phthalazinylhydrazones **63** afforded *s*-triazolo[3,4-*a*]phthalazines **64** by reaction with thianthrene cation radical perchlorate <99SC583>. An efficient synthesis of 6-substituted 2-(2*H*-[1,2,4]triazol-3-ylmethyl)-1,2,3,4-tetrahydroisoquinolines using a bis-alkylation process has been described <99SC645>. Reactions of *N*-alkyl-*N*-formyl hydrazines **65**, with *in situ* generated imidoyl chlorides **66**, efficiently afforded previously inaccessible 3,4-disubstituted 1-alkyl-4*H*-1,2,4-triazol-1-ium salts **67** <99SC1>. A novel synthesis of chiral 5-aryltriazolo[3,4-*b*]-3-α-phenylethyl-2,4-2*H*-1,3,5-thiadiazines was achieved from an intramolecular bis-Mannich reaction of 3-aryl-5-mercapto-1,2,4-triazole, formalin, and *S*-(-)-α-phenylethylamine in the presence of acid <99SC2027>. [1,2,4]Triazolo[1,5-*a*]pyrimidinium–2-amides were obtained from the reaction of 4-alkyl-3,5-diamino-1,2,4-triazoles with pentane-2,4-dione <99JCS(P1)1527>. Reactions of 6-alkoxy-2-aryl-4*H*-1,3-oxazin-4-ones with hydrazine or phenylhydrazine gave 1,2,4-triazole-5-acetic acid esters <99S483>. A simple one-step synthesis of new 3,5-disubstituted-4-amino-1,2,4-triazoles **69** by the reaction of aromatic nitriles **68** with hydrazine dihydrochloride in ethylene or diethylene glycol was reported <99JHC149>. Three-component condensation of ethyl trifluoroacetate (**70**) and amidines in the presence of sodium hydroxide gave 3-trifloromethyl-5-substituted-1,2,4-triazoles **71** <99JCR(S)300>. Reaction of diphenyldiazomethane with *N*-methyloxy- and *N*-ethyloxycarbonyl-*N*-(2,2,2-trichloroethylidene)amines afforded Δ^3-1,2,4-triazolines <99EJOC1541>. Oxidation of thiosemicarbazones of 4-piperidones furnished spiropiperidino-1,2,4-triazoles <99HC313>. Bis(4-amino-5-mercapto-1,2,4-triazol-3-yl)alkanes were prepared from aliphatic dicarboxylic acids and thiocarbohydrazide <99JCR(S)170>. The synthesis and reactivity of 3-alkylthio-5-cyanomethyl-4-phenyl-1,2,4-triazoles from phenyl isothiocyanate and 2-cyanoacetohydrazide was studied <99JCR(S)76>. A 1,3-dipolar cycloaddition of 1-(chloroalkyl)-1-aza-2-azoniaallene salts and nitriles gave 1,2,4-triazolium salts <99T751>. Reactions of ferrocenylmethylidenehydrazones with dipolarophiles led to 1,2,4-triazole cycloadducts <99T5441>. Intramolecular condensation of 4-substituted 1,2,4-triazol-1-ium 4-substituted benzoyl 2,4,6-trinitrophenylmethylides gave 1*H*-1,2,4-triazolo[5,1-*a*]isoindoles <99H(51)2213>. The preparation of 1,2,4-triazolo[1,5-*c*]quinazolines by cyclocondensation of 3-amino-4-imino-2-thioxo-1,2,3,4-tetrahydroquinazoline with carboxylic acids and anhydrides has been investigated <99JHC1327>.

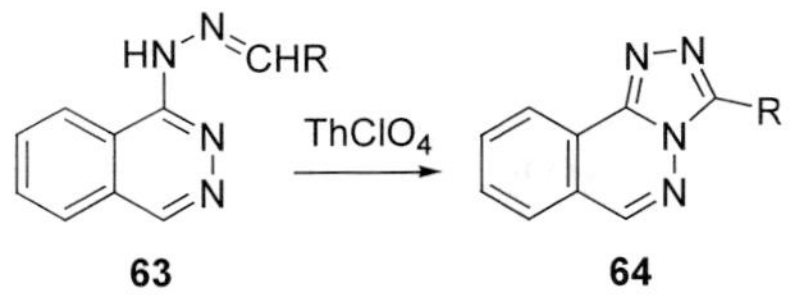

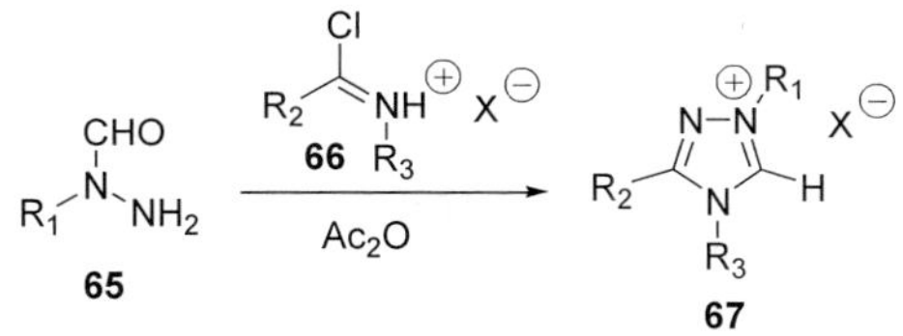

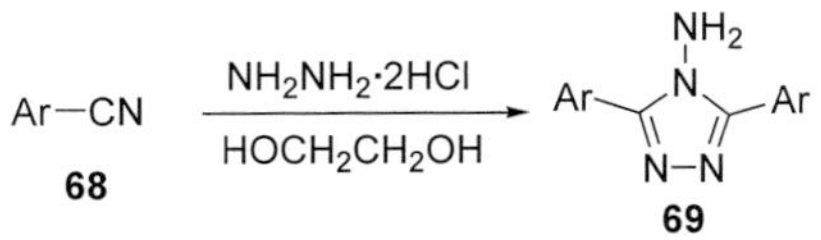

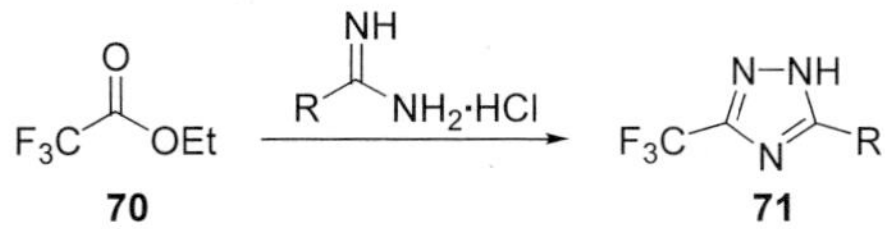

The synthesis of 1,2,4-triazole-functionalized solid support **72** and its use in the solid-phase synthesis of various trisubstituted 1,2,4-triazoles **73** was reported <99OL1189>. A solid-phase library synthesis of triazolopyridazines **76** was achieved by [4+2] cycloaddition of diene amides **74** with 4-substituted urazines **75** <99TL619>.

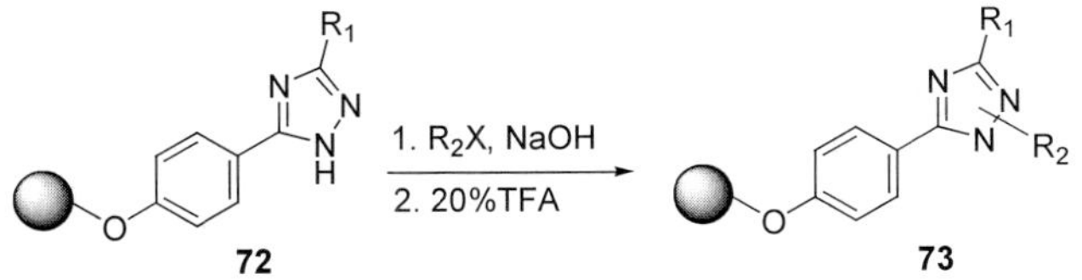

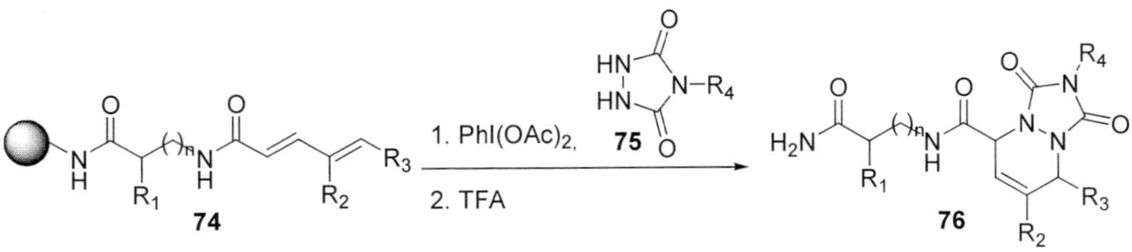

A highly effective method for the selective removal of chlorine from the 7-position of the 1,2,4-triazolo[1,5-*a*]pyrimidine ring in the presence of a chlorine at the 5- or 6-positions involved treatment with zinc-copper in the presence of acetic acid in methanol-tetrahydrofuran <99JHC183>. Cyclocondensation of 3-amino-1,2,4-triazole with substituted methyl cinnamates led selectively to the formation of 7-aryl-6,7-dihydro[1,2,4]triazolo[1,5-*a*]pyrimidin-5(4*H*)-ones <99JHC205>. The regio- and diastereoselective ene reaction of 4-phenyl-1,2,4-triazoline-3,5-dione with chiral allylic alcohols and their derivatives was investigated <99JOC2194>. 3-Propynylthio-1,2,4-triazoles were employed in the synthesis of 2-cyanoamidothiazoles <99H(51)475>.

4,5-Dihydro-1,2,4-triazolo[1,5-*a*]quinoxalin-4-ones bearing different substituents on the benzofused ring and at the 2-position were synthesized and evaluated for their binding at glycine/NMDA and AMPA receptors <99JMC2478>. Pyrazolo[4,3-*e*]-1,2,4-triazolo[1,5-*c*]pyrimidine derivatives were found to be highly potent and selective human A_3 adenosine receptor antagonists <99JMC4473>. 5-(1,2,4-Triazol-4-yl)-3-(piperazinylpropyl)indoles, 4-hdroxy-1-[3-(5-(1,2,4-triazol-4yl)-1*H*-indol-3-yl)propyl]piperidines, and 5-(1,2,4-triazol-4-

yl)-3-[2-(pyrrolidin-1-yl)ethyl]indoles were synthesized and evaluated to be potent agonists for the H_5-HT_{1D} receptor in the treatment of migraine headaches <99JMC677, 99JMC691, 99JMC2087, 99BML3369>. The synthesis and anticonvulsant activity of 3-[3-[(dimethylamino)methyl]-5-methyl-4*H*-1,2,4-triazol-4-yl]-4-(*o*-chlorobenzoyl)pyridine <99JHC201> and 5-(2-chlorophenyl)-7*H*-pyrido[4,3-*f*][1,2,4]triazolo[4,3-*a*][1,4]diazepines <99JHC377> were studied. Heterocycle-peptide hybrid compounds containing the 1,2,4-triazole moiety have been tested as agonists of the thrombin receptor PAR-1 <99BML1423>. 5-Phenyl-10-methyl-7*H*-pyrimido[4,5-*f*][1,2,4]triazolo[4,3-*a*][1,4]diazepine was prepared and found to be a potent anticonvulsant agent <99CJC216>. Asymmetric synthesis of four optically pure D- and L-1,3-dioxolanyl triazole *C*-nucleosides were accomplished and were evaluated for activity against HIV and hepatitis B viruses <99T9073>. The synthesis of 5,7-dimethylpyrazolo[3,4:4,5]thiazolo-[2,3-*c*]-1,2,4-triazole, an analogue of tricyclazole, was published <99H(51)829>. Boc-protected Phe-Gly dipeptidomimetics containing 1,2,4-triazole ring systems were synthesized <99JMC4331>.

5.4.6 TETRAZOLES AND RING-FUSED DERIVATIVES

2-Hydroxytetrazole, a new novel and efficient acylation catalyst in peptide synthesis, was prepared from the sodium salt of ethyl tetrazole-5-carboxylate <99TL6093>. Synthesis of 1-(5-methylisoxazole-3-yl)-5-aryltetrazoles and its pyrolysis to 3-aryl-6-acetyl(1,2,4)triazines was reported <99SC2847>. *tris*(2-Perfluorohexylethyl)tin azide has been developed as a new reagent for the preparation of 5-substituted tetrazoles **78** from nitriles **77** with purification by fluorous/organic liquid-liquid extraction <99T8997>. Oxidative cyclization by lead(IV) acetate of 1-(4-methoxyphenyl)-4-(tetrazol-5-ylmethyl)azetidin-2-one to 3-methoxy-9,9a-dihydroazeto[1,2-*a*]tetrazolo[5,1-*d*][1,5]benzodiazepin-11(10*H*)-ones has been reported <99T8457>. Photochemical conversion of 5-azido-1,3-diaryltetrazolium salts **79** to novel tricyclic meso-ions **80** with a tetrazolo[1,5-*a*]benzimidazole skeleton has been disclosed <99JHC863>. Tetrazolo[1,5-*a*][1,4]benzodiazepin-6-ones were prepared by intramolecular azide cycloadditions onto the cyano group <99H(51)1295>. The synthesis and structural features of 11*H*-tetrazolo[1,5-*c*][2,3]-benzodiazepines were examined <99H(51)1303>. Nucleophilic aromatic substitution reactions of novel 5-(2-methoxyphenyl)tetrazole derivatives with organolithium reagents were investigated <99JOC9301>.

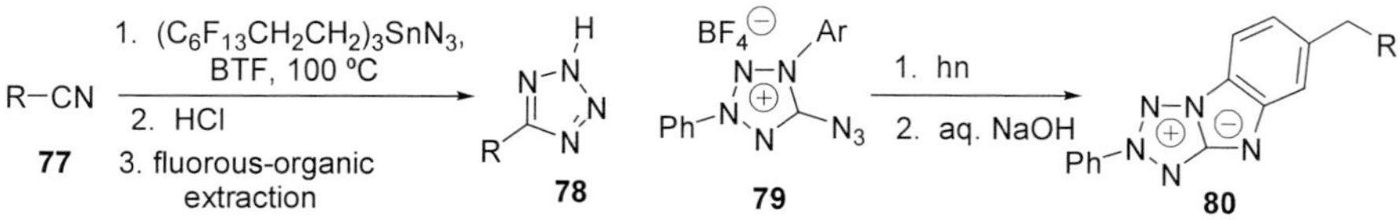

77 78 79 80

The structure-activity relationships of biphenyl tetrazoles **81** as metallo-β-lactamase inhibitors were studied <99BML2741>. Biaryl tetrazoles **82** incorporating amino acid side chains were synthesized and evaluated for *in vitro* growth hormone release <99BML3237>. Tetrazole enol ethers **83** were examined as group 2 metabotropic glutamate receptor antagonists <99BML2173>. Tetrazoles of manno- and rhamno-pyranoses <99T4489> and tetrazoles of manno- and rhamno-furanoses <99T4501> were prepared and evaluated for inhibitory activity towards glycosidases.

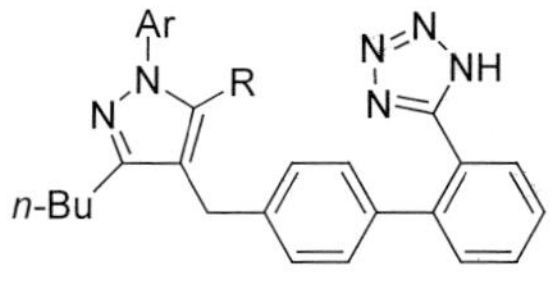

81

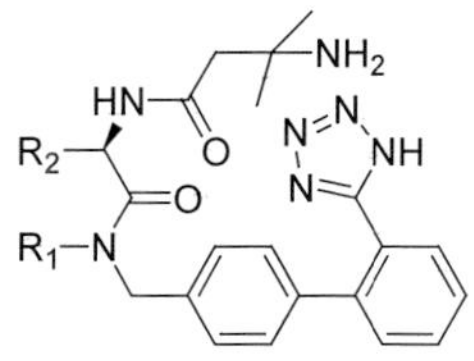

82

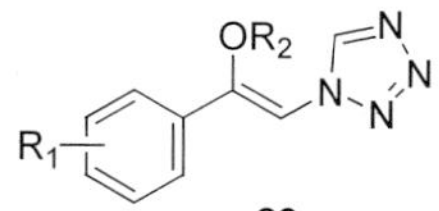

83

5.4.7 REFERENCES

<99BCSJ1327> H. Tomoda, T. Hirano, S. Saito, T. Mutai, K. Araki, *Bull. Chem. Soc. Jpn.* **1999**, *72*, 1327.

<99BCSJ1629> Y. Murakami, T. Yamamoto, *Bull. Chem. Soc. Jpn.* **1999**, *72*, 1629.

<99BMC481> Y. Wang, G. Inguaggiato, M. Jasamai, M. Shah, D. Hughes, M. Slater, C. Simons, *Bioorg. Med. Chem.* **1999**, *7*, 481.

<99BMC1059> O. Vitse, F. Laurent, T.M. Pocock, V. Bénézech, L. Zanik, K.R.F. Elliot, G. Subra, K. Portet, J. Bompart, J.-P. Chapat, R.C. Small, A. Michel, P.-A. Bonnet, *Bioorg. Med. Chem.* **1999**, *7*, 1059.

<99BMC1151> H. Miyachi, H. Kiyota, H. Uchiki, M. Segawa, *Bioorg. Med. Chem.* **1999**, *7*, 1151.

<99BMC1425> V.S. Parmar, A. Kumar, A.K. Prasad, S.K. Singh, N. Kumar, S. Mukherjee, H.G. Raj, S. Goel, W. Errington, M.S. Puar, *Bioorg. Med. Chem.* **1999**, *7*, 1425.

<99BMC1617> A. Martínez, A.I. Esteban, A. Herrero, C. Ochoa, G. Andrei, R. Snoeck, J. Balzarini, E.D. Clercq, *Bioorg. Med. Chem.* **1999**, *7*, 1617.

<99BMC1913> B.G. Wachall, M. Hector, Y. Zhuang, R.W. Hartmann, *Bioorg. Med. Chem.* **1999**, *7*, 1913.

<99BMC1941> Y. Lee, P. Martasek, L.J. Roman, B.S.S. Masters, R.B. Silverman, *Bioorg. Med. Chem.* **1999**, *7*, 1941.

<99BMC2271> M.L. López-Rodríguez, B. Benhamú, A. Viso, M.J. Morcillo, M. Murcia, L. Orensanz, M.J. Alfaro, M.I. Martin, *Bioorg. Med. Chem.* **1999**, *7*, 2271.

<99BMC2373> J.C. Bussolari, R.P. Panzica, *Bioorg. Med. Chem.* **1999**, *7*, 2373.

<99BMC2705> S. Selleri, F. Bruni, C. Costagli, A. Costanzo, G. Guerrini, G. Ciciani, B. Costa, C. Martini, *Bioorg. Med. Chem.* **1999**, *7*, 2705.

<99BMC3003> I.J.P. De Esch, A. Gaffar, W.M.P.B. Menge, H. Timmerman, *Bioorg. Med. Chem.* **1999**, *7*, 3003.

<99BML251> P. G. Baraldi, P. Cozzi, C. Geroni, N. Mongelli, R. Romagnoli, G. Spalluto, *Bioorg. Med. Chem. Lett.* **1999**, *9*, 251.

<99BML647> H. Zarrinmayeh, D.M. Zimmerman, B.E. Cantrell, D.A. Schober, R.F. Bruns, S.L. Gackenheimer, P.L. Ornstein, P.A. Hipskind, T.C. Britton, D.R. Gehlert, *Bioorg. Med. Chem. Lett.* **1999**, *9*, 647.

<99BML961> T. Kaiya, S. Aoyama, K. Kohda, *Bioorg. Med. Chem. Lett.* **1999**, *9*, 961.

<99BML1267> S.A. Everett, M.A. Naylor, K.B. Patel, M.R.L. Stratford, P. Wardman, *Bioorg. Med. Chem. Lett.* **1999**, *9*, 1267.

<99BML1423> D.F. McComsey, M.J. Hawkins, P. Andrade-Gordon, M.F. Addo, D. Oksenberg, B.E. Maryanoff, *Bioorg. Med. Chem. Lett.* **1999**, *9*, 1423.

<99BML2173> S. Kolcqewski, G. Adam, H. Stadler, V. Mutel, J. Wichmann, T. Woltering, *Bioorg. Med. Chem. Lett.* **1999**, *9*, 2173.

<99BML2205> E. Couloigner, D. Cartier, R. Labia, *Bioorg. Med. Chem. Lett.* **1999**, *9*, 2205.

<99BMK2233> M. Kanyonyo, S.J. Govaerts, E. Hermans, J.H. Poupaert, D.M. Lambert, *Bioorg. Med. Chem. Lett.* **1999**, *9*, 2233.

<99BML2339> M.L. López-Rodriguez, A. Viso, B. Benhamú, J. L. Rominguera, M. Murcia, *Bioorg. Med. Chem. Lett.* **1999**, *9*, 2339.

<99BML2525> L. Garuti, M. Roberti, C. Cermelli, *Bioorg. Med. Chem. Lett.* **1999**, *9*, 2525.

<99BML2741> J.H. Toney, K.A. Cleary, G. Hammond, X. Yuan, W.J. May, S.M. Hutchins, W.T. Ashton, E.E. Vanderwall, *Bioorg. Med. Chem. Lett.* **1999**, *9*, 2741.

<99BML2953> R.N. Atkinson, S.B. King, *Bioorg. Med. Chem. Lett.* **1999**, *9*, 2953.

<99BML3003> H. Miyachi, H. Kiyota, M. Segawa, *Bioorg. Med. Chem. Lett.* **1999**, *9*, 3003.

<99BML2815> J.L. Wright, T.F. Gregory, P.A. Boxer, L.T. Meltzer, K.A. Serpa, L.D. Wise, S. Hong-Bae, J.C. Huang, C.S. Konkoy, R.B. Upasani, E.R. Whitemore, R.M. Woodward, K.C. Yang, Z.-L. Zhou, *Bioorg. Med. Chem. Lett.* **1999**, *9*, 2815.

<99BML3217> M. Patel, J.D. Rodgers, R.J. McHugh, Jr., B.L. Johnson, B.C. Cordova, R.M. Klabe, L.T. Bacheler, S. Erickson-Viitanen, S.S. Ko, *Bioorg. Med. Chem. Lett.* **1999**, *9*, 3217.

<99BML3237> P. Lin, J.M. Pisano, W.R. Schoen, K. Cheng, W.W.-S. Chan, B.S. Butler, R.G. Smith, M.H. Fisher, M.J. Wyvratt, *Bioorg. Med. Chem. Lett.* **1999**, *9*, 3237.

<99BML3301> C.J. Dinsmore, T.M. Williams, T.J. O'Neill, D. Liu, E. Rands, J.C. Culberson, R.B. Lobell, K.S. Koblan, N.E. Kohl, J.B. Gibbs, A.I. Oliff, S.L. Graham, G.D. Hartman, *Bioorg. Med. Chem. Lett.* **1999**, *9*, 3301.

<99BML3369> S. Bourrain, J.G. Neduvelil, M.S. Beer, J.A. Stanton, G.A. Showell, A.M. MacLeod, *Bioorg. Med. Chem. Lett.* **1999**, *9*, 3369.

<99CJC216> O.A. Phillips, K.S.K. Murthy, C.Y. Fiakpui, E.E. Knaus, *Can. J. Chem.* **1999**, *77*, 216.

<99CJC1005> A. Khan, A.A. Neverov, A.K. Yatsimirsky, R.S. Brown, *Can. J. Chem.* **1999**, *77*, 1005.

<99EJOC893> J. Streith, H. Rudyk, T. Tschamber, C. Tarnus, C. Strehler, D. Deredas, A. Frankowski, *Eur. J. Org. Chem.* **1999**, 893.

<99EJOC1541> S. Jacquot, A. Belaissaoui, G. Schmitt, B. Laude, M.M. Kubicki, O. Blacque, *Eur. J. Org. Chem.* **1999**, 1541.

<99H(50)767> A.R. Katritzky, A.E.S. Ferwanah, S.N. Denisenko, *Heterocycles* **1999**, *50*, 767.

<99H(50)799> W. Holzer, K. Mereiter, B. Plagens, *Heterocycles* **1999**, *50*, 799.

<99H(50)1081> H. Salgado-Azmora, E. Campos, R. Jimenez, E. Sanchez-Pavon, H. Cervantes, *Heterocycles* **1999**, *50*, 1081.

<99H(51)475> D.K. Bates, M. Xia, M. Aho, H. Mueller, R.R. Raghavan, *Heterocycles* **1999**, *51*, 475.

<99H(51)481> A.M.S. Silva, J.S. Vieira, J.A.S. Cavaleiro, T. Patonay, A. Levai, J. Elguero, *Heterocycles* **1999**, *51*, 481.

<99H(51)599> C.P. Hadjiantoniou-Maroulis, *Heterocycles* **1999**, *51*, 599.

<99H(51)751> R.M. Claramunt, I. Forfar, P. Cabildo, J. Lafuente, J. Barbera, R. Gimenez, J. Elguero, *Heterocycles* **1999**, *51*, 751.

<99H(51)829> C.B. Vicentini, M. Manfrini, D. Mares, A.C. Veronese, *Heterocycles* **1999**, *51*, 829.

<99H(51)1295> G. Broggini, L. Garanti, G. Molteni, G. Zecchi, *Heterocycles* **1999**, *51*, 1295.

<99H(51)1303> A. Chimirri, M. Zappala, R. Gitto, S. Quartarone, F. Bevacqua, *Heterocycles* **1999**, *51*, 1303.

<99H(51)1661> J. Milhavet, A. Gueiffier, L. Bernal, J. Teulade, *Heterocycles* **1999**, *51*, 1661.

<99H(51)1807> S. Freeze, P. Norris, *Heterocycles* **1999**, *51*, 1807.

<99H(51)1877> A.R. Katritzky, M.V. Voronkov, A. Pastor, D. Tatham, *Heterocycles* **1999**, *51*, 1877.

<99H(51)2171> P. Sanna, A. Carta, G. Paglietti, *Heterocycles* **1999**, *51*, 2171.

<99H(51)2213> G.G. Surpateanu, G. Vergoten, A. Elass, G. Surpateanu, *Heterocycles* **1999**, *51*, 2213.

<99HC303> Z. Wang, Z. Li, J. Ren, H. Chen, *Heteroatom Chem.* **1999**, *10*, 303.

<99HC313> D.B. Reddy, A.S. Reddy, A. Padmaja, *Heteroatom Chem.* **1999**, *10*, 313.

<99HC391> A.A. Tolmachev, S.I. Dovgopoly, A.N. Kostyuk, E.S. Kozlov, A.O. Pushechnikov, W. Holzer, *Heteroatom Chem.* **1999**, *10*, 391.

<99HC446> R. Brugidou, J.P. Bazureau, J. Hamelin, Z. Dahmani, M. Rahmouni, *Heteroatom Chem.* **1999**, *10*, 446.

<99HC508> A.O. Abdelhamid, H.F. Zohdi, G.S. Mohamed, *Heteroatom Chem.* **1999**, *10*, 508.

<99CC1461> T. Nagamatsu, T. Fujita, *J. Chem. Soc., Chem. Commun.* **1999**, 1461.

<99JACS308> M.D. Erion, S.R. Kasibhatla, B.C. Bookser, P.D. van Poelje, M.R. Reddy, H.E. Gruber, J.R. Appleman, *J. Am. Chem. Soc.* **1999**, *121*, 308.

<99JACS1459> T.A. Gadosy, R.A. McClelland, *J. Am. Chem. Soc.* **1999**, *121*, 1459.

<99JACS2329> C.D. Abernethy, J.A.C. Clyburne, A.H. Cowley, R.A. Jones, *J. Am. Chem. Soc.* **1999**, *121*, 2329.

<99JACS2621> A. Varrot, M. Schülein, M. Pipelier, A. Vasella, G.J. Davies, *J. Am. Chem. Soc.* **1999**, *121*, 2621.

<99JACS2647> J.K. Huang, E.D. Stevens, S.P. Nolan, J.L. Petersen, *J. Am. Chem. Soc.* **1999**, *121*, 2674.

<99JACS4536> J.R. Perera, M.J. Heeg, H.B. Schlegel, C.H. Winter, *J. Am. Chem. Soc.* **1999**, *121*, 4536.

<99JACS4961> Z.-F. Tao, T. Fujiwara, I. Saito, H. Sugiyama, *J. Am. Chem. Soc.* **1999**, *121*, 4961.

<99JACS8106> M. Kawano, T. Sano, J. Abe, Y. Ohashi, *J. Am. Chem. Soc.* **1999**, *121*, 8106.

<99JACS9758> E.S. Schmidt, A. Jockisch, H. Schmidbaur, *J. Am. Chem. Soc.* **1999**, *121*, 9758.

<99JACS9889> J. Huang, S.P. Nolan, *J. Am. Chem. Soc.* **1999**, *121*, 9889.

<99JCC173> A.R. Katritzky, S.A. Belyakov, D.O. Tymoshenko, *J. Comb. Chem.* **1999**, *1*, 173.

<99JCC195> A. Nefzi, J.M. Ostresh, M. Giulianotti, R.A. Houghten, *J. Comb. Chem.* **1999**, *1*, 195.

<99JCC317> A. Paio, A. Zaramella, R. Ferritto, N. Conti, C. Marchioro, P. Seneci, *J. Comb. Chem.* **1999**, *1*, 317.

<99JCC368> J.M. Smith, J. Gard, W. Cummings, A. Kanizsai, V. Krchnak, *J. Comb. Chem.* **1999**, *1*, 368.

<99JCR(S)76> R.A. Mekheimer, R.M. Shaker, *J. Chem. Research (S)* **1999**, 76.

<99JCR(S)162> A.R. Katritzky, J. Cobo-Domingo, B. Yang, P.J. Steel, *J. Chem. Research (S)* **1999**, 162.

<99JCR(S)170> P.-F. Xu, X.-W. Sun, L.-M. Zhang, Z.-Y. Zhang, *J. Chem. Research (S)* **1999**, 170.

<99JCR(S)230> A.R. Katritzky, D.A. Monteux, D.O. Tymoshenko, S.A. Belyakov, *J. Chem. Research (S)* **1999**, 230.

<99JCR(S)274> M.W. Branco, R.Z. Cao, L.Z. Liu, G. Ege, *J. Chem. Research (S)* **1999**, 274.

<99JCR(S)300> K. Funabiki, N. Noma, G. Kuzuya, M. Matsui, K. Shibata, *J. Chem. Research (S)* **1999**, 300.

<99JCS(P1)615> A.K. Sharma, G. Hundal, S. Obrai, M.P. Mahajan, *J. Chem. Soc., Perkin Trans. 1* **1999**, 615.

<99JCS(P1)629> G. Tennant, C. J. Wallis, G.W. Weaver, *J. Chem. Soc., Perkin Trans. 1* **1999**, 629.

<99JCS(P1)693> S. Yamazaki, M. Hanada, Y. Yanase, C. Fukumori, K. Ogura, T. Saeki, S. Umetani, *J. Chem. Soc., Perkin Trans. 1* **1999**, 693.

<99JCS(P1)817> G. Tennant, C. J. Wallis, G.W. Weaver, *J. Chem. Soc., Perkin Trans. 1* **1999**, 817.

<99JCS(P1)1153> F. Esser, P. Ehrengart, H.P. Ignatow, *J. Chem. Soc., Perkin Trans. 1* **1999**, 1153.

<99JCS(P1)1339> N. Abe, K. Odagiri, M. Otani, E. Fujinaga, H. Fujii, A. Kakehi, *J. Chem. Soc., Perkin Trans. 1* **1999**, 1339.

<99JCS(P1)1527> B.C. Bishop, H. Marley, P.N. Preston, S.H.B. Wright, *J. Chem. Soc., Perkin Trans. 1* **1999**, 1527.

<99JCS(P1)2183> R.A. Mekheimer, *J. Chem. Soc., Perkin Trans. 1* **1999**, 2183.

<99JCS(P1)2429> M.J. Crossley, J.A. McDonald, *J. Chem. Soc., Perkin Trans. 1* **1999**, 2429.

<99JHC11> J. Quiroga, B. Insuasty, A. Hormaza, D. Gamenara, L. Dominguez, J. Saldaña, *J. Heterocyclic Chem.* **1999**, *36*, 11.

<99JHC45> H.G. Bonacorso, S.R.T. Bittencourt, A.D. Wastowski, A.P. Wentz, N. Zanatta, M.A.P. Martins, *J. Heterocyclic Chem.* **1999**, *36*, 45.

<99JHC117> R. Rezner, W. Kramer, R. Neidlein, H. Suschitzky, *J. Heterocyclic Chem.* **1999**, *36*, 117.

<99JHC149> F. Bentiss, M. Lagrenee, M. Traisnel, B. Mernari, H. Elattari, *J. Heterocyclic Chem.* **1999**, *36*, 149.

<99JHC183> W.T. Monte, W.A. Kleschick, J. Bordner, *J. Heterocyclic Chem.* **1999**, *36*, 183.

<99JHC201> V.K. Arora, E.E. Knaus, *J. Heterocyclic Chem.* **1999**, *36*, 201.

<99JHC205> S.M. Desenko, V.V. Lipson, O.V. Shishkin, S.A. Komykhov, V.D. Orlov, E.E. Lakin, V.P. Kuznetsov, H. Meier, *J. Heterocyclic Chem.* **1999**, *36*, 205.

<99JHC217> M.A.P. Martins, R.A. Freitag, A. da Rosa, A.F.C. Flores, N. Zanatta, H.G. Bonacorso, *J. Heterocyclic Chem.* **1999**, *36*, 217.

<99JHC321> K. Makino, H.S. Kim, Y. Kurasawa, *J. Heterocyclic Chem.* **1999**, *36*, 321.

<99JHC371> A.R. Katritzky, X. Cui, Q. Long, *J. Heterocyclic Chem.* **1999**, *36*, 371.

<99JHC377> C.Y. Fiakpui, O.A. Phillips, K.S.K. Murthy, E.E. Knaus, *J. Heterocyclic Chem.* **1999**, *36*, 377.

<99JHC473> A.R. Katritzky, S. Agamy, B. Yang, G. Qiu, *J. Heterocyclic Chem.* **1999**, *36*, 473.

<99JHC549> F. Guerrera, L. Salerno, M.C. Sarvà, M.A. Siracusa, A. Corsaro, V. Pistarà, R. Capasso, G.M. Raso, *J. Heterocyclic Chem.* **1999**, *36*, 549.

<99JHC589> J. Andersch, D. Sicker, *J. Heterocyclic Chem.* **1999**, *36*, 589.

<99JHC595> L. Infantes, C. Foces-Foces, R.M. Claramunt, C. López, J. Elguero, *J. Heterocyclic Chem.* **1999**, *36*, 595.

<99JHC607> S. Zupancic, J. Svete, B. Stanovnik, *J. Heterocyclic Chem.* **1999**, *36*, 607.

<99JHC635> B. Insuasty, H. Insuasty, J. Quiroga, C. Saitz, C. Jullian, *J. Heterocyclic Chem.* **1999**, *36*, 635.

<99JHC697> X.-L. Li, Y.-M. Wang, T. Matsura, J.-B. Meng, *J. Heterocyclic Chem.* **1999**, *36*, 697.

<99JHC767> H. Morita, K. Harada, Y. Okamoto, K. Takagi, *J. Heterocyclic Chem.* **1999**, *36*, 767.

<99JHC771> N. Su, J.S. Bradshaw, X.X. Zhang, P.B. Savage, K.E. Krakowiak, R.M. Izatt, *J. Heterocyclic Chem.* **1999**, *36*, 771.

<99JHC777> A.R. Katritzky, A. Pastor, M.V. Voronkov, *J. Heterocyclic Chem.* **1999**, *36*, 777.

<99JHC799> A. Preseren, J. Svete, B. Stanovnik, *J. Heterocyclic Chem.* **1999**, *36*, 799.

<99JHC813> T. Zimmermann, *J. Heterocyclic Chem.* **1999**, *36*, 813.

<99JHC863> S. Araki, H. Hattori, N. Shimizu, K. Ogawa, H. Yamamura, M. Kawai, *J.* Heterocyclic *Chem.* **1999**, *36*, 863.

<99JHC889> E. Díez-Barra, A. de la Hoz, A. Sánchez-Migallón, J. Elguero, *J. Heterocyclic Chem.* **1999**, *36*, 889.

<99JHC927> A.R. Katritzky, W. Du, Y. Matsukawa, I. Ghiviriga, S.N. Denisenko, *J. Heterocyclic Chem.* **1999**, *36*, 927.

<99JHC1143> J.M. Bakke, J. Riha, *J. Heterocyclic Chem.* **1999**, *36*, 1143.

<99JHC1231> A.J. Angel, A.E. Finefrock, A.R. Williams, J.D. Townsend, T.-H.V. Nguyen, D.R. Hurst, F.J. Heldrich, C.F. Beam, I.T. Badejo, *J. Heterocyclic Chem.* **1999**, *36*, 1231.

<99JHC1291> A. Bendaas, M. Hamdi, N. Sellier, *J. Heterocyclic Chem.* **1999**, *36*, 1291.

<99JHC1311> J. Quiroga, M. Alvarado, B. Insuasty, R. Moreno, E. Ravina, I Estevez, R.H. de Almeida, *J. Heterocyclic Chem.* **1999**, *36*, 1311.

<99JHC1327> W.-D. Pfeiffer, A. Hetzheim, P. Pazdera, A. Bodtke, J. Mücke, *J. Heterocyclic Chem.* **1999**, *36*, 1327.

<99JMC50> C. Hamdouchi, J. de Blas, M. del Prado, J. Gruber, B.A. Heniz, L. Vance, *J. Med. Chem.* **1999**, *42*, 50.

<99JMC593> A. Sasse, K. Kiec-Kononowics, H. Stark, M. Motyl, S. Reidemeister, C.R. Ganellin, X. Ligneau, J.-C. Schwartz, W. Schunack, *J. Med. Chem.* **1999**, *42*, 593.

<99JMC677> F. Sternfeld, A.R. Guiblin, R.A. Jelley, V.G. Matassa, A.J. Reeve, P.A. Hunt, M.S. Beer, A. Heald, J.A. Stanton, B. Sohal, A.P. Watt, L.J. Street, *J. Med. Chem.* **1999**, *42*, 677.

<99JMC691> M.S. Chambers, L.J. Street, S. Goodacre, S.C. Hobbs, P. Hunt, R.A. Jelley, V.G. Matassa, A.J. Reeve, F. Sternfeld, M.S. Beer, J.A. Stanton, D. Rathbone, A.P. Watt, A.M. MacLeod, *J. Med. Chem.* **1999**, *42*, 691.

<99JMC769> R. Lan, Q. Liu, P. Fan, S. Lin, S.R. Fernando, D. McCallion, R. Pertwee, A. Makriyannis, *J. Med. Chem.* **1999**, *42*, 769.

<99JMC779> A. Akahane, H. Katayama, T. Mitsunaga, T. Kato, T. Kinoshita, Y. Kita, T. Kusunoki, T. Terai, K. Yoshida, Y. Shiokawa, *J. Med. Chem.* **1999**, *42*, 779.

<99JMC833> R.J. Chorvat, R. Bakthavatchalam, J.P. Beck, P.J. Gilligan, R.G. Wilde, A.J. Cocuzza, R.W. Hobbs, R.S. Cheeseman, M. Curry, J.P. Rescinito, P. Krenitsky, D. Chidester, J.A. Yarem, J.D. Klaczkiewicz, C. N. Hodge, P.E. Aldrich, Z.R. Wasserman, C.H. Fernandez, R. Zaczek, L.W. Fitzgerald, S.-M. Huang, H. L. Shen, Y.N. Wong, B.M. Chien, C.Y. Quon, A. Arvanitis, *J. Med. Chem.* **1999**, *42*, 833.

<99JMC903> S.M. Ali, C.E. Tedford, R. Gregory, M.K. Handley, S.L. Yates, W.W. Hirth, J.G. Phillips, *J. Med. Chem.* **1999**, *42*, 903.

<99JMC1115> I.J.P. De Esch, R.C. Vollinga, K. Goubitz, H. Schenk, U. Appelberg, U. Hacksell, S. Lemstra, O.P. Zuiderveld, M. Hoffman, R. Leurs, W.M.P.B. Menge, H. Timmerman, *J. Med. Chem.* **1999**, *42*, 1115.

<99JMC1123> E.J. Jacobsen, L.S. Stelzer, R.E. TenBrink, K.L. Belonga, D.B. Carter, H.K. Im, W.B. Im, V.H. Sethy, A.H. Tang, P.F. VonVoigtlander, J.D. Petke, W.-Z. Zhong, J.W. Mickelson, *J. Med. Chem.* **1999**, *42*, 1123.

<99JMC1193> J.T. Kovalainen, J.A.M. Christiaans, S. Kotisaari, J.T. Laitinen, P.T. Männistö, L. Tuomisto, J. Gynther, *J. Med. Chem.* **1999**, *42*, 1193.

<99JMC1333> A. Vasudevan, Y. Qian, A. Vogt, M.A. Blaskovich, J. Ohkanda, S.M. Sebti, A.D. Hamilton, *J. Med. Chem.* **1999**, *42*, 1333.

<99JMC2046> B. Zacharie, M. Lagraoui, M. Dimarco, C.L. Penney, L. Gagnon, *J. Med. Chem.* **1999**, *42*, 2046.

<99JMC2087> M.B. van Niel, I. Collins, M.S. Beer, H.B. Broughton, S.K.F. Cheng, S.C. Goodacre, A. Heald, K.L. Locker, A.M. MacLeod, D. Morrison, C.R. Moyes, D. O'Connor, A. Pike, M. Rowley, M.G.N. Russell, B. Sohal, J.A. Stanton, S. Thomas, H. Verrier, A.P. Watt, J.L. Castro, *J. Med. Chem.* **1999**, *42*, 2087.

<99JMC2180> N.J. Liverton, J.W. Butcher, C.F. Claiborne, D.A. Claremon, B.E. Libby, K.T. Nguyen, S.M. Pitzenberger, H.G. Selnick, G.R. Smith, A. Tebben, J.P. Vacca, S.L. Varga, L. Agarwal, K. Dancheck, A. J. Forsyth, D.S. Fletcher, B. Frantz, W.A. Hanlon, C.F. Harper, S.J. Hofsess, M. Kostura, J. Lin, S. Luell, E.A. O'Neill, C.J. Orevillo, M. Pang, J. Parsons, A. Rolando, Y. Sahly, D.M. Visco, S.J. O'Keefe, *J. Med. Chem.* **1999**, *42*, 2180.

<99JMC2218> A. Costanzo, G. Guerrini, G. Ciciani, F. Bruni, S. Selleri, B. Costa, C. Martini, A. Lucacchini, P.M. Aiello, A. Ipponi, *J. Med. Chem.* **1999**, *42*, 2218.

<99JMC2373> B.D. Palmer, A.J. Kraker, B.G. Hartl, A.D. Panopoulos, R.L. Panek, B.L. Batley, G. H. Lu, S. Trumpp-Kallmeyer, H.D.H. Showalter, W.A. Denny, *J. Med. Chem.* **1999**, *42*, 2373.

<99JMC2478> D. Catarzi, V. Colotta, F. Varano, L. Cecchi, G. Filacchioni, A. Galli, C. Costagli, *J. Med. Chem.* **1999**, *42*, 2478.

<99JMC2706> R.W. Carling, K.W. Moore, C.R. Moyes, E.A. Jones, K. Bonner, F. Emms, R. Marwood, S. Patel, S. Patel, A.E. Fletcher, M. Beer, B. Sohal, A. Pike, P.D. Leeson, *J. Med. Chem.* **1999**, *42*, 2706.

<99JMC2737> W. Quaglia, P. Bousquet, M. Pigini, A. Carotti, A. Carrieri, M. Dontenwill, F. Gentili, M. Giannella, F. Maranca, A. Piergentili, L. Brasili, *J. Med. Chem.* **1999**, *42*, 2737.

<99JMC3324> W.A. Craigo, B.W. LeSueur, E.B. Skibo, *J. Med. Chem.* **1999**, *42*, 3324.

<99JMC3356> N.J. Anthony, R.P. Gomez, M.D. Schaber, S.D. Mosser, K.A. Hamilton, T.J. O'Neil, K.S. Koblan, S.L. Graham, G.D. Hartman, D. Shah, E. Rands, N.E. Kohl, J.B. Gibbs, A.I. Oliff, *J. Med. Chem.* **1999**, *42*, 3356.

<99JMC3934> G. Trapani, M. Franco, A. Latrofa, L. Ricciardi, A. Carotti, M. Serra, E. Sanna, G. Biggio, G. Liso, *J. Med. Chem.* **1999**, *42*, 3934.

<99JMC4269> A. Sasse, H. Stark, S. Reidemeister, A. Huls, S. Elz, X. Ligneau, C.R. Ganellin, J.-C. Schwartz, W. Schunack, *J. Med. Chem.* **1999**, *42*, 4269.

<99JMC4331> S. Borg, R.C. Vollinga, M. Labarre, K. Payze, L. Terenius, K. Luthman, *J. Med. Chem.* **1999**, *42*, 4331.

<99JMC4473> P.G. Baraldi, B. Cacciari, R. Romagnoli, G. Spalluto, K.-N. Klotz, E. Leung, K. Varani, S. Gessi, S. Merighi, P.A. Borea, *J. Med. Chem.* **1999**, *42*, 4473.

<99JMC5020> M.L. López-Rodríguez, B. Benhamú, M.J. Morcillo, I.D. Tejada, L. Orensanz, M.J. Alfaro, M.I. Martín, *J. Med. Chem.* **1999**, *42*, 5020.

<99JOC313> T.V. Hughes, M.P. Cava, *J. Org. Chem.* **1999**, *64*, 313.

<99JOC488> J.H. Ahn, M.J. Jound, N.M. Yoon, D.C. Oniciu, A.R. Katritzky, *J. Org. Chem.* **1999**, *64*, 488.

<99JOC1065> M.A. Shalaby, H. Rapoport, *J. Org. Chem.* **1999**, *64*, 1065.

<99JOC1331> X.-T. Zhou, Y.-R. Lin, L.-X. Dai, J. Sun, L.-J. Xia, M.-H. Tang, *J. Org. Chem.* **1999**, *64*, 1331.

<99JOC1859> Z. Song, A. DeMarco, M. Zhao, E.G. Corley, A.S. Thompson, J. McNamara, Y. Li, D. Rieger, P. Sohar, D.J. Mathre, D.M. Tschaen, R.A. Reamer, M.F. Huntington, G.-J. Ho, F.-R. Tsay, K. Emerson, R. Shuman, E.J.J. Grabowski, P.J. Reider, *J. Org. Chem.* **1999**, *64*, 1859.

<99JOC2194> A.-H. Gau, G.-L. Lin, B.-J. Uang, F.-L. Liao, S.-L. Wang, *J. Org. Chem.* **1999**, *64*, 2194.

<99JOC2258> R. Lenaršic, M. Kocevar, S. Polanc, *J. Org. Chem.* **1999**, *64*, 2558.

<99JOC2540> P. Molina, P.M. Fresneda, M.A. Sanz, *J. Org. Chem.* **1999**, *64*, 2540.

<99JOC2814> N. Svenstrup, K.B. Simonsen, N. Thorup, J. Brodersen, W. Dehaen, J. Becher, *J. Org. Chem.* **1999**, *64*, 2814.

<99JOC2941> T.H. Kim, G.-J. Lee, *J. Org. Chem.* **1999**, *64*, 2941.

<99JOC3332> A.R. Katritzky, D.O. Tymoshenko, S.A. Belyakov, *J. Org. Chem.* **1999**, *64*, 3332.

<99JOC4169> J.-L. Girardet, L.B. Townsend, *J. Org. Chem.* **1999**, *64*, 4169.

<99JOC4196> J. Felding, J. Kristensen, T. Bjerregaard, L. Sander, P. Vedsø, M. Begtrup, *J. Org. Chem.* **1999**, *64*, 4196.

<99JOC4492> D. Gao, Y.-K. Pan, *J. Org. Chem.* **1999**, *64*, 4492.

<99JOC4972> K. Schiemann, H.D.H. Showalter, *J. Org. Chem.* **1999**, *64*, 4972.

<99JOC5366> T. Balle, P. Vedsø, M. Begtrup, *J. Org. Chem.* **1999**, *64*, 5366.

<99JOC5499> M. Béres, G. Hajós, Z. Riedl, T. Soós, G. Timári, A. Messmer, *J. Org. Chem.* **1999**, *64*, 5499.

<99JOC6071> A.R. Katritzky, A.A.A. Abdel-Fattah, D.O. Tymoshenko, S.A. Belyakov, I. Ghiviriga, P.J. Steel, *J. Org. Chem.* **1999**, *64*, 6071.

<99JOC6076> A.R. Katritzky, A. Denisenko, M. Arend, *J. Org. Chem.* **1999**, *64*, 6076.

<99JOC6566> G. Krishnamoorthy, S.K. Dogra, *J. Org. Chem.* **1999**, *64*, 6566.

<99JOC6984,6989> T. Isobe, T. Ishikawa, *J. Org. Chem.* **1999**, *64*, 6984, 6989.

<99JOC7618> A.R. Katritzky, G. Qui, B. Yang, H.-Y. He, *J. Org. Chem.* **1999**, *64*, 7618.

<99JOC7622> A.R. Katritzky, Y. Fang, A. Silina, *J. Org. Chem.* **1999**, *64*, 7622.

<99JOC7625> A.R. Katritzky, Z. Huang, Y. Fang, *J. Org. Chem.* **1999**, *64*, 7625.

<99JOC8084> G.J. Griffiths, M.B. Hauck, R. Imwinkelried, J. Kohr, C.A. Roten, G.C. Stucky, *J. Org. Chem.* **1999**, *64*, 8084.

<99JOC8608> S. Harusawa, T. Imazu, S. Takashima, L. Araki, H. Ohishi, T. Kurihara, Y. Sakamoto, Y. Yamamoto, A. Yamatodani, *J. Org. Chem.* **1999**, *64*, 8608.

<99JOC9301> D.P.G. Norman, A.E. Bunnell, S.R. Stabler, L.A. Flippin, *J. Org. Chem.* **1999**, *64*, 9301.

<99O1216> H.M.J. Wang, C.Y.L. Chen, I.J.B. Lin, , *Organometallics* **1999**, *18*, 1216.

<99O2370> J. Huang, H.-J. Schanz, E.D. Stevens, S.P. Nolan, *Organometallics* **1999**, *18*, 2370.

<99O4270> C. Churlaud, J. Pornet, D.C. Oniciu, A.R. Katritzky, *Organometallics* **1999**, *18*, 4270.

<99OL249> P.W. Baures, *Organic Lett.* **1999**, *1*, 249.

<99OL577> A.R. Katritzky, D.A. Monteux, D.O. Tymoshenko, *Organic Lett.* **1999**, *1*, 577.

<99OL949> J.P. Collman, M. Zhong, Z. Wang, *Organic Lett.* **1999**, *1*, 949.

<99OL953> M. Scholl, S. Ding, C.W. Lee, R.H. Grubbs, *Organic Lett.* **1999**, *1*, 953.

<99OL961> J.E. McCusker, C.A. Grasso, A.D. Main, L. McElwee-White, *Organic Lett.* **1999**, *1*, 961.

<99OL1035> E. Alcalde, S. Ramos, L. Pérez-Garcia, *Organic Lett.* **1999**, *1*, 1035.

<99OL1189> A.R. Katritzky, M. Qi, D. Feng, G. Zhang, M.C. Griffith, K. Watson, *Organic Lett.* **1999**, *1*, 1189.

<99OL1307> J. Huang, G. Grasa, S.P. Nolan, *Organic Lett.* **1999**, *1*, 1307.

<99OL1351> M. Gu, M. Fernandez, M.L. Smith, J.A. Flygare, *Organic Lett.* **1999**, *1*, 1351.

<99OL1521> B. Batanero, J. Escudero, F. Barba, *Organic Lett.* **1999**, *1*, 1521.

<99OPRD104> G.P. Baker, I.A. Bourne, M.J. Ford, R.W.G. Foster, T.H. Jackson, R.W. Pannell, M.W. Whitmore, *Org. Proc. Res. Dev.* **1999**, *3*, 104.

<99OPRD260> J.H. Cohen, C.A. Maryanoff, S.M. Stefanick, K.L. Sorgi, F.J. Villani, Jr., *Org. Proc. Res. Dev.* **1999**, *3*, 260.

<99S157> M. Moreno-Mañas, R.M. Sebastián, A. Vallribera, F. Carini, *Synthesis* **1999**, 157.

<99S453> P.G. Baraldi, B. Cacciari, R. Romagnoli, G. Spalluto, *Synthesis* **1999**, 453.

<99S483> A.V. Prep'yalov, I.P. Yakovlev, V.E. Zakhs, *Synthesis* **1999**, 483.

<99S588> S. Caron, E. Vazquez, *Synthesis* **1999**, 588.

<99S878> S.M. Sondhi, V.K. Sharma, R.P. Verma, N. Singhal, R. Shukla, R. Raghubir, M.P. Dubey, *Synthesis* **1999**, 878.

<99S947> T. Mimura, N. Kato, T. Sugaya, M. Ikuta, S. Kato, Y. Kuge, S. Tomioka, M. Kasai, *Synthesis* **1999**, 947.

<99S1437> A.R. Katritzky, A.A.A. Abdel-Fattah, D.O. Tymoshenko, S.A. Belyakov, *Synthesis* **1999**, 1437.

<99S1613> M. Heras, M. Ventura, A. Linden, J.M. Villalgordo, *Synthesis* **1999**, 1613.

<99S1961> P. Grosche, A. Höltzel, T.B. Walk, A.W. Trautwein, G. Jung, *Synthesis* **1999**, 1961.

<99S2114> A.R. Katritzky, A.A.A. Abdel-Fattah, D.O. Tymoshenko, S.A. Essawy, *Synthesis* **1999**, 2114.

<99SC1> J.H. Teles, K. Breuer, D. Enders, H. Gielen, *Synth. Commun.* **1999**, *29*, 1.

<99SC289> N.B. Ambati, V.N.S.R. Babu, V. Anand, P. Hanumanthu, *Synth. Commun.* **1999**, *29*, 289.

<99SC311> P. Seneci, M. Nicola, M. Inglesi, E. Vanotti, G. Resnati, *Synth. Commun.* **1999**, *29*, 311.

<99SC387> A.E. Kihel, M. Benchidmi, E.M. Essassi, R. Danion-Bougot, *Synth. Commun.* **1999**, *29*, 387.

<99SC495> M.E. Rampey, D.R. Hurst, A. Sood, S.L. Studer-Martinez, C.F. Beam, *Synth. Commun.* **1999**, *29*, 495.

<99SC583> K.H. Park, K. Jun, S.R. Shin, S.W. Oh, *Synth. Commun.* **1999**, *29*, 583.

<99SC645> F.J. Urban, R. Breitenbach, *Synth. Commun.* **1999**, *29*, 645.

<99SC655> G. Sabitha, R. Srividya, B. Archana, J.S. Yadav, *Synth. Commun.* **1999**, *29*, 655.

<99SC763> B. Özgün, N. Degirmenbasi, *Synth. Commun.* **1999**, *29*, 763.

<99SC803> X. Huang, H. Qian, *Synth. Commun.* **1999**, *29*, 803.

<99SC1057> H.B. Lazrek, A. Rochdi, Y. Kabbaj, M. Taourirte, S. Sebti, *Synth. Commun.* **1999**, *29*, 1057.

<99SC1171> M.-W. Ding, Z.-F. Xu, T.-J. Wu, *Synth. Commun.* **1999**, *29*, 1171.

<99SC1217> C.-H. Zhou, X.-R. Gu, R.-G. Xie, M.-S. Cai, *Synth. Commun.* **1999**, *29*, 1217.

<99SC1741> S.S. Chaudhari, K.G. Akamanchi, *Synth. Commun.* **1999**, *29*, 1741.

<99SC2027> S. Haijian, W. Zhongyi, S. Haoxin, *Synth. Commun.* **1999**, *29*, 2027.

<99SC2053> L. Tao, Y.-G. Wang, C. Ma, B. Zheng, Y.-Z. Chen, *Synth. Commun.* **1999**, *29*, 2053.

<99SC2355> Z. Wang, Z. Li, J. Ren, H. Chen, *Synth. Commun.* **1999**, *29*, 2355.

<99SC2365> K.V.V. Reddy, P.S. Rao, D. Ashok, *Synth. Commun.* **1999**, *29*, 2365.

<99SC2435> A.E. Kihel, M. Benchidmi, E.M. Essassi, P. Bauchat, R. Danion-Bougot, *Synth. Commun.* **1999**, *29*, 2435.

<99SC2547> J.S. Lee, Y.S. Oh, J.K. Lim, W.Y. Yang, I.H. Kim, C.W. Lee, Y.H. Chung, S.J. Yoon, *Synth. Commun.* **1999**, *29*, 2547.

<99SC2617> M.S. Khajavi, K. Rad-Moghadam, H. Hazarkhani, *Synth. Commun.* **1999**, *29*, 2617.

<99SC2847> K.K. Vasu, K.K. Reddy, , *Synth. Commun.* **1999**, *29*, 2847.

<99SC3025> X. Chen, X. Wang, H. Wang, H. Lian, Y. Pan, Y. Shi, *Synth. Commun.* **1999**, *29*, 3025.

<99SC3959> D. Csányi, G. Timári, G. Hajós, *Synth. Commun.* **1999**, *29*, 3959.

<99SC4087> J.P. Jayachandran, M.-L. Wang, *Synth. Commun.* **1999**, *29*, 4087.

<99SL299> N. Almirante, A. Benicchio, A. Ceri, G. Fedrizzi, G. Marazzi, M. Santagostino, *Synlett* **1999**, 299.

<99SL307> C.G. Blettner, W.A. König, G. Rühter, W. Stenzel, T. Schotten, *Synlett* **1999**, 307.

<99SL765> A. Torrens, J.A. Castrillo, A. Claparols, J. Redondo, *Synlett* **1999**, 765.

<99SL801> P. Vanelle, K. Benakli, L. Giraud, M.P. Crozet, *Synlett* **1999**, 801.

<99SL810> C.-M. Yeh, C.-M. Sun, *Synlett* **1999**, 810.

<99T345> H.G. Bonacorso, M.R. Oliveira, A.P. Wentz, A.D. Wastowski, A.B. de Oliveira, M. Höerner, N. Zanatta, M.A.P. Martins, *Tetrahedron* **1999**, *55*, 345.

<99T353> R.P. Subrayan, P.G. Rasmussen, *Tetrahedron* **1999**, *55*, 353.

<99T449> N. Boukamcha, R. Gharbi, M.-T. Martin, A. Chiaroni, Z. Mighri, A. Khemiss, *Tetrahedron* **1999**, *55*, 449.

<99T541> C. Hamdouch, J. de Blas, J. Ezquerra, *Tetrahedron* **1999**, *55*, 541.

<99T751> N.A. Al-Masoudi, Y.A. Al-Soud, A. Geyer, *Tetrahedron* **1999**, *55*, 751.

<99T1417> P. Molina, A. Tárraga, D. Curiel, C.R. de Arellano, *Tetrahedron* **1999**, *55*, 1417.

<99T2317> J.A. Vega, J.J. Vaquero, J. Alvarez-Builla, J. Ezquerra, C. Hamdouchi, *Tetrahedron* **1999**, *55*, 2317.

<99T2811> E.C. Coad, H. Liu, P.G. Rasmussen, *Tetrahedron* **1999**, *55*, 2811.

<99T3489> A.R. Katritzky, G. Qiu, B. Yang, P.J. Steel, *Tetrahedron* **1999**, *55*, 3489.

<99T3987> B.P. Medaer, G.J. Hoornaert, *Tetrahedron* **1999**, *55*, 3987.

<99T4109> F. Aldabbagh, W.R. Bowman, *Tetrahedron* **1999**, *55*, 4109.

<99T4271> Y.H. Kang, K. Kim, *Tetrahedron* **1999**, *55*, 4271.

<99T4489> B.G. Davis, T.W. Brandstetter, L. Hackett, B.G. Winchester, R.J. Nash, A.A. Watson, R.C. Griffiths, C. Smith, G.W.J. Fleet, *Tetrahedron* **1999**, *55*, 4489.

<99T4501> B.G. Davis, R.J. Nash, A.A. Watson, C. Smith, G.W.J. Fleet, *Tetrahedron* **1999**, *55*, 4501.

<99T4889> P. de la Cruz, E. Espildora, J.J. Garcia, A. de la Hoz, F. Langa, N. Martin, L. Sánchez, *Tetrahedron* **1999**, *55*, 4889.

<99T5441> A. Abrán, A. Csámpai, Z. Böcskei, P. Sohár, *Tetrahedron* **1999**, *55*, 5441.

<99T6167> F.Fabis, S. Jolivet-Fouchet, S. Rault, *Tetrahedron* **1999**, *55*, 6167.

<99T6917> V. Collot, P. Dallemagne, P.R. Bovy, S. Rault, *Tetrahedron* **1999**, *55*, 6917.

<99T7279> A. Szöllósy, T. Tischer, I. Kádas, L. Tóke, G. Tóth, *Tetrahedron* **1999**, *55*, 7279.

<99T7645> K. Kishore, K.R. Reddy, J.R. Suresh, H. Ila, H. Junjappa, *Tetrahedron* **1999**, *55*, 7645.

<99T8111> F. Aldabbagh, W.R. Bowman, E. Mann, A.M.Z. Slawin, *Tetrahedron* **1999**, *55*, 8111.

<99T8457> L.T. Giang, J. Fetter, M. Kajtár-Peredy, K. Lempert, F. Bertha, G.M. Keserú, G. Czira, T. Czuppon, *Tetrahedron* **1999**, *55*, 8457.

<99T8997> D.P. Curran, S. Hadida, S.-Y. Kim, *Tetrahedron* **1999**, *55*, 8997.

<99T9073> F. Qu, J.H. Hong, J. Du, M.G. Newton, C.K. Chu, *Tetrahedron* **1999**, *55*, 9073.

<99T9623> J.R. Carrillo, A. Díaz-Ortiz, A. de la Hoz, M. J. Gómez-Escalonilla, A. Moreno, P. Prieto, *Tetrahedron* **1999**, *55*, 9623.

<99T10351> S. Kuroda, A. Akahane, H. Itani, S. Nishimura, K. Durkin, T. Kinoshita, I. Nakanishi, K. Sakane, *Tetrahedron* **1999**, *55*, 10351.

<99T11619> M.G. Siegel, M.O. Chaney, R.F. Bruns, M.P. Clay, D.A. Schober, A.M.V. Abbema, D.W. Johnson, B.E. Cantrell, P.J. Hahn, D.C. Hunden, D.R. Gehlert, H. Zarrinmayeh, P.L. Ornstein, D.M. Zimmerman, G.A. Koppel, *Tetrahedron* **1999**, *55*, 11619.

<99T11641> M.G. Johnson, D.D. Bronson, J.E. Gillespie, D.S. Gifford-Moore, K. Kalter, M.P. Lynch, J.R. McCowan, C.C. Redick, D.J. Sall, G.F. Smith, R.J. Foglesong, *Tetrahedron* **1999**, *55*, 11641.

<99T11903> D.P.M. Pleynet, J.K. Dutton, A.P. Johnson, *Tetrahedron* **1999**, *55*, 11903.

<99T11927> J.K. Dutton, D.P.M. Pleynet, A.P. Johnson, *Tetrahedron* **1999**, *55*, 11927.

<99T12881> B. Abarca, R. Ballesteros, M. Elmasnouy, *Tetrahedron* **1999**, *55*, 12881.

<99T12957> E. Hasegawa, A. Yoneoka, K. Suzuki, T. Kato, T. Kitazume, K. Yanagi, *Tetrahedron* **1999**, *55*, 12957.

<99T14451> F. Palacios, A.M.O. de Retana, J. Pagalday, *Tetrahedron* **1999**, *55*, 14451.

<99T14523> A.J. Arduengo, III, R. Krafczyk, R. Schmutzler, H.A. Craig, J.R. Goerlich, W.J. Marshall, M. Unverzagt, *Tetrahedron* **1999**, *55*, 14523.

<99T14791> W.H. Midura, J.A. Krysiak, M. Mikolajcyzk, *Tetrahedron* **1999**, *55*, 14791.

<99TA255> A.R. Katritzky, J. C.-Domingo, B. Yang, P.J. Steel, *Tetrahedron: Asymm.* **1999**, *10*, 255.

<99TA487> G. Broggini, L. Garanti, G. Molteni, G. Zecchi, *Tetrahedron: Asymm.* **1999**, *10*, 487.

<99TA2203> G. Broggini, L. Garanti, G. Molteni, T. Pilati, A. Ponti, G. Zecchi, *Tetrahedron: Asymm.* **1999**, *10*, 2203.

<99TA3011> J.G. Fernández-Bolaños, E. Zafra, O. López, I. Robina, J. Fuentes, *Tetrahedron: Asymm.* **1999**, *10*, 3011.

<99TA3873> G. Molteni, T. Pilati, *Tetrahedron: Asymm.* **1999**, *10*, 3873.

<99TA4071> M. Avalos, R. Babiano, P. Cintas, F.J. Higes, J.L. Jiménez, J.C. Palacios, G. Silvero, *Tetrahedron: Asymm.* **1999**, *10*, 4071.

<99TA4447>	G. Broggini, G. Casalone, L. Garanti, G. Molteni, T. Pilati, G. Zecchi, *Tetrahedron: Asymm.* **1999**, *10*, 4447.
<99TL53>	Y.F. Yong, J.A. Kowalski, J.C. Thoen, M. A. Lipton, *Tetrahedron Lett.* **1999**, *40*, 53.
<99TL117>	S.Y. Kim, N.-D. Sung, J.-K. Choi, S.S. Kim, *Tetrahedron Lett.* **1999**, *40*, 117.
<99TL619>	A.M. Boldi, C.R. Johnson, H.O. Eissa, *Tetrahedron Lett.* **1999**, *40*, 619.
<99TL809>	G. Sabatino, M. Chelli, S. Mazzucco, M. Ginanneschi, A.M. Papini, *Tetrahedron Lett.* **1999**, *40*, 809.
<99TL883>	D. Bourissou, G. Bertrand, *Tetrahedron Lett.* **1999**, *40*, 883.
<99TL1103>	J.J. Perkins, A.E. Zartman, R.S. Meissner, *Tetrahedron Lett* **1999**, *40*, 1103.
<99TL1587>	P.de la Cruz, A. D.-Ortiz, J.J. Garcia, M.J. Gomex-Escalonilla, A. de la Hoz, F. Langa, *Tetrahedron Lett.* **1999**, *40*, 1587.
<99TL1623>	A.P. Combs, S. Saubern, M. Rafalski, P.Y.S. Lam, *Tetrahedron Lett.* **1999**, *40*, 1623.
<99TL1739>	C. Seipelt, P. López, T. Kirschgen, A. Dölle, M.D. Zeidler, I. Fonseca, F. H. Cano, P. Ballesteros, *Tetrahedron Lett.* **1999**, *40*, 1739.
<99TL2247>	M. Scholl, T.M. Trnka, J.P. Morgan, R.H. Grubbs, *Tetrahedron Lett.* **1999**, *40*, 2247.
<99TL2439>	I.A. Barrett, M.A. Kerr, *Tetrahedron Lett.* **1999**, *40*, 2439.
<99TL2541>	M. Kawase, M. Hirabayashi, S. Saito, K. Yamamoto, *Tetrahedron Lett.* **1999**, *40*, 2541.
<99TL2561>	S. Harusawa, T. Imazu, S. Takashima, L. Araki, H. Ohishi, T. Kurihara, Y. Yamamoto, A. Yamatodani, *Tetrahedron Lett.* **1999**, *40*, 2561.
<99TL2657>	A. Kiyomori, J.-F. Marcoux, S.L. Buchwald, *Tetrahedron Lett.* **1999**, *40*, 2657.
<99TL2665>	W. Huang, R.M. Scarborough, *Tetrahedron Lett.* **1999**, *40*, 2665.
<99TL2669>	R.A. Batey, C. Y.-Ishii, S.D. Taylor, V. Santhakumar, *Tetrahedron Lett.* **1999**, *40*, 2669.
<99TL2825>	S. Eleftheriou, D. Gatos, A. Panagopoulos, S. Stathopoulos, K. Barlos, *Tetrahedron Lett.* **1999**, *40*, 2825.
<99TL3891>	O.A. Attanasi, P. Filippone, C. Fiorucci, F. Mantellini, *Tetrahedron Lett.* **1999**, *40*, 3891.
<99TL4035>	P.I. Dalko, *Tetrahedron Lett.* **1999**, *40*, 4035.
<99TL4119>	A.S. Kiselyov, *Tetrahedron Lett.* **1999**, *40*, 4119.
<99TL4787>	L. Ackermann, A. Fürstner, T. Weskamp, F.J. Kohl, W.A. Hermann, *Tetrahedron Lett.* **1999**, *40*, 4787.
<99TL5459>	B.V. Meprathu, S. Diltz, P.J. Walsh, J.D. Protasiewicz, *Tetrahedron Lett.* **1999**, *40*, 5459.
<99TL6093>	R.G. Giles, N.J. Lewis, P.W. Oxley, J.K. Quick, *Tetrahedron Lett.* **1999**, *40*, 6093.
<99TL6185>	D. Tumelty, M.K. Schwarz, K. Cao, M.C. Needels, *Tetrahedron Lett.* **1999**, *40*, 6185.
<99TL6193>	D. Tzalis, C. Koradin, P. Knochel, *Tetrahedron Lett.* **1999**, *40*, 6193.
<99TL6443>	P.-C. Pan, C.-M. Sun, *Tetrahedron Lett.* **1999**, *40*, 6443.
<99TL6875>	C.H. Senanayake, Y. Hong, T. Xiang, H.S. Wilkinson, R.P. Bakale, A.R. Jurgens, M.F. Pippert, H.T. Butler, S.A. Wald, *Tetrahedron Lett.* **1999**, *40*, 6875.
<99TL7247>	C.-M. Yeh, C.-M. Sun, *Tetrahedron Lett.* **1999**, *40*, 7247.
<99TL7289>	F. Messina, M. Botta, F. Corelli, C. Mugnaini, *Tetrahedron Lett.* **1999**, *40*, 7289.
<99TL7399>	B. Liu, M.-X. Wang, Z.-T. Huang, *Tetrahedron Lett.* **1999**, *40*, 7399.
<99TL7633>	J.M. Smith, V. Krchnák, *Tetrahedron Lett.* **1999**, *40*, 7633.
<99TL7655>	R.S. Varma, D. Kumar, *Tetrahedron Lett.* **1999**, *40*, 7665.
<99TL7925>	C. Hulme, L. Ma, J. Romano, M. Morrissette, *Tetrahedron Lett.* **1999**, *40*, 7925.
<99TL8097>	S. Jouneau, J.P. Bazureau, *Tetrahedron Lett.* **1999**, *40*, 8097.
<99TL8157>	A. Taher, A.M.Z. Slawin, G.W. Weaver, *Tetrahedron Lett.* **1999**, *40*, 8157.
<99TL8163>	A. Guirado, R. Andreu, J. Gálvez, *Tetrahedron Lett.* **1999**, *40*, 8163.
<99TL8701>	L. De Luca, M. Falorni, G. Giacomelli, A. Porcheddu, *Tetrahedron Lett.* **1999**, *40*, 8701.
<99TL8849>	K.-I. Washizuki, K. Nagai, S. Minakata, I. Ryu, M. Komatsu, *Tetrahedron Lett.* **1999**, *40*, 8849.
<99TL9029>	J. Lachheb, M.-T. Martin, A.-K. Khemiss, *Tetrahedron Lett.* **1999**, *40*, 9029.
<99TL9277>	O.A. Attanasi, P. Filippone, B. Guidi, T. Hippe, F. Mantellini, L.F. Tietze, *Tetrahedron Lett.* **1999**, *40*, 9277.

Chapter 5.5

Five-Membered Ring Systems: With N & S (Se) Atoms

Paul A. Bradley and David J. Wilkins
Knoll Pharmaceuticals, Research and Development Department, Nottingham, England
E-mail, david.wilkins@knoll.co.uk; paul.bradley@knoll.co.uk

5.5.1 ISOTHIAZOLES

Various 2,5-disubstituted furans **1** have been converted into 5-acyl-3-substituted isothiazoles **2** by treatment with ethyl carbamate, thionyl chloride and pyridine. The reactive species produced by this combination of reagents may be the highly reactive thiazyl chloride, NSCl, or a related species such as its trimer $(NSCl)_3$. This mixture of reagents was reported to be much more convenient to use than $(NSCl)_3$ and in many cases produced higher yields of isothiazole <99S757>.

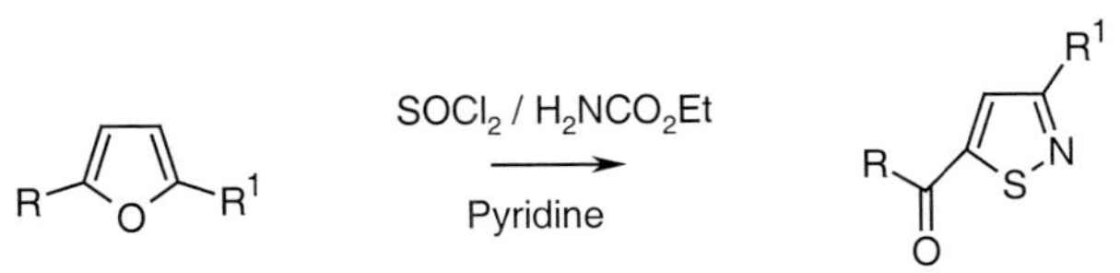

1 **2**

R = Ph, *p*-tolyl, ᵗ-Bu, 4-MeOC$_6$H$_4$
R¹ = Ph, *p*-tolyl, ᵗ-Bu, 4-O$_2$NC$_6$H$_4$

Zajic has described an extremely convenient and high yielding preparation of *N*-bromo-saccharin, which is an excellent source of electrophilic bromine, by treatment of saccharin with $KBrO_3$ and sulfuric acid in aqueous acetic acid <99SC1779>.

N-fluoro-3-cyclohexyl-3-methyl-2,3-dihydobenzo[1,2-*d*]isothiazole 1,1-dioxide **5** which was synthesised by manipulation of the imine **3** by the sequence outlined in scheme 1 has proved an efficient reagent for electrophilic asymmetric fluorination of enolates. Thus, alkylation of **3** with cyclohexylmagnesium bromide gave racemic **4** which on optical resolution with (-) menthoxyacetyl chloride and subsequent separation of the resulting diastereoisomers and removal of auxillary gave enantiomerically pure **4**. Fluorination of the pure enantiomers with 15% F_2/He in the presence of spray dried KF then gave **5**. As an example of the application of **5** in the electrophilic asymmetric fluorination of enolates, reaction with 2-benzyl-1-tetralone gave an excellent yield of the 2-fluoro derivative **6** in 88 % ee (scheme 1) <99JOC5708>.

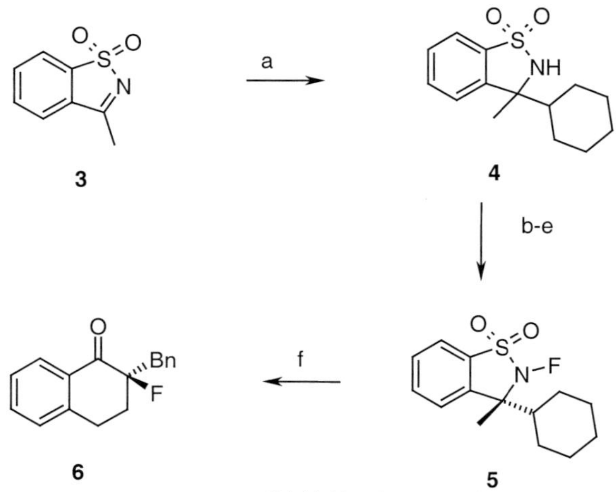

(a) c-C_6H_{11}MgBr, THF (b) (-) Menthoxyacetyl chloride, NaH, THF,

(c) Diastereoisomeric separation, (d) 2M LiOH, (e) F_2/He, spray-dried KF

(f) 2-Benzyl-1-tetralone

Scheme 1

A series of bicyclic isothiazole-*S*-oxides **8** were prepared by reaction of fused 1,2-dithole-*S*-oxides **7** with a range of primary amines *via* an S/N exchange reaction <99JHC161>.

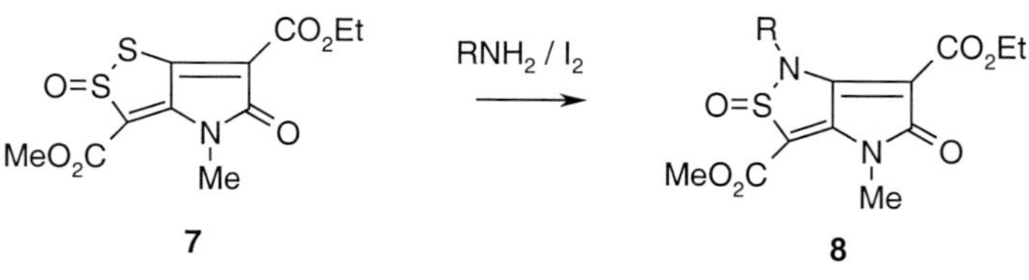

R = Me, Bn, Ad,^{t}-Bu,H

Michael addition of various sulfur, oxygen and nitrogen nucleophiles to the isothiazole **9** at C-5 gave 4,5-dihydro-derivatives, for example, treatment of **9** with methylthiolate (MeSNa) in acetonitrile gave **10** in 40% yield. However, the presence of a 5-bromo substituent in the isothiazole ring allowed regeneration of the 4,5-double bond (e.g. reaction of **11** with MeSNa gave **12** in 58% yield) <99T2001>.

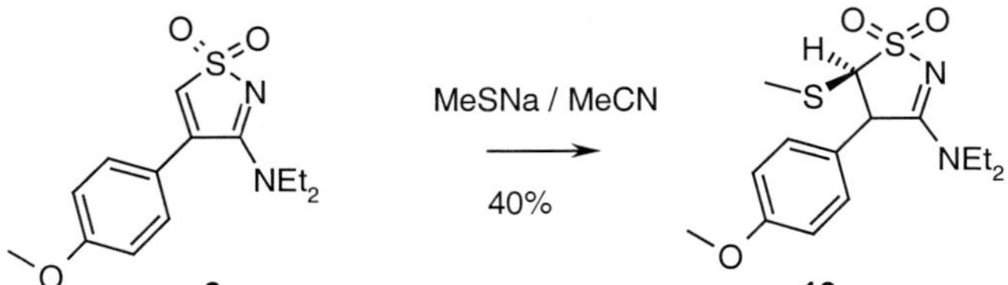

9 **10**

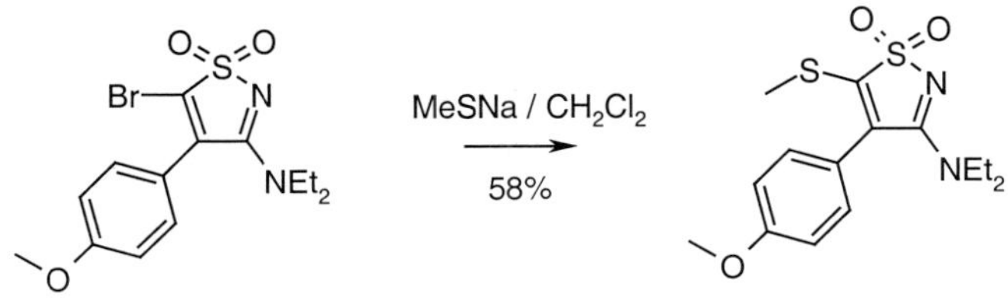

11 **12**

Hartung and co-workers have reported that oxidation of isothiazolium salts such as **13** with H_2O_2 in AcOH led to stable 3-hydroperoxy-2-phenylhexahydro-1,2-benzisothiazole 1-oxides **14** which could be isolated in fair-to-good yields (38-70%). 3-Hydroperoxy and 3-hydroxy sultams were observed as over oxidation products in some cases and was dependent on the substituent present in the 2-aryl ring <99HCA685>.

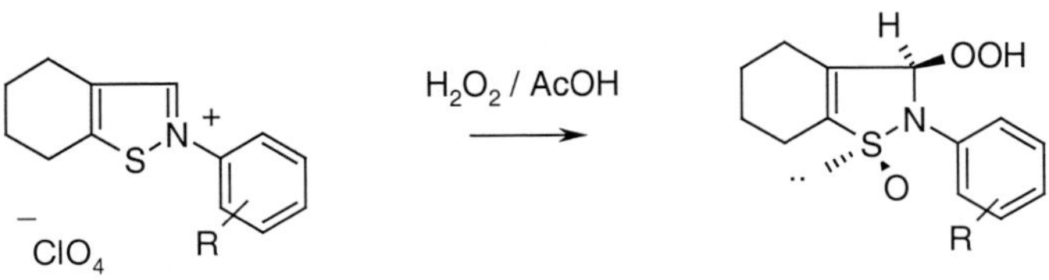

13 **14**

Clerici *et al.* studied the reaction of 3-diethylamino-4-(4-methoxyphenyl)-5-vinyl-isothiazole 1,1-dioxides **15** with nitrile oxides and with munchnones. These reactions produced cycloadducts such as **16** and **17** which underwent thermal rearrangement to α, β-unsaturated nitriles <98T11285>.

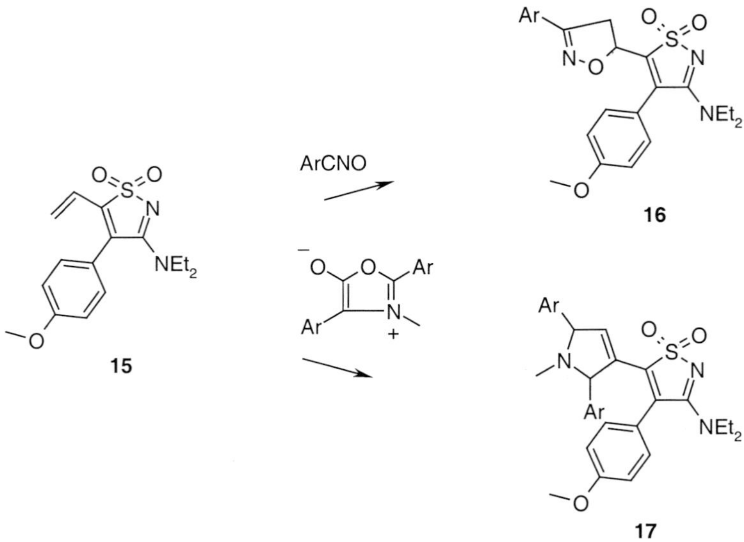

Jurczak and Kiegiel reported that additions of allylmagnesium chloride and allyl bromide in the presence of Zn to *N*-methyl and *N*-phenylglyoxyloyl-(2R)-borane-10,2-sultam occurred in a diastereoselective manner. Similarly, the Lewis acid mediated addition of allyltrimethylsilane also gave good diastereoselectivity and in the case with $TiCl_4$ a change of direction of asymmetric induction was observed <99TL1009>.

5.5.2 THIAZOLES

The synthesis of novel pyrano[2,3-*d*]thiazole derivatives has been reported. The reaction of the hydrazinothiazolone **18** with cinnamonitrile derivatives such as **19** yielded the pyrano[2,3-*d*]thiazole **20** <98JCR(S)730>.

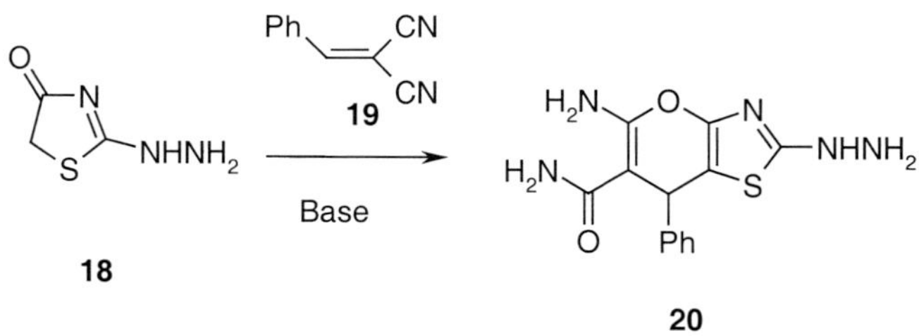

A number of papers have reported Stille couplings of either 2- or 5-stannylated thiazoles. The 2-stannylated thiazole **21** undergoes Stille couplings with the triflate **22** to give the aryldifluorophosphonate **23** which is useful as a building block for the synthesis of phosphotyrosine mimics such as (difluorophosphonomethyl)-phenylalanine **24** <99TL2601>.

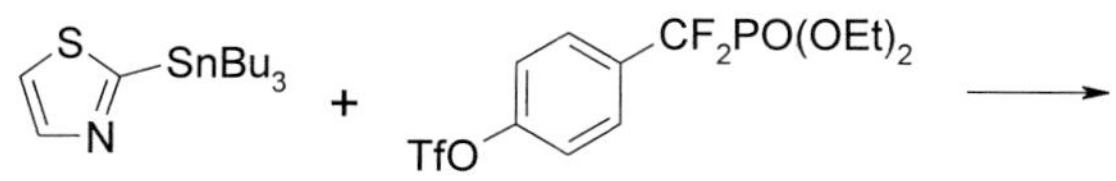

21 **22**

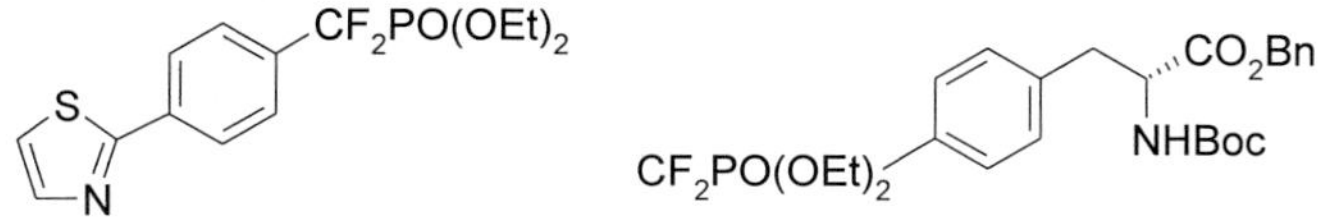

23 **24**

5-Stannylated thiazole derivatives **25** undergo Stille cross coupling reactions with 6-haloindoles such as **26** to give 6-heteroarylindoles such as **27** in high yield. The corresponding Suzuki couplings using heterocyclic boronic acids derivatives gave 6-heterocyclic indoles in only poor to moderate yields <99TL5701>.

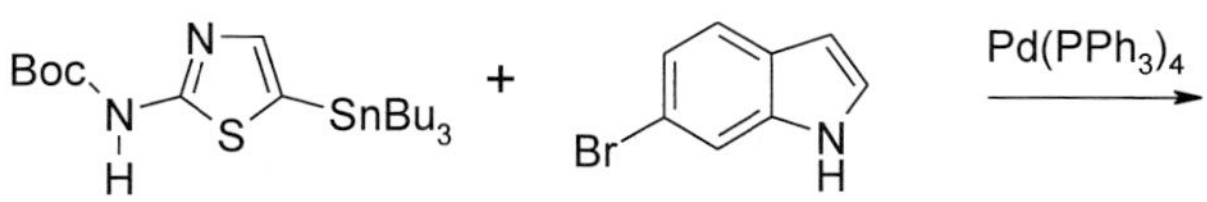

25 **26**

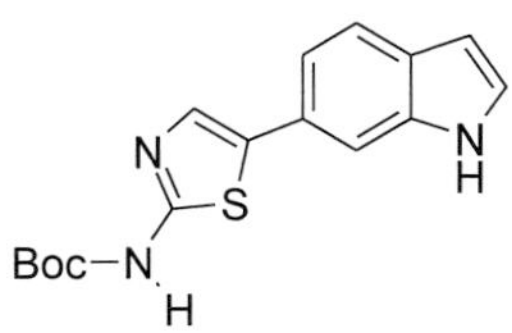

27

The first efficient Stille coupling of heteroaromatic cations to tributyl stannyl derivatives has been reported. 2-Tributylstannylthiazole **21** was coupled with the bromoquinolizinium salt **28**, in the presence of Pd(0) / CuI catalysis, to afford the quinolizinium salt **29** in moderate yield. This method provides a good alternative to nucleophilic substitution of quinolizinium derivatives, which are usually very unstable in the presence nucleophilic species <99OL545>.

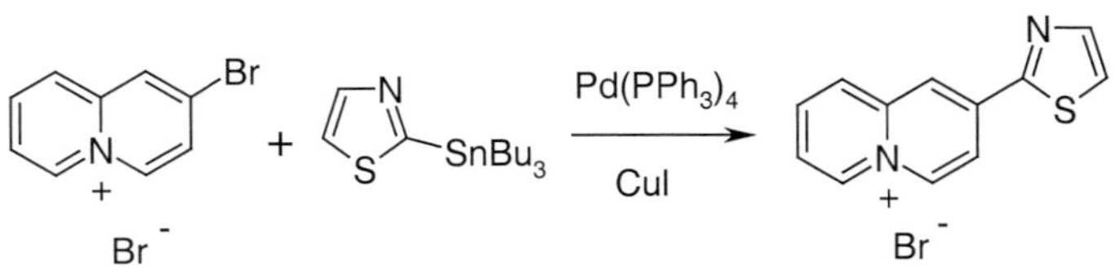

28 **21** **29**

A Stille type coupling strategy has been utilised to complete a total synthesis of epothilone E . The vinyl iodide **30** and the thiazole stannane **31** were coupled to give the macrolactone **32** which is a precursor to natural epithilone E. The thiazole stannane **31** was prepared from 4-bromo-2-hydroxymethylthiazole *via* treatment of the lithiated protected 4-bromo-2-hydroxymethylthiazole with tributylstannyl chloride. This Stille coupling approach was also used to prepare a range of epothilone B analogues <99BMC665>.

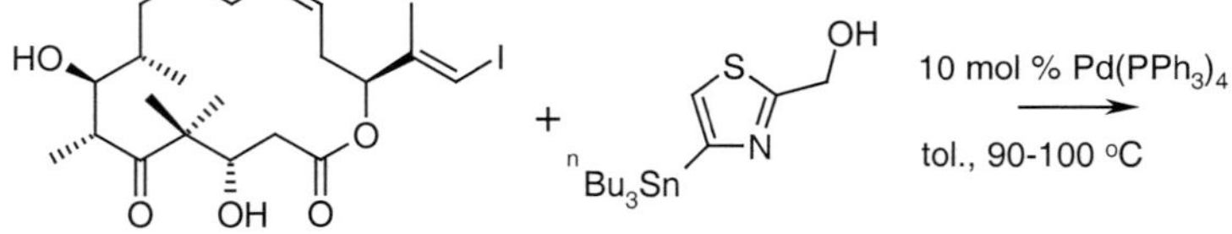

30 **31**

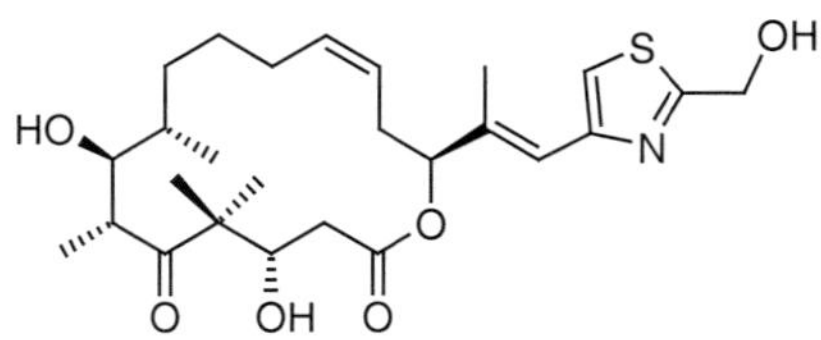

32

The synthesis of epothilone A and C using an antibody catalysed aldol reaction has been reported <98M14603>. The synthesis of 12,13-desoxyepothilone has also been reported <99JA7050>.

2-Aminothiazolines are usually prepared by the acid catalysed cyclization of *N*-(2-hydroxyethyl)thioureas or the cyclization of the hydrogen sulfate of thioureas in aqueous basic conditions. These methods give low yields of 2-aminothiazolines and are not suitable for acid sensitive or racemization prone substrates. Mitsunobu reaction of thioureas such as **33** afforded 2-methylaminothiazolines **34** in good to excellent yields <99TL3125>.

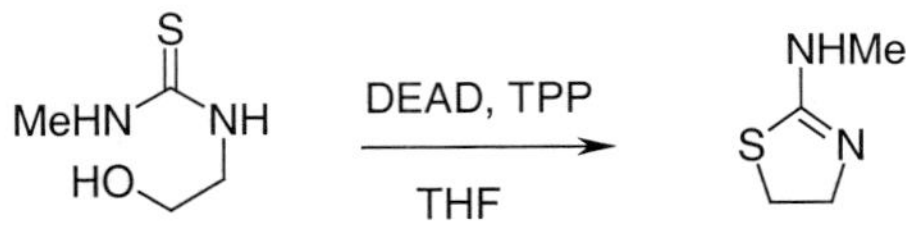

33 **34**

Oxidative nucleophilic substitution of hydrogen with tertiary carbanions is a useful method for introducing carbon substituents into heterocyclic nitroarene rings. 2-Nitrothiazole **35** reacts with the carbanion of 2-phenylpropanenitrile **36** generated from sodamide in liquid ammonia at –70 °C to give the adduct **37**, after oxidation of the intermediate σ^H complex with potassium permanganate, in moderate yield. 2-Nitrothiazole also reacted with 2,3,3-triphenylpropanenitrile under similar conditions to give the corresponding adduct in good yield <98S1631>.

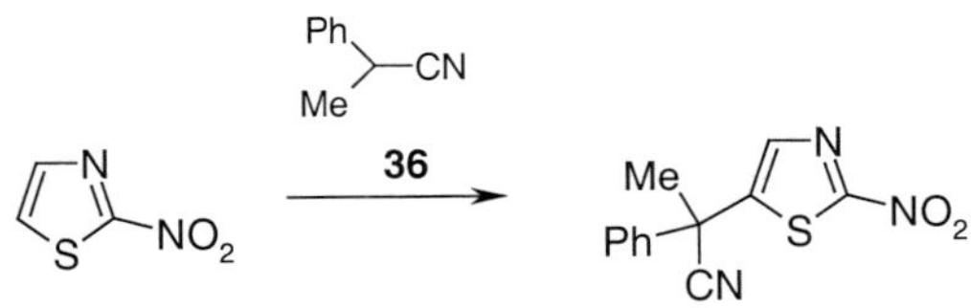

35 **37**

The generation and Diels Alder reaction of 4,5-bis(bromomethylene)-4,5-dihydrothiazole **39** has been investigated. **39** is a heterocyclic analogue of *ortho*-quinodimethanes and is generated by treating 4,5-bis(dibromomethyl)thiazole **38** with sodium iodide in dimethylformamide. **39** can be trapped *in situ* with symmetrical dienophiles to give substituted benzothiazoles such as **40** <98EJOC2047>.

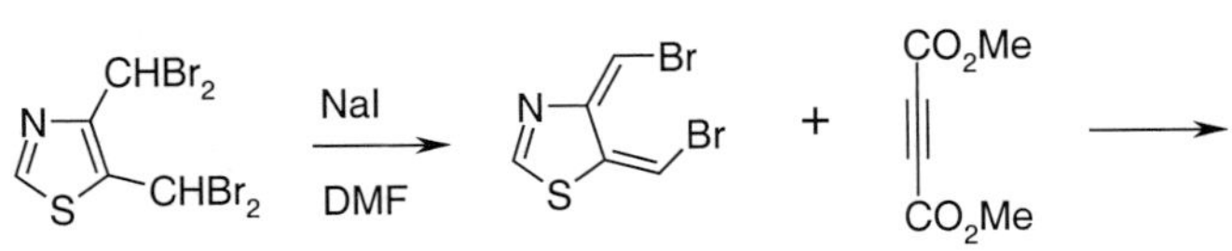

38 **39**

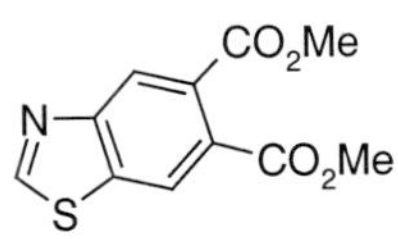

40

When **39** was trapped with unsymmetrical dienophiles such as acrylonitrile a mixture of regioisomers was obtained with some selectivity for 6-substituted benzothiazoles **41** over 5-substituted benzothiazoles **42** <98EJOC2047>.

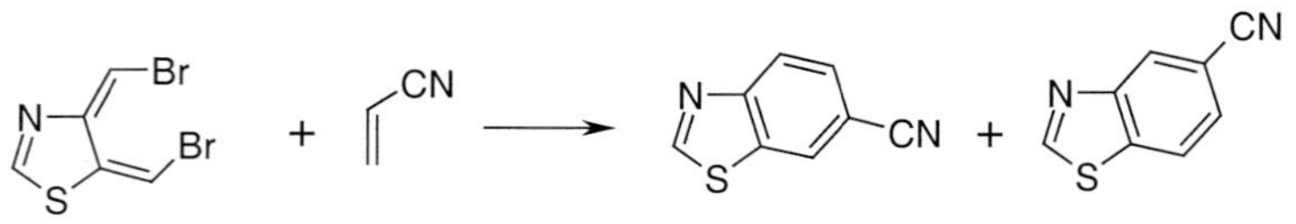

39

67 : 33

41 42

4-Oxoalkyl and 4-iminoalkyl-5-azidothiazoles such as **43** undergo a ring transformation with loss of nitrogen at relatively low temperatures to afford 4-cyanooxazoles such as **46** and 4-cyanoimidazoles, respectively. The mechanism is thought to involve an initial ring opening to give a thiocarbonyl intermediate **44** which then undergoes a 1,6-electrocyclization to an unstable oxathiazine **45** which then extrudes sulfur to form the oxazole **46** <99T1977>.

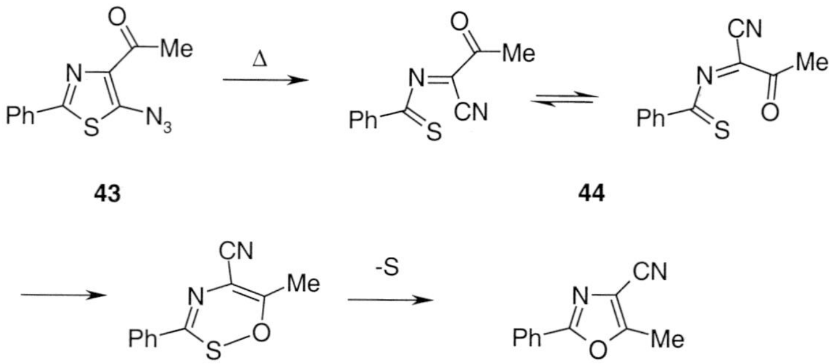

43 44

45 46

5-Aminothiazolium salts such as **47**, when treated with base generate mesoionic thiazoles **48** which are potential cyclic azomethine ylids. They undergo intramolecular 1,3-dipolar cycloaddition reactions across the pendant olefin group attached to *N*-3 to give a mixture of regioisomeric *N*-bridged thiazoloquinolines **49** and **50** in good yield <98EJOC2631>.

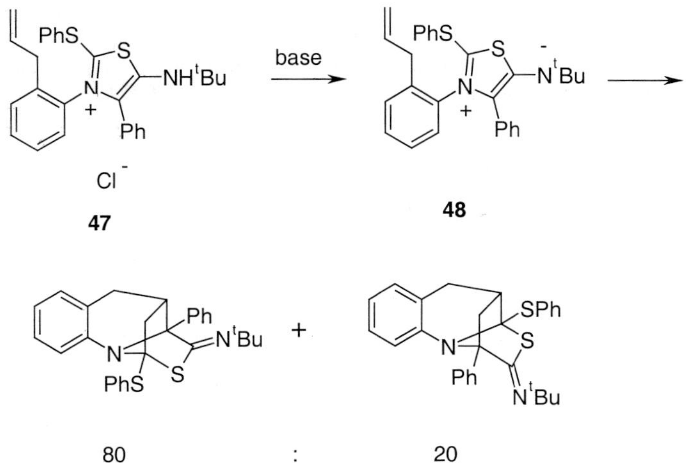

47 48 49 50

N-arylimino-1,2,3-dithiazole derivatives such as **51** prepared from the corresponding anilines can be converted in high yield into 2-cyanobenzothiazoles such as **52**. The conversion is achieved either by vigorous heating (200-250 °C for 1 to 2 mins.) or by focussed microwave irradiation in pyridine containing cuprous iodide. The mechanism is thought to proceed *via* an electrocyclization and fragmentation process which is facilitated by halogen complexation with copper <98JCS(P1)3925>.

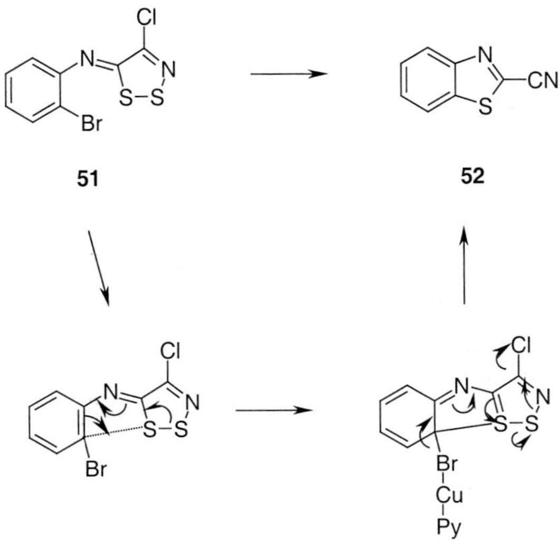

51 52

Thiazolium ion based ionic liquids (OIL) have been used to promote the benzoin condensation of benzaldehyde. 4- And 5-methylthiazoles are readily alkylated with *n*-butyl bromide to give the corresponding bromide salt. Anion exchange with sodium tetrafluoroborate gave the tetrafluoroborate salt **53** as a stable yellow orange oil. When activated with a small quantity of triethylamine (5 mol%) the oil promotes the coupling of benzaldehyde to benzoin <99TL1621>.

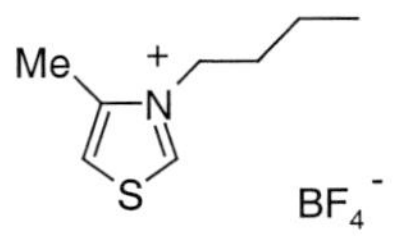

53

3-(2-propynylthio)triazoles **54** undergo thermal rearrangement to give 5-substituted-2-cyanoamidothiazoles **55** in good yields. The rearrangement is thought to proceed *via* the bicyclic intermediate **56** <99H475>.

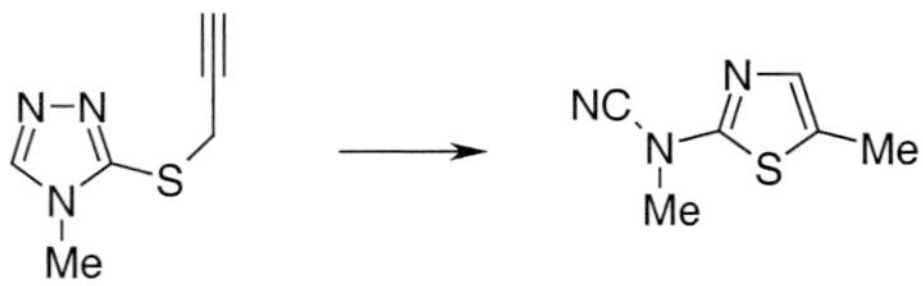

54 **55**

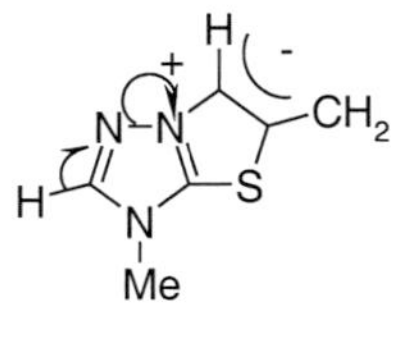

56

The hydroboration of alkynylchlorosilanes gave chlorodimethylsilyldiethylborylalkenes such as **57**. The alkene possess two electrophilic centres on the silyl and boryl groups and they react with 2-lithiated thiazoles **58** to give the zwitterionic compound **59** which can drawn as resonance structures **59a** or **59b** <99JOM98>.

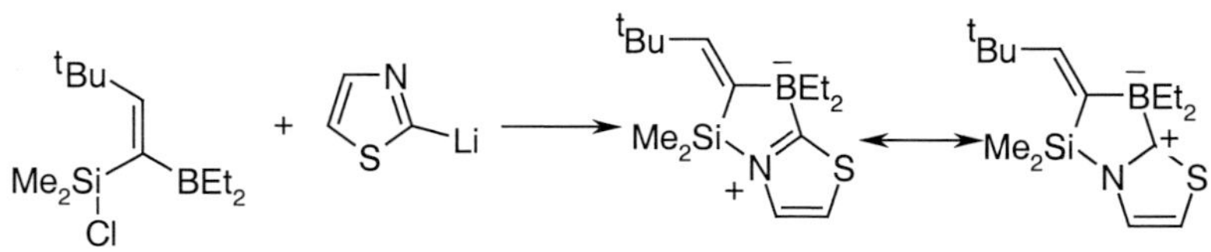

57 **58** **59a** **59b**

5.5.3 THIADIAZOLES

5.5.3.1 1,2,3-Thiadiazoles

During 1999, numerous references appeared in the literature describing the synthesis of the 1,2,3-thiadiazole ring system. As expected, the majority of these references involved the Hurd-Mori cyclization of a semicarbazone with thionyl chloride, producing various fused systems (e.g: <99HAC17, HC285, IJC308 and SC667>). In addition, Porco Jr *et al* used the Hurd-Mori reaction in the parallel synthesis of 1,2,3-thiadiazoles <99JOC1049>.

Alkylation of 4,5-diaryl-1,2,3-thiadiazoles **60** and 1,2,3-benzothiadiazoles with trimethylsilylmethyl trifluoromethanesulfonate occurred at *N*-3 giving compound **61**. Subsequent treatment of **61** with CsF then produced new 1,2,3-thiadiazol-3-ium-3-methane 1,3-dipoles **62** which underwent *in situ* cycloaddition-rearrangement reactions with alkyne and alkene dipolarophiles producing new vinylthioethenylpyrazole systems **63** and **64**. A second molecule of the dipolarophile had been added at the thiol SH which was generated by opening of the thiadiazole ring <99JCS(P1)1415>.

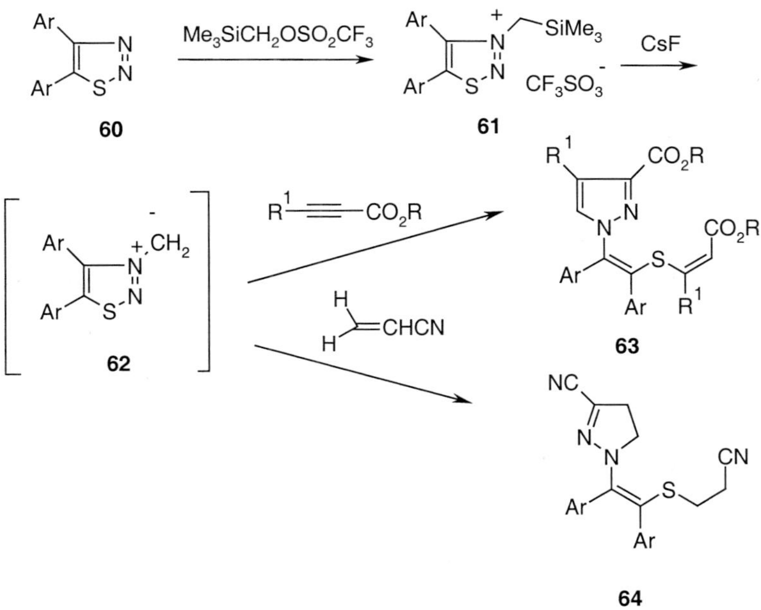

Dahaen and co-workers studied the reaction of 5-chloro-1,2,3-thiadiazoles **65** with organolithium and Grignard reagents and discovered a novel ring cleavage reaction with loss of nitrogen and chloride ion, giving alkynyl sulfides of type **66** <99JCS(P1)1473>.

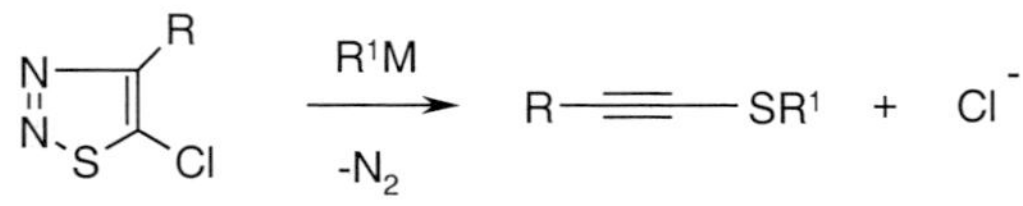

65 66

R = Ph, t-Bu; R^1 = n-Bu, Ph, t-Bu, Et, Me, Ph—≡

Dehaen and Smeets also described the smooth conversion of *meso*-(1,2,3-thiadiazol-4-yl)porphyrins into the corresponding ethynyl derivatives using potassium *tert*-butoxide as base <98TL9841>.

A greatly improved experimental procedure for the synthesis of thieno[2,3-*d*]-1,2,3-thiadiazole carboxylates **68** was reported by Stanetty *et al.* and involved diazotisation of aminothiophene derivatives **67** <99JHC761>. In these systems, substituents could be introduced into the 5-position by nucleophilic displacement of a chlorine atom or by metallation of the unsubstituted compound (**68**; R = H) and subsequent electrophilic quenching <99JPR391>.

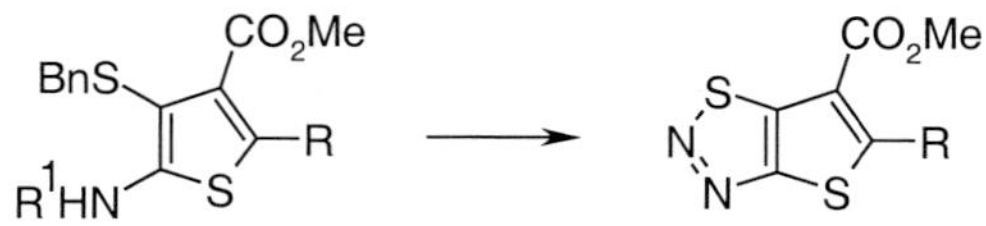

67 68

5.5.3.2 1,2,4-Thiadiazoles

Reaction of 4,5-dichloro-1,2,3-dithiazolium chloride **69** with benzamidine gave 5-cyano-3-phenyl-1,2,4-thiadiazole **70**. Benzamidoxime **71a** reacted in a similar manner to give the corresponding 4-oxide **72** as a minor product (8%), this being the first 1,2,4-thiadiazole *N*-oxide ever reported. A series of O-acylbenzamidoximes **71b-e** gave the same *N*-oxide **72** in somewhat higher yields (20-30%). Yields were again similar (21-29%) when electron releasing substituents were present in the benzene ring of the benzamidoxime. The structures of the *N*-oxides were determined by NMR analysis, mass spectra of ^{15}N-labelled and unlabelled products and x-ray analysis of the derived carboxamide. A mechanism in which the amidoxime reacts with **69** *via* their oximino, rather than amino, nitrogen atom was suggested by the authors <99JCS(P1)2243>.

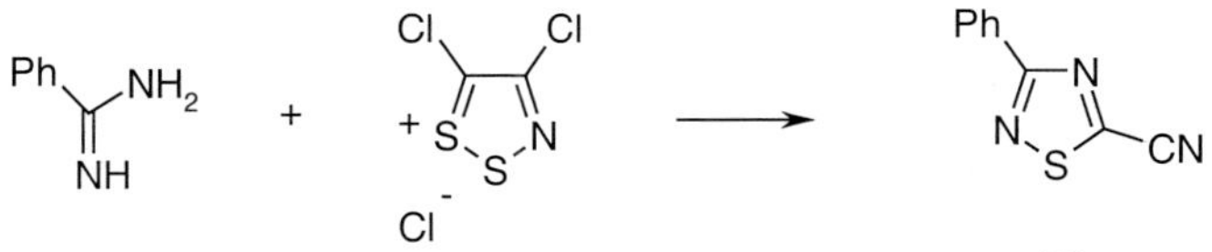

69 **70**

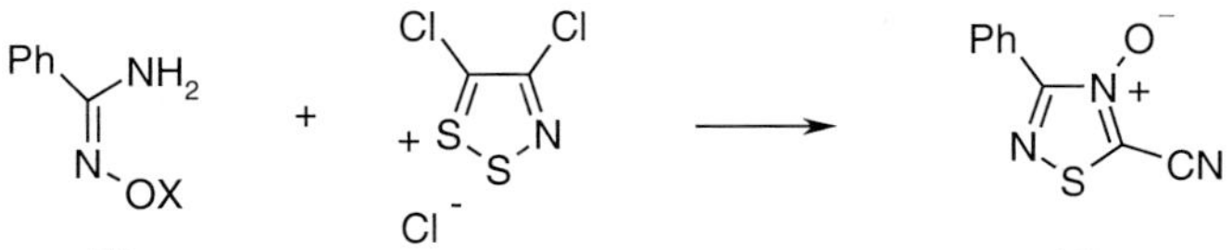

71 **69** **72**

X = (a) H, (b) Ac, (c) Bz, (d) 4-ClBz, (e) CONHMe

3-Phenyl-5-ethoxy-1,2,4-thiadiazole **74** has been synthesised as a by-product (13%) in the reaction of phenylbromodiazirine **73** with potassium ethyl xanthate and was suggested to have arisen by a radical mechanism. As expected, the major product formed in the reaction (87%) was benzonitrile, arising by reduction of the bromodiazirine <99TL29>.

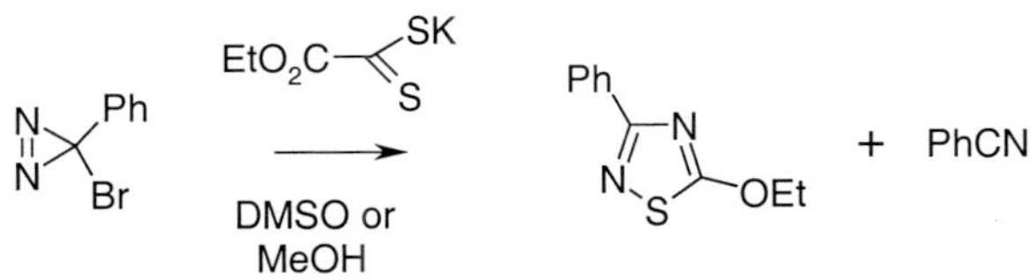

73 **74**

A series of 3-substituted 5-cyano-1,2,4-thiadiazoles **76** have been prepared in varying yields (21-92%) *via* reaction of nitrogen nucleophiles with 4-chloro-5-isoxazolylimino-5*H*-1,2,3-dithiazoles **75** <99H811>.

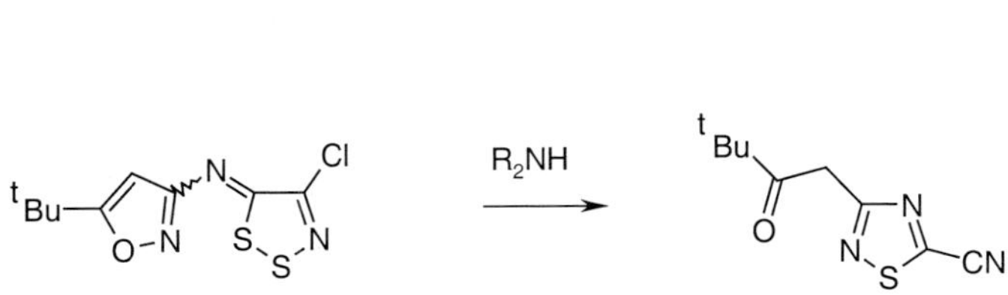

75 **76**

Butler and co-workers studied the quaternisation of 3,5-diaryl-1,2,4-thiadiazoles **77** with trimethylsilylmethyl triflate at 40 °C and observed reaction at *N*-2 to give salt **78**. Desilyation of **78** with caesium fluoride resulted in ring expansion to 2*H*-1,3,5-thiadiazines **79** which on heating in ethanolic sodium ethoxide gave 2,4-

disubstituted imidazoles **80** *via* sulfur extrusion and ring contraction <99JCS(P1)1709>.

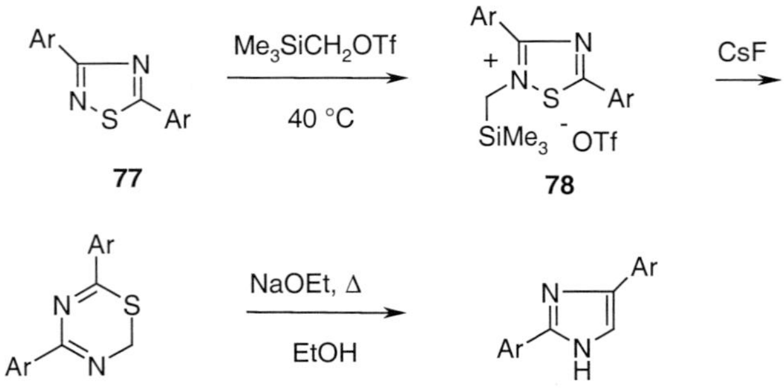

77 78 79 80

5.5.3.3 1,2,5-Thiadiazoles

A series of 3-acyl and 3-aroyl-4-substituted 1,2,5-thiadiazoles **82** have been synthesised by reaction of 3,5-disubstituted isoxazoles **81** with tetrasulfur tetranitride antimony pentachloride ($S_4N_4.SbCl_5$) in toluene at 90 °C to reflux temperature. Compounds **82** are produced regioselectively and a plausible mechanism for their formation discussed. Under the same conditions, 3,4-dialkyl and 5-alkyl(aryl)-isoxazoles furnished chloroketones of type **83** <98JCS(P1)2175>.

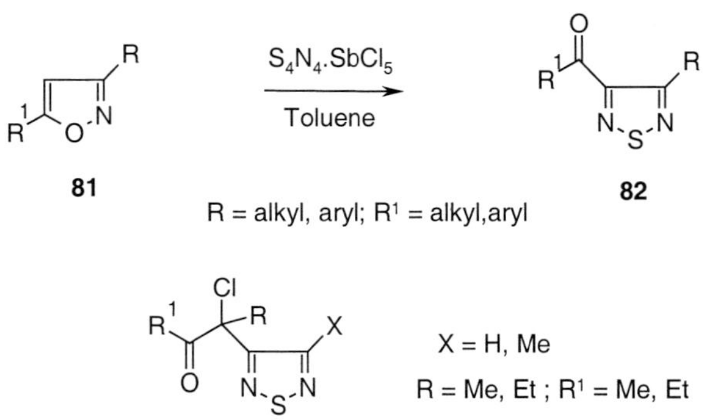

81 82

R = alkyl, aryl; R¹ = alkyl,aryl

83

X = H, Me

R = Me, Et ; R¹ = Me, Et

Similarly, ($S_4N_4.SbCl_5$) reacted with alkyl methyl ketoximes **84** in aromatic solvents (e.g benzene and toluene) to give 3-alkyl-4-methyl-1,2,5-thiadiazoles **85**, albeit in low yields (3-37%). A mechanism for the formation of **85** was proposed and the regioselective formation of **85** ascribed to the stability of an enamine intermediate. Suprisingly, this appears to be only the second example of a synthesis of a 3,4-dialkyl-1,2,5-thiadiazole that has been reported in the literature <99H147>.

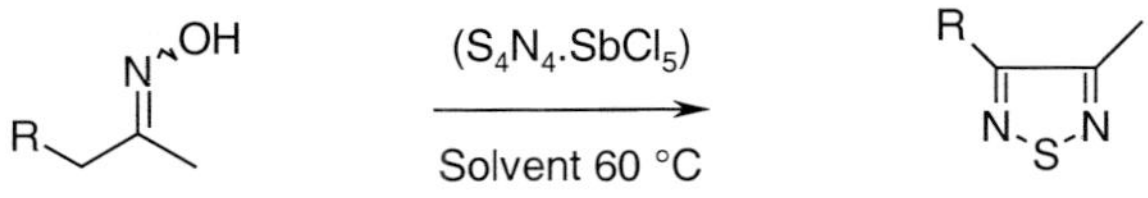

84 **85**

5.5.3.4 1,3,4-Thiadiazoles

Reaction of the amino-1,3,4-thiadiazole **86** with a series of benzaldehydes gave the arylidene amines **87** which when treated with arylacetyl chlorides and triethylamine gave 5-substituted 1,3,4-thiadiazolo[3,2-*b*]pyrimidin-6-ones **88** in good yields (75-95%). The reaction was thought to proceed by a (4+2) cycloaddition reaction between **87** and the ketene which was produced *in situ* by the interaction of arylacetyl chlorides and triethylamine <99JCR(S)36>.

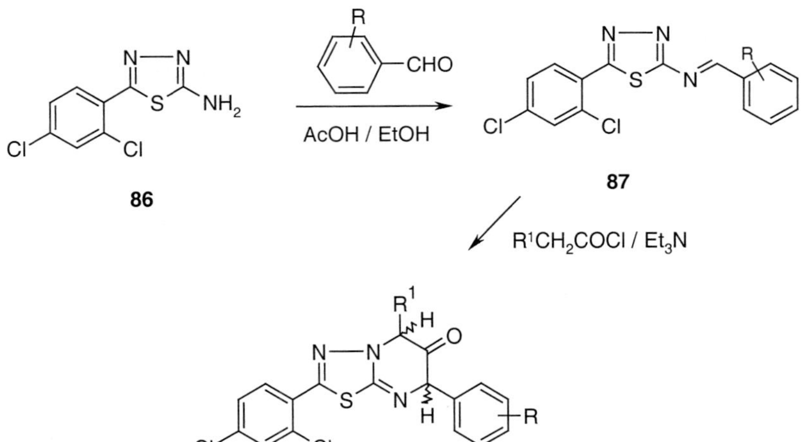

86 **87** **88**

The novel heterocyclic system **92** has been prepared by reaction of 2-amino-1,3,4-thiadiazoles **91** with either 1-(haloalkyl)pyridinium halides **89** or N,N'-methylenebis(pyridinium) dihalides **90**. A mechanism for the formation of **92** was proposed and involved a series of specific proton migrations, bond-breaking and bond-forming processes <98EJOC2923>.

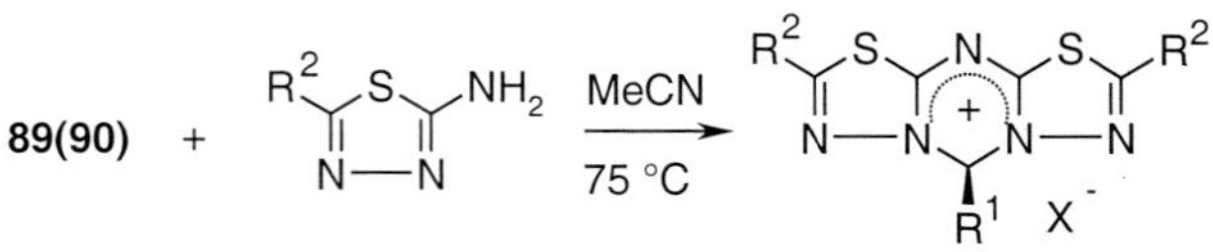

91 **92**

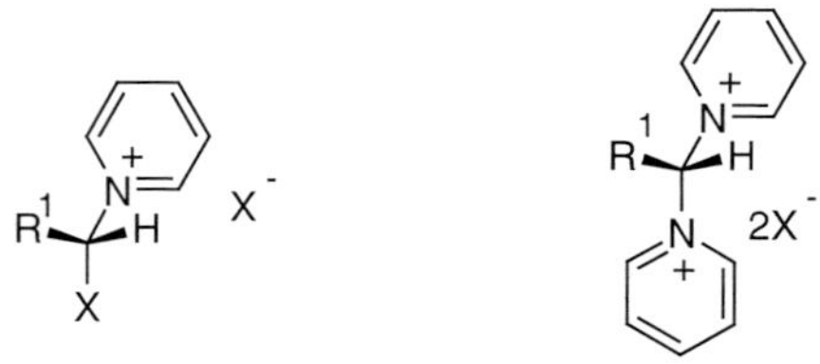

89 **90**

X = Cl, Br

Treatment of the bromide **93** with the potassium salt of 1,1-dicyanothioacetanide **94** gave the dihydrothiadiazole **95** *via* initial nucleophilic displacement of the bromine by the sulfur atom in **94**, cyclization and then loss of malononitrile <99JCR(S)184>.

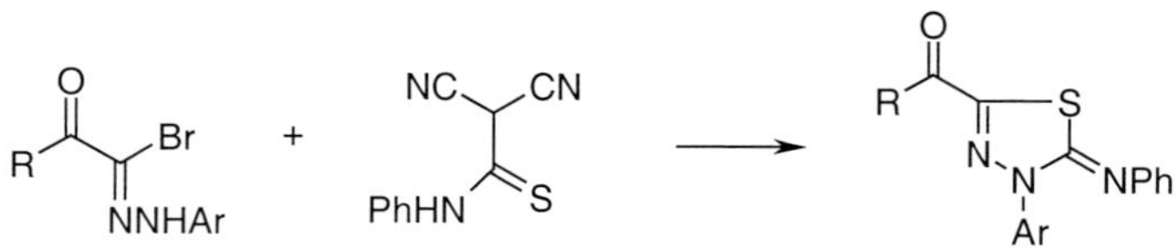

93 **94** **95**

A new and convenient synthesis of 1,3,4-thiadiazoles was reported and involved the direct conversion of 1,3,4-oxadiazoles using thiourea as the thionating agent <98SC4611>. Courtois *et al.* have described an efficient monohydroxyarylation (or alkylation) and symmetrical bis-hydroxyarylation (or alkylation) of 2,5-dimethyl-1,3,4-thiadiazole using LDA and the appropriate carbonyl compound <99SC145>.

5.5.4 SELENAZOLES AND SELENADIAZOLES

4-substituted 1,2,3-selenadiazoles are usually easily decomposed with the liberation of nitrogen to afford alkynyl selenoates under the action of strong bases such as organo lithiums or potassium ethoxide. When 4-(2-hydroxyphenyl)-1,2,3-selenadiazole **97**, formed by treating the semi-carbazone of *o*-hydroxyacetophenone **96** with selenium dioxide, reacts with potassium carbonate the extra functional group on the aryl ring leads to the formation of the 2-benzofuranselenoate **100**. The mechanism is thought to involve initial loss of nitrogen to give the alkyne selenoate **98** which then undergoes an intramolecular proton shift to give the reactive selenoketene **99** which cyclises to give the selenolate **100**. The alkyne selenoate **98** can be trapped if methyl iodide is added during the decomposition process to give the dimethylated product **101**. The selenoate **100** can readily be alkylated with methyl iodide or benzyl chloride <99TL3903>.

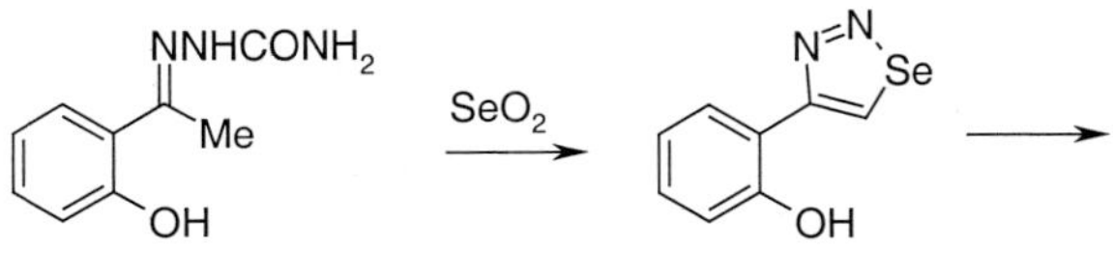

96 **97**

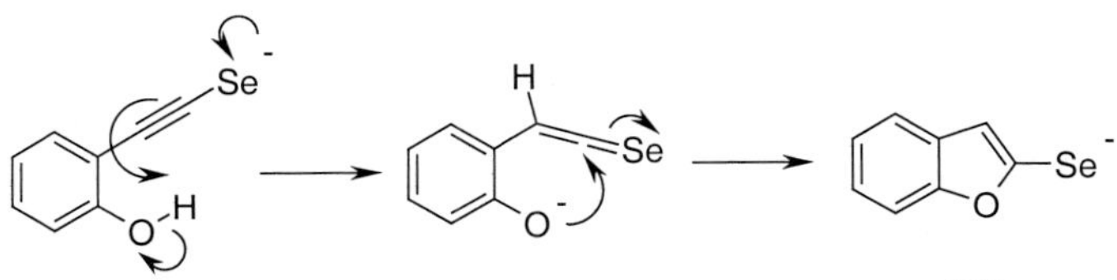

98 **99** **100**

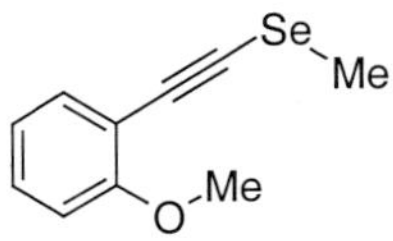

101

When 1,2,3-selenadiazoles such as **102** are treated with a catalytic amount of tributyl tin hydride and AIBN a vinyl radical species **103** is formed. This radical species can add to electron deficient olefins to give intermediates such as **104** which then undergo intramolecular cyclization to give the dihydroselenophenes such as **105** in moderate to good yields <99TL6293>.

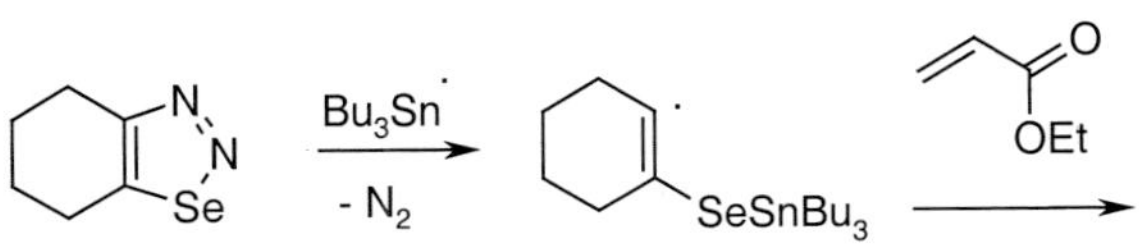

102 **103**

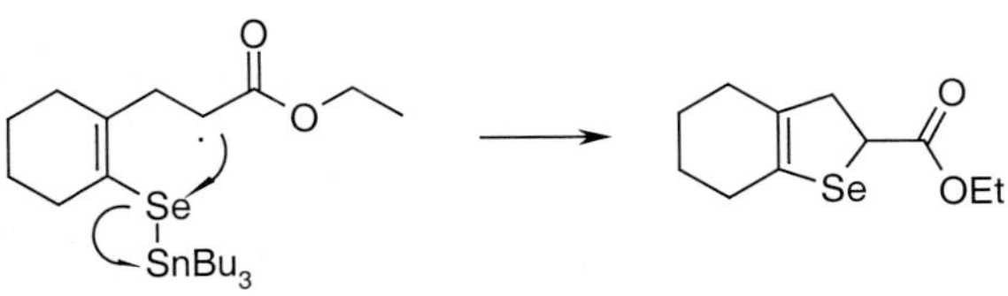

104 **105**

5.5.5 REFERENCES

98EJOC2047 K. Jouve, F. Pautet, M. Domard and H. Fillion, *Eur. J. Org. Chem.*, **1998**, 2047.

98EJOC2631 G. Morel, E. Marchand, A. T. Benjelloun, S. Sinbandhit, O. Guillou and P. Gall, *Eur. J. Org. Chem.*, **1998**, 2631.

98EJOC2923 E. Anders, K. Werman, B. Wiedel, W. Gunther and H. Gorls, *Eur. J. Org. Chem.*, **1998**, 2923.

98JCR(S)730 S. M. Eldin, *J. Chem. Res. (S)*, **1998**, 730.

98JCS(P1)2175 K.J. Kim and K. Kim, *J. Chem. Soc., Perkin Trans. 1*, **1998**, 2175.

98JCS(P1)3925 T. Besson, M-J. Dozias, J. Guillard and C. W. Rees, *J. Chem. Res. (S)*, **1998**, 3925.

98M14603 S. C. Sinha, C. F. Barbas III and R. A. Lerner, *Proc. Natl. Acad. Sci, USA*, **1998**, 14603.

98S1631 M. Makosza and K. Stalinski, *Synthesis*, **1998**, 1631.

98SC4611 N. Linganna and K. M. Lokanatha Rai, *Synth. Commun.*, **1998**, 28, 4611.

98T11285 F. Clerici, M. L. Gelmi, R. Soave and M. Valle, *Tetrahedron*, **1998**, 54, 11285.

98TL9841 S. Smeets and W. Dehaen, *Tetrahedron Lett.*, **1998**, 39, 9841.

99BMC665 K. C. Nicolaou, N. P. King, M. R. V. Finlay, Y. He, F. Roschangar, D. Vourloumis, H. Vallberg, F. Sarabia, S. Ninkovic and D. Hepworth, *Biorg. and Med. Chem. Lett.*, **1999**, 665.

99H147 K. J. Kim and K. Kim, *Heterocycles*, **1999**, 50, 147.

99H475 D. K. Bates, M. Xia, M. Aho, H. Mueller and R. R. Raghavan, *Heterocycles*, **1999**, 475.

99H811 T. Iwakawa and A. Murabayashi, *Heterocycles*, **1999**, 51, 811.

99HAC17 D. B. Reddy, M. V. R. Reddy and V. Padmavathi, *Heteroatom. Chemistry*, **1999**, 10, 17.

99HC285 D. B. Reddy, A. Balaiah, V. Padmavathi and A. Padmaja, *Heterocycl. Commun.*, **1999**, 5, 285.

99HCA685 C. Hartung, K. Illgen, J. Sieler, B. Schneider and B. Schulze, *Helv. Chim. Acta.*, **1999**, 82, 685.

99IJC308 V. Padmavathi, A. Padmaja and D. B. Reddy, *Ind. J. Chem. Incl. Med. Chem.*, **1999**, 38B, 308.

99JA7050 C. R. Harris, S. D. Kuduk, A. Balog, K. Savin, P. W. Glunz and S. J. Danishefsky, *J. Amer. Chem. Soc.*, **1999**, 7050.

99JCR(S)36 B. C. Dutta, K. K. Das and B. N. Goswani, *J. Chem. Res (S)*, **1999**, 36.

99JCR(S)184 A. O. Abdelhamid, H. F. Zohdi and N. M. Rateb, *J. Chem. Res (S)*, **1999**, 184 and (M), 0920.

99JCS(P1)1415 R. N. Butler, M. O. Cloonan, P. McArdle and D. Cunningham, *J. Chem. Soc., Perkin Trans 1*, **1999**, 1415.

99JCS(P1)1473 M. Voets, M. Smet and W. Dehaen, *J. Chem. Soc., Perkin Trans 1*, **1999**, 1473.

99JCS(P1)1709 R. N. Butler, M. O. Cloonan, J. McMahon and L. A. Burke, *J. Chem. Soc., Perkin Trans 1*, **1999**, 1709.

99JCS(P1)2243 L. S. Konstantinova, O. A. Rakitin, C. W. Rees, T. Torroba, A. J. P T. White and D. J. Williams, *J. Chem. Soc., Perkin Trans 1*, **1999**, 2243.

99JHC161 J. E. Schachtner, T. Zoukas, H. D. Stachel, K. Polborn and H. Noth, *J. Heterocycl. Chem.*, **1999**, 36, 161.

99JHC761 P. Stanetty, E. Gorner amd M. D. Mihovilovic, *J. Heterocycl. Chem.*, **1999**, 36, 761.

99JOC1049 Y. Hu, S. Baudart and J. A. Porco Jr, *J. Org. Chem.*, **1999**, 64, 1049.

99JOC5708 Y. Takeuchi, T. Suzuki, A. Satoh, T. Shiragami and N. Shibata, *J. Org. Chem.*, **1999**, 64, 5708.

99JOM98 B. Wrackmeyer, A. Badshah, E. Molla and A. Mottalib, *J. Orgmetal. Chem.*, **1999**, 98.

99JPR391 P. Stanetty, M. Jaksits and M. D. Mihovilovic, *J. Prakt. Chem.*, **1999**, 341, 391.

99OL545 B. M. Barchin, J. Valenciano, A. M. Cuadro, J. Alvarez-Builla and J. J. Vaquero, *Org. Letts*, **1999**, 545.

99S757 S. M. Laaman, O. Meth-Cohn and C. W. Rees, *Synthesis*, **1999**, 757.

99SC145 P. Cousin, G. Anselme, G. Courtois and D. Mesnard, *Synth. Commun.*, **1999**, 29, 145.

99SC667 D. B. Reddy, M. V. R. Reddy and V. Padmavathi, *Synth. Commun.*, **1999**, 29, 667.

99SC1779 B. Zajic, *Synth. Commun.*, **1999**, 29, 1779.

99T1977 E. Ceulemans, L. K. Dyall and W. Dehaen, *Tetrahedron*, **1999**, *55*, 1977.
99T2001 E. M. Beccalli, F. Clerici and M. L. Gelmi, *Tetrahedron*, **1999**, *55*, 2001.
99TL29 X. Creary, *Tetrahedron Lett.*, **1999**, *40*, 29.
99TL1009 K. Kiegel and J. Jurczak, *Tetrahedron Lett.*, **1999**, *40*, 1009.
99TL1621 J. H. Davis Jr and K. J. Forrester, *Tetrahedron Lett.*, **1999**, *40*, 1621.
99TL2601 G. S. Cockerill, H. J. Easterfield and J. M. Percy, *Tetrahedron Lett.*, **1999**, *40*, 2601.
99TL3125 T. H. Kim and M-H. Cha, *Tetrahedron Letts.*, **1999**, *40*, 3125.
99TL3903 M. L. Petrov, M. A. Abramov, W. Dehaen and S. Toppet, *Tetrahedron Lett.*, **1999**, *40*, 3903.
99TL5701 R. Benhida, F. Lecubin, J-L, Fourrey, L. R. Castellanos and L. Quintero, *Tetrahedron Lett.*, **1999**, *40*, 5701.
99TL6293 Y. Nishiyama, Y. Hada, M. Anjiki, S. Hanita and N. Sonoda, *Tetrahedron Lett.*, **1999**, *40*, 6293.

Chapter 5.6

Five-Membered Ring Systems: With O & S (Se, Te) Atoms

R. Alan Aitken
University of St. Andrews, UK
(e-mail: raa@st-and.ac.uk)

5.6.1 1,3-DIOXOLES AND DIOXOLANES

New catalysts for reaction of epoxides with acetone to give 2,2-dimethyl-1,3-dioxolanes include $TiO(CF_3CO_2)_2$ and $TiCl_3(CF_3SO_3)$ <98JCR(S)466> and montmorillonite clay catalyses the reaction of aldehydes and ketones with catechol to give benzodioxoles <98JCS(P1)3561>. A process for reaction of glyoxylate esters with 1,2-diols to give dioxolane esters **1** has been patented <99JAP11189591>. Treatment of **2** with MCPBA provides a new approach to the dioxolane aldehydes **3** <99SL303> and reaction of **4** with LDA results in an unusual rearrangement to give chiral dioxolanes **5** <99TL1583>. Radical addition of 2,2-dimethyl-1,3-dioxolane to perfluoroalk-1-enes, $R_FCF{=}CF_2$, followed by deprotection gives the

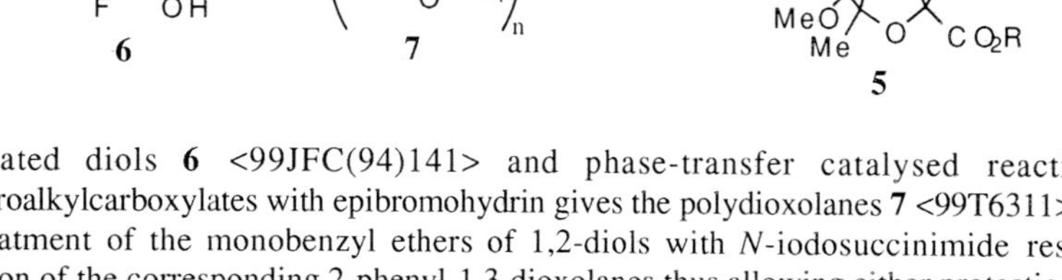

fluorinated diols **6** <99JFC(94)141> and phase-transfer catalysed reaction of perfluoroalkylcarboxylates with epibromohydrin gives the polydioxolanes **7** <99T6311>.

Treatment of the monobenzyl ethers of 1,2-diols with *N*-iodosuccinimide results in formation of the corresponding 2-phenyl-1,3-dioxolanes thus allowing either protection of the alcohol or deprotection of the ether <98AG(E)3177>. The carbohydrate derivative **8** reacts with silver carbonate in acetone to give iminodioxolane **9** <99CC591>. Reaction of substituted thiobenzophenones and α-hydroxy acids with silver nitrate and triethylamine provides a new synthesis of 2,2-diaryl-1,3-dioxolan-4-ones <99JAP11209367> and new minor products identified in the reaction of butanedione with conc. HCl include **10** and **11** <99T5867>.

New methods for the deprotection of 2,2-disubstituted-1,3-dioxolanes to give the corresponding carbonyl compounds include ceric ammonium nitrate either in stoichiometric

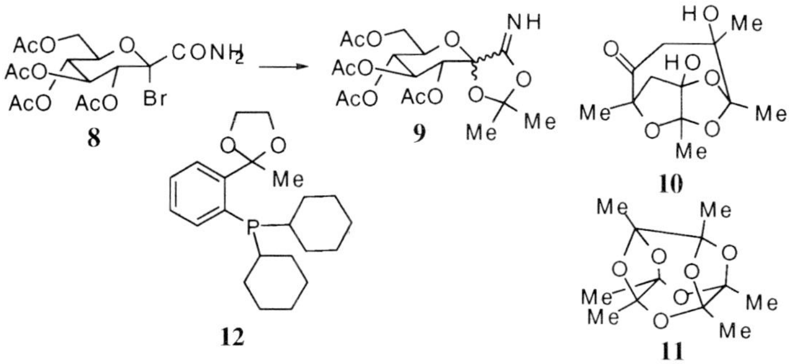

<99TL1799> or catalytic <99AG(E)3207> amount, and a polymer supported dicyanoketene acetal <99SL1960> while treatment with Oxone® and wet alumina results in oxidative cleavage of 2-substituted dioxolanes to give the 2-hydroxyethyl esters, $RC(=O)OCH_2CH_2OH$ <99SL777>. Bromination of both 2-methyl- and 2,2-dimethyldioxolane results in bromination of a methyl group <98JGU914> and the dioxolane-containing phosphine **12** is an effective ligand for Pd-catalysed Suzuki coupling <99JOC6797>.

There have again been a large number of studies on preparation of chiral dioxolanes and their use in asymmetric synthesis. A reliable procedure for preparation of diols **14** from isopropylidenethreitol **13** has appeared <99OS101> and the C_2 symmetric benzoquinone monoketals **15** can be prepared in enantiomerically pure form <99T7907> while large scale kinetic resolution of dioxolanemethanols **16** is possible using esterases <99TA3747> or lipases <99MI447>. Full details of the behaviour of chiral 2-methylene-1,3-dioxolanes as dienophiles have appeared <99T12907> and dioxolanes such as **17** <99TA2749> and **18** <99JOC6443> derived from quinic acid are effective catalysts for asymmetric epoxidation of double bonds using dimethyldioxirane. Several methods for asymmetric cyclopropanation of double bonds directed by an adjacent chiral dioxolane function have appeared <99TA4245, 99MI285,

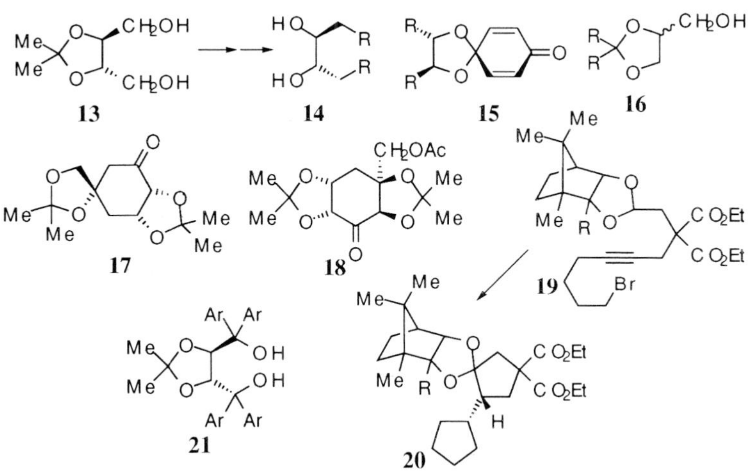

tertamethylTTF <99SC2953>. The structure of the cobalt salt of TTF tetracarboxylate in relation to its extent of hydration has been examined by X-ray methods <98AG(E)3158> and two new methods for synthesis of unsymmetrically substituted TTFs have appeared <99T9979, 99T13029>.

New results on annulated TTF derivatives include an improved synthesis of **46** <99S577>, the preparation of **47**/**48** <98MI1945>, and synthesis of new donors such as **49** <98MCLC251> and the pentathiepine-fused compound **50** <99AM758>. The structure and properties of a complex between bis(ethylenedithio)TTF **51** and chromium^{III} oxalate <99CC513> and also a compound of its tetraselena analogue with Cu_5I_6 <99MI687> have been reported, and complexes of **52**, an isomer of **51** have been examined similarly

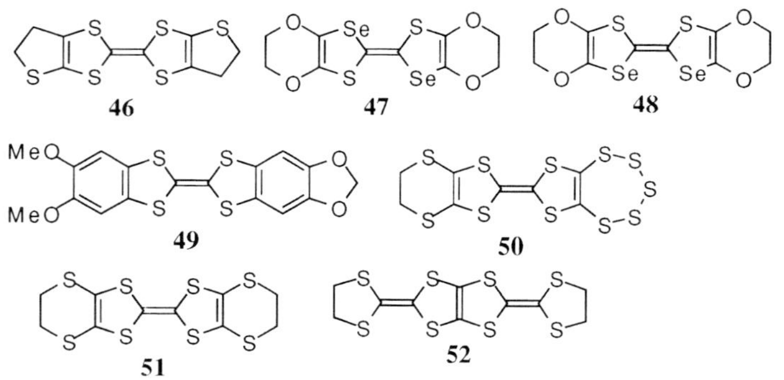

46 **47** **48** **49** **50** **51** **52**

<99AG(E)810>. A simple TTF monoamide differs markedly in its formation of charge transfer complexes according to whether the anion is AsF_6^- or RuO_4^- <99CEJ2971>. Further donors whose preparation and properties have been described include **53** <99TL6635>, **54**

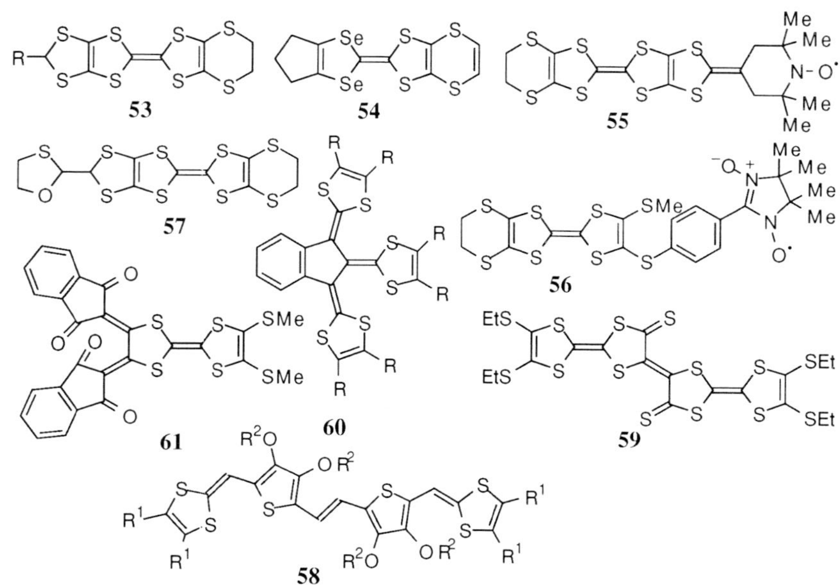

53 **54** **55** **57** **56** **61** **60** **59** **58**

<99MI693>, and the radical-containing donors **55** <99CC2417> and **56** <99TL5027>. The more extended systems **57** <98TL7709>, **58** <99JOC4267>, **59** <99CL987> and **60** <99T9915> have also been prepared and **61** shows a useful near IR absorption at 800 nm <99JAP11116567>.

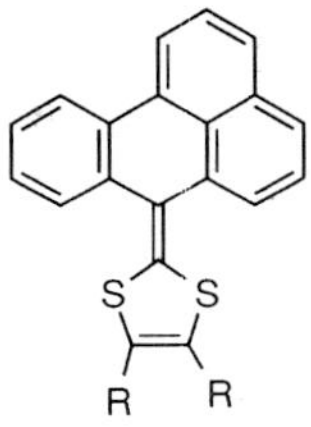

62

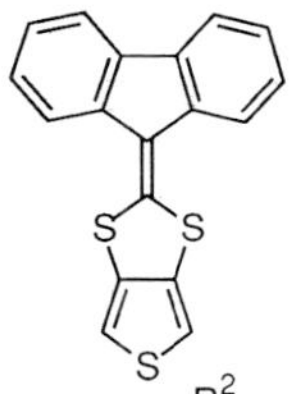

63

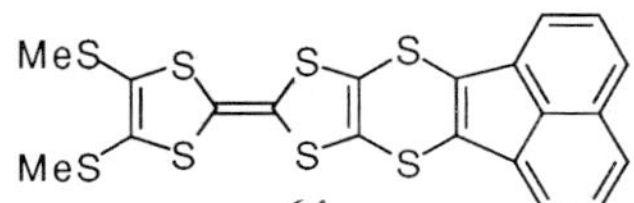

64

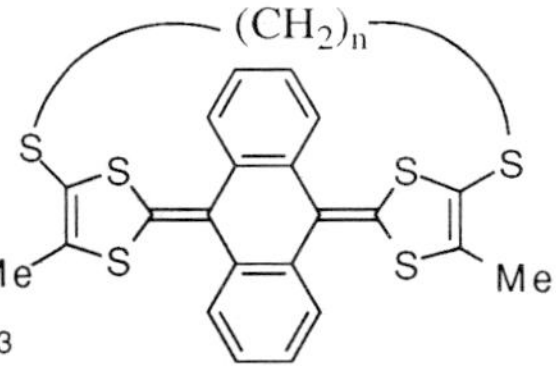

65 n = 5,6

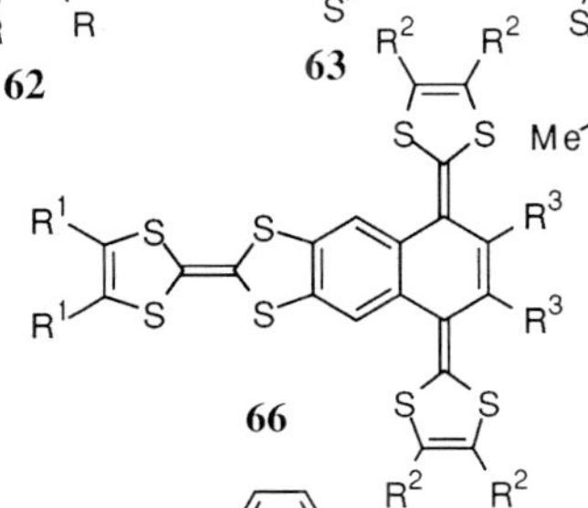

66

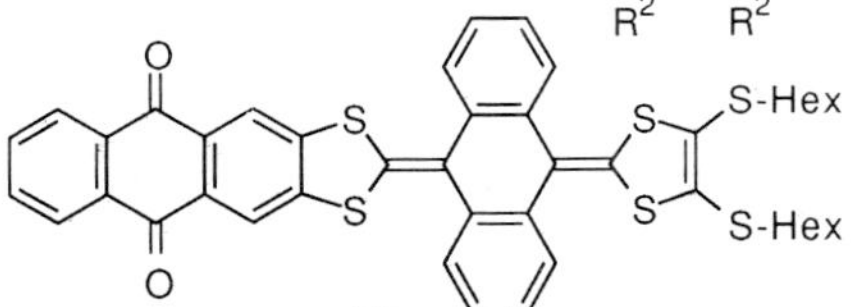

67

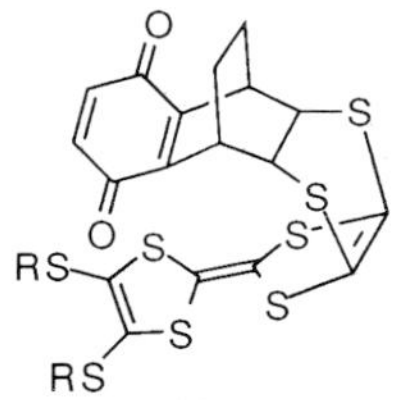

68

New aromatic fused dithiole systems reported include **62** <99MI1239>, **63** <99JCS(P2)505>, **64** <99TL801>, **65** <99CC1835> and **66** <99TL5997>. The preparation and X-ray structures of **67** <99CC2433> and **68** <99CC1125> containing donor and acceptor groups have appeared and in the latter case the molecule is bent round as shown to give an intramolecular charge transfer interaction. Precursors for synthesis of TTF-containing chlorinated benzoquinones have been reported <98CHE907> and new studies on TTF–C_{60} systems have appeared <99MI323, 99JCS(P2)657>. Further developments in TTF-containing supramolecular systems have been reported including the effect of self-assembled monolayers of a TTF thiol on metal electrodes <99CC737>, the use of crown and thiacrown ether TTFs as ion sensors <99CC1417, 99CC1493>, formation of TTF containing cyclophanes <98AM1360>, macrobicyclic cages <98MI1743> and "molecular belts" <99AG(E)1417>. TTF-containing dendrimers <98CC2565> and polymers <99CC515, 99CC1407> have also been described.

New applications of dithioles and dithiolanes include evaluation of compounds such as **69** and **70** as flavour components <98MI177>, **71** for antimycotic activity <99MIP33826>, **72** for antiviral activity <99CC1245>, **73** for antifungal and antibacterial activity <99MI6> and use of **74** in prevention and treatment of asthma complications <99EUP909758>.

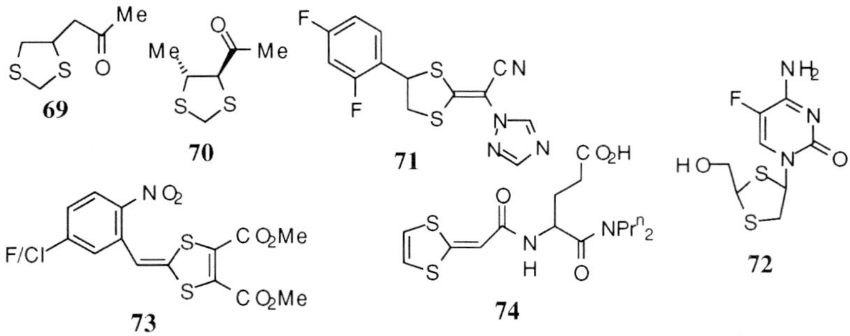

69 70 71 72 73 74

5.6.3 1,3-OXATHIOLES AND OXATHIOLANES

Reaction of thiazoline-5-thiones with epoxycycloalkenes to give spiro thiazoline/oxathiolane products **75** has been described <99HCA1458> and intramolecular hetero Diels-Alder reactions based on 1,3-oxathiolane *S*-oxides such as **76** going to **77** have been reported <98MI2733>.

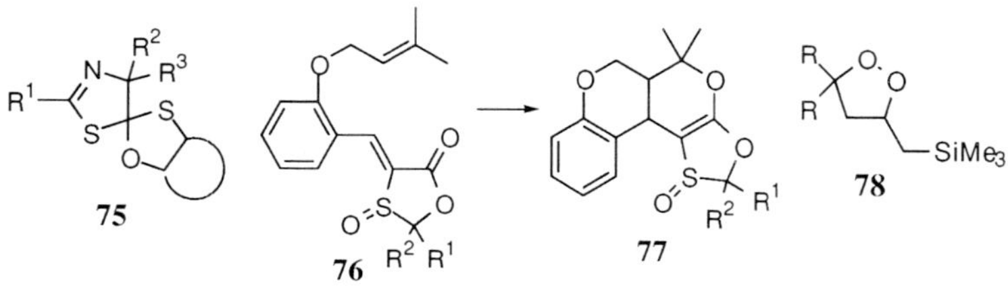

75 76 77 78

5.6.4 1,2-DIOXOLANES

Lewis acid treatment of 1,2,4-trioxolanes gives metallated carbonyl oxides which may be trapped by cycloaddition to allylsilanes to give 1,2-dioxolanes **78** <99TL6553>.

5.6.5 1,2-DITHIOLES AND DITHIOLANES

A mild method for preparation of 1,2-dithiolanes involves treatment of 1,3-dithiocyanates with $Bu^n{}_4N^+ F^-$ <99TL6489>. The 1,2-benzodithiolium salt **79** has been prepared and its X-ray structure determined <99CC1891>. Treatment with sodium results in reduction to the corresponding radical which can be observed by ESR. A variety of mixed dichalcogen dications **80** (X = S, Se, Te; Y = S, Se) have been prepared <99CL723>. A new improved synthesis of the "Beaucage sulfurising agent" **81** has appeared <99S43>. Several new reactions involving S_2Cl_2 which lead to 1,2-dithiole products have been described <99JOC4376, 99JCS(P1)1023> and the synthesis and reactivity of monosulfoxides of pyrrolodithioles has been examined <99JHC161>. The enatiomers of **82** have been separated on a chiral column and their racemisation studied <99ACS710>. The 1,2-ditellurole derivative **83** forms a conducting thiocyanate whose electrical and magnetic properties have been examined <99CL845>.

New applications for 1,2-dithiole and dithiolane compounds include evaluation of dithiolanes such as **84** and **85** as flavour components <98MI177>, **86** as a new indigo chromophore <99T14429>, **87** as a fungicide and insecticide <99JAP11279179>, **88** as a glutathione reductase enhancer <98EUP869126> and **89** as an antitumour, antiproliferative and antiinflammatory agent <99MIP43313>.

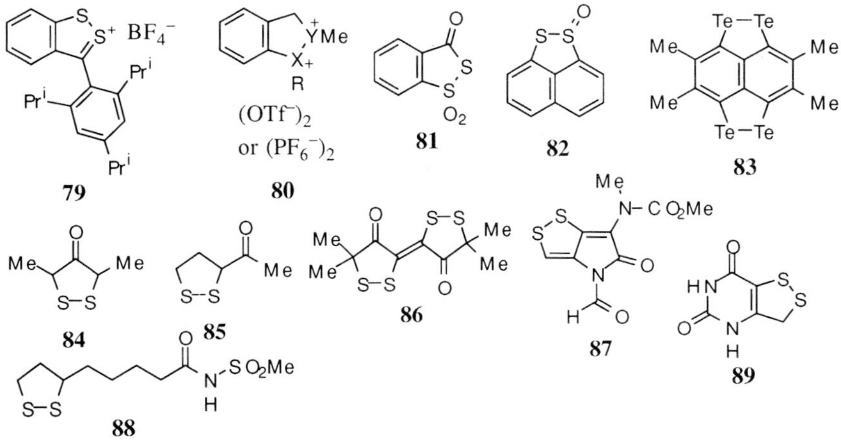

79 **80** **81** **82** **83** **84** **85** **86** **87** **88** **89**

5.6.6 1,2-OXATHIOLES AND OXATHIOLANES

Treatment of 2,3-epoxysulfonyl chlorides **90** with Et_3N results in formation of the 1,2-oxathiole dioxides **91** whose reactivity has been studied <98SL1411> and reaction of an iron cyclopropyl complex with SO_2 to give **92** has been reported <98OM5534>. Synthesis of the spiro piperidine–oxathiole dioxides **93** has been reported <99T7625> and reaction of *p*-$MeOC_6H_4TeCl_3$ with HC≡C–CH_2OH gives the 1,2-oxatellurole **94** <99OM803>. A convenient synthesis of **95** and **96** has been described and **95** may be converted into **96** by treatment with HCl <99TA4183>. Several spiro oxaselenolanes and oxatellurolanes **97** have been prepared and their X-ray structures show trigonal bipyramidal geometry at the chalcogen atom <98TA3303>. The intramolecular Diels-Alder reaction of furans tethered to a dienophile by a sulfinate link affords products such as **98** <99MI487>.

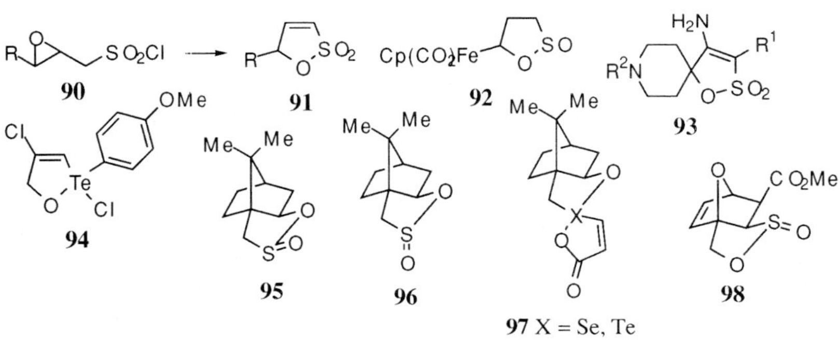

90 **91** **92** **93** **94** **95** **96** **98**

97 X = Se, Te

5.6.7 THREE OR FOUR HETEROATOMS

The gas phase formation of 1,2,4-trioxolanes (secondary ozonides) has been studied <99JCS(P2)239> and reaction of photochemically generated dimesitylsilylene with carbonyl compounds gives products **99** <99CC1857>. The bis-dioxaborole and dithiaborole compounds **100** form charge transfer complexes with acceptors such as TCNQ and TCNE and their X-ray structures have been determined <99JCS(D)2127>. Theoretical calculations on the strain energy of 1,3,2-dioxathiolane and its mono and dioxides **101** have been reported <99ACS1003> and a convenient synthetic approach to 4-methylene-1,3,2-dioxathiolane 2-oxides **102** has been described <99T10845>. Either enantiomer of the cyclic sulfite **103** may be obtained by enzymatic resolution <99TA4755> and carbohydrate-derived cyclic sulfates have been used in the synthesis of 3,6-anhydro sugars <99T14649>. The chiral cyclic sulfates **104** and **105** have been used as key intermediates in the synthesis of (*R*)-(–)-mevalonolactone <99TA4349> and (*S*)-massoialactone <99T13445> respectively.

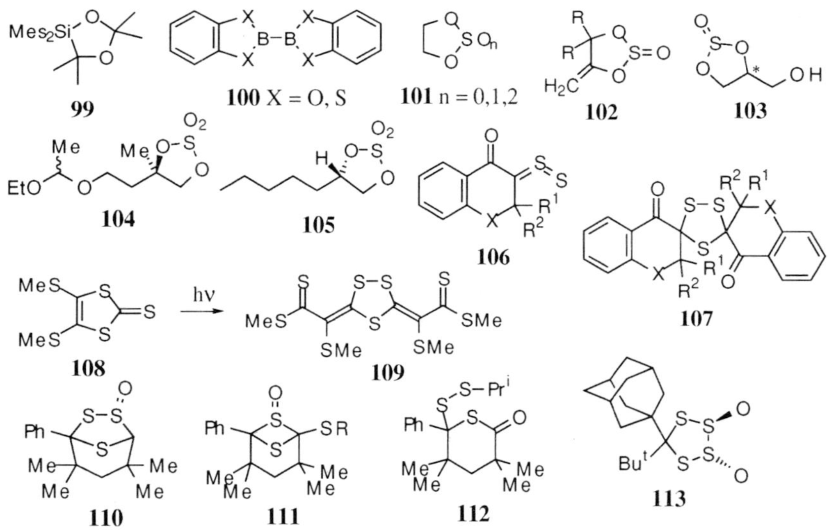

Generation of the thiocarbonyl *S*-sulfides (thiosulfines) **106** (X = O, S, CH_2) leads to formation of the spiro 1,2,4-trithiolane products **107** <98JOC9840, 99ACS133>. Photochemical reaction of the simple dithiolethione **108** leads to formation of the 1,2,4-trithiolane product **109** whose X-ray structure was determined <99JHC823>. The carbanion derived from **110** by treatment with LDA undergoes a variety of interesting and complex reaction sequences as illustrated by formation of **111** with methyl or ethyl iodide but **112** with isopropyl bromide <99T10341>. Finally treatment of the corresponding tetrathiolane with dimethyldioxirane at low temperature has allowed preparation of the first tetrathiolane 2,3-dioxide **113** whose NMR spectra and X-ray structure have been reported <99JA7959>.

5.6.8 REFERENCES

98AG(E)3158 C. J. Kepert, D. Hesek, P. D. Beer and M. J. Rosseinsky, *Angew. Chem., Int Ed. Engl.*, **1998**, *37*, 3158.

98AG(E)3177 J. Madsen and M. Bols, *Angew. Chem., Int. Ed. Engl.*, **1998**, *37*, 3177.

98AM1360 A. S. Batsanov, D. E. John, M. R. Bryce and J. A. K. Howard, *Adv. Mater. (Weinheim, Ger.)*, **1998**, *10*, 1360.

98CC2565 M. R. Bryce, P. de Miguel and W. Devonport, *Chem. Commun.*, **1998**, 2565.

98CHE738 L. V. Timokhina, G. M. Panova and M. G. Voronkov, *Chem. Heterocycl. Compd. (Engl. Transl.)*, **1998**, *34*, 738 [*Chem. Abstr.*, **1999**, *130*, 209620].

98CHE907 G. A. Karlivan, R. E. Valter and Yu. V. Gulbis, *Chem. Heterocycl. Compd. (Engl. Transl.)*, **1998**, *34*, 907 [*Chem. Abstr.*, **1999**, *130*, 311717].

98EUP869126 T. Fujita and T. Yokoyama, *Eur. Pat.* 869 126 (1998) [*Chem. Abstr.*, **1998**, *129*, 302641].

98H(48)2003 Y. Gareau and A. Beauchemin, *Heterocycles*, **1998**, *48*, 2003.

98JA10027 J. Nakayama, I. Akiyama, Y. Sugihara and T. Nishio, *J. Am. Chem. Soc.*, **1998**, *120*, 10027.

98JAP10298183 F. Toda, *Jpn. Pat.* 10 298 183 (1998) [*Chem. Abstr.*, **1999**, *130*, 25060].

98JCR(S)452 A. S. Gajare, M. S. Shingare and B. P. Bandgar, *J. Chem. Res. (S)*, **1998**, 452.

98JCR(S)466 N. Iranpoor and B. Zeynizadeh, *J. Chem. Res. (S)*, **1998**, 466.

98JCS(P1)2771 V. K. Aggarwal, Z. Gültekin, R. S. Grainger, H. Adams and P. L. Spergo, *J. Chem. Soc., Perkin Trans. 1*, **1998**, 2771.

98JCS(P1)3561 T.-S. Li, L.-J. Li, B. Lu and F. Yang, *J. Chem. Soc., Perkin Trans. 1*, **1998**, 3561.

98JGU914 S. A. Kotlyar and S. M. Pluzhnik-Gladyr', *Russ. J. Gen. Chem.*, **1998**, *68*, 914 [*Chem. Abstr.*, **1999**, *130*, 95491].

98JOC9840 F. A. G. El-Essawy, S. M. Yassin, I. A. El-Sakka, A. F. Khattab, I. Søtofte, J. Ø. Madsen and A. Senning, *J. Org. Chem.*, **1998**, *63*, 9840.

98MCLC251 T. Inayoshi, K. Sato, I. Miyazaki, S. Yokoyama and I. Ono, *Mol. Cryst. Liq. Cryst. Sci. Technol., Sect. A*, **1998**, *309*, 251 [*Chem. Abstr.*, **1999**, *130*, 182417].

98MI21 H. Li, S. Yue, J. Sun and Q. Jia, *Jingxi Huagong*, **1998**, *15*, 21 [*Chem. Abstr.*, **1999**, *130*, 125010].

98MI103 J. Irurre, A. Fernandez-Serrat, M. Altayo and M. Riera, *Enantiomer*, **1998**, *3*, 103 [*Chem. Abstr.*, **1999**, *130*, 24993].

98MI147 N. N. Romanova, A. G. Gravis, I. F. Leshcheva and Yu. G. Bundel', *Mendeleev. Commun.*, **1998**, 147.

98MI177 P. Weyerstahl and T. Oldenburg, *Flavour Fragrance J.*, **1998**, *13*, 177 [*Chem. Abstr.*, **1998**, *129*, 275883].

98MI1011 H.-J. Lee, Y.-Y. Kim and D.-Y. Noh, *Bull. Korean Chem. Soc.*, **1998**, *19*, 1011 [*Chem. Abstr.*, **1998**, *129*, 302585].

98MI1743 P. Blanchard, N. Svenstrup, J. Rault-Berthelot, A. Riou and J. Becher, *Eur. J. Org. Chem.*, **1998**, 1743.

98MI1945 T. Imakubo and K. Kobayashi, *J. Mater. Chem.*, **1998**, *8*, 1945.

98MI2733 L. F. Tietze, T. Pfeiffer and A. Schuffenhauser, *Eur. J. Org. Chem.*, **1998**, 2733.

98OM5534 B. L. Hayes and M. E. Welker, *Organometallics*, **1998**, *17*, 5534.

98S1615 C. Wang, A. S. Batsanov, M. R. Bryce and J. A. K. Howard, *Synthesis*, **1998**, 1615.

98S1710 C. E. Keefer, S. T. Purrington and R. D. Bereman, *Synthesis*, **1998**, 1710.

98SL1411 B. F. Bonini, G. Kemperman, S. T. H. Willems, M. Fochi, G. Mazzanti and B. Zwanenburg, *Synlett*, **1998**, 1411.

98T10779 D. Stien, D. Crich and M. P. Bertrand, *Tetrahedron*, **1998**, *54*, 10779.

98TA3303 J. Zhang, S. Takahashi, S. Saito and T. Koizumi, *Tetrahedron: Asymmetry*, **1998**, *9*, 3303.

98TL7709 J. Yamada, R. Oka, H. Anzai, H. Nishikawa, I. Ikemoto and K. Kikuchi, *Tetrahedron Lett.*, **1998**, *39*, 7709.

99ACS133 M. I. Hegab, F. M. E. Abdel-Megeid, F. A. Gad, S. A. Shiba, I. Søtpfte, J. Møller and A. Senning, *Acta Chem. Scand.*, **1999**, *53*, 133.

99ACS710 J. Oxelbark, L. Thunberg, A. Andersson and S. Allenmark, *Acta Chem. Scand.*, **1999**, *53*, 710.

99ACS1003 J. F. Liebman, K. N. Petersen and P. N. Skancke, *Acta Chem. Scand.*, **1999**, *53*, 1003.

99AG(E)810 J. Yamada, M. Watanabe, H. Anzai, H. Nishikawa, I. Ikemoto and K. Kikuchi, *Angew. Chem., Int. Edn.*, **1999**, *38*, 810.

99AG(E)1417 K. B. Simonsen, N. Svenstrup, J. Lau, N. Thorup and J. Becher, *Angew. Chem., Int. Edn.*, **1999**, *38*, 1417.

99AG(E)1918 H. Sellner and D. Seebach, *Angew. Chem., Int. Edn.*, **1999**, *38*, 1918.

99AG(E)1946 A. Krief, L. Provins and W. Dumont, *Angew. Chem., Int. Edn.*, **1999**, *38*, 1946.

99AG(E)3207 I. E. Markó, A. Ates, A. Gautier, B. Leroy, J.-M. Plancher, Y. Quesnel and J.-C. Vanherck, *Angew. Chem., Int. Ed.*, **1999**, *38*, 3207.

99AM758 E. Ojima, H. Fujiwara and H. Kobayashi, *Adv. Mater. (Weinheim, Ger.)*, **1999**, *11*, 758.

99CC513 L. Martin, S. S. Turner, P. Day, K. M. A. Malik, S. J. Coles and M. B. Hursthouse, *Chem. Commun.*, **1999**, 513.

99CC515 T. Shimizu and T. Yamamoto, *Chem. Commun.*, **1999**, 515.

99CC591 L. Somsák, L. Kocács, V. Gyóllai and E. Õsz, *Chem. Commun.*, **1999**, 591.

99CC737 H. Fujihara, H. Nakai, M. Yoshihara and T. Maeshima, *Chem. Commun.*, **1999**, 737.

99CC1125 E. Tsiperman, T. Regev, J. Y. Becker, J. Bernstein, A. Ellern, V. Khodorkovsky, A. Shames and L. Shapiro, *Chem. Commun.*, **1999**, 1125.

99CC1245 N. Nguyen-Ba, W. L. Brown, L. Chan, N. Lee, L. Brasili, D. Lafleur and B. Zacharie, *Chem. Commun.*, **1999**, 1245.

99CC1407 H. Müller, F. Salhi, B. Divisia-Blohorn, F. Genoud, T. Narayanan, M. Lorenzen and C. Ferrero, *Chem. Commun.*, **1999**, 1407.

99CC1417 F. LeDerf, M. Mazari, N. Mercier, E. Llevillain, P. Richomme, J. Becher, J. Garín, J. Orduna, A. Gorgues and M. Sallé, *Chem. Commun.*, **1999**, 1417.

99CC1493 H. Liu, S. Liu and L. Echegoyen, *Chem. Commun.*, **1999**, 1493.

99CC1673 R. A. Aitken, L. Hill and N. J. Wilson, *Chem. Commun.*, **1999**, 1673.

99CC1835 T. Finn, M. R. Bryce, A. S. Batsanov and J. A. K. Howard, *Chem. Commun.*, **1999**, 1835.

99CC1857 N. Sakai, T. Fukushima, S. Minakata, I. Ryu and M. Komatsu, *Chem. Commun.*, **1999**, 1857.

99CC1891 S. Ogawa, M. Kikuchi, Y. Kawai, S. Niizuma and R. Sato, *Chem. Commun.*, **1999**, 1891.

99CC2417 H. Fujiwara and H. Kobayashi, *Chem. Commun.*, **1999**, 2417.

99CC2433 C. A. Christensen, M. R. Bryce, A. S. Batsanov, J. A. K. Howard, J. O. Jeppesen and J. Becher, *Chem. Commun.*, **1999**, 2433.

99CEJ2971 K. Heuzé, M. Fourmigué, P. Batail, E. Canadell and P. Auban-Senzier, *Chem. Eur. J.*, **1999**, *5*, 2971.

99CL723 H. Naka, M. Shindo, T. Maruyama, S. Sato and N. Furukawa, *Chem. Lett.*, **1999**, 723.

99CL845 E. Ojima, B. Zh. Narymbetov, H. Fujiwara, H. Kobayashi, A. Kobayashi, K. Takimiya, T. Otsubo and F. Ogura, *Chem. Lett.*, **1999**, 845.

99CL987 K. Ueda, Y. Kamata, T. Kominami, M. Iwamatsu and T. Sugimoto, *Chem. Lett.*, **1999**, 987.

99EUP909758 F. Ishii, K. Mogi, H. Eto, S. Sato and H. Matsuda, *Eur. Pat.* 909 758 (1999) [*Chem. Abstr.*, **1999**, *130*, 296687].

99HCA1096 D. K. Heldmann and D. Seebach, *Helv. Chim. Acta*, **1999**, *82*, 1096.

99HCA1458 M. Blagoev, A. Linden and H. Heimgartner, *Helv. Chim. Acta*, **1999**, *82*, 1458.

99HCA1829 D. Seebach, I. M. Lyapkalo and R. Dahinden, *Helv. Chim. Acta*, **1999**, *82*, 1829.

99JA7959 A. Ishii, M. Nakabayashi and J. Nakayama, *J. Am. Chem. Soc.*, **1999**, *121*, 7959.

99JAP11116567 T. Sigimoto and S. Komatsu, *Jpn. Pat.* 11 116 567 (1999) [*Chem. Abstr.*, **1999**, *130*, 352274].

99JAP11189591 M. Nonokuchi, H. Baba and N. Saito, *Jpn. Pat.* 11 189 591 (1999) [*Chem. Abstr.*, **1999**, *131*, 73641].

99JAP11209367 I. Shibuya, Y. Gama and M. Shimizu, *Jpn. Pat.* 11 209 367 (1999) [*Chem. Abstr.*, **1999**, *131*, 129984].

99JAP11279179 S. Kawata, H. Yoshida and M. Kobayashi, *Jpn. Pat.* 11 279 179 (1999) [*Chem. Abstr.*, **1999**, *131*, 257566].

99JCR(S)50 M. Lourak, S. Becker, Y. Fort and P. Caubère, *J. Chem. Res. (S)*, **1999**, 50.

99JCS(D)2127 T. B. Marder, N. C. Norman, A. G. Orpen, M. J. Quayle and C. R. Rice, *J. Chem. Soc., Dalton Trans.*, **1999**, 2127.

99JCS(D)3007 N. Bricklebank, P. J. Skabara, D. E. Hibbs, M. B. Hursthouse and K. M. A. Malik, *J. Chem. Soc., Dalton Trans.*, **1999**, 3007.

99JCS(P1)1023 L. S. Konstantinova, O. A. Rakitin, C. W. Rees, L. I. Souvorova and T. Torroba, *J. Chem. Soc., Perkin Trans. 1*, **1999**, 1023.

99JCS(P2)239 R. Fajgar, J. Vítek, Y. Haas and J. Pola, *J. Chem. Soc., Perkin Trans. 2*, **1999**, 239.

99JCS(P2)505 P. J. Skabara, I. M. Serebryakov and I. F. Perepichka, *J. Chem. Soc., Perkin Trans. 2*, **1999**, 505.

99JCS(P2)657 K. B. Simonsen, V. V. Konovalov, T. A. Konovalova, T. Kawai, M. P. Cava, L. D. Kispert, R. M. Metzger and J. Becher, *J. Chem. Soc., Perkin Trans 2*, **1999**, 657.

99JCS(P2)1405 I. M. Serebryakov, P. J. Skabara and I. F. Perepichka, *J. Chem. Soc., Perkin Trans. 2*, **1999**, 1405.

99JFC(94)141 V. Církva and O. Paleta, *J. Fluorine Chem.*, **1999**, *94*, 141.

99JHC161 J. E. Schachtner, T. Zoukas, H.-D. Stachel, K. Polborn and H. Noth, *J. Heterocycl. Chem.*, **1999**, *36*, 161.

99JHC823 M. R. Bryce, A. K. Lay, A. Batsanov and J. A. K. Howard, *J. Heterocycl. Chem.*, **1999**, *36*, 823.

99JOC4267 S. Akoudad, P. Frère, N. Mercier and J. Roncali, *J. Org. Chem.*, **1999**, *64*, 4267.

99JOC4376 C. W. Rees, O. A. Rakitin, C. F. Marcos and T. Torroba, *J. Org. Chem.*, **1999**, *64*, 4376.

99JOC6443 Z.-X. Wang, S. M. Miller, O. P. Anderson and Y. Shi, *J. Org. Chem.*, **1999**, *64*, 6443.

99JOC6797 X. Bei, H. W. Turner, W. H. Weinberg, A. S. Guram and J. L. Petersen, *J. Org. Chem.*, **1999**, *64*, 6797.

99MI6 H. Hamedifar and A. Shafiee, *Daru, J. Fac. Pharm., Tehran Univ. Med. Sci.*, **1999**, *7*, 6 [*Chem. Abstr.*, **1999**, *131*, 73592].

99MI23 A. Morikami, K. Takimiya, Y. Aso and T. Otsubo, *Org. Lett.*, **1999**, *1*, 23.

99MI55 D. Seebach, A. Pichota, A. K. Beck, A. B. Pinkerton, T. Litz, J. Karjalainen and V. Gramlich, *Org. Lett.*, **1999**, *1*, 55.

99MI89 P. Angeli, L. Brasili, S. Franchini, D. Giardina, U. Gulin and G. Marucci, *Med. Chem. Res.*, **1999**, *9*, 89 [*Chem. Abstr.*, **1999**, *130*, 352203].

99MI285 D. Ma, Y. Cao, Y. Yang and D. Cheng, *Org. Lett.*, **1999**, *1*, 285.

99MI323 M. M. Alam, O. Ito, N. Sakurai and H. Moriyama, *Res. Chem. Intermed.*, **1999**, *25*, 323 [*Chem. Abstr.*, **1999**, *130*, 102738].

99MI447 T. Miyazawa, S. Kurita, H. Sakamoto, T. Otomatsu, K. Hirose and T. Yamada, *Biotechnol. Lett.*, **1999**, *21*, 447 [*Chem. Abstr.*, **1999**, *131*, 336960].

99MI487 A. L. Schwan, J. L. Snelgrove, M. L. Kalin, R. D. J. Froese and K. Morokuma, *Org. Lett.*, **1999**, *1*, 487.

99MI617 J. Yamada, H. Nishikawa and K. Kikuchi, *J. Mater. Chem.*, **1999**, *9*, 617.

99MI687 N. D. Kushch, O. A. Dyachenko, V. V. Gritsenko, L. I. Buravov, V. A. Tkacheva, E. B. Yagubskii, M. G. Kaplunov, E. V. Golabev, T. G. Togonidze, A. Kobayashi and H. Kobayashi, *J. Mater. Chem.*, **1999**, *9*, 687.

99MI693 H. Nishikawa, T. Sato, T. Kodama, I. Ikemoto, K. Kikuchi, H. Anzai and J. Yamada, *J. Mater. Chem.*, **1999**, *9*, 693.

99MI1081 J. Zhu, Z.-Y. Zhou, F.-M. Fu, J.-G. Deng, A.-Q. Mi, Y.-Z. Jiang and T.-Y. Chau, *Gaodeng Xuexiao Huaxue Xuebao*, **1999**, *20*, 1081 [*Chem. Abstr.*, **1999**, *131*, 271824].

99MI1239 N. Martín, E. Ortí, L. Sánchez, P. M. Viruela and R. Viruela, *Eur. J. Org. Chem.*, **1999**, 1239.

99MIP16766 E. Oshima, H. Nakasato, K. Yanagawa and H. Manabe, *PCT Int. Appl.* WO 16 766 (1999) [*Chem. Abstr.*, **1999**, *130*, 252350].

99MIP33826 T. Kagawa, H. Kodama, Y. Goto, Y. Niwano and M. Yoshida, *PCT Int. Appl.* WO 33 826 (1999) [*Chem. Abstr.*, **1999**, *131*, 87917].

99MIP43313 J. A. Dunn, T. J. Bardos and H. F. Kung, *PCT Int. Appl.* WO 43 313 (1999) [*Chem. Abstr.*, **1999**, *131*, 199701].

99OM803 G. Zeni, A. Chieffi, R. L. O. R. Cunha, J. Zukerman-Schpector, H. A. Stefani and J. V. Comasseto, *Organometallics*, **1999**, *18*, 803.

99OS12 A. K. Beck, P. Gysi, L. LaVecchia and D. Seebach, *Org. Synth.*, **1999**, *76*, 12.

99OS101 M. A. Robbins, P. N. Devine and T. Oh, *Org. Synth.*, **1999**, *76*, 101.

99S43 V. Marchán, M. Gibert, A. Messeguer, E. Pedroso and A. Grandas, *Synthesis*, **1999**, 43.

99S577 A. Pérez-Benítez, J. Tarrés, E. Ribera, J. Veciana and C. Rovira, *Synthesis*, **1999**, 577.

99SC1889 P. T. Kaye and W. E. Molema, *Synth. Commun.*, **1999**, *29*, 1889.

99SC2953 C. A. Merlic, A. Baur, D. J. Tantillo and S. E. Brown, *Synth. Commun.*, **1999**, *29*, 2953.

99SL303 C. Wattenbach, M. Maurer and H. Frauenrath, *Synlett*, **1999**, 303.

99SL319 H. Firouzabadi, N. Iranpoor and B. Karimi, *Synlett*, **1999**, 319.

99SL413 H. Firouzabadi, N. Iranpoor and B. Karimi, *Synlett*, **1999**, 413.

99SL415 R. V. Anand, P. Saravanan and V. K. Singh, *Synlett*, **1999**, 415.

99SL777 M. Curini, F. Epifano, M. C. Marcotullio and O. Rosati, *Synlett*, **1999**, 777.

99SL1811 A. Alexakis, C. Benhaïm, X. Fournioux, A. van den Heuvel, J.-M. Levêque, S. March and S. Rosset, *Synlett*, **1999**, 1811.

99SL1814 R. Hilgraf and A. Pfalz, *Synlett*, **1999**, 1814.

99SL1936 A. Krief, L. Provins and A. Froidbise, *Synlett*, **1999**, 1936.

99SL1960 N. Tanaka and Y. Masaki, *Synlett*, **1999**, 1960.

99T2353 S. Braverman, M. Cherinsky, E. Nov and M. Sprecher, *Tetrahedron*, **1999**, *55*, 2353.

99T5867 I. Alexandropoulou, T. A. Crabb, A. V. Patel and J. Hudec, *Tetrahedron*, **1999**, *55*, 5867.

99T6311 A. Kameyama, N. Kijima, H. Hashikawa and T. Nishikubo, *Tetrahedron*, **1999**, *55*, 6311.

99T7625 J. L. Marco, S. T. Ingate and P. M. Chinchón, *Tetrahedron*, **1999**, *55*, 7625.

99T7907 P. de March, M. Figueredo, J. Font and J. Medrano, *Tetrahedron*, **1999**, *55*, 7907.

99T9915 M. R. Bryce, M. A. Coffin, A. J. Moore, A. S. Batsanov and J. A. K. Howard, *Tetrahedron*, **1999**, *55*, 9915.

99T9979 R. D. McCullough, M. A. Petruska and J. A. Belot, *Tetrahedron*, **1999**, *55*, 9979.

99T10341 A. Ishii, T. Nakaniwa, K. Umezawa and J. Nakayama, *Tetrahedron*, **1999**, *55*, 10341.

99T10845 M. Shipman, H. R. Thorpe and I. R. Clemens, *Tetrahedron*, **1999**, *55*, 10845.

99T11475 G. Mloston, R. Huisgen and K. Polborn, *Tetrahedron*, **1999**, *55*, 11475.

99T12907 R. Hayes, K.-D. Li, P. Leeming, T. W. Wallace and R. C. Williams, *Tetrahedron*, **1999**, *55*, 12907.

99T13029 H. J. Cristeau, F. Darviche, M.-T. Babouneau, J.-M. Fabre and E. Torreilles, *Tetrahedron*, **1999**, *55*, 13029.

99T13445 G. C. G. Pais, R. A. Fernandes and P. Kumar, *Tetrahedron*, **1999**, *55*, 13445.

99T14429 R. Gerke, L. Fitjer, P. Müller, I. Usón, K. Kowski and P. Rademacher, *Tetrahedron*, **1999**, *55*, 14429.

99T14649 M. P. Molas, M. I. Matheu, S. Castillón, J. Isac-García, F. Hernández-Mateo, F. G. Calvo-Flores and F. Santoyo-González, *Tetrahedron*, **1999**, *55*, 14649.

99TA2749 W. Adam, C. R. Saha-Möller and C.-G. Zhao, *Tetrahedron: Asymmetry*, **1999**, *10*, 2749.

99TA3747 M. Jungen and H.-J. Gais, *Tetrahedron: Asymmetry*, **1999**, *10*, 3747.

99TA3859 J. Cossy, S. Bouzbouz and J. C. Caille, *Tetrahedron: Asymmetry*, **1999**, *10*, 3859.

99TA4183 R. Kawecki, *Tetrahedron: Asymmetry*, **1999**, *10*, 4183.

99TA4245 J. Rifé and R. M. Ortuño, *Tetrahedron: Asymmetry*, **1999**, *10*, 4245.

99TA4349 R. A. Fernandes and P. Kumar, *Tetrahedron: Asymmetry*, **1999**, *10*, 4349.

99TA4755 M. Lemaire and J. Bolte, *Tetrahedron: Asymmetry*, **1999**, *10*, 4755.

99TL801 M. R. Bryce, A. K. Lay, A. S. Batsanov and J. A. K. Howard, *Tetrahedron Lett.*, **1999**, *40*, 801.

99TL1583 M. T. Barros, A. J. Burke and C. D. Maycock, *Tetrahedron Lett.*, **1999**, *40*, 1583.

99TL1779 P. C. Ray and S. M. Roberts, *Tetrahedron Lett.*, **1999**, *40*, 1779.

99TL1799 A. Ates, A. Gautier, B. Leroy, J.-M. Plancher, Y. Quesnel and I. E. Markó, *Tetrahedron Lett.*, **1999**, *40*, 1799.

99TL2927 P. Hudhomme, S. G. Liu, D. Kreher, M. Cariou and A. Gorgues, *Tetrahedron Lett.*, **1999**, *40*, 2927.

99TL5027 J. Nakazaki, M. M. Matsushita, A. Izuoka and T. Sugawara, *Tetrahedron Lett.*, **1999**, *40*, 5027.

99TL5997	N. Gautier, N. Mercier, A. Riou, A. Gorgues and P. Hudhomme, *Tetrahedron Lett.*, **1999**, *40*, 5997.
99TL6489	C. J. Burns, L. D. Field, J. Morgan, D. D. Ridley and V. Vignevich, *Tetrahedron Lett.*, **1999**, *40*, 6489.
99TL6553	P. H. Dussault and X. Liu, *Tetrahedron Lett.*, **1999**, *40*, 6553.
99TL6571	A. Krief and L. Defrère, *Tetrahedron Lett.*, **1999**, *40*, 6571.
99TL6635	J. Yamada, K. Aoki, S. Nakatsuji, H. Nishikawa, I. Ikemoto and K. Kikuchi, *Tetrahedron Lett.*, **1999**, *40*, 6635.

Chapter 5.7

Five-Membered Ring Systems with O & N Atoms

Thomas L. Gilchrist
The University of Liverpool, UK

email: tlg57@liverpool.ac.uk

5.7.1 ISOXAZOLES

Cycloaddition reactions of nitrile oxides continue to be the source of new isoxazoles and dihydroisoxazoles. The oxidation of α-hydroxyimino acids **1** by ammonium hexanitratocerium(IV) has provided a new method of generation of benzonitrile oxide and other nitrile oxides; on the other hand the oxime **2** is simply dehydrogenated by this oxidant to produce benzoylnitrile oxide <99BSJ2277> . Manganese dioxide has also been used to oxidise aldoximes to nitrile oxides; the reaction is most efficient with methyl (hydroxyimino)acetate <99TL5605> .

1

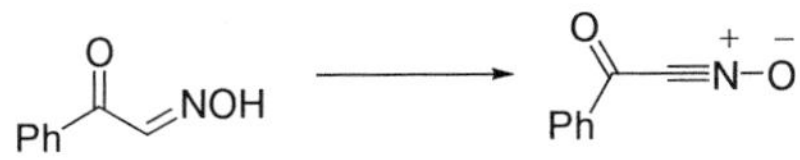

2

Several isoxazoles and dihydroisoxazoles bearing a diethylphosphonyl substituent have been produced by the cycloaddition of nitrile oxides to unsaturated phosphonates; an example is the preparation of the isoxazole **3** <99S2027, 99SC3621> . These derivatives have then be used in further reactions; for example, as precursors to 5-alkenylisoxazoles. Cycloaddition of nitrile oxides having the general structure **4** (P = protecting group) to enamines of β-ketoesters produces 4-(1-aminoalkyl)isoxazole-4-carboxylic acid esters **5**; these are precursors to 5-substituted 3-acyltetramic acids <99SL873> . In this route the roles of 1,3-dipole and dipolarophile used in earlier syntheses of 3-acyltetramic acids <99JCS(P1)765> are reversed. Protected (β-aminoalkyl)isoxazolines have also been synthesised by nitrile oxide cycloaddition <99JOC9297, 99TL4085> and the fused isoxazolines **6** <99S1569> and **7** <99TL3535> were produced by intramolecular cycloaddition of nitrile oxides to alkynes.

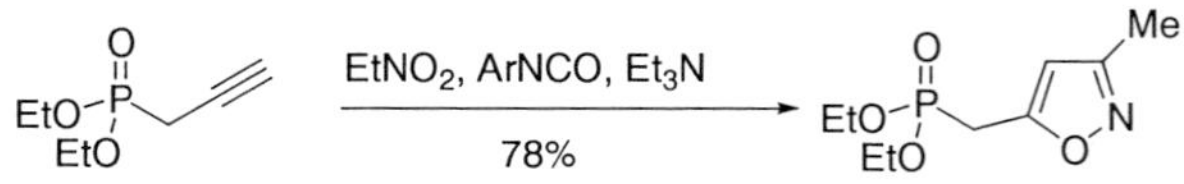

3

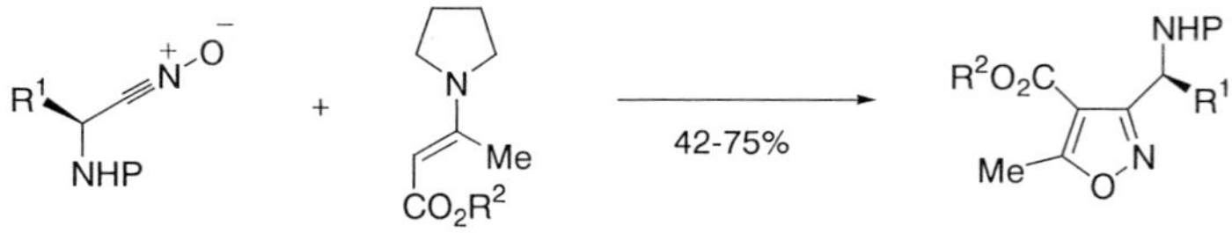

4 **5**

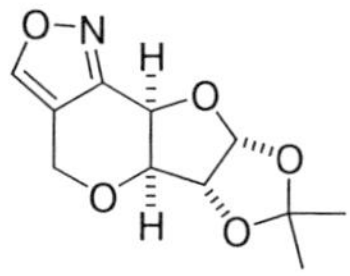

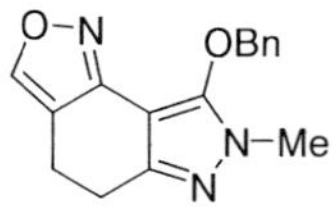

6 **7**

The nitroisoxazolone salt **8** has been converted into a series of 4-substituted 3,5-dicyanoisoxazoles **9** by successive reaction with pyrrolidine, an aldehyde, and acetyl chloride. The sequence, which can be carried out in one pot, goes by way of an intermediate dianion **10** which is *O*-acylated and cyclised <99JOC2160> . The nitro group of the imidazoles **11** also provides the N-O fragment of the isoxazole ring in the imidazo[4,5-*c*]isoxazoles **12** that are produced by thermolysis of **11** <99JCS(P1)817> . A ketene intermediate, produced by loss of ethanol from the ethyl ester, gives the isoxazole ring after further loss of carbon dioxide. The trifluoromethylisoxazole **13** was produced from the ketone **14** by reaction with an excess of hydroxylamine followed by dehydration of an intermediate isoxazoline with thionyl chloride <99H(50)377> .

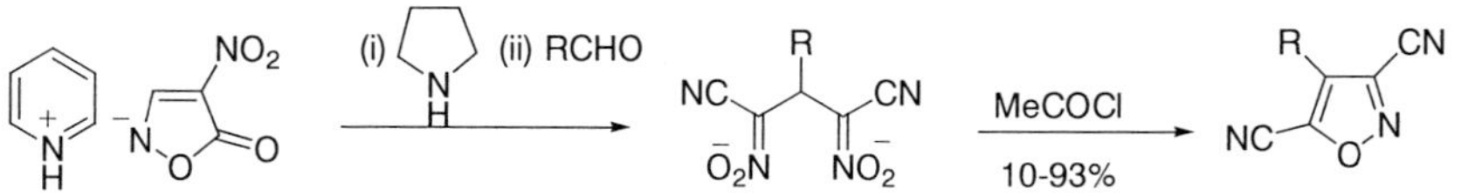

8 **10** **9**

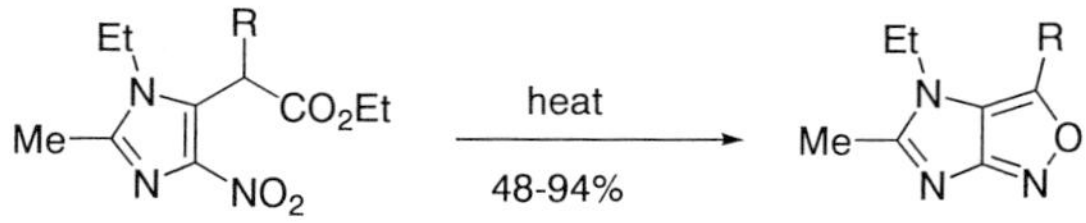

11 **12**

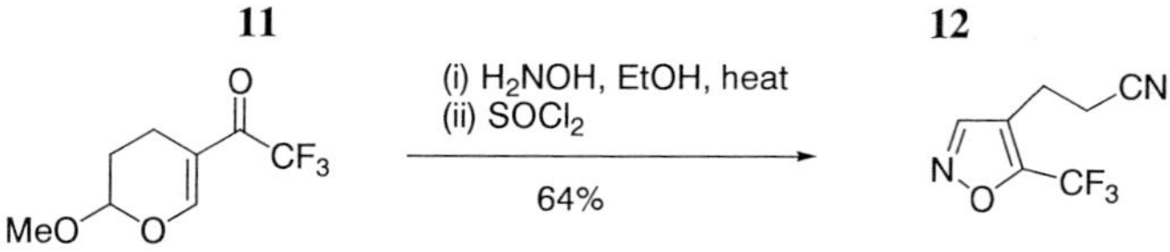

14 **13**

A high yielding synthesis of anthranil from 2-nitrobenzaldehyde makes use of zinc and allyl bromide as the reducing agent. The method has also been used for the preparation of 3-substituted anthranils from 2-nitrophenyl ketones <99H(51)1921> . A solid phase synthesis of 3-aminobenzisoxazoles **15** is based on the displacement of fluoride from 2-fluorobenzonitriles by the Kaiser oxime resin and subsequent hydrolysis of the C=N bond <99JOC4547> . A synthesis of 3-(2-dialkylaminoethyl)benzisoxazoles from oximes of 2-hydroxyphenyl ketones has also been described <99H(51)2139> .

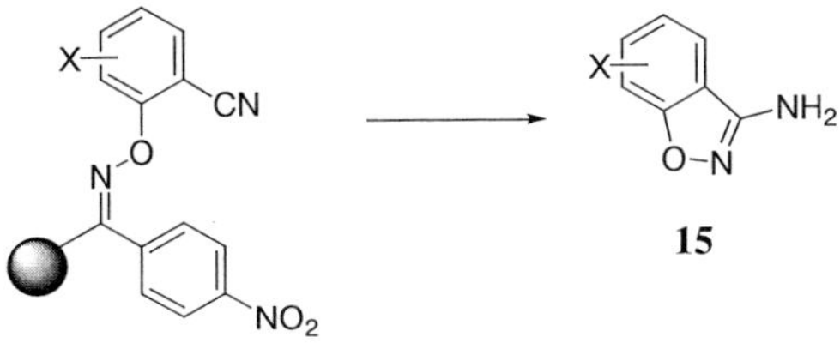

15

5.7.2 ISOXAZOLINES

The first isoxazoline ligands to be used in asymmetric synthesis have been constructed by double intramolecular addition of nitrile oxides produced from the oximes **16** and sodium hypochlorite. Spirobis(isoxazolines) **17** were formed; the pure enantiomers were separated by chiral stationary phase chromatography and were used as ligands for copper(II) <99OL1795> . Spiroisoxazolines are also formed by the cycloaddition of nitrile oxides to α-methylenecycloalkanones **18** <99SC877> and the spiroisoxazoline structure **19** present in psammaplysin-A has been constructed by an intramolecular oxime cyclisation reaction <99TL1291> . Isoxazolines that are produced by the addition of nitrile oxides to the optically active trimethylsilyl substituted allylic alcohols **20** are almost diastereomerically pure <99TL4349> .

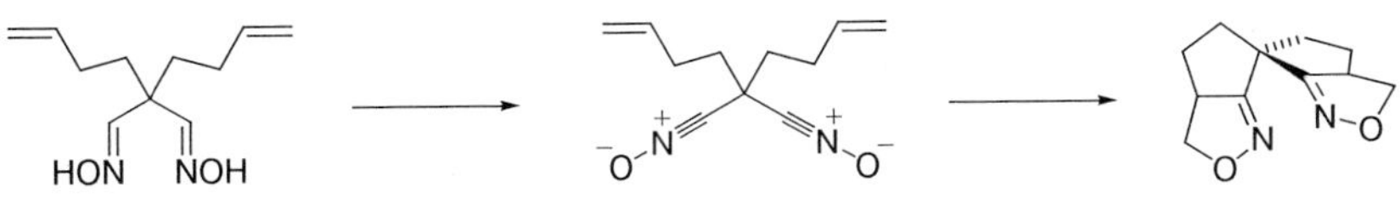

16 **17**

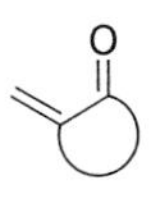

18

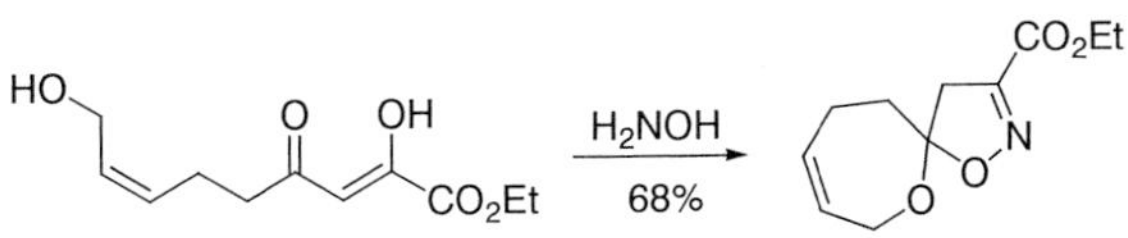

19

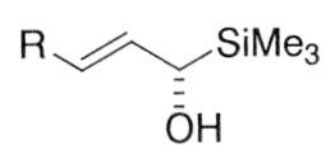

20

Other nitrile oxide cycloadditions have been reported including the addition of benzonitrile oxide to alkenes catalysed by a chiral vinyl dioxazaborocine catalyst (the *ee*'s are moderate) <98TL8513> and the reaction of *N*-phenylpyrazolylnitrile oxide with C_{60} <99TL489> .

The variety of mechanisms by which isoxazol-5(2*H*)-ones are cleaved by bases and nucleophiles have been reviewed <99H(51)3013> ; the cleavage of the salt **8** is an example of a reaction of this type <99JOC2160> . 2-Thioacyl derivatives of this ring system are converted into 1,3-oxazinones by reaction with triphenylphosphine <98JCS(P1)3245> . The addition of diallylzinc to 5-phenylisoxazoline **21** is diastereoselective, giving predominantly the *trans*-isoxazolidine **22**; the reaction with other 5-substituted isoxazolines shows similar diastereoselectivity <99SL798> . The products obtained from the reductive cleavage of 5-(bromomethyl)isoxazolines **23** by reaction with magnesium in methanol are temperature dependent <99SC3165> .

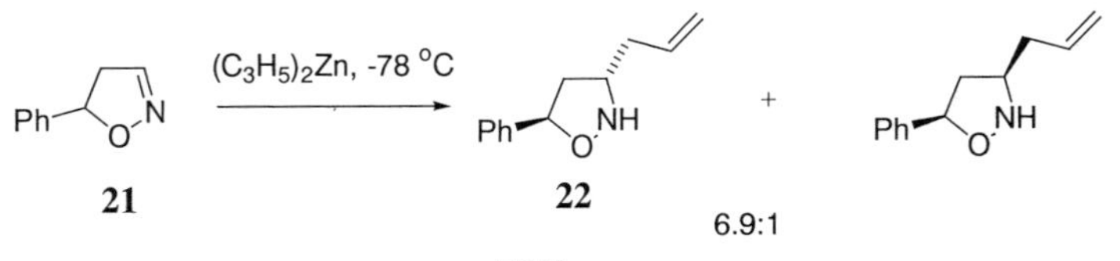

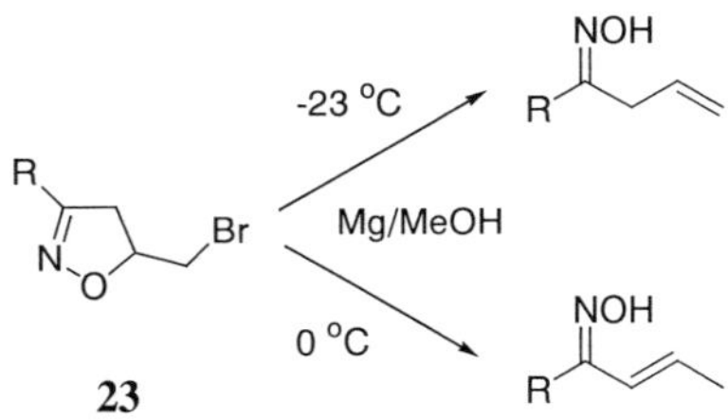

5.7.3 ISOXAZOLIDINES

The cycloaddition of nitrones to enol ethers can be achieved with high enantioselectivity when chiral binaphthol-aluminium catalysts are used. For example, the cycloaddition of *C,N*-diphenylnitrone to *tert*-butyl vinyl ether gave mainly the *exo* isomer **24** in 89% *ee* <99JA3845> . A polymeric version of the catalyst has also been described; this enables the ligand to be recovered and re-used <99CC811> . The asymmetric addition of nitrones to enol ethers has also been achieved using chiral copper-bis(oxazoline) catalysts <99JOC2353> . The synthesis of enantiomerically pure isoxazolidines from chiral precursors is exemplified by the intramolecular addition of the nitrone **25** <99H(51)37> and **26** <99JCS(P1)3349> . The nitrone **26** is generated by rearrangement of an oxime **27**; similarly, thermal rearrangement of the oximes **28** leads to the formation of the isoxazolidines **29** by intramolecular addition of transient nitrones <99JCS(P1)185> .

24

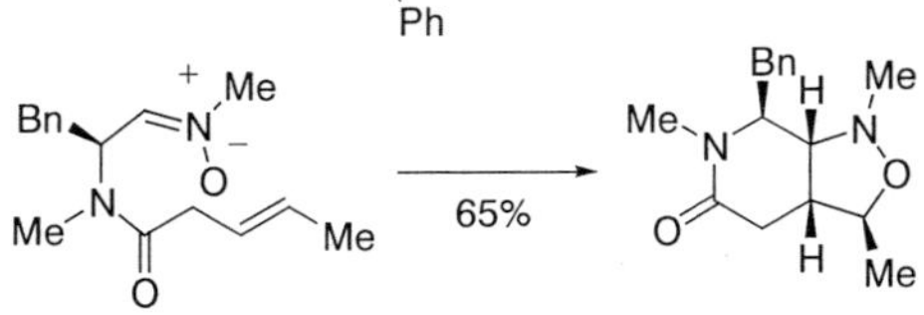

25

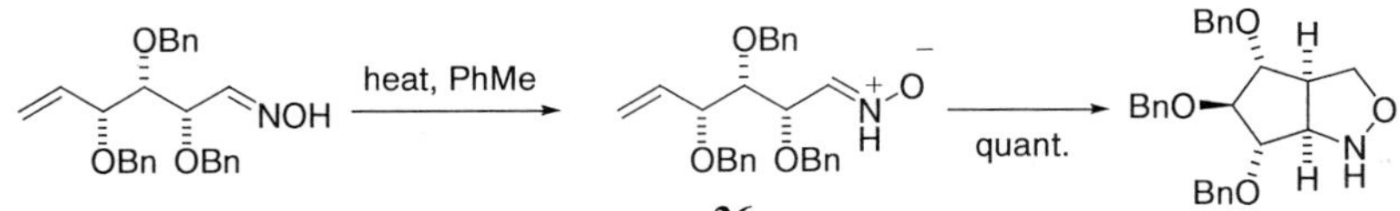

27 **26**

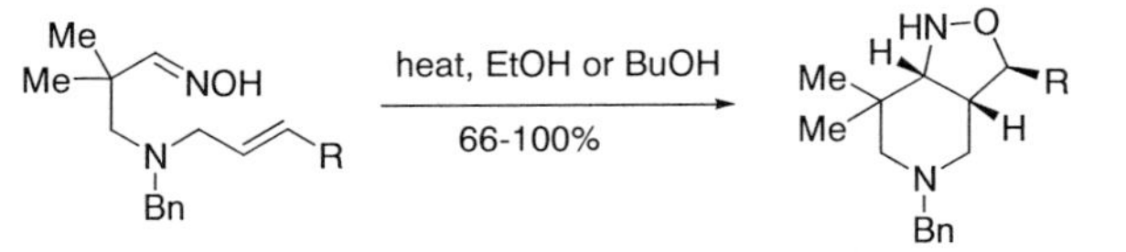

28 **29**

Both the rate and the stereoselectivity of the addition of nitrones to allylic and homoallylic alcohols are improved in the presence of 1 equivalent of magnesium dibromide etherate. For example, the addition of *N*-methyl-*C*-phenylnitrone to allyl alcohol is slow and unselective in the absence of this Lewis acid, but gives the isoxazolidine **30** stereoselectively in its presence <99JCS(P1)3337> . Nitronates have also been added to the magnesium salts of allylic alcohols to give isoxazolidinones in high yield <98TL8865> . Selective *endo* addition of the nitrone **31** to the alkene **32** gave the isoxazolidine **33** which was then converted into the β-lactone (–)-tetrahydrolipstatin <99OL753> . The uncatalysed addition of the nitrone **34** to alkenes is also highly regioselective and diastereoselective <99SL653> . Isoxazolidinones have been prepared in enantiomerically pure form by cyclisation and elimination from chiral oxiranes such as **35** <98JCS(P1)2923> . The radical induced rearrangement of 5-bromomethylisoxazolidines **36** also results in the formation of isoxazolidinones; the reaction involves N-O bond cleavage and recyclisation <99SL79> .

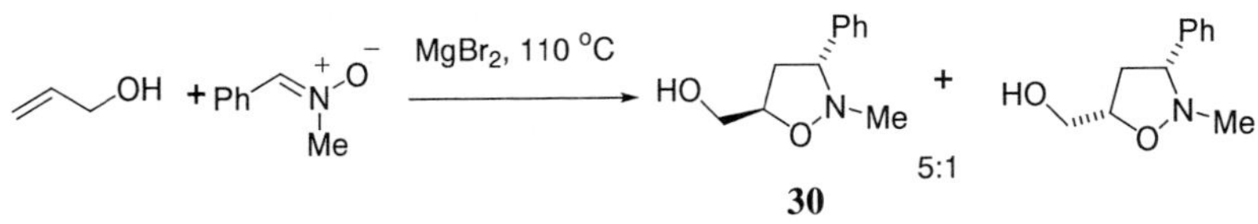

30

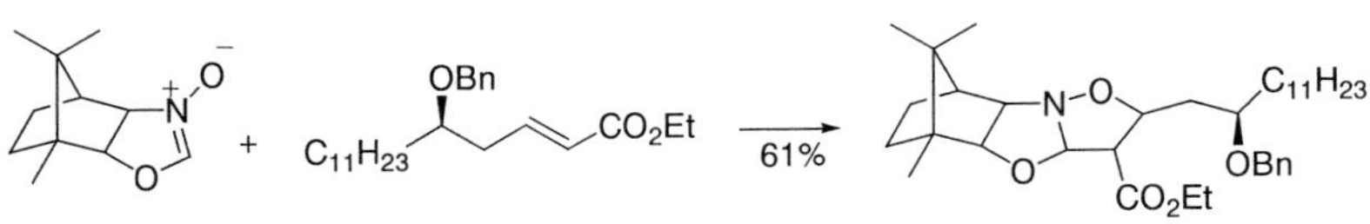

31 **32** **33**

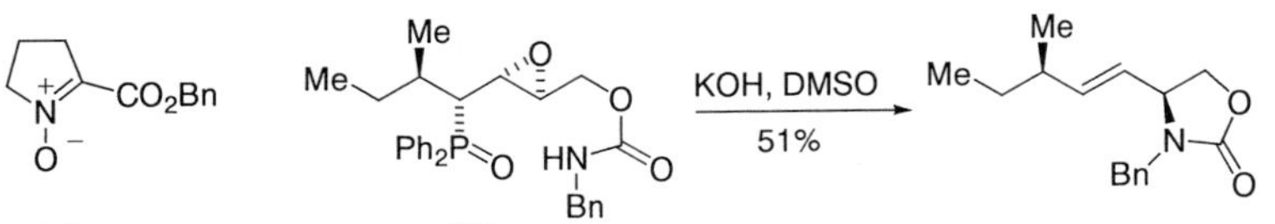

34 **35**

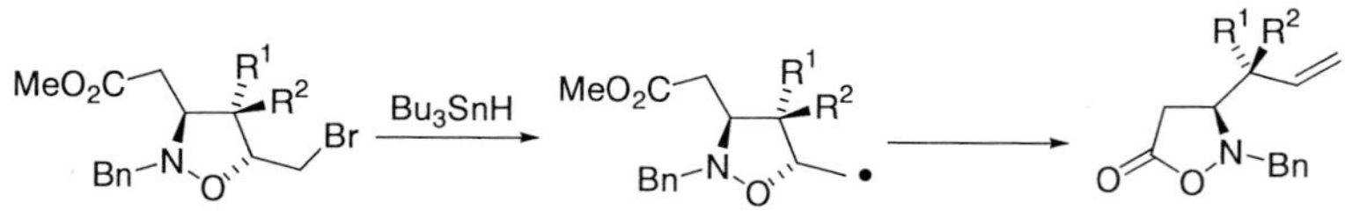

36

5.7.4 OXAZOLES

Isocyanides are useful reagents for the preparation of oxazoles unsubstituted at C-2. For example, TosMIC was used to generate the oxazole ring in a synthesis of phthoxazolin A **37** <99OL1137> . However in the reaction of *N*-tosyl aldimines with the isocyanide **38** the oxazoles that were isolated were 2:1 adducts **39** that had been further substituted at C-2 <98T12445> .

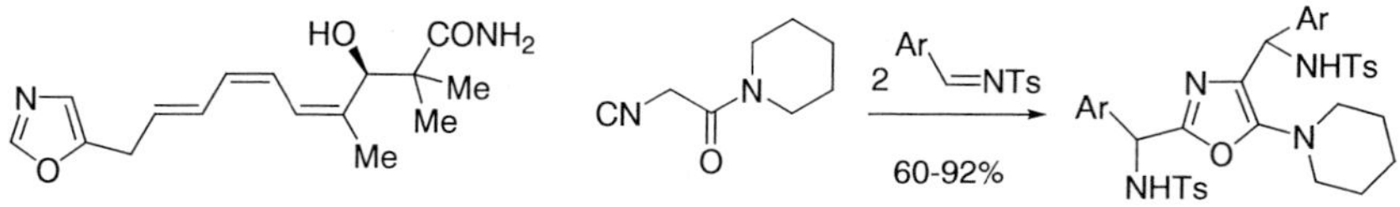

37 **38** **39**

A new variant of the Robinson-Gabriel synthesis of oxazoles involves the use of the Burgess reagent to dehydrate acylaminocarbonyl compounds **40**; the reaction takes place rapidly under microwave irradiation <99SL1642> . Mercury(II) chloride was used to promote the cyclisation of a thioimidate **41** to a 5-aminooxazole in the synthesis of a fungal depsipeptide <99SL893> . 2-Dimethylaminino-5-aryloxazoles **42** and the corresponding 3-methyloxazolium salts have been prepared from α-aminoacetophenones and dichloroiminium chloride <99HCA1981> and several 2-aryl-5-aroyloxazoles have been isolated in moderate yield from the reaction of α-hydroxyiminoacetophenones with S_4N_4 <99JHC911> .

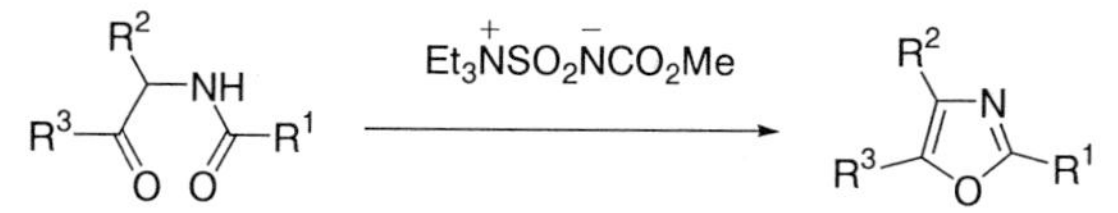

40

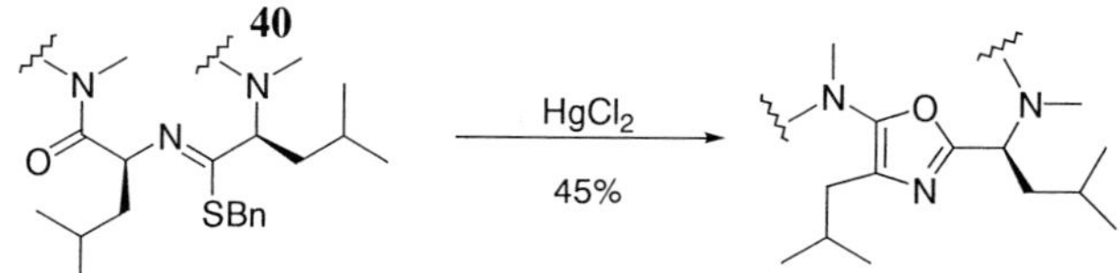

41

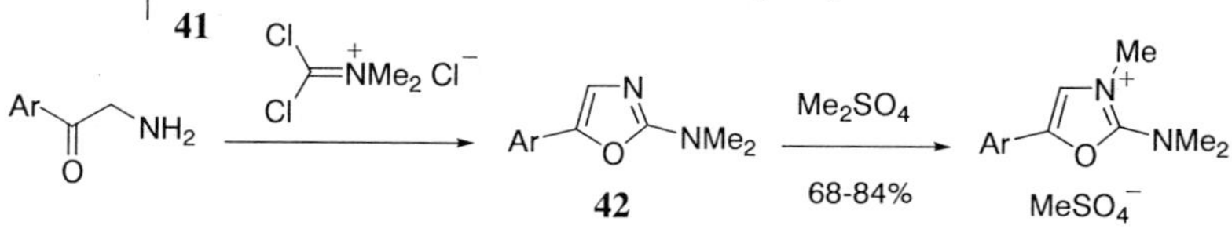

42

Lithiation of 2-methyloxazoles is usually unselective in giving both the 5-oxazolyllithium derivatives (the kinetic products) and the 2-lithiomethyloxazoles. Lithium diethylamide has been shown to give products of substitution at the methyl group with greater than 95:5 selectivity since diethylamine acts as a proton source to mediate the equilibration <99OL87>.

4-Nitrooxazoles have been found to undergo a variety of addition reactions with nucleophiles. With ethyl vinyl ether, for example, the oxazoles undergo a formal 4 + 2 cycloaddition reaction similar to that observed with some nitroalkenes; the final products are 2:1 adducts such as compound **44** from the nitro ester **43** <99T13809> . The known *O*- to *C*-acyl rearrangement of *O*-acyloxazoles **45** can be achieved with high enantioselectivity when a chiral catalyst, a ferrocene derivative bearing a 4-(pyrrolidino)pyridine unit, is used <98JA11532> . The reaction of 2-aryl-5-methoxyoxazoles **46** with benzaldehyde, catalysed by an aluminium(III)-BINOL catalyst, results in ring opening and ring closure to give methyl oxazoline-4-carboxylates **47** with high enantioselectivity. The rearrangement may go by way of a chiral intermediate **48** <99JOC7040> .

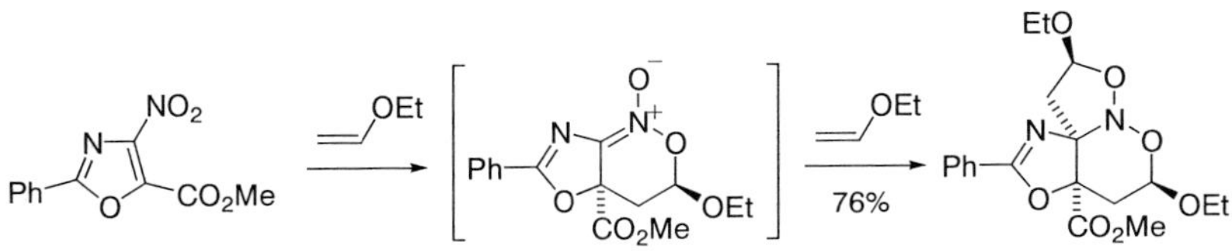

43 **44**

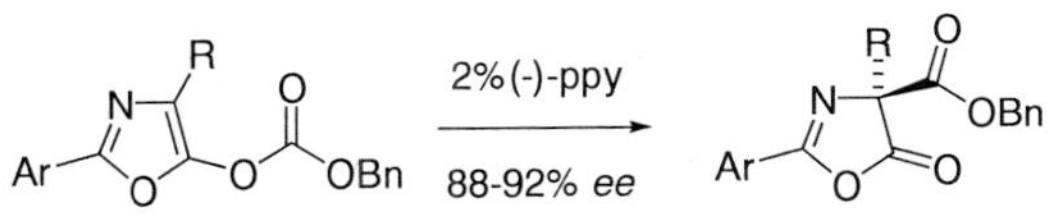

45

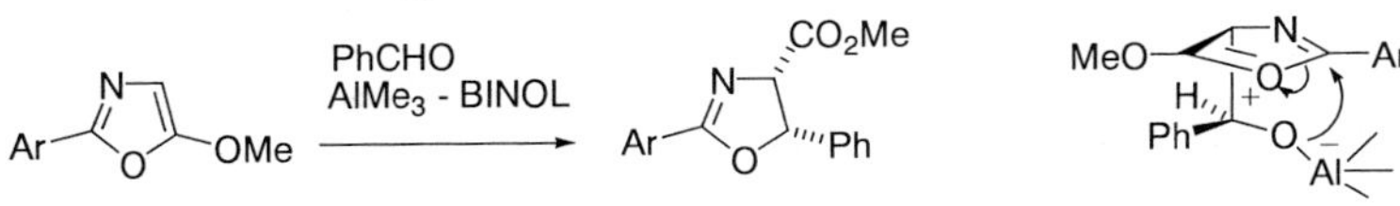

46 **47** **48**

Mesoionic oxazolium-5-oxides **49** react with aminomalonic ester to give pyrrolidinones **50** as the major or exclusive products <99H(50)71> and the oxazolamine **51** is converted by sodium acetate in acetic acid into the hydantoin **52** <99JHC283> . The intramolecular Diels-Alder cycloaddition of the oxazole **53** and related compounds has been used as a route to substituted isoquinolines <99JOC3595> .

The furo[3,4-*d*]oxazole **54**, constructed by cyclisation of a rhodium carbenoid, is a useful compound for the production of unusual benzoxazoles; for example, cycloaddition of *N*-phenylmaleimide gave the benzoxazole **55** <98JOC7680> . Simpler benzoxazole syntheses include the base catalysed cyclizsation (with loss of the trifluoromethyl anion) of the imines **56** <99TL4119> and the acid catalysed cyclisation of diacylated aminophenols **57** <99H(51)979> .

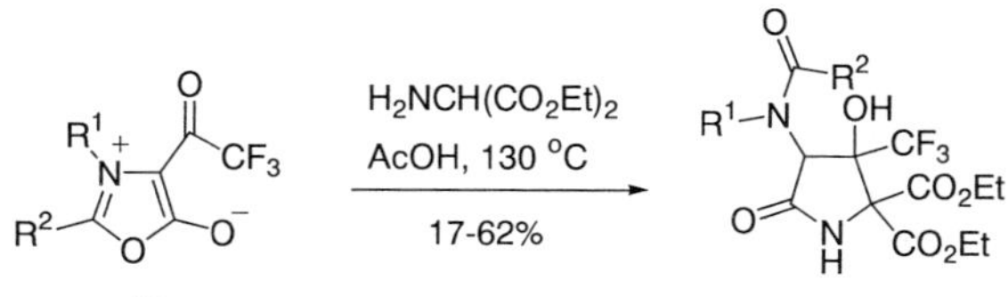

49 50

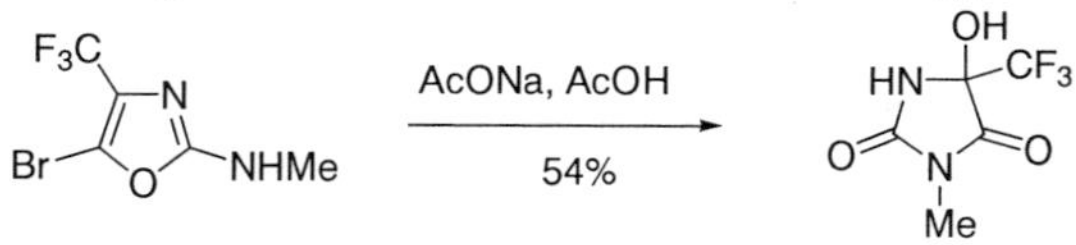

51 52

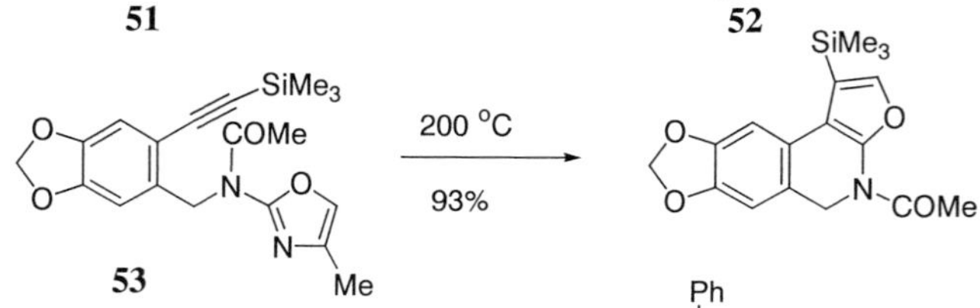

53

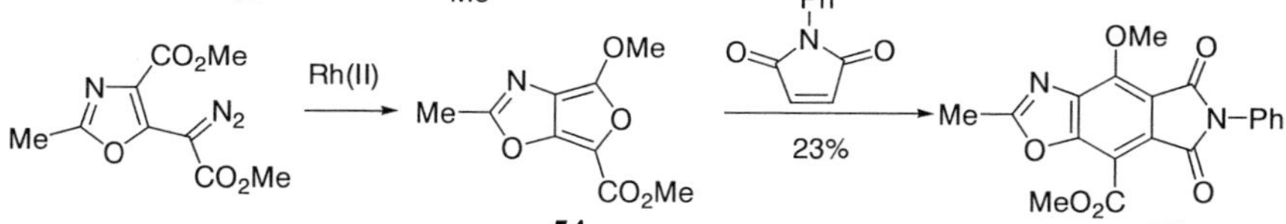

54 55

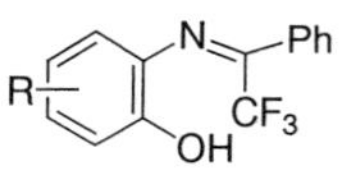

56

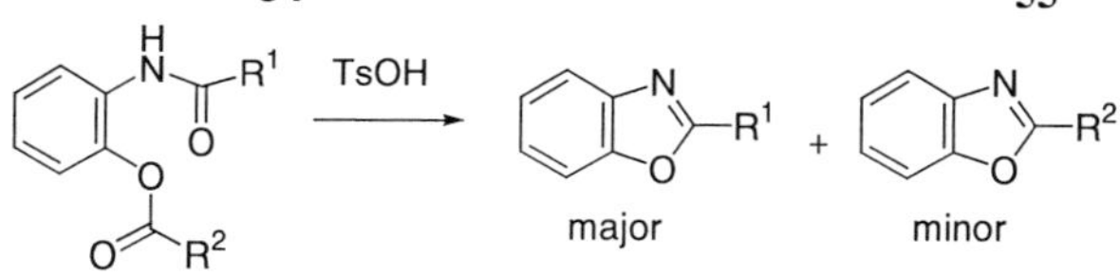

57

Naturally occurring benzoxazoles of marine origin are rare, but two new structurally related compounds of this type have been isolated. One of these, pseudopteroxazole **58**, is a potent antimycobacterial agent <99OL527> .

The benzoxazoles **59** and **60** can act as azadienes in Diels-Alder reactions with nucleophilic alkenes such as vinyl ethers <98JCS(P1)3389> .

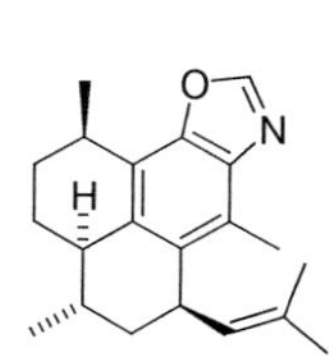

58

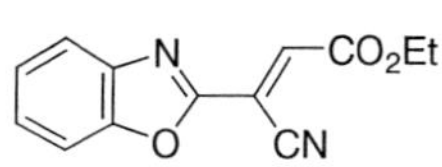

59

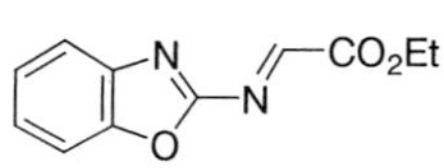

60

5.7.5 OXAZOLINES

Chiral oxazoline ligands continue to be among those most widely used for asymmetric catalysis. The phosphines **61** are superior to other oxazoline based phosphines for palladium catalysed asymmetric allylations <99OL1745> . New naphthyl substituted oxazoline ligands include the compounds **62** <99JOC1620> and **63** <98T15721> .

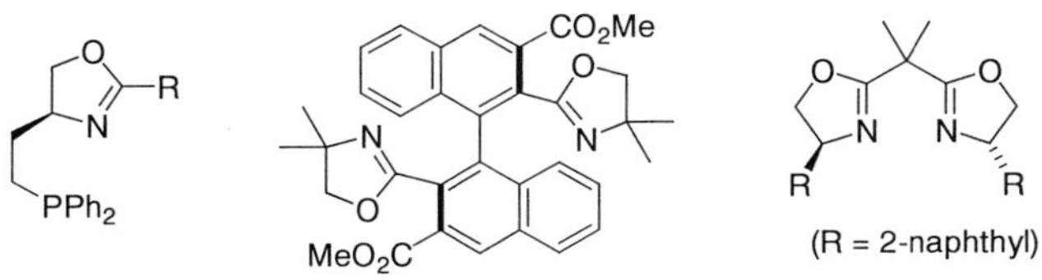

61 **62** **63**

2-Chloromethyloxazolines such as **64** have been synthesised from chiral 2-amino alcohols and ortho esters <99H(51)373> . The palladium(0) catalysed cyclisation of the chiral allylic benzamide **65** gave the *trans*-oxazoline **66** <99JOC9450> . The oxazoline **67** has been used as a source of acyloxiranes; it is selectively deprotonated by *sec*-butyllithium and , after addition of an electrophile, the oxazoline function is converted into an acyl group by reaction with a Grignard reagent <99EJO409> . The oxazoline **68** and related compounds are cleaved by LDA and chlorotrimethylsilane to give ketenimines such as **69**; these can subsequently be converted into ketones with a chiral quaternary α-carbon atom <99TL4765> .

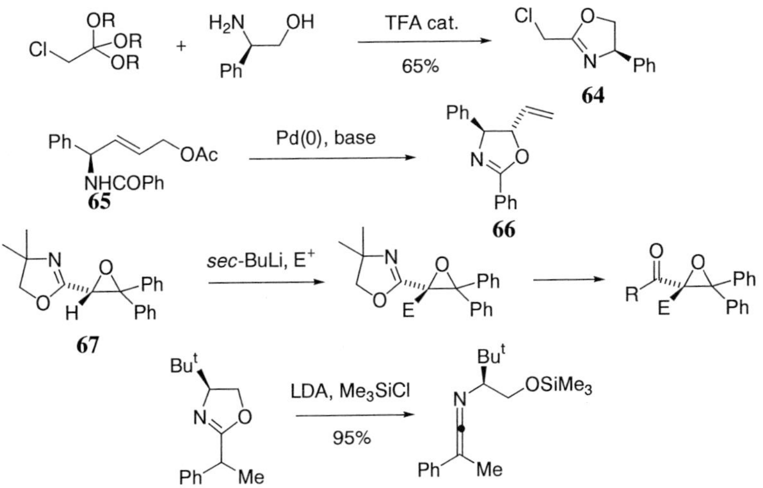

68 **69**

Alkyl oxazoline-5-carboxylates **71**, precursors of β-amino-α-hydroxycarboxylic acids, have been produced by iodocyclisation of alkyl 3-benzamidocarboxylates **70**. The oxazolines can be resolved enzymatically <99SL1727> . The amides **72** are cyclised to *N*-aryloxazolium salts **73** by fluoroboric acid <99EJO297>.

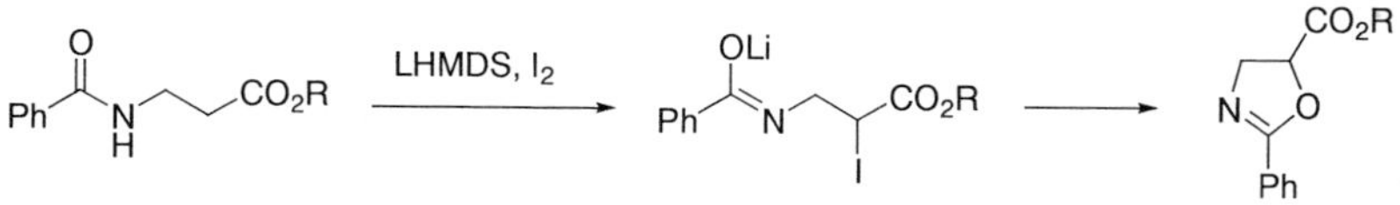

70 71

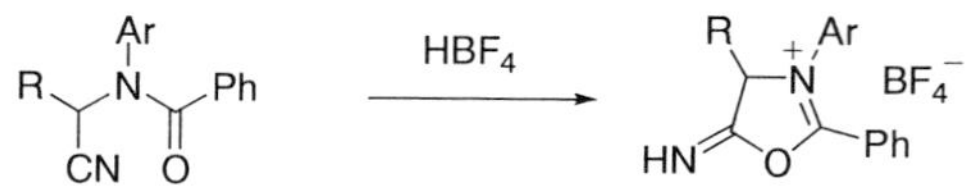

72 73

5.7.6 OXAZOLIDINES

Oxazolidines such as **75** are produced by the cycloaddition of non stabilised azomethine ylides across the aldehydic C=O bond of the pyranone **74** <99JCS(P1)1167> . The reaction of 1*S*,2*R*-1-amino-2-indanol and other chiral amino alcohols with formaldehyde in excess leads to the formation of bis(oxazolidinyl)methanes such as **76** in high yield <99SC43> . The reaction of the oxazolidinium salt **77** with potassium tert-butoxide leads to the selective formation of the rearranged isoxazolidine **78**. Since the salt **77** is produced by selective *N*-methylation the overall process represents a double transfer of chirality, from carbon to nitrogen and then from nitrogen to carbon <99OL31> . Chirality is also transferred in the conjugate addition of lithium cuprates to the oxazolidine imine **79** <99SL727, <99SL1838> and in the formation of the azetidinone **80** <99JOC8461> .

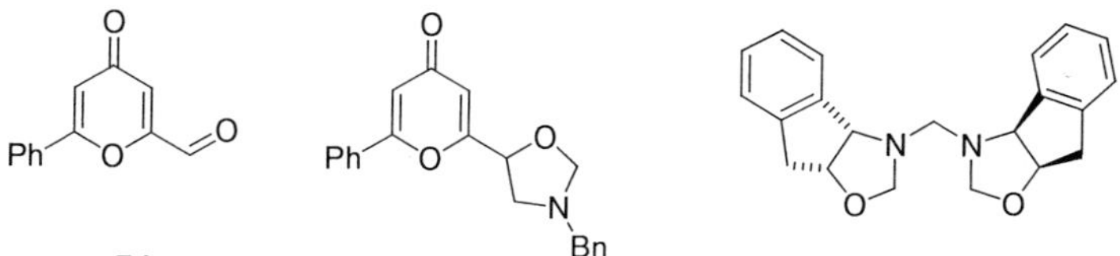

74 75 76

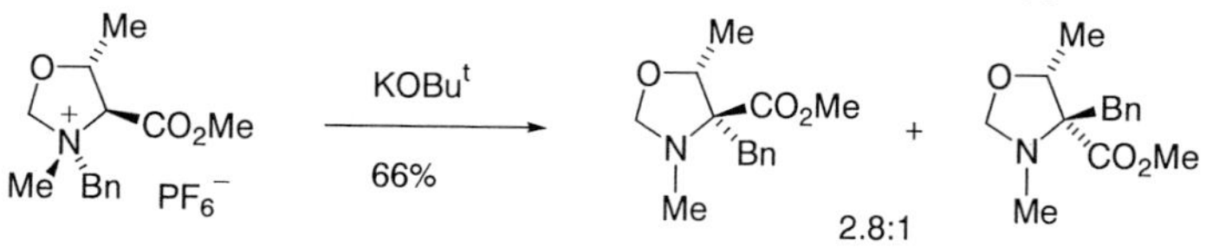

77 78

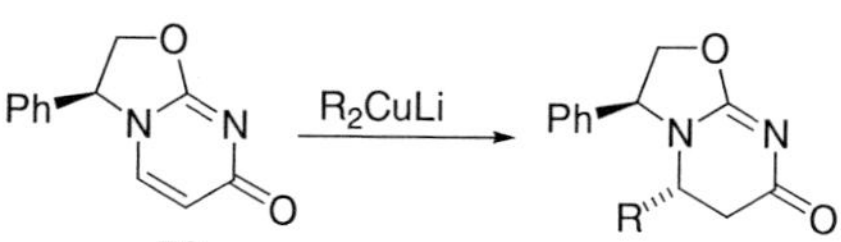

79

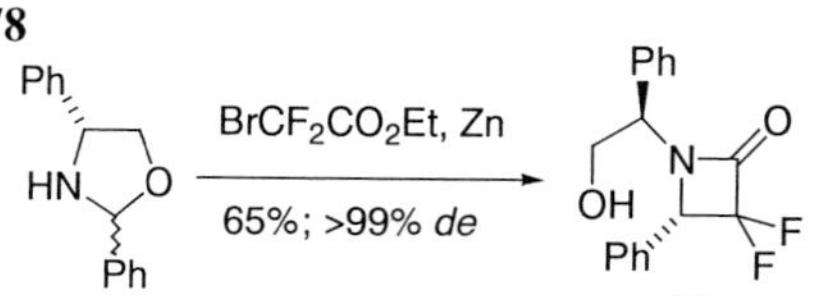

80

N-Protected oxazolidin-5-ones **81** can be rapidly prepared from protected amino acids and paraformaldehyde under microwave irradiation <99SC4017> . They undergo several useful transformations <99CC317> including conversion into protected hydroxamic acids by reaction with hydroxylamine <99SC3613> and reduction to protected amino alcohols with sodium borohydride in excess <99TL2653> . 4-Isopropyl-5,5-diphenyloxazolidin-2-one **82** has been prepared (in both enantiomeric forms) and offers some advantages over the Evans chiral oxazolidinone auxiliary <98HCA2093> . It can be acylated at 0 °C and the lithium enolates of the acylated species can be generated with butyllithium because the carbonyl group is sterically protected from nucleophilic attack. A synthesis of the oxazolidin-2-one (–)-cytoxazone **83** has been described <99SL1915> and several other syntheses of oxazolidin-2-ones have been reported. These include the palladium catalysed reaction of the dicarbonate **84** with anilines bearing electron withdrawing groups to give compounds **85** <99EJO181> , the Lewis acid catalysed ring expansion of aziridines **86** to oxazolidinones **87** <99OL2153> , the synthesis of *N*-alkyloxazolidinones **88** from *N*-alkyl-2-aminoethanol using nitric acid to activate the alcohol <98TL8211> and the cyclisation of chiral 2-aminoalcohols with di-*tert*-butyl dicarbonate to give the oxazolidinones **89** <98TL9407> . 4-Vinyloxazolidines **90** have been prepared (as a mixture of diastereoisojmers) by the palladium catalysed addition of vinyloxiranes to *N*-tosylaldimines <99TL1053> .

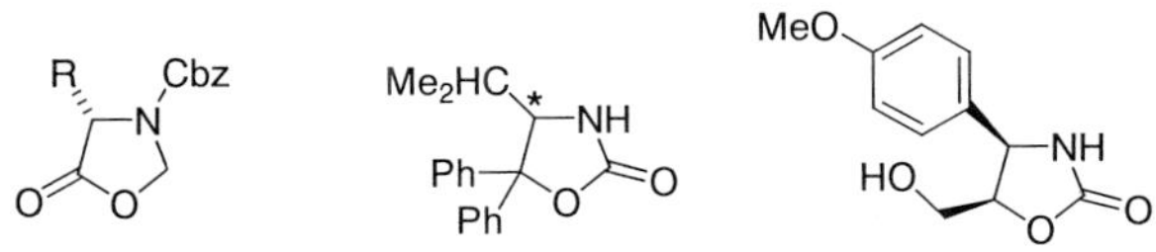

81 **82** **83**

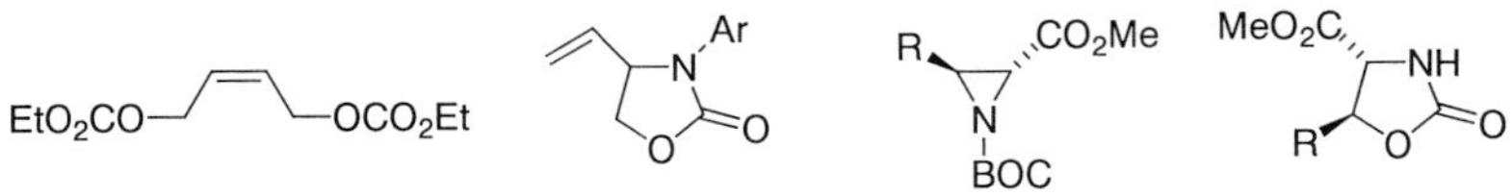

84 **85** **86** **87**

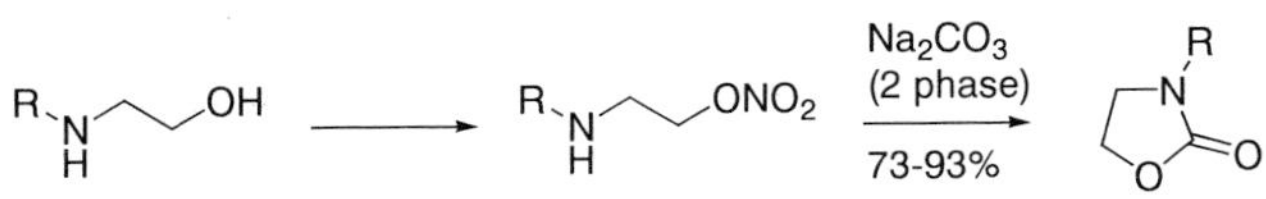

88

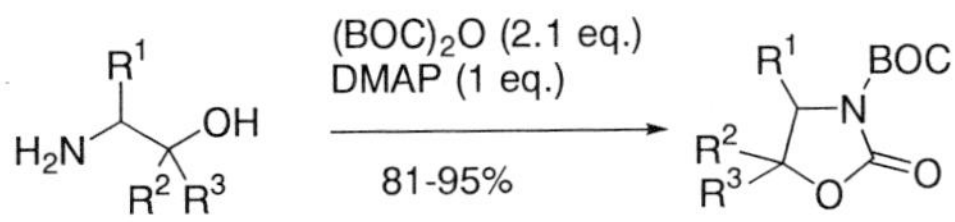

89

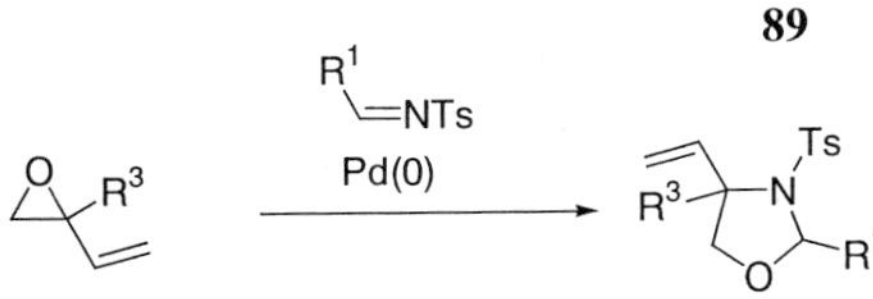

90

Chiral 4-phenyloxazolidin-2-one has been *N*-alkylated with nitroalkenes, and the products subsequently converted into chiral amino acids and diamines <99EJO2583> . *N*-acylation of this oxazolidinone by amino acids can be achieved in one pot using pivaloyl chloride and triethylamine <98TL9369> Silylcupration of the chiral oxazolidine **91** produced **92** which, after deprotection and oxidation gave silicon substituted amino acids <99JOC9211> . The bicyclic oxazolidinone **93** which was derived from proline and chloral was steresoselectively alkylated and then cleaved to give C-2 substituted proline derivatives <99SL33>

The oxazolidinediones **94** are crystalline, stable solids that are a convenient source of dipeptides when heated in THF with amino acids. The compounds are prepared from *N*-protected amino acids and phosgene or triphosgene <99JOC2532> . The chiral oxazolidinediones **95** act as Friedel-Crafts reagents with aromatic hydrocarbons and aluminium chloride, giving the ketones **96** in moderate yield but with high *ee* <99S423> .

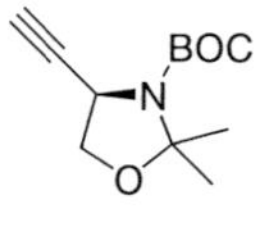

91

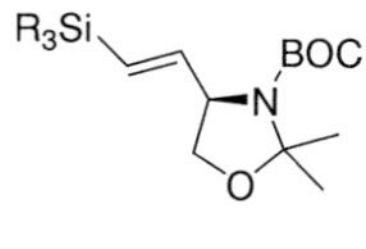

92

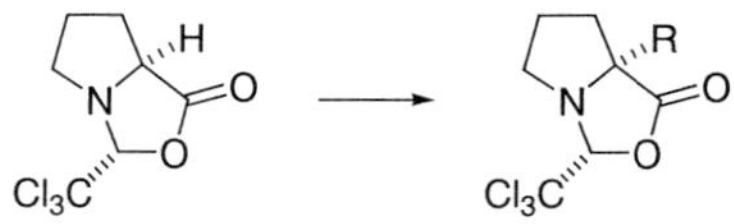

93

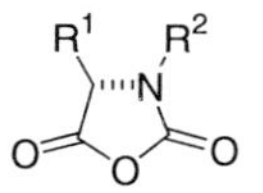

94 (R^2 = trityl or phenylfluorenyl)

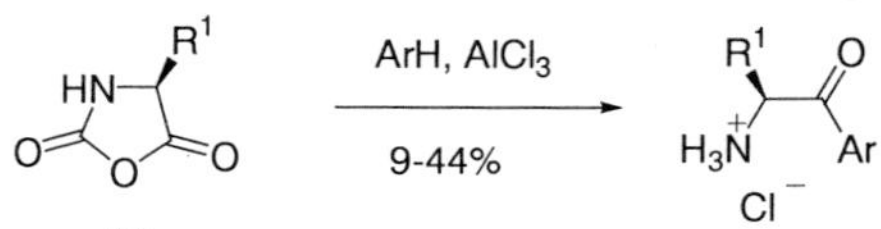

95 **96**

5.7.7 OXADIAZOLES

The standard route to sydnones is the cyclodehydration of *N*-nitroso-β-amino acids. 2-Chloro-4,5-dihydro-1,3-dimethylimidazolium chloride **97** has been shown to be an excellent reagent for this conversion, as, for example, in the synthesis of 3-phenylsydnone **98** <99JOC6989> . Functional group manipulations of fused sydnones have led to the isolation of several new compounds including **99** and **100** <99H(51)1433> . 4-Acetylsydnones react with hydrazine at room temperature to give 2,4-dihydropyrazol-3-ones **101** <99H(51)95> .

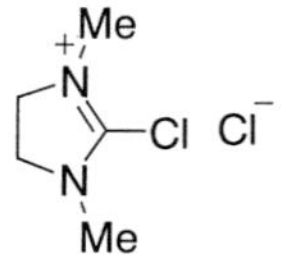

97

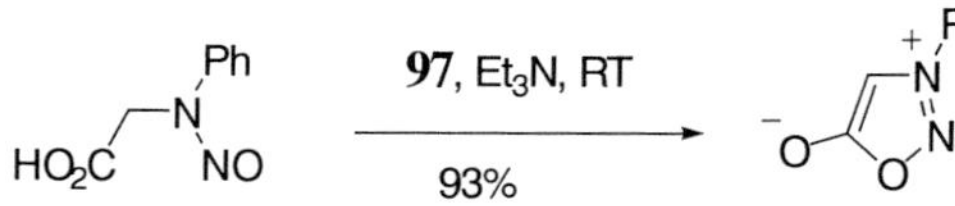

98

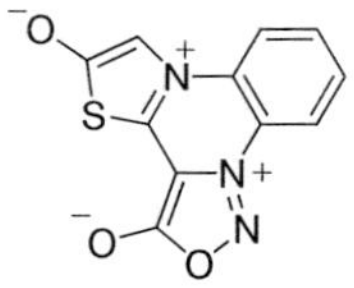

99

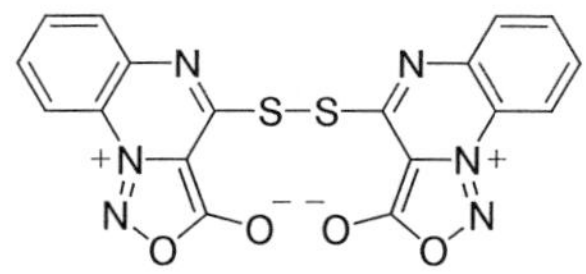

100

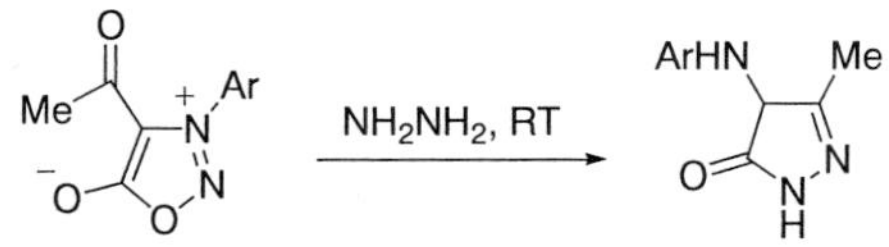

101

The standard preparative route to 1,2,4-oxadiazoles, the condensation of an amidoxime with a carboxylic acid using DCCI or a similar activating agent, has been extended in a number of ways. It has been used to make the thienylmethyloxadiazole **102** <99SC1437> and it has been applied to several parallel and solid phase syntheses of 1,2,4-oxadiazoles. 1,1-Carbonylbis(imidazole) has been shown to be an excellent reagent for the cyclodehydration and its use facilitates purification in parallel synthesis <99BMC209> . Syntheses on solid supports have been carried out with supported amidoximes and *N*-protected amino acid anhydrides <99TL8547> and with amidoximes and supported acyl fluorides, which were generated from the carboxylic acids and 2,4,6-trifluoro-1,3,5-triazine **103** <99TL9359> . The 5-(4-piperidinyl)oxadiazoles **104** were synthesised on a solid support from carboxylic acid esters and amidoximes under basic conditions at room temperature <99BMC2101>

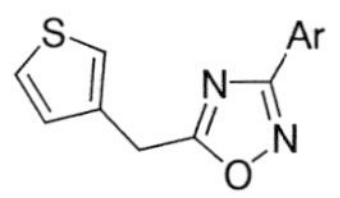

102

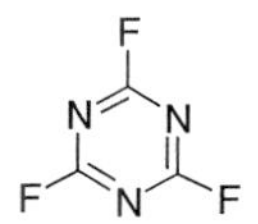

103

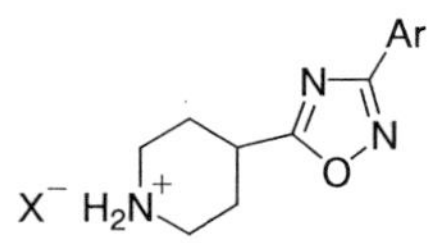

104

Dihydrooxadiazoles **106** have been syntheised in moderate to high diastereomeric excess by the addition of aromatic nitrile oxides across the C=N bond of the hydrazones **105**. The N-N bond can subsequently be cleaved with formic acid, and the chiral auxiliary recycled <99H(50)995> . The oxadiazolone **108** was produced (56%) from the oxime **107** by heating it with phenyl isocyanate <99SC3889> .

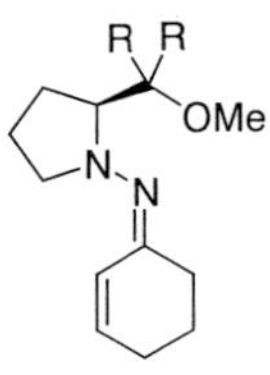

105

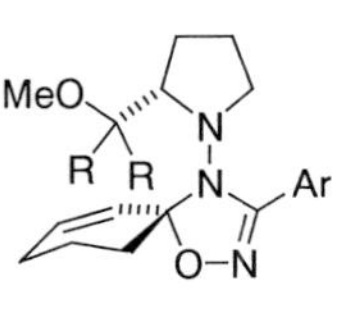

106

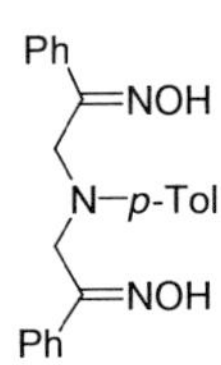

107

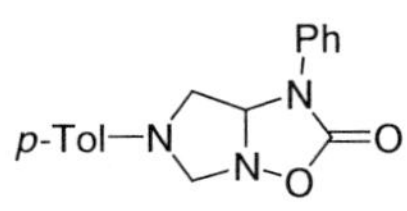

108

A new method of generation of acylnitroso compounds is the photochemical cleavage of

1,2,4-oxadiazole-4-oxides **109** <99TL797> . The sensitised photochemical rearrangement of 1,2,4-oxadiazoles **110** leads to the production of quinazolin-4-ones **111** <99JOC7028> .

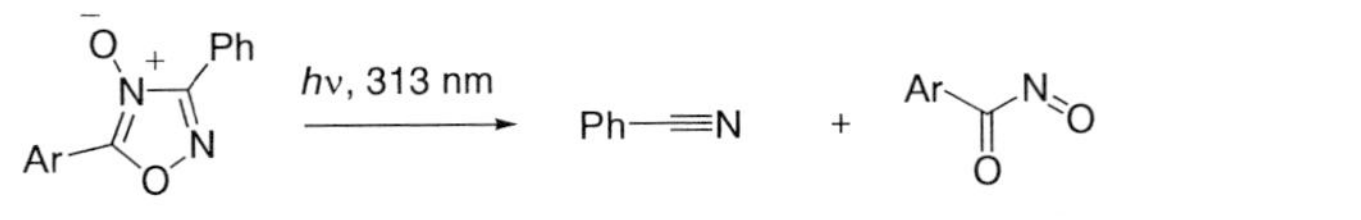

109

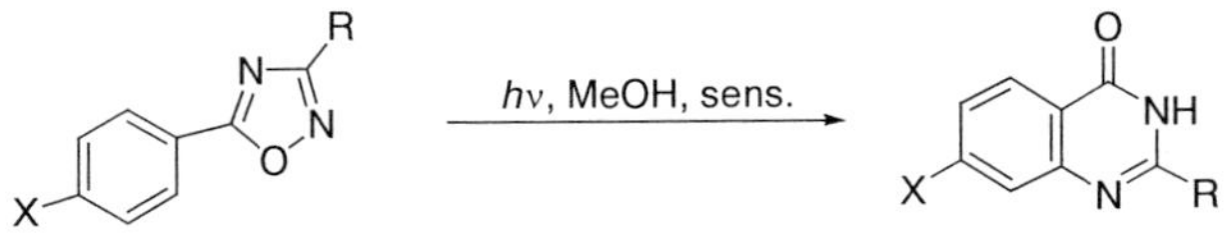

110 **111**

Trifluoromethyl substituted 1,2,5-oxadiazoles **113** were prepared by heating the bis(oximes) **112** with dry silica <99H(51)627> . 1,2,5-Oxadiazole-2-oxides (furoxans) are often produced by dimerisation of nitrile oxides and some novel examples of the method, which produces 3,4-disubstituted derivatives bearing medium and large rings, have been described <99JOC8428> . Two examples of compounds prepared in this way from bis(nitrile oxides) are the furoxans **114** (84%) and **115** (86%).

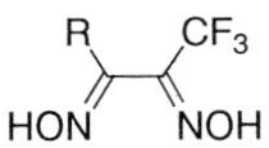

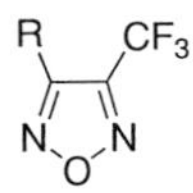

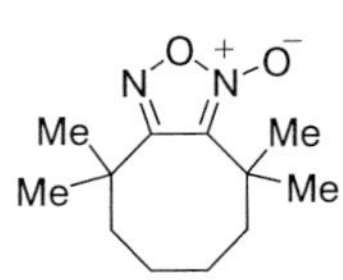

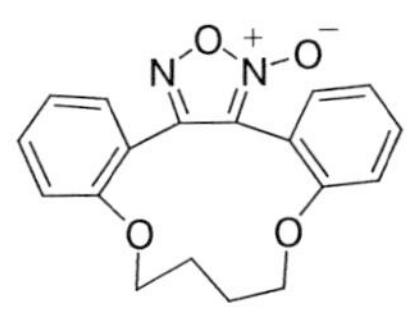

112 **113** **114** **115**

The dinitrobenzofuroxan **116** has previously been shown to be able to act as a dienophile in Diels-Alder reactions and further examples of such reactions, with cylopentadiene and with 1,3-cyclohexadiene, have been described <99JOC9254> . Cyclopentadiene acts both as a dienophile and as a diene, the final product being the adduct **117**. An attempt to convert the formyl group of the furoxan **118** into a nitrile function by reaction with ammonia, oxygen and a copper catalyst led instead to cleavage of the ring, indicating that 3-substituted furoxans are susceptible to ring cleavage even under mild conditions <99JOC8748> .

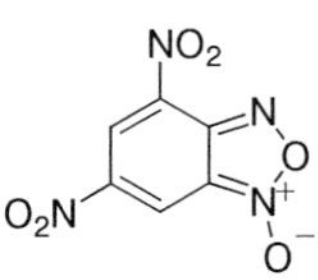

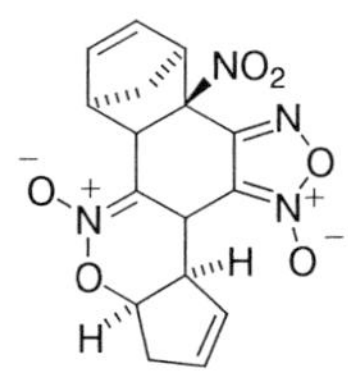

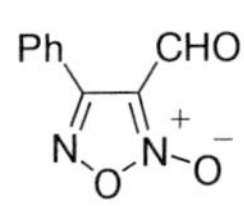

116 **117** **118**

The preparation of symmetrical 2,5-diaryl-1,3,4-oxadiazoles **119** from aromatic carboxylic acids and hydrazine is often carried out in polyphosphoric acid. An improved procedure has been suggested: the reaction is carried out in 85% aqueous orthophosphoric acid to which 3 equivalents of phosphorus pentoxide and 1 equivalent of phosphorus oxychloride are added slowly <99JHC1029>. Bis(1,3,4-oxadiazolyl)ethylenes **120** have been formed from tetrazoles and fumaroyl chloride; the reaction sequence involves acylation of the tetrazoles followed by thermal degradation of the acyltetrazoles <99S999>. Unsymmetrical 2,5-diaryl-1,3,4-oxadiazoles such as **121** were formed from aryltrichloromethanes and benzoylhydrazine <99KGS557>.

Several examples of the mesoionic 3-aryl-1,2,3,4-oxatriazole-5-oxide ring system **123** have been prepared by cyclisation of the arylhydrazones **122**; these in turn are prepared from aryldiazonium salts and bromonitromethane <99KGS413>.

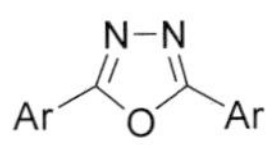

119

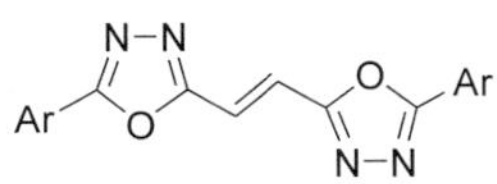

120

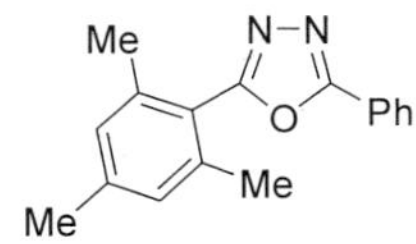

121

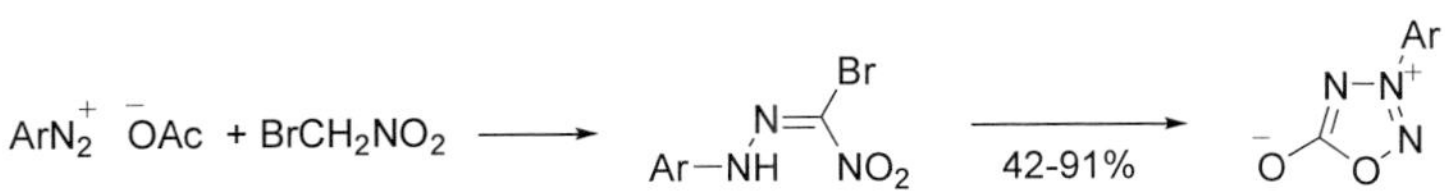

122 **123**

Acknowledgement: A preliminary literature search for this review was carried out by Professor G. V. Boyd, who usually produces the chapter but was unable to do so this year.

5.7.8 REFERENCES

98HCA2093	T. Hintermann and D. Seebach, *Helv. Chim. Acta*, **1998**, *81*, 2093.
98JA11532	J. C. Ruble and G. C. Fu, *J. Am. Chem. Soc.*, **1998**, *120*, 11532.
98JCS(P1)2923	J. Clayden and S. Warren, *J. Chem. Soc., Perkin Trans. 1*, **1998**, 2923.
98JCS(P1)3245	D. S. Millan and R. H. Prager, *J. Chem. Soc., Perkin Trans. 1*, **1998**, 3245.
98JCS(P1)3389	M. Sakamoto, K. Satoh, M. Nagano, and O. Tamura, *J. Chem. Soc., Perkin Trans. 1*, **1998**, 3389.
98JOC7680	S. Reck and W. Friedrichsen, *J. Org. Chem.*, **1998**, *63*, 7680.
98T12445	X. T. Zhou, Y. R. Lin, L. X. Dai, and J. Sun, *Tetrahedron*, **1998**, *54*, 12445.
98T15721	S. Crosignani, G. Desimoni, G. Faita, and P. Righetti, *Tetrahedron*, **1998**, *54*, 15721.
98TL8211	M. E. Sitzmann, J. A. Kenar, and N. J. Trivedi, *Tetrahedron Lett.*, **1998**, *39*, 8211.
98TL8513	C. D. Davies, S. P. Marsden, and E. S. E. Stokes, *Tetrahedron Lett.*, **1998**, *39*, 8513.
98TL8865	S. Kanemasa, S. Kaga, and E. Wada, *Tetrahedron Lett.*, **1998**, *39*, 8865.
98TL9369	M. Prashad, H. Y. Kim, D. Har, O. Repic, and T. J. Blacklock, *Tetrahedron Lett.*, **1998**, *39*, 9369.
98TL9407	H. J. Knölker and T. Braxmeier, *Tetrahedron Lett.*, **1998**, *39*, 9407.
99BMC209	T. L. Deegan, T. J. Nitz, D. Cebzanov, D. E. Pufko, and J. A. Porco, *Bioorg. Med.*

	Chem. Lett., **1999**, *9*, 209.
99BMC2101	G. B. Liang and X. X. Qian, *Bioorg. Med. Chem. Lett.*, **1999**, *9*, 2101.
99BSJ2277	N. Arai, M. Iwakoshi, K. Tanabe, and K. Narasaka, *Bull. Chem. Soc. Jpn.*, **1999**, *72*, 2277.
99CC317	G. V. Reddy, G. V. Rao, and D. S. Iyengar, *Chem. Commun.*, **1999**, 317.
99CC811	K. B. Simonsen, K. A. Jørgensen, Q. S. Hu, and L. Pu, *Chem. Commun.*, **1999**, 811.
99EJO181	M. Moreno-Mañas, L. Morral, R. Pleixats, and S. Villarroya, *Eur. J. Org. Chem.*, **1999**, 181.
99EJO297	S. Perrin, K. Monnier, B. Laude, M. Kubicki, and O. Blacque, *Eur. J. Org. Chem.*, **1999**, 297.
99EJO409	S. Florio, V. Capriati, S. DiMartino, and A. Abbotto, *Eur. J. Org. Chem.*, **1999**, 409.
99EJO2583	D. Lucet, S. Sabelle, O. Kostelitz, T. LeGall, and C. Mioskowski, *Eur. J. Org. Chem.*, **1999**, 2583.
99H(50)71	M. Kawase, H. Miyamae, and S. Saito, *Heterocycles*, **1999**, *50*, 71.
99H(50)377	E. Okada, H. Okumura, Y. Nishida, and T. Kitahora, *Heterocycles*, **1999**, *50*, 377.
99H(50)995	D. Enders, I. Meyer, J. Runsink, and G. Raabe, *Heterocycles*, **1999**, *50*, 995.
99H(51)37	U. Chiacchio, A. Corsaro, and A. Piperno, *Heterocycles*, **1999**, *51*, 37.
99H(51)95	H. J. Tien, S. T. Lin, and M. L. Yang, *Heterocycles*, **1999**, *51*, 95.
99H(51)373	K. Kamata, H. Sato, E. Takagi, I. Agata, and A. I. Meyers, *Heterocycles*, **1999**, *51*, 373.
99H(51)627	Y. Kamitori, *Heterocycles*, **1999**, *51*, 627.
99H(51)979	M. R. DeLuca, I. B. Taraporewala, and S. M. Kerwin, *Heterocycles*, **1999**, *51*, 979.
99H(51)1433	C. W. Lo, W. L. Chan, Y. S. Szeto, and C. W. Yip, *Heterocycles*, **1999**, *51*, 1433.
99H(51)1921	B. H. Kim, T. K. Kim, J. W. Cheong, S. W. Lee, Y. M. Jun, W. Baik, and B. M. Lee, *Heterocycles*, **1999**, *51*, 1921.
99H(51)2139	E. Comanita, I. Popovici, G. Roman, G. Robertson, and B. Comanita, *Heterocycles*, **1999**, *51*, 2139.
99H(51)3013	R. H. Prager and C. M. Williams, *Heterocycles*, **1999**, *51*, 3013.
99HCA1981	T. Moschny and H. Hartmann, *Helv. Chim. Acta*, **1999**, *82*, 1981.
99JA3845	K. B. Simonsen, P. Bayon, R. G. Hazell, K. V. Gothelf, and K. A. Jørgensen, *J. Am. Chem. Soc.*, **1999**, *121*, 3845.
99JCS(P1)185	M. Noguchi, H. Okada, S. Nishimura, Y. Yamagata, S. Takamura, M. Tanaka, A. Kakehi, and H. Yamamoto, *J. Chem. Soc., Perkin Trans. 1*, **1999**, 185.
99JCS(P1)765	R. C. F. Jones, G. Bhalay, P. A. Carter, K. A. M. Duller, and S. H. Dunn, *J. Chem. Soc., Perkin Trans. 1*, **1999**, 765.
99JCS(P1)1167	M. Rudas, I. Fejes, M. Nyerges, A. Szöllõsy, L. Tõke, and P. W. Groundwater, *J. Chem. Soc., Perkin Trans. 1*, **1999**, 1167.
99JCS(P1)3337	A. D. Jones, D. W. Knight, and S. R. Thornton, *J. Chem. Soc., Perkin Trans. 1*, **1999**, 3337.
99JCS(P1)3349	P. J. Dransfield, S. Moutel, M. Shipman, and V. Sik, *J. Chem. Soc., Perkin Trans. 1*, **1999**, 3349.
99JHC283	J. P. Lawson and K. A. VanSant, *J. Heterocycl. Chem.*, **1999**, *36*, 283.
99JHC911	Y. C. Kong and K. Kim, *J. Heterocycl. Chem.*, **1999**, *36*, 911.
99JHC1029	F. Bentiss and M. Lagrenée, *J. Heterocycl. Chem.*, **1999**, *36*, 1029.
99JOC1620	Y. Uozumi, H. Kyota, K. Kato, M. Ogasawara, and T. Hayashi, *J. Org. Chem.*, **1999**,

64, 1620.
99JOC2160 N. Nishiwaki, T. Nogami, C. Tanaka, F. Nakashima, Y. Inoue, N. Asaka, Y. Tohda, and M. Ariga, *J. Org. Chem.*, **1999**, *64*, 2160.
99JOC2353 K. B. Jensen, R. G. Hazell, and K. A. Jørgensen, *J. Org. Chem.*, **1999**, *64*, 2353.
99JOC2532 T. B. Sim and H. Rapoport, *J. Org. Chem.*, **1999**, *64*, 2532.
99JOC3595 A. Padwa, M. A. Brodney, B. Liu, K. Satake, and T. Wu, *J. Org. Chem.*, **1999**, *64*, 3595.
99JOC4547 S. D. Lepore and M. R. Wiley, *J. Org. Chem.*, **1999**, *64*, 4547.
99JOC6989 T. Isobe and T. Ishikawa, *J. Org. Chem.*, **1999**, *64*, 6989.
99JOC7028 S. Buscemi, A. Pace, N. Vivona, T. Caronna, and A. Galia, *J. Org. Chem.*, **1999**, *64*, 7028.
99JOC7040 H. Suga, K. Ikai, and T. Ibata, *J. Org. Chem.*, **1999**, *64*, 7040.
99JOC8428 N. Maugein, A. Wagner, and C. Mioskowski, *J. Org. Chem.*, **1999**, **64**, 8428.
99JOC8461 S. Marcotte, X. Pannecoucke, C. Feasson, and J. C. Quirion, *J. Org. Chem.*, **1999**, *64*, 8461.
99JOC8748 A. R. Butler, P. Lightfoot, and D. M. Short, *J. Org. Chem.*, **1999**, *64*, 8748.
99JOC9211 G. Reginato, A. Mordini, M. Valacchi, and E. Grandini, *J. Org. Chem.*, **1999**, *64*, 9211.
99JOC9254 P. Sepulcri, J. C. Hallé, R. Goumont, D. Riou, and F. Terrier, *J. Org. Chem.*, **1999**, *64*, 9254.
99JOC9297 K. H. Park and M. J. Kurth, *J. Org. Chem.*, **1999**, *64*, 9297.
99JOC9450 K. Y. Lee, Y. H. Kim, M. S. Park, C. Y. Oh, and W. H. Ham, *J. Org. Chem.*, **1999**, *64*, 9450.
99KGS413 S. A. Shevelev, I. L. Dalinger, V. I. Gulevskaya, T. I. Cherkasova, V. M. Vinogradov, and B. I. Ugrak, *Khim. Geterotsikl. Soedin.*, **1999**, 413.
99KGS557 L. I. Belenskii, S. I. Luiksaar, and M. M. Krayushkin, *Khim. Geterotsikl. Soedin.*, **1999**, 557.
99OL31 K. W. Glaeske and F. G. West, *Org. Lett.*, **1999**, *1*, 31.
99OL87 D. A. Evans, V. J. Cee, T. E. Smith, and K. J. Santiago, *Org. Lett.*, **1999**, *1*, 87.
99OL527 A. D. Rodríguez, C. Ramírez, I. I. Rodríguez and E. González, *Org. Lett.*, **1999**, *1*, 527.
99OL753 O. Dirat, C. Kouklovsky, and Y. Langlois, *Org. Lett.*, **1999**, *1*, 753.
99OL1137 N. Hénaff and A. Whiting, *Org. Lett.*, **1999**, *1*, 1137.
99OL1745 D. R. Hou and K. Burgess, *Org. Lett.*, **1999**, *1*, 1745.
99OL1795 M. A. Arai, T. Arai, and H. Sasai, *Org. Lett.*, **1999**, *1*, 1795.
99OL2153 C. Tomasini and A. Vecchione, *Org. Lett.*, **1999**, *1*, 2153.
99S423 O. Itoh and A. Amano, *Synthesis*, **1999**, 423.
99S999 H. Detert and D. Schollmeier, *Synthesis*, **1999**, 999.
99S1569 A. Pal, A. Bhattacharjee, and A. Bhattacharjya, *Synthesis*, **1999**, 1569.
99S2027 S. Y. Lee, B. S. Lee, and D. Y. Oh, *Synthesis*, **1999**, 2027.
99SC43 C. Bolm, T. H. Chuang, G. Raabe, and J. M. Fang, *Synth. Commun.*, **1999**, *29*, 43.
99SC877 H. M. S. Kumar, S. Anjaneyulu, and J. S. Yadav, *Synth. Commun.*, **1999**, *29*, 877.
99SC1437 R. M. Srivastava, F. J. S. Oliveira, D. S. Machado, and R. M. SoutoMaior, *Synth. Commun.*, **1999**, *29*, 1437.
99SC3165 S. J. Ha, G. H. Lee, I. K. Yoon, and C. S. Pak, *Synth. Commun.*, **1999**, *29*, 3165.

99SC3613	G. V. Reddy, *Synth. Commun.*, **1999**, *29*, 3613.
99SC3621	S. Y. Lee, B. S. Lee, C. W. Lee, and D. Y. Oh, *Synth. Commun.*, **1999**, *29*, 3621.
99SC3889	N. Coşkun, F. T. Tat, and Ö. Ö. Güven, *Synth. Commun.*, **1999**, *29*, 3889.
99SC4017	G. V. Reddy, G. V. Rao, and D. S. Iyengar, *Synth. Commun.*, **1999**, *29*, 4071.
99SL33	H. Wang and J. P. Germanas, *Synlett*, **1999**, 33.
99SL79	M. Jurczak, D. Socha, and M. Chmielewski, *Synlett*, **1999**, 79.
99SL653	R. C. Bernotas, J. S. Sabol, L. Sing, and D. Friedrich, *Synlett*, **1999**, 653.
99SL727	C. Agami, S. Cheramy, L. Dechoux, and C. KadouriPuchot, *Synlett*, **1999**, 727.
99SL798	F. J. F. Castro, M. M. Vila, P. R. Jenkins, M. L. Sharma, G. Tustin, J. Fawcett, and D. R. Russell, *Synlett*, **1999**, 798.
99SL873	R. C. F. Jones, C. E. Dawson, and M. J. O'Mahony, *Synlett*, **1999**, 873.
99SL893	B. Oberhauser, K. Baumann, B. Grohmann, and H. Sperner, *Synlett*, **1999**, 893.
99SL1642	C. T. Brain and J. M. Paul, *Synlett*, **1999**, 1642.
99SL1727	G. Cardillo, L. Gentilucci, A. Tolomelli, and C. Tomasini, *Synlett*, **1999**, 1727.
99SL1838	C. Agami, S. Cheramy, and L. Dechoux, *Synlett*, **1999**, 1838.
99SL1915	O. Miyata, H. Asai, and T. Naito, *Synlett*, **1999**, 1915.
99T13809	R. Nesi, S. Turchi, D. Giomi, and A. Danesi, *Tetrahedron*, **1999**, *55*, 13809.
99TL797	P. Quadrelli, M. Mella, and P. Caramella, *Tetrahedron Lett.*, **1999**, *40*, 797.
99TL1053	J. G. Shim and Y. Yamamoto, *Tetrahedron Lett.*, **1999**, *40*, 1053.
99TL1291	M. Smietana, V. Gouverneur, and C. Mioskowski, *Tetrahedron Lett.*, **1999**, *40*, 1291.
99TL2653	G. V. Reddy, G. V. Rao, and D. S. Iyengar, *Tetrahedron Lett.*, **1999**, *40*, 2653.
99TL3535	D. E. Kizer, R. B. Miller, and M. J. Kurth, *Tetrahedron Lett.*, **1999**, *40*, 3535.
99TL4085	R. C. F. Jones, C. E. Dawson, M. J. O'Mahony, and P. Patel, *Tetrahedron Lett.*, **1999**, *40*, 4085.
99TL4119	A. S. Kiselyov, *Tetrahedron Lett.*, **1999**, *40*, 4119.
99TL4349	A. Kamimura, Y. Kaneko, A. Ohta, A. Kakehi, H. Matsuda, and S. Kanemasa, *Tetrahedron Lett.*, **1999**, *40*, 4349.
99TL4765	M. P. Dwyer, D. A. Price, J. E. Lamar, and A. I. Meyers, *Tetrahedron Lett.*, **1999**, *40*, 4765.
99TL4889	P. de la Cruz, E. Espildora, J. J. Garcia, A. de la Hoz, F. Langa, N. Martin, and L. Sánchez, *Tetrahedron Lett.*, **1999**, *40*, 4889.
99TL5605	J. Kiegiel, M. Poplawska, J. Jóźwik, M. Kosior, and J. Jurczak, *Tetrahedron Lett.*, **1999**, *40*, 5605.
99TL8547	N. Hébert, A. L. Hannah, and S. C. Sutton, *Tetrahedron Lett.*, **1999**, *40*, 8547.
99TL9359	C. K. Sams and J. Lau, *Tetrahedron Lett.*, **1999**, *40*, 9359.

Chapter 6.1

Six-Membered Ring Systems: Pyridines and Benzo Derivatives

Robert D. Larsen and Jean-Francois Marcoux
Department of Process Research, Merck Research Laboratories, Merck & Co., Inc.
Rahway, NJ, USA
rob_larsen@merck.com

6.1.1 INTRODUCTION

Continuing with the approach of this chapter from previous years metal-mediated reactions, cycloadditions, radical processes and asymmetric applications will be highlighted. Syntheses using traditional approaches will not be covered, unless improvements are reported. Due to the volume of publications concerning pyridines and associated heterocycles many subject areas could not be covered. Combinatorial or solid-phase synthesis will not represented since the area is rather specialized and many of the processes utilize existing methodology. The synthesis and reactions of polyaza-fused systems of the pyridine class will also not be included in this review.

6.1.2 PYRIDINES

6.1.2.1 Preparation of Pyridines

Recent advances in the large-scale synthesis of pyridines using conventional methods and gas phase synthesis over ZSM-5 zeolites were reviewed <98M1>.

Cycloaddition reactions are very efficient in setting up substituted pyridines (Scheme 1). Treatment of the diazoimide **1** with a catalytic quantity of rhodium(II) acetate <99JOC8648> or the Pummerer reaction of the imidosulfoxide **2** <99JOC2038> produces an intermediate isomünchnone, which in the presence of dipolarophiles forms 3-oxy-2-pyridones. Stannylated bi- and terpyridines can be obtained from 2,6-(1,2,4-triazinyl)pyridine in the presence of tributyl(ethynyl)tin <99EJOC313>. Arenecarbothioamides **3** undergo a photochemical cyclization with hexadienal, hexadienone or *o*-vinylbenzaldehyde to the corresponding 2-arylsubstituted pyridines <99CC2371>.

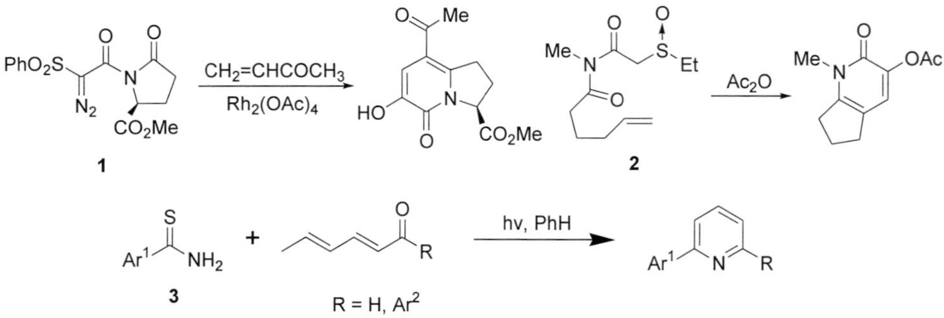

Scheme 1

Condensation reactions are the most frequently used methods for the synthesis of the pyridine ring. The condensation of enamines with aldehydes or ketones affords pyridines and 1,4-dihydropyridines <99JCR(S)536, 99JPR147, 99S1216, 99JOC6076, 99JCR(S)208>. Benzotriazolyl ketones react with α,β-unsaturated ketones to give various 2,4,6-trisubstituted pyridines <99S2114>. An efficient two-step synthesis of polysubstituted symmetric terpyridines has been carried out by condensation of a *bis*-pyridinium iodide derived from 2,6-diacetylpyridine with α, β-unsaturated aldehydes and ammonium acetate in formamide <99S815>. A 2-pyridone was obtained as a side product from the treatment of a 4-pyronecarboxylate with an excess of 4-methoxyaniline <99JHC493>. Malonate derivatives are common building blocks for the preparation of carboxy-substituted pyridines <99ZN(B)1205, 99H(51)1855, 99JOC9493, 99H(51)2093, 99JHC541>. The reaction of lithium dienediolates of α,β-unsaturated carboxylic acids and nitriles yields 4,6- and 3,4,6-substituted pyridinones <99SL1088>. Silylated azaallyl anions combine with trifluoroenones or 1,3-diones to provide 2- and 4-trifluoromethylpyridines <99JCS(P1)2803>.

The cyanoenamino ketoester **4** reacts with hydrazines to form the pyridone derivative **5** <99H(50)791>. *N*-Arylsulfonylamino-2-pyridones have been prepared via the reaction of acetohydrazides with arylidenecyanoacetate derivatives <99JCR(S)6>.

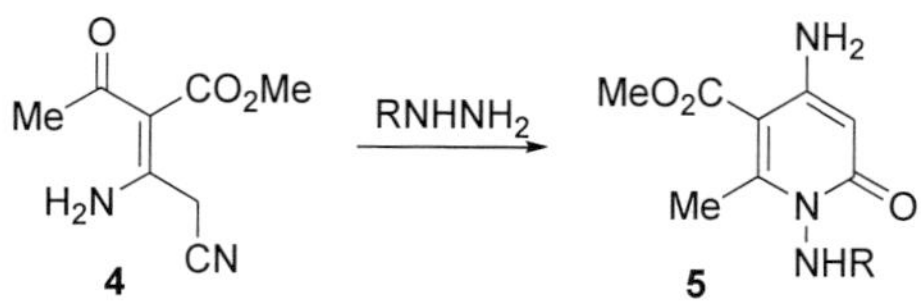

Several functionalized pyridines and dihydropyridines are accessible through the reaction of phosphazenes **6** and 2-azadienes **7** <99JOC6239>.

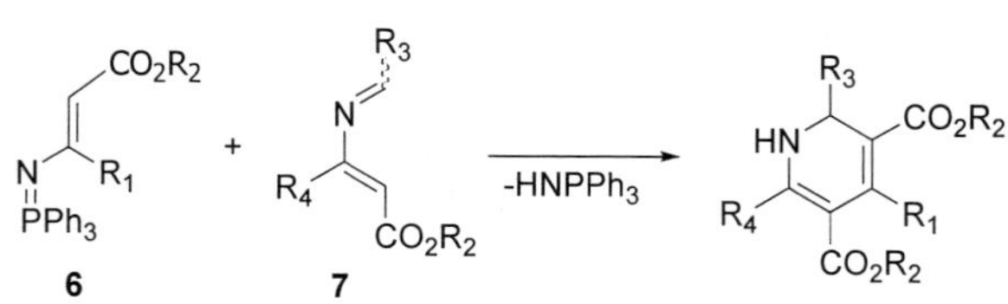

Often the oxidation of reduced forms of pyridine is necessary to prepare the ring. The mild oxidation of Hantzsch 1,4-dihydropyridines with manganese triacetate gives the corresponding pyridines in high yields without dealkylation at the 4-position <99TL21>. The aromatization of Hantzsch 1,4-dihydropyridines can also be accomplished in the presence of Magtrieve™ CrO_2 <99TL3207>. The mechanism of aromatization of Hantzsch esters has been studied <99JOC8980>.

The enzymatic conversion of 1-deoxy-D-xylulose-5-phosphate to pyridoxol phosphate (Vitamin B6) in *E. coli* was demonstrated <99JA7722>.

6.1.2.2 Reactions of Pyridines

Metal-mediated reactions are a significant advance in the available substitution reactions of pyridines. Palladium-catalyzed reactions have been particularly effective. Many examples of the Stille approach employing trialkylstannyl intermediates were reported <99SL342, 99S683, 99S754, 99S779, 99OL1027, 99JHC(36)869, 99T5047, 99JCR(S)636, 99EJOC313>. The Suzuki-cross coupling is an alternative that avoids the use of tin reagents <99SL45, 99SC103, 99TL4069, 99ZN(B)559>. An interesting example of the cross-coupling of the cyclopropylboronic acid **8** with bromopyridines and other *N*-heterocyclic bromides was reported <99SC2477>. A comparison of the coupling of pyridine substrates through the Stille or Suzuki methodology was carried out as part of the synthetic studies of streptonigrin <99H(51)721>. Linear bipyridyl ligands containing alkynyl spacers can be prepared using a modified Sonogashira reaction <99TL5413, 99S306>. The enantioselective reductive Heck-type heteroarylation of an azabicycle has been explored with various optically active ligands to give *N*-protected epibatidine with enantiomeric excesses up to 81% <99SL804>. Selectivity and reactivity in palladium-catalyzed carbonylation reactions were evaluated <99OL745, 99TL3717, 99T393, 99H(51)2589>. The cyanation of 2-amino-5-bromo-6-picoline to the 5-cyano derivative has been accomplished using DPPF as a ligand <99TL8193>.

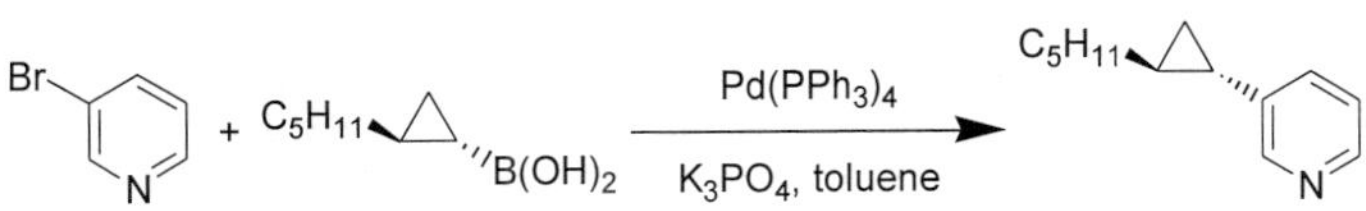

8

The Ni-catalyzed coupling of 2-chloropyridine with amines provides the 2-aminopyridines effectively <99T12829>. The homocoupling of 2-chloro- and 2-bromopyridines to substituted 2,2'-bipyridines has been achieved in the presence of $Ni(CO)_2(PPh_3)_2$ or nickel dichloride using zinc metal as the reducing agent <99TL4243, 99SC3341>. In addition to the carbon-carbon coupling of the ring, *N*-arylation of various 2-pyridones and 3-pyridazinones with 4-methylphenylboronic acid can proceed in moderate yields in the presence of a stoichiometric amount of copper(II) acetate <99T12757>.

Lithiation of the pyridine ring offers new avenues for substitution <99H(50)341, 99JA3539, 99TL5483>. Chiral (*p*-tolylsulfinyl)pyridines **9** undergo diastereoselective reactions with aldehydes through aromatic *ortho*-directed metalation <99JOC4512>. Bromopyridines can be converted to the corresponding pyridylmagnesium chlorides at room temperature by treatment with *i*-PrMgCl <99TL4339>.

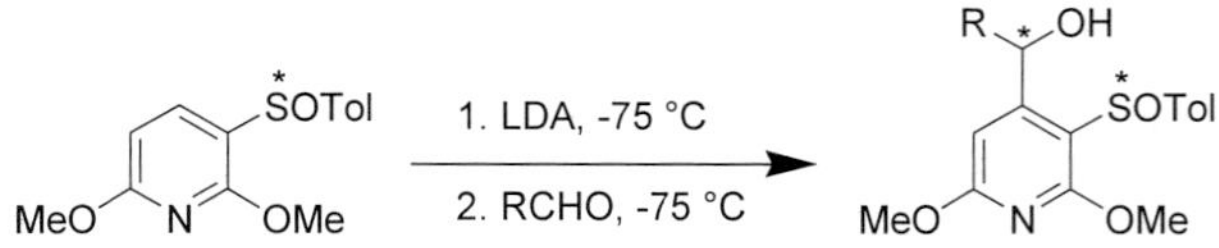

9 (*R* or *S*: >99%ee)

R = Me, d.e. 36%;
R = Ph, d.e. >99%

Hydroxymethylpyridines are reduced directly to alkylpyridines in the presence of samarium diiodide <99TL8823>. The electroreduction of pyridinedicarboxylic acid derivatives to 1,2- and 1,4-dihydropyridines has been achieved in good yield <99TL8587>.

Substitution reactions are of continuing importance to pyridine synthesis. Derivatives of 2-hydroxypyridine are halogenated to the corresponding 2-bromo- and 2-chloropyridines using triphenylphosphine and NBS or NCS <99TL7477>. The selective fluorination at the 2- <99JCS(P1)803> or 3-position <99JOC1007> has been reported. Pyridines are nitrated with N_2O_5 to 3-nitropyridines <99ACS141, 99ACS356>.

New nucleophilic reactions of pyridines were disclosed. Pyridylsulfides can be accessed by the K-selectride-promoted nucleophilic displacement of the corresponding 2- and 3-halopyridines with mercaptans <99TL5565>. Pyridinosulfilimines are obtained by nucleophilic aromatic substitution of 2-chloronitropyridines or pentachloropyridine <99T10243, 99JCR(S)520>. The nucleophilic substitution of pyridines with amines is an increasingly attractive alternative to the reduction of nitro groups <99SL1559, 99SL1747, 99H(50)843, 99H(51)157>. *N*-Arylation <99TL8759> and *N*-alkylation <99SC4051> of pyridones with activated halides were reported. The protonation of 2-pyridyl diazotate salts was utilized as a method for carbocation formation at the 2-position, which led to the formation of acyl-, hydroxyl- and pyrrolyl-substituted pyridines <99OL1957>.

New examples of the photochemistry of the pyridine ring were reported. The photochemical reactions of pyridines with furan have been investigated <99JCS(P1)171>. The [4+4] photocycloaddition of 4-methoxy-2-pyridone **10** and *N*-butyl-2-pyridone **11** affords the mixed product **12a** specifically <99JOC950, 99JOC954, 99TL4007>. Several 2-pyridone derivatives also undergo [4+4] photocycloaddition with 1,3-dienes <99TL3527, 99OL1775>.

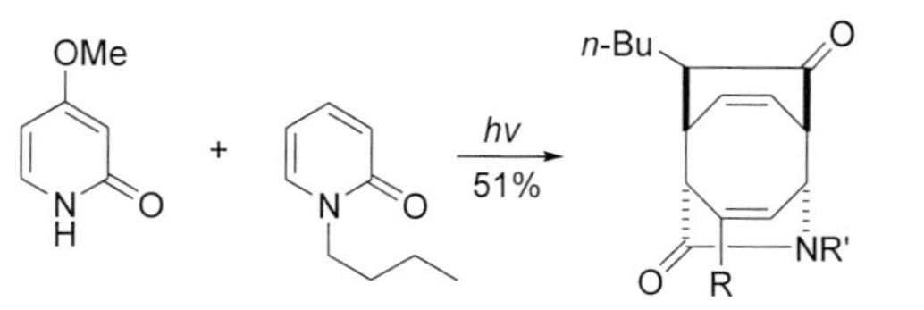

10 **11**

7 : 1

12a/12b 13:1

12a: R = OMe, R' = H
12b: R = H, R' = Bu

6.1.2.3 Pyridine *N*-oxides and Pyridinium

The oxidation of picolinaldehydes to the corresponding *N*-oxides with dimethyldioxirane proceeds in good yield without the need to protect the aldehyde function <99T12557>. The urea-hydrogen peroxide complex oxidizes pyridines to pyridine *N*-oxides <99OL189>.

The microwave-assisted *N*-alkylation of 2-chloropyridines proceeds more rapidly and in better yields compared to the corresponding thermal reactions <99T2317>.

Pyridinium intermediates readily undergo addition reactions. 2-Fluoropyridinium salts react with various enamines to give the corresponding 2-pyridylenamines and acetamides <99TL6661>. Stabilized carbon nucleophiles add to 1-alkyl-3-acylpyridinium intermediates <99TL3961>. Further applications of chiral *N*-acylpyridinium intermediates to the asymmetric synthesis of natural products were reported <99TL217, 99OL1941, 99JOC2184>. Either antipode of the *N*-acylpyridinium intermediate containing the (+)- or (-)-*trans*-2-(α-cumyl)-cyclohexanol (TCC) carbamate can be used to afford either enantiomer of the piperidine. The addition of benzylmagnesium chloride to the (+)-TCC-substituted pyridinium ion **13** was used in the asymmetric synthesis of benzomorphans <1999OL657>.

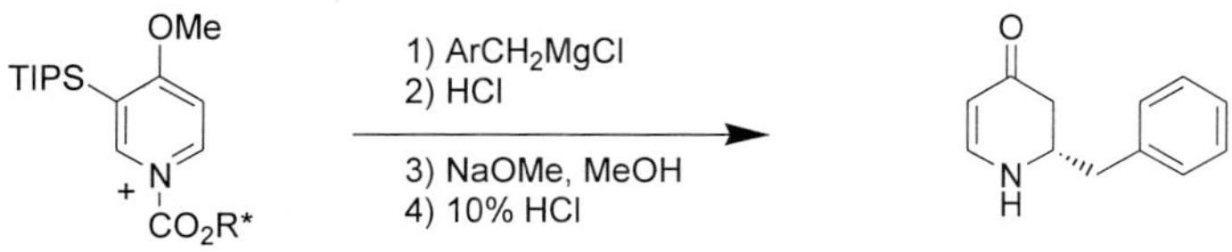

13: R* = (+)-TCC

Cycloaddition reactions of pyridinium species can easily produce more complex polycyclic products. Cycloaddition of 3-oxidopyridinium betaines with cyclopentadiene leads to the tricyclic core of Sarain A <99OL2017>. The dipolar additions of 3-oxidopyridinium betaines **14** with activated olefins give the tricyclic tropane homologues **15** <99H(50)929>. New syntheses of indolizines from pyridinium *N*-ylides were developed <99JOC7618, 99S166, 99JCR(S)136, 99JCS(P1)1571>.

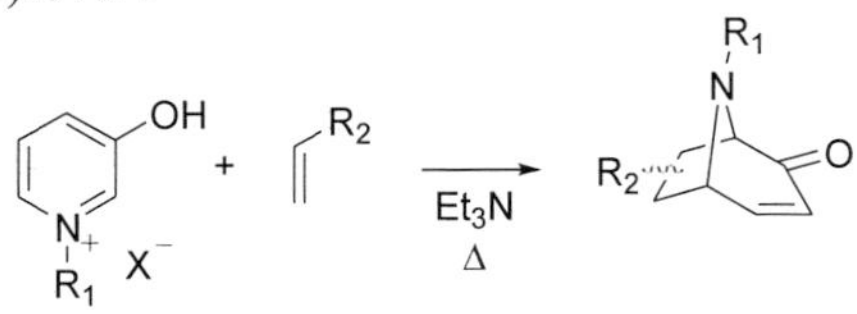

14 **15**

Photosolvolysis of 3-alkoxypyridinium tetrafluoroborates **16** under basic aqueous conditions generates the substituted cyclopentenone ketals **17** <99SL93, 99T6183>.

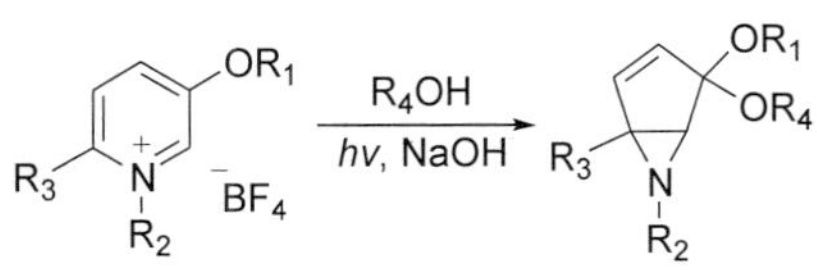

16 **17**

6.1.3 QUINOLINES

6.1.3.1 Preparation of Quinolines

Only a few applications of organometallics to quinoline synthesis (Scheme 2) were reported in 1999. The palladium-catalyzed transfer hydrogenation/heterocyclization of the ynone **18** offers an interesting means to prepare the quinoline ring <99SL401>. Alternatively, aryl-*N* bond formation can be used to prepare the *N*-acyltetrahydroisoquinoline or quinolin-2-one by intramolecular coupling of the amine or amide, respectively, with an aryl bromide <99OL35>. The ruthenium-catalyzed synthesis of 2-ethyl-3-methylquinoline **19** from anilines and triallylamine was reported <99TL1499>. The reaction of anilines with tricarbonyl iron-complexed cyclohexadienyl cations affords spiro-substituted tetrahydroquinolines. A revised mechanism for the reaction was proposed <99TL4331>.

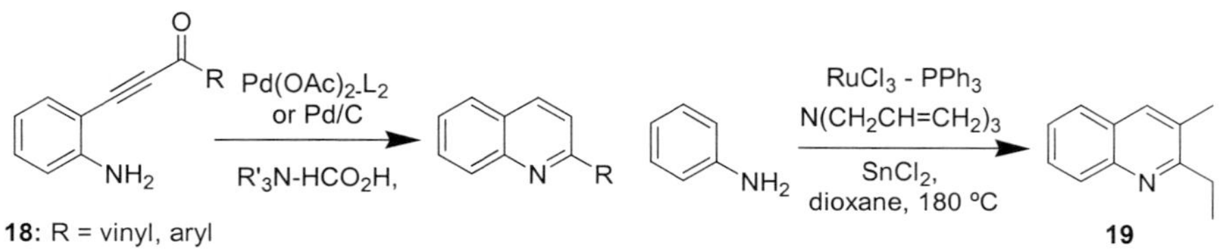

Scheme 2

Inter- and intramolecular aza-Diels-Alder reactions of *o*-azaxylylenes can be used to prepare hydroquinolines **20** <99AG(E)1928>. This methodology was a key step in the synthesis of virantmycin <99OL823>. An *o*-azaxylylene reacted with fullerene to afford the fullerotetrahydroquinoline <99SL207>. The tricyclic core of martinelline and martinellic acid was prepared through a [3+2] azomethine ylide cycloaddition reaction <99TL3339, 99TL2079>. The Diels-Alder reaction between 1,2,3-benzotriazines and enamines of acetophenones affords 2-arylquinolines <99CPB1038>.

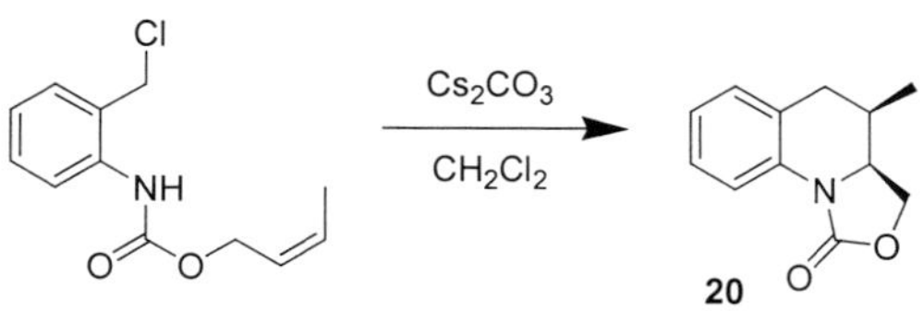

New methods for the direct cyclization of anilines to quinolines continue to be explored (Scheme 3). A procedure, which had previously failed, has been developed for the preparation of 4-arylquinolines using directed *ortho*-lithiation of *N*-pivaloylanilines <99S1335>. Reaction of anilines with the iminium triflate **21** yields the 2-trifluoromethylquinolines <99EJOC937>. Reaction of lithiated *beta*-enaminophosphonates of aniline and isocyanates, followed by cyclization of the amides with triphenylphosphine/triethylamine affords the 3-phosphonyl-4-aminoquinolines **22** <99T5947>. Tetrahydroquinolines can be prepared from allyl silanes and *N*-aryl benzotriazolemethanamines, which generate iminium species upon addition of tin chloride <99JHC371>. The reaction of anilines with *beta*-diketones <99H(51)2171> or ethoxymethylenemalonate <99JFC7> is useful for the preparation of quinolines.

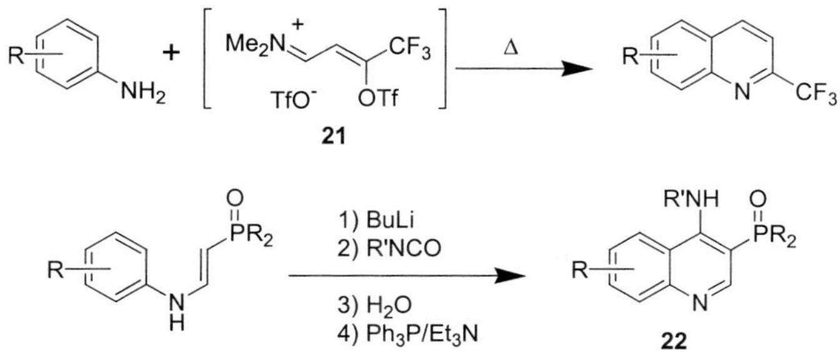

Scheme 3

Aniline derivatives substituted at the 2-position are frequently employed as starting materials for quinoline synthesis (Scheme 4). Nucleophilic addition onto the *o*-alkynyl isocyanobenzene **23** or aniline **24** induces ring closure to substituted quinolines <99OL1977, 99T13233>. 2-Aminophenylketones are used most often in the Friedlander synthesis <99H(50)479, 99SC4403> and produce 1,2-dihydroquinoline-3-carboxylates when reacted with acrylates <99JCS(P1)1547>. Anthranilic acid derivatives afford 4-quinolinones. The acetonyl or phenacyl esters form 2-substituted-3-hydroxy-4-quinolinones under acid-catalyzed conditions with PPA or simply in refluxing NMP <99JHC141>. Benzoxazin-4-ones **25** react with ketene silyl acetals or other active methylene compounds to provide the 3,3-disubstituted quinolin-2,4-diones **26** <99OL1953> or 3-substituted 4-hydroxyquinolin-2-ones **27** <99H(51)1543>, respectively. The reaction of *N*-formyl toluidines, containing an electron-withdrawing group, with base and diethyloxalate provides the corresponding 3-hydroxyquinolin-2-one <99ACS616>.

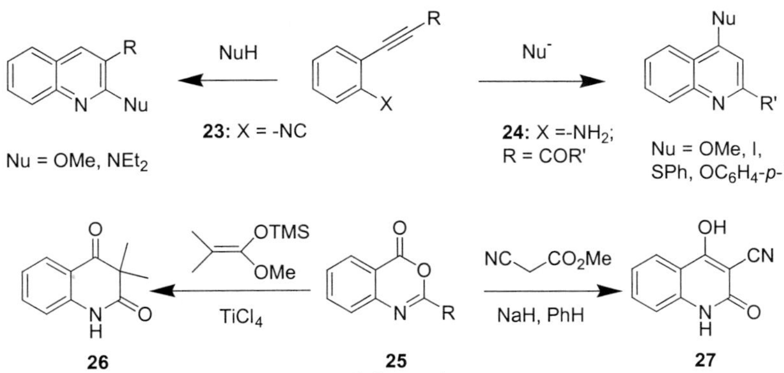

Scheme 4

Since most anilines are derived from nitro intermediates the reduction of the aromatic nitro group followed by cyclization of the aniline *in situ* has offered a direct approach to the synthesis of quinolines. The *ortho*-nitro cinnamic acid derivatives **28** undergo cyclization, where R corresponds to R', respectively, when treated with zinc in near-critical water at 250 °C

<99NJC641>. Similarly, the reaction of nitroarenes with TiO_2, as a photocatalyst, in the presence of alcohols leads to the formation of the tetrahydroquinoline <99TL1145>.

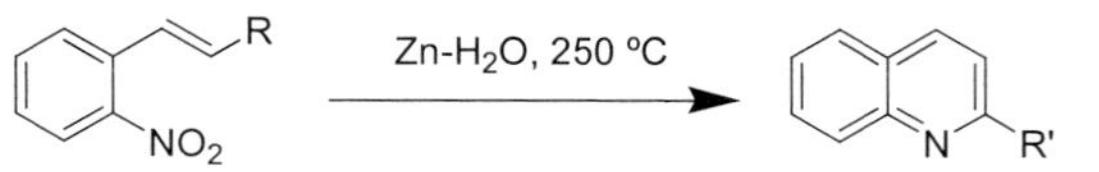

28: R =CHO, COMe, CO_2H, CO_2Me R' =H, Me, OH, OH

An interesting new method to cyclize phenylalkylazides to the quinone imine **29** by oxidative cyclization with phenyliodine (III) bis(trifluoroacetate) (PIFA) was reported <99CPB241>.

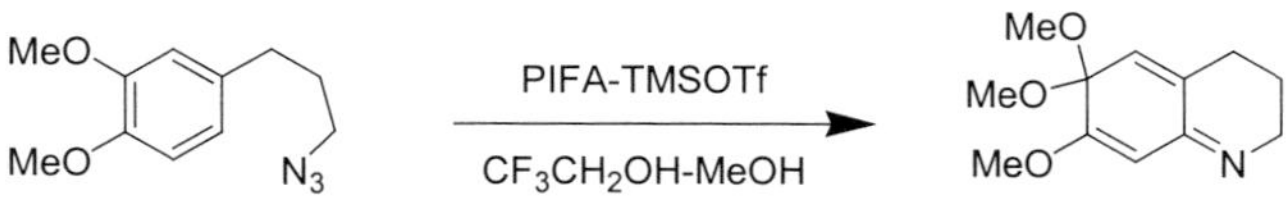

29

Pyrolysis of some *N*-aryl five-membered-ring heterocycles causes ring opening and rearrangement to the quinoline ring <99H(51)2377, 99JOC3608>.

6.1.3.2 Reactions of Quinolines

The selective addition reactions of quinolines are important to the preparation of more complex products. Tetrahydroquinoline undergoes anodic oxidation with incorporation of cyanide in the 2-position to prepare the 2-cyanoquinolines <99EJOC2645>.

Conditions for controlling the regiochemistry of the addition reactions between benzyl zinc reagents and 2,4-dichloroquinoline under palladium-catalyzed conditions were developed <99JOC453>. Similarly, the regiochemistry of the palladium-catalyzed carbonylation of 4,7-dichloroquinoline was evaluated <99TL3719>.

Frequently, it is easier to add nucleophiles through the quinolinium species (Scheme 5). The reaction of Grignard reagents with quinolinium salts formed by *N*-silylation showed that as the silyl group becomes larger, addition at the 4-position is favored. A unique redox reaction between the 1,4 and 1,2-dihydroquinoline species was also observed <99JOC6911>. Quinoline is suitably activated as the *N*-trifluoromethylsulfonamide for addition of phosphites at the 2- and 4- positions yielding the heteroarylphosphonates <99S2071>. Quinoline *N*-oxides undergo selective addition at the 2-position when treated with alkyl and aryllithium reagents. In the presence of an oxidant the addition product is oxidized back to the quinoline <99H(51)2385>. The sulfate salts of 1-methyl- or 1-ethyl-4-amino-3-aminosulfonylquinolinium are hydrolyzed at the 4-position to yield the 1,4-dihydro-4-oxoquinolinesulfonamides. The 4-thioxo analogue can also be prepared by displacement with sodium sulfide <99H(51)2111>. By conversion of the chiral-3-quinolinyl aminal **30** to an *N*-acylisoquinolinium intermediate asymmetric alkylation at the 2- or 4-position can be carried out <99TL6241>. Aryl and alkenyl boronates add stereoselectively to the *N*-acyliminium ion intermediate generated from **31** by addition of a Lewis acid to form the substituted tetrahydroquinoline **32** <99JA5075>.

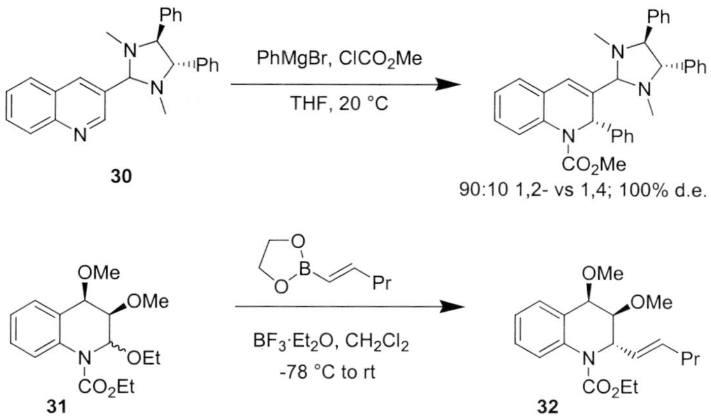

Scheme 5

The selective reduction of a quinoline to the tetrahydroquinoline is an important reaction. With $NiCl_2$-lithium arene <99T14479>, $NiCl_2$-$NaBH_4$ <99EJOC1173>, or LiH_3BNMe_2 <99OL799> tetrahydroquinolines are obtained cleanly. Interestingly, when the *N*-alkyl group is an aminal, such as methoxymethyl, the ring undergoes a 3-aza-Grob fragmentation with $NaBH_4$ to provide the ring opened *o*-methylaminophenylpropyl alcohol <99JOC5725>.

The *N*- vs *O*-alkylation of quinolinone derivatives was studied <99TL6999>.

6.1.4 ISOQUINOLINES

6.1.4.1 Preparation of Isoquinolines

Reviews of the total synthesis of naphthylisoquinoline <98M2> and isoquinoline <99NPR367> alkaloids were published.

Examples of palladium-catalyzed heterocyclizations continue to expand (Scheme 6). A palladium-catalyzed reaction of the iodobenzaldimine **33** with alkynes <99OL553> or allenes <99TL4255> affords isoquinolines. An intramolecular version of the alkyne coupling with a benzamide moiety similarly leads to the 3-substituted isoquinolin-1-one <99S1145>. Reaction of 2-bromobenzylamines with allenes affords 4-alkylidene-1,2,3,4-tetrahydroisoquinolines in low yields <99H(50)83>. Palladium-catalyzed coupling of 10-dimethylamino-1-iodophenanthrene with alkynes gives the aporphine-related heterocycles <99EJOC1957>.

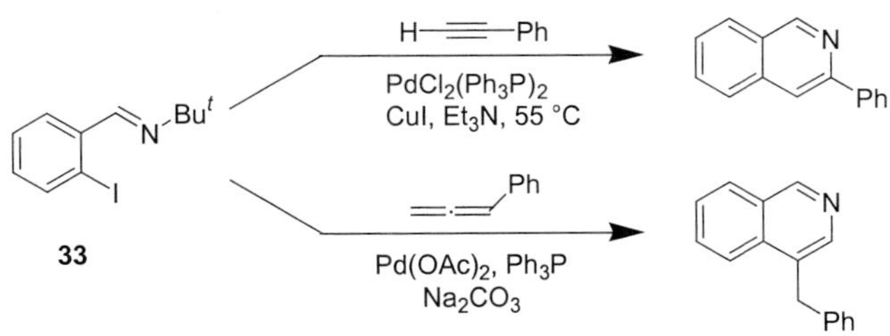

Scheme 6

Cycloaddition reactions can construct either the benzo or hetero ring of isoquinolines (Scheme 7). The piperidine-fused zirconocyclopentadiene **34** formed from *N*-propargyl-*N*-homopropargylbenzylamine undergoes alkyne coupling to yield the tetra-substituted isoquinoline **35** <99JA11093>. A similar intramolecular cycloaddition of the vinyl group and the aminofuran of **36** can be used in the synthesis of the Amaryllidaceae class of alkaloids <99JOC3595>.

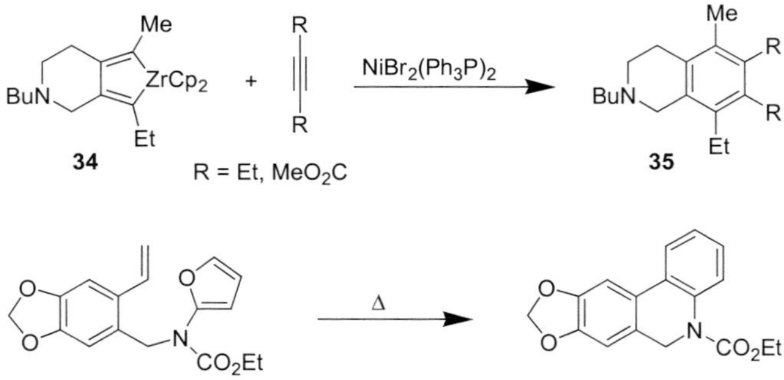

Scheme 7

Radical cyclization is an effective approach to the synthesis of isoquinolines (Scheme 8). In some cases these offer an alternative to the palladium-catalyzed reactions with aryl halide intermediates <99EJOC1925, 99TL1125>. For example, the radical cyclization of the iodide **37** onto the vinylsulfide moiety was followed by a cascade cyclization to form the benzo[*a*]quinolizidine system <99TL1149>. In some cases the radical cyclization can take place without the need for a halo intermediate. The reactive intermediate of **38** was formed on the nitrogen as an amidyl radical, which underwent tandem cyclizations to the lycorane system <99TL2125, 99SL441>.

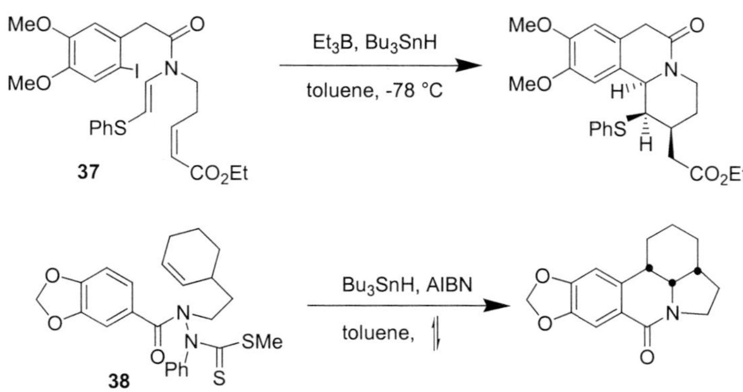

Scheme 8

Ring transformations are an interesting means to prepare isoquinolines. The cyclobutene **39** undergoes ring expansion and subsequent cyclization to form the isoquinolinedione <99JOC5979, 99JOC6881>. The pyrylium ring is a precursor to pyridines through substitution with ammonia. This strategy has been applied to the synthesis of isoquinolines <99JOC6499>. Reaction of piperidinones with pyranones affords the tetrahydroisoquinoline framework <99S467>.

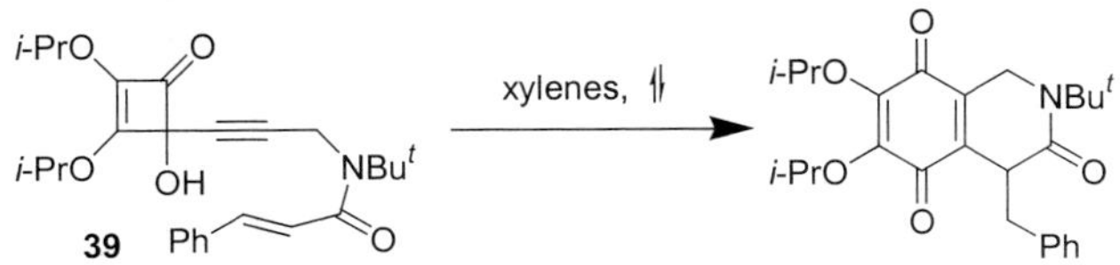

39

Improved methods for preparing isoquinolines through ring closure at the C4-aryl bond were reported (Scheme 9). The Pummerer reaction induces cyclization of the *N*-formylbenzylamine sulfoxide **40** by treatment with TFAA or BF_3•OEt_2/TFAA <99H(51)119>. The unactivated titanium-mediated cyclization of the Schiff base **41** to 6-methoxyisoquinoline is effected through intermediacy of the *N*-ethylcarbamate-aminophosphate derivative **42** <99SC1617>.

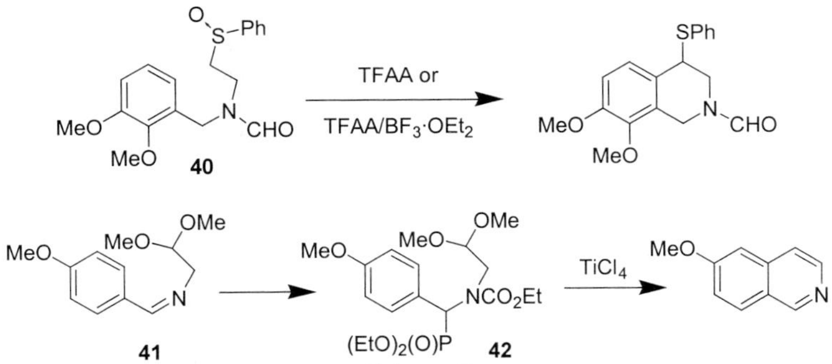

40 **41** **42**

Scheme 9

New synthetic methods for the preparation of chiral tetrahydroisoquinolines are available. The chiral ligand-mediated addition of an organolithium onto the imine **43** <99TA221> and reaction of the chiral anion of **44** with a benzylidene <99EJOC503> provide entries into chiral 1-substituted-tetrahydroisoquinolines (Scheme 10). Chiral *N*-benzyl-3-substituted-tetrahydroisoquinolines are prepared from chiral dibenzylaminols by intramolecular Friedel-Crafts cyclization of the tosylate <99OL877>. Lithiated *o*-toluamides undergo alkylation with the chiral benzylidene-*p*-toluenesulfinamides followed by intramolecular amidation to afford the 3-aryltetrahydroisoquinolines <99JOC8627>. Asymmetric syntheses of chiral Amaryllidaceae alkaloids have utilized the Banwell-modification of the Bischler-Napieralski reaction to complete the ring closure of the isoquinoline ring <99TL3077, 99TL3081>. A general strategy to the asymmetric synthesis of 4-alkyl-3-aryl-1,2,3,4-tetrahydroisoquinolines was reported <99JOC4610>.

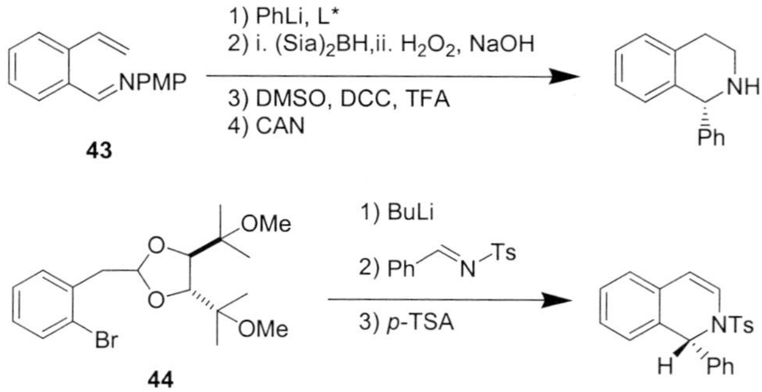

Scheme 10

The direct anionic cyclization of 2-alkynylbenzonitriles **45** was applied to the synthesis of isoquinolinones <99T13193> and phenanthridines <99OL767>. A *bis*-alkylation process of an amine with a *bis*-mesylate afforded the tetrahydroisoquinoline ring system <99SC645>.

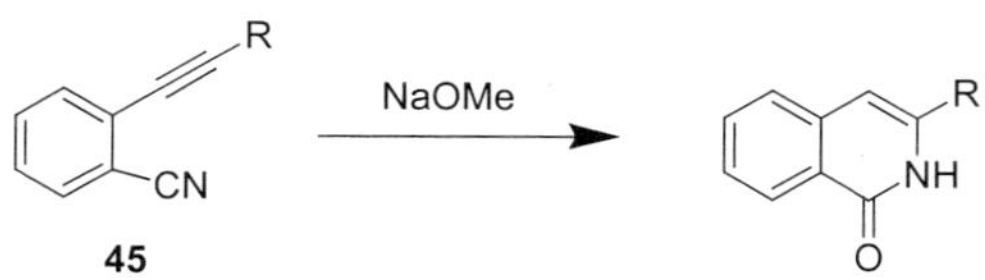

The intermediacy of aryne intermediates generated by dehydrohalogenation of the corresponding aryl bromide is useful in preparing polycyclic isoquinolines with the phenanthridine skeleton <99T5195, 99OL985>.

Improvements in the traditional aryl C-1 cyclization reaction were reported (Scheme 11). Generally, condensation reactions of phenethylamine derivatives do not perform well with electron-withdrawing groups. Reaction of sulfamoyl-β-phenethylamines **46**, even containing an aryl nitro group, with chloro(methylthio)acetate <99H(51)103> or α-chloro(phenylseleno)acetate <99TL4969> in the presence of a Lewis acid gives the isoquinolines **47** in good yield. A study of the Pictet-Spengler reaction with superacids revealed the involvement of a dicationic species in the cyclization of the prototypic *beta*-phenethylamine <99JOC611>. Aerobic oxidation of dopamine afforded the 6,7-dihydroxy-1,2,3,4-tetrahydroisoquinoline via a Pictet-Spengler reaction, where the formaldehyde was generated from oxidation of a second dopamine substrate <99TL2833>. An asymmetric variant of the Pictet-Spengler reaction will afford chiral 3-arylisoquinolines <99JOC1115>. Morphine derivatives are prepared via *N*-acyliminium ion-olefin cyclization reactions <99CCC203>. The benzotriazole group of the substituted phenylalaninol **48** is employed to generate the *N*-acyliminium ion in the intramolecular Friedel-Crafts cyclization to the tetrahydroisoquinolines <99JCS(P1)179, 99TA255>. Enamides of *N*-formyl-*beta*-phenethylamines undergo ring closure to the tetrahydroisoquinoline upon treatment with 9-BBN triflate or triflic acid <99T4481>.

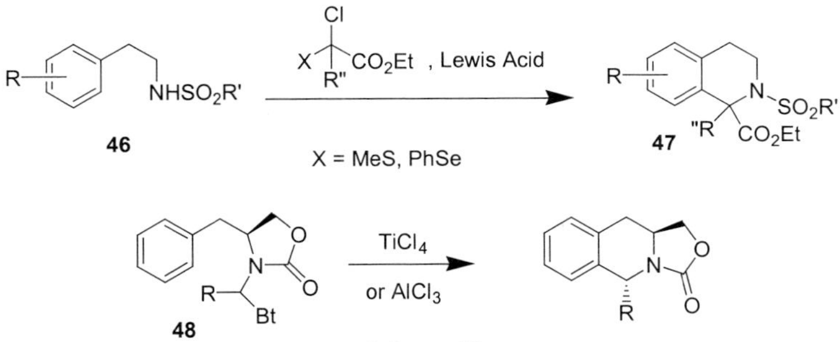

Scheme 11

Reaction of isophthaldehyde with phosphonoglycine derivatives affords the 3-carboxyisoquinoline **49** <99TL7935>.

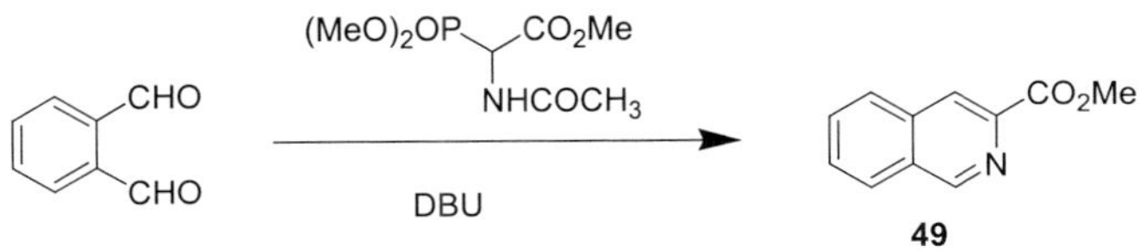

6.1.4.2 Reactions of Isoquinolines

New processes for setting chirality at the C-1 position by hydrogenation of a 3,4-dihydroisoquinoline intermediate were published. In the hydrogenation of 3,4-dihydroisoquinolines with Rh or Ru-catalysis, chiral *mono*-sulfonamide diamine ligands are effective for preparing chiral 1-substituted-1,2,3,4-tetrahydroisoquinoline <99EJOC2315, 99OL841, 99JOC6724>. The MBA amide of the 1-carboxy-3,4-dihydroisoquinoline **50** can be reduced to either diastereomer depending on the reagent <99TA3371>.

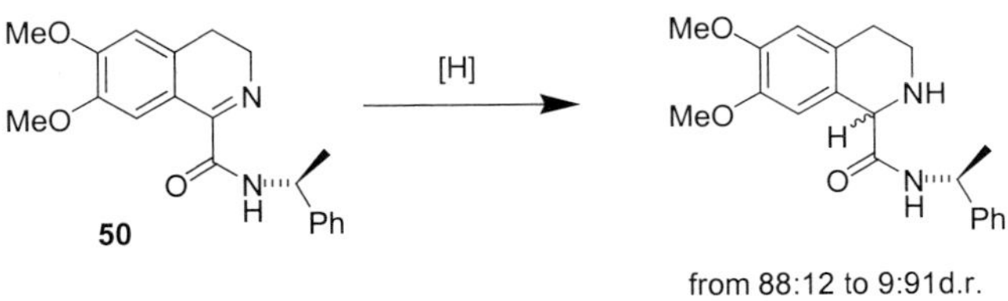

4-Arylisoquinolines can be prepared by Suzuki coupling of the 4-bromoderivatives <99JOC1407>. This methodology was used in the synthesis of dinapsoline. A synthesis of 4-alkylcarbonylmethylidene isoquinolines was reported using a modified Jones oxidation and alkene acylation conditions <99H(51)2311>.

Substitution of the 1-position is an important transformation in isoquinoline chemistry (Scheme 12). Isoquinolinium intermediates, analogous to the pyridinium examples (see Sect. 6.1.2.3), are frequently used. A series of reports studied the reaction of isoquinolinium species with hydroxylamine, which resulted in the formation of ring-opened products or the *N*-oxide through rearrangement pathways <99ZN(B)225, 99ZN(B)532, 99ZN(B)913>. *N*-Acylisoquinolinium species undergo enantioselective allylation (>90% e.e.) with a chiral allylsilane <99CC2213>. Silyl enol ethers add to the isoquinolinium species **51** activated with the chiral acyl group providing d.e.'s >80% <99SL1154>. The *alpha*-diazo amide **52** of the tetrahydroisoquinoline is effectively oxidized under rhodium catalysis to the 1-oxo product <99TL8269>. Anodic oxidation of a benzyl tetrahydroisoquinoline in the presence of sodium cyanide brings about effective 1-cyanation. Upon treatment with $ZnCl_2$ the intermediate produces an *N*-acyliminium precursor as part of the preparation of the quinazoline ring system <99SL1383>. Peroxyiodanes oxidize tetrahydroisoquinolines to the 1-oxo compounds <99TL5541>. 1,3-Isoquinolinediones undergo oxidation with singlet oxygen to the 1,3,4,-trione, as well as the rearranged isobenzofuranone <99T9185>.

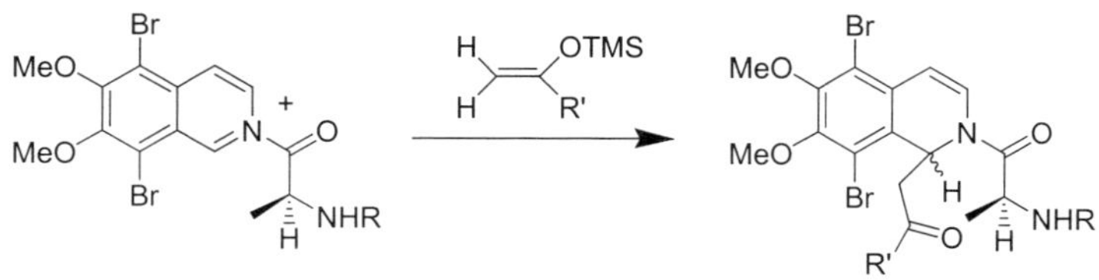

51: R = *p*-nitrophenylsulfonyl

d.e. 83-96%

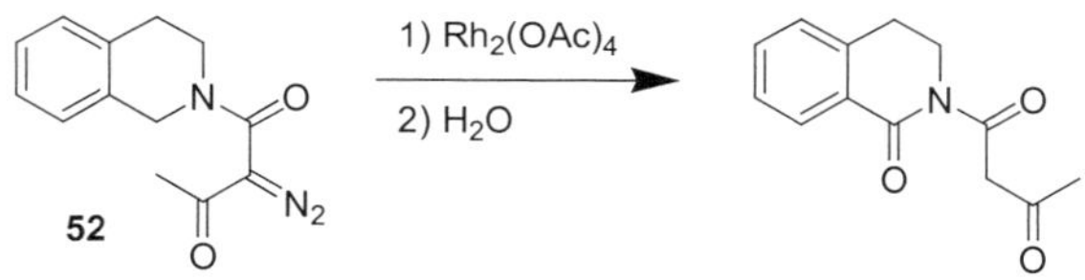

52

Scheme 12

The cycloaddition reactions of isoquinolinium species produce fused isoquinoline products. The *N*-ylide of **53**, formed with base addition, couples with alkenes <99S51> or imines <99T7279> to afford tricyclic products, such as **54**. Pyrrole-fused isoquinolines result from the reaction between münchnone imine intermediates and *α,β*-ethylenic esters <99EJOC297>. *N*-Arylimides undergo 1,3-dipolar cycloaddition with strained *trans*-cyclooctenes, as opposed to common cycloalkenes, to afford the pyrazolidine-fused ring system <99H(50)353>.

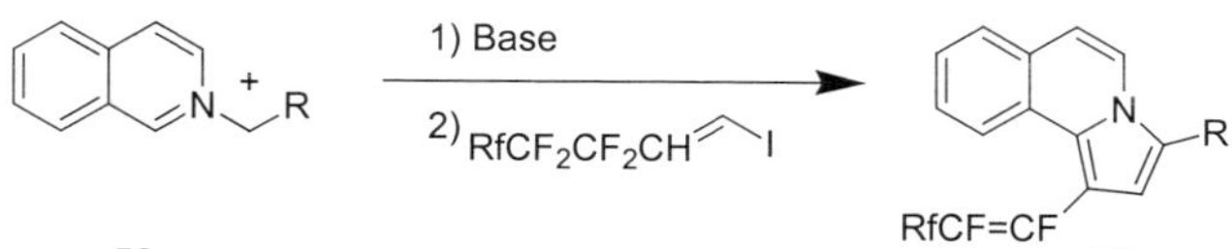

53 **54**

6.1.5 ACRIDINES

Condensation reactions are a common approach to the synthesis of acridines. The Bernthsen reaction can be applied to the synthesis of 9-acridinylalkanoic acids. Diphenylamine is condensed with dicarboxylic acids in the presence of $ZnCl_2$ to afford a mixture of the *bis*-acridinyl alkanes and acridinylalkanoic acids depending on the stoichiometry <99SC4007>. The Friedlander reaction under Fehnel conditions <99TL4097> and Ullmann condensation/Friedel-Crafts acylation <99SL641, 99S947> are commonly used to prepare the acridine and acridinone rings, respectively. Nafion-H is a useful acid catalyst for the intramolecular Friedel-Crafts acylation of *N*-(2'-carboxyphenyl)aniline to afford the acridinone <99SL1067>. Amination of benzyne with the *o*-aminobenzoate **55** followed by Friedel-Crafts acylation provides the acridinone <99TL7003>. A diimine was reported to add to benzyne in a [2+2] fashion forming a benzazetidine followed by electrocyclization and aromatization to produce the acridine ring system <99T1111>.

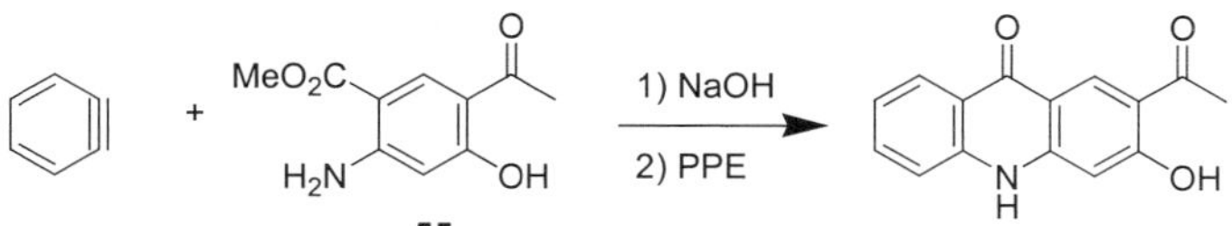

55

Photocyclization of the condensed adduct **56** between 2-methyl-4-oxoquinoline and cinnamaldehyde gives the acridine **57** <99ZN(B)1337>. This approach is notable for its application to the synthesis of the hitherto unknown 1-phenyl and 1-naphthacridones.

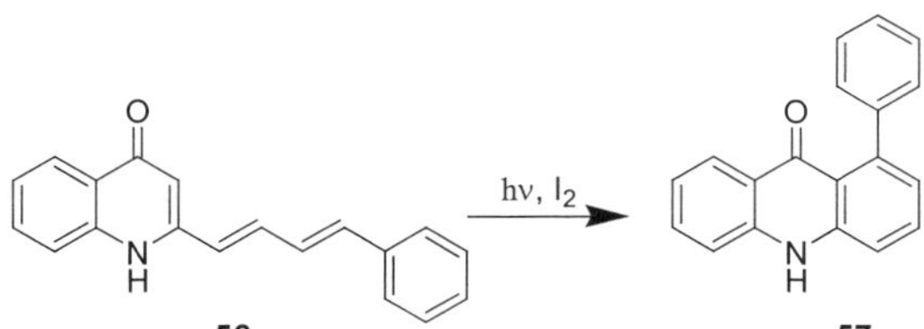

56 **57**

Substitution of the nine position is a common transformation for acridines. An optimized method for preparing the 9-carboxamides uses BOP/DMF <99SC4341>. Reaction of 9-isothioacridines with the sodium anion of diethylmalonate is followed by alkylation with bromoacetate to afford the spiro[dihydroacridine-9(10*H*)-thiazolidines] <99H(51)137>.

N-Methyl-9-*t*-butylacridine undergoes oxidation to acridine with loss of the methyl and *t*-butyl groups by treatment with PhIO <99TL5425>. Acridine can be converted to the diols and tetraols under biocatalysis conditions <99CC1201>.

Acridine-4,5-diol was converted to the corresponding 18-crown-6 ligand by alkylation of the hydroxy groups with tetraethylene glycol di-*p*-tosylate <99T1491>.

6.1.6 PIPERIDINES

6.1.6.1 Preparation of Piperidines

A review was published this year covering the literature of saturated nitrogen heterocycles over 1998 with a section devoted to piperidine synthesis <99JCS(P1)2553>.

A few synthetic applications of palladium catalysis appeared this year. The palladium-catalyzed cyclization of amino allenes **58** occurs with coupling of aryl iodides or vinyl triflates at the 3-position <99OL717, 99SL324>. The cyclization can also proceed by the 4-*exo-trig* pathway, but under suitable reaction conditions the piperidine **59** is prepared selectively. The intramolecular cyclization of amines onto *N*-allylbenzotriazoles similarly affords piperidines <99JOC6066>.

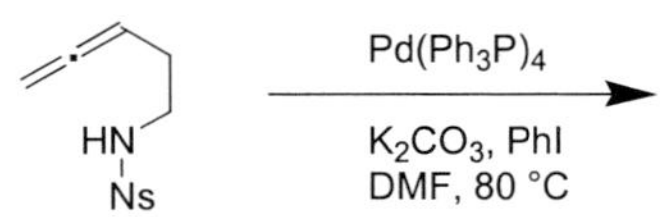

58 59

The ring-closing metathesis (RCM) reaction continues to be an excellent avenue for constructing the piperidine ring. A review describes the application of these reactions to the synthesis of azasugars and alkaloids <99EJOC959>. In one example (*S*)-(+)-*N*-Boc-coniine was prepared <99TL5581>. An interesting combination of RCM and ring-opening metathesis afforded the precursor **60** to (-)-halosaline <99T8179>.

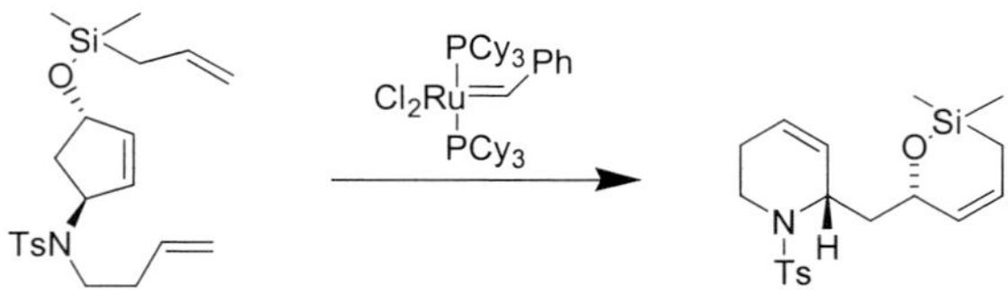

60

The enantiopure tricarbonyl(dienal)iron complex **61** suitably transfers chirality in the piperidine ring formation. Condensation to the Schiff base is followed by the intramolecular Mannich reaction catalyzed with *p*-TSA. The piperidine was converted to dienomycin C (**62**) in five additional steps <99EJOC1517>.

61 62

Radical cyclizations are often used in ring formations and are an effective methodology in the synthesis of piperidines. The intramolecular cyclization of an oxime ether, such as **63** onto an aldehyde or ketone gives a new entry into cyclic amino alcohols <99JOC2003, 99H(51)2711>. Similarly, reaction of a terminal acetylene with Bu_3SnH generates a vinyl radical, which will cyclize with an imine moiety to give 3-methylenepiperidine <99TL1515>. The indolizidine alkaloid ipalbidine was prepared by a sulfur-controlled 6-*exo*-selective radical cyclization of an *alpha*-phenylthio amide <99H(50)31>.

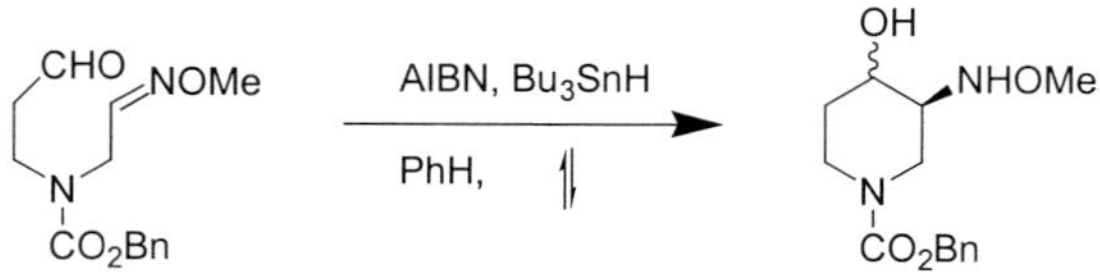

63

1.3:1 t/c

The preparation of piperidines by cycloaddition can use either an imino (Scheme 13) or azadiene (Scheme 14) substrate. Highly functionalized tetrahydropiperidines were prepared by a high-pressure aza-Diels-Alder reaction <99TL1877>. The hetero [4+2] cycloaddition of vinyl ketenes with imines provides α,β-unsaturated δ-valerolactams <99OL641>. Activations of the imine – diene hetero-Diels-Alder reaction with 5.0 M $LiClO_4$-diethyl ether <99JOC6041>, HBF_4 in aqueous media <99TL7831>, or indium triflate <99TL5621> were reported. Two examples of chiral imines provided asymmetric syntheses of the heterocycle. High diastereoselectivities are achieved using the planar chiral Cr-complex **64** <99SL626, 99OL1997>. With the benzylimine of glyceraldehyde high diastereoselectivities were reported as well <99T7601>. Alternatively, a chiral 2-aminodiene reacted with *N*-trimethylsilylbenzaldimine to afford the piperidine in high optical purity after hydrolysis of the resultant enamine to the ketone <99JOC3736>. A zirconium-catalyzed aza-Diels-Alder reaction provided piperidinones with high enantioselectivity using chiral binaphthol ligands <99JOC4220>. A formal [3+3] cycloaddition was achieved by reaction of a vinylogous amide of cyclohexan-1,3-dione with an α,β-unsaturated iminium intermediate <99OL509>.

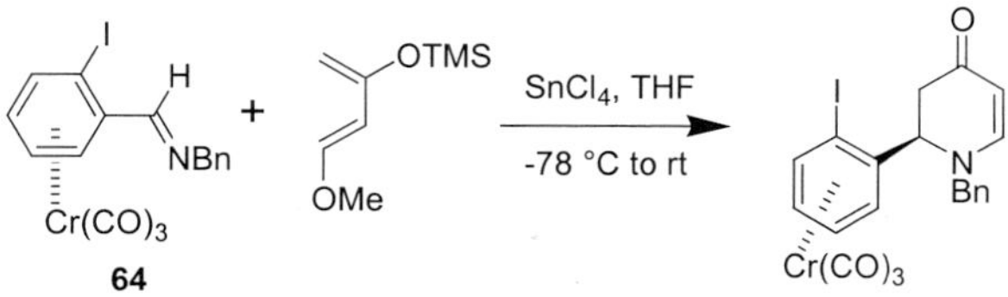

64

Scheme 13

The reaction of diphenylketene with an azadiene was reported to produce piperidinones <99SL1379>. An intramolecular hetero-Diels-Alder of the activated azadiene **65** was carried out either by heating or with catalysis <99TL7211,99TL7215>.

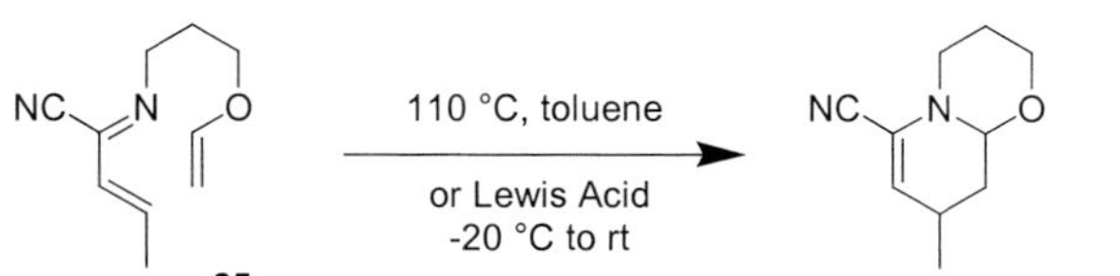

65

Scheme 14

Other cycloadditions were reported. The intramolecular cycloaddition of alkenylnitrones was applied to the synthesis of piperidines <99TL1397, 99JCS(P1)185>. Cycloaddition of an alkenyl azide afforded piperidines after reduction of the bicyclo triazole <99T1043, 99EJOC1407>. Similar to the cyclization of the diazo imide **2** in section 6.1.2.1, isomünchnone intermediates can rearrange to functionalized piperidines <99JOC556>.

The photocyclization of the γ-ketoamide **66** provides a diastereoselective ring closure to the piperidine <99TL3137>. A similar photocyclization of a γ-ketoamide of proline affords the indolizinone ring <99TL5987>.

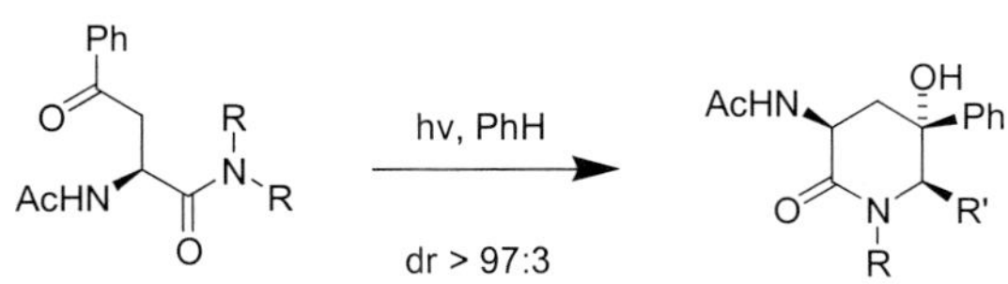

66: R = Bn or CH_2CO_2Me R' = Ph or CO_2Me

Furan-substituted ethanol amines are easily oxidized affording chiral 2-hydroxymethylpyridinones **67** as precursors to azasugars <99TA3649> or piperidine alkaloids <99S1889, 99TL6869>.

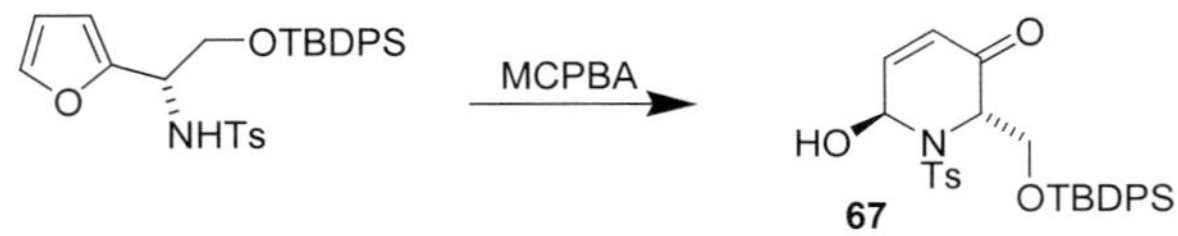

67

Ring expansions of pyrrolidine precursors often produce substituted piperidines. The chiral pyrrolidine **68** derived from serine undergoes ring expansion to the piperidine precursor **69** to aza-sugars <99TL5331>. Optically active 3-hydroxypiperidines are prepared by ring expansion of pyrrolidine methanol derivatives <99EJOC1693>. 5-Spirocyclopropane-isoxazolidines thermally rearrange to tetrahydropyridinones <99JOC7846>, where an *N*-phenyl substituent significantly reduces the temperature necessary for the reaction <99TL6657>. Enantiomerically pure 6,6-disubstituted-piperidinones are prepared by Beckmann rearrangement of chiral 1,1-disubstituted-cyclopentanones <99SL1127>. The ring fragmentation of the tricyclo intermediate **70** provides an asymmetric synthesis of piperidine alkaloids <99T2911>.

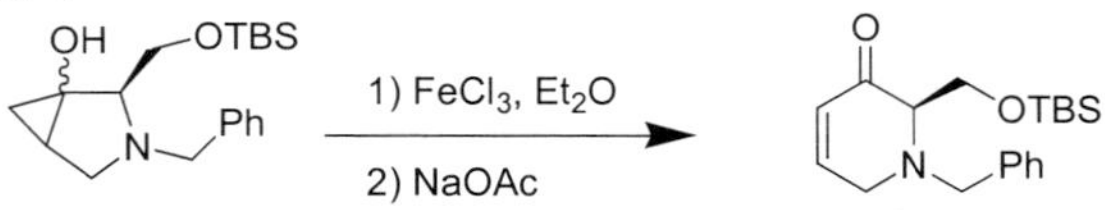

68 **69**

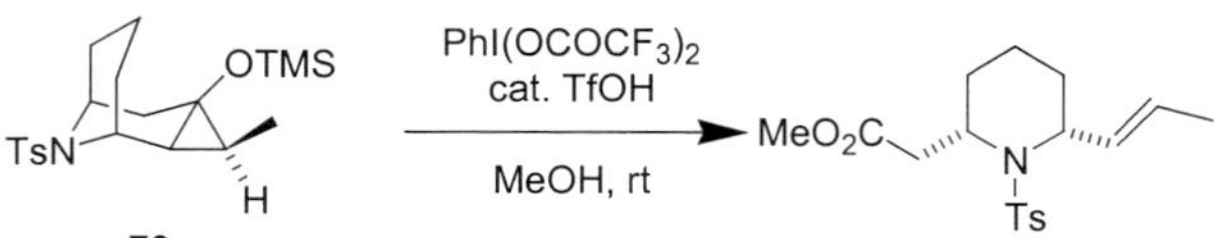

70

N-Acyliminium ions are useful intermediates for preparing heterocycles in general. This methodology was applied to the synthesis of functionalized pipecolic acids <99EJOC1127>. *N*-But-3-enyl-*N*-styrylformamides undergo cyclization to the tetrahydropyridine when treated with 9-BBN triflate <99T4481>.

The reaction of enol ethers with allenesulfonamides produces tetrahydropyridines with a novel 1,3-sulfonyl shift <99AG(E)121>.

6.1.6.2 Reactions of Piperidines

Regio- and stereoselective alkylation of the piperidine ring is a major focus in the literature (Scheme 15). Using the piperidinone intermediates discussed in section 6.1.2.3 further elaboration of the piperidine ring can be carried out. An interesting new application of the Mukaiyama-Michael reaction allows the regio- and stereoselective synthesis of *cis*-2,6-disubstituted tetrahydropyridines **71** <99OL1031>. Without the phenylsulfide(selenide) the diastereoselectivity was poor. Similarly, the stereoelectronic effects in the 1,4-addition of cuprates to *N*-*t*-butoxy-6-oxo-1,2,3,6-tetrahydropyridine-2-carboxylates were evaluated as part of the synthesis of substituted adipic acid derivatives <99CC683>. *DL*-Febrifugine and isofebrifugine were synthesized via an unusual Claisen rearrangement of allylenol ether **72** to the corresponding allylpiperidone **73** <99S1814>. The palladium-catalyzed allylic alkylation of acetoxy tetrahydropyridines was studied. With the appropriate chiral ligand either enantiomer can be produced in >95% e.e. <99EJOC2515>. *N*-Boc-2-methylpiperidine is converted to homo-chiral *trans*-2,6-dimethylpiperidine using a metalation/alkylation approach <99SC1747>. Piperidinones are converted to the *gem*-diaryl compounds in the presence of the super acid triflic acid <99JOC6702>. *N*-Alkynyl amino chromium carbene complexes of piperidinones undergo cyclization to quinolizidines <99EJOC2825>.

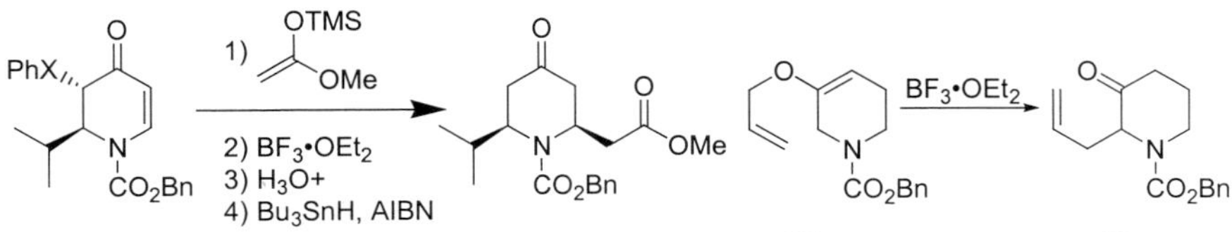

X = S, Se 71 72 73

Scheme 15

N-Acyl *N,O*-acetals of piperidines are frequently used for alkylation at the 2-position through the *N*-acyliminium species. Titanium enolates of chiral *N*-propionyl-oxazolidinones add in moderate to good diastereoselectivity <99TL2891, 99OL175>. The piperidinone **74** undergoes high diastereoselective allylation <99TL739, 99SL37>. Alkenyl and aryl boronates are mild nucleophiles for the stereoselective addition reactions <99JA5075> (see section 6.1.3.2). An enantiomeric synthesis of (-)-porantheridine was carried out by addition of a *beta*-ketoester to such an *N*-acyliminium intermediate <99JOC8402>.

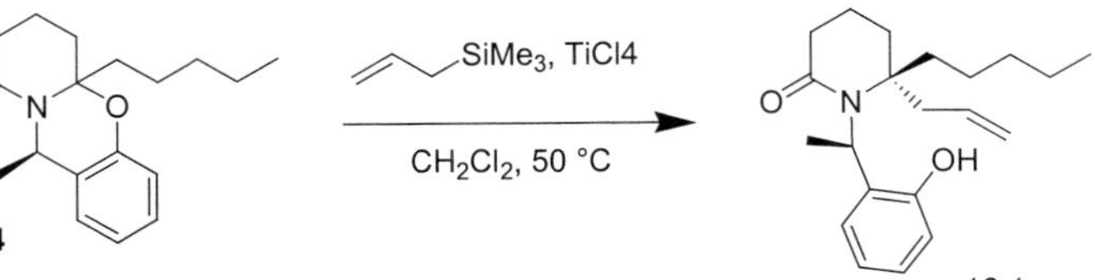

74

46:1 16:1

Similarly, the 2-cyano-6-oxazolopiperidine **75** (Scheme 16) can be used to provide a variety of substituted piperidines <99TL3731, 99H(51)2065>. Conversion to the enamide **76** provides a means to introduce C-3 alkyl groups by Michael reaction <99TL3699>. Electrochemical *bis*-bromination and dehydrohalogenation affords the vinyl bromide **77**, which can undergo substitution at the 4-position by the addition of nucleophiles as simple as water <99T8931>.

Chiral piperidinones have been prepared and used to diastereoselectively substitute the piperidine ring <99S258, 99T10173, 99JOC4914>. 3-Carboxypiperidin-4-ones undergo regioselective deprotonation and substitution reactions. After dehydrogenation with phenylselenyl chloride a Sakurai allylation was carried out on the menthyl ester to afford low to excellent facial selectivity <99SL405>. In another case the dianion underwent regioselective alkylation as part of the synthesis of hexopyranose mimics <99TL367>.

Clavepictines A and B were prepared using a variety of effective reactions on the piperidine ring, such as a silver-promoted cyclization of an aminoallene intermediate, diastereoselective alkylation, and cross coupling of an enol triflate <99JA10012>.

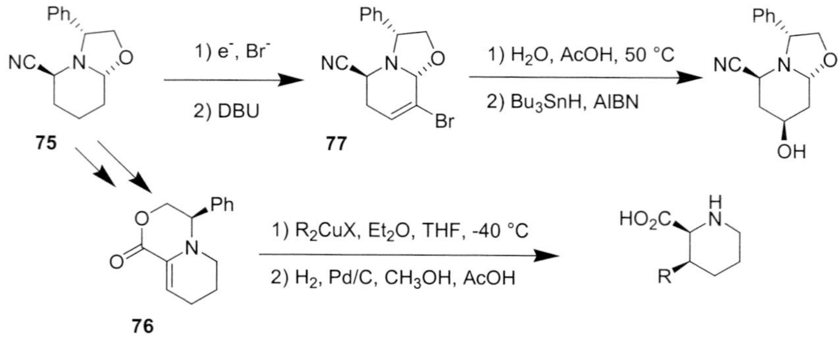

Scheme 16

The carbene insertion of the aryldiazoacetate **78** onto the 2-position of *N*-Boc-piperidine is highly selective with Rh catalysis <99JA6509, 99JA6511>. The enantiomeric excess could be easily increased to >95% by recrystallization.

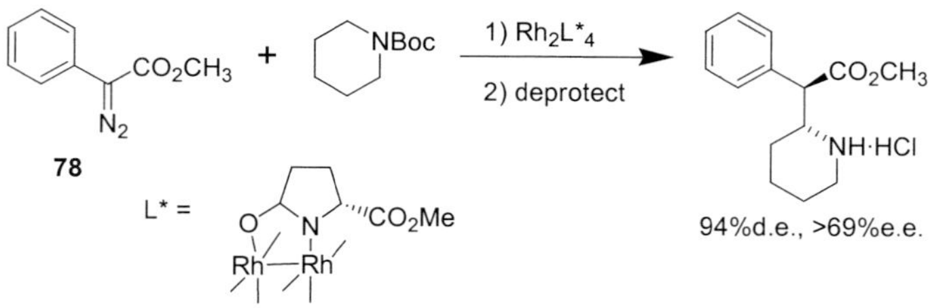

Asymmetric desymmetrization is effective for converting *meso*-substituted piperidines to chiral products. The asymmetric reduction of *meso*-imide **79** was used in the preparation of the antipode of a nojirimycin analogue <99TA657>. Enzymatic desymmetrization of 2,6-*bis*(acetoxymethyl)piperidines affords chiral intermediates as part of the synthesis of 6-hydroxymethylpipecolic acids <99JOC3178> and indolizidine alkaloids <99TA3117, 99H(51)593>. *N*-Benzyl-2,6-dicarbomethoxypiperidine undergoes symmetry-breaking enolization and alkylation with a chiral *bis*-lithium amide base in high diastereo- and enantioselectivity <99SL1292>.

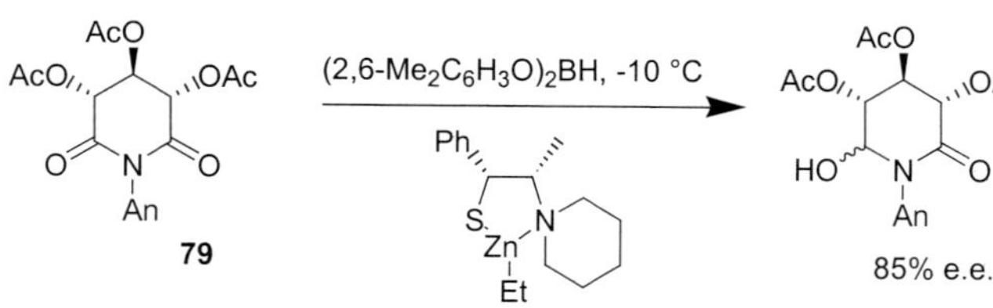

Oxidation and dimerization of *N*-arylpiperidines with mercury-EDTA reagents was reported <99ZN(B)214>. The synthesis of 1-cyclopropyl-1,4-dihydro-4-oxo-3-pyridinecarboxylic acid is accomplished via the double dehydrogenation of the corresponding 4-piperidinone <99H(50)867>. Base addition frequently causes ring contractions of halogenated piperidines to pyrroli(di)nes accordingly <99CC47, 99S1309, 99TL5495>.

6.1.7 REFERENCES

98M1	S. Shimizu, N. Abe, A. Iguchi, *Catalysis Surveys Jpn* **1998**, *2*, 71.
98M2	M.A. Rizzacasa; Atta-ur-Rahman, *Ed.*, *Studies in Natural Products Chem.* **1998**, *20*, 407.
99ACS141	J.M. Bakke, E. Ranes, J. Riha, H. Svensen, *Acta Chem. Scand.* **1999**, *53*, 141.
99ACS356	J.M. Bakke, J. Riha, *Acta Chem. Scand.* **1999**, *53*, 356.
99ACS616	J. Bergman, T. Brimert, *Acta Chem. Scand.* **1999**, *53*, 616.
99AG(E)121	Y. Horino, M. Kimura, Y. Wakamiya, T. Okajima, Y. Tamaru, *Angew. Chem. Int. Ed. Engl.* **1999**, *38*, 121.
99AG(E)1928	H. Steinhagen, E.J. Corey, *Angew. Chem. Int. Ed. Engl.* **1999**, *38*, 1928.
99CC47	R.E. Banks, M.K. Bescheesh, N.J. Lawrence, R.G. Pritchard, D.J. Tovell, *Chem. Commun.* **1999**, 47.
99CC683	M. Muller, A. Schoenfelder, B. Didier, A. Mann, C.-G. Wermuth, *Chem. Commun.* **1999**, 683.
99CC1201	D.R.Boyd, N.D. Sharma, J.G. Carroll, C.C.R. Allen, D.A. Clarke, D.T. Gibson, *Chem.Commun.* **1999**, 1201.
99CC2213	R. Yamaguchi, M. Tanaka, T. Matsuda, K.-i. Fujita, *Chem. Commun.* **1999**, 2213.
99CC2371	K. Oda, R. Nakagami, N. Nishizono, M. Machida, *Chem. Commun.* **1999**, 2371.
99CCC203	P. Bottari, M.A.A. Endoma, T. Hudlický, I. Ghiviriga, K.A. Abboud, *Collect. Czech. Chem. Commun.* **1999**, *64*, 203.
99CPB241	Y. Kita, M. Egi, M. Ohtsubo, T. Saiki, A. Okajima, T. Takada, H. Tohma, *Chem. Pharm. Bull.* **1999**, *47*, 241.
99CPB1038	J. Koyama, I. Toyokuni, K. Tagahara, *Chem. Pharm. Bull.* **1999**, *47*, 1038.
99EJOC297	S. Perrin, K. Monnier, B. Laude, M.Kubicki, O. Blacque, *Eur. J. Org. Chem.* **1999**, 297.
99EJOC313	J. Sauer, D.K. Heldmann, G.R. Pabst, *Eur. J. Org. Chem.* **1999**, 313.
99EJOC503	B. Wünsch, S. Nerdinger, *Eur. J. Org. Chem.* **1999**, 503.
99EJOC937	I.L. Baraznenok, V.G. Nenajdenko, E.S. Balenkova, *Eur. J. Org. Chem.* **1999**, 937.
99EJOC959	U.K. Pandit, H.S. Overkleeft, B.C. Borer, H. Bieräugel, *Eur. J. Org. Chem.* **1999**, 959.
99EJOC1127	F.P.J.T. Rutjes, J.J.N. Veerman, W.J.N. Meester, H. Hiemstra, H.E. Schoemaker, *Eur. J. Org. Chem.* **1999**, 1127.
99EJOC1173	M. Alvarez, M.A. Bros, G. Gras, W. Ajana, J.A. Joule, *Eur. J. Org. Chem.* **1999**, 1173.
99EJOC1407	C. Herdeis, J. Telser, *Eur. J. Org. Chem.* **1999**, 1407.
99EJOC1517	I. Ripoche, J.-L. Canet, J. Gelas, Y. Troin, *Eur. J. Org. Chem.* **1999**, 1517.
99EJOC1693	J. Cossy, C. Dumas, D.G. Pardo, *Eur. J. Org. Chem.* **1999**, 1693.
99EJOC1925	J. Cossy, L. Tresnard, D.G. Pardo, *Eur. J. Org. Chem.* **1999**, 1925.
99EJOC1957	A-E. Gies, M. Pfeffer, C. Sirlin, J. Spencer, *Eur. J. Org. Chem.* **1999**, 1957.
99EJOC2315	G.J. Meuzelaar, M.C.A. van Vliet, L. Maat, R.A. Sheldon, *Eur. J. Org. Chem.* **1999**, 2315.
99EJOC2515	S. Schleich, G. Helmchen, *Eur. J. Org. Chem.* **1999**, 2515.
99EJOC2645	E. Le Gall, J.-P. Hurvois, S. Sinbandhit, *Eur. J. Org. Chem.* **1999**, 2645.
99EJOC2825	H. Rudler, A. Parlier, S. Bezennine-Lafollée, J. Vaissermann, *Eur. J. Org. Chem.* **1999**, 2825.
99H(50)31	M. Ikeda, J. Shikaura, N. Maekawa, K. Daibuzono, H. Teranishi, Y. Teraoka, N. Oda, H. Ishibashi, *Heterocycles* **1999**, *50*, 31.
99H(50)83	F. Kato, Y. Hiratsuka, T. Mitsui, T. Watanabe, K. Hiroi, *Heterocycles* **1999**, *50*, 83.
99H(50)341	J. Uenishi, T. Ueno, S. Hata, K. Nishiwaki, T. Tanaka, S. Wakabayashi, O. Yonemitsu, S. Oae *Heterocycles* **1999**, *50*, 341.
99H(50)353	R. Huisgen, F. Palacios-Gambra, K. Polborn, D. Boeckh, *Heterocycles* **1999**, *50*, 353.
99H(50)479	M. Suzuki, K. Tanikawa, R. Sakoda, *Heterocycles* **1999**, *50*, 479.
99H(50)791	C.B. Vicentini, M. Manfrini, M. Mazzanti, A.C. Veronese, *Heterocycles* **1999**, *50*, 791.
99H(50)843	S. Tandel, E.R. Biehl, *Heterocycles* **1999**, *50*, 843.
99H(50)867	H.R. Pfaendler, W. Jenni, *Heterocycles* **1999**, *50*, 867.
99H(50)929	F. Estour, S. Rézel, D. Fraisse, J. Métin, V. Gaumet, C. Lartigue, G. Miscoria, A. Gueiffier, Y. Blache, J.C. Teulade, O. Chavignon, *Heterocycles* **1999**, *50*, 929.
99H(51)103	H. Kohno, K. Yamada, *Heterocycles* **1999**, *51*, 103.
99H(51)119	T. Shinohara, A. Takeda, J. Toda, M. Kohno, T. Sano, *Heterocycles* **1999**, *51*, 119.
99H(51)137	J. Bernát, I. Chomca, P. Kristian, K. Pihlaja, K. D. Klika, J. Imrich, *Heterocycles* **1999**, *51*, 137.
99H(51)157	W.J. Smith III, J.S. Sawyer, *Heterocycles* **1999**, *51*, 157.
99H(51)593	R. Chênevert, G.M. Ziarani, M. Dasser, *Heterocycles* **1999**, *51*, 593.
99H(51)721	R. Crous, C. Dwyer, C.W. Holzapfel, *Heterocycles* **1999**, *51*, 721.

99H(51)1543 C. Mitsos, A. Zografos, O. Igglessi-Markopoulou, *Heterocycles* **1999**, *51*, 1543.
99H(51)1855 Y. Yamaguchi, K. Takagi, Y. Okamoto, K. Harada, Y. Kurasawa, *Heterocycles* **1999**, *51*, 1855.
99H(51)2093 D.-L. Wang, K. Imafuku, *Heterocycles* **1999**, *51*, 2093.
99H(51)2111 L. Skrzypek, *Heterocycles* **1999**, *51*, 2111.
99H(51)2065 O. Froelich, F. Cossart, M. Bonin, J.-C. Quirion, H.-P. Husson, *Heterocycles* **1999**, *51*, 2065.
99H(51)2171 P. Sanna, A. Carta, G. Paglietti, *Heterocycles* **1999**, *51*, 2171.
99H(51)2311 I. Moreno, R. SanMartin, I.Tellitu, E. Domínguez, *Heterocycles* **1999**, *51*, 2311.
99H(51)2377 K. Mohri, A. Kanie, Y. Horiguchi, K. Isobe, *Heterocycles* **1999**, *51*, 2377.
99H(51)2385 Y. Tagawa, M. Nomura, H. Yamashita, Y. Goto, M. Hamana, *Heterocycles* **1999**, *51*, 2385.
99H(51)2589 Y. Bessard, R. Crettaz, *Heterocycles* **1999**, *51*, 2589.
99H(51)2711 T. Kiguchi, M. Okazaki, T. Naito, *Heterocycles* **1999**, *51*, 2711.
99JA3539 Y. Kondo, M. Shilai, M. Uchiyama, T. Sakamoto, *J. Am. Chem. Soc.* **1999**, *121*, 3539.
99JA5075 R.A. Batey, D.B. MacKay, V. Santhakumar, *J. Am. Chem. Soc.* **1999**, *121*, 5075.
99JA6509 H.M.L. Davies, T. Hansen, D.W. Hopper, S.A. Panaro, *J. Am. Chem. Soc.* **1999**, *121*, 6509.
99JA6511 J.M. Axten, R. Ivy, L. Krim, J.D. Winkler, *J. Am. Chem. Soc.* **1999**, *121*, 6511.
99JA7722 D.E. Cane, S. Du, J.K. Robinson, Y. Hsiung, I.D. Spenser, *J. Am. Chem. Soc.* **1999**, *121*, 7722.
99JA10012 J.D. Ha, J.K. Cha, *J. Am. Chem. Soc.* **1999**, *121*, 10012.
99JA11093 T. Takahashi, F.-Y. Tsai, Y. Li, K. Nakajima, M. Kotora, *J. Am. Chem. Soc.* **1999**, *121*, 11093.
99JCR(S)6 G.H. Elgemeie, A.H.H. Elghandour, H.A. Ali, H.M. Abdel-Azzez, *J. Chem. Research (S)* **1999**, 6.
99JCR(S)136 J. Zhou, Y. Hu, H. Hu, *J. Chem. Research (S)* **1999**, 136.
99JCR(S)208 G.H. Elgemeie, N.H. Metwally, *J. Chem. Research (S)* **1999**, 208.
99JCR(S)520 R.P. Claridge, R.W. Millar, J.P.B. Sandall, C. Thompson, *J. Chem. Research (S)* **1999**, 520.
99JCR(S)536 I.S.A. Hafiz, E.S. Darwish, F.F. Mahmoud, *J. Chem. Research (S)* **1999**, 536.
99JCR(S)636 C.-m. Liu, Q.-h. Xu, Y.-m. Liang, Y.-x. Ma, *J. Chem. Research (S)* **1999**, 636.
99JCS(P1)171 M. Sakamoto, A. Kinbara, T. Yagi, M. Takahashi, K. Yamaguchi, T. Mino, S. Watanabe, T. Fujita, *J. Chem. Soc., Perkin Trans. 1* **1999**, 171.
99JCS(P1)179 C. Locher, N. Peerzada, *J. Chem. Soc., Perkin Trans. 1* **1999**, 179.
99JCS(P1)185 M. Noguchi, H. Okada, S. Nishimura, Y. Yamagata, S. Takamura, M. Tanaka, A. Kakehi, H. Yamamoto, *J. Chem. Soc., Perkin Trans 1* **1999**, 185.
99JCS(P1)803 R.D. Chambers, M. Parsons, G. Sandford, C.J. Skinner, M.J. Atherton, J.S. Moilliet, *J. Chem. Soc., Perkin Trans. 1* **1999**, 803.
99JCS(P1)1547 K. Kobayashi, R. Nakahashi, A. Shimizu, T. Kitamura, O. Morikawa, H. Konishi, *J. Chem. Soc., Perkin Trans 1* **1999**, 1547.
99JCS(P1)1571 B. Wang, X. Zhang, J. Li, X. Jiang, Y. Hu, H. Hu, *J. Chem. Soc., Perkin Trans. 1* **1999**, 1571.
99JCS(P1)2553 A. Mitchinson, A. Nadin, *J. Chem. Soc., Perkin Trans 1* **1999**, 2553.
99JCS(P1)2803 T. Konakahara, M. Hojahmat, S. Tamura, *J. Chem. Soc., Perkin Trans. 1* **1999**, 2803.
99JFC7 E. Leyva, E. Monreal, A. Hernández, *J. Fluorine Chem.* **1999**, *94*, 7.
99JHC141 P. Hradil, J. Hlavác, K. Lemr, *J. Heterocyclic Chem.* **1999**, *36*, 141.
99JHC371 A.R. Katritzky, X. Cui, Q. Long, *J. Heterocyclic Chem.* **1999**, *36*, 371.
99JHC493 A. Nagl, A. Hergold-Brundic, *J. Heterocyclic Chem.* **1999**, *36*, 493.
99JHC541 B. Schnell, *J. Heterocyclic Chem.* **1999**, *36*, 541.
99JHC869 Y.-H. Fan, J. Haseltine, *J. Heterocyclic Chem.* **1999**, *36*, 869.
99JOC453 T. Shiota, T. Yamamori, *J. Org. Chem.* **1999**, *64*, 453.
99JOC556 M.A. Brodney, A. Padwa, *J. Org. Chem.* **1999**, *64*, 556.
99JOC611 A. Yokoyama, T. Ohwada, K. Shudo, *J. Org. Chem.* **1999**, *64*, 611.
99JOC950 S. McN. Sieburth, C.-H. Lin, D. Rucando, *J. Org. Chem.* **1999**, *64*, 950.
99JOC954 S. McN. Sieburth, D. Rucando, C.-H. Lin, *J. Org. Chem.* **1999**, *64*, 954.
99JOC1007 A.G. de Viedma, V. Martinez-Barrasa, C. Burgos, M.L. Izquierdo, J. Alvarez-Builla, *J. Org. Chem.* **1999**, *64*, 1007.
99JOC1115 L. Carrillo, D. Badía, E. Domínguez, E. Anakabe, I. Osante, I. Tellitu, J.L. Vicario, *J. Org. Chem.* **1999**, *64*, 1115.
99JOC1407 A.M. Qandil, D. Ghosh, D.E. Nichols, *J. Org. Chem.* **1999**, *64*, 1407.
99JOC2003 T. Naito, K. Nakagawa, T. Nakamura, A. Kasei, I. Ninomiya, T. Kiguchi, *J. Org. Chem.* **1999**, *64*, 2003.
99JOC2038 A. Padwa, T.M. Heidelbaugh, J.T. Kuethe, *J. Org. Chem.* **1999**, *64*, 2038.
99JOC2184 D.L. Comins, A.H. Libby, R.S. Al-awar, C.J. Foti, *J. Org. Chem.* **1999**, *64*, 2184.
99JOC3178 R. Chênevert, M.-P. Morin, *J. Org. Chem.* **1999**, *64*, 3178.
99JOC3595 A. Padwa, M.A. Brodney, B. Liu, K. Satake, T. Wu, *J. Org. Chem.* **1999**, *64*, 3595.
99JOC3608 C. Wentrup, V.V.R. Rao, W. Frank, B.E. Fulloon, D.W.J. Moloney, T. Mosandl, *J. Org. Chem.* **1999**, *64*,

3608.
99JOC3736 J. Barluenga, F. Aznar, C. Ribas, C. Valdés, *J. Org. Chem.* **1999**, *64*, 3736.
99JOC4220 S. Kobayashi, K.-i. Kusakabe, S. Komiyama, H. Ishitani, *J. Org. Chem.* **1999**, *64*, 4220.
99JOC4512 P. Pollet, A. Turck, N. Plé, G. Quéguiner, *J. Org. Chem.* **1999**, *64*, 4512.
99JOC4610 J.L. Vicario, D. Badía, E. Domínguez, L. Carrillo, *J. Org. Chem.* **1999**, *64*, 4610.
99JOC4914 N. Toyooka, Y. Yoshida, Y. Yotsui, T. Momose, *J. Org. Chem.* **1999**, *64*, 4914.
99JOC5725 J.-J. Wang, W.-P. Hu, *J. Org. Chem.* **1999**, *64*, 5725.
99JOC5979 A.R. Hergueta, H.W. Moore, *J. Org. Chem.* **1999**, *64*, 5979.
99JOC6041 P.A. Grieco, M.D. Kaufman, *J. Org. Chem.* **1999**, *64*, 6041.
99JOC6066 A.R. Katritzky, J. Yao, B. Yang, *J. Org. Chem.* **1999**, *64*, 6066.
99JOC6076 A.R. Katritzky, A. Denisenko, M. Arend, *J. Org. Chem.* **1999**, *64*, 6076.
99JOC6239 F. Palacios, E. Herran, G. Rubiales, *J. Org. Chem.* **1999**, *64*, 6239.
99JOC6499 J.D. Tovar, T.M. Swager, *J. Org. Chem.* **1999**, *64*, 6499.
99JOC6702 D.A. Klumpp, M. Garza, A. Jones, S. Mendoza, *J. Org. Chem.* **1999**, *64*, 6702.
99JOC6724 E. Vedejs, P. Trapencieris, E. Suna, *J. Org. Chem.* **1999**, *64*, 6724.
99JOC6881 P. Wipf, C.R. Hopkins, *J. Org. Chem.* **1999**, *64*, 6881.
99JOC6911 N.S. Mani, P. Chen, T.K. Jones, *J. Org. Chem.* **1999**, *64*, 6911.
99JOC7618 A.R. Katritzky, G. Qiu, B. Yang, H.-Y. He, *J. Org. Chem.* **1999**, *64*, 7618.
99JOC7846 C. Zorn, B. Anichini, A. Goti, A. Brandi, S.I. Kozhushkov, A. de Meijere, L. Citti, *J. Org. Chem.* **1999**, *64*, 7846.
99JOC8402 M. David, H. Dhimane, C. Vanucci-Bacqué, G. Lhommet, *J. Org. Chem.* **1999**, *64*, 8402.
99JOC8627 F.A. Davis, Y.W. Andemichael, *J. Org. Chem.* **1999**, *64*, 8627.
99JOC8648 A. Padwa, S.M. Sheehan, C.S. Straub, *J. Org. Chem.* **1999**, *64*, 8648.
99JOC8980 X.-Q. Zhu, Y.-C. Liu, J.-P. Cheng, *J. Org. Chem.* **1999**, *64*, 8980.
99JOC9493 A. Alberola, L.A. Calvo, A.G. Ortega, M.C.S. Ruiz, P. Yustos, *J. Org. Chem.* **1999**, *64*, 9493.
99JPR147 A.W. Erian, *J. Prakt. Chem.* **1999**, *341*, 147.
99NJC641 C. Boix, J.M. de la Fuente, M. Poliakoff, *New J. Chem.* **1999**, *23*, 641.
99NPR367 K.W. Bentley, *Natural Products Reports* **1999**, *16*, 367.
99OL35 B.H. Yang, S.L. Buchwald, *Organic Lett.* **1999**, *1*, 35.
99OL175 Y. Matsumura, Y. Kanda, K. Shirai, O. Onomura, T. Maki, *Organic Lett.* **1999**, *1*, 175.
99OL189 R.S. Varma, K.P. Naicker, *Organic Lett.* **1999**, *1*, 189.
99OL509 R.P. Hsung, L.-L. Wei, H.M. Sklenicka, C.J. Douglas, M.J. McLaughlin, J.A. Mulder, L.J. Yao, *Organic Lett.* **1999**, *1*, 509.
99OL553 K.R. Roesch, R.C. Larock, *Organic Lett.* **1999**, *1*, 553.
99OL641 D.M. Bennet, I. Okamoto, R.L. Danheiser, *Organic Lett.* **1999**, *1*, 641.
99OL657 D.L. Comins, Y.-m. Zhang, S.P. Joseph, *Organic Lett.* **1999**, *1*, 657.
99OL717 F.P.J.T. Rutjes, K.C.M.F. Tjen, L.B. Wolf, W.F.J. Karstens, H.E. Schoemaker, H. Hiemstra, *Organic Lett.* **1999**, *1*, 717.
99OL745 G.G. Wu, Y. Wong, M. Poirier, *Organic Lett.* **1999**, *1*, 745.
99OL767 M.-J. Wu, C.-F. Lin, S.-H. Chen, *Organic Lett.* **1999**, *1*, 767.
99OL799 J.M. Flaniken, C.J. Collins, M. Lanz, B. Singaram, *Organic Lett.* **1999**, *1*, 799.
99OL823 H. Steinhagen, E.J. Corey, *Organic Lett.* **1999**, *1*, 823.
99OL841 J. Mao, D.C. Baker, *Organic Lett.* **1999**, *1*, 841.
99OL877 S. Chandrasekhar, P.K. Mohanty, K. Harikishan, P.K. Sasmal, *Organic Lett.* **1999**, *1*, 877.
99OL985 T. Nakanishi, M. Suzuki, *Organic Lett.* **1999**, *1*, 985.
99OL1027 U.S. Schubert, C. Eschbaumer, *Organic Lett.* **1999**, *1*, 1027.
99OL1031 J.T. Kuethe, D.L. Comins, *Organic Lett.* **1999**, *1*, 1031.
99OL1775 S. McN. Sieburth, K.F. McGee, Jr., *Organic Lett.* **1999**, *1*, 1775.
99OL1941 D.L. Comins, A.B. Fulp, *Organic Lett.* **1999**, *1*, 1941.
99OL1953 A.L. Zografos, C.A. Mitsos, O. Igglessi-Markopoulou, *Organic Lett.* **1999**, *1*, 1953.
99OL1957 F. Song, V.R. St. Hilaire, E.H. White, *Organic Lett.* **1999**, *1*, 1957.
99OL1977 M. Suginome, T. Fukuda, Y. Ito, *Organic Lett.* **1999**, *1*, 1977.
99OL1997 H. Ratni, E.P. Kündig, *Organic Lett.* **1999**, *1*, 1997.
99OL2017 M.J. Sung, H.I. Lee, Y. Chong, J.K. Cha, *Organic Lett.* **1999**, *1*, 2017.
99S51 X.-c. Zhang, W.-y. Huang, *Synthesis* **1999**, 51.
99S166 J. Zhou, Y. Hu, H. Hu, *Synthesis* **1999**, 166.
99S258 P. Forns, M.M. Fernández, A. Diez, M. Rubiralta, M.P. Cherrier, M. Bonin, J.-C. Quirion, *Synthesis* **1999**, 258.

99S306	A. Numata, Y. Kondo, T. Sakamoto, *Synthesis* **1999**, 306.
99S467	V.J. Ram, A. Goel, *Synthesis* **1999**, 467.
99S683	O. Henze, U. Lehmann, A.D. Schlüter, *Synthesis* **1999**, 683.
99S754	P. Gros, Y. Fort, *Synthesis* **1999**, 754.
99S779	U.S. Schubert, C. Eschbaumer, G. Hochwimmer, *Synthesis* **1999**, 779.
99S815	I. Sasaki, J.C. Daran, G.G.A. Balavoine, *Synthesis* **1999**, 815.
99S947	T. Mimura, N. Kato, T. Sugaya, M. Ikuta, S. Kato, Y. Kuge, S. Tomioka, M. Kasai, *Synthesis* **1999**, 947.
99S1145	H. Sashida, A. Kawamukai, *Synthesis* **1999**, 1145.
99S1216	E. Gräf, R. Troschütz, *Synthesis* **1999**, 1216.
99S1309	S.W. Stork, M.W. Makinen, *Synthesis* **1999**, 1309.
99S1335	J.I. Úbeda, M. Villacampa, C. Avendaño, *Synthesis* **1999**, 1335.
99S1814	Y. Takeuchi, M. Hattori, H. Abe, T. Harayama, *Synthesis* **1999**, 1814.
99S1889	S.D. Koulocheri, S.A. Haroutounian, *Synthesis* **1999**, 1889.
99S2071	M. Haase, W. Günther, H. Görls, E. Anders, *Synthesis* **1999**, 2071.
99S2114	A.R. Katritzky, A.A.A. Abdel-Fattah, D.O. Tymoshenko, S.A. Essawy, *Synthesis* **1999**, 2114.
99SC103	M. Wasgindt, E. Klemm, *Synth. Commun.* **1999**, *29*, 103.
99SC645	F.J. Urban, R. Breitenbach, *Synth. Commun.* **1999**, *29*, 645.
99SC1617	R. Kucznierz, J. Dickhaut, H. Leinert, W. von der Saal, *Synth. Commun.* **1999**, *29*, 1617.
99SC1747	M.F.A. Adamo, V.K. Aggarwal, M.A. Sage, *Synth. Commun.* **1999**, *29*, 1747.
99SC2477	H.-r. Ma, X.-h. Wang, M.-z. Deng, *Synth. Commun.* **1999**, *29*, 2477.
99SC3341	C. Janiak, S. Deblon, H.-P. Wu, *Synth. Commun.* **1999**, *29*, 3341.
99SC4007	N.V. Eldho, M. Saminathan, D. Ramaiah, *Synth. Commun.* **1999**, *29*, 4007.
99SC4051	W.R. Bowman, C.F.Bridge, *Synth. Commun.* **1999**, *29*, 4051.
99SC4403	G. Sabitha, R.S. Babu, B.V.S. Reddy, J.S. Yadav, *Synth. Commun.* **1999**, *29*, 4403.
99SC4341	A. Kossanyi, B. Mestre, M. Perrée-Fauvet, *Synth. Commun.* **1999**, *29*, 4341.
99SL37	N. Yamazaki, T. Ito, C. Kibayashi, *Synlett* **1999**, 37.
99SL45	O. Lohse, P. Thevenin, E. Waldvogel, *Synlett* **1999**, 45.
99SL93	C.S. Penkett, I.D. Simpson, *Synlett* **1999**, 93.
99SL207	M. Ohno, H. Sato, S. Eguchi, *Synlett* **1999**, 207.
99SL324	S.-K. Kang, T.-G. Baik, A.N. Kulak, *Synlett* **1999**, 324.
99SL342	U.S. Schubert, C. Eschbaumer, C.H. Weidl, *Synlett* **1999**, 342.
99SL401	S. Cacchi, G. Fabrizi, F. Marinelli, *Synlett* **1999**, 401.
99SL405	S. Brocherieux-Lanoy, H. Dhimane, C. Vanucci-Bacqué, G. Lhommet, *Synlett* **1999**, 405.
99SL441	A.J. Clark, R.P. Filik, J.L. Peacock, G.H. Thomas, *Synlett* **1999**, 441.
99SL626	H. Ratni, B. Crousse, E.P. Kündig, *Synlett* **1999**, 626.
99SL641	S. Issmaili, G.Boyer, J.-P. Galy, *Synlett* **1999**, 641.
99SL804	J.C. Namyslo, D.E. Kaufmann, *Synlett* **1999**, 804.
99SL1067	G.A. Olah, T. Mathew, M. Farnia, G.K.S. Prakash, *Synlett* **1999**, 1067.
99SL1088	E.M. Brun, S. Gil, R. Mestres, M. Parra, *Synlett* **1999**, 1088.
99SL1127	N. Diedrichs, B. Westermann, *Synlett* **1999**, 1127.
99SL1154	T. Itoh, K. Nagata, M. Miyazaki, A. Ohsawa, *Synlett* **1999**, 1154.
99SL1292	N.J. Goldspink, N.S. Simpkins, M. Beckmann, *Synlett* **1999**, 1292.
99SL1379	M.C. Elliott, A.E. Monk, E. Kruiswijk, D.E. Hibbs, R.L. Jenkins, D.V. Jones, *Synlett* **1999**, 1379.
99SL1383	E. Le Gall, R. Malassene, L. Toupet, J.-P. Hurvois, C. Moinet, *Synlett* **1999**, 1383.
99SL1559	L. Schio, G. Lemoine, M. Klich, *Synlett* **1999**, 1559.
99SL1747	D. Heber, E.V. Stoyanov, *Synlett* **1999**, 1747.
99T393	Y. Bessard, J.P. Roduit, *Tetrahedron* **1999**, *55*, 393.
99T1043	C. Herdeis, T. Schiffer, *Tetrahedron* **1999**, *55*, 1043.
99T1111	A.A. Aly, N.K. Mohamed, A.A. Hassan, A.-F. E. Mourad, *Tetrahedron* **1999**, *55*, 1111.
99T1491	P. Huszthy, E. Samu, B.Vermes G. Mezey-Vándor, M. Nógrádi, J.S. Bradshaw, R.M. Izatt, *Tetrahedron* **1999**, *55*, 1491.
99T2317	J.A. Vega, J.J. Vaquero, J. Alvarez-Builla, J. Ezquerra, C. Hamdouchi, *Tetrahedron* **1999**, *55*, 2317.
99T2911	M. Kirihara, T. Nishio, S. Yokoyama, H. Kakuda, T. Momose, *Tetrahedron* **1999**, *55*, 2911.
99T4481	G.J. Meuzelaar, L. Maat, R.A. Sheldon, *Tetrahedron* **1999**, *55*, 4481.
99T5047	G.R. Pabst, O.C. Pfüller, J. Sauer, *Tetrahedron* **1999**, *55*, 5047.
99T5195	C. González, E. Guitián, L. Castedo, *Tetrahedron* **1999**, *55*, 5195.
99T5947	F. Palacios, A.M. O. de Retana, J. Oyarzabal, *Tetrahedron* **1999**, *55*, 5947.
99T6183	C.S. Penkett, I.D. Simpson, *Tetrahedron* **1999**, *55*, 6183.

99T7279 A. Szöllösy, T. Tischer, I. Kádas, L. Töke, G. Tóth, *Tetrahedron* **1999**, *55*, 7279.
99T7601 R. Badorrey, C. Cativiela, M.D. Díaz-de-Villegas, J.A. Gálvez, *Tetrahedron* **1999**, *55*, 7601.
99T8179 R. Stragies, S. Blechert, *Tetrahedron* **1999**, *55*, 8179.
99T8931 A.I. Meyers, C.J. Andres, J.E. Resek, C.C. Woodall, M.A. McLaughlin, P.H. Lee, D.A. Price, *Tetrahedron* **1999**, *55*, 8931.
99T9185 K.-Q. Ling, J.-H. Ye, X.-Y. Chen, D.-J. Ma, J.-H. Xu, *Tetrahedron* **1999**, *55*, 9185.
99T10173 M. A. Estiarte, M.V.N. de Souza, X. del Rio, R.H. Dodd, M. Rubiralta, A. Diez, *Tetrahedron* **1999**, *55*, 10173.
99T10243 R.P. Claridge, R.W. Millar, J.P.B. Sandall, C. Thompson, *Tetrahedron* **1999**, *55*, 10243.
99T12557 G. Dyker, B. Hölzer, *Tetrahedron* **1999**, *55*, 12557.
99T12757 W.W.K.R. Mederski, M. Lefort, M. Germann, D. Kux, *Tetrahedron* **1999**, *55*, 12757.
99T12829 E. Brenner, R. Schneider, Y. Fort, *Tetrahedron* **1999**, *55*, 12829.
99T13193 M.-J. Wu, L.-J. Chang, L.-M. Wei, C.-F. Lin, *Tetrahedron* **1999**, *55*, 13193.
99T13233 A. Arcadi, F. Marinelli, E. Rossi, *Tetrahedron* **1999**, *55*, 13233.
99T14479 G. Radivoy, F. Alonso, M. Yus, *Tetrahedron* **1999**, *55*, 14479.
99TA221 D. Taniyama, M. Hasegawa, K. Tomioka, *Tetrahedron: Asymm.* **1999**, *10*, 221.
99TA255 A.R. Katritzky, J. Cobo-Domingo, B. Yang, P.J. Steel, *Tetrahedron: Asymm.* **1999**, *10*, 255.
99TA657 J. Kang, C.W. Lee, G.J. Lim, B.T. Cho, *Tetrahedron: Asymm.* **1999**, *10*, 657.
99TA3117 R. Chênevert, G.M. Ziarani, M.P. Morin, M. Dasser, *Tetrahedron: Asymm.* **1999**, *10*, 3117.
99TA3371 M. Ziólkowski, Z. Czarnocki, A. Leniewski, J.K. Maurin, *Tetrahedron: Asymm.* **1999**, *10*, 3371.
99TA3649 L.-X. Liao, Z.-M. Wang, H.-X. Zhang, W.-S. Zhou, *Tetrahedron: Asymm.* **1999**, *10*, 3649.
99TL21 R.S. Varma, D. Kumar, *Tetrahedron Lett.* **1999**, *40*, 21.
99TL217 D.L. Comins, G.M. Green, *Tetrahedron Lett.* **1999**, *40*, 217.
99TL367 P. Bach, A. Lohse, M. Bols, *Tetrahedron Lett.* **1999**, *40*, 367.
99TL739 N. Yamazaki, T. Ito, C. Kibayashi, *Tetrahedron Lett.* **1999**, *40*, 739.
99TL1125 J. Cossy, L. Tresnard, D.G. Pardo, *Tetrahedron Lett.* **1999**, *40*, 1125.
99TL1145 K.H. Park, H.S. Joo, S.W. Kim, M.S. Park, P.S. Shin, *Tetrahedron Lett.* **1999**, *40*, 1145.
99TL1149 H. Ishibashi, M. Inomata, M. Ohba, M. Ikeda, *Tetrahedron Lett.* **1999**, *40*, 1149.
99TL1397 C.-H. Tan, T. Stork, N. Feeder, A.B. Holmes, *Tetrahedron Lett.* **1999**, *40*, 1397.
99TL1499 C.S. Cho, B.H. Oh, S.C. Shim, *Tetrahedron Lett.* **1999**, *40*, 1499.
99TL1515 I. Ryu, S.-I. Ogura, S. Minakata, M. Komatsu, *Tetrahedron Lett.* **1999**, *40*, 1515.
99TL1877 S.C. Schürer, S. Blechert, *Tetrahedron Lett.* **1999**, *40*, 1877.
99TL2079 C.J. Lovely, H. Mahmud, *Tetrahedron Lett.* **1999**, 40, 2079.
99TL2125 X. Hoang-Cong, B. Quiclet-Sire, S.Z. Zard, *Tetrahedron Lett.* **1999**, *40*, 2125.
99TL2833 A. Napolitano, A. Pezzella, G. Prota, *Tetrahedron Lett.* **1999**, *40*, 2833.
99TL2891 R.A. Pilli, C. de F. Alves, M.A. Bockelmann, Y.P. Mascarenhas, J.G. Nery, I. Vencato, *Tetrahedron Lett.* **1999**, *40*, 2891.
99TL3077 D. Gonzalez, T. Martinot, T. Hudlicky, *Tetrahedron Lett.* **1999**, *40*, 3077.
99TL3081 H. Akgün, T. Hudlicky, *Tetrahedron Lett.* **1999**, *40*, 3081.
99TL3137 A.G. Griesbeck, H. Heckroth, H. Schmickler, *Tetrahedron Lett.* **1999**, *40*, 3137.
99TL3207 K.-Y. Ko, J.-Y. Kim, *Tetrahedron Lett.* **1999**, *40*, 3207.
99TL3339 B.B. Snider, Y. Ahn, B.M. Foxman, *Tetrahedron Lett.* **1999**, *40*, 3339.
99TL3527 S. McN. Sieburth, F. Zhang, *Tetrahedron Lett.* **1999**, *40*, 3527.
99TL3699 A. Zaparucha, M. Danjoux, A. Chiaroni, J. Royer, H.-P. Husson, *Tetrahedron Lett.* **1999**, *40*, 3699.
99TL3717 S. Couve-Bonnaire, J.-F. Carpentier, Y. Castanet, A. Mortreux, *Tetrahedron Lett.* **1999**, *40*, 3717.
99TL3719 D. Najiba, J.-F. Carpentier, Y. Castanet, C. Biot, J. Brocard, A. Mortreux, *Tetrahedron Lett.* **1999**, *40*, 3719.
99TL3731 F. Billon-Souquet, T. Martens, J. Royer, *Tetrahedron Lett.* **1999**, *40*, 3731.
99TL3961 M.-L. Bennasar, E. Zulaica, C. Juan, L. Llauger, J. Bosch, *Tetrahedron Lett.* **1999**, *40*, 3961.
99TL4007 S. McN. Sieburth, K.F. McGee, Jr., T.H. Al-Tel, *Tetrahedron Lett.* **1999**, *40*, 4007.
99TL4069 I. Collins, J.L. Castro, *Tetrahedron Lett.* **1999**, *40*, 4069.
99TL4097 M. del Mar Blanco, J.A. de la Fuente, C. Avendaño, J.C. Menéndez, *Tetrahedron Lett.* **1999**, *40*, 4097.
99TL4243 N.E. Leadbeater, S.M. Resouly, *Tetrahedron Lett.* **1999**, *40*, 4243.
99TL4255 J.J.H. Diederen, R.W. Sinkeldam, H.-W. Frühauf, H. Hiemstra, K. Vrieze, *Tetrahedron Lett.* **1999**, *40*, 4255.
99TL4331 C. Guillou, F. Bintein, J.-P. Biron, C. Thal, *Tetrahedron Lett.* **1999**, *40*, 4331.
99TL4339 F. Trécourt, G. Breton, V. Bonnet, F. Mongin, F. Marsais, G. Quéguiner, *Tetrahedron Lett.* **1999**, *40*, 4339.

99TL4969 C.C. Silveira, C.R. Bernardi, A.L. Braga, T.S. Kaufman, *Tetrahedron Lett.* **1999**, *40*, 4969.
99TL5331 M. Shirai, S. Okamoto, F. Sato, *Tetrahedron Lett.* **1999**, *40*, 5331.
99TL5413 N.R. Champness, A.N. Khlobystov, A.G. Majuga, M. Schröder, N.V. Zyk, *Tetrahedron Lett.* **1999**, *40*, 5413.
99TL5425 E. Baciocchi, A. Lapi, *Tetrahedron Lett.* **1999**, *40*, 5425.
99TL5483 F. Mongin, F. Trécourt, G. Quéguiner, *Tetrahedron Lett.* **1999**, *40*, 5483.
99TL5495 T. Giard, M.-C. Lasne, J.-C. Plaquevent, *Tetrahedron Lett.* **1999**, *40*, 5495.
99TL5541 M. Ochiai, D. Kajishima, T. Sueda, *Tetrahedron Lett.* **1999**, *40*, 5541.
99TL5565 E. Takashiro, Y. Nakamura, K. Fujimoto, *Tetrahedron Lett.* **1999**, *40*, 5565.
99TL5581 E. Jo, Y. Na, S. Chang, *Tetrahedron Lett.* **1999**, *40*, 5581.
99TL5621 T. Ali, K.K. Chauhan, C.G. Frost, *Tetrahedron Lett.* **1999**, *40*, 5621.
99TL5987 P. Wessig, *Tetrahedron Lett.* **1999**, *40*, 5987.
99TL6241 F. Rezgui, P. Mangeney, A. Alexakis, *Tetrahedron Lett.* **1999**, *40*, 6241.
99TL6661 C.-Y. Yu, D.L. Taylor, O. Meth-Cohn, *Tetrahedron Lett.* **1999**, *40*, 6661.
99TL6657 F.M. Cordero, I. Barile, F. De Sarlo, A. Brandi, *Tetrahedron Lett.* **1999**, *40*, 6657.
99TL6869 S.D. Koulocheri, S.A. Haroutounian, *Tetrahedron Lett.* **1999**, *40*, 6869.
99TL6999 Z.-X. Guo, A.N. Cammidge, A. McKillop, D.C. Horwell, *Tetrahedron Lett.* **1999**, *40*, 6999.
99TL7003 M. Rudas, M. Nyerges, L. Töke, B. Pete, P.W. Groundwater, *Tetrahedron Lett.* **1999**, *40*, 7003.
99TL7211 I.A. Motorina, D.S. Grierson, *Tetrahedron Lett.* **1999**, *40*, 7211.
99TL7215 I.A. Motorina, D.S. Grierson, *Tetrahedron Lett.* **1999**, *40*, 7215.
99TL7477 O. Sugimoto, M. Mori, K.-i. Tanji, *Tetrahedron Lett.* **1999**, *40*, 7477.
99TL7831 T. Akiyama, J. Takaya, H. Kagoshima, *Tetrahedron Lett.* **1999**, *40*, 7831.
99TL7935 J. Hiebl, H. Kollmann, S.H. Levinson, P. Offen, S.B. Shetzline, R. Badlani, *Tetrahedron Lett.* **1999**, *40*, 7935.
99TL8193 P.E. Maligres, M.S. Waters, F. Fleitz, D. Askin, *Tetrahedron Lett.* **1999**, *40*, 8193.
99TL8269 G. Chelucci, N. Culeddu, A. Saba, R. Valenti, *Tetrahedron Lett.* **1999**, *40*, 8269.
99TL8587 Y. Kita, H. Maekawa, Y. Yamasaki, I. Nishiguchi, *Tetrahedron Lett.* **1999**, *40*, 8587.
99TL8759 F. You, R.J. Twieg, *Tetrahedron Lett.* **1999**, *40*, 8759.
99TL8823 Y. Kato, T. Mase, *Tetrahedron Lett.* **1999**, *40*, 8823.
99ZN(B)214 H. Möhrle, J. Mehrens, *Z. Naturforsch.* **1999**, *54b*, 214.
99ZN(B)225 H. Möhrle, R. Niessen, *Z. Naturforsch.* **1999**, *54b*, 225.
99ZN(B)532 H. Möhrle, R. Niessen, *Z. Naturforsch.* **1999**, *54b*, 532.
99ZN(B)559 M. Schmittel, A. Ganz, W.A. Schenk, M. Hagel, *Z. Naturforsch.* **1999**, *54b*, 559.
99ZN(B)913 H. Möhrle, R. Niessen, *Z. Naturforsch.* **1999**, *54b*, 913.
99ZN(B)1205 M.M. Mashaly, M. Hammouda, *Z. Naturforsch.* **1999**, *54b*, 1205.
99ZN(B)1337 S. Thamaraiselvi, P.S. Mohan, *Z. Naturforsch.* **1999**, *54b*, 1337.

Chapter 6.2

Six-Membered Ring Systems: Diazines and Benzo Derivatives

Brian R. Lahue
Boston University, Boston, MA, USA
blahue@bu.edu

John K. Snyder
Boston University, Boston, MA, USA
jsnyder@chem.bu.edu

6.2.1 INTRODUCTION

In recent years, diazines and their derivatives have become extremely important to the field of chemistry as well as to the general population in terms of their invaluable biological activities. In 1999 alone, there were hundreds of publications on their syntheses as well as important reactions of these heterocycles. This review is comprised of the most significant of these reports.

6.2.2 PYRIMIDINES

6.2.2.1 Preparations of Pyrimidines

The most common method for synthesizing the fully aromatized pyrimidine skeleton is the condensation of an amidine-containing substrate with an α,β-unsaturated carbonyl compound. For example, the aza-Wittig reaction of **1** with a variety of aldehydes **2** was reported by Rossi and co-workers to produce pyrimidines **3** <99SL1265>.

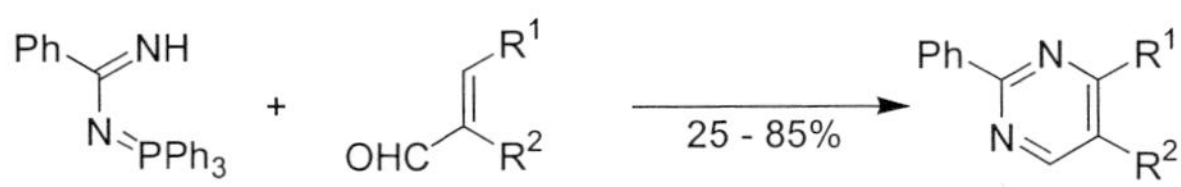

1 **2** **3**

Similar transformations using α,β-unsaturated ketones activated with a trifluoromethyl group also proved to be highly efficient (*e.g.*, **4** → **5** <99SL756>, **6** → **7** <99TL2541>) for the preparation of medicinally and agriculturally important trifluoromethyl-containing pyrimidines.

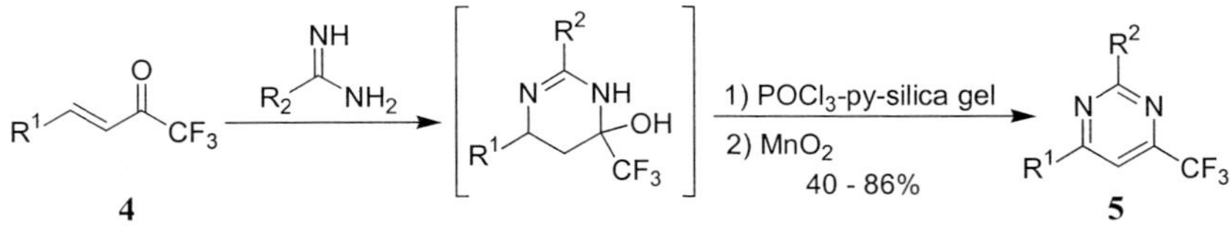

4 5

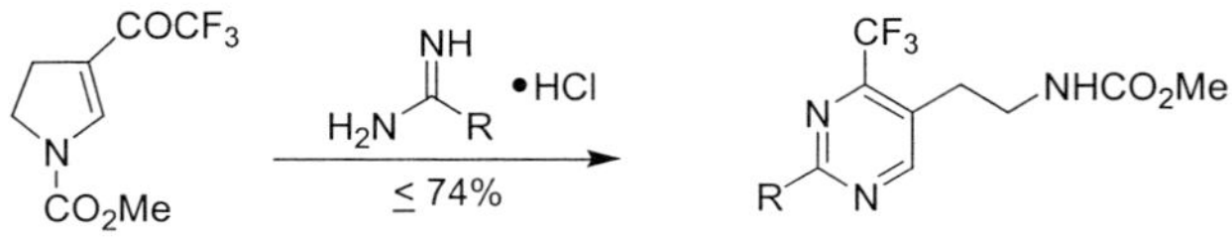

6 7

Aminopyrimidines were prepared in analogous fashion beginning with guanidine instead of amidines. For example, the reaction of **8** with guanidinium nitrate produced aminopyrimidine **9** <99H2445>, while a similar condensation of **10** with guanidine gave **11** <99JCR(S)88>.

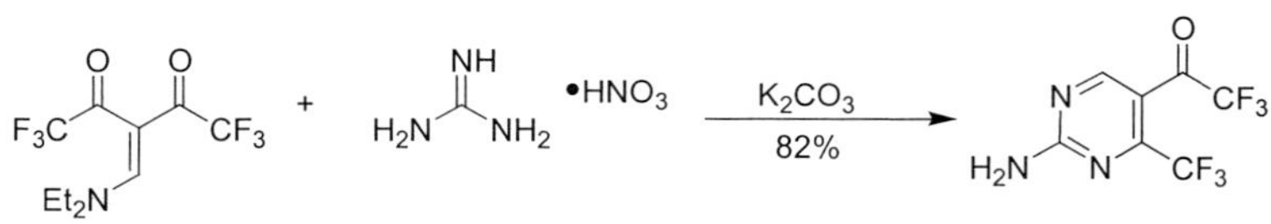

8 9

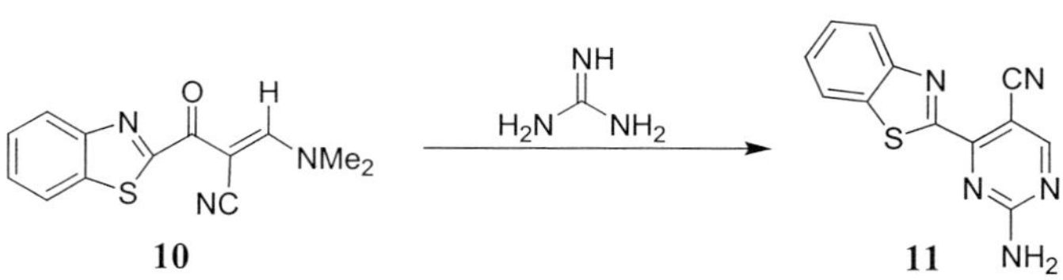

10 11

Nucleophilic attack on a nitrile rather than a carbonyl has also provided aminopyrimidines as reported by Hassanien and co-workers in their efforts to discover new sulfonamide drugs <99JCR(S)8>. The reactions of sulfonamides **12** with a variety of nitrogen-based nucleophiles produced aminopyrimidines **13**.

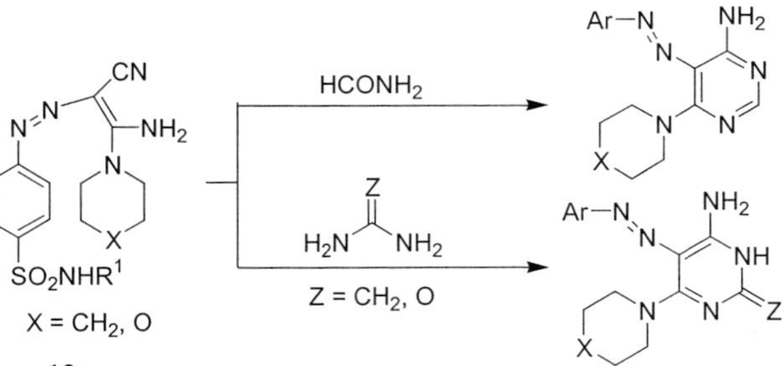

12 **13**

A variety of 4-alkoxypyrimidines **16** were synthesized by the condensation and cyclization of numerous esters **14** with 2 equivalents of nitriles **15** <99T4825>. This methodology is an extension of other work by Fernandez and co-workers with ketones and nitriles <92JOC1627>.

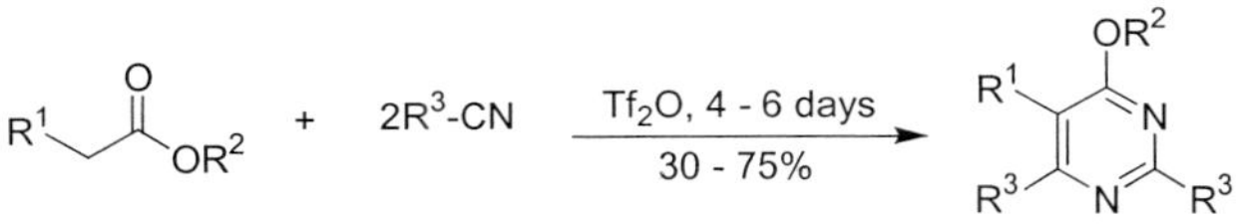

14 **15** **16**

In an effort to explore the chemistry of pyrrolodiazines and their quaternized salts (see Section 6.2.2.2), Alvarez-Builla and co-workers prepared a series of pyrrolo[1,2-*c*]pyrimidines *via* methodology developed in their laboratory <99JOC7788>. Cyclocondensation of tosylmethyl isocyanide with substituted pyrrole-2-carboxaldehydes **17** produced pyrimidine derivatives **18** after removal of the tosyl group. The key to this procedure was the use of tosylmethyl isocyanide, which provided a relatively easily removed tosyl group in comparison to the more problematic decarboxylation of a carboxylic acid functionality.

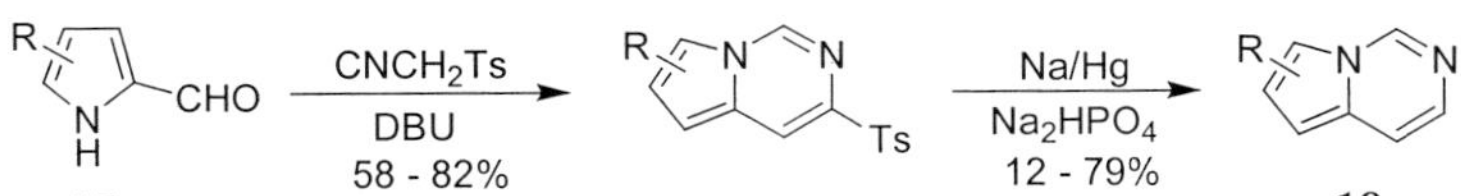

17 **18**

The reaction of acetophenone (**19**) with formamide is known to produce **21** after reduction of the imine and hydrolysis of the formate group. This is accompanied by a trace of pyrimidine **22** in the reaction mixture. Lejon and co-workers have optimized the production of **22** by adding CuCl, which is thought to oxidize the formate ion produced from the reaction of water with formamide, thereby minimizing the reduction of **20** and allowing the cyclocondensation with a second equivalent of formamide <99H611>.

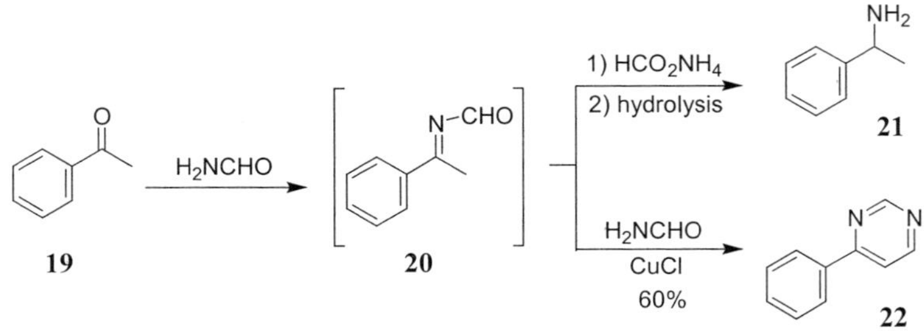

19 **20** **21** **22**

6.2.2.2 Reactions of Pyrimidines

Nucleophilic substitution reactions (S_NAr) are among the most common transformations of pyrimidines. Direct displacements of a variety of leaving groups have been reported, such as the reactions of **23** with heteroaromatic nucleophiles which produced 2-substituted pyrimidines **24** <99JCS(P1)1325>.

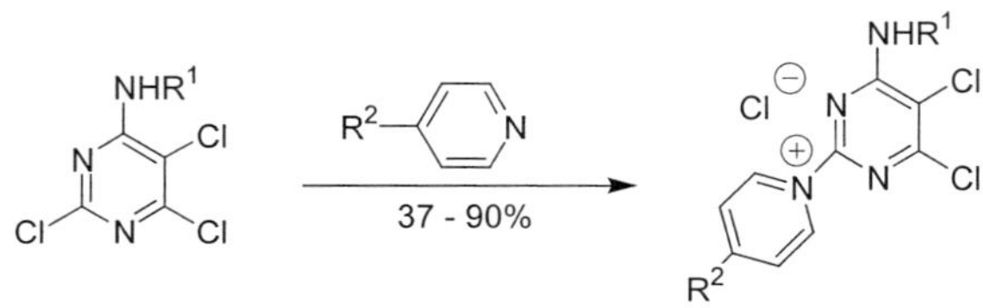

23 **24**

This reaction exemplified the difference between the reactivity of polychloropyrimidines with heteroaromatic and that with aliphatic nucleophiles, which predominantly yield 4-substituted pyrimidines. For example, a series of trichloropyrimidines **25** reacted with various Grignard, lithium, sodium, and thiolate reagents (R_2M) to produce mainly **26**, along with occasional, minor amounts of the competitive products **27** <99SC1503>.

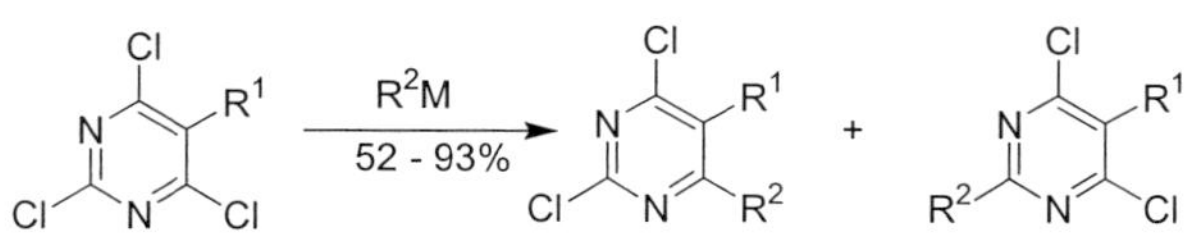

25 **26** **27**

In the same vein, the selective hydrolysis of the 4-fluoro substituent in trifluoropyrimidine **28** was realized by the reaction with [Ni(cod)$_2$] in the presence of triethylphosphine <99AC(E)3326>. Hydrolysis of the isolable metal-bound pyrimidine resulted in the production of **29**.

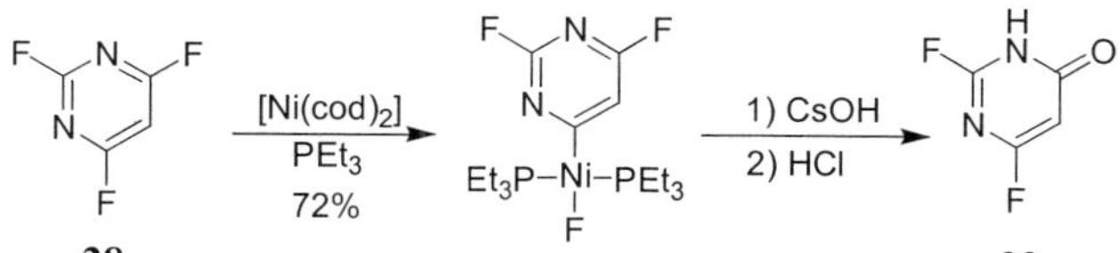

28 29

Nagamatsu and co-workers also reported the reaction of **30** with hydrazine to produce **31**, a key intermediate in the synthesis of potential xanthine oxidase inhibitors **32** <99CC1461>. The regioselectivity of this hydrazine displacement thus paralleled that observed with carbanionic nucleophiles.

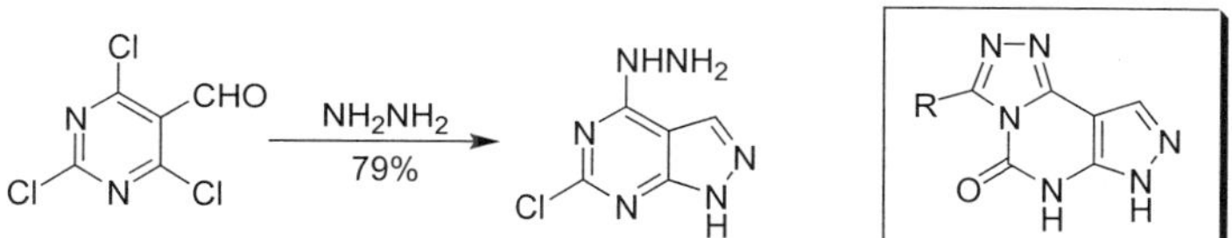

30 31 32

On a similar note, Chorvat and co-workers reported progress toward understanding the pharmacokinetic properties of various pyrimidine derivatives <99JMC833>. They noted that while the reaction sequence **33** → **34** → **35** could theoretically produce the same product in the reverse order of addition, the aryl amine added prior to the alkyl amine, the former sequence was necessary to avoid hydrolysis of the remaining chloro substituent after the first step. This could be attributed to the greater electron donation from the alkyl amines, which inhibited hydrolysis in comparison to aryl amines.

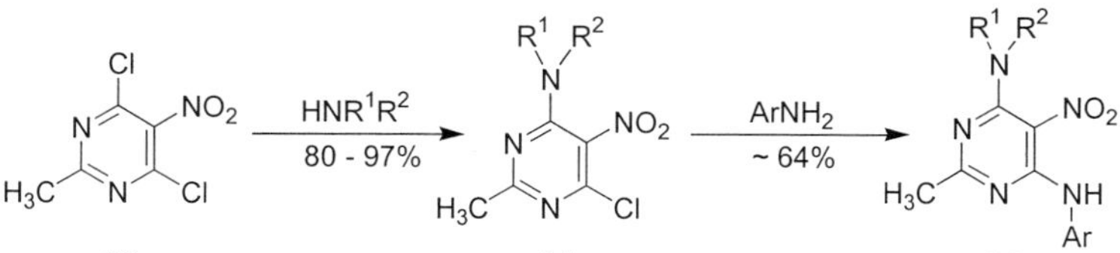

33 34 35

From what was planned as a straightforward displacement of the chloride atom in **36** with hydrazine followed by a condensation with 2-tetralone and Fischer indolization to produce **39**, dihydrazone **38** was isolated as an intermediate, resulting from dihydrazine **37** <99JHC441>. Subsequent Fischer indole cyclization and aminolysis of **38** produced **39**; a mono-hydrazone intermediate (as opposed to **38**) was ruled out by the authors on the basis of ^{1}H NMR.

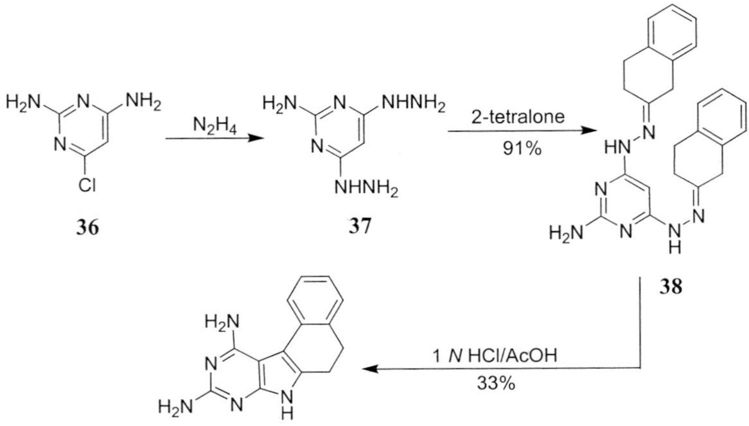

36 37 38 39

Unexpected products also arose from the reactions of **40** with excess (6-14 equivalents) hydroxylamine hydrochloride <99JHC787>. Unless R was very small (*i.e.*, H or Me), this reaction provided the pyrimidine-opened **42** exclusively; the oxime products **41** could not be isolated. With small substituents (*i.e.*, R = H or Me), the normal oximes **41** were the sole product.

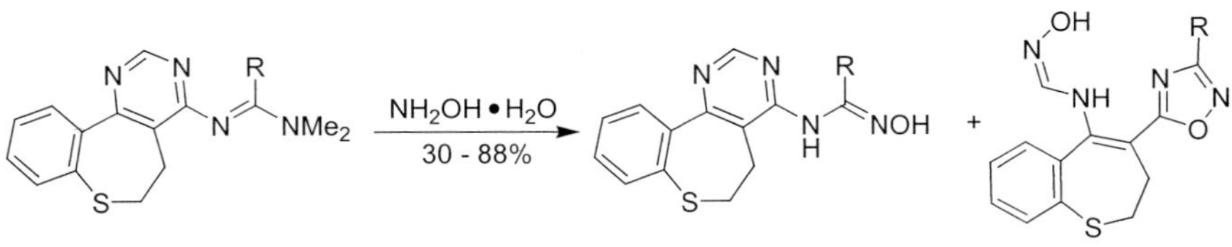

40 41 42

The formation of pyrimidine Grignard reagents (or equivalents thereof) followed by their reactions with electrophiles was also a widely reported topic. For example, the cerium derivative **43**, as well as the Grignard and lithium pyrimidine analogs of **43**, were produced from the bromo precursor and allowed to react with a range of ketones and aldehydes to yield **44** after removal of the *tert*-butyl protecting groups <99S495>.

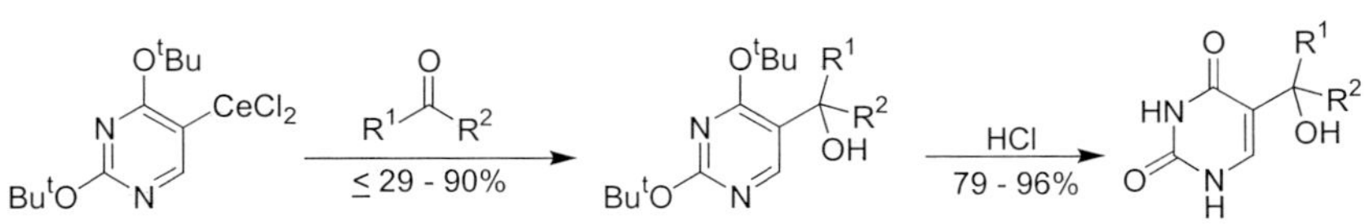

43 44

Similarly, 5-bromopyrimidine (**45**) was converted to its lithium derivative and was allowed to react with chiral sulfinate esters to give chiral sulfoxides **46** with high enantioselectivities <99JOC4512>.

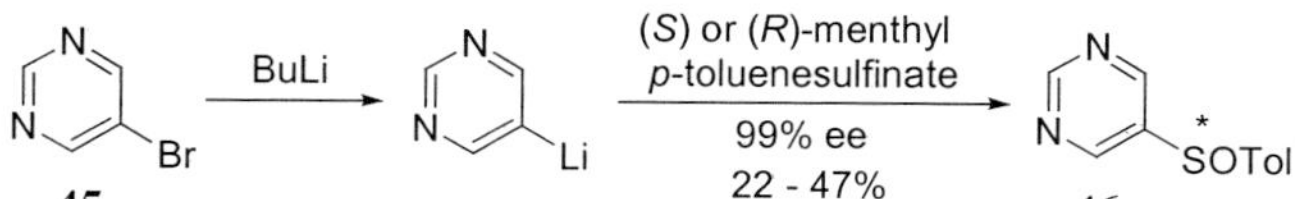

45 **46**

The recent popularity of palladium-catalyzed cross coupling reactions has been extended into the field of pyrimidines as well. For example, the palladium-catalyzed couplings of arylthiols and 2-bromopyrimidine (**47**) to produce **48** were reported by Thorarensen and co-workers <99SL1579>.

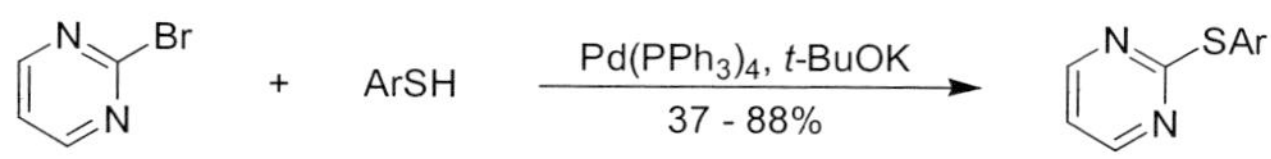

47 **48**

Traditional Stille-type (**49** + **50** → **51** <99S615>) and Heck-type couplings (**52** → **53** <99JHC145>) of halopyrimidines were also well represented. The latter reaction was utilized enroute to a new class of dihydrofolate reductase inhibitors **53**.

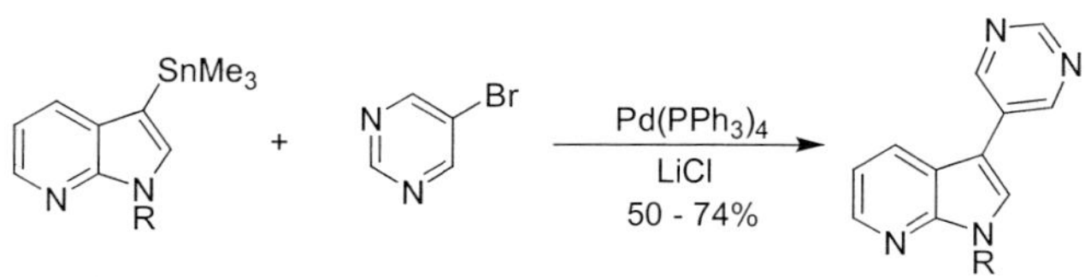

49 **50** **51**

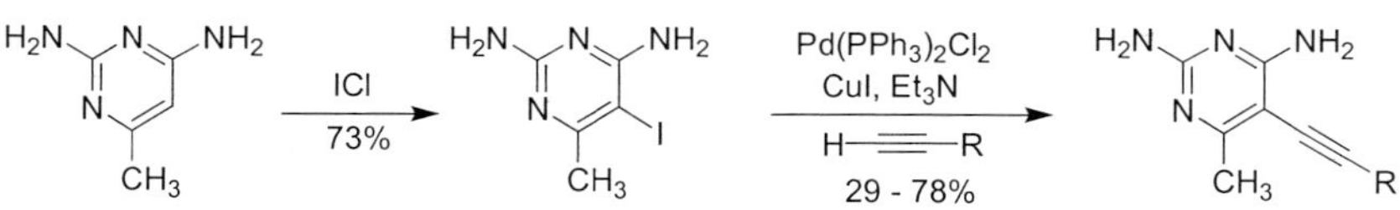

52 **53**

In a similar fashion, palladium-catalyzed alkoxycarbonylation of **54** was effective in producing pyrimdine esters **55** <99T405>. It was noted that dppf along with the use of the alcohols as solvents (rather than solely as reagents) was required for optimal conversion.

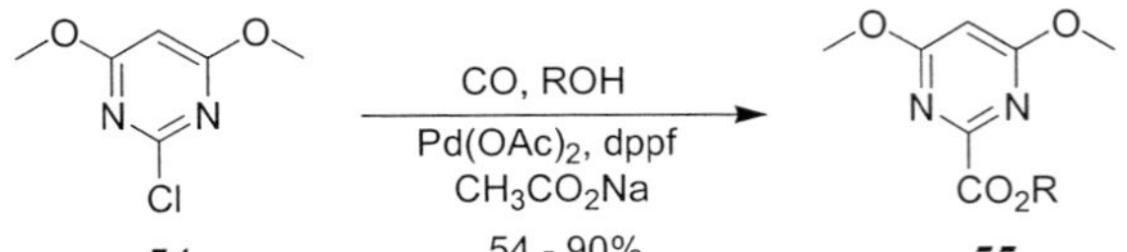

54 55

Despite the overall electron deficiency of nitrogen-containing heterocycles like pyrimidines, the use of electron-donating substituents enables pyrimidines to undergo reactions in which the ring nitrogens act as nucleophiles. For example, tosylated 2-aminopyrimidines **56** were alkylated to form mainly **57** along with traces of **58** <99S2124>. The authors reported that both products were converted to a single regioisomer **59** upon treatment with trifluoroacetic anhydride.

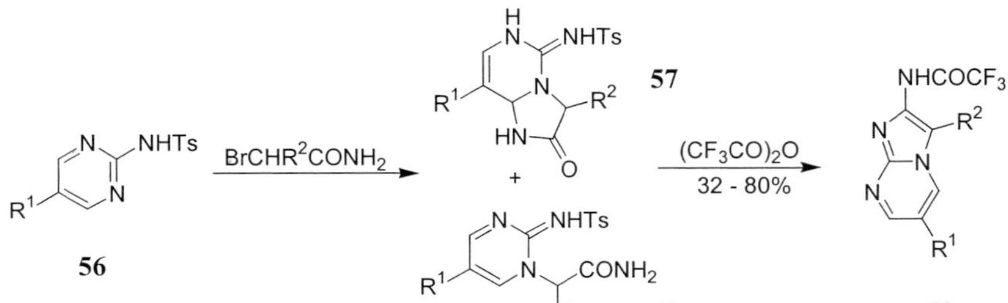

56 57 58 59

Similarly, Kumar and co-workers reported the coupling of **60** and **61** in the presence of iodine to yield **62** as the sole regioisomer in a single step <99JOC7717>. This was a key intermediate in their syntheses of heterocalixarenes which were used in subsequent biological cation binding studies.

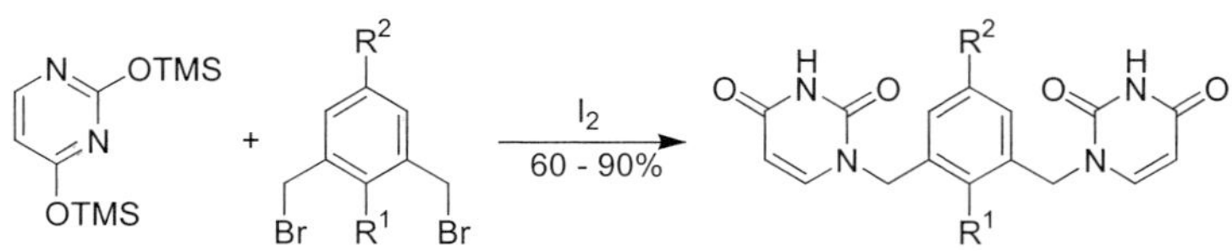

60 61 62

Additional exploitation of the nucleophilicity of activated pyrimidine ring-nitrogens was reported by Oberdorfer and co-workers in the conversion of **64** to **65** (R = H); trace amounts of acylated **65** (R = Ac) were deprotected upon silica gel chromatography <99S2057>. The authors did not address the issue of a direct transformation of **63** to **65** in a single step.

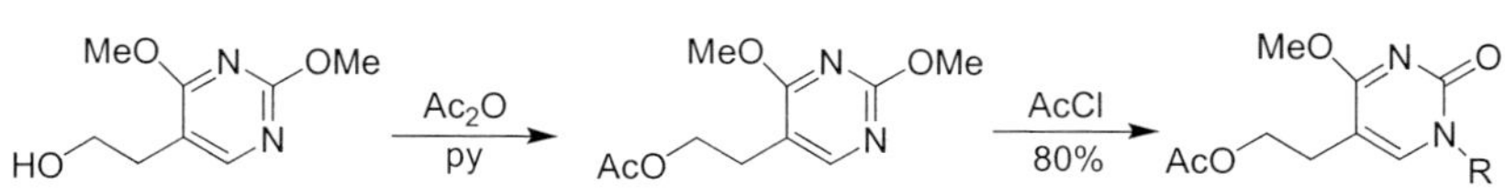

63 64 65

In an analogous fashion, pyrrolo[1,2-*c*]pyrimidine (**66**), the preparation of which was discussed in Section 6.2.2.1, was converted to the corresponding quaternized salt **67** <99JOC7788>. This heterocycle was shown to undergo 1,3-dipolar additions with a variety of dipolarophiles such as alkynes to produce **68** after oxidation with DDQ.

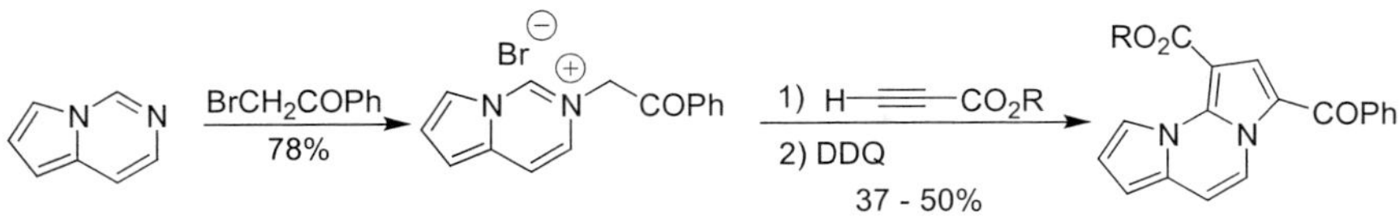

66 **67** **68**

An alternative form of reactivity of pyrimidines also assisted by electron donating substituents is the nucleophilicity of a ring carbon atom towards various electrophiles. For example, the reaction of **69** with aryl aldehydes **70** in the presence of ethyl cyanoacetate produced **71**, presumably from the conjugate addition of **69** to the *in situ* generated double bond formed in the Knoevenagel condensation of **70** and ethyl cyanoacetate <99JHC113>.

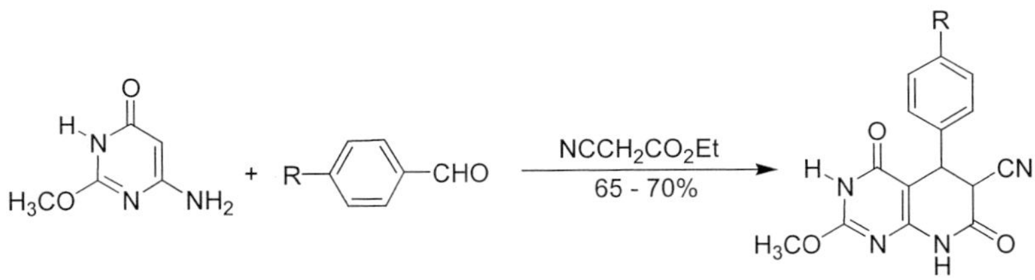

69 **70** **71**

In a similar transformation, **74** was synthesized from the reaction of **72** with **73** <99JHC501>. It was assumed that this reaction also proceeds *via* the Michael addition of **72** to the *in situ* generated double bond produced by the elimination of dimethylamine hydrochloride from **73**.

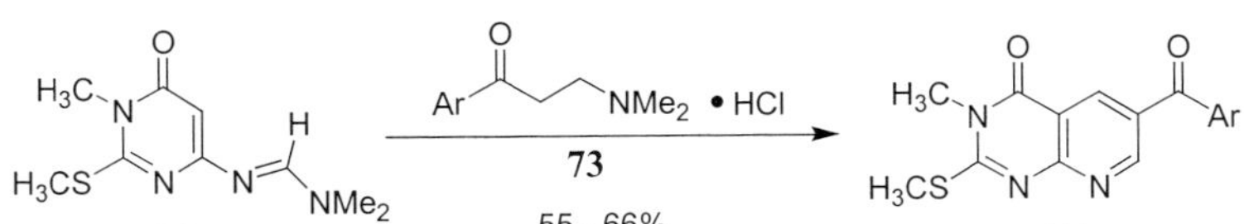

72 **74**

The reaction of **75** with **76** in the presence of nitromethane to ultimately produce **77** is thought to proceed by an analogous mechanism <99TL4023, 99TL4027>. The authors noted that this tandem Nef reaction/Michael addition produced **77** in a single step with sonication.

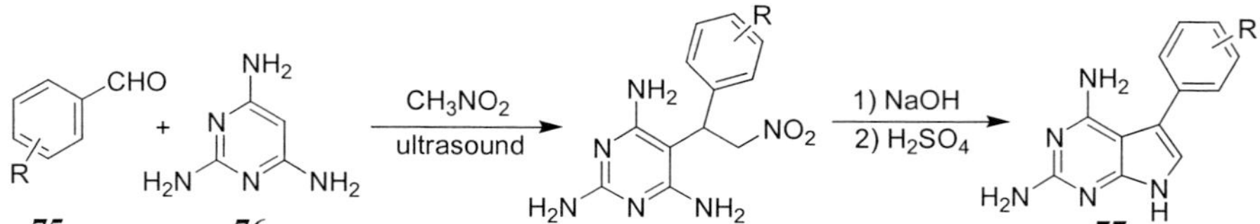

75 **76** **77**

Dauzonne and co-workers reported an interesting mechanistic investigation into the reduction of **81**, a compound formed through a Michael addition-initiated sequence, similar to those just discussed (*i.e.*, **78** + **79** → **80**) <99CPB156>. Hydrogenation of **81** in the presence of 5% Pd-C provided **84** while the use of 10% Pd-C produced **82**. The authors offered the following mechanism to explain the *a priori* unexpected furan ring-opening that produced **82**. While pathways *a* and *c* lead to product **84**, further hydrogenation of intermediate **83** in pathway *b* would lead to the ring-opened alcohol **82**, thus explaining the production of this product with the increased presence of Pd (10% Pd-C).

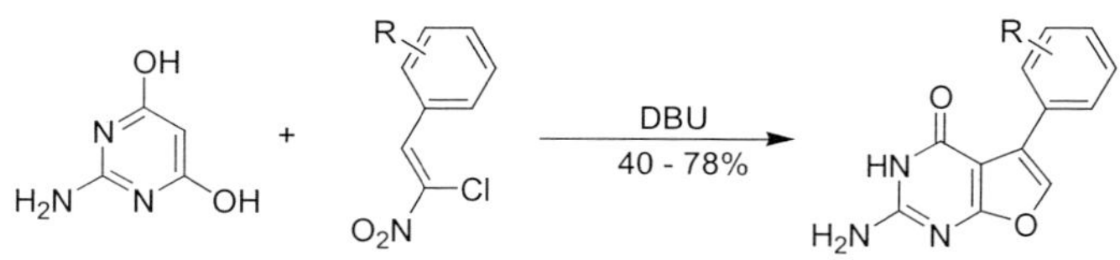

78 **79** **80**

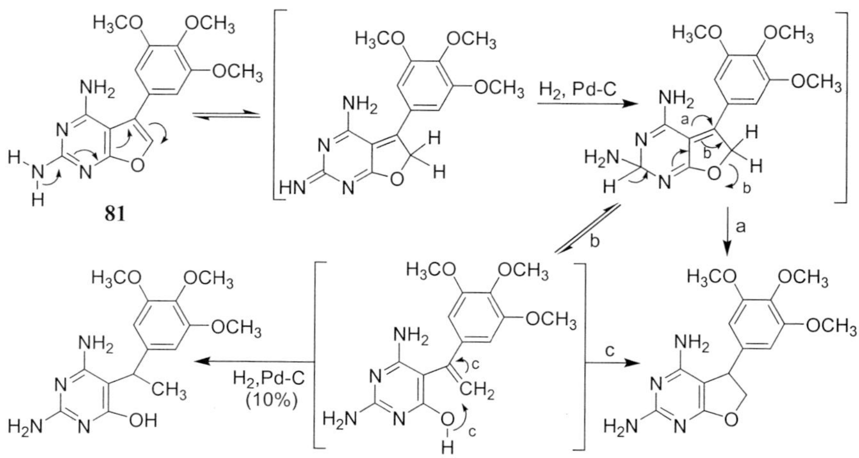

82 **83** **84**

Control over the site of nucleophilic addition of aminopyrimidines by solvent choice and reaction conditions was demonstrated by Vasudevan and co-workers in their report of the reaction of **85** with **86** <99JOC634>. Either **87** (if aqueous sodium acetate was used) or **88** (if a polar aprotic solvent like DMF was used) was formed regioselectively in modest yields. This trend follows literature precedence, which argues that hydrogen bonding between the 4/6-amino groups and water (in an aqueous solvent) allows the N1 nitrogen to react with the more reactive chloro-containing carbon. In aprotic solvents with no hydrogen donor capabilities, the 4-amino group was argued to react with the chloro-containing carbon.

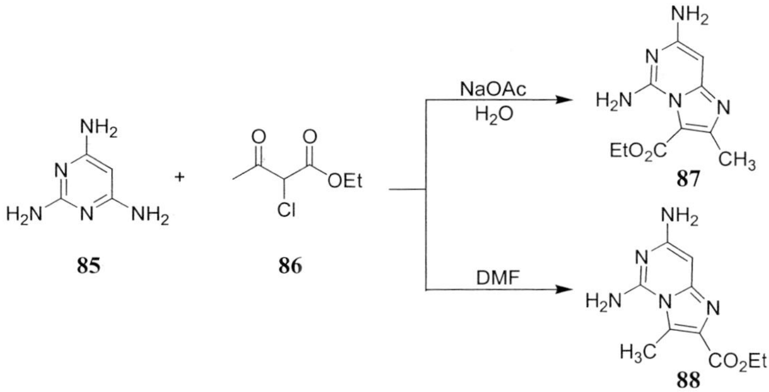

85 **86**

Dominguez and co-workers reported the intramolecular coupling of the two phenyl rings in **89** to produce phenanthro[9,10-*d*] fused pyrimidines **90** <99TL3479>.

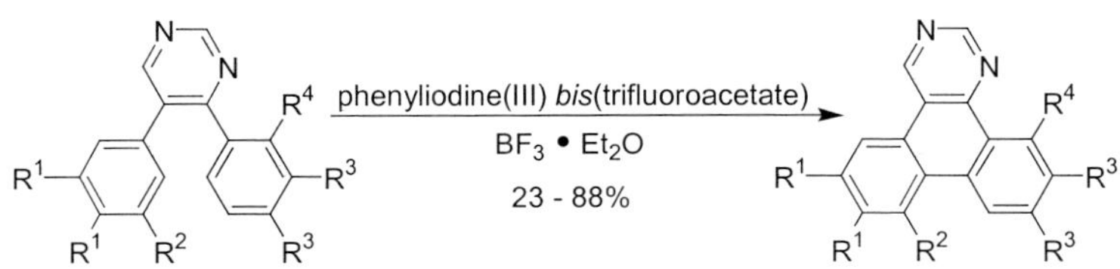

89 **90**

6.2.3 QUINAZOLINES

6.2.3.1 Preparations of Quinazolines

Quinazolines, the benzo derivatives of pyrimidines, were prepared in a variety of ways, from methods analogous to those for synthesizing pyrimidines to vastly different condensation schemes. Following the popular condensation routes using amidines, Kotsuki and co-workers reported the condensation of various amidines **92** with 2-fluorobenzaldehydes **91** which yielded quinazolines **93** after intramolecular S_NAr closure <99SL1993>.

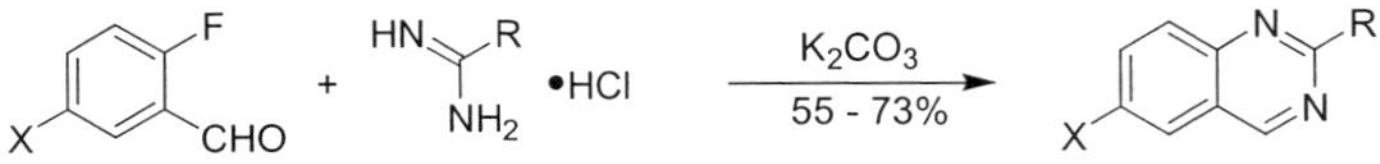

91 **92** **93**

X = CN, NO_2

In analogous fashion, quinazolines **96** were synthesized through *in situ* formation of the corresponding amidine by the reaction of ammonia with imines **95** <99JCS(P1)421>. While this reaction occurs in a single step beginning with triazolines **94**, intermediates **95** could be isolated by heating **94** in the absence of the ammonia source. Subsequent condensation of **95** with ammonia gave rise to **96**.

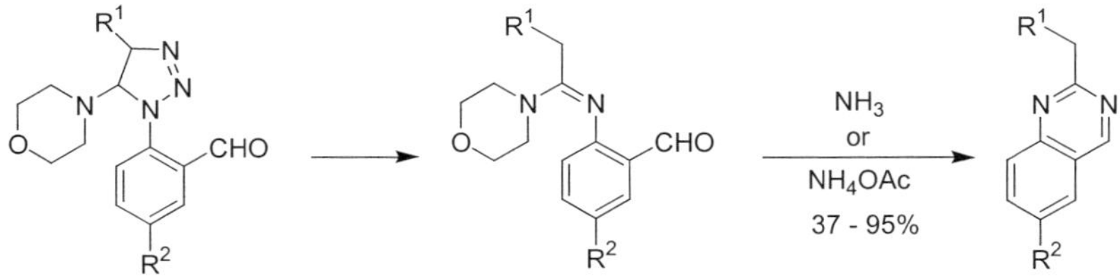

94 **95** **96**

The one-pot preparation of **99** through the reactions of various isocyanates with amide anions **98**, which were generated *in situ* from **97**, proved to be a useful method for the syntheses of these potentially biologically useful molecules <99BCJ1071>.

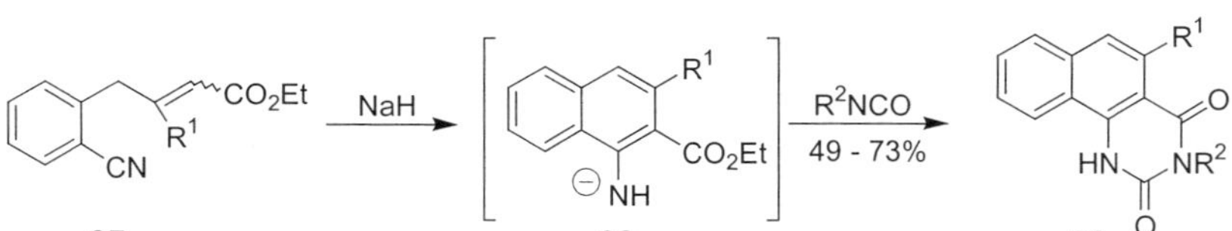

97 **98** **99**

In similar transformations, agriculturally and medicinally significant fluorine-containing quinazolines **101**, isolated as the hydrates, were produced in good yields from **100** after condensation with various aldehydes in the presence of ammonia followed by oxidation with DDQ <99H2471>.

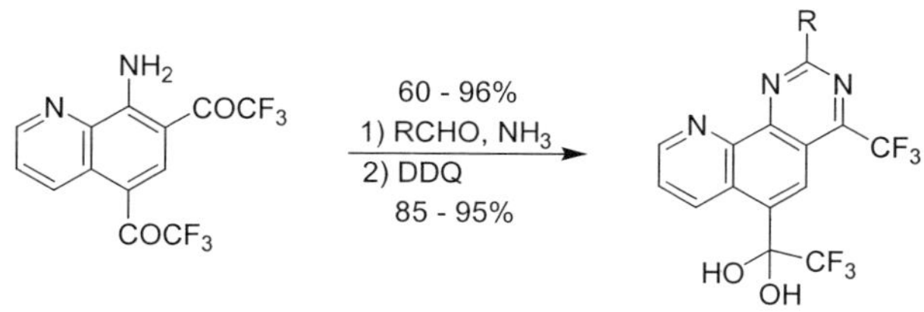

Ortho esters (**102 → 103** <99SC2617>) and anhydrides (**104** + **105 → 106** <99H1883>) were also used as cyclization reactants to form the pyrimidine ring of various quinazolines.

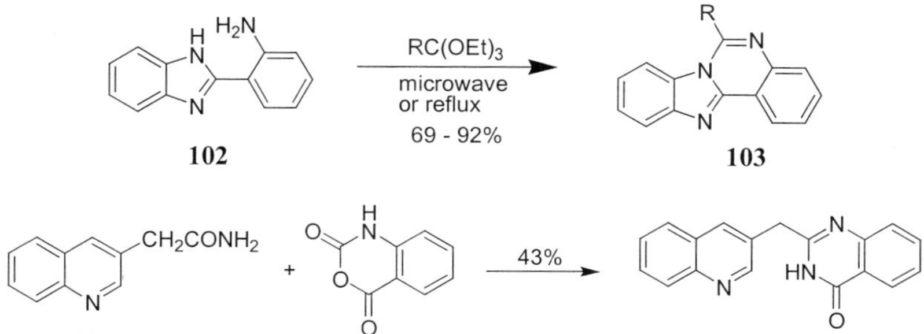

Several similar ring-closing strategies have also been published, such as the *in situ* reduction of the nitro group in **107** followed by condensation of the resulting amino group with the acetyl carbonyl to produce quinazoline **108** in 46% yield <99H2193>. The acetyl transfer product **109** was also produced (32%).

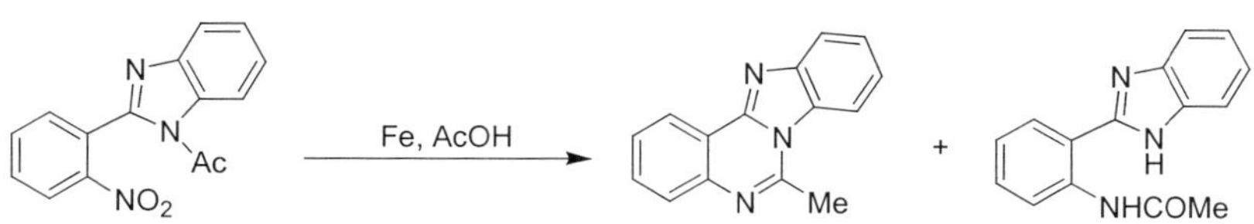

In a report from Sashida and co-workers, the unexpected ring-contracted products **112** were produced from the treatment of **111** with NaOMe at room temperature <99H2407>. This tandem ring-expansion (**110 → 111**) ring-contraction provided a facile route to quinazolines **112**.

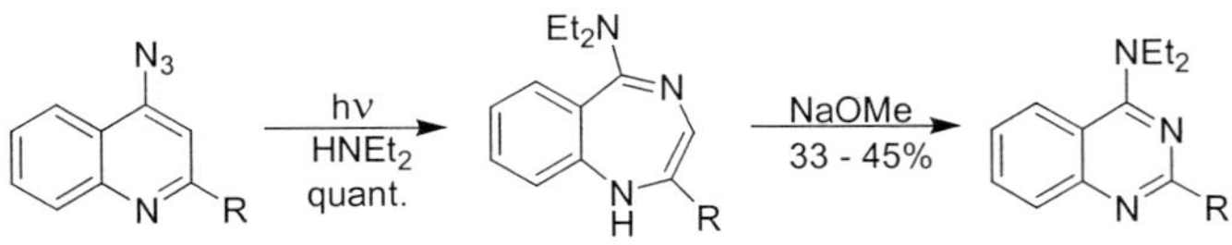

110 **111** **112**

The solid phase synthesis of quinazoline **114** was reported by Abell and co-workers, in which a traceless linker was utilized <99TL1045>. The key step in this procedure was the removal of the desired quinazoline from the resin with concurrent decarboxylation to produce **114** in 69% yield from **113**.

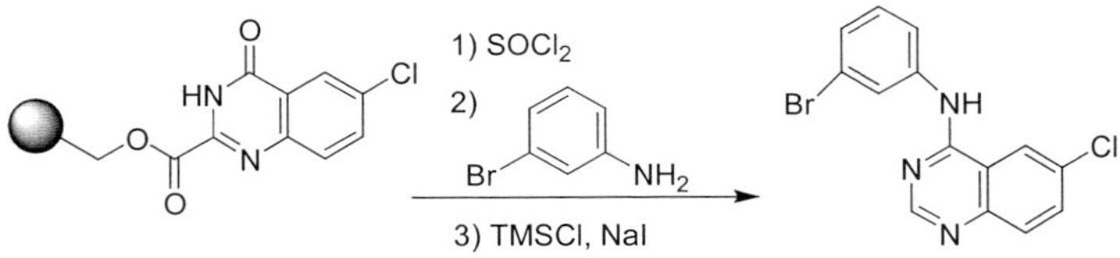

113 69% **114**

6.2.3.2 Reactions of Quinazolines

Quinazolines take part in the same types of reactions as pyrimidines, but because of their additional benzene ring, the products of these reactions may have the added feature of hindered rotation. An example of this is the synthesis of 2-phenyl-Quinazolinap by Guiry and co-workers <99TA2797>. Suzuki coupling of 4-chloro-2-phenylquinazoline (**115**) with boronic acids **116** led to **117** (R = OMe). These intermediates were parlayed into phosphinamines **117** (R = PPh_2) and then subjected to chiral resolution to produce new chiral phosphinamine ligands for asymmetric catalysis.

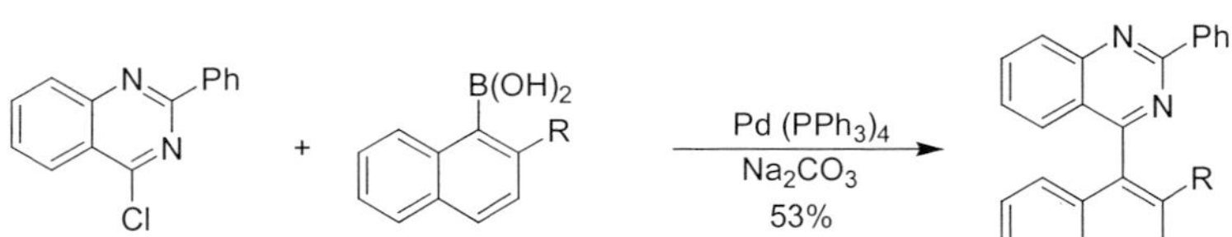

115 **116** **117**

6.2.4 PYRIDAZINES

6.2.4.1 Preparations of Pyridazines

Novel synthetic approaches to pyridazines, isomers of the popular pyrimidines already discussed, were significantly lacking in the number of publications. Nonetheless, Elassar reported the Japp-Klingemann-type reaction of aryldiazonium salts with **118** to produce pyridazine derivatives **119** after cyclization, though no yields were given <99JCR(S)96>.

118 **119**

A more classical method of introducing the ring nitrogens of pyridazines is the reaction of hydrazine with a 1,4-dicarbonyl compound. This was illustrated by Hafez and co-workers in which the reactions of **120** with hydrazine produced pyridazine derivatives **122** <99JCR(S)360>. It was noted that depending on the substituents on pyrazoles **120**, this reaction may not proceed at all (**120 → 121**).

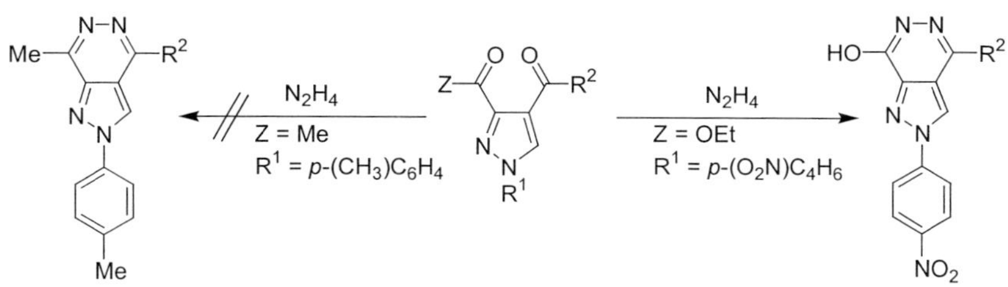

121 **120** **122**

Aromatization of tetrahydropyridazines is another method of synthesizing the aromatic pyridazine ring, although this route is sometimes met with difficulty. Ravina and co-workers reported that the oxidation of tetrahydropyridazine **123** produced **124** in 45% yield in the course of the preparation of a series of 5-substituted pyridazines <99JHC985>. The synthetically useful bromo derivative **125** was prepared either from aromatic alcohol **124** or in a single step from **123**.

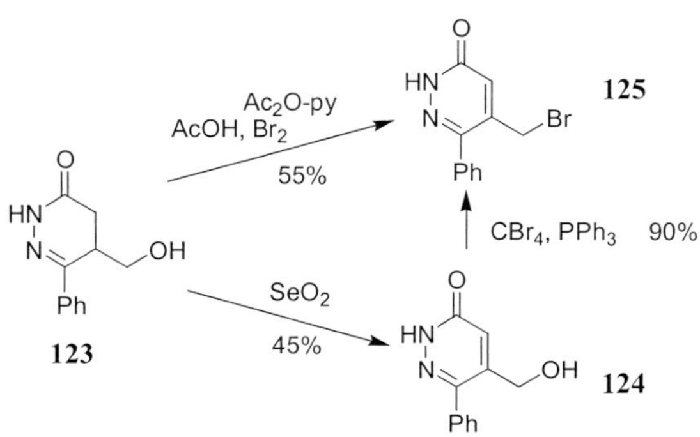

The [4 + 2] hetero Diels-Alder reaction of *in situ*-generated chlorodiazadienes **127** with various electron rich dienophiles (such as enamines) yielded a series of substituted pyridazines **128** after aromatization <99JHC301>. In this publication, South noted that the use of trichlorohydrazones **126** (X = Cl) gave rise to chloro-substituted pyridazines **128**, although not through the [4 + 2] mechanism.

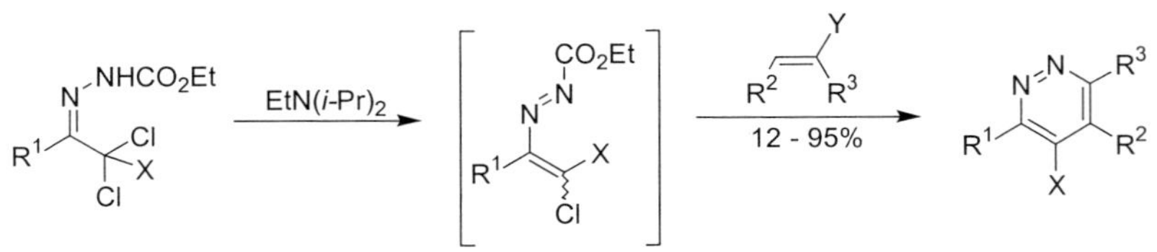

6.2.4.2 Reactions of Pyridazines

The inverse electron demand Diels-Alder of pyridazines continued to be a commonly explored topic. The adjacent nitrogen atoms of pyridazines not only help create an electron-deficient heteroaromatic diene, but also function as a good leaving group in a subsequent retro Diels-Alder reaction. This was illustrated by Haider and co-workers in their preparation of drug intermediates **131** through the reactions of **129** with enamines **130** <99SC1577>.

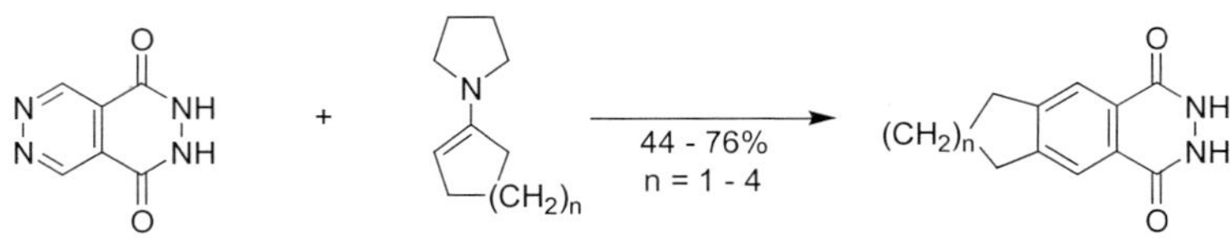

Direct lithiation of pyridazine **132** followed by trapping with chiral sulfinate esters produced chiral sulfoxides **133**, analogous to the pyrimidine reaction covered in Section 6.2.2.2 <99JOC4512>. Queguiner and co-workers demonstrated that a second lithiation/trapping sequence can provide fully substituted pyridazines **134** with high diastereoselectivities.

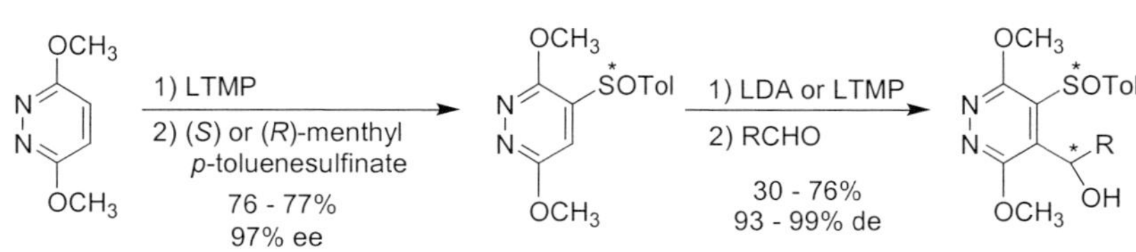

Lehn and Romero-Salguero <99TL859> reported the Stille coupling of chloropyridazine **135** with **136** to produce **137**, an intermediate in the preparation of various bidentate and tetradentate ligands.

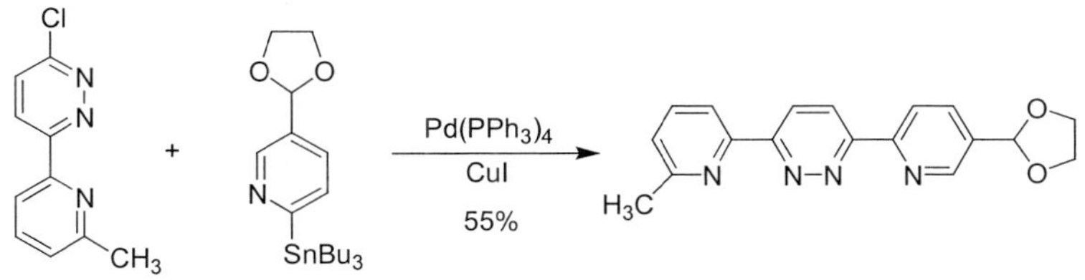

135 **136** **137**

6.2.5 CINNOLINES

6.2.5.1 Preparations of Cinnolines

Cinnolines, one of the two benzo derivatives of pyridazines, have been primarily prepared through condensations of hydrazine derivatives with carbonyl compounds followed by ring closures of various sorts. For example, boron-containing estrogen mimic **139** was prepared through the condensation of aldehyde **138** with 2-hydrazino-6-methoxypyridine followed by selective O-demethylation <99AX(C)1701>. A hydrogen bonding interaction between the B-OH and the pyridine ring nitrogen in **139** provides a "virtual six-membered ring" which corresponds to the basic steroid tetracyclic structure.

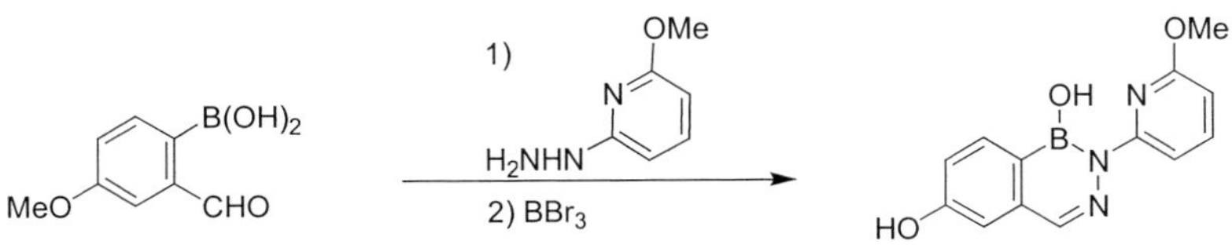

138 44% **139**

Kiselyov and Dominguez reported the formation of aminocinnolines **141** from the reaction of NaHMDS with **140**, products of aldehyde hydrazine condensations <99TL5111>. It was noted that purification of the cinnolines could be simplified by using a resin-bound aryl aldehyde and performing a solid-phase extraction. The ring formation was thought to proceed with the loss of two sequential fluoride leaving groups and subsequent displacement of the third fluoride with HMDS. Hydrolysis then produced **141**.

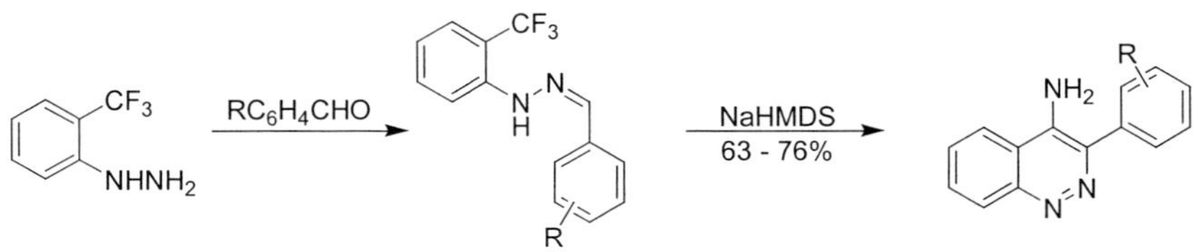

140 **141**

In the first reported solid-phase Richter reaction, cinnolines **143** were prepared through a two step sequence of a Heck-type coupling to give intermediates **142** followed by intramolecular ring formation and release of cinnolines **143** <99TL6201>.

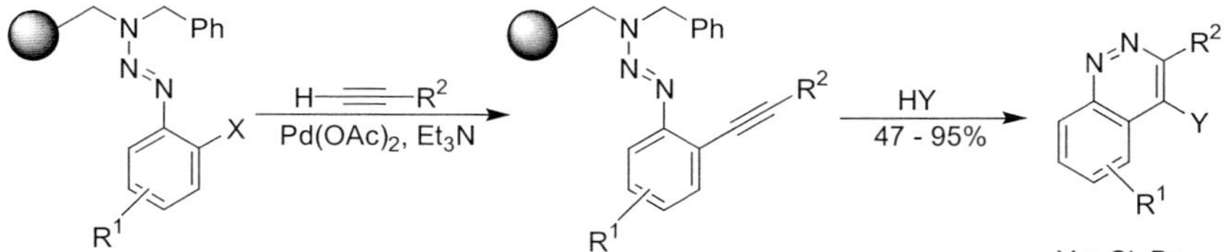

142

Y = Cl, Br

143

Cirrincione and co-workers reported the serendipitous discovery of another intramolecular ring-forming reaction in the generation of cinnolines **145** <99JMC2561>. Diazotization of the aniline ring in derivatives **144** was followed by intramolecular ring-closure not to the indole nitrogen, but instead to the indole C3 in a Japp-Klingemann-type reaction with the loss of a bromonium ion.

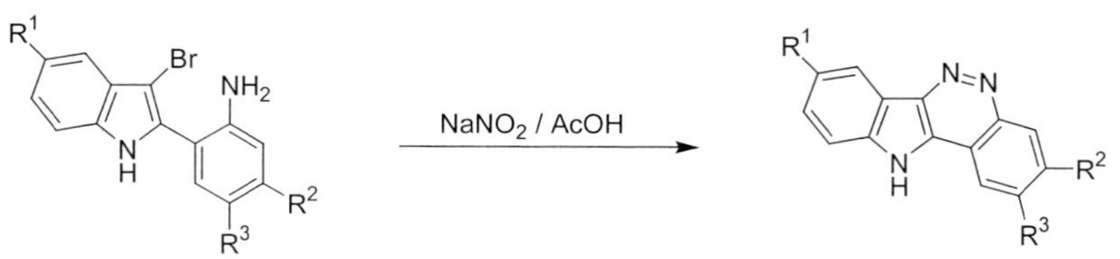

144 **145**

6.2.5.2 Reactions of Cinnolines

Due to the fact that many nitrogen-containing heterocyclic rings like those of cinnolines are effective pharmacophores, these compounds tend to be targets of syntheses rather than reactants in subsequent steps. An interesting example of a reaction of cinnoline derivative **146** was reported by Murakami and co-workers in their studies of Fischer indolizations <99CPB791>. The reduction of dihydrocinnoline **146** was followed by ring contraction to give **147** after loss of ammonia.

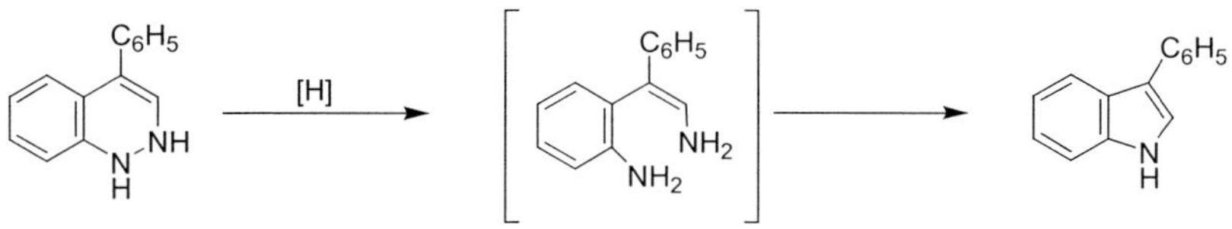

146 **147**

6.2.6 PHTHALAZINES

6.2.6.1 Preparations of Phthalazines

As with cinnolines, phthalazines were also prepared most frequently through condensations of hydrazine derivatives and carbonyl-containing compounds. For example, Monneret and co-workers reported the condensation of dialdehyde **148** with hydrazine to produce phthalazine derivative **149**, an advanced intermediate in the preparation of anticancer analogs of etoposide <99T12805>.

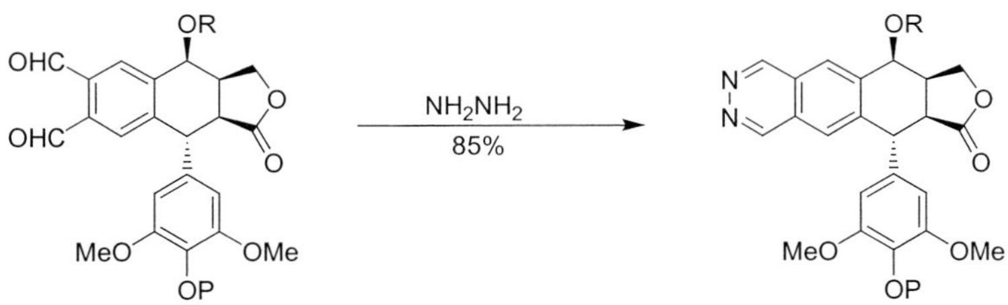

148 **149**

In a similar manner, aryl acid hydrazides **150** were condensed with benzaldehydes **151** <99SC3503>. Intermediates **152** underwent cyclodehydration in the presence of polyphosphate ester (PPE) to provide phthalazines **153** in good yields.

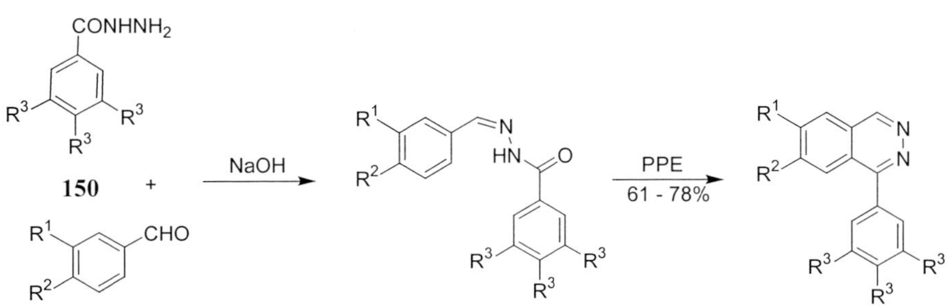

151 **152** **153**

As previously noted, Haider and co-workers reported inverse electron demand Diels-Alder reactions of various enamines **130** with an appropriately substituted pyridazine **129** as a method for phthalazine synthesis as well (see section 6.2.4.3) <99SC1577>.

6.2.6.2 Reactions of Phthalazines

Phthalazines are commonly used as ligands in transition metal cataysis since the structure provides a planar backbone with coordinating nitrogens. One of the most prevalent phthalazine-based ligands is known as $(DHQD)_2PHAL$ (**154**) <94CR2483>. A recent example of the use of **154** was in the catalytic asymmetric dihydroxylation by osmium tetroxide with air as the ultimate oxidant reported by Krief and co-worker <99TL4189>.

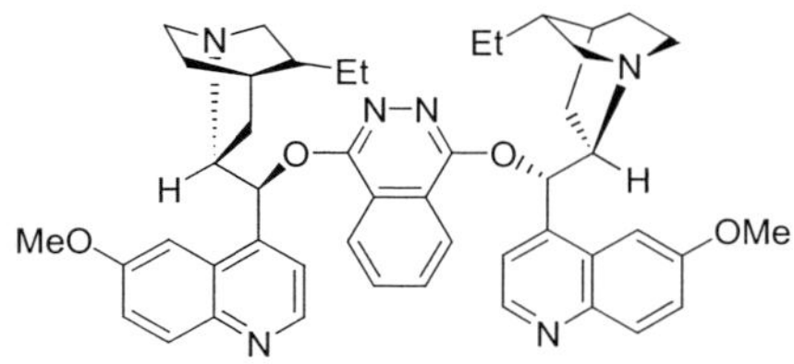

154

6.2.7 PYRAZINES

6.2.7.1 Preparations of Pyrazines

The pyrazine ring structure warrants the use of methodology analogous to that of pyridazines for their preparation. Condensation of diaminoethane with 1,2-dicarbonyl compounds **155** provides non-symmetrical pyrazines **156** after aromatization <99SL1203>.

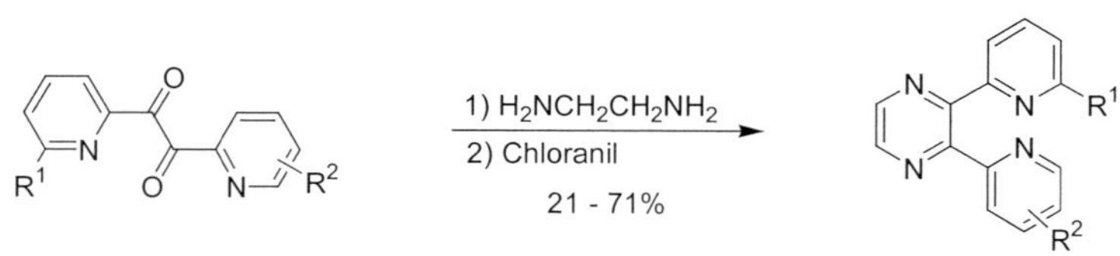

155 **156**

6.2.7.2 Reactions of Pyrazines

Pyrazines undergo nearly all of the same reactions as pyrimidines, from nucleophilic substitution (S_NAr) to palladium-catalyzed cross coupling reactions. Displacement of the chlorides *via* S_NAr reactions with nitrogen (**157 → 158**) and sulfur-based nucleophiles (**158 → 159**) was the methodology employed by Oakley and co-workers in the course of the preparation of neutral π-radical conductors <99JA969>.

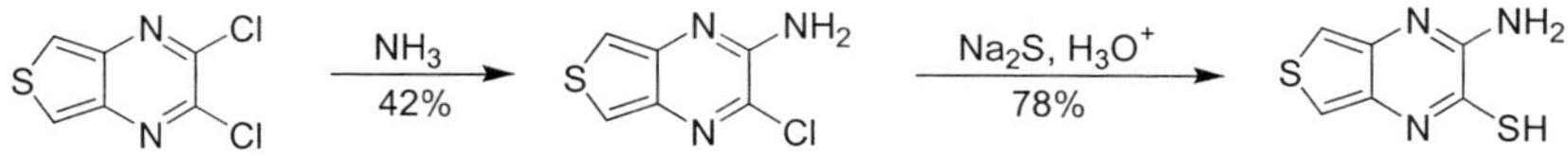

157 **158** **159**

Sato and Narita provided an improved synthesis of various halopyrazines in which hydroxypyrazines **160** were activated with TMSCl to give silyl ethers **161** <99JHC783>. Subsequent treatment of **161** with the appropriate phosphorus-based halogen source provided halopyrazines **162** in 46-94% overall yield. This two-step process was accomplished without isolation of intermediate **161** and provides a milder, more convenient approach than the traditional heating of hydxoxypyrazines with PX_n directly.

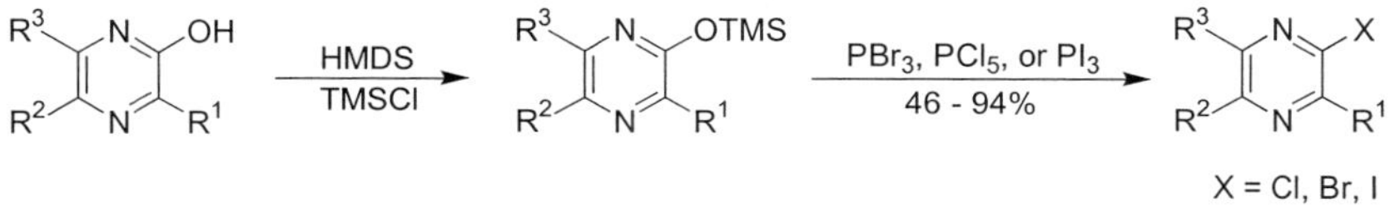

160 **161** **162**

Lithiation and subsequent trapping of pyrazine carbanions can be directed to the *para* position of pyrazines with electron withdrawing groups as with the thioamides **163**. In the event, Queguiner and co-workers noted that this reaction with pyrazines **163** to produce **164** could be accomplished either in a single step where the lithiated pyrazine was trapped *in situ*, or in a two step procedure depending on the electrophile <99H2349>. It was also noted that under certain conditions (*i.e.*, R^1=R^2 = *i*-Pr, electrophile = TMSCl, 2.2 eq LTMP), the *ortho* product could be selectively generated.

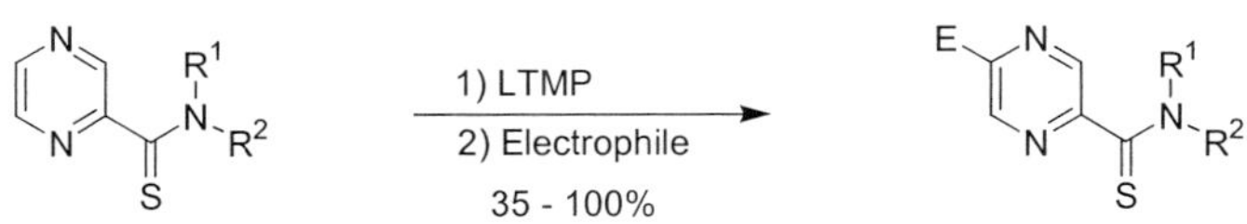

163 **164**

Similarly, Tour and Zhang reported a lithiation/trapping sequence to produce two different pyrazines which would ultimately be coupled together to form pyrazine polymers <99JA8783>. Diiodopyrazines **166** were prepared by the reaction of **165** with LTMP followed by trapping with iodine. The Stille coupling partner **168** was synthesized through a similar reaction with Boc-protected diaminopyrazine **167**. The authors noted that the first two steps (deprotonation with NaH and trapping with Bu_3SnCl) were necessary to further protect the exocyclic nitrogens.

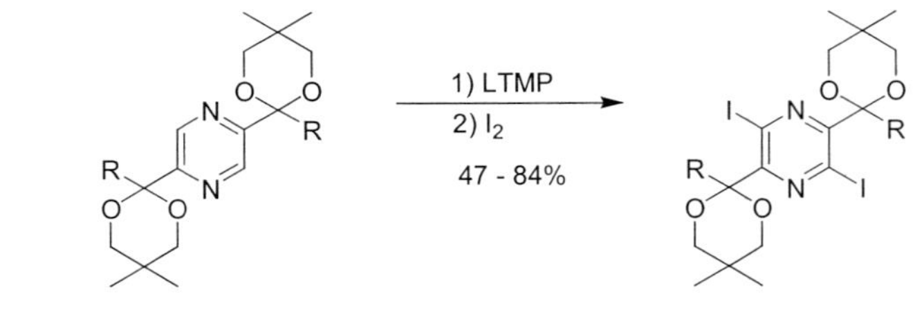

165 **166**

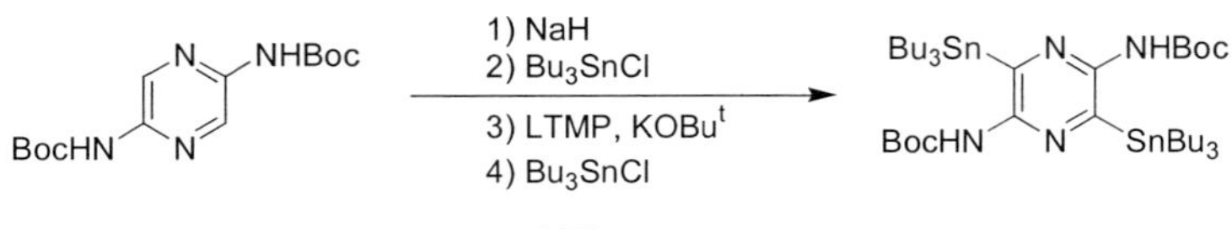

167 **168**

The Stille-type coupling of these intermediates (**166** + **168**) led to pyrazine polymer **169** in good yields.

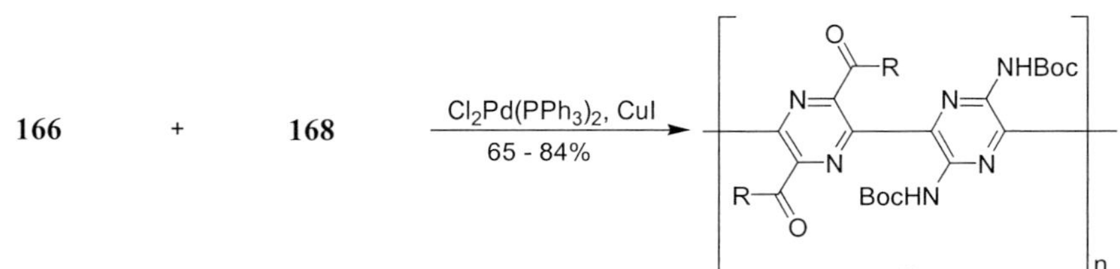

169

6.2.8 PHENAZINES

6.2.8.1 Preparations of Phenazines

Phenazines and their derivatives are known to be biologically significant molecules, especially in the field of photodynamic therapy. A recent example of the preparation of red shifted azine dyes potentially for photodynamic therapy was reported by Gloster and co-workers <99JHC25>. The synthesis of phenazine **171** from **170** was thought to take place through the following mechanism.

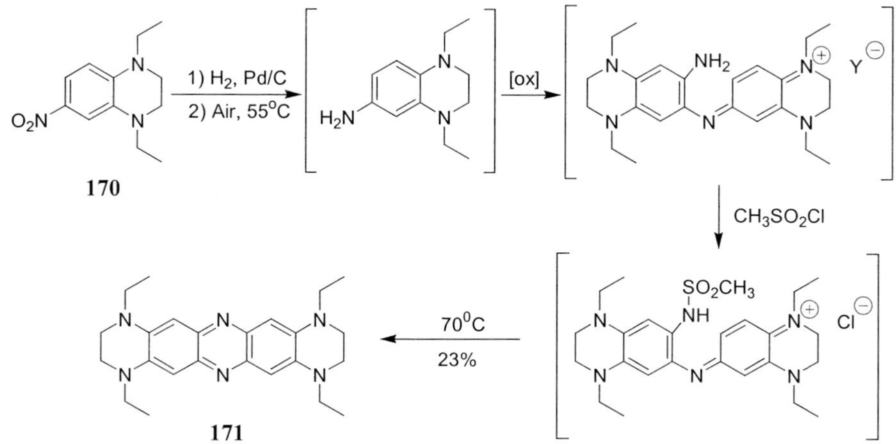

In studies of the role of electronic effects in the Bergman cyclization, Russell and Kim reported the preparation of phenazines **173** from heterodiyne **172** <99TL3835>. The authors noted the significant rate dependence on solvent in these reactions.

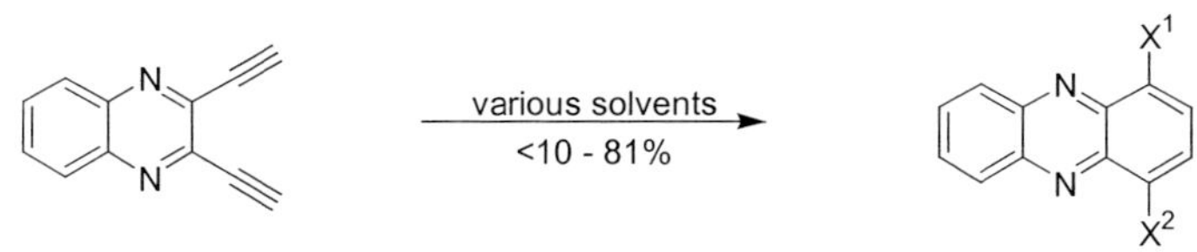

172 **173**

Another cyclization thought to proceed *via* a free radical mechanism was the aromatization of **174** to produce phenazine **175** along with varying amounts of ketones **176** <99JHC1057>. It was noted that each product could be selectively formed depending on the nitrogen substituents.

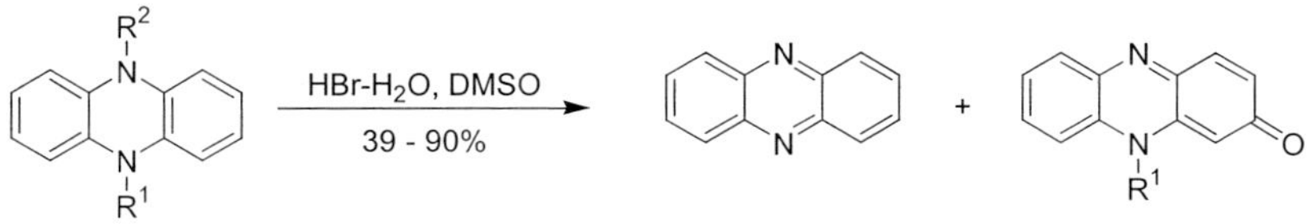

174 **175** **176** 0

6.2.8.2 Reactions of Phenazines

The preparation of phenazine crown ether derivatives was reported by Huszthy and co-workers in their investigation of the use of these heterocycles as enzyme mimics <99T1491,

99TA2775>. The reaction of phenazine diol **177** with various chiral ditosylates **178** gave rise to phenazino-crown ethers **179** in enantiopure form.

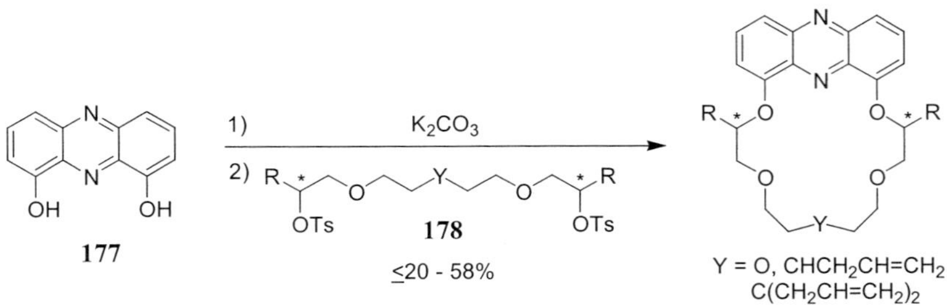

177 **178** **179**

6.2.9 QUINOXALINES

6.2.9.1 Preparations of Quinoxalines

Quinoxalines have received a significant amount of attention due to their potential use in fighting various pathophysiological conditions like epilepsy, Parkinson's, and Alzheimer's diseases. Preparatory methodologies range from straightforward condensation reactions to complex rearrangements. The desulfurization of isothiazoles **180** produced quinoxaline-N-oxides **181** after rearrangement, albeit in low yields <99HCA238>.

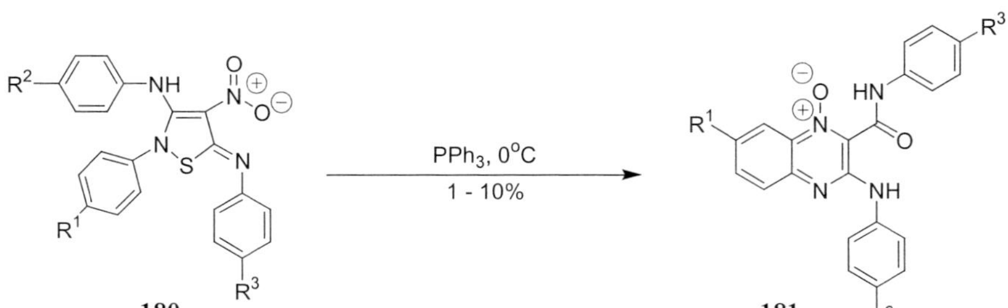

180 **181**

Another quinoxaline-yielding rearrangement, reported by Hanaineh-Abdelnour and co-workers, entailed treating imides **182** with sodium azide to produce quinoxalines **183** in moderate to good yields <99H2931, 99T11859>. The reaction presumably proceeds by a nitrene insertion to close the pyrazine ring.

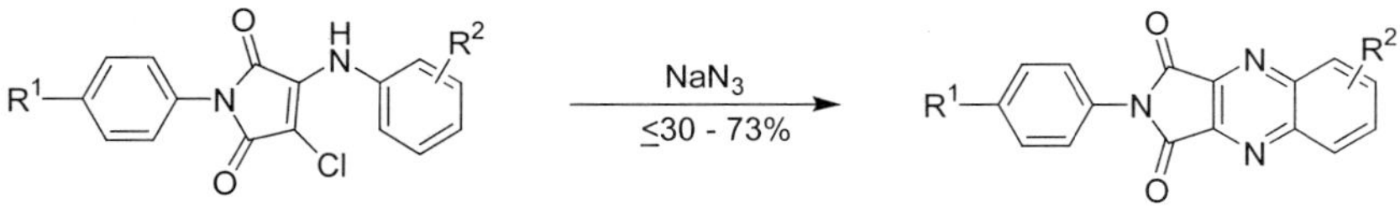

182 **183**

From a more traditional standpoint, 1,2-diaminobenzenes were condensed with a variety of 1,2-dielectrophiles ranging from α-keto esters (**184** + **185** → **186** <99H1213>) and 1,2-diketones (**187** + **188** → **189** <99JA10438>) to α-keto imines (**190** + **191** → **192** <99JCS(P1)1789>) and halohydrins (**193** + **194** → **195** <99SC3459>). Regioselectivity in the production of highly substituted quinoxalines **192** favored R[1] adjacent to the lactam nitrogen in most cases, except when R[2] was a strong withdrawing group like CF_3, then the other regioisomer was predominant. The reaction of **193**, the ring-opened product of the corresponding epoxide with HCl, produced quinoxalines **195** exclusively, but the authors noted that if the epoxides were condensed directly with 1,2-diaminobenzene (**194**), the analogous dihydroquinoxalines were the sole products.

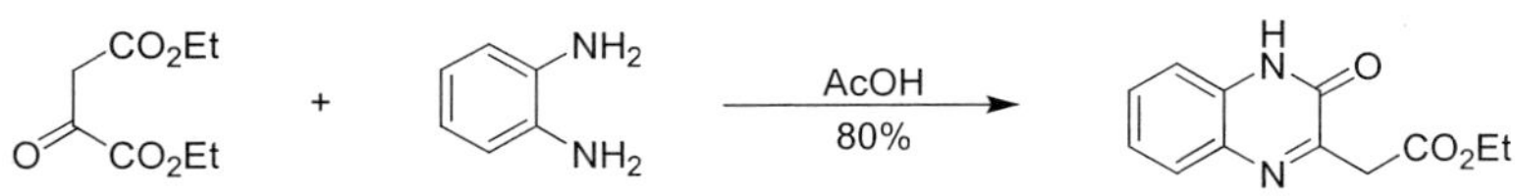

184 **185** **186**

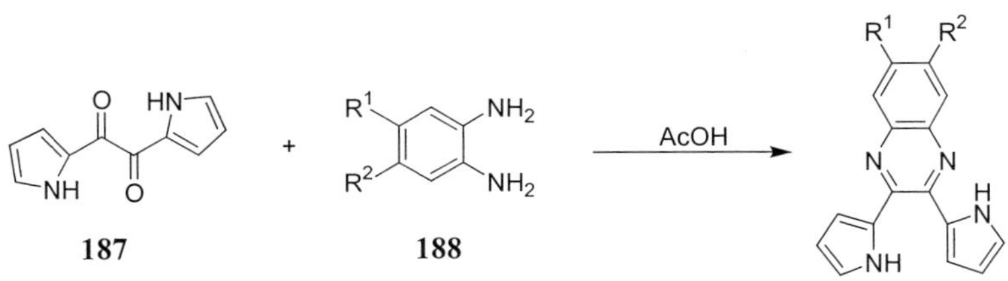

187 **188**

189

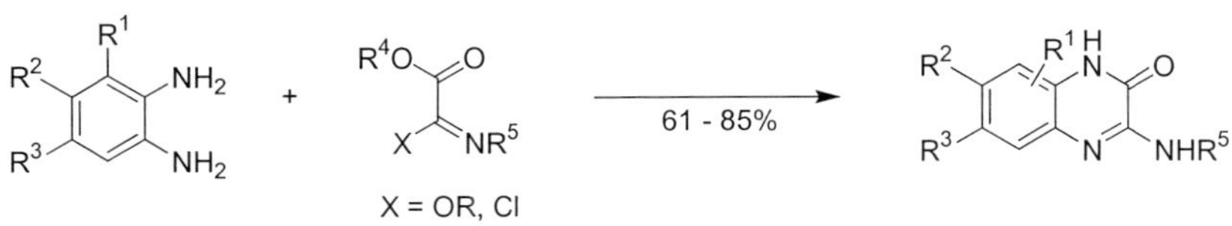

190 **191** **192**

193 **194** **195**

Gawinecki and co-workers reported the structural determination of isomeric products from a similar reaction, the condensation of 1,2-diaminobenzenes **197** with 1,2-dicarbonyl compound **196** <99T8475>. The two different regioisomeric quinoxalines **198**, which were produced in nearly equal amounts, were distinguished through the use of advanced NMR techniques including 2D *z*-gradient selected H^1, N^{15} HMBC.

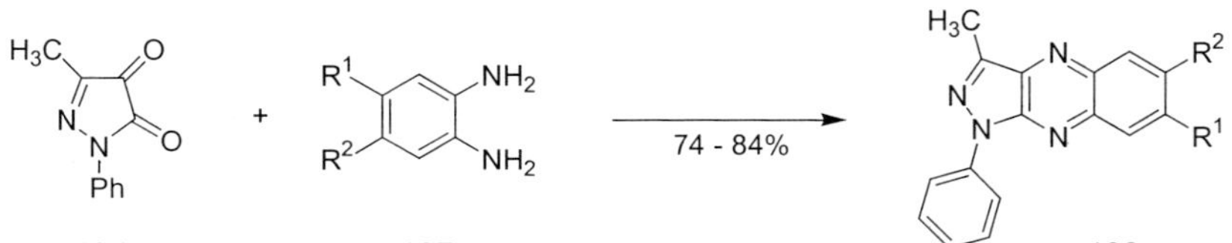

196 **197** **198**

A plethora of quinoxalines were synthesized for the production of chemical libraries in search of biologically active derivatives. Nikam and co-workers reported the condensation of 1,2-diaminobenzene **199** (or an analog thereof) with oxalic acid in the preparation of various analogs of 5-aminomethyl quinoxaline **200** <99JMC2266>.

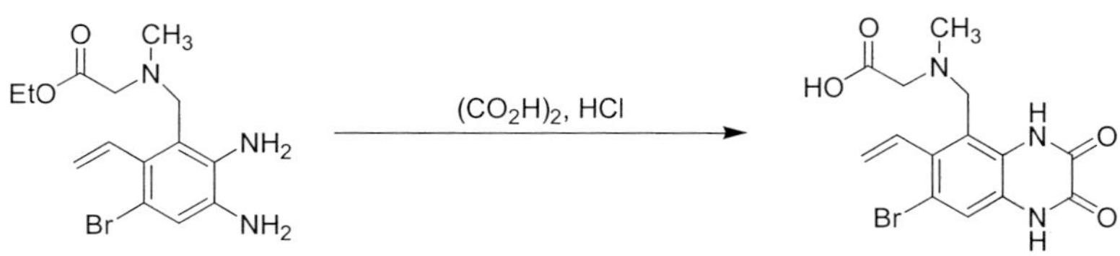

199 **200**

The analogous reaction with 1,2-diaminobenzene **201** was reported by Kornberg and co-workers in another quinoxaline library production <99JHC1271>. Quinoxaline **202** was a key intermediate in this methodology.

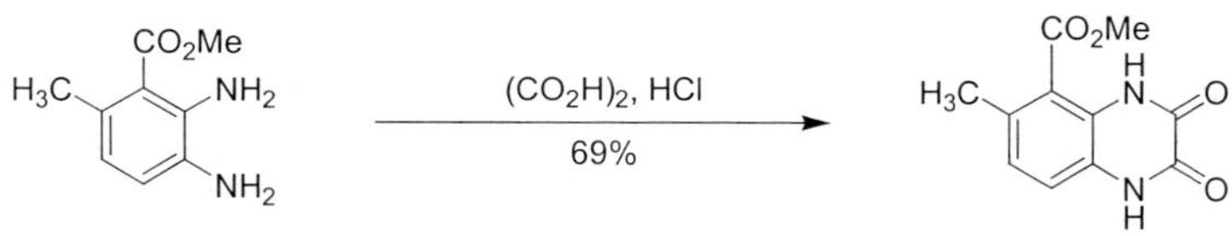

201 **202**

6.2.9.2 Reactions of Quinoxalines

Due to the significant biological interest of quinoxalines, several methodologies for derivatization of these heterocyclic skeleta have been reported. The majority of the reactions involving quinoxalines are analogous to those of the other heterocycles discussed previously including S_NAr substitution, lithiation, and transition metal catalyzed cross coupling reactions.

Electrophilic aromatic substitution (S_EAr) of 1H,4H-2,3-dione precursors has also been demonstrated. For example, intermediate **202** prepared by Kornberg and co-workers (see previous section) was converted to nitro derivative **203** through S_EAr chemistry, which was subsequently used to prepare a library of derivatives for biological testing, including the synthetically useful dichloro derivative **204** <99JHC1271>.

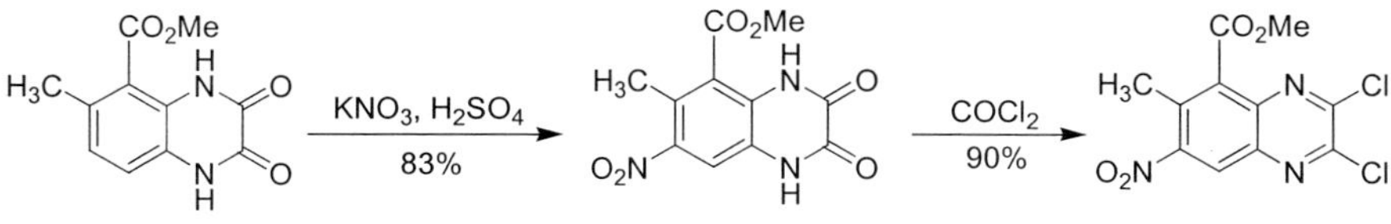

202 **203** **204**

Similar dicholoro precursor **205** was utilized by Katoh and co-workers in the discovery of a new fluorescence derivatizing agent for fatty acids <99H299>. The conversion of **205** to fatty acid-coupled **206** proceeded through the displacement of both chloro groups with morpholine in several high-yielding steps.

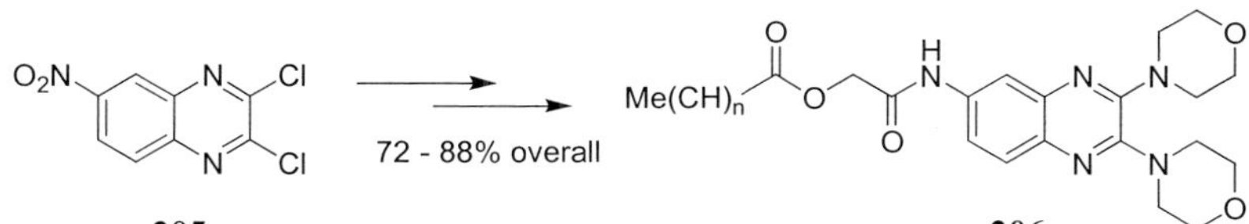

205 **206**

The coupling of quinoxaline **207** with phenolates **208** was reported to proceed rapidly in the presence of Ag^+ to produce **209** in good yields <99SC1393>. The authors noted that in the absence of silver ions, the reaction proceeded at a significantly lower rate.

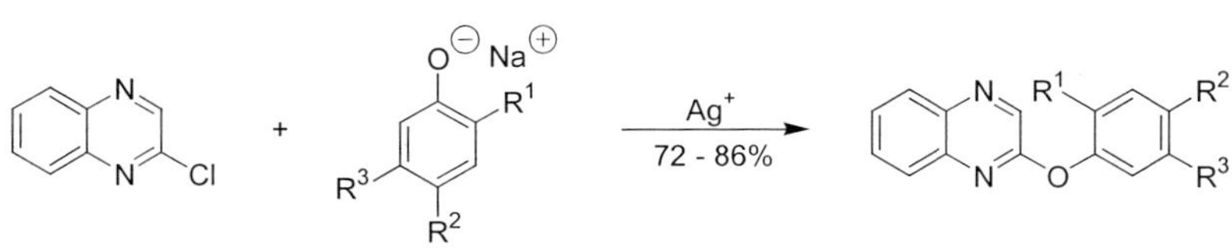

207 **208** **209**

In a similar fashion, one of the halogens in quinoxaline **210** was displaced with allyl amines **211** in order to prepare intermediates **212** <99JOC8425>. These monohaloquinoxalines were subjected to intramolecular palladium-catalyzed couplings to provide **213**.

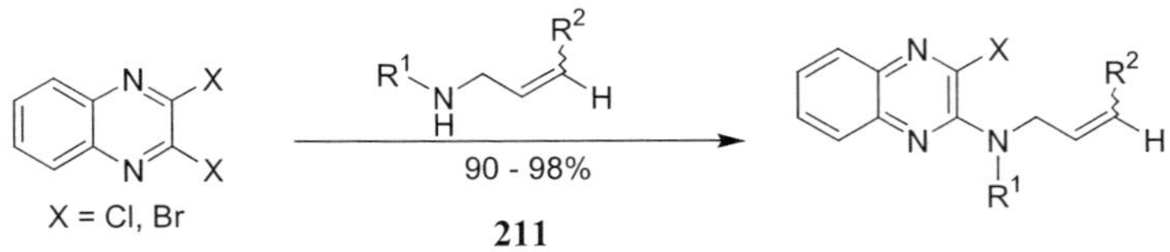

210 **211** **212**

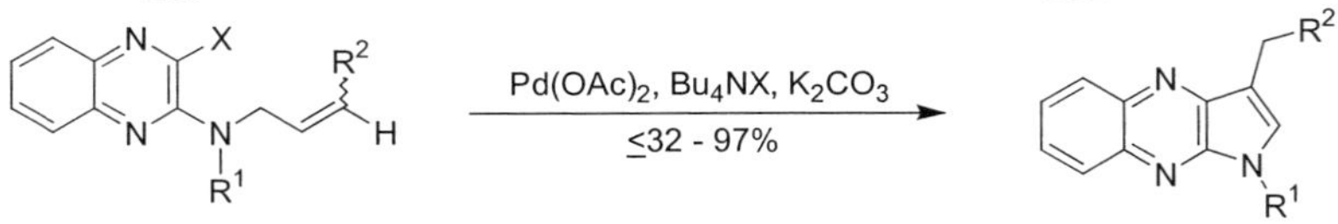

212 **213**

Li and Yue also reported the intermolecular palladium catalyzed cross coupling reactions of bromo quinoxalines **214** and **216** with aryl boronic acids and heterocyclic stannanes, respectively <99TL4507>. The Suzuki couplings (*i.e.*, **214** → **215**) required the use of a strong base for the reaction to take place which also limited the scope of the reaction due to the instability of the heterocyclic boronic acids, amides, and esters to these harsh conditions. This difficulty was overcome by the use of Stille reaction conditions (*i.e.*, **216** → **217**), which allowed for these various functionalities to be present.

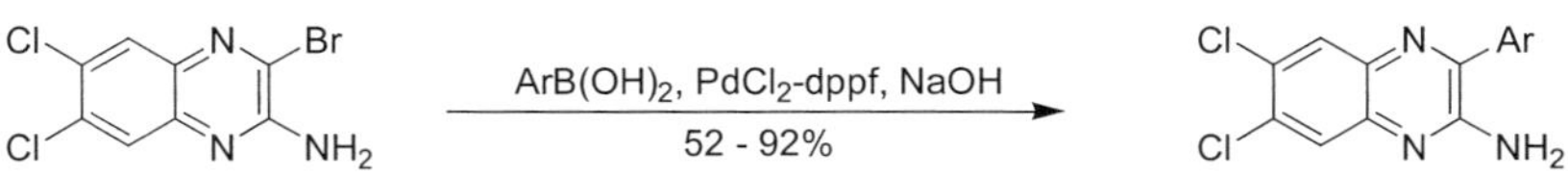

214 **215**

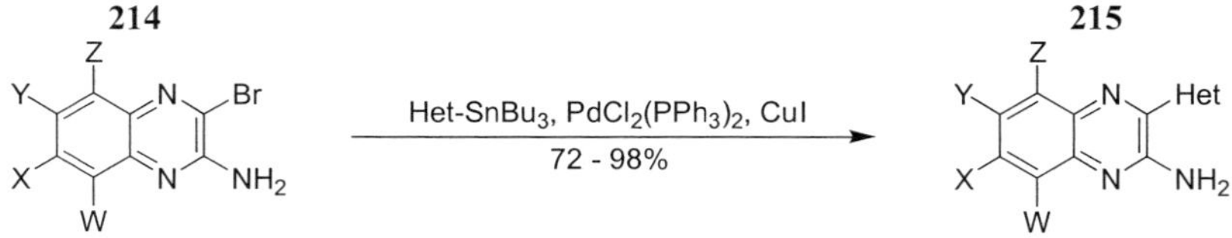

216 **217**

An alternative form of reactivity of quinoxalines is lithiation followed by electrophilic trapping of the anion. An example of this was reported by Queguiner and co-workers in the regioselective lithiation of quinoxaline **218** and subsequent trapping with various electrophiles <99T5389>. It was noted that regioisomer **219** predominated presumably due to the adjacent electron-withdrawing ring-nitrogens.

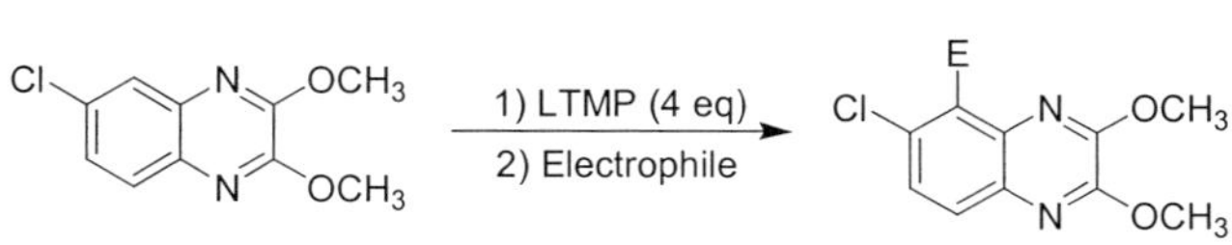

218 **219**

Chambers and co-workers reported similar results from the fluorination of disubstituted quinoxalines **220** <99JCS(P1)803>. Chemoselectivity could be controlled simply through the use of 1.5 equivalents (for **221**) or 2 equivalents (for **222**) of fluorine.

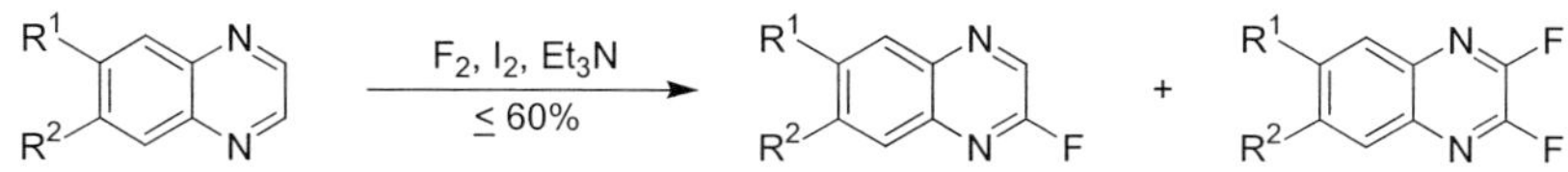

220 **221** **222**

The 1,3-dipolar addition of ylide **223** with various dipolarophilic alkenes to produce **224** after aromatization of the adducts was reported to proceed significantly more selectively in the presence of MnO_2 than in the original reaction where TPCD (tetrakis(pyridine) cobalt(II) dichromate) was used as the oxidant in which **224** were minor by-products <99JCR(S)552>.

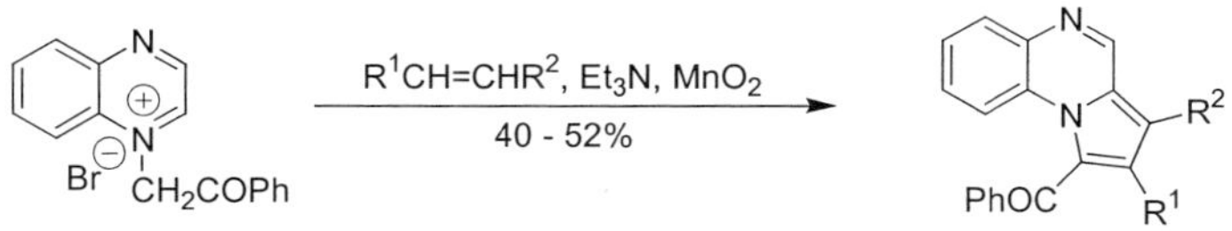

223 **224**

Another interesting example of a reaction of quinoxalines is the preparation of phenazines **173** from quinoxaline **172** which was covered in Section 6.2.8.1.

6.2.10 REFERENCES

92JOC1627	A.G. Martinez, A.H. Fernandez, F.M. Jimenez, A.G. Fraile, L.R. Subramanian, M. Hanack, *J. Org. Chem.* **1992**, *57*, 1627.
94CR2483	H.C. Kolb, M.S. VanNieuwenhze, K.B. Sharpless, *Chem. Rev.* **1994**, *94*, 2483.
99AC(E)3326	T. Braun, S.P. Foxon, R.N. Perutz, P.H. Walton, *Angew. Chem. Int. Ed.* **1999**, *38*, 3326.
99AX(C)1701	P.D. Robinson, M.P. Groziak, *Acta Crystallogr., Sect. C* **1999**, *C55*, 1701.
99BCJ1071	K. Kobayashi, H. Tanaka, H. Takabatake, T. Kitamura, R. Nakahashi O. Morikawa, H. Konishi, *Bull. Chem. Soc. Jpn.* **1999**, *72*, 1071.
99CC1461	T. Nagamatsu, T. Fujita, *J. Chem. Soc., Chem. Commun.* **1999**, 1461.
99CPB156	F. Wahid, C. Monneret, D. Dauzonne, *Chem. Pharm. Bull.* **1999**, *47*, 156.
99CPB791	Y. Murakami, H. Yokoo, Y. Yokoyama, T. Watanabe, *Chem. Pharm. Bull.* **1999**, *47*, 791.
99H299	A. Katoh, T. Fujimoto, M. Takahashi, J. Ohkanda, *Heterocycles* **1999**, *50*, 299.
99H611	I. Helland, T. Lejon, *Heterocycles* **1999**, *51*, 611.
99H1213	D.W. Rangnekar, V.R. Kanetkar, G.S. Shankarling, J.V. Malanker, C.R. Shanbhag, *J. Heterocycl. Chem.* **1999**, *36*, 1213.
99H1883	Z.-Z. Ma, Y. Hano, T. Nomura, Y.-J. Chen, *Heterocycles* **1999**, *51*, 1883.
99H2193	J.A. Vladerrama, H. Pessoa-Mahana, G. Sarras, R. Tapia, *Heterocycles* **1999**, *51*, 2193.
99H2349	C. Fruit, A. Turck, N. Ple, G. Queguiner, *Heterocycles* **1999**, *51*, 2349.
99H2407	M. Kaname, T. Tsuchiya, H. Sashida, *Heterocycles* **1999**, *51*, 2407.
99H2445	M. Soufyane, S. van der Broek, L. Khamliche, C. Mirand, *Heterocycles* **1999**, *51*, 2445.
99H2471	E. Okada, M. Tsukushi, *Heterocycles* **1999**, *51*, 2471.

99H2931	L. Hanaineh-Abdelnour, B.A. Salameh, *Heterocycles* **1999**, *51*, 2931.
99HCA238	D.M. Argilagos, A. Linden, H. Heimgartner, *Helv. Chim. Acta* **1999**, *82*, 238.
99JA969	T.M. Barclay, A.W. Cordes, R.C. Haddon, M.E. Itkis, R.T. Oakley, R.W. Reed, H. Zhang, *J. Am. Chem. Soc.* **1999**, *121*, 969.
99JA8783	C. Zhang, J.M. Tour, *J. Am. Chem. Soc.* **1999**, *121*, 8783.
99JA10438	C.B. Black, B. Andrioletti, A.C. Try, C. Ruiperez, J.L. Sessler, *J. Am. Chem. Soc.* **1999**, *121*, 10438.
99JCR(S)8	A.Z.A.E.-B. Hassanien, I.S.A. Hafiz, M.H. Elnagdi, *J. Chem. Res. (S)* **1999**, 8.
99JCR(S)88	K.M. Dawood, A.M. Farag, Z.E. Kandeel, *J. Chem. Res. (S)* **1999**, 88.
99JCR(S)96	A.A. Elassar, *J. Chem. Res. (S)* **1999**, 96.
99JCR(S)360	K.M. Al-Zaydi, E.A.A. Hafez, *J. Chem. Res. (S)* **1999**, 360.
99JCR(S)552	J. Zhou, L. Zhang, Y. Hu, H. Hu, *J. Chem. Res. (S)* **1999**, 552.
99JCS(P1)421	E. Erba, D. Pocar, M. Valle, *J. Chem. Soc., Perkin Trans. 1* **1999**, 421.
99JCS(P1)803	R.D. Chambers, M. Parsons, G. Sandford, C.J. Skinner, M.J. Atherton, J.S. Moilliet, *J. Chem. Soc., Perkin Trans. 1* **1999**, 803.
99JCS(P1)1325	A. Schmidt, M. Nieger, *J. Chem. Soc., Perkin Trans. 1* **1999**, 1325.
99JCS(P1)1789	E. Csikos, C. Gonczi, B. Podanyl, G. Toth, I. Hermecz, *J. Chem. Soc., Perkin Trans. 1* **1999**, 1789.
99JHC25	D.F. Gloster, L. Cincotta, J.W. Foley, *J. Heterocycl. Chem.* **1999**, *36*, 25.
99JHC113	J. Quiroga, M. Alvarado, B. Insuasty, *J. Heterocycl. Chem.* **1999**, *36*, 113.
99JHC145	M.L. Jones, D.P. Baccanari, R.L. Tansik, C.M. Boytos, S.K. Rudolph, L.F. Kuyper, *J. Heterocycl. Chem.* **1999**, *36*, 145.
99JHC301	M.S. South, *J. Heterocycl. Chem.* **1999**, *36*, 301.
99JHC441	A. Gangjee, L. Chen, *J. Heterocycl. Chem.* **1999**, *36*, 441.
99JHC501	J. Quiroga, M. Alvarado, B. Insuasty, M. Nogueras, A. Sanchez, J. Cobo, , *J. Heterocycl. Chem.* **1999**, *36*, 501.
99JHC783	N. Sato, N. Narita, *J. Heterocycl. Chem.* **1999**, *36*, 783.
99JHC787	Y.-X. Zhang, K. Saski, T. Hirota, *J. Heterocycl. Chem.* **1999**, *36*, 787.
99JHC985	E. Sotelo, E. Ravina, I. Estevez, *J. Heterocycl. Chem.* **1999**, *36*, 985.
99JHC1057	A. Sugimoto, Y. Yoshino, R. Watanabe, K. Mizuno, K. Uehara, *J. Heterocycl. Chem.* **1999**, *36*, 1057.
99JHC1271	B.E. Kornberg, S.S. Nakim, M.F. Rafferty, *J. Heterocycl. Chem.* **1999**, *36*, 1271.
99JMC833	R.J. Chorvat, R. Bakthavatchalam, J.P. Beck, P.J. Gilligan, R.G. Wilde, A.J. Cocuzza, F.W. Hobbs, R.S. Cheeseman, M. Curry, J.P. Rescinito, P. Krenitsky, D Chidester, J.A. Yarem, J.D. Klaczkiewicz, C.N. Hodge, P.E. Aldrich, Z.R. Wasserman, C.H. Fernandez, R. Zaczek, L.W. Fitzgerald, S.-M. Huang, H.L. Shen, Y.N. Wong, B.M. Chien, C.Y. Quon, A. Arvanitis, *J. Med. Chem.* **1999**, *42*, 833.
99JMC2266	S.S. Nikam, J.J. Cordon, D.F. Ortwine, T.H. Heimbach, A.C. Blackburn, M.G. Vartanian, C.B. Nelson, R.D. Schwarz, P.A. Boxer, M.F. Rafferty, *J. Med. Chem.* **1999**, *42*, 2266.
99JMC2561	G. Cirrincione, A.M. Almerico, P. Barraja, P. Diane, A. Lauria, A. Passannanti, C. Musiu, A. Pani, P. Murtas, C. Minnei, M.E. Marongiu, P.L. Colla, *J. Med. Chem.* **1999**, *42*, 2561.
99JOC634	A. Vasudevan. F. Mavandadi, L. Chen, A. Gangjee, *J. Org. Chem.* **1999**, *64*, 634.
99JOC4512	P. Pollet, A. Turck, N. Ple, G. Queguiner, *J. Org. Chem.* **1999**, *64*, 4512.
99JOC7717	S. Kumar, G. Hundal, D. Paul, M.S. Hundal, H. Singh, *J. Org. Chem.* **1999**, *64*, 7717.
99JOC7788	J.M. Minguez, J.J. Vaquero, J. Alvarez-Builla, O. Castano, J.L. Andres, *J. Org. Chem.* **1999**, *64*, 7788.
99JOC8425	J.J. Li, *J. Org. Chem.* **1999**, *64*, 8425.
99S495	A. Shimura, A. Momotake, H. Togo, M. Yokoyama, *Synthesis* **1999**, 495.
99S615	M. Alvarez, D. Fernandez, J.A. Joule, *Synthesis* **1999**, 615.
99S2057	C.-S. Yu, F. Oberdorfer, *Synthesis* **1999**, 2057.
99S2124	A. Acero-Alarcon, T. Armero-Alarte, J.M. Jorda-Gregori, C. Rojas-Argudo, E. Zaballos-Garcia, J. Server-Carrio, F.Z. Ahjyaje, J. Sepulveda-Arques, *Synthesis* **1999**, 2124.

99SC1393	A. Cuenca, S.E. Lopez, I. Garces, A. Aranda, *Synth. Commun.* **1999**, *29*, 1393.
99SC1503	Y.S. Lee, Y.H. Kim, *Synth. Commun.* **1999**, *29*, 1503.
99SC1577	N. Haider, E. Mavrokordatou, A. Steinwender, *Synth. Commun.* **1999**, *29*, 1577.
99SC2617	M.S. Khajavi, K. Rad-Moghadam, H. Hazarkhani, *Synth. Commun.* **1999**, *29*, 2617.
99SC3459	G. Abderrazak, S. Albdelaziz, C. Gerard, *Synth. Commun.* **1999**, *29*, 3459.
99SC3503	S. Shashikanth, S.K. Ahmad, G.L. Hegde, K.M.L. Rai, *Synth. Commun.* **1999**, *29*, 3503.
99SL756	K. Funabiki, H. Nakamura, M. Matsui, K. Shibata, *Synlett.* **1999**, *6*, 756.
99SL1203	F.R. Heirtzler, *Synlett.* **1999**, 1203.
99SL1265	E. Rossi, G. Abbiati, E. Pini, *Synlett.* **1999**, *6*, 1265.
99SL1579	M.S. Harr, A.L. Presley, A. Thorarensen, *Synlett.* **1999**, 1579.
99SL1993	H. Kotsuki, H. Sakai, H. Morimoto, H. Suenaga, *Synlett.* **1999**, 1993.
99T405	Y. Bessard, R. Crettaz, *Tetrahedron* **1999**, *55*, 405.
99T1491	P. Huszthy, E. Samu, B. Vermes, G. Mezey-Vandor, M. Nogradi, J.S. Bradshaw, R.M. Izatt, *Tetrahedron* **1999**, *55*, 1491.
99T4825	A.G. Martinez, A.H. Fernandez, R.M. Alvarez, M.D.M. Vilchez, M.L.L. Gutierrez, L.R. Subramanian, *Tetrahedron* **1999**, *55*, 4825.
99T5389	V.G. Chapoulaud, I. Salliot, N. Ple, A. Turck, G. Queguiner, *Tetrahedron* **1999**, *55*, 5389.
99T8475	E. Kolehmainen, Z. Kucybala, R. Gawinecki, J. Paczkowski, A. Kacala, *Tetrahedron* **1999**, *55*, 8475.
99T11859	L. Hanaineh-Abdelnour, S. Bayyuk, R. Theodorie, *Tetrahedron* **1999**, *55*, 11859.
99T12805	P. Meresse, E. Bertounesque, T. Imbert, C. Monneret, *Tetrahedron* **1999**, *55*, 12805.
99TA2775	E. Samu, P. Huszthy, L. Somogyi, M. Hollsi, *Tetrahedron: Asymmetry* **1999**, *10*, 2775.
99TA2797	M. McCarthy, R. Goddard, P.J. Guiry, *Tetrahedron: Asymmetry* **1999**, *10*, 2797.
99TL859	F.J. Romero-Salguero, J. Lehn, *Tetrahedron Lett.* **1999**, *40*, 859.
99TL1045	J.M. Cobb, M.T. Fiorini, C.R. Goddard, M.-E. Theoclitou, C. Abell, *Tetrahedron Lett.* **1999**, *40*, 1045.
99TL2541	M. Kawase, M. Hirabayashi, S. Saito, K. Yamamoto, *Tetrahedron Lett.* **1999**, *40*, 2541.
99TL3479	R. Olivera, R. SanMartin, S. Pascual, M. Herrero, E. Dominguez, *Tetrahedron Lett.* **1999**, *40*, 3479.
99TL3835	C.-S. Kim, K.C. Russell, *Tetrahedron Lett.* **1999**, *40*, 3835.
99TL4023	E.C. Taylor, B. Lui, *Tetrahedron Lett.* **1999**, *40*, 4023.
99TL4027	E.C. Taylor, B. Lui, *Tetrahedron Lett.* **1999**, *40*, 4027.
99TL4189	A. Krief, C. Colaux-Castillo, *Tetrahedron Lett.* **1999**, *40*, 4189.
99TL4507	J.J. Li, W.S. Yue, *Tetrahedron Lett.* **1999**, *40*, 4507.
99TL5111	A.S. Kiselyov, C. Dominguez, *Tetrahedron Lett* **1999**, *40*, 5111.
99TL6201	S. Brase, S. Dahmen, J. Heuts, *Tetrahedron Lett.* **1999**, *40*, 6201.

Chapter 6.3

Six-Membered Ring Systems: Triazines, Tetrazines and Fused Ring Polyaza Systems

Carmen Ochoa and Pilar Goya
Instituto de Química Médica (CSIC), Madrid, Spain.
e-mail: carmela@iqm.csic.es, iqmg310@iqm.csic.es

6.3.1 TRIAZINES

A comparison of chemical and photochemically induced reduction of some 2(4),5-dihydro-1,2,4-triazines and aromatic triazines has been investigated <99EJOC685>. Metallic complexes of 3-mercapto-1,2,4-triazin-5-ones with Co-II, Ni-II, Ru-II and Ru-III <99JIC239> and Ti-IV and Zr-IV <99IJC(A)956> have been described. Structure and magnetic properties of the organic triradical with triazine skeleton 2,4,6-tris[P-(*N*-oxy-*N*-*tert*-butylamino)phenyl]-1,3,5-triazine have been studied <99CL545>. Formation of 24 cooperative hydrogen bonds drives the spontaneous assembly of a rigid bifunctional trimelamine and bis-barbituric acid to give selectively the [2x2] hydrogen-bonded grid <99CC1311>. Noncovalent assembly of a 15-component hydrogen-bonded nanostructure, in which 1,3,5-triazines are involved, has been reported <99AG(E)933>. 2-Methoxy-4,6-dichloro-1,3,5-triazine has been used as a probe for the determination of the absolute configuration of (+)-1-(9-anthryl)ethylamine by circular dichroism <99JOC5754>. The photochemistry and photophysics of 2-(4'-methoxynaphthyl)-4,6-bis(trichloromethyl)-1,3,5-triazine have been investigated <99JA6167>.

6.3.1.1 Synthesis

A novel type of heterocyclization involving the [3+2] cycloaddition of *N,N*-dialkylamino-substituted thioisomünchnones **1** with azodicarboxylate **2** gives rise to 1,2,4-triazine derivatives **3** after a selective fragmentation pathway of the transient cycloadducts <99TL8675>.

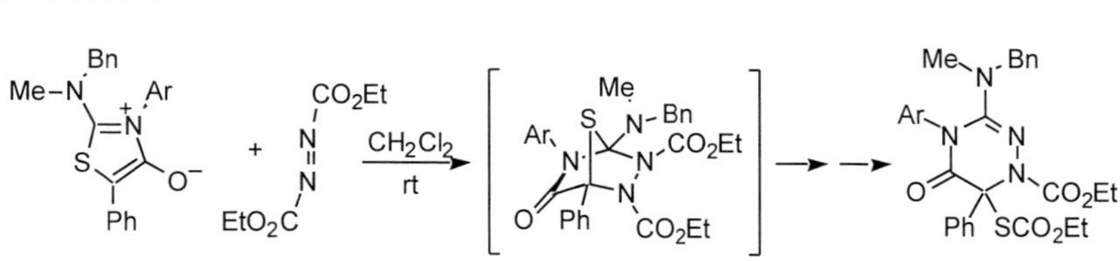

1a, Ar = Ph
1b, Ar = $4\text{-}MeOC_6H_4$

2

3a, Ar = Ph (71% yield)
3b, Ar = $4\text{-}MeOC_6H_4$ (50% yield)

Novel synthesis of 4,5-dihydro-1,2,4-triazin-6(1*H*)-ones has been developed by use of imidoyl chlorides <99AJC379>. New 5,5'-dithiophen-1,2,4-triazines have been synthesized by cyclocondensation of amidrazones with appropriate 1,2-dicarbonyl compounds <99T5047>. The biological activity of new 1,2,4-triazine-5-ones, obtained in the reaction of *N*-3-substituted amidrazones with dimethyl acetylenedicarboxylate, has been investigated <99PHA503>. New 3-alkylamino substituted 1,2,4-triazines have been synthesized as acetylcholinesterase inhibitors <99JMC730>. The first example of a triaryl-1,4-dihydro-1,2,4-triazine has been reported <99EJOC685>. 5-Arylidene-2-(2',4'-dinitrophenyl)-1,2,4-triazines have been synthesized from 4-arylidene-2-phenyl-5-oxazolones and dinitrophenylhydrazine followed by cyclocondensation in alkaline medium <99IJC(B)445>. The reaction of pyrido[1,2-*a*]pyrazines **4** with nitroso compounds **5** provides pyridyl substituted 1,2,4-triazinones **6** *via* a domino reaction which involves a cycloaddition and a ring transformation reaction <99JHC627>.

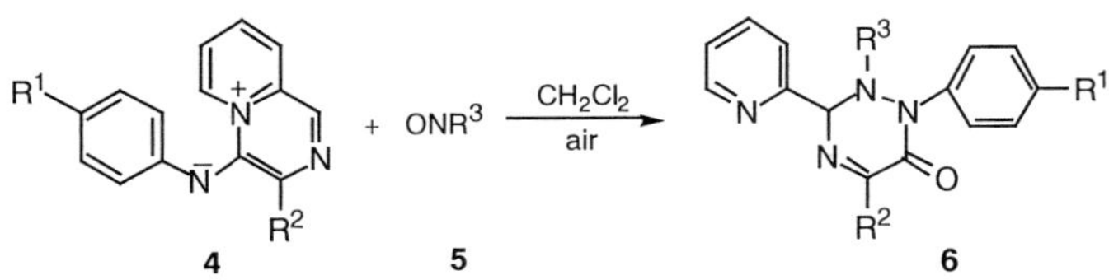

R^2 = NH-(*p*-R^1Ph)

The synthesis of a series of dichloroamino-1,3,5-triazines and dihalosulfonamido-1,3,5-triazines and investigation of their hindered rotation and stereodynamic behaviour by NMR spectroscopy have been reported <99JCS(P2)1231>. Combinatorial synthesis of 64 dihydrophenyl-1,3,5-triazines has been performed and the compounds screened as antifolates <99BMC1255>. A multidirectional cleavage procedure of 2-alkylsulfinyl intermediates with different nucleophiles forms highly substituted 1,3,5-triazines of type **9** and **10**. This methodology has successfully been transferred onto solid support using the polymer-bound thioronium salt **7** to yield the resin-bound thiol **8** or triazines **9** and **10**. On the other hand, the use of resin-bound thiol **8** and cyanuric chloride has chemioselectively generated on solid support 1,3,5-triazines libraries <98H2489>.

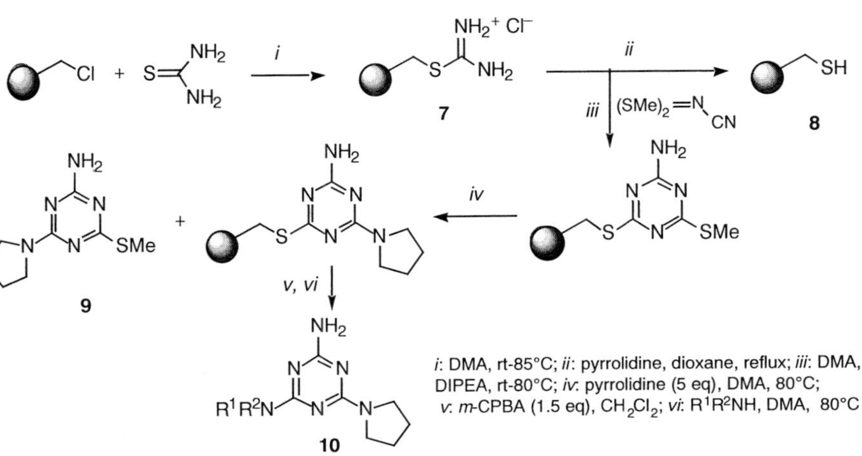

i: DMA, rt-85°C; *ii*: pyrrolidine, dioxane, reflux; *iii*: DMA, DIPEA, rt-80°C; *iv*: pyrrolidine (5 eq), DMA, 80°C; *v*: *m*-CPBA (1.5 eq), CH_2Cl_2; *vi*: R^1R^2NH, DMA, 80°C

4-Aryl-1-(4-tolyl)-2-phenyl- and 1-benzyl-2,4-diphenyl-1,3-diaza-1,3-butadienes are almost quantitatively transformed into the corresponding 1,3,5-triazines when allowed to stand at room temperature in benzene solution (five examples). The mechanism of the reaction is discussed <99H1401>. 2-(1-Methylhydrazino)-4,6-dimethyl-1,3,5-triazine (**12**) has been prepared by generating the free base from 1-methyl-1-aminoguanidine sulfate and treating it with ethyl *N*-acetylacetimidate (**11**). Reaction of the hydrazine derivative **12** with the appropriate isothiocyanate derivative gives the corresponding triazinyl thiosemicarbazides **13**. Attempts to cyclize triazines **13** to obtain triazinium betaine **14** have been unsuccessful, and triazine derivative **15a** (R=Ph) has been identified by X-ray analysis. The authors considered that triazines **15a** and **15b** could arise from nucleophilic attack of the N(8) of the thiosemicarbazide **13** to the C(5) of the putative triazinium betaine **14** <99JCS(P1)1517>.

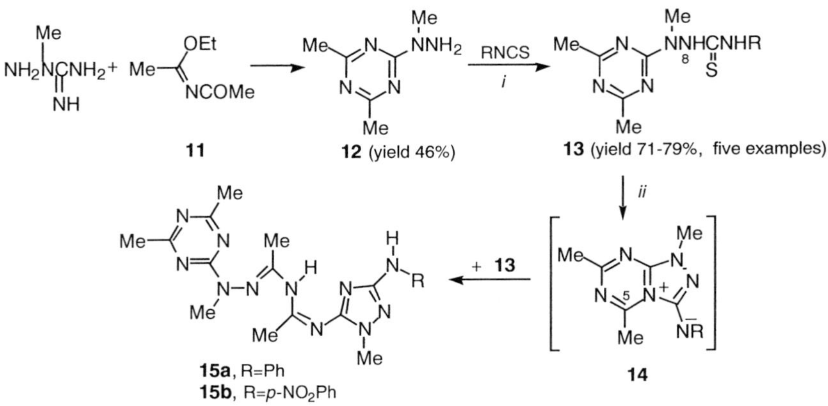

i: tert-butyl methyl ether; *ii:* DCC, r t

Reaction of hydroxyalkylbiuret derivatives **16** and diethyl carbonate yields 1-(hydroxyalkyl)-1,3,5-triazin-2,4,6-triones **17**, which lead to the corresponding 1-(carboxyalkyl)-1,3,5-triazin-2,4,6-triones **18**. Triazinones **18** afford 1:1 complexes with a series of aminoalkyl substituted pyrimidine-2,4,6-triamines, which form hydrogen-bonded molecular ribbons <99CEJ381>.

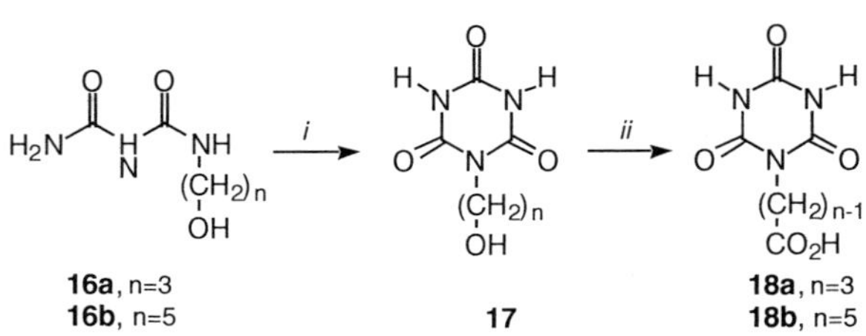

i: $(EtO)_2CO$, NaOEt, EtOH, reflux; *ii:* CrO_3/H_2SO_4, aqueous HOAc (**18a**); $NaIO_4/RuCl_3$, aqueous acetone (**18b**)

6.3.1.2 Reactions

Several examples of a new and simple "LEGO" system to obtain thienyl substituted 2,6-oligopyridines <99T5047>, 4-stannyl-, 4-bromo-oligopyridines and branched oligopyridines <99T5067>, stannylated bipyridines and terpyridines <99EJOC313> from

1,2,4-triazines *via* [4+2] cycloadditions have been reported. Synthesis and biological evaluation of novel 1,2,4-triazine nucleoside analogs containing 2,3-epoxypropyl or 3-amino-2-hydroxypropyl moieties have been reported <99EJM405>.

The synthesis of the novel [1,2]thiaphospholo[4,5-*e*][1,2,4]triazine ring system, exemplified by derivatives **20**, has been accomplished by the action of Lawesson's reagent on 1,2,4-triazin-6-ones **19** <99T13457>.

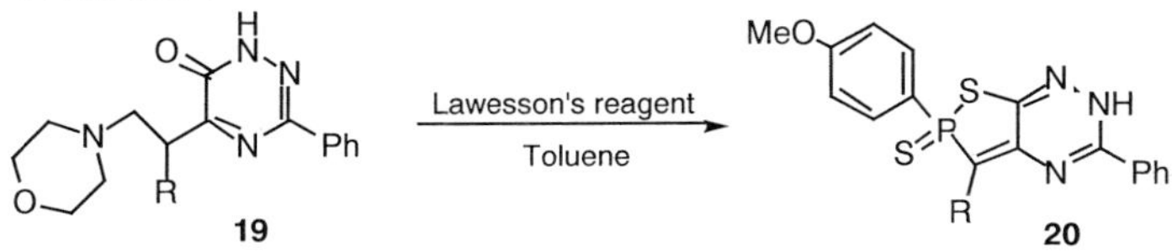

A series of papers dealing with transformations of 1,2,4-triazines under the actions of nucleophiles have been published <98MI388>, <98MI400>. The reaction of 3-pyrrolidino-1,2,4-triazine-4-oxide with ammonia leads to the product of *tele*-substitution of pyrrolidine-5-amino-1,2,4-triazine-4-oxide <99TL6099>. Reaction of 3-amino-5,6-disubstituted-1,2,4-triazines **21** with some cyclic and acyclic compounds yields new 3,5-disubstituted-6-methyl-1,2,4-triazines among which compounds **22** are an example. Compounds synthesized have been tested as anti-HIV and anticancer drugs <99PHA347>, <99PHA667>.

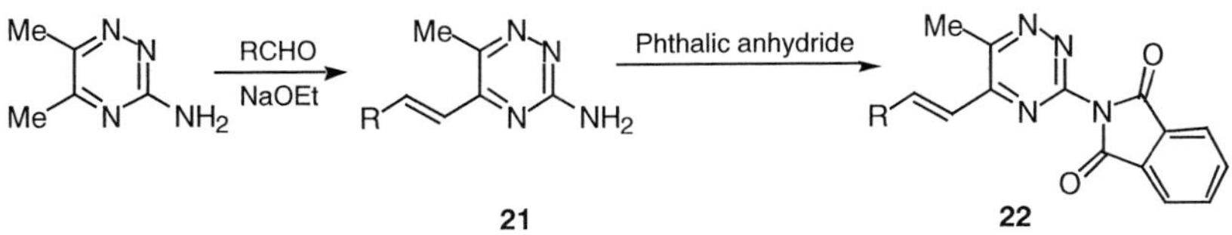

By reactions of different 1,2,4-triazine derivatives, C-ribosyl imidazo[2,1-*f*][1,2,4]triazines <99JCS(P1)2929> and C-ribosyl 1,2,4-triazolo[3,4-*f*][1,2,4]triazines <99JCS(P1)2937> have been synthesized as inhibitors of adenosine and AMP deaminases. Catalytic asymmetric aminohydroxylation with amino substituted 1,2,4-triazine and 1,3,5-triazine derivatives, as nitrogen sources, has been described <99AG(E)1080>.

From 1,3,5-triazines **23** and 5-amino-4-imidazole-carboxylic acids **24** a variety of purines and purine nucleosides **25** have been prepared *via* an inverse electron demand Diels-Alder reaction <99JA5833>.

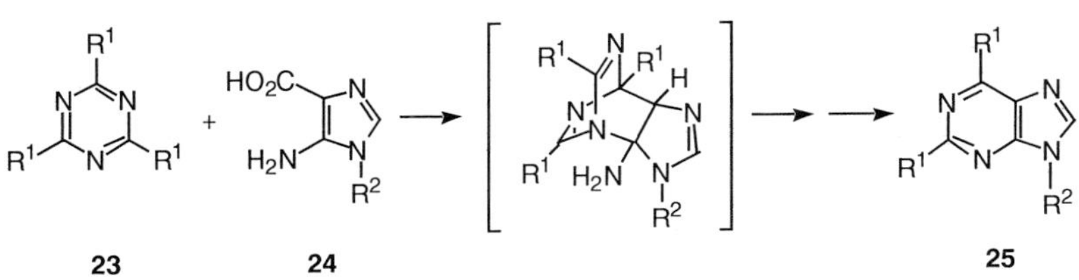

R^1 = H, CO_2Et, $CONH_2$; R^2 = Bn, β-D-ribofuranosyl, 2,3,5-tri-*O*-acetyl-β-D-ribofuranosyl

New reagents, 2-acyloxy-4,6-dimethoxy-1,3,5-triazines, have been used as acylating agents for the synthesis of esters from primary, secondary and tertiary alcohols. Because of mild acylation conditions the method could be applied to esterification of labile alcohols with aromatic and aliphatic, also α–branched, acids in good yields <99S593>.

Cyanuric chloride (**27**) has been loaded on different types of NH_2-functionalized resins **26** to give a new supported reagent **28**. This reagent has been employed for the solution-phase synthesis of different amides and dipeptides **29** <99OL1355>.

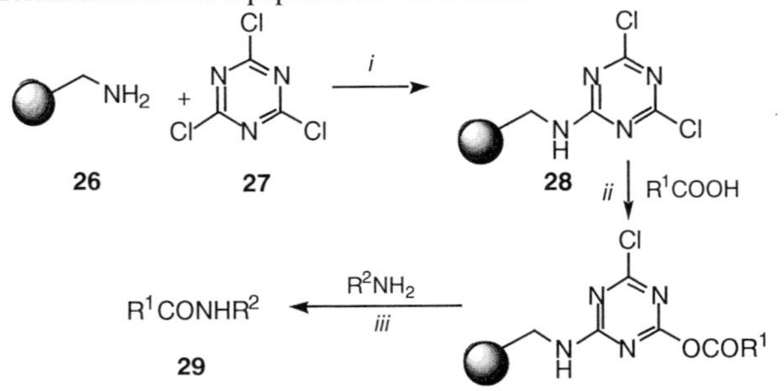

i: DIPEA,THF; *ii*:NMM, THF; *iii*:NMM,THF; fifteen examples, 60-87% yields

Many reactions of cyanuric chloride to give different 1,3,5-triazine derivatives have been reported including a series of cycloalkyl derivatives with airways smooth muscle relaxant properties <99BMC509>. Reaction of cyanuric chloride with different amines to give novel inhibitors of methyltransferases has been reported <99JMC3852>. Condensation of cyanuric chloride or other 2,4-di-chloro-6-(substituted-hydrazino)-1,3,5-triazines with isoniazide, pyrazinamide and 2-aminobutanol gave some new 1,3,5-triazine derivatives which were evaluated as antitubercular agents <99IJC(B)508>. Reaction of cyanuric chloride with alkylaminoethynyltributyltin afforded the corresponding 2-aminoacetylene-4,6-dichloro-1,3,5-triazines which were used to study their capability to form stacked "push-pull"-acetylenes <99HCA326>. Cyanuric chloride gave 2-anilino-1,3,5-triazines which were synthesized as potential nonpeptide corticotropin-releasing hormone antagonists <99JMC805>. Reaction of cyanuric chloride with 1-benzylpyrazole under solvent-free conditions and microwave irradiation yielded, within ten minutes, tris-2,4,6-(pyrazol-1-yl)-1,3,5-triazine by a quaternization-dequaternization procedure <98JHC1263>. Starting from cyanuric chloride, the synthesis of reactive 1,3,5-triazines bearing a cage system derived from adamantane has been reported. These compounds are precursors of hexamethyleneamine analogues <99H1891>.

Condensation of 1-[4,6-bis(allylamino)-1,3,5-triazin-2-yl]-4-piperidone (**32**) with amines **30** and **31** afforded the verapamil-like analogs **33** and **34** which were synthesized as potential drugs able to revert multidrug resistance (MDR). Both compounds show chemosensitizing activity but maintain some cardiovascular action <99JMC1687>.

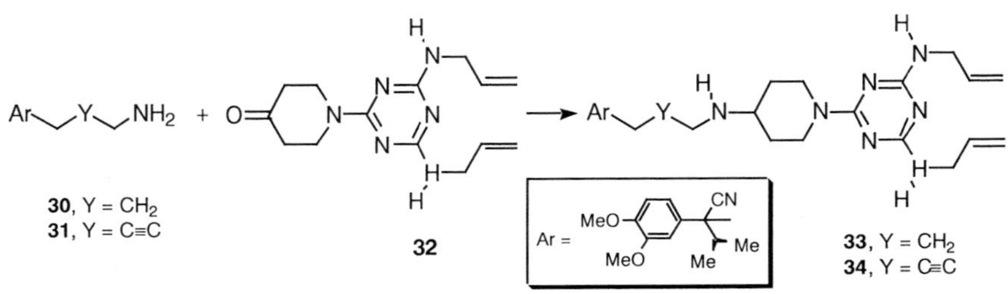

Nucleophilic attack of *N*-lithio-*S,S*-diphenylsulfilimine on 2,4-dichloro-6-propoxy-1,3,5-triazine affords the corresponding *N*-aryl-*S,S*-diphenylsulfilimine 2-mono (9% yield) and 2,6-disubstituted derivatives (3% yield) <99T10243>. Coupling of 2-chloro-4,6-dimethoxy-1,3,5-triazine and *N*-methylmorpholine, in THF, quantitatively yields 4-(4,6-dimethoxy-1,3,5-triazin-2-yl)-4-methylmorpholinium chloride (DMTMM). Condensation of carboxylic acids and amines by DMTMM proceeds effectively in THF to give the corresponding amides in good yield <99TL5327>. In the presence of catalytic amounts of (Ph3P)4Pd, 2-chloro-4,6-dimethoxy-1,3,5-triazine reacts with alk-1-ynes to give the corresponding 2-(alk-1'-ynyl)-derivatives in satisfactory yields <99TL8419>.

Reaction of 2,4,6-trifluoro-1,3,5-triazine and hexafluoropropene gives perfluoro-(isopropyl)-1,3,5-triazines which react with a range of oxygen nucleophiles. The application of this chemistry to the surface treatment of cellulose has been described <99JFC69>. Various routes and improved experimental conditions for the synthesis of 1,3,5-triazine-2,4,6-trisulfenyltrichloride derivatives from 1,3,5-triazine-2,4,6-trithiol have been reported and their reactions with water, alcohols, ammonia, and acetone investigated <99ZN(B)609>. The amination, transamination and disproportionation reactions of some hexahydro-1,3,5-triazines have been reported <99H2079>.

1,3,5-Triazine containing polysulfides having high molecular weights have been readily synthesized by the phase-transfer catalyzed polycondensation of 6-diethylamino or dibutylamino-1,3,5-triazine-2,4-dithiols with 1,10-dibromodecane in the presence of cetyl trimethyl ammonium bromide in a nitrobenzene-aqueous alkaline solution system <99MI294>.

The conversion of 2,4-bis(4-cyanophenyl)-6-dimethylamino-1,3,5-triazine into the corresponding diaryl cyclic diamidine *via* reaction with an excess of 1,2-diaminopropane saturated with hydrogen sulfide has been reported. 2,4-Bis(4-cyanophenyl)-1,3,5-triazine is resistant to cyclization in reacting with 1,4-diaminobutane under similar conditions and affords 2,4-bis[4-[*N*-(4-aminobutyl)thiocarbamoyl]phenyl]-1,3,5-triazine <99TL2841>.

2-Azido-4,6-dichloro-1,3,5-triazine **35**, matrix isolated in argon at 10 K, yields triplet nitrene **36** and the strained cyclic carbodiimide **37** upon photolysis <99CC2113>.

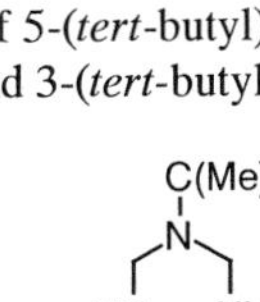

35

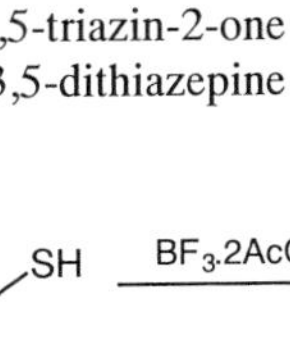

36

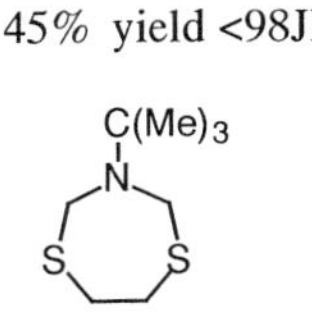

37

Reaction of 5-(*tert*-butyl)perhydro-1,3,5-triazin-2-one (**38**) and 1,2-ethanedithiol affords the new compound 3-(*tert*-butyl)perhydro-1,3,5-dithiazepine **39**, in 45% yield <98JHC1531>.

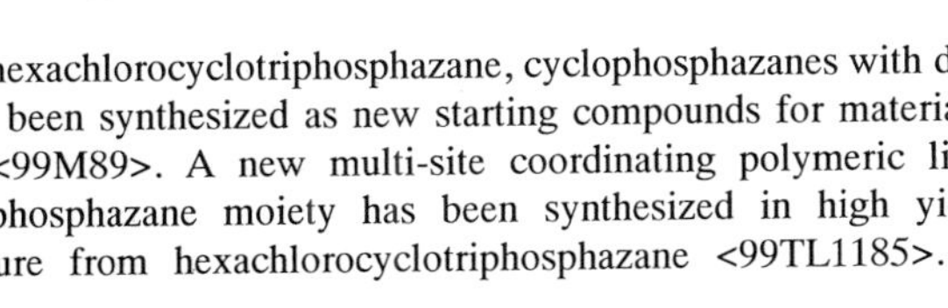

38 **39**

Starting from hexachlorocyclotriphosphazane, cyclophosphazanes with different alkoxysilyl substituents have been synthesized as new starting compounds for material synthesis *via* the sol-gel process <99M89>. A new multi-site coordinating polymeric ligand containing a pendant cyclotriphosphazane moiety has been synthesized in high yield by a simple synthetic procedure from hexachlorocyclotriphosphazane <99TL1185>. The synthesis of cyclic phosphazanes bearing diphenylphosphine groups has been reported <99POL2853>. Monochloropentaphenoxyphosphazane has been converted into alkoxy, aryloxy, and dialkylamino phosphazane azides <99JA884>. A regioselective route to new polytopic receptors by diaminolysis of chlorocyclotriphosphazatriene containing crown ethers has been reported <99JOC7299>. The reaction of cyclotriphosphazane **40** with two equivalents of the chelating diols 2,2'-dihydroxybiphenyl and 2',2''-dihydroxy-1',1''-bi(2-naphthyl) has been investigated, only the formation of the meso-compounds [(*R*,*S*)-(O,O)(2)Cl2P3N3], such as **41**, has been found <99MI1673>.

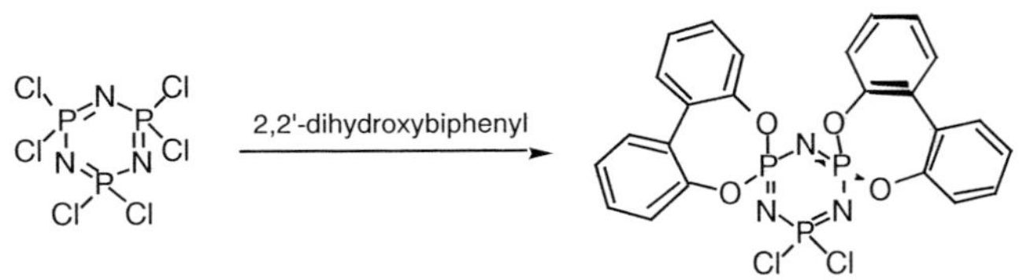

40 **41**

6.3.2 TETRAZINES

Coordination complexes derived from 3,6-di(2-pyridyl)-1,4-dihydro-1,2,4,5-tetrazines (DPDHT) and trimethylgallium have been described <98CJC1800>.

6.3.2.1 Synthesis

The stereoselective normal electron demand Diels-Alder reaction of chiral 1,2-diaza-1,3-butadienes **42**, derived from acyclic carbohydrates, with diethyl azodicarboxylate **2** yields the corresponding functionalized 1,2,3,6-tetrahydro-1,2,3,4-tetrazines **43**. The observed stereoselectivity is markedly dependent on the relative stereochemistry at C-1',2'. Reactions proceed slowly in benzene solution at room temperature, but are greatly accelerated by microwave irradiation <99JOC6297>.

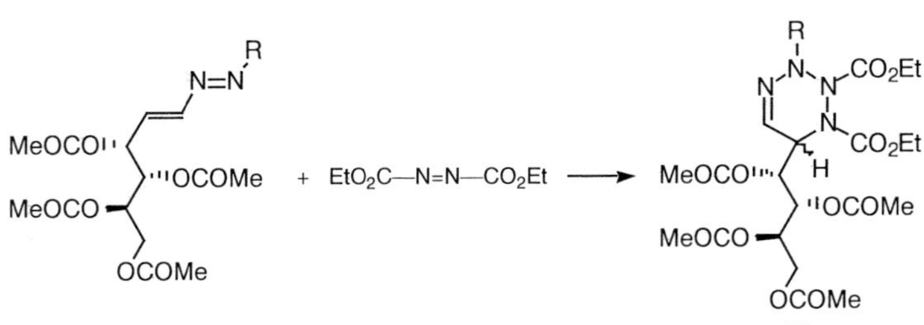

42 **2** **43**

6.3.2.2 Reactions

Appropriately substituted azolyldieneamines were found to undergo double inverse electron demand Diels-Alder reactions with tetrazine derivatives, yielding azolylpyridazines and dihydropyridazines as products <99TL6313>. The inverse electon demand Diels-Alder reactions of 3,6-di(trifluoromethyl)-1,2,4,5-tetrazine (**44**) with some 2-deoxy-α and β-D-riboses and proline derivatives have been used to obtain 1,2,4-triazine-C-nucleosides **45** <99ZN(B)549>, and hitherto unknown chiral 5-(2'-pyrrolidinyl-)1,2,4-triazines **46** <99TA573>. Both, 1,2,4-triazine derivatives **45** and **46**, are newly used in inverse electron demand Diels-Alder reactions to give pyridine substituted and isoquinoline substituted C-nucleosides of 2-deoxy-D-ribose and novel optically active nicotine analogs, respectively.

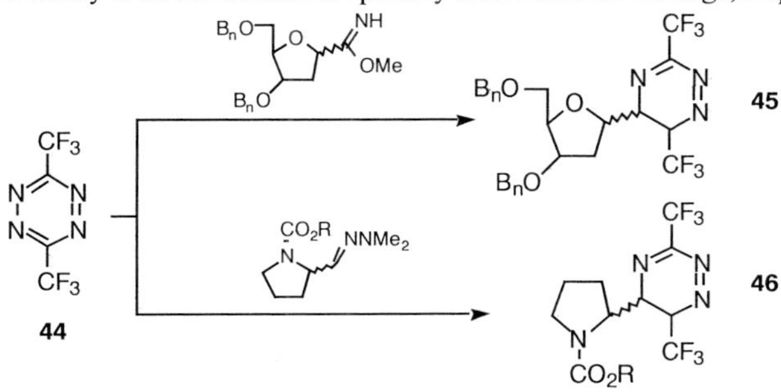

Treatment of 3-hydrazino-1,2,4,5-tetrazine (**47**), obtained following the pathway shown below, with diethoxymethyl acetate yields the unknown parent ring system 1,2,4-triazolo [4,3-*b*][1,2,4,5]tetrazine (**48**), some 3,6-disubstituted derivatives being also described <98JHC1329>.

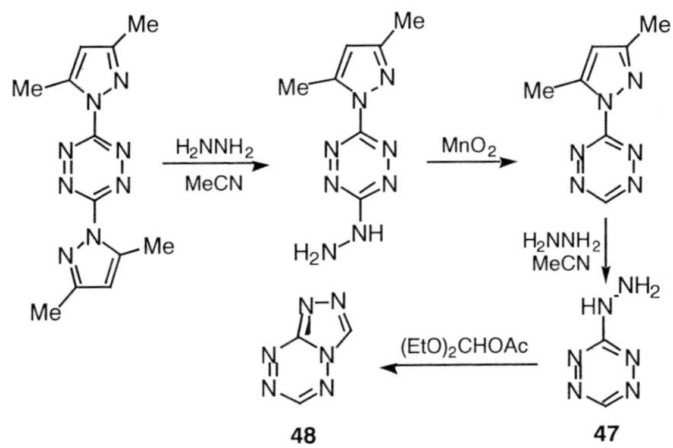

Dihydro-1,2,4-tetrazine **49** reacts with trimethylaluminium to produce mono **50a** and diketones **50b** depending upon the reaction conditions. Borohydride reduction of **50a** gives alcohol **50c**. Aromatization of **50a-c** by exposure to nitrous gases affords tetrazines **51a-c** which have proved to be very good electron-defficient heteroatomic azadienes for inverse electron demand Diels-Alder chemistry. Numerous examples are described with symmetric and nonsymmetric electron rich dienophiles <98JOC10063>.

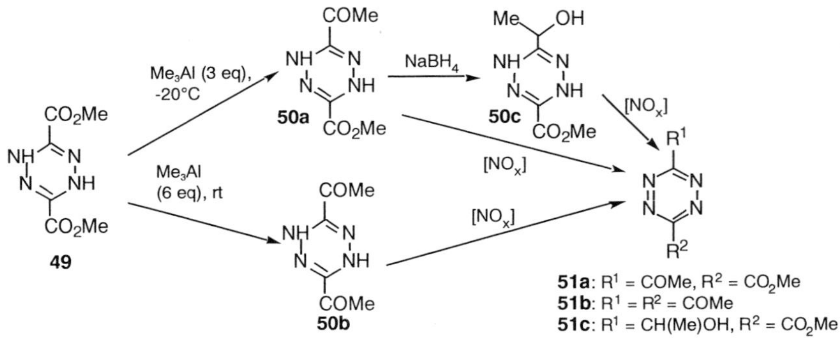

7-Arylidene-3,3-diethyl-3,4,6,7-tetrahydro-2*H*-thiazolo[3,2-*b*][1,2,4,5]-tetrazin-6-ones **53** have been synthesized in a single step by the condensation of tetrahydro-1,2,4,5-tetrazin-3-thione **52** with ethyl chloroacetate and aldehydes in the presence of pyridine and piperidine. Condensation of **53** with hydrazine afforded fused derivatives **54** <98IJC(B)819>.

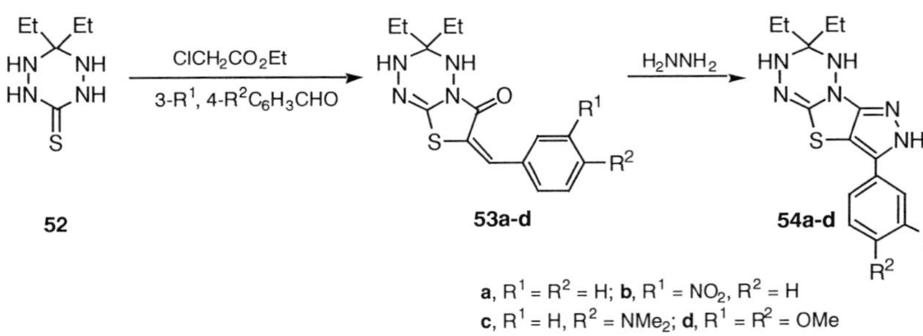

6.3.3 FUSED [6]+[5] POLYAZA SYSTEMS

A new purine derivative, 1,3-dimethylguanine, has been isolated from the ascidian *Botrylloides leachi*. Its structure has been elucidated by analysis of spectroscopic data by comparison with the regioisomer 1,3-dimethylisoguanine and by hydrolysis to theophylline <99MI638>.

6.3.3.1 Synthesis

Some new bactericidal and fungicidal derivatives of purine and pyrazolo[1,2,4]triazine have been synthesized by the reaction of 4-arylidene-2-phenyl-5-oxazolones with different nucleophilic reagents <99IJC(B)445>. A variety of pyrazolo[3,2-*c*][1,2,4]-triazin-3-yl **55**, **59** and 1,2,4-triazolo[3,2-*c*][1,2,4]-triazin-3-yl-phosphonic acid dialkyl esters **56**, **60** have been synthesized from diazobetaines of pyrazoles **57** and triazoles **58** and monocarbanions of certain phosphonates, respectively <99H513>.

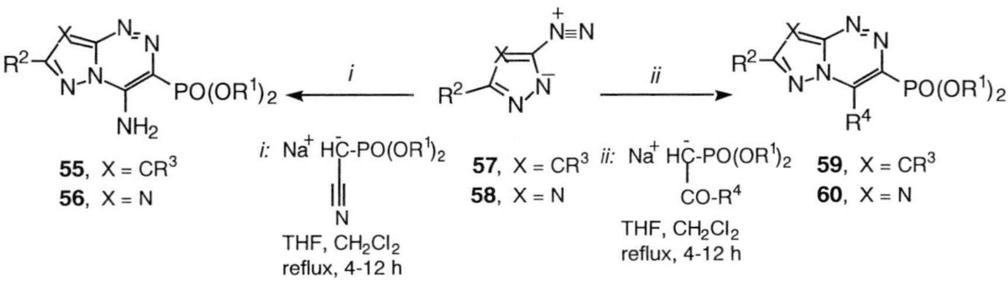

55, X = CR[3]
56, X = N
57, X = CR[3]
58, X = N
59, X = CR[3]
60, X = N

Methyl (4*R*)-3-(2-diazo-3-oxobutanoyl)-1,1-dioxo-1λ[6],3-thiazolidine-4-carboxylate (**61**) undergoes a base-induced cyclization to give methyl (8a*S*)-3-acetyl-4,7,7-trioxo-1,4,6,7,8,8a-hexahydro-7λ[6]-[1,3]thiazolo[4,5-*c*][1,2,4]triazine-8a-carboxylate (**62**), in high enantiomeric purity <99JCS(P1)1067>

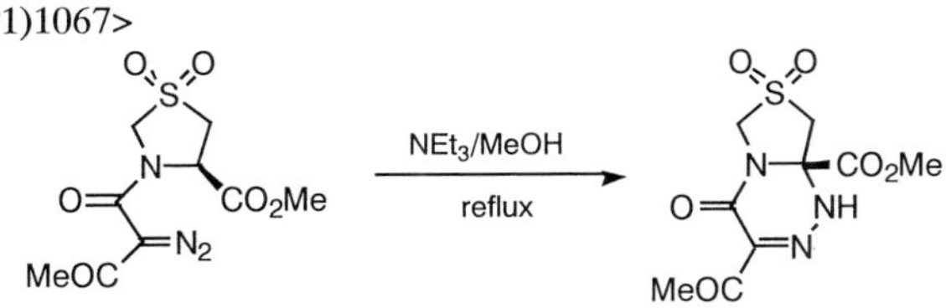

61 **62** (86% yield)

A series of potential receptor molecules based on the dipurinyl-2,6-pyridinecarboxamide motif has been prepared and the intramolecular hydrogen bonding characterized by ^{1}H-NMR <99T549>. The synthesis of [1,3,6,7-^{15}N,8-^{13}C] adenine using [1,3,7-^{15}N,8-^{13}C] xanthine, as starting labelled material, has been reported. The experimental procedure is an adaptation of the synthetic methods for the corresponding unlabelled compounds <99MI377>. A new and more efficient method for the synthesis of 8-^{13}C enriched adenine has been presented <99MI23>. Synthesis, corticotropin-releasing factor receptor binding affinity, and pharmacokinetic properties of a new series of *N*-aryl-purines, -triazolopyrimidines, -pyrrolopyrimidines, and -pyrrolopyridines have been described <99JMC833>.

The Schiff base derivative **65**, prepared from the respective 5,6-diaminouracil **63** and aldehydes **64**, can be converted into the C-8 substituted xanthines **66** through an oxidative cyclization with *m*-chloroperbenzoic acid <99H29>.

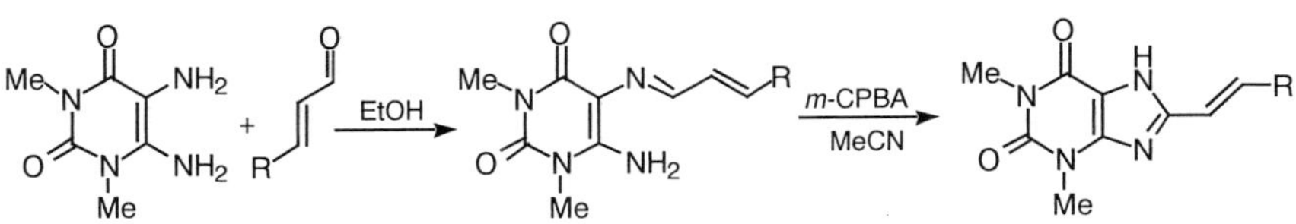

63 **64** **65** **66**

New pyrazolo[1,5-*a*]pyrimidine derivatives have been synthesized. These compounds are potent angiotensin-II receptor antagonists <99CPB928>. Pyrazolo[3,4-*d*]dihydropyridazinone derivatives have been obtained by the reaction of 5-methyl-4-methoxycarbonyl-3-acetyl-1-phenylpyrazole with different hydrazines <99TL3891>. A direct synthesis of pyrazolo [3,4-*b*]pyridines **69** from pyrazole **67** and benzothiazole **68**, through a Friendlander condensation, has been described <99SC655>.

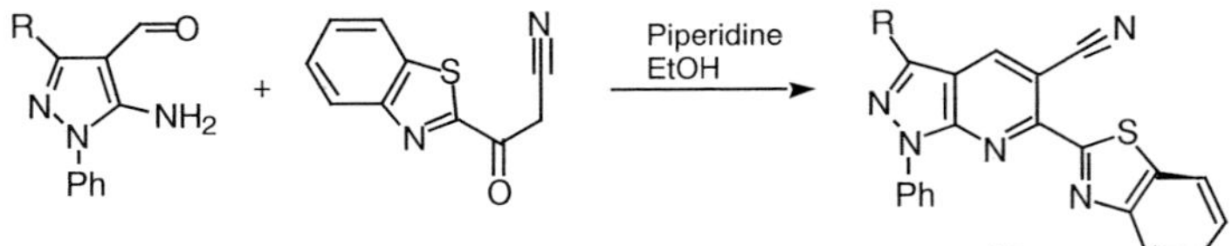

67 **68** **69**

The 2-hydrazino derivative of 6-perfluorohexylpyrimidines undergoes a rearrangement to give some new fluorinated 1,2,4-triazolo[1,5-*a*]pyrimidines <99JFC51>. New 1,2,4-triazolopyrimidinones have been synthesized from 5-cyano-6-(4-pyridyl)-2-thiouracil <99IJC(B)173>. Trifluoromethyl substituted triazolopyrimidines have been prepared by condensation of new 4-trifluoroacetyl-2,3-dihydropyrroles with hydrazines, as bifunctional *N*-nucleophiles, with opening of the dihydropyrrole ring <99TL2541>. Cyclocondensation of 3-amino-1,2,4-triazoles with substituted methyl cinnamates leads selectively to the formation of 7-aryl-6,7-dihydro[1,2,4]triazolo[1,5-*a*]pyridine-5(4*H*)-ones <99JHC205>. Several 1,2,4-triazolo[1,5-*a*]pyrimidino-2-sulfonanilides have been synthesized and their herbicidal activities tested <98MI521>. Triazolo[1,5-*a*]pyrimidinium 2-amines **71** have been synthesized from the reaction of 4-alkyl-3,5-diamino-1,2,4-triazoles **70** with pentane-2,4-dione or 1,1,3,3-tetramethoxypropane <99JCS(P1)1527>.

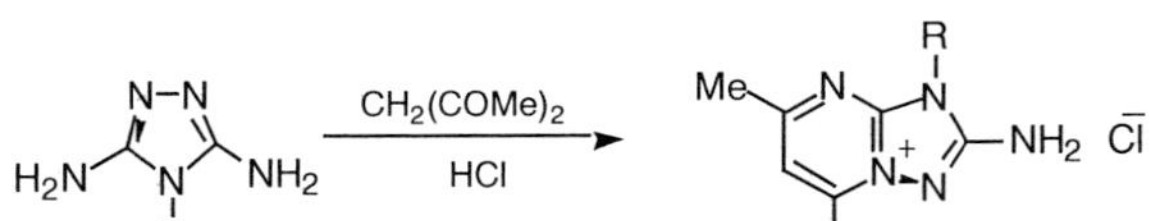

a, R = Me
b, R = $(CH_2)_2Me$
c, R = CH_2Ph

70 **71**

Solid-phase library synthesis of triazolopyridazines *via* a [4+2] cycloaddition strategy has been accomplished <99TL619>. Intramolecular bis-Mannich reaction of 3-aryl-5-mercapto-1,2,4-triazole, formaldehyde and α-phenylethylamine yields chiral 5-aryltriazolo[3,4-*b*]-[1,3,5]thiadiazine derivatives. These compounds have been screened for antibacterial activities and some of them show potent biological activity <99SC2027>.

6.3.3.2 Reactions

An unprecedented nitrogen elimination reaction of 4-amino-7-benzylpyrrolo [2,3-*d*][1,2,3]triazine-5-carbonitrile (**72**) to give the pyrrole derivative **74** has been described. The following mechanism, presumably *via* a a retro Diels-Alder reaction of the imino tautomer **73**, has been proposed <99OL537>.

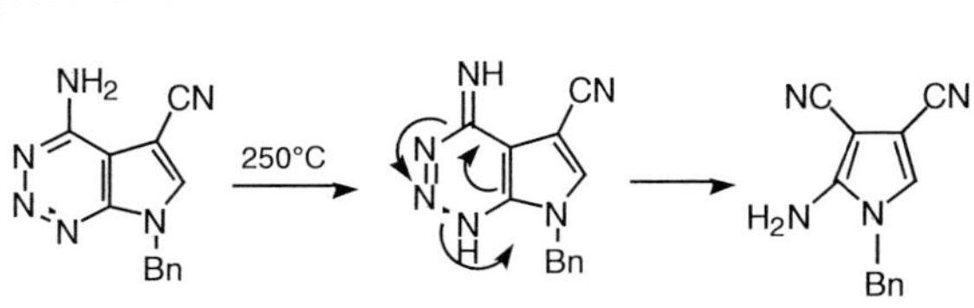

72 **73** **74**

The Suzuki-Miyaura cross-coupling reaction of 9-benzyl-6-chloropurine with boronic acids gives 6-alkylated purines in moderate to excellent yields. The reaction has been successfully applied for the synthesis of 6-phenylpurine bases and nucleosides <99SL1145>. Reaction of 6-chloropurine and (*Z*)-4-amino-2(and 3)-methylbut-2-en-1-ols yields (*Z*)-6-(4-hydroxy-3-methylbut-2-en-1-ylamino)purine, (Z)-zeatin, and the isomeric (*Z*)-isozeatin, respectively, both free of the *E*-isomers <99CCC696>. Analogs of the cytokinins E-zeatin and benzylaminopurine have been prepared by Heck coupling on 6-vinylpurines or Sonogashira coupling on 6-halopurines as key-steps <99T211>. Starting from 2,6-dichloropurine new 2,9-disubstituted 6-benzylaminopurines have been synthesized as cyclin-dependent kinase inhibitors <99AP187>. Mitsunobu reaction of 2,6-dichloropurine with 2,6-difluorobenzyl alcohol gives 2,6-dichloro-9-(2',6'-difluorobenzyl)-9*H*-purine and its 7-benzylated isomer in 73% and 10% yield, respectively <99CPB574>. The regiochemistry in the Pd-mediated coupling between 6,8-diaholopurines and organometallic reagents has been examined. In the case of 6,8-dichloropurines highly selective coupling occurs at the purine 6-position and, the regiochemical outcome is completely reverse when a better leaving group (Br or I) is introduced at C-8 <99ACS366>.

Readily accessible *N*-2-acetyl-*O*-6-(2-(*p*-nitrophenyl)ethyl)guanine undergoes Mitsunobu reactions with either a primary or secondary alcohol to generate only 9-substituted derivatives of the starting guanine <99SC3003>. Condensation of 8-bromotheophylline with arylalkyl and aryloxyalkyl bromides and subsequent aminolysis with glycine to obtain *N*-(7-substituted-8-theophyllyl)-glycines has been reported <99PJC783>. A variety of caffeine analogs, in which methyl groups at the 1 and 7 positions have been replaced with alkyl chains containing different functional groups, have been synthesized from the appropriate purines, in order to increase the known ability to induce Ca^{+2} release from intracellular stores of caffeine itself <99JMC2527>. A bacterial consortium consisting of strains belonging to the genus *Klebsiella* and *Rhodococcus* quantitatively converts 1-, 3-, and 7-substituted xanthines to their respective 8-oxo derivatives <99JCS(P1)677>.

Methylation of adenine 3-oxide derivatives with methyl iodide in *N,N*-dimethylacetamide affords the correspoding 3-methoxyadenines which undergo hydroxy-ion attack at the 2 position to give isoguanine derivatives in 38% yied <99CPB554>. Nucleophilic addition and cycloaddition to 2- and 8-vinylpurines have been reported <99ACS269>. The chemistry, physicochemical properties and biological activities of different *N*-oxygenated adenines have been described <99H1971>. Lithiation at the 2-position of purine ring has been accomplished, for the first time, by using a riboside of 8-triisopropylsilylpurine derivative as substrate and lithium 2,2,6,6-tetramethylpiperidide (LTMP) as lithiating agent <99JOC7773>.

Remarkable acceleration has been observed for *N*-sulfopropylation of heterocyclic compounds such as adenine using 1,3-propane sulfone (**75**) under microwave irradiation affording the *N*-sulfopropyl derivative **76**, in 30 seconds and 95% yield <98TL9587>.

75 **76**

A series of 2-amino-6-fluoro-9-(1,3-dihydroxy-2-propoxymethyl)purine mono and diesters has been synthesized, as potential prodrugs of ganciclovir, from 2-amino-6-chloro-9-(1,3-dihydroxy-2-propoxymethyl)purine <99JMC324>. 2-Amino-6-arylsulfanylpurines **78** can easily be prepared from guanine (**77**) in what is essentially a one-pot reaction <99T5239>.

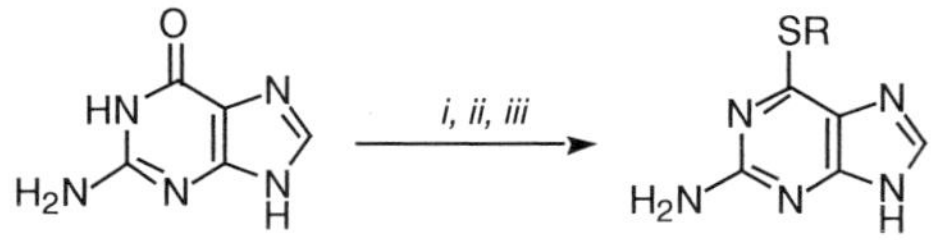

77

78a, R = *p*-ClPh (88% yield)
78b, R = *p*-MePh (76% yield)
78c, R = Ph (75% yield)

i: $(CF_3CO)_2O$, C_5H_5N, 0°C, 35 min.
ii: *p*-chlorothiophenol for **78a**, *p*-methylthiophenol for **78b** and thiophenol and MeCN for **78c**, rt, 2 h
iii: **a**, aq. NH_3 (d 0.88), **b**, 27% H_2O_2, rt, for **78a** and **78b** or $MeNH_2$ in EtOH for **78c**

Sandmeyer reaction of 3-nitropyrazolo[3,4-*c*]pyridines yields the corresponding 3-cyano derivatives <99H1661>. Reactions from pyrazolo[3,4-*d*]pyrimidine nucleosides afford oligonucleotide derivatives <99JCS(P1)479>. Lithiation of 1*H*-pyrazolo[3,4-*d*]pyrimidine derivatives using lithium alkanetellurolate has been reported <99TL2139>. Selective removal of chlorine at the 7-position on 1,2,4-triazolo[1,5-*a*]pyrimidine ring has been described <99JHC183>. 1,2,4-Triazolo[4,3-*c*]pyrimidin-5-ones give rise to a novel Dimroth-type rearrangement leading to 1,2,4-triazolo[1,5-*c*]pyrimidin-5-ones <99JCS(P1)1333>.

6.3.4 FUSED [6]+[6] POLYAZA SYSTEMS

6.3.4.1 Synthesis

Diazotization of 2-amino-N-cyanomethylbenzamide (**79**) and subsequent treatment with sodium azide yields 3-cyanomethyl-1,2,3-benzotriazin-4-one **80** <99H1295>.

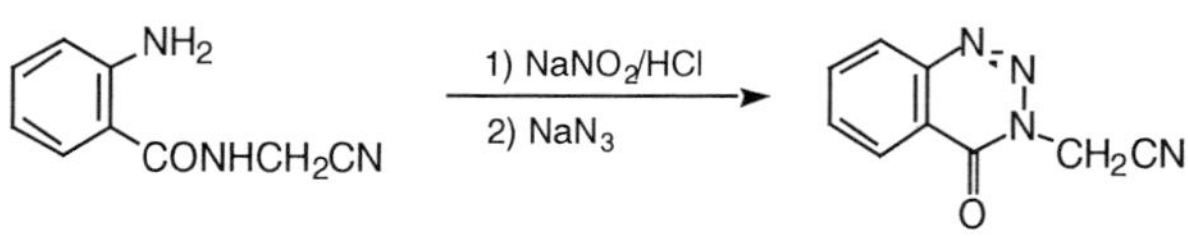

79 **80**

New 4-thiadiazolylmethyl substituted 1,2,4-benzotriazines have been prepared as anticonvulsant agents <99IJC(B)623>. Reductive cyclization of carbohydrate 2-nitrophenylhydrazones to chiral functionalized 1,2,4-benzotriazines has been reported <99JHC589>. Some new heterobicyclic nitrogen systems bearing the 1,2,4-triazine moiety (**81, 82** and **83**) have been achieved by treatment of 3-amino-5,6-disubstituted-1,2,4-triazines with some cyclic and acyclic oxygen compounds followed by heterocyclization. Anti-HIV and anticancer *in vitro* activities have been observed for some members <99PHA347>.

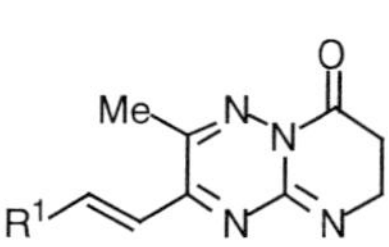

81

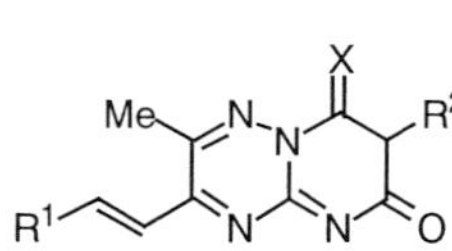

82

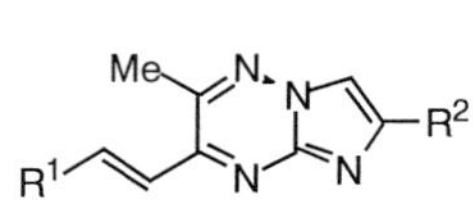

83

Pyrimido[1,2,4]triazine derivatives **84** and **85** have been synthesized from 2-thio-4-phenyl-6-pyrimidinone and their antibacterial and antifungal activities tested <99ZN(B)788>.

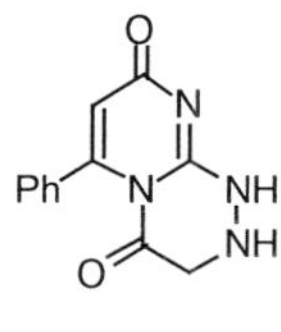

84

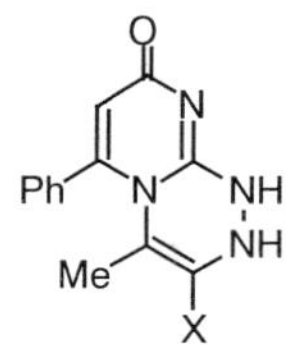

85

Treatment of 6-benzylidenehydrazinouracils **86** with potassium nitrate and sulfuric acid in acetic acid affords fervenulin-4-oxides **87** <99M819>.

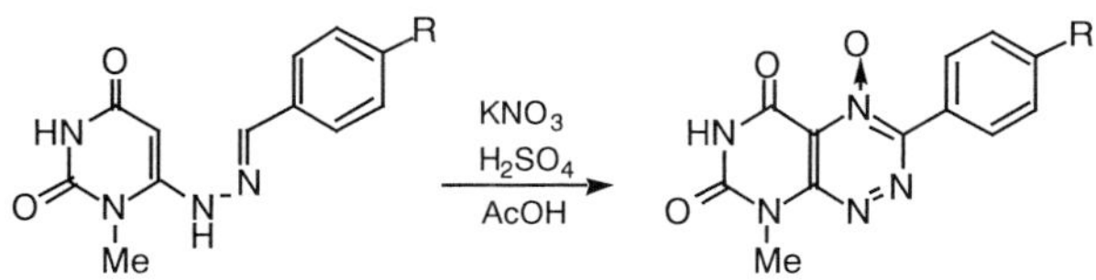

86, R = H, Cl **87**, R = H, Cl

New derivatives of 4-amino and 2,4-diaminopteridines have been synthesized and their capability to inhibit neuronal nitric oxide synthase evaluated <99JMC4108>. The synthesis of folic acid multiply labeled with stable isotopes, for bioavailability studies in human nutrition, has been reported <99JCS(P1)1311>. Synthesis and antiviral evaluation of several 6-(methylenecarbomethoxy)pteridine-4,7-diones have been described <99JHC435>. Synthesis and biochemical evaluation of bis(6,7-dimethyl-8-D-ribityllumazines) as potential bisubstrate analog inhibitors of riboflavin synthase have been reported <99JOC4635>. Synthesis and cyclization of novel lumazine-enediyne chimeras have been reported <99H13>.

A new route to pyrido[1,2-*b*]pyridazinium inner salts has been achieved *via* a 1,3-dipolar cycloaddition-ring expansion process <99TL763>. The synthesis, in several steps, of new 6-aroylpyrido[2,3-*d*]pyrimidines <99JHC501> and novel 5-aryl-6-cyano-pyrido [2,3-*d*]pyrimidinediones <99JHC113> from 6-aminopyrimidin-4-ones have been reported. Reactions of 6-aminopyrimidines with biselectrophiles to give pyrido and pyrimidopyrimidines have been carried out <99JOC634>. An easy access to bicyclic peptides with an octahydropyrazino[1,2-*a*]pyrazine skeleton has been reported <99EJOC1345>. The reaction of 6-hydrazinouracils and acetylenedicarboxylates, at room temperature, affords tetrahydro-pyrimido[4,5-*c*]pyridazines in excellent yield <99TL1793>. Novel pyrazino [2,3-*c*][1,2,6] thiadiazine 2,2-dioxides **89** have been prepared from triaminothiadiazine 1,1-dioxide **88** and carbonyl compounds. Regioselectivity at the 7-position can be achieved using the α-hydroxyiminoketone as carbonyl compound. Different *N*-substituted derivatives **90** have been prepared and tested as inhibitors of platelet aggregation <99JMC1698> and their activity has been optimized by means of a QSAR study using LDF *ab initio* calculations <99JMC3279>.

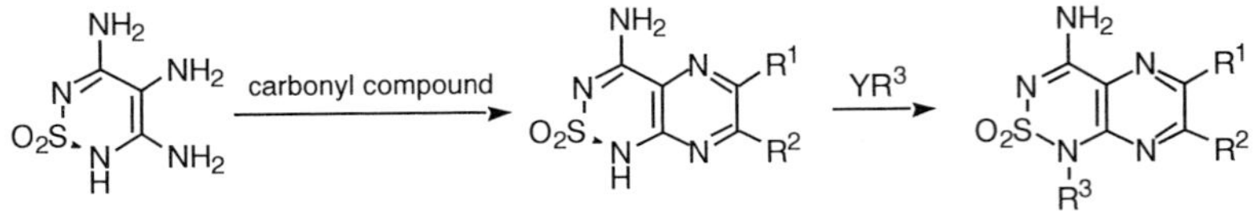

88 **89** **90**

Synthesis and structural studies of new 3-alkylamino-pyrido[4,3-*e*][1,2,4]thiadiazine 1,1-dioxides have been reported <99T5419>.

6.3.4.2 Reactions

Diels-Alder reaction of 1,2,3-benzotriazines with enamines yields 2-arylquinoline and 2-aryl-4-quinolone alkaloids <99CPB1038>. Refluxing in acetic acid 3-isoxazolyl-1,2,3-benzotriazin-4-one derivatives and potassium iodide the corresponding *N*-isoxazolyl-2-iodobenzamides have been obtained <99FES90>. *N*-[*O*-1,2,3-Benzotriazin-4(3*H*)one-yl]-3-(2-pyridyldithio)-propionate (**92**) has been synthesized, from hydroxybenzotriazinone **91**. Benzotriazinone **92** reacts with secondary amines to give high yields of acylated products **93** (five examples) <99TL1107>.

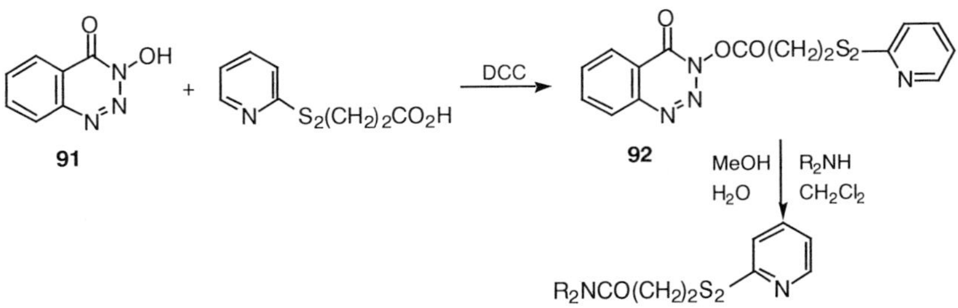

91 **92** **93**

R = Me, Pr, $(CH_2)_{17}Me$, $(CH_2)_2OH$, CH_2COONa

A novel method to introduce carbon substituents into the pteridine skeleton has been described <99H117>. The Wittig reaction has been applied to 6-formylpterin and 6-formyl-7,7-dihydro-7,7-dimethylpterin, in order to provide access to highly functionalized pterins as potential inhibitors of enzymes in the folate biosynthesis pathway <99JCS(P1)163>.

S-Galactosyl derivatives of 6,7-diaryl-2-thiolumazines have been synthesized from the corresponding thiolumazines <98MI197>. 6-Acetyl and 7-methyl groups of 6-acetyl-1,3,7-trimethyllumazine have been oxidized to carboxylic acid by nitric acid. The oxidation of starting lumazine affords 1,3,7-trimethyllumazine-6-carboxylic acid or 1,3-dimethyllumazine-6,7-dicarboxylic acid depending on the concentration of nitric acid <99H857>.

Oxidation of pyrido[2,3-*e*][1,2,4]thiadiazines **94** with sodium hypochlorite and, separately, *m*-chloroperbenzoic acid affords the 1,1-dioxide derivatives **95** and the 5-oxide derivatives **97**, respectively. Oxidation of 1,1-dioxide derivatives yields the novel 1,1,5-trioxides **96** <98T13645>.

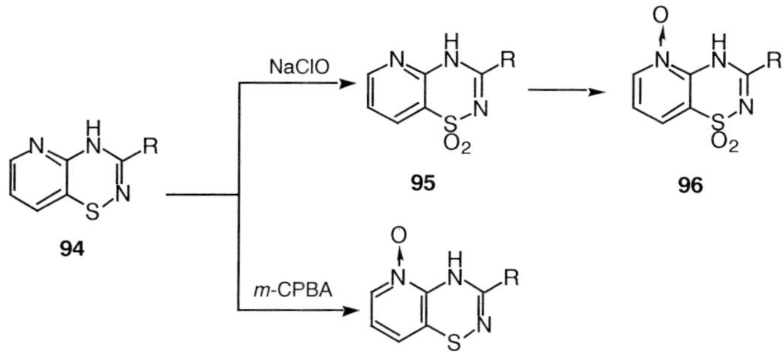

94 95 96 97

A series of 6- and 7-acrylamide derivatives of 4-(phenylamino)-quinazoline and 4-(phenylamino)-pyridopyrimidine, classes of epidermal growth factor receptor (EGFR) inhibitors, have been prepared from the corresponding amino compounds by reaction with acroyl chloride/base <99JMC1803>. Reaction of thionyl chloride with hexahydro-7-methyl-pyrimido[1,6-*a*]-pyrimidine-6,8-dione yields the corresponding 9,9'-thiobispyrimidopyrimidine derivative <99JHC453>.

6.3.5 MISCELLANEOUS FUSED RING POLYAZA SYSTEMS

Many reports have dealt with structures which can be included under this heading and so, only examples bearing triazine and tetrazine rings will be highlighted.

6.3.5.1 Synthesis

The novel heterocyclic system **102** with the phenantrene type skeleton, in which the benzene ring is annulated with two 1,2,3,4-tetrazine 1,3-dioxide rings has been reported. The step-by-step synthetic approach involves treatment of **98** with N_2O_5, resulting in the first 1,2,3,4-tetrazine 1,3-dioxide ring formation. Displacement of the chlorine atom at 6-position of **99** with ammonia yields the mixture of isomers **100** and **101** which are separated by HPLC. Subsequent treatment of **101** with N_2O_5 results in the formation of the second 1,2,3,4-tetrazine 1,3-dioxide ring. System **102** is rather stable and comparatively much more stable than 1,2,3,4-tetrazines lacking the N-oxide oxygens, thus **102** undergoes nucleophilic attack to give **103**, the tetrazinobenzotetrazine skeleton being left intact <99OL721>.

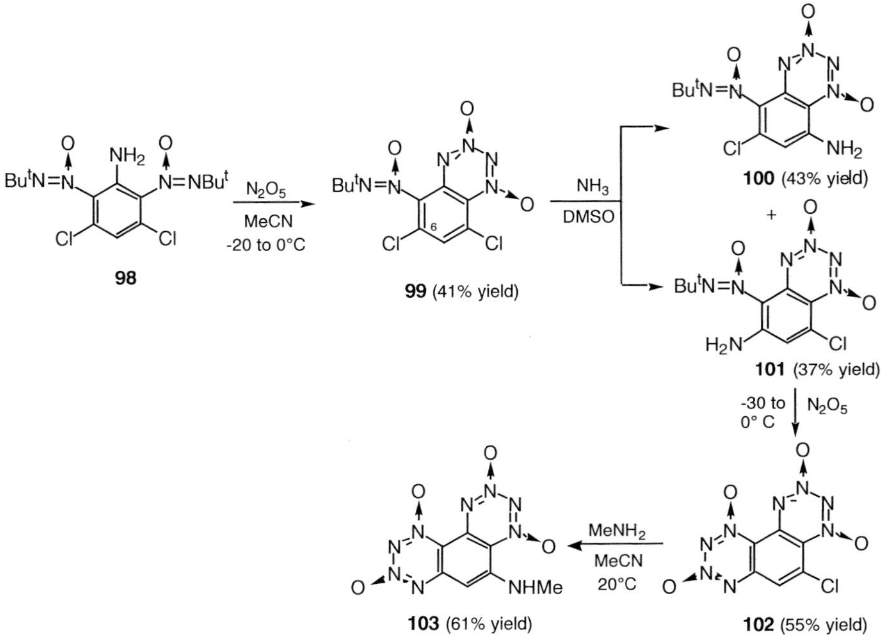

Reaction of 2-phenyl-3,1-benzoxazin-4*H*-one (**104**) with thiocarbohydrazide yields thioxotetrazinoquinazoline **105** when the reaction mixture is melted in an oil bath at 160 °C <99IJC(B)850>.

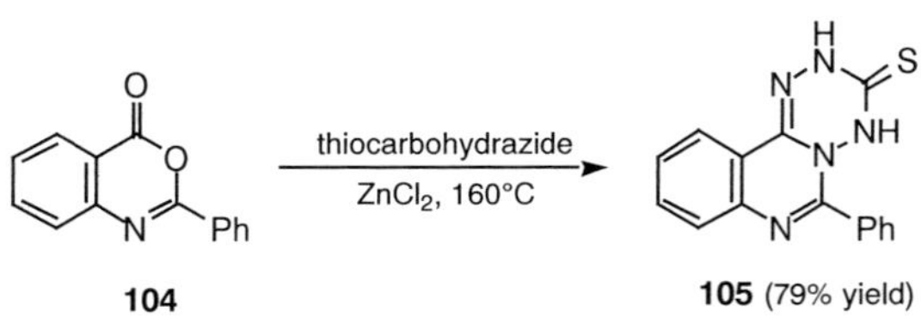

The synthesis of new substituted pyrazolo[5,1-*c*][1,2,4]benzotriazines 5-oxides **106**, their binding activities at the central benzodiazepine receptor <99JMC2218>, <99FES375> and their antimicrobial activity <99MCR223> have been reported. Derivatives of the new ring system indolo[1,2-*c*]benzo[1,2,3]triazine **107** have been synthesized by diazotization of substituted 2-(2'-aminophenyl)indoles followed by an intramolecular coupling reaction of the diazonium group with the indole nitrogen. Some derivatives show potent antitumor and antimicrobial activities <99JMC2561>.

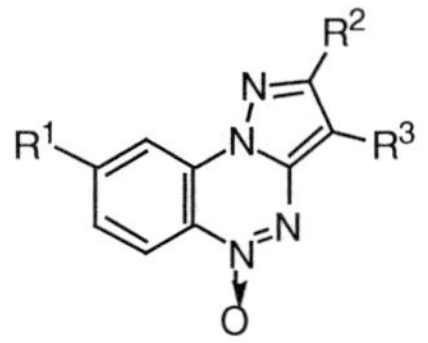

106

R^1 = H, Cl
R^2 = H, Ph, 2- thienyl, 3- thienyl, 2- furyl
2-MeOC_6H_4; 2-MeO,5-Br- C_6H_3, 3-Br-2-furyl
R^3 = H, Br, Ph, 2-FC_6H_4, 3-FC_6H_4, 4-FC_6H_4

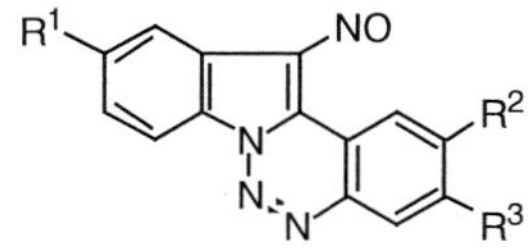

107

R^1 = H, MeO
R^2 = H, Cl
R^3 = H, Cl

Straightforward synthesis of some substituted naphtho[1,2,4]triazines **108** and quinolino[1,2,4]triazines **109-112**, *via* the cyclocondenstion of nitronaphthalenes and nitroquinolines with guanidine have been reported <99JOC3361>.

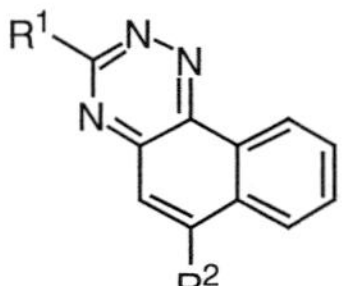

108

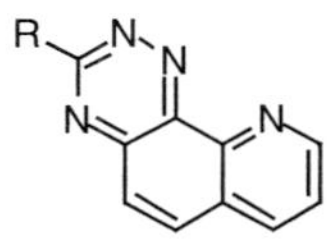

109

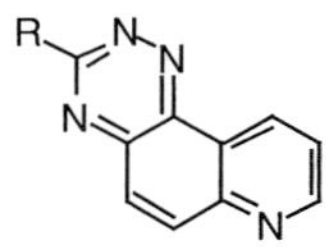

110

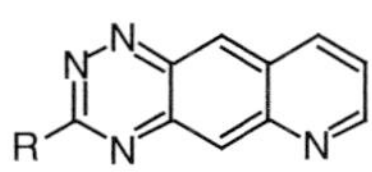

111

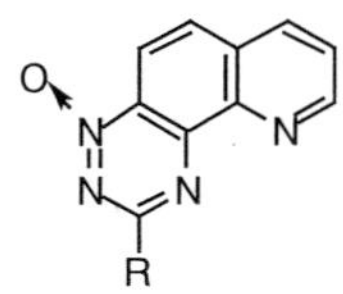

112

A convenient approach to the synthesis of 2-aryl-4-alkyl-1,3,5-triazino[1,2-*a*]benzimidazoles **113** from 2-aminobenzimidazole has been reported <98IJC(B)1283>. New 8-cyanopyridothieno[1,2,3]triazines **114** as inhibitors of nitric oxide and eicosanoid biosynthesis have been described <99JMC4720>. Synthesis of pyridodithienotriazines **115**, from dithienopyrimidine derivatives, and their antihistaminic and cytotoxic activities have been reported <98EJM887>.

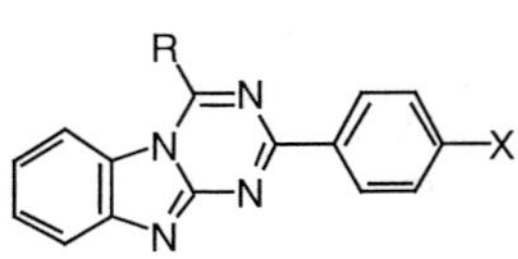

113

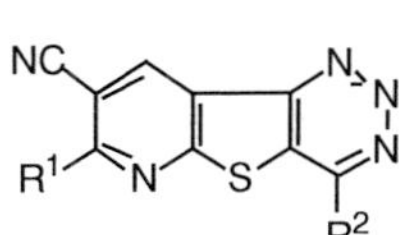

114

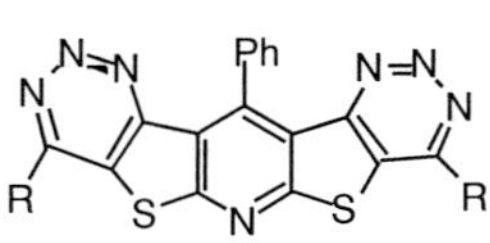

115

6.3.5.2 Reactions

Sterically controlled regiospecific heterocyclization of aldehydo sugars (5-methyl-1,2,4-triazino[5,6-*b*]indol-3-yl)hydrazones **116** to 3-(alditol-1-yl)-10-methyl[1,2,4]triazolo-[4',3':2,3][1,2,4]triazino[5,6-*b*]indoles **117** has been carried out and the antimicrobial activity of some representative memberes tested <99PHA580>.

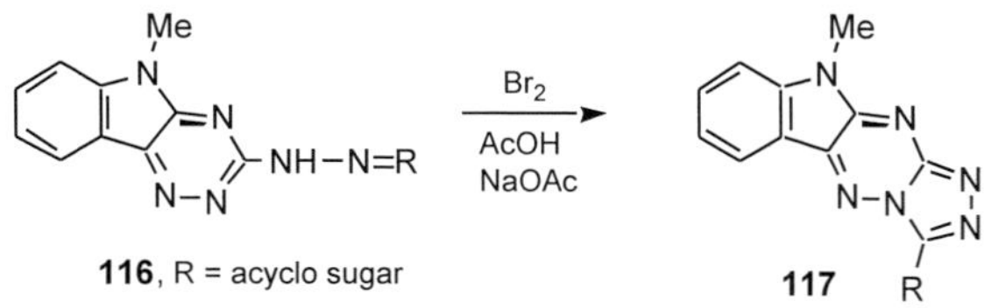

6.3.6 REFERENCES

98CJC1800	P. N. Preston, S. J. Rettig, A. Storr, J. Trotter, *Can. J. Chem.*, **1998**, *76*, 1800.
98EJM887	J. M. Quintela, C. Peinador, M. C. Veiga, L. M. Botana, A. Alfonso, R. Riguera, *Eur. J. Med. Chem.*, **1998**, *33*, 887.
98H2489	T. Masquelin, N. Meunier, F. Gerber, G. Rosse, *Heterocycles*, **1998**, *48*, 2489.
98IJC(B)819	J. Mohan, S. Kataria, *Indian J. Chem (B)*, **1998**, *37*, 819.
98IJC(B)1283	B. S. Reddy, T. Sambaiah, K. K. Reddy, *Indian J. Chem (B)*, **1998**, *37* , 1283.
98JHC1263	I. Almena, E. Díez-Barra, A. de la Hoz, J. Ruiz, A. Sánchez-Migallón, J. Elguero, *J. Heterocyclic Chem.*, **1998**, *35*, 1263.
98JHC1329	D. E. Chavez, M. A. Hiskey, *J. Heterocyclic Chem.*, **1998**, *35*, 1329.
98JHC1531	U. Wellmar, *J. Heterocyclic Chem.*, **1998**, *35*, 1531.
98JOC10063	M. Girardot, R. Nomak, J. K. Snyder, *J. Org. Chem.*, **1998**, *63*, 10063.
98MI197	M. A. M. Alho, C. Ochoa, A. Chana, N. B. D'Accorso, *An. Asoc. Quim. Argent.*, **1998**, *86*, 197.
98MI388	O. N. Chupakhin, V. N. Kozhevnikov, D. N. Kozhevnikov, V. L. Rusinov, *Russ. J. Org. Chem*, **1998**, *34*, 388. *Chem. Abstr.*, **1999**, *130*, 25045j.
98MI400	V. L. Rusinov, D. N. Kozhevnikov, E. N. Ulomskii, O. N. Chupakhin, G. G. Aleksandrov, H. Neunhoeffer, *Russ. J. Org. Chem*, **1998**, *34*, 400. *Chem. Abstr.*, **1999**, *130*, 25046k.
98MI521	G. F. Yang, H. Y. Liu, X. F. Yang, H. Z. Yang, *Chinese J. Chem.*, **1998**, *16*, 521.
98T13645	C. G. Neill, P. N. Preston, R. H. Wightman, *Tetrahedron*, **1998**, *54*, 13645.
98TL9587	M. Adamczyk, S. Rege, *Tetrahedron Lett.*, **1998**, *39*, 9587.
99ACS269	F. S. Liu, B. Dalhus, L. L. Gundersen, F. Rise, *Acta Chem. Scand.*, **1999**, *53*, 269.
99ACS366	J. M. J. Nolsoe, , L. L. Gundersen, F. Rise, *Acta Chem. Scand.*, **1999**, *53*, 366.
99AG(E)933	K. A. Jolliffe, P. Timmerman, D. N. Reinhoudt, *Angew. Chem. Int. Ed*, **1999**, *38*, 933.
99AG(E)1080	L. J. Goossen, H. Liu, K. R. Dress, K. B. Sharpless, *Angew. Chem. Int. Ed*, **1999**, *38*, 1080.
99AJC379	D. J. Collins, T. C. Hughes, W. M. Johnson, *Aust. J. Chem.*, **1999**, *52*, 379.
99AP187	C. H. Oh, S. C. Lee, K. S. Lee, E. R. Woo, C. Y. Hong, B. S. Yang, D. J. Baek, J. H. Cho, *Arch. Pharm.*, **1999**, *332*, 187.
99BMC509	F. Leroux, B. J. van Keulen, J. Daliers, N. Pommery , J. P. Hénichart, *Bioorg. Med. Chem.*, **1999**, 509.
99BMC1255	H.-K. Lee, W.-K. Chui, *Bioorg. Med. Chem.*, **1999**, 1255.
99CC1311	P. Lipkowski, A. Bielejewska, H. Kooijman, A. L. Spek, P. Timmerman, D. N. Reinhoudt, *J. Chem. Soc., Chem. Commun.*, **1999**, 1311.
99CC2113	G. Bucher, F. Siegler, J. J. Wolff, *J. Chem. Soc., Chem. Commun.*, **1999**, 2113.
99CCC696	V. Tolman, J. Hanus, P. Sedmera, *Collect. Czech. Chem. Commun.*, **1999**, *64*, 696.
99CEJ381	M. Mascal, J. Hansen, P. S. Fallon, A. J. Blake, B. R. Heywood, M. H. Moore, J. P. Turkenburg, *Chem.-Eur. J*, **1999**, *5*, 381.
99CL545	S. Hayami, K. Inoue, *Chem. Lett.*, **1999**, 545.
99CPB548	M. Nagaoka, E. Nagaseva, S. Sato, M. Numazawa, *Chem. Pharm. Bull.*, **1999**, *47*, 548.
99CPB554	T. Itaya, T. Kanai, Y. Takada, M. Kaneko, K. Yasuhara, T. Fujii, *Chem. Pharm. Bull.*, **1999**, *47*, 554.

99CPB574 S. Kozai, T. Maruyama, *Chem. Pharm. Bull.*, **1999**, *47*, 574.

99CPB928 T. Shiota, T. Yamamori, K. Sakai, M. Kiyokawa, T. Honma, M. Ogawa, K. Hayashi, N. Ishizka, K. Matsumura, M. Hara, M. Fujimoto, T. Kawabata, S. Nakajima, *Chem. Pharm. Bull.*, **1999**, *47*, 928.

99CPB1038 J. Koyama, I. Toyokuni, K. Tagahara, *Chem. Pharm. Bull.*, **1999**, *47*, 1038.

99EJOC313 J. Sauer, D. K. Heldmann, G. R. Pabst, *Eur. J. Org. Chem.*, **1999**, 313.

99EJOC685 J. Nagy, A. Horvath, A. Szollosy, J. Nyitrai, *Eur. J. Org. Chem.*, **1999**, 685.

99EJOC1345 V. N. Belov, C. Funke, T. Labahn, M. Essayed, A. Demeijere, *Eur. J. Org. Chem.*, **1999**, 1345.

99EJM405 S. Raicmalic, M. Grdisa, K. Pavelic, M. Mintas, *Eur. J. Med. Chem.*, **1999**, *34*, 405.

99FES90 D. Raffa, G. Daidone, B. Maggio, D. Schillaci, F. Plescia, L. Torta, *Farmaco*, **1999**, *54*, 90.

99FES375 G. Guerrini, A. Costanzo, F. Bruni, G. Ciciani, S. Selleri, P. Gratteri, B. Costa, C. Martini, A. Lucacchini, *Farmaco*, **1999**, *54*, 375.

99H13 N. Choy, K. C. Russell, *Heterocycles*, **1999**, *51*, 13.

99H29 A. D. Araujo, E. Bacher, F. W. J. Demnitz, D. A. Santos, *Heterocycles*, **1999**, *51*, 29.

99H117 S. Muruta, M. Kujime, T. Sugimoto, K. Murakami, C. Seo, *Heterocycles*, **1999**, *50*, 117.

99H513 T. Ankenbrand, R. Neidlein, *Heterocycles*, **1999**, *51*, 513.

99H857 Y. Kim, J. Kim, S. Jang, Y. Kang, *Heterocycles*, **1999**, *51*, 857.

99H1295 G. Broggini, L. Garanti, G. Molteni, G. Zecchi, *Heterocycles*, **1999**, *51*, 1295.

99H1401 E. Rossi, G. Abbiati, D. Nava, *Heterocycles*, **1999**, *51*, 1401.

99H1661 J. C. Milhavet, A. Gueiffier, L. Bernal, J. C. Teulade, *Heterocycles*, **1999**, *51*, 1661.

99H1891 M. Hedayatullah, C. Lion, A. B. Slimane, L. Da Conceiçao, I. Nachawati, *Heterocycles*, **1999**, *51*, 1891.

99H1971 T. Fujii, T. Itaya, *Heterocycles*, **1999**, *51*, 1971.

99H2079 A. Flores-Parra, S. A. Sánchez-Ruiz, *Heterocycles*, **1999**, *51*, 2079.

99HCA326 D. Berger, P. Wilhelm, M. Neuenschwander, *Helv. Chim. Acta*, **1999**, *82*, 326.

99IJC(A)956 S. K. Sengupta, O. P. Pandey, A. K. Srivastava, R. Rai, K. D. Mishra, *Indian J. Chem (A)*, **1999**, *38*, 956.

99IJC(B)173 D. N. Upadhyay, V. J. Ram, *Indian J. Chem (B)*, **1999**, *38*, 173.

99IJC(B)445 H. A. Allimony, H. A. Saad, F. A. A. El Mariah, *Indian J. Chem (B)*, **1999**, *38*, 445.

99IJC(B)508 C. V. Shabadi, B. A. Shelar, A. R. Shelar, *Indian J. Chem (B)*, **1999**, *38*, 508.

99IJC(B)623 S. D. Srivastavas, T. R. Rawat, *Indian J. Chem (B)*, **1999**, *38*, 623.

99IJC(B)850 M. A. Abdo, I. F. Zeid, G. A. Elhiti, O. E. Mahmoud, *Indian J. Chem (B)*, **1999**, *38*, 850.

99JA884 M. B. Mcintosh, T. J. Hartle, H. R. Allcock, *J. Am. Chem. Soc*, **1999**, *121*, 884.

99JA5833 Q. Dang, Y. Liu, M. D. Erion, *J. Am. Chem. Soc*, **1999**, *121*, 5833.

99JA6167 G. Pohlers, J. C. Scaiano, E. Step, R. Sinta, *J. Am. Chem. Soc*, **1999**, *121*, 6167.

99JCS(P1)163 C. L. Gibson, A. Lang, K. Ohta, C. J. Suckling, *J. Chem. Soc. Perkin Trans 1*, **1999**, 163.

99JCS(P1)479 F. Seela, M. Zulauf, *J. Chem. Soc. Perkin Trans 1*, **1999**, 479.

99JCS(P1)677 K. M. Madyastha, G. R. Sridhar, *J. Chem. Soc. Perkin Trans 1*, **1999**, 677.

99JCS(P1)1067 M. J. Betts, R. G. Pritchard, A. Schofield, R. J. Stoodley, S. Vohra, *J. Chem. Soc. Perkin Trans 1*, **1999**, 1067.

99JCS(P1)1311 P. Maunder, P. M. Finglas, A. I. Mallet, F. A. Mellon, M. A. Razzaque, B. Ridge, L. Vahteristo, C. Wittoft, *J. Chem. Soc. Perkin Trans 1*, **1999**, 1311.

99JCS(P1)1333 D. Loakes, D. M. Brown, S. A. Salisbury, *J. Chem. Soc. Perkin Trans 1*, **1999**, 1333.

99JCS(P1)1517 D. L. Crabb, K. J. McCullough, P. N. Preston, G. M. Rosair, B. C. Bishop, S. H. B. Wright, W. Clegg, S. Coles, *J. Chem. Soc. Perkin Trans 1*, **1999**, 1517.

99JCS(P1)1527 B. C. Bishop, H. Marley, P. N. Preston, S. H. B. Wright, *J. Chem. Soc. Perkin Trans 1*, **1999**, 1527.

99JCS(P1)2929 P. J. Dudfield, V. D. Le, S. D. Lindell, C. W. Rees, *J. Chem. Soc. Perkin Trans 1*, **1999**, 2929.

99JCS(P1)2937 P. J. Dudfield, V. D. Le, S. D. Lindell, C. W. Rees, *J. Chem. Soc. Perkin Trans 1*, **1999**, 2937.

99JCS(P2)1231 S. A. Brewer, H. T. Burnell, I. Holden, B. G. Jones, C. R. Willis, *J. Chem. Soc. Perkin Trans 2*, **1999**, 1231.

99JFC51 L. M. Popova, A. U. Trishina, S. V. Vershilov, A. I. Ginak, B. N. Maksimov, *J. Fluorine Chem.*, **1999**, *96*, 51.

99JFC69 R. D. Chambers, C. Magron, G. Sandford, J. A. K. Howard, D. S. Yufit, *J. Fluorine Chem.*, **1999**, *97*, 69.

99JHC113 J. Quiroga, M. Alvarado, B. Insuasty, M. Nogueras, A. Sánchez M. D. López, *J. Heterocyclic Chem.*, **1999**, *36*, 113.

99JHC183 W. T. Monte, W. A. Kleschick, J. Bordner, *J. Heterocyclic Chem.*, **1999**, *36* , 183.

99JHC205 S. M. Desenko, V. V. Lipson, O. V. Shishkin, S. A. Komykhov, V. D. Orlov, E. E. Lakin, V. P. Kuznetsov, H. Meier, *J. Heterocyclic Chem.*, **1999**, *36*, 205.

99JHC435 S. Molina, J. Cobo, A. Sánchez, M. Nogueras, E. De Clercq, *J. Heterocyclic Chem.*, **1999**, *36*, 435.

99JHC453 O. Simo, A. Ribar, J. Alfoldi, V. Patoprsty, *J. Heterocyclic Chem.*, **1999**, *36*, 453.

99JHC501 J. Quiroga, G. Viveros, B. Insuasty, M. Nogueras, A. Sánchez, J. Cobo, *J. Heterocyclic Chem.*, **1999**, *36*, 501.

99JHC589 J. Andersch, D. Sicker *J. Heterocyclic Chem.*, **1999**, *36*, 589.

99JHC627 T. Billert, R. Beckert, P. Fehling, M. Doring, J. Brandenburg, H. Gorls, P. Langer, *J. Heterocyclic Chem.*, **1999**, *36*, 627.

99JIC239 B. Ramachandra, B. Narayana, *J. Indian Chem. Soc.*, **1999**, *76*, 239.

99JMC324 D. K. Kim, K. Y. Chang, G. J. Im, H. T. Kim, N. K. Lee, K. H. kim *J. Med. Chem.*, **1999**, *42*, 324.

99JMC730 J. M. Contreras, Y. M. Rival, S. Chayer, J. J. Bourguignon, G. C. Wermuth, *J. Med. Chem.*, **1999**, *42*, 730.

99JMC805 A. G. Arvanitis, P. J. Gilligan, R. J. Chorvat, R. S. Cheeseman, T. E. Christos, R. Bakthavatchalam, J. P. Beck, A. J. Cocuzza, F. W. Hobbs, R. G. Wilde, C. Arnold, D. Chidester, M. Curry, L. Q. He, A. Hollis, J. Klaczkiewicz, P. J. Krenitsky, J. P. Rescinito, E. Scholfield, S. Culp, E. B. Desouza, L. Fitzgerald, D. Grigoriadis, S. W. Tam, Y. N. Wong, S. M. Huang, H. L. Shen, *J. Med. Chem.*, **1999**, *42*, 805.

99JMC833 R. J. Chorvat, R. Bakthavatchalam, J. P. Beck, P. J. Gilligan, R. G. Wilder, A. J. Cocuzza, F. W. Hobbs, R. S. Cheeseman, M. Curry, J. P. Rescinito, P. Krenitsky, D. Chidester, J. A. Yarem, J. D. Klaczkiewicz, C. N. Hodge, P. E. Aldrich, Z. R. Wasserman, C. H. Fernández, R. Zaczek, L. W. Fitgerald, S. M. Huang, H. L. Shen, Y. N. Wong, B. M. Chien, C. Y. Quon, A. Arvanitis, *J. Med. Chem.*, **1999**, *42*, 833.

99JMC1687 E. Teodori, S. Dei, P. Quidu, R. Budriesi, A. Chiarini, A. Garnier-Suillerot, F. Gualteri, D. Manetti, M. N. Romanelli, S. Scapecchi, *J. Med. Chem.*, **1999**, *42*, 1687.

99JMC1698 N. Campillo, C. García, P. Goya, J. A. Páez, E. Carrasco, M. Grau, *J. Med. Chem.*, **1999**, *42*, 1698.

99JMC1803 J. B. Smaill, B. D. Palmer, G. W. Rewcastle, W. A. Denny, D. J. McNamara, E. M. Dobrusin, A. J. Bridges, H. R. Zhou, H. D. H. Showalter, R. T. Winters, W. R. Leopold, D. W. Fry, J. M. Nelson, V. Slintak, W. L. Elliot, B. J. Roberts, P. W. Vicent, S. J. Patmore, *J. Med. Chem.*, **1999**, *42*, 1803.

99JMC2218 A. Costanzo, G. Guerrini, G. Ciciani, F. Bruni, S. Selleri, B. Costa, C. Martini, A. Lucacchini, P. M. Aiello, A. Ipponi, *J. Med. Chem.*, **1999**, *42*, 2218.

99JMC2527 R. A. Cavallaro, L. Filocamo, A. Galuppi, A. Galione, M. Brufani, A. A. Ganazzani, *J. Med. Chem.*, **1999**, *42*, 2527.

99JMC2561 G. Cirrincione, A. M. Almerico, P. Barraja, P. Diana, A. Lauria, A. Passannanti, C. Musiu, A. Pani, P. Murtas, C. Minnei, M. E. Marongui, P. Lacolla, *J. Med. Chem.*, **1999**, *42*, 2561.

99JMC3279 N. Campillo, P. Goya, J. A. Páez, *J. Med. Chem.*, **1999**, *42*, 3279.

99JMC3852 P. J. Hajduk, J. Dinges, J. M. Schkeryantz, D. Janowick, M. Kaminski, M. Tufano, D. J. Augeri, A. Petros, V. Nienaber, P. Zhong, R. Hammond, M. Coen, B. Beutel, L. Katz, S. W. Fesik, *J. Med. Chem.*, **1999**, *42*, 3852.

99JMC4108 L. G. Fröhlich, P. Kotsonis, H. Traub, S. Taghavi-Moghadam, N. Al-Masoudi, H. Hofmann, H. Strobel, H. Matter, W. Pfleiderer, H. H. W. Schmidt, *J. Med. Chem.*, **1999**, *42*, 4108.

99JMC4720 J. M. Quintela, C. Peinador, L. M. González, R. Riguera, I. Rioja, M. C. Terencio, A. Ubeda, M. J. Alcaraz, *J. Med. Chem.*, **1999**, *42*, 4720.

99JOC634 A. Vasudevan, F. Mavandadi, L. H. Chen, A. Gangjee, *J. Org. Chem.*, **1999**, *64*, 634.

99JOC3361 H. Suzuki, T. Kawakami, *J. Org. Chem.*, **1999**, *64*, 3361.

99JOC4635 M. Cushman, F. Mavandadi, D. L. Yang, K. Kugelbrey, K. Kis, A. Bacher, *J. Org. Chem.*, **1999**, *64*, 4635.

99JOC5754 A. Iuliano, I. Voir, P. Salvadori, *J. Org. Chem.*, **1999**, *64*, 5754.

99JOC6297 M. Avalos, R. Babiano, P. Cintas, F. R. Clemente, J. L. Jiménez, J. C. Palacios, J. B. Sánchez, *J. Org. Chem.*, **1999**, *64*, 6297.

99JOC7299 K. Brandt, I. Porwolikczomperlik, M. Siwy, T. Kupka, R. A. Shaw, S. Ture, A. Clayton, D. B. Davies, M. B. Hursthouse, G. D. Sykara, *J. Org. Chem.*, **1999**, *64*, 7299.

99JOC7773 H. Kumamoto, H. Tanaka, R. Tsukioka, Y. Ishida, A. Nakamura, S. Kimura, H. Hayakawa, K. Kato, T. Miyasaka, *J. Org. Chem.*, **1999**, *64*, 7773.

99M89 A. Schneider, S. Kairies, K. Rose, *Monatsh Chem.*, **1999**, *130*, 89.

99M819 S. Youssif, *Monatsh Chem.*, **1999**, *130*, 819.

99MCR223 A. Costanzo, G. Ciciani, G. Guerrini, F. Bruni, S. Selleri, R. Donato, C. Sacco, C. Musiu, C. Milia, S. Longu, C. Minnei, P. La Colla, *Med. Chem. Res.*, **1999**, *9*, 223.

99MI23 C. J. Lafrancois, J. Fujimoto and L. C. Sowers, *Nucleosides&Nucleotides*, **1999**, *18*, 23.

99MI294 Y. Oishi, J. J. Kim, M. Nakamura, H. Hirahara, K. Mori, *Macromol. Rapid Commun.*, **1999**, *20*, 294.

99MI377 M. Chiriac, D. Axente, N. Palibroda, C. T. Craescu, *J. Label. Compound Radiopharm.*, **1999**, *42*, 377.

99MI638 B. S. Lindsay, C. N. Battershill , B. R. Copp, *J. Nat. Prod.*, **1999**, *62*, 638.

99MI1673 I. Dez, J. Levaloismitjaville, H. Grutzmacher, V. Gramlich, R. Dejaeger, *Eur. J. Inorg. Chem.*, **1999**, 1673.

99OL537 M. T. Migawa, L. B. Townsend, *Org. Lett.*, **1999**, *1*, 537.

99OL721 A. E. Frumkin, A. M. Churakov, Y. A. Strelenko, V. V. Kachala, V. A. Tartakovsky, **1999**, *Org. Lett.*, *1*, 721.

99OL1355 S. Masala, M. Taddei, *Org. Lett.*, **1999**, *1*, 1355.

99PHA347 R. M. Abdelrahman, J. M. Morsy, F. Hanafy, H. A. Amene, *Pharmazie,* **1999**, *54*, 347.

99PHA503 B. Modzelewskabanachiewicz, J. Kalabun, *Pharmazie,* **1999**, *54*, 503.

99PHA580 M. A. E. Shaban, A. Z. Nasr, A. E. A. Morgaan, *Pharmazie,* **1999**, *54*, 580.

99PHA667 R. M. Abdelrahman, J. M. Morsy, S. Eledfawy, H. A. Amine, *Pharmazie,* **1999**, *54*, 667.

99PJC783 A. Drabczynska, J. Karolakwojciechoska, K. Kieckononowicz, *Polish. J. Chem.*, **1999**, *73*, 783.

99POL2853 G. A. Carriedo, F. J. G. Alonso, P. A. González, P. Gomezelipe, *Polyhedron*, **1999**, *18*, 2853.

99S593 J. E. Kaminska, Z. J. Kaminski , J. Gora, *Synthesis*, **1999**, 593.

99SC655 G. Sabitha, R. Srividya, B. Archana, J. S. Yadav, *Syn. Commun.*, **1999**, *29*, 655.

99SC2027 H. J. Shi, Z. Y. Wang, H. X. Shi, *Syn. Commun.*, **1999**, *29*, 2027.

99SC3003 J. L. Zhou, J. Y. Tsai, K. Bouhadir, P. B. Shevlin, *Syn. Commun.*, **1999**, *29*, 3003.

99SL1145 M. Havelkova, M. Hocek, M. Cesnek, D. Dvorak, *Synlett*, **1999**, 1145.

99T211 A. Brathe, L. L. Gundersen, E. Rise, A. B. Eriksen, A. V. Vollsnes, L. N. Wang, *Tetrahedron* , **1999**, *55*, 211.

99T549 G. Crisp, Y. L. Jiang, *Tetrahedron*, **1999**, *55*, 549.

99T5047 G. R. Pabst, P. Fuller, J. Sauer, *Tetrahedron*, **1999**, *55*, 5047.

99T5067 G. R. Pabst, J. Sauer, *Tetrahedron* , **1999**, *55*, 5067.

99T5239	B. Brand, C. B. Reese, Q. Song, C. Visintin, *Tetrahedron*, **1999**, *55*, 5239.
99T5419	P. Detullio, R. Quedraogo, L. Dupont, F. Somers, S. Boverie, J. M. Dogne, J. Delarge, B. Pirotte, *Tetrahedron*, **1999**, *55*, 5419.
99T10243	R. P. Claridge, R. W. Millar, J. P. B. Sandall, C. Thompson, *Tetrahedron* , **1999**, *55*, 10243.
99T13457	Y. A. Ibrahim, A. M. Kadri, M. R. Ibrahin, J. N. Lisgarten, B. S. Potter, R. A. Palmer, *Tetrahedron*, **1999**, *55*, 13457.
99TA573	D. Q. Che, J. Siegl, G. Seitz, *Tetrahedron -Asymmetry*, **1999**, *10*, 573.
99TL619	A. M. Boldi, C. R. Johnson, H.O. Eissa, *Tetrahedron Lett.*, **1999**, *40*, 619.
99TL763	J. Valenciano, A. M. Cuadro, J. J. Vaquero, J. Alvarez-Builla, *Tetrahedron Lett.*, **1999**, *40*, 763.
99TL1107	V. Janout, M. Lanier, S. L. Regen, *Tetrahedron Lett.*, **1999**, *40*, 1107.
99TL1185	V. Chandrasekhar, A. P. Athimoolam, K. Vivekanandan, S. Nagendran, *Tetrahedron Lett.*, **1999**, *40*, 1185.
99TL1793	P. J. Bhuyan, K. C. Lekhok, J. S. Sandhu, *Tetrahedron Lett.*, **1999**, *40*, 1793.
99TL2139	O. Sugimoto, M. Sudo, K. Taji, *Tetrahedron Lett.*, **1999**, *40*, 2139.
99TL2541	M. Kawase, M. Hirabayashi, S. Saito, K. Yamamoto, *Tetrahedron Lett.*, **1999**, *40*, 2541.
99TL2841	J. Spychala, *Tetrahedron Lett.*, **1999**, *40*, 2841.
99TL3891	O. A. Attanasi, P. Filippone, C. Fiorucci, F. Mantellini *Tetrahedron Lett.*, **1999**, *40*, 3891.
99TL5327	M. Kunishima, C. Kawachi, F. Iwasaki, K. Terao, S. Tani, *Tetrahedron Lett.*, **1999**, *40*, 5327.
99TL6099	O. N. Chupakhin, V. N. Kozhevnikov, D. N. Kozhevnikov, V. L. Rusinov, *Tetrahedron Lett.*, **1999**, *40*, 6099.
99TL6313	A. Kotschy, Z. Novak, Z. Vincze, D. H. Smith, G. Hajos, *Tetrahedron Lett.*, **1999**, *40*, 6313.
99TL8419	R. Menicagli, S. Samaritani, S. Gori, *Tetrahedron Lett.*, **1999**, *40*, 8419.
99TL8675	M. J. Arévalo, M. Avalos, R. Babiano, P. Cintas, M. B. Hursthouse, J. L. Jiménez, M. E. Light, I. López, J. C. Palacios, *Tetrahedron Lett.*, **1999**, *40*, 8675.
99ZN(B)549	G. Seitz, J. Lachmann, *Z. Naturforsch. (B)*, **1999**, *54*, 549.
99ZN(B)609	R. Tripolt, S. Schmuck, E. Nachbaur, *Z. Naturforsch. (B)*, **1999**, *54*, 609.
99ZN(B)788	F. A. Attaby, S. M. Eldin, *Z. Naturforsch. (B)*, **1999**, *54*, 788.

Chapter 6.4

Six-Membered Ring Systems: With O and/or S Atoms

John D. Hepworth
University of Hull, Hull, UK
B. Mark Heron
Department of Colour Chemistry,
University of Leeds, Leeds, UK
Email: ccdbmh@leeds.ac.uk

Introduction

Interest in marine natural products continues with syntheses of the macrolides bryostatin 2 <99JA7540> and (+)-miyakolide <99JA6816>, the pyran fragment of the altohyrtins <99TL7135> and of the tetracyclic puupehedione <99T15181>. In the polycyclic ether group, syntheses of brevetoxin A <99CEJ599, 618, 628, 646>, members of the cephalostatin family <99JA2056, 2071>, 7-deoxyokadaic acid <99AG(E)2258>, 18-*O*-methyl mycalamide B <99S2087> and the triterpene hippospongic acid A <99JCS(P1)489, 2271> have been described, whilst studies towards the synthesis of ciguatoxin have been published <99CC1063, 2035, 99JOC37, 9399, 9416>. Ningalin, a di(benzopyrano)pyrrole, and various related compounds have been synthesised using Diels-Alder methodology <99JA54>.

Routes to fused bi- and tri- cyclic ethers include the use of a protected glucal <99SL1945>, a SmI_2-induced reductive intramolecular cyclisation <99TL2811, 8859>, the CAN oxidation of 3-oxabicyclo[3.1.0]hexyl sulfides <99CC2515>, a sequential ring-closing metathesis (RCM) and hydroboration of enol ethers <99T8231>, a RCM of diene and enyne carbohydrate derivatives <99T8253> and an aldol reaction between a tetrahydropyranyl aldehyde and a dithioacetal *S*-oxide <99SL1037>.

The synthesis of bis-spiroacetal ring systems has been discussed <99T7661>. An enantioselective synthesis of reveromycin B has been achieved <99JA456> and fungal metabolites containing a 1,8-dihydroxynaphthalene derived spiroacetal unit have been synthesised <99JCS(P1)1073, 99JOC1092>.

The Handbook of Natural Flavonoids <B-99MI1> is a major resource for this large group of natural products, whilst reviews on saturated and unsaturated lactones <99JCS(P1)1377>, lactone-bridged biaryls <99S525>, artemisinin <99H(51)1681>, synthetic applications of the non-classical Wittig reaction <99JCS(P1)3049> and *o*-quinodimethanes <99CRV3199> also contain material pertinent to this chapter.

6.4.1 HETEROCYCLES CONTAINING ONE OXYGEN ATOM

6.4.1.1 Pyrans

The conditions necessary to achieve enantioselectivity in the hetero-Diels-Alder (hDA) reaction between ketomalonates and dienes have been established and the mechanism investigated. The product from cyclohexadiene offers access to optically active cyclohexene-diols *via* the chiral CO_2-synthon **1** (Scheme 1) <99JOC6677>. Application of the hDA reaction to 6-oxocyclohexene-1-carbaldehydes using ketene acetals as the 2π-electron component is the key step in a stereoselective approach to a range of substituted cyclohexenes <99T12907>. Studies of the hDA reaction between crotonyl phosphate and vinyl ethers have shown that the enantiomeric adducts can be obtained by changing the substituent on the Cu(II)-oxazoline catalyst <99TL2879>. Chiral polymeric binaphthol-Al complexes are efficient catalysts for hDA reactions <99JOC299>.

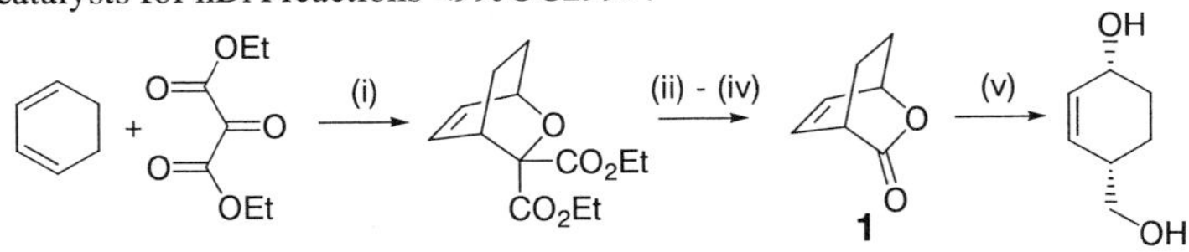

Reagents: (i) 10 mol % $Zn(OTf)_2$, bisoxazoline ligand, Et_2O, rt, 144 h. (94 %, 91 % ee); (ii) aq. KOH, 6h.; (iii) aq. HCl; (iv) CAN, aq. MeCN, 15 min.; (v) LAH, THF, 18 h. (42 % over steps (ii) - (v))

Scheme 1

Novel 2-amino-substituted- 3-methylenedihydropyrans **2** have been obtained from inverse electron demand Diels-Alder reactions of allenamides with heterodienes <99TL6903> and the reaction of styrenes with α,β-unsaturated 2-oxonitriles yields predominantly the *cis*-2-aryldihydropyran-6-carbonitriles **3** <99HCA1122>.

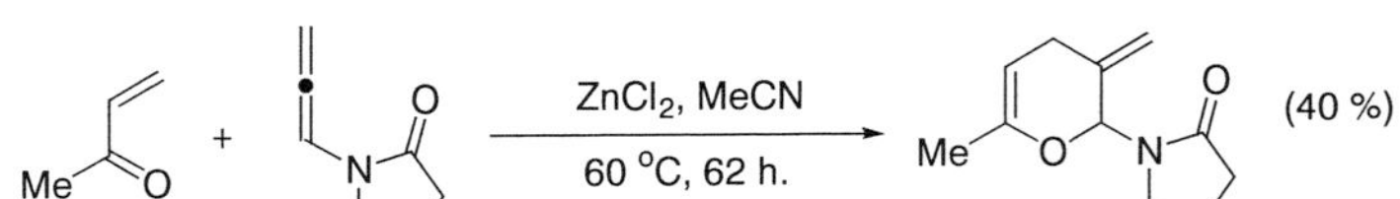

2

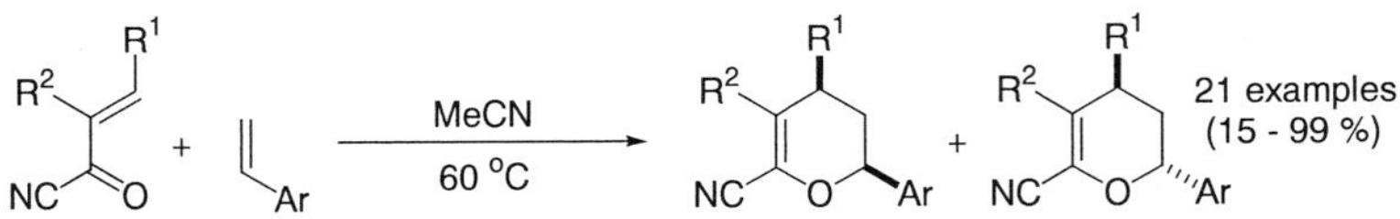

3

Tetraacetylethene yields dihydropyrans in hDA reactions with ethyl vinyl ether and cyclopentadiene, but 1,2-dimethylenecyclohexane affords both the hDA and normal Diels-Alder adducts, the latter presumably arising *via* ring opening of the former adduct to a zwitterionic intermediate (Scheme 2) <99EJO3343>.

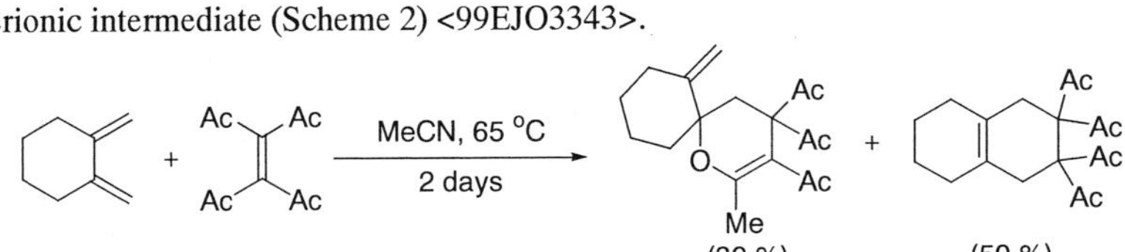

Scheme 2

An unusual cycloaddition, described as a [2+1+1+1] process, results when ethene is lost from the cyclopropane moiety of the Cr-complex **4**. The product is a mixture of the fused dihydropyrans **5** and **6** <99OL16>. This methodology has been extended to complexes **4** in which a chiral centre features in the tether <99JOC1291>. A RCM reaction converts allyl homoallyl ethers into dihydropyrans **7** from which 2,3-dihydropyran-4-ols **8** can be obtained by way of rearrangement of the derived oxiranes. Diastereomeric mixtures of *rac*-**8** undergo a Lewis acid-mediated addition of allyltrimethylsilane to give 2-allyl-5,6-dihydropyrans **9** as a single diastereoisomer (Scheme 3) <99TL4319>.

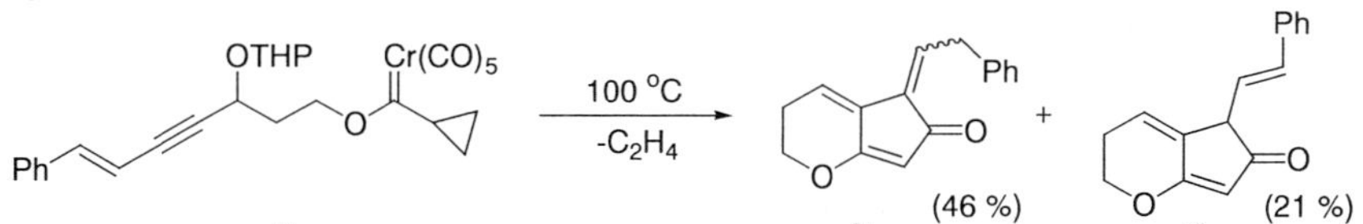

4 **5** **6**

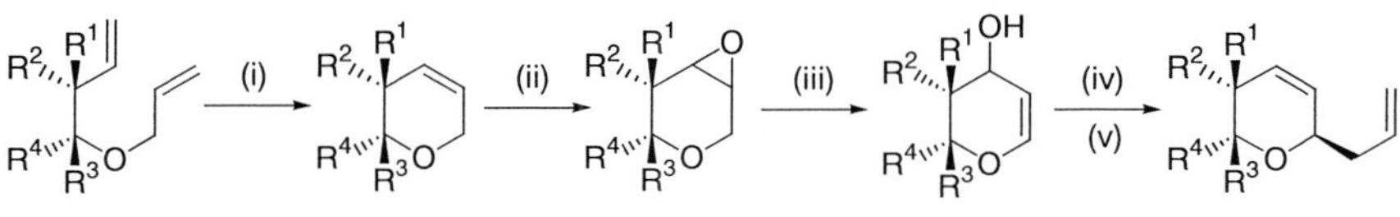

7 **8** **9**

Reagents: (i) 2 mol % $Cl_2P(Cy_3)_2Ru$=CH-CH=CPh_2, CH_2Cl_2; (ii) *m*-CPBA, CH_2Cl_2, rt; (iii) LDA, THF, 65 °C; (iv) AcCl, Et_3N, DMAP, CH_2Cl_2, 0 °C; (v) allyltrimethylsilane, $BF_3.OEt_2$, CH_2Cl_2, -78 °C

Scheme 3

2,6-Disubstituted dihydropyrans are produced with high *anti*-selectivity when 2-phenyl-4-(4-tolylsulfonyl)-3,4-dihydro-2*H*-pyrans are treated with Al-based Lewis acids <99SL132>. Tetraenes **10**, derived from dienes *via* their epoxides, undergo a double RCM reaction under Ru-catalysis to yield polycyclic ethers **11** in which the dihydropyran units can be joined by a variable number of carbon atoms <99JOC3354>. Continued work on the use of dispiroketals in synthesis has led to an improved route to the enantiomers of bi(dihydropyrans) **12** <99JCS(P1)1639>.

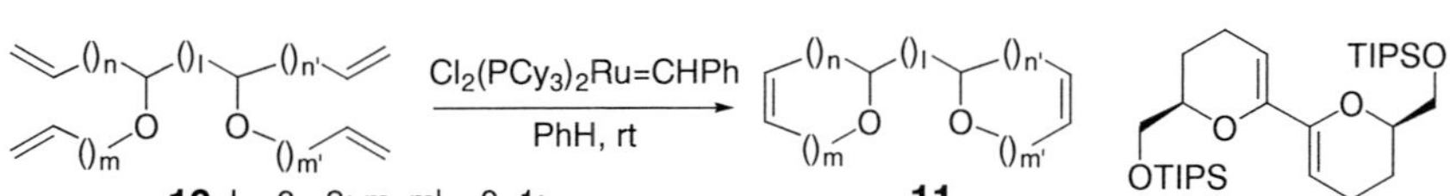

10 l = 0 - 2; m, m' = 0, 1; n, n' = 0 - 2

11 (50 - 94 %)

12

Hexahydro-4*H*-chromenes are formed from 1,5-dienes in a one-pot process in which a Rh-complex catalyses sequential hydroformylation, carbonyl ene reaction, a second hydroformylation, cyclisation to a lactol and dehydration <99TL7455>.

The oxonium ion generated when dihydropyran is coupled with ethyl glyoxylate in the presence of $TiCl_4$ can be trapped by nucleophiles offering a useful route to 2,3-disubstituted tetrahydropyrans <99TL1083, 4751>.

5-Methylenetetrahydropyrans are formed regioselectively in good yields from the Ru-catalysed reaction of prop-2-yn-1-ols with allyl alcohol <99JOC3524>, whilst 2-vinyltetrahydropyrans result from a Pd-catalysed intramolecular hydrocarbonation of alkoxyallenes <99TL1747>. Various hepta-5,6-dien-1-ols **13** undergo a Ru-catalysed alkylative cyclisation with vinyl ketones to give both simple and fused 2-substituted tetrahydropyrans *e.g.* **14**. Reduction of the carbonyl function in the side chain allows the synthesis of spiroketals **15** (Scheme 4) <99JA10842>.

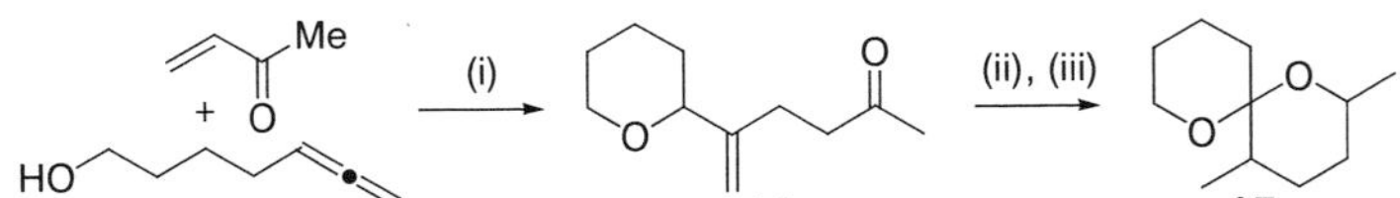

13 **14** **15**

Reagents: (i) Ru-catalyst, DMF, 60 °C (74 %); (ii) $NaBH_4$, MeOH, (83 %); (iii) 4-TsOH, PhH, (78 %)

Scheme 4

But-3-en-1-ols react with aldehydes in a $Sc(OTf)_3$-catalysed Prins reaction which leads to tetrahydropyran-4-ols and their ethers <99CC291> and 4,4-dichlorotetrahydropyrans are formed with high diastereoselectivity when $InCl_3$ mediates the cross-coupling reaction involving 3-chlorobut-3-en-1-ols <99SL717>. All *cis*-2,4,6-trisubstituted tetrahydropyrans are the products when 2-methylene-1,4-diols react with aldehydes in an acid-promoted Prins cyclisation-pinacol rearrangement sequence <99JA1092>. An intramolecular Lewis acid mediated cyclisation of the hydrazones of γ-alkoxyallylstannanes **16** leads to the *trans* 3-amino-tetrahydropyrans **17**. Asymmetric syntheses of the heterocycles can be achieved using a chiral imine auxiliary and by means of a chiral Lewis acid <99JOC4901>.

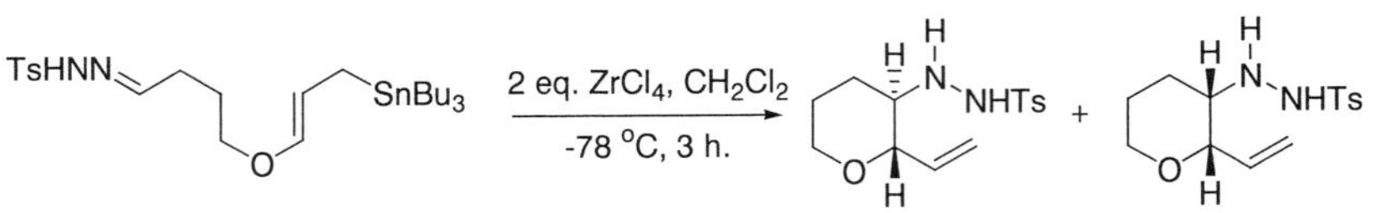

16 **17** (81 %) *trans* : *cis* = >95 : 5

Chloromethanesulfonate is an efficient leaving group for the synthesis of 2,3-*trans*-disubstituted tetrahydropyrans by the rearrangement and ring expansion of tetrahydrofurans (Scheme 5) <99TL2145>.

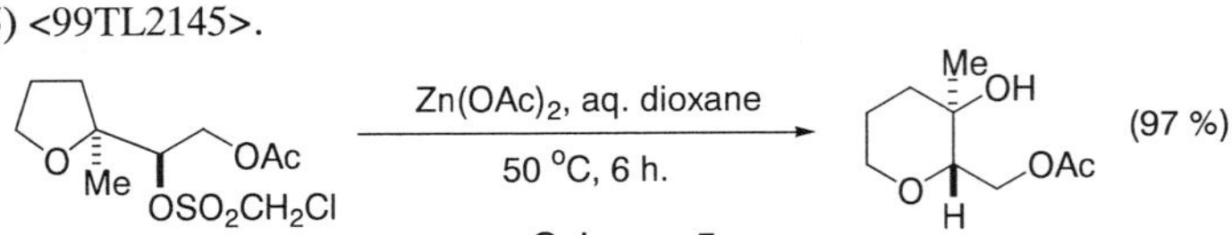

Scheme 5

The *endo*-cyclisation of the diol-cyclic sulfite **19** obtained by oxidation of the dihydropyran **18** affords a *trans*-fused bipyran <99TL2235> and this system is also accessible from the sulfonyl-stabilised oxirane **20** (Scheme 6) <99TL8019>.

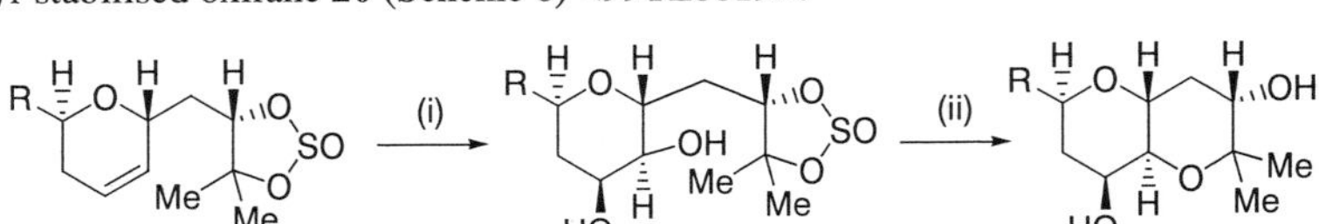

18 **19**

Reagents: (i) $RuCl_3.3H_2O$, $NaIO_4$, EtOAc, aq. MeCN; (ii) 4-TsOH, aq. MeCN, heat

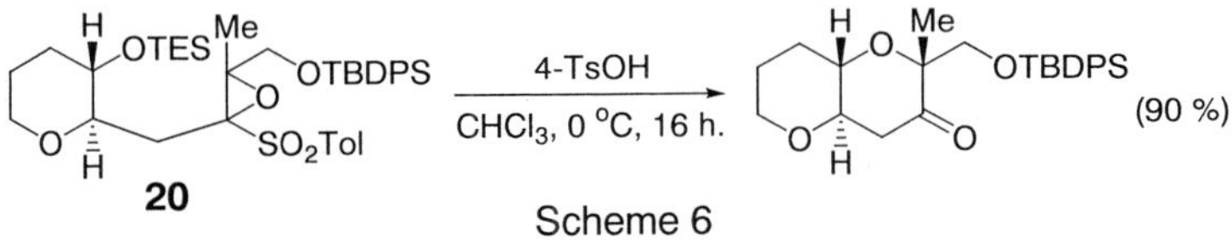

Scheme 6

6.4.1.2 Benzopyrans (Chromenes)

Chiral 2*H*-1-benzopyrans can be obtained by a Ru-catalysed RCM in which Ti(O-*i*Pr)$_4$ is added to compete with chelation involving an *o*-ester function which otherwise reduces the efficacy of the Ru-catalyst. A facile photoracemisation at C-2 is noted for a 2-cyclopropylchromene that presumably proceeds through ring opening to the dienone <99JOC5321>.

The formation of the chromene carbamate **21** from resorcinol dicarbamate involves directed *ortho* metallation and an intramolecular *O* → *O* carbamoyl transfer. Further manipulation utilising directed metallation and transition metal catalysed reactions allows the synthesis of plicadin, a naturally occurring coumestan <99AG(E)1435>.

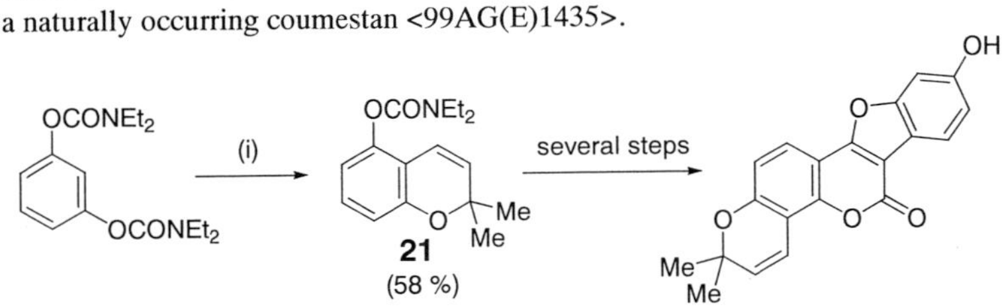

Reagents (i) 1.1 eq. tBuLi, THF, -78 °C, 30 min.; then 1.3 eq. Me$_2$C=CHCHO, -78 °C - rt, 1h. then 3 eq. AcOH, 0 - 25 °C, 1h.

A phenylboronic acid-mediated synthesis of chromenes from naphthols and 3-methylbut-2-enal features in two short routes to β-lapachone <99S1875>.

6-Dimethylaminofulvene undergoes a [6+3] cycloaddition with benzoquinones to give cyclopenta[*c*]chromenes, offering a new approach to the 11-oxasteroid system (Scheme 7) <99CC2125>.

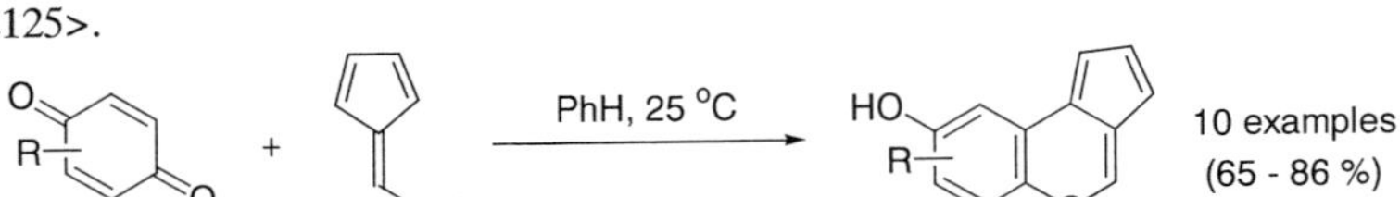

Scheme 7

6.4.1.3 Dihydro[1]benzopyrans (Chromans)

The aryl iodide **22** undergoes a Pd-catalysed cross-coupling with *C*-, *N*- and *O*-nucleophiles to yield the substituted tricyclic chromans **23** with good diastereoselectivity (Scheme 8) <99JOC1875>.

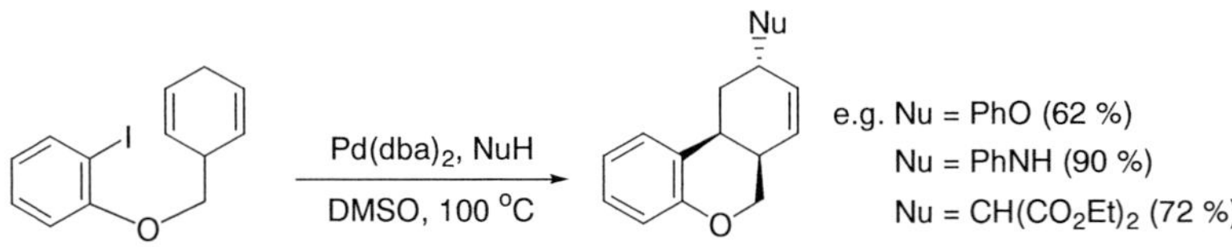

22 **23**

Scheme 8

A high yield and good enantiomeric excess characterise a synthesis of the vitamin E core in which the key feature is the discrimintaion of enantiotopic faces of a prochiral alkene during a Pd-catalysed allylic alkylation <99S1491>. Alternative enantioselective approaches to precursors of the vitamin involve a Pd-catalysed Sonogashira alkynol coupling and a stereoselective ketone allylation <99EJO1075>. A photochemical benzannulation of the chromium complex **24** offers a novel route to trisubstituted chroman-6-ols (Scheme 9) <99SL231>.

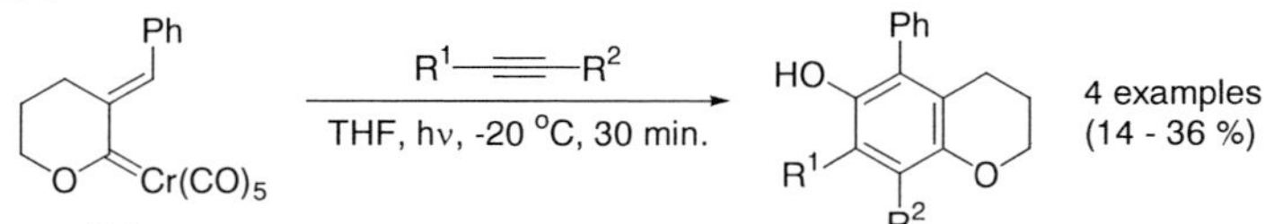

24

Scheme 9

Chromans with defined stereochemistry are accessible by manipulation of dihydrochromeno[3,2-*b*]azete-2,8-diones **26** which result from the cyclisation of azetidine-2-carboxylic chlorides **25** <99T5567> and also from chroman-4-ones *via* an initial enantioselective reduction by BH_3 in the presence of Corey's oxazaborolidine <99T7555>.

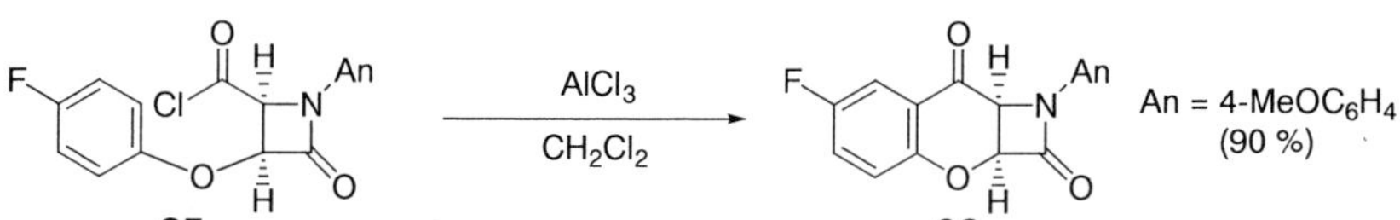

25 **26**

There are a number of examples of the synthesis of chromans using *o*-quinone methides as the heterodiene in a hDA reaction. Both pyrano[3,2-*c*]-benzopyrans and cyclopenta[*c*][1]benzopyrans result from an intramolecular cycloaddition of a substituted *o*-quinonemethide generated under mild conditions. In the former case, salicylaldehyde and an unsaturated alcohol yield the *trans*-fused tetrahydropyranobenzopyran (Scheme 10) <99JOC9507>. However, the latter synthesis (Scheme 11) is less selective <99BCJ73>.

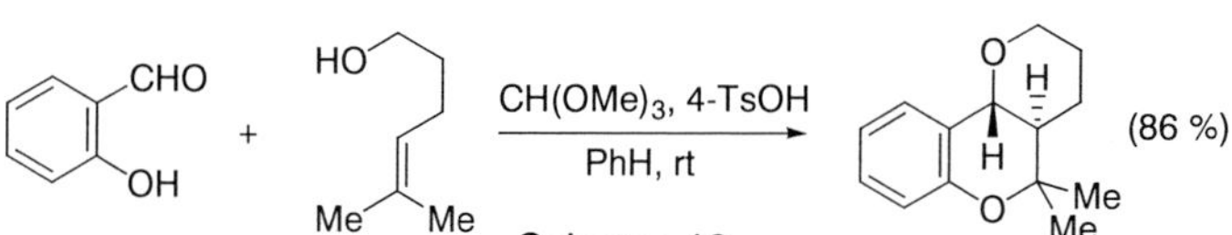

Scheme 10

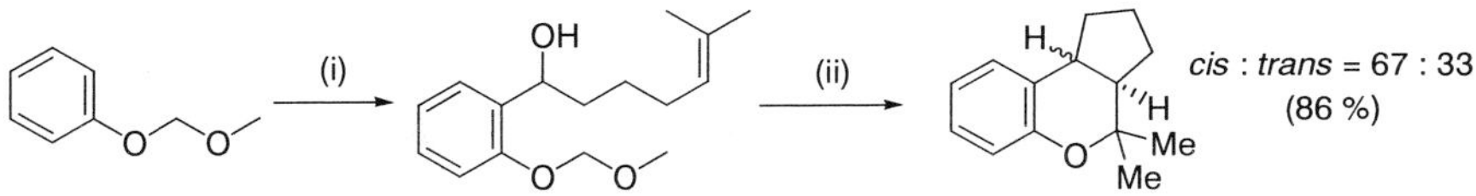

Reagents: (i) nBuLi, TMEDA, hexane 0 °C, then 6-methyl-5-heptenal; (ii) c.HCl, MeOH, heat, 1h.

Scheme 11

The flavonoid sideroxylonal B has been synthesised from 3,5-dimethoxyphenol through initial conversion to the 2-(3'-methylbutan-1'-ol) derivative. Reaction with EtMgBr simultaneously generates the quinone methide and the 2-isopentenylphenol which react together to produce the flavan (Scheme 12) <99TL1925>.

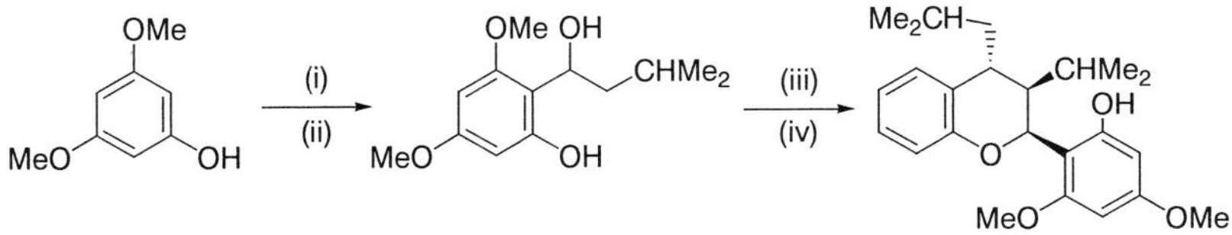

Reagents: (i) isovaleric acid, $BF_3.OEt_2$, 80 °C, (84 %); (ii) LAH/THF, Et_2O, 0 °C, (82 %); (iii) EtMgBr/THF, Et_2O; (iv) PhH, heat, 29 h., (74 %).

Scheme 12

Application of the intramolecular Pauson-Khand reaction to enynes derived from salicylaldehyde leads to the cyclopentenone **27** (Scheme 13) <99TL2817>. Intramolecular Diels-Alder reactions feature in syntheses of the bipyridyl **28** <99CC793> and benzopyrano[4,3-*b*]quinolines <99JCR254>.

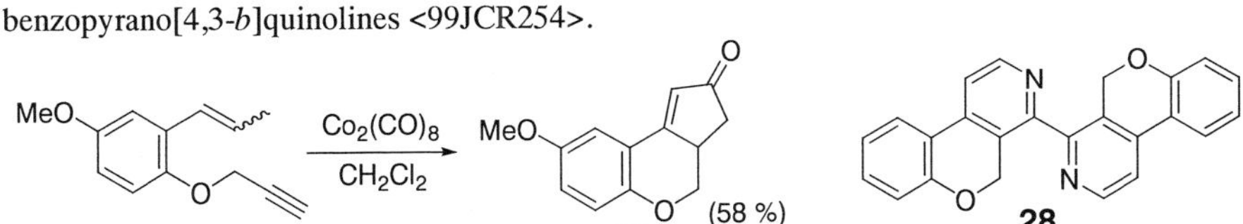

Scheme 13 **27** **28** (60 - 87 % crude)

6.4.1.4 Dihydro[2]benzopyrans (Isochromans)

Alkenyl derivatives of benzyl alcohol and related naphthalene substrates undergo a rapid Pd-catalysed cyclisation to 1*H*-2-benzopyrans and the benzologues (Scheme 14) <99TL4871>.

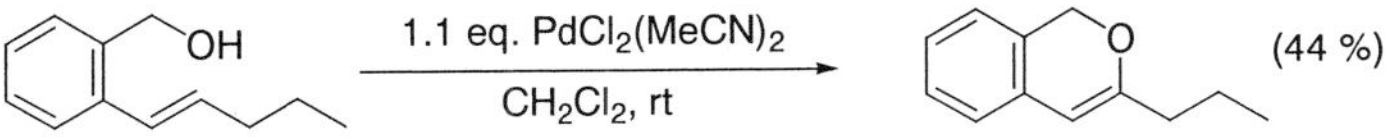

Scheme 14

An intramolecular arylation of the asymmetric aldehyde **29** occurs on treatment with $Ti(O\text{-}iPr)_4$ and sonication giving the isochroman with complete diastereoselection, unfortunately not replicated with its diastereoisomer, and subsequent oxidation provides an asymmetric synthesis of the 2-benzopyrano-5,8-quinone **30** (Scheme 15) <99JCS(P1)3039>. Two naturally occurring naphthoquinones containing the isochroman system have been obtained by the OsO_4-$NaIO_4$ oxidative cyclisation of 2-allylbenzyl alcohols followed by sequential oxidative demethylation and dehydration <99JOC1173>. A tetracyclic naphthoquinone and other isochromenes and isocoumarins result from reaction of 2-(1,2-dihydroxyethyl)naphthoquinone with acylated pyridinium ylides <99JOC438>.

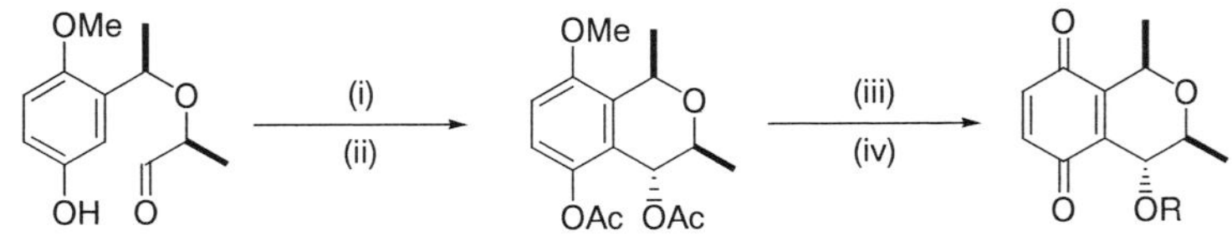

29

30

Reagents: (i) $Ti(O\text{-}iPr)_4$, CH_2Cl_2, ultrasound, 5 h., ~20 °C; (ii) py, Ac_2O, 20 h.; (iii) LAH, Et_2O; (iv) AgO, dioxane, aq. HNO_3

Scheme 15

The four stereoisomers of galaxolide, a synthetic musk widely used in perfumery, have been obtained by the Friedel-Crafts alkylation of 1,1,2,3,3-pentamethylindane with (*R*)- and (*S*)- methyloxirane. Acid-catalysed reaction with $(CH_2O)_n$ afforded the epimeric isochromans which were separated through their tricarbonylchromium complexes <99HCA1656>. The $Cr(CO)_3$ complex of the allyl ether **31** undergoes a Pd-catalysed cyclisation to the 4-methylene-isochroman **32** (Scheme 16) <99TL6757>. The acylpalladium(II) intermediate formed during the Pd-catalysed carbonylation of aryl iodides can be trapped with *N*-BOC-*O*-benzylhydroxylamine and by resin-bound hydroxylamines giving, after deprotection, isochromanylhydroxamic acids **33** <99TL7709>.

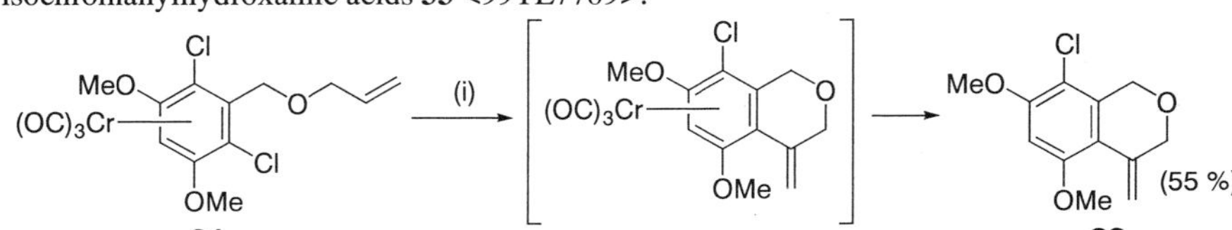

31

32

Reagents: (I) $Pd(OAc)_2$, PPh_3, Et_3N, DMF, 80 °C, 24 h.

Scheme 16

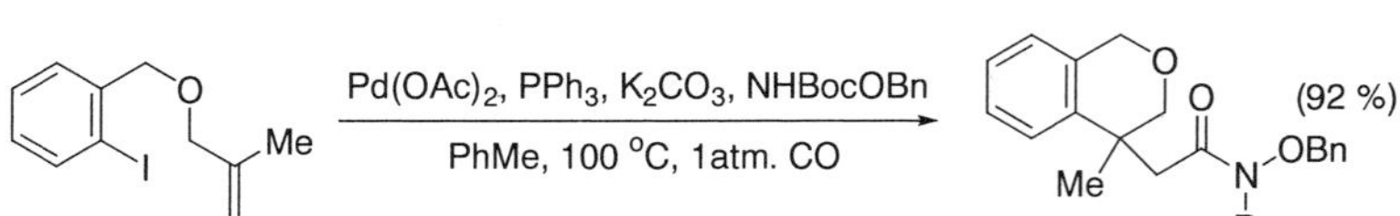

33

6.4.1.5 Pyranones

5,6-Disubstituted pyran-2-ones are formed through the Ni-catalysed coupling of (*Z*)-3-iodopropenoates with alkynes; the corresponding bromo compound yields cyclopentadienes <99T4969>.

β-Keto acids **34** can be converted into 5,6-dialkyl-4-hydroxypyran-2-ones by thermolysis of the Meldrum's acid derivatives **35**. A Fries rearrangement of the derived enol acylates and reduction of the resulting 3-acylpyranone with Et_3SiH provides a route to the 3,6-dialkylpyranones **36** (Scheme 17) <99T4783>.

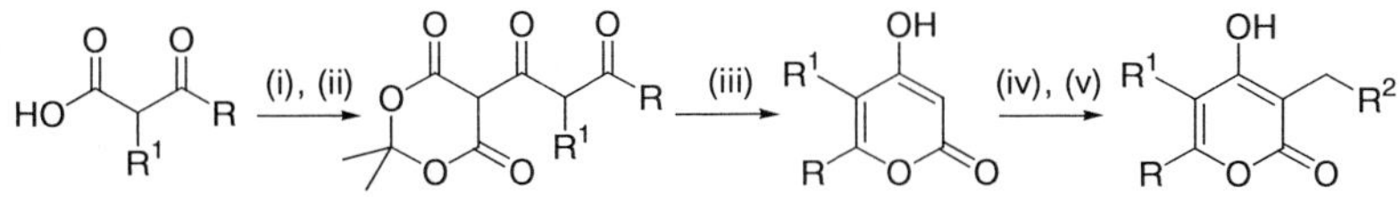

34 **35** **36**

Reagents: (i) DCC, 0.3 eq. DMAP, Et_3N, CH_2Cl_2; (ii) aq. HCl; (iii) PhMe, heat; (iv) R^2COCl, TFA, heat; (v) Et_3SiH, TFA, 0.01 eq. $LiClO_4$

Scheme 17

Two one-pot syntheses of highly substituted pyran-2-ones have been published. One involves the reaction between t-BuNC, dialkyl acetylenedicarboxylates and bromomalonates <99JCR368>. In the other, cyclobutenediones are treated with *O*-silylated cyanohydrins to yield a 4-acylcyclobutenone by a 1,4-silyl migration and cyanide displacement which rearranges to the pyranone (Scheme 18) <99JOC2145>.

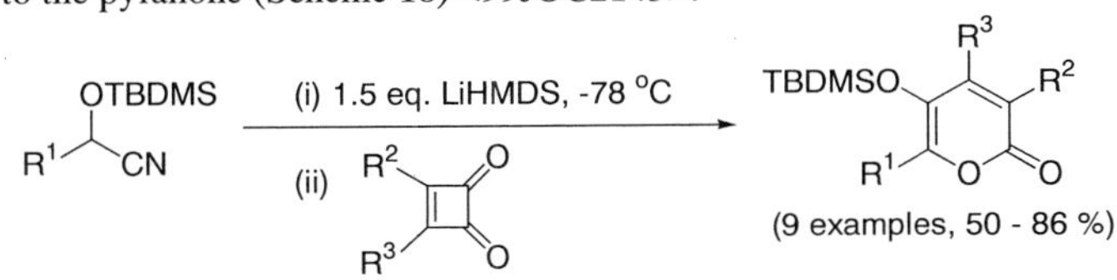

Scheme 18

A combination of a chiral auxiliary group and a chiral Lewis acid catalyst enable the steric steering of a hDA reaction involving electron-rich dienes to be achieved; high diastereoisomeric ratios usually result (Scheme 19) <99EJO329>. Cr(III)-catalysis of the hDA reaction between aldehydes and various moderately nucleophilic 2-alkoxy-1,3-dienes ensures almost complete diastereoselectivity in favour of *endo* cyclisation, affording the all *cis* tetrahydropyran-4-one. The good enantioselectivity is enhanced when the reaction is carried out in acetone (Scheme 20) <99AG(E)2398>.

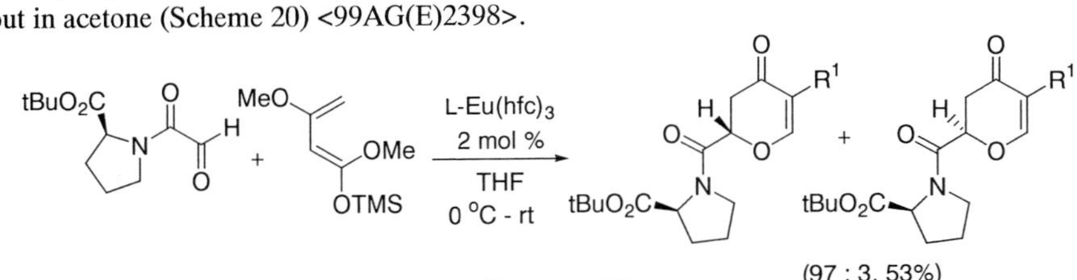

Scheme 19

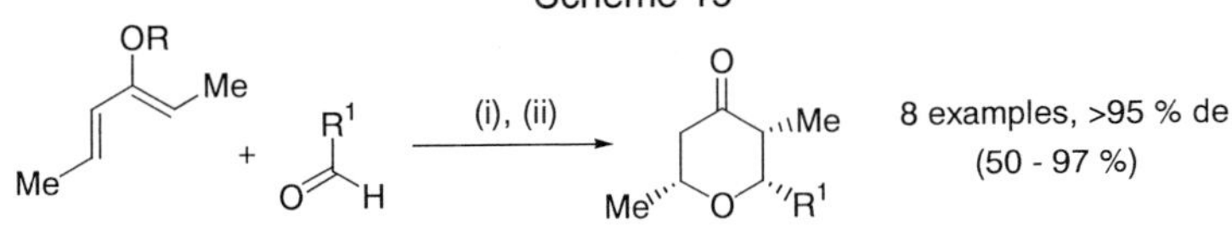

Reagents: (i) 3 mol % Cr(III) Schiff base complex, 4A sieves, rt, 16 - 40 h.; (ii) TBAF, AcOH, THF

Scheme 20

Good yields of the bridged tetrahydropyran-3-one **38** are obtained when the α-diazoketones **37** are decomposed by chiral Rh(II)-catalysts in the presence of DMAD. It is proposed that an enantioselective intermolecular 1,3-dipolar cycloaddition follows the generation of a carbonyl ylide which is bound to the rhodium (Scheme 21) <99JA1417>.

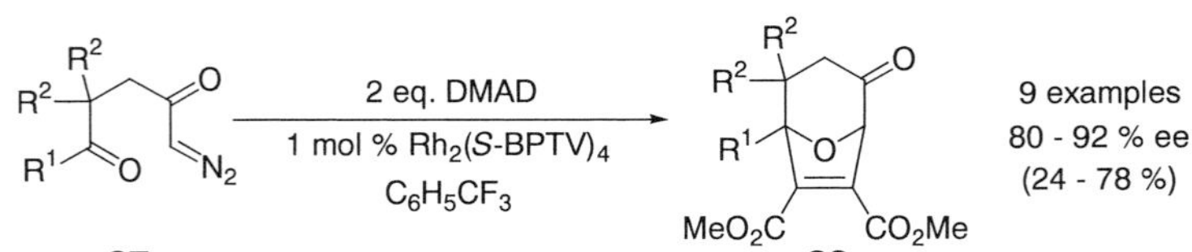

37 **38**

Scheme 21

At low temperatures, the Zn enolate derived from dimethyl 3-methylpent-2-endioate **39** reacts with aldehydes in a one-pot aldolisation and cyclisation to yield the *syn*-dihydropyran-2-one **40**. At the higher temperatures necessary to achieve reaction with α-aminoaldehydes, the *anti*-products predominate indicating thermodynamic control (Scheme 22) <99T7847>. An aldol condensation features in the asymmetric synthesis of phomalactone. The key step is the reaction of the enolate of the vinylogous urethane **41** with crotonaldehyde which occurs with 99% *syn*-diastereoselectivity and in 99% ee (Scheme 23) <99TL1257>.

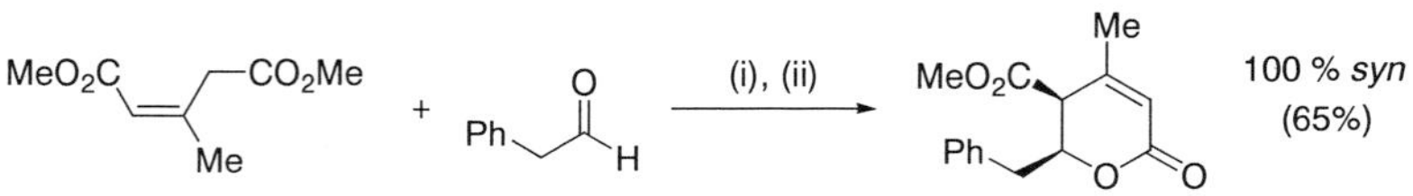

39 **40**

Reagents: (i) LiHMDS, $ZnCl_2$, THF, -65 °C; (ii) aq. NH_4Cl

Scheme 22

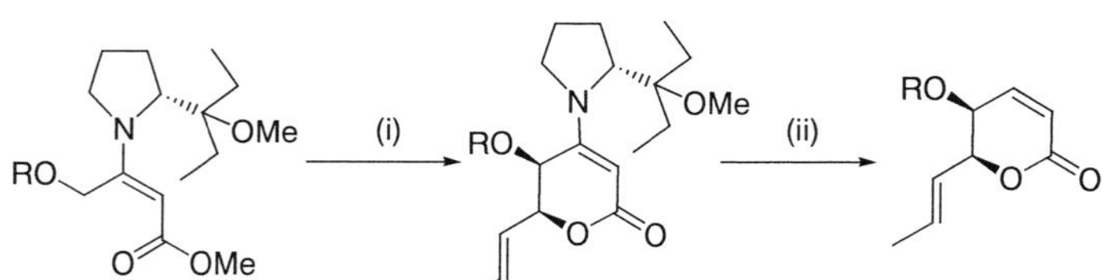

41

Reagents: (i) 1.5 eq. LDA, 4.0 eq. crotonaldehyde, -78 - 0 °C, THF, (51%); (ii) 10 eq. $NaCNBH_3$, AcOH, then 1.3 eq. *m*-CPBA, -78 °C - rt, CH_2Cl_2, (71 %)

Scheme 23

Both α,β-unsaturated iminium species and enals react with 4-hydroxypyran-2-ones to give pyrano[4,3-*b*]pyranones in a formal [3+3] cycloaddition (Scheme 24) <99JOC690>. In the presence of butadienes, the malononitrile derivative **42** obtained from a 3-hydroxypyran-4-one undergoes a one-pot sequential intramolecular [5+2] pyranone - alkene cycloaddition and a Diels-Alder reaction to give the *O*-bridged tricyclic system **43** (Scheme 25) <99JOC966>.

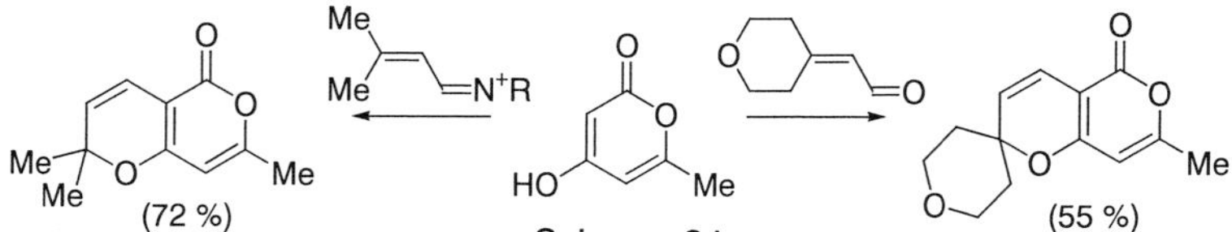

Scheme 24

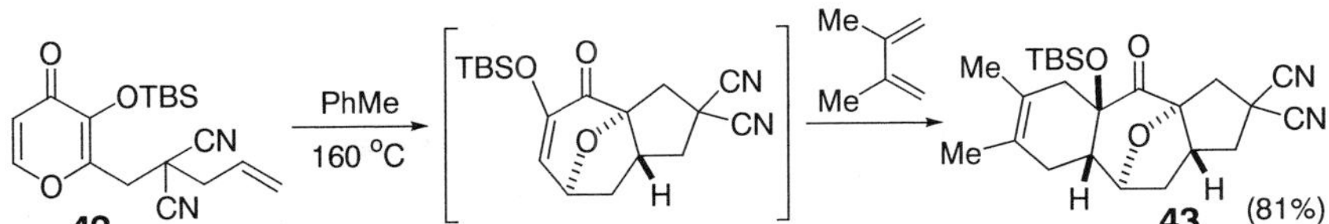

42 **43**

Scheme 25

6.4.1.6 Coumarins

The Pechmann and Knoevenagel reactions have been widely used to synthesise coumarins and developments in both have been reported. Activated phenols react rapidly with ethyl acetoacetate, propenoic acid and propynoic acid under microwave irradiation using cation-exchange resins as catalyst <99SL608>. Similarly, salicylaldehydes are converted into coumarin-3-carboxylic acids when the reaction with malonic acid is catalysed by the montmorillonite KSF <99JOC1033>. In both cases the use of a solid catalyst has environmentally friendly benefits. Methyl 3-(3-coumarinyl)propenoate **44**, prepared from dimethyl glutaconate and salicylaldehyde, is a stable electron deficient diene which reacts with enamines to form benzo[*c*]coumarins. An inverse electron demand Diels-Alder reaction is followed by elimination of a secondary amine and aromatisation (Scheme 26) <99SL477>.

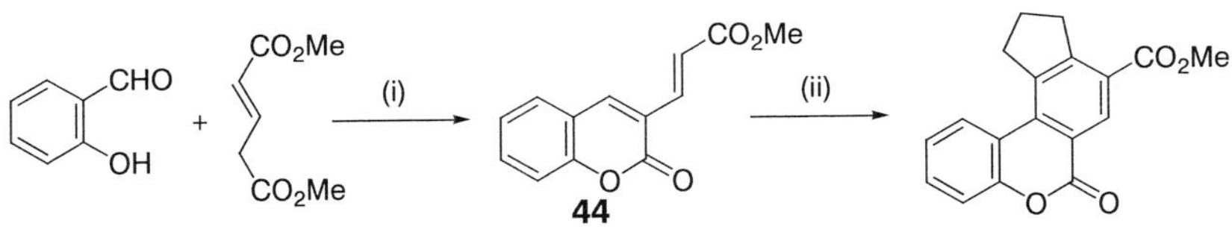

44

Reagents: (i) piperidine, THF, heat, 3 h, 69%; (ii) 1-pyrrolidinocyclopentene, rt, 3 h, 43%

Scheme 26

2-Phenylbenzoic acid is converted into benzo[*c*]coumarin in the Suárez system, that is irradiation when mixed with trivalent iodine compounds and iodine; the 2-iodo derivative may also be formed <99JCS(P1)1713>.

Various crown ethers have been constucted onto the 3- and 4- positions of coumarin by way of 3-diazo-2*H*-1benzopyran-2,4(3*H*)-dione. Carbenoid generation under Rh-catalysis is followed by formation of an oxonium ion by reaction with the ether solvent and a sigmatropic rearrangement completes the sequence <99T6577>.

The liberation of umbelliferone, 7-hydroxycoumarin, which is readily detected by its fluorescence, forms the basis of methods for the detection of NAD(P)H <99CC1637> and lipase catalytic antibodies <99HCA400>.

The asymmetric catalytic hydrogenation of 4-arylcoumarins which leads to chiral 3,4-dihydrocoumarins proceeds through opening of the lactone ring <99TL3293>. 3-Alkyl-3-iodo-3,4-dihydrocoumarins undergo an enantioselective radical reduction to the 3-alkyldihydrocoumarin in the presence of a chiral Lewis acid <99T10295>.

The Cu(I)-catalysed asymmetric conjugate addition of dialkyl zinc reagents to 3-nitrocoumarins **45** gives high yields of 3,4-dihydrocoumarins in a pH-dependent diastereoisomeric ratio. Subsequent decarboxylation gives optically active β-aryl nitroalkanes

<99TL5803>. Libraries of 2-substituted phenylpropyl amides and amines have been obtained from dihydrocoumarins <99TL1241>.

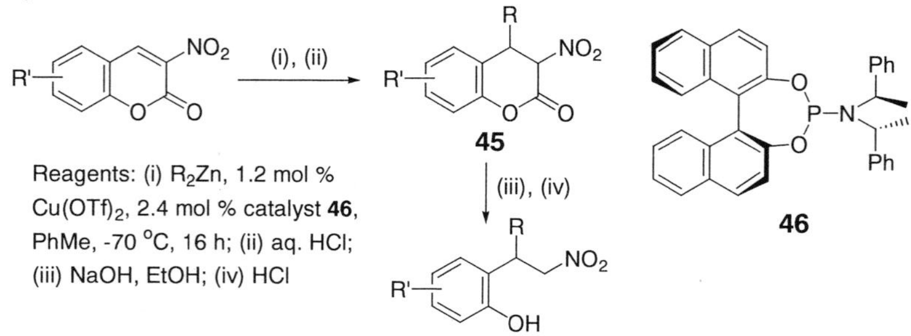

Reagents: (i) R_2Zn, 1.2 mol % $Cu(OTf)_2$, 2.4 mol % catalyst **46**, PhMe, -70 °C, 16 h; (ii) aq. HCl; (iii) NaOH, EtOH; (iv) HCl

A facile synthesis of 3,4-disubstituted isocoumarins and their benzologues is based on the Pd-catalysed annulation of internal alkynes by 2-iodobenzoate esters (Scheme 27). Tri- and tetra- substituted pyran-2-ones are also available by this route <99JOC8770>. 3-Substituted isocoumarins result from the Pd-catalysed *6-endo-dig* cyclisation of 2-ethynylbenzoic acids. The *5-exo-dig* products, phthalides, are also formed when there is a terminal bulky group on the alkyne moiety <99S1145>.

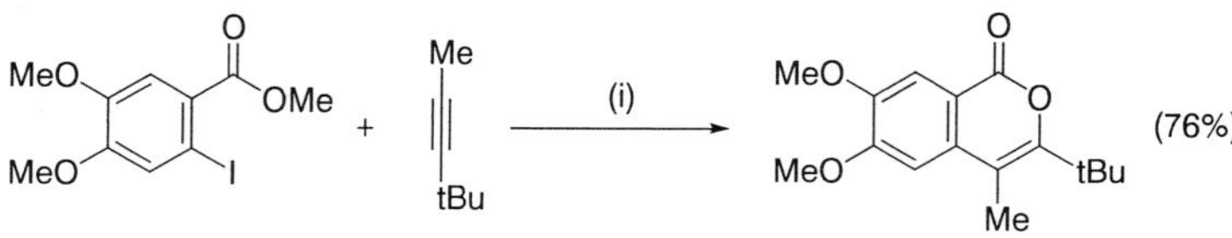

Reagents: (i) 5 mol % $Pd(OAc)_2$, 1 eq. Na_2CO_3, 1 eq. LiCl, DMF, 100 °C

Scheme 27

The oxidation of 3-arylisochromans by dimethyloxirane offers an attractive route to both 3-arylisocoumarins and their 3,4-dihydro derivatives <99T14719>. Although 2'-substituents in 2-oxopropylbenzenes generally inhibit enzymatic reduction, a 2'-cyano group is tolerated. Thus, reduction of **47** with NAD^+ and a formate dehydrogenase followed by a hydrolytic cyclisation gives the dihydroisocoumarin **48** in high yield and with excellent ee (Scheme 28) <99S2045>.

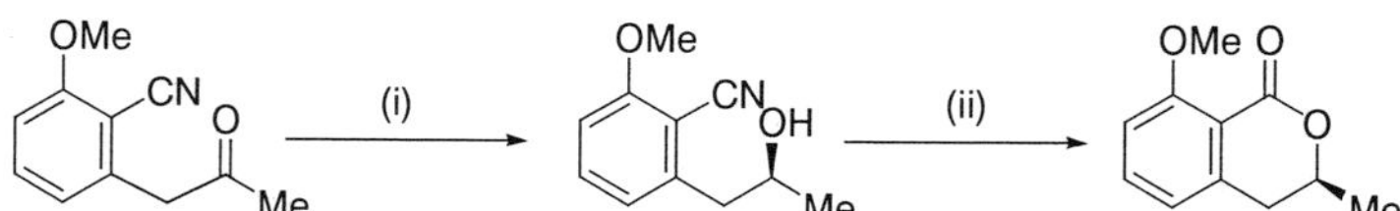

47 **48**

Reagents: (i) NAD^+, HCO_2Na, FDH-, *C*PCR-soln, $N(CH_2CH_2OH)_3$, pH 7, (90 %); (ii) HCl, MeOH, heat (94%, ee > 99.5 %)

Scheme 28

Good stereoselectivity is observed in the synthesis of 3,4-dihydroisocoumarins from the oxazoline **49** and the protected leucinal **50** (Scheme 29). Further reaction with an azetidinone introduces a dipeptide unit at C-3 <99JCS(P1)1083>.

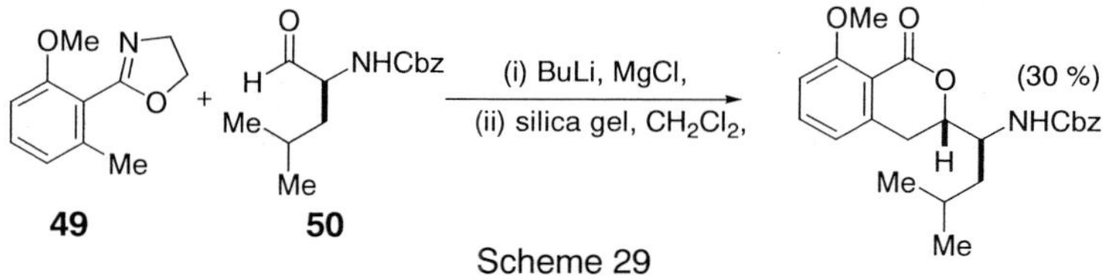

49 **50**

Scheme 29

6.4.1.7 Chromones

Alkynones *e.g.* **51** are efficiently converted into chromones *via* the enaminoketones in a one-pot reaction offering an attractive route to the heterocycle from salicylic acids by way of a chlorination-Sonogashira coupling sequence which lends itself to combinatorial chemistry (Scheme 30) <99TL2469>.

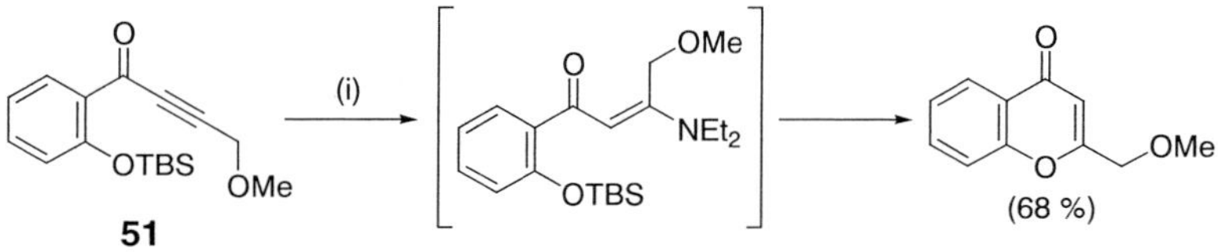

51

Reagents: (i) 10 eq. Et_2NH, EtOH, heat, 24 h.

Scheme 30

5-Hydroxyflavones can be obtained directly from 2,6-dihydroxyacetophenone by reaction with aroyl chlorides in the presence of K_2CO_3, a simplification of the usual Baker-Venkataraman synthesis <99SL1480>. Interest continues in styryl chromones with a facile synthesis of 3-styrylflavones from arylacetaldehydes and substituted propane-1,3-diones (Scheme 31) <99TL6761> and the construction of 2-[2-(3-arylnaphthyl)]chromones **52** from a Diels-Alder reaction between 2-styrylchromones and *o*-benzoquinodimethanes <99EJO135>. 7-Vinylflavones, available from a Pd-catalysed cross-coupling of flavone triflates with tetravinyltin, are a source of 7-aminoflavones through reaction with benzophenone imine and hydrogenolysis <99EJO2683>. Syntheses of 4-oxochromone-3-carbonitriles <99JOC8736>, deuterated isoflavonoids <99T3445> and the anodic fluorination of flavones <99JOC3346> have been described.

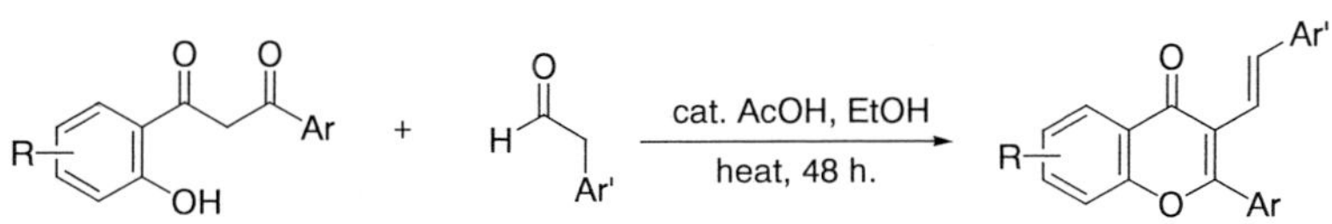

Scheme 31

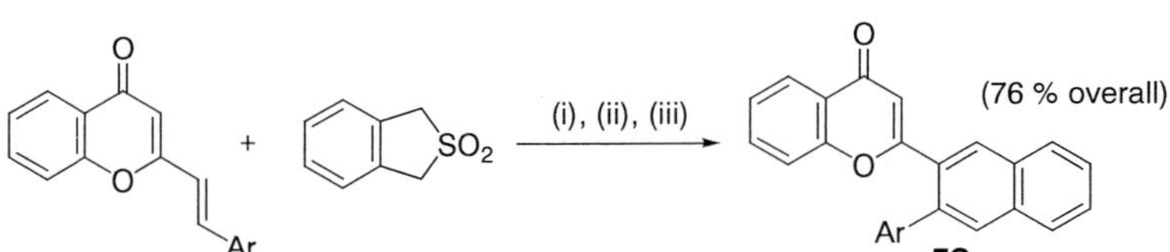

52

Reagents: (i) 1,2,4-trichlorobenzene, heat; (ii) NBS, DBP, CCl_4, heat; (iii) Et_3N, heat

The *cis*-2,3-dimethylchroman-4-one **53** is obtained with fair enantioselectivity through an asymmetric Michael addition in the presence of (-)-quinine (Scheme 32) <99TL3777>. Directed metallation of protected phenols and subsequent reaction of the Li derivative with enantiopure Weinreb amides of glycidic acids feature in a route to stereoisomers of 2-alkyl-3-hydroxychroman-4-ones (Scheme 33) <99JOC3489>.

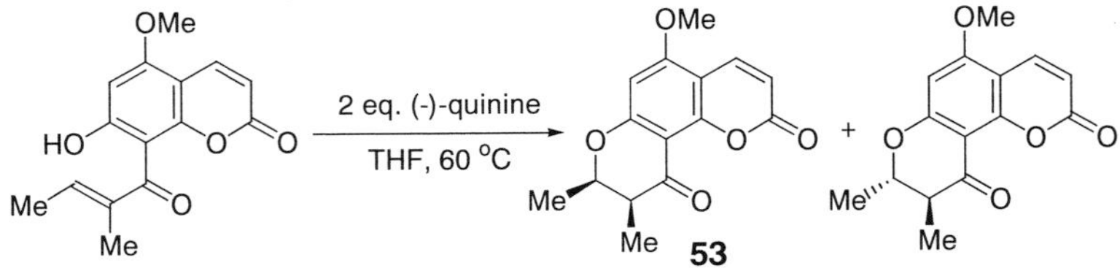

ee 75 %   cis : trans = 1:1

Scheme 32

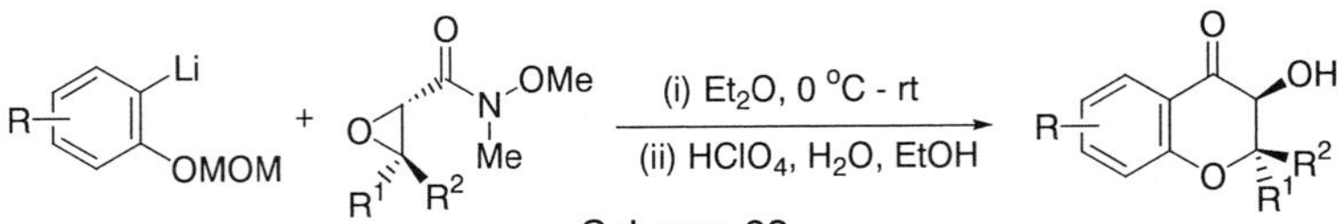

Scheme 33

6.4.1.8 Xanthones

N-Alkylnitrilium salts, prepared from benzonitriles and $SbCl_5$, acylate activated aromatic compounds under mild conditions. The resulting benzophenone imines can be cyclised and hydrolysed providing a good route to substituted xanthones (Scheme 34) <99 JOC4050>.

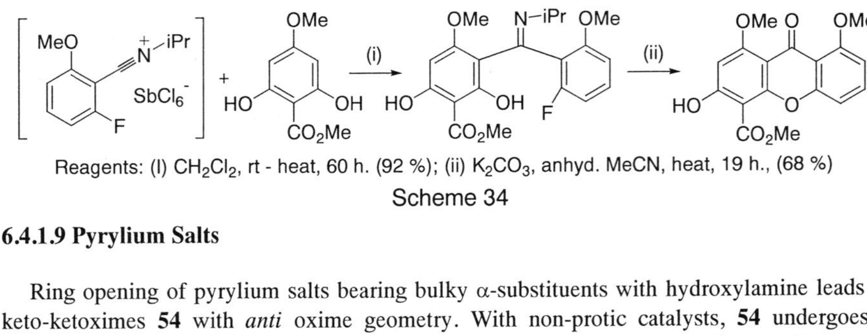

Reagents: (I) CH_2Cl_2, rt - heat, 60 h. (92 %); (ii) K_2CO_3, anhyd. MeCN, heat, 19 h., (68 %)

Scheme 34

6.4.1.9 Pyrylium Salts

Ring opening of pyrylium salts bearing bulky α-substituents with hydroxylamine leads to keto-ketoximes **54** with *anti* oxime geometry. With non-protic catalysts, **54** undergoes a Beckmann rearrangement in which the carbonyl oxygen atom behaves as the terminator and the ensuing cyclisation yields 2-aminopyrylium salts. These in turn act as precursors of 2*H*-pyran-2-imines (Scheme 35) <99T15011>.

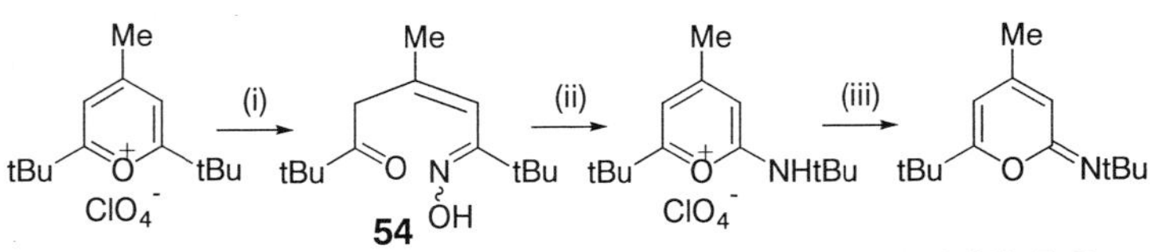

Reagents: (i) $NH_2OH.HCl$, NaOH, Et_2O, rt; (ii) $SOCl_2$, CCl_4; (iii) NH_4OH, Et_2O, rt

Scheme 35

Chromeno[4,3,2-*gh*]phenanthridine **55** is formed when the anion derived from the addition product from the reaction of benzotriazole and the xanthylium ion is oxidised with copper iodide (Scheme 36). Thioxanthylium cations behave in a similar manner <99JHC927>.

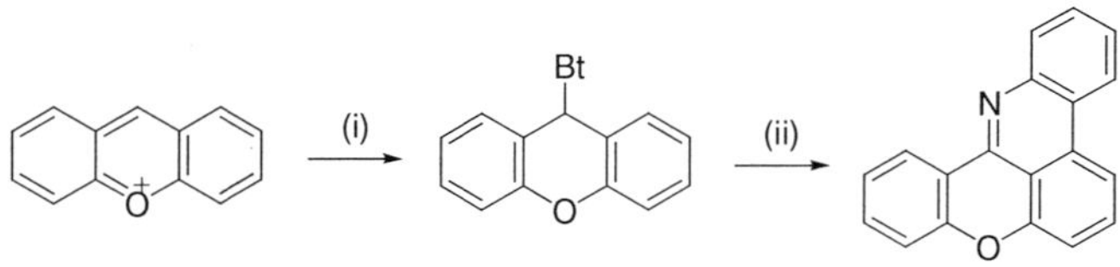

55

Reagents: (i) benzotriazole (Bt), PhH, 4-TsOH, heat, 8 h.; (ii) nBuLi, THF, -78 °C, then CuI, heat (22 %)

Scheme 36

6.4.2 HETEROCYCLES CONTAINING ONE SULFUR ATOM

6.4.2.1 Thiopyrans and analogues

Application of cycloadditions to the synthesis of the thiopyran system has continued. Acetylenic thioketones and thioaldehydes are accessible from β-silyl-protected ynones by reaction with HMDST and are readily trapped by 1,3-dienes to give 5,6-dihydrothiopyrans. Desilylation is facile and leads to 6-ethynyl derivatives. The reaction between cyclohexadiene and a thioaldehyde shows *exo*-selectivity <99SL1739>. The thermolysis of heteroaryl-substituted phenacylsulfoxides generates sulfines which cycloadd to butadiene to give 5,6-dihydrothiopyran 1-oxides. The stereoselectivity of the sequence which usually gives the *cis* adduct is thought to be controlled by the elimination of the heterocyclic moiety which leads to the sulfine. An external base not only accelerates the reaction but also influences the *cis/trans* ratio, favouring the latter <99JOC6730>.

Thiabutadienes undergo highly enantioselective hDA reactions in the presence of homochiral bis(oxazoline) and bis(imine) complexes with Cu and Ni (Scheme 37) <99CC1001>. Homochiral camphor-based thiabutadienes show good *exo* selectivity and give rise to bornene ring-fused dihydrothiopyrans (Scheme 38) <99TL8383>.

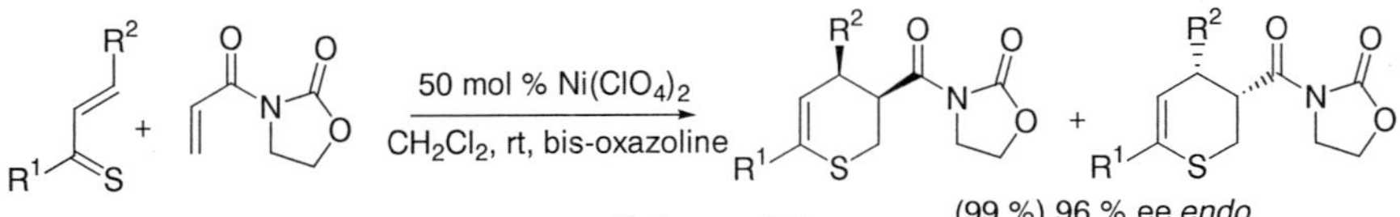

Scheme 37

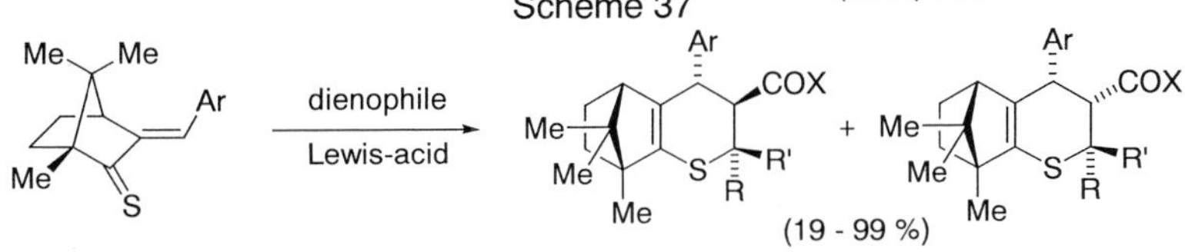

Scheme 38

Dihydrobenzothiopyran-3-carboxylic acid derivatives are formed when 2-mercapto-benzophenone reacts with α,β-unsaturated carboxylates and nitriles in the presence of magnesium bis(diisopropyl)amide. A facile dehydration leads to the 2*H*-1-benzothiopyrans <99JCS(P1)1547>. Alkyl 4-oxo-1*H*-2-benzothiopyran-1-carboxylates result from the Friedel-Crafts cyclisation of the sulfides **56** <99JMC751>.

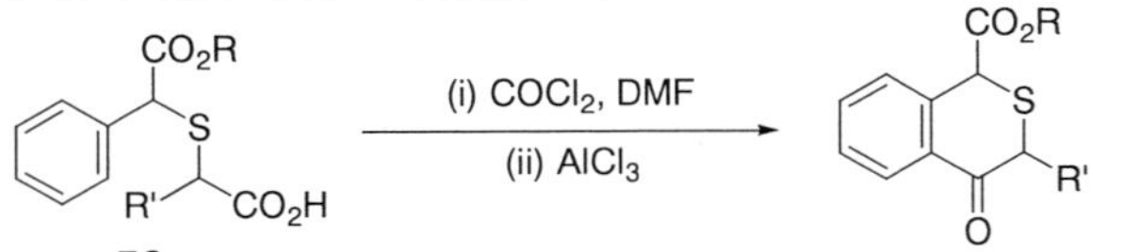

56

3-*tert*-Butyl-1*H*-2-benzotelluropyran undergoes a facile Li-Te exchange on reaction with BuLi, providing access to a (*E*)-2-(2'-lithiovinyl)benzyllithium and thus to dihydrometala-naphthalenes (Scheme 39) <99S1866>.

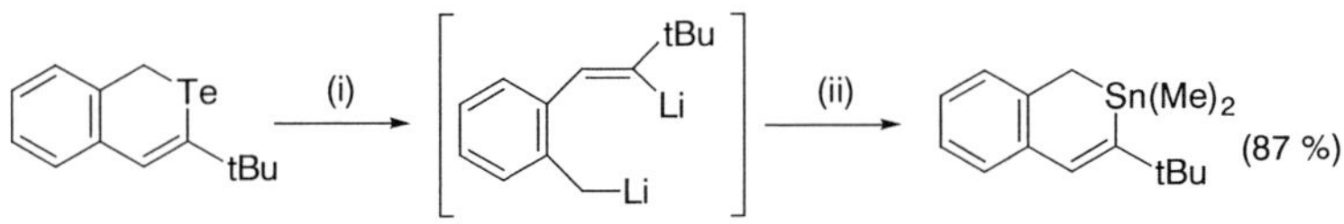

Reagents: (i) nBuLi, anhyd. THF, -80 °C, 10 min.; (ii) Me_2SnCl_2

Scheme 39

The enol ethers generated *in situ* by the addition of ethanol to 1,4-pentadiyn-3-ones are cyclised to the 4*H*-chalcogenopyran-4-ones on treatment with disodium chalcogenides (Scheme 40) <99JHC707>. The 2,6-diphenyl derivatives are useful catalysts for the Baylis-Hillman reaction <99TL3741>.

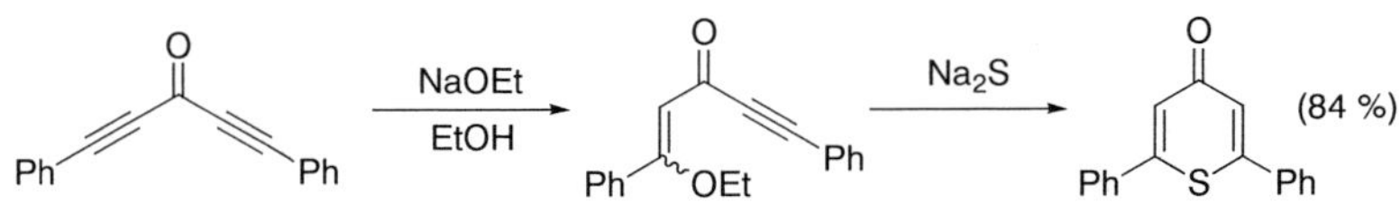

Scheme 40

Thiochroman-4-ones result from the Pd-catalysed reaction of iodothiophenols with allenes and carbon monoxide <99JOC9646> and various thioflavonoids are accessible from acetoquinone *via* derived chalcones <99JOC359>.

The 9-phenylthioxanthyl (*S*-pixyl) group is a photocleavable 5'-OH protecting group for deoxyribonucleosides <99TL7911>.

The ability of a range of 2,4,6-triaryl thio-, seleno- and telluro- pyrylium dyes to generate singlet oxygen and hence to function as sensitisers for photodynamic therapy has been examined <99JMC3942, 3953>.

The synthesis of 2-benzoselenopyrylium and 2-substituted 1-benzotelluropyrylium salts have been achieved and their reactions with nucleophiles investigated <99H(51)17, 99JCS(P1)1665>.

6.4.3 HETEROCYCLES CONTAINING TWO OR MORE OXYGEN ATOMS

6.4.3.1 Dioxins

The photooxygenation of both (*E,E*)- and (*E,Z*)-1-aryl-1,3-pentadienes sensitised by tetraphenylporphine leads almost exclusively to *cis*-3-aryl-6-methyl-1,2-dioxines. A photoisomerisation of the (*E,Z*)-diene is involved and the addition of singlet oxygen occurs only to the (*E,E*)-isomer <99JOC493>. Chiral dienol ethers provide access to alkoxy-1,2-dioxines under similar conditions; the extent of competing reactions and of diastereoselectivity is dependent on the substitution and stereochemistry of the diene <99T11437>.

Electron-withdrawing groups occupy axial and electron-donating groups equatorial positions in the chair conformation of 2,2-di(4-substituted phenyl)-1,3-dioxanes <99JOC1436>. 1,3-Dioxane derivatives function as a source of γ-oxy-substituted benzyllithium derivatives <SYN664>, γ-amino acid derivatives <99CL879>and of 1,3-dioxane-4-carbaldehydes <99SL303>.

Functionalised 2,3-dihydro-1,4-dioxins can be synthesised in a three step-sequence from β-keto esters. The key step is the insertion of a Rh-carbenoid derived from an α-diazo-β-keto ester into an O-H bond of a 1,2-diol <99H(51)1073>. The reaction of 2-(1,4-dioxenyl)alkanols with silyl enol ethers yields 2,3-disubstituted 1,4-dioxanes. When 1,2-bis(trimethylsilyloxy)-cyclobut-1-ene is used, the expected cyclobutanone products are accompanied by a spirocyclopropane derivative <99TL863>. 1,4-Dioxane-monochloroborane **57** is a highly reactive hydroborating reagent <99OL315>.

Whereas 2-substituted 1,4-benzodioxanes are attacked predominantly at the 7-position in $AlCl_3$-DMA-promoted Friedel-Crafts reactions, the corresponding benzodioxin gives mainly the C-6 acylated product <99TL3523>.

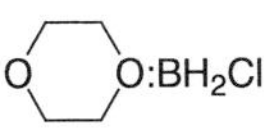

57

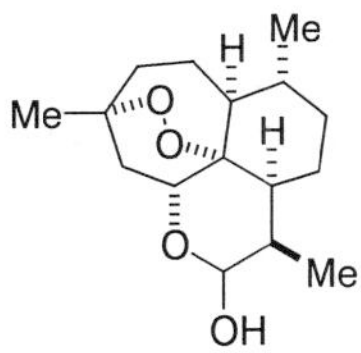

58

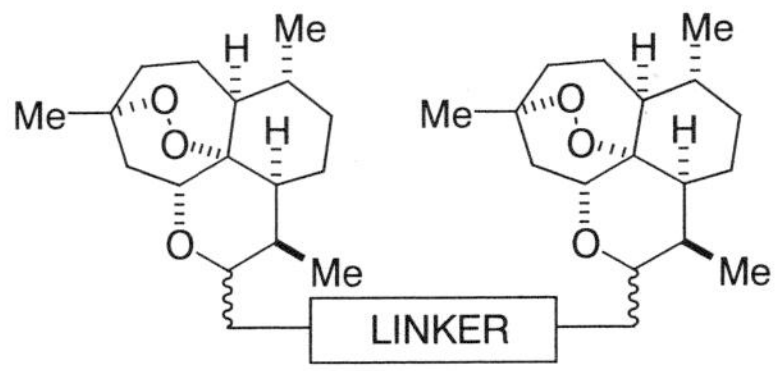

59

6.4.3.2 Trioxins

Developments in the chemistry and biological activity of artemisinin and related antimalarials have been reviewed <99H(51)1681>. Dihydroartemisinin **58**, readily obtained by reduction of the lactone function of artemisinin, is a source of 10-aryloxy <99TL8543>, 10-aryl, -heteroaryl and -alkynyl <99JMC300, 99T3625>, 10-naphthyl <99JCS(P1)1827> and 10-alkyl <99TL8543> derivatives, all of which appear to offer advantages over artemisinin itself. Two artemisinin moieties have been joined through various linker units and **59** exhibit good biological activities <99JMC4275>. Detailed assignments of the ^1H and ^{13}C NMR spectra of artether have been made and correlated with a conformational model <99JCS(P2)2089>. Simpler 1,2,4-trioxanes related to artemisinin have been synthesised from cyclohexanone involving a Co(II)-catalysed oxygenation of an allylic alcohol <99TL8391> and an asymmetric $MgCl_2$-promoted Michael addition of an enamine to acrylonitrile <99TL9133> as key steps.

6.4.3.3 Tetraoxanes

Antimalarial activity is also shown by 1,2,4,5-tetraoxanes and these compounds have been synthesised by the acid-catalysed dimerisation of ozonides <99JMC2604>. Ozonolysis of vinyl ethers and cyclocondensation of the derived bis(hydroperoxides) with carbonyl compounds is an alternative approach <99JCS(P1)1867> and dispiro-linked tetraoxanes are accessible from cyclohexanones by treatment with hydrogen peroxide <99JMC1477>.

6.4.4 HETEROCYCLES CONTAINING TWO OR MORE SULFUR ATOMS

6.4.4.1 Dithianes

Cyclodecasulfur, S_{10}, an allotrope obtained by reaction of titanocene pentasulfide with sulfuryl chloride, reacts with 1,3-dienes to give 3*H*,6*H*-1,2-dithiins. A radical mechanism and cycloelimination of S_8 is proposed, with the process offering a diatomic sulfur transfer <99TL7961>. Titanocene derivatives also feature in a general synthesis of 1,2-dichalcogenins from alkynes and butadienes (Scheme 41) <99AG(E)1604>.

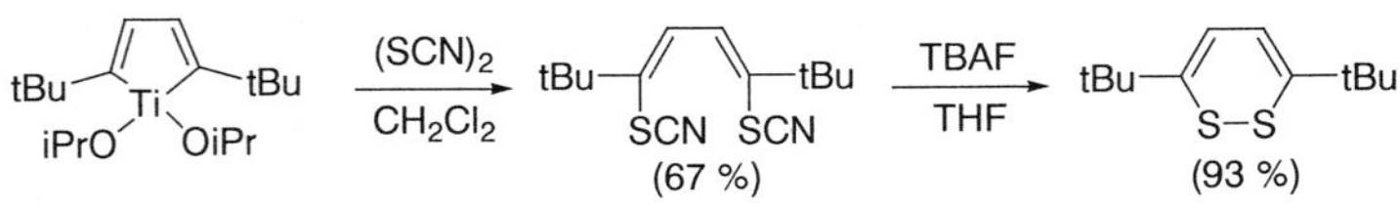

Scheme 41

1-Adamantyl-*tert*-butyltetrathiolane 2,3-dioxide is an efficient precursor of S_2O which can be trapped by 1,3-dienes to give 1,2-dithiin 1-oxides <99JA7959>.

The stereospecific formation of 5,6-dihydro-1,4-dithiins from the reaction of the 1,2-dithiete **60** with alkenes has been shown to proceed through its valence isomer, 1,2-bis(methoxycarbonyl)ethane-1,2-dithione (Scheme 42) <99JOC8489>.

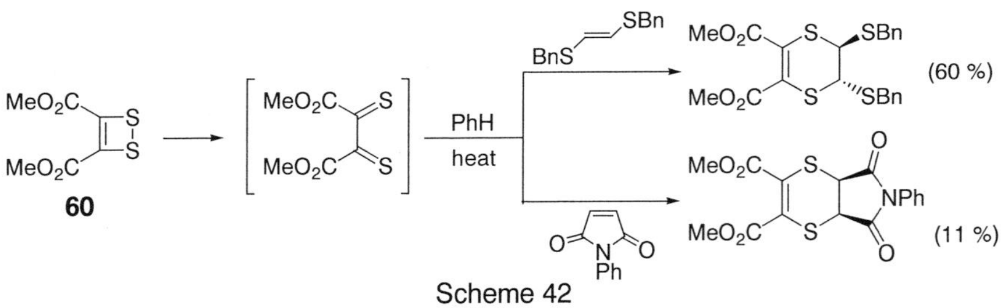

Scheme 42

Expansion of the dithiolane ring is involved in the stereoselective formation of 2-azido-1,4-dithianes (Scheme 43) <99T801> and of the spiro-linked dithiane **61** <99CC1673>.

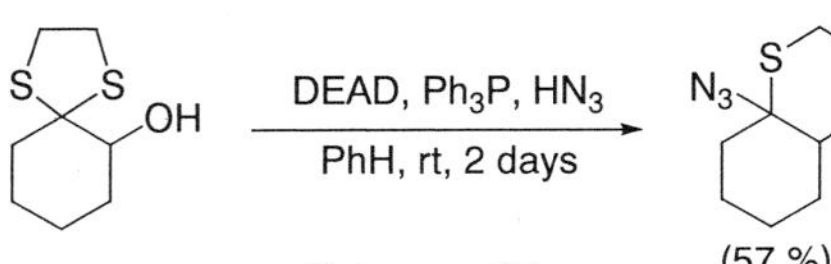

Scheme 43

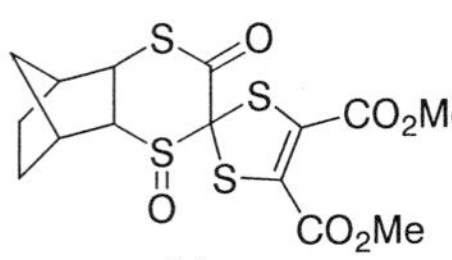

61

Thermodynamic data has been published for 1,3- and 1,4- dithianes <99JCT635, 99JOC9328, 99T359>. Both theoretical and experimental evidence indicates that the 1,4-dithiin dication is aromatic <99CC777>.

6.4.5 HETEROCYCLES CONTAINING BOTH OXYGEN AND SULFUR IN THE SAME RING

6.4.5.1 Oxathiines

1,2-Oxathiane 2-oxides are formed by the oxidative ring expansion of 2-alkylthio-2-benzylthiolane 1-oxides brought about by [bis(trifluoroacetoxy)iodo]benzene. That the reaction is only successful with the (1R*,2S*)-diastereoisomers is attributed to chelation between the nucleophilic S and O atoms and the hypervalent iodine <99EJO943>. A diazo-mediated thiolane ring expansion is the key step in a synthesis of the acenaphtho-[1,2-*b*][1,4]oxathiine system <99JCS(P2)755>.

The use of *o*-hydroxyphenylthiophthalimides as a source of *ortho*-thioquinones has been extended to an atropisomeric biphenol **62** and thus to the synthesis of various 1,4-oxathiin derivatives including the crown ethers **63** <99TL4421>. Similar methodology enables enantiomerically pure 4'-thiaspiroacetals to be obtained using substituted exoenitols in inverse electron demand Diels-Alder reactions <99JOC6490>.

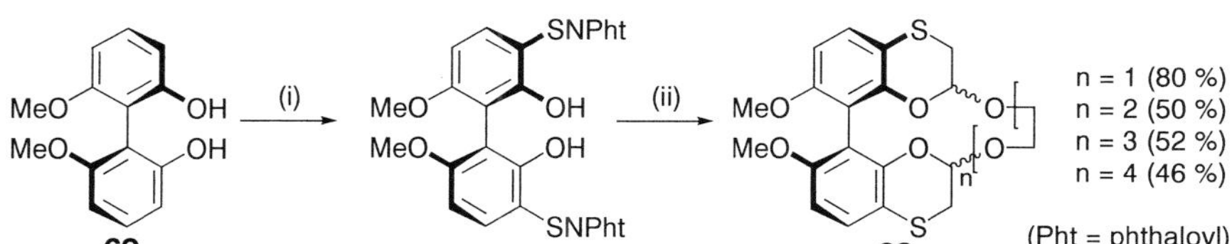

62 **63**

Reagents: (i) PhtNSCl, $CHCl_3$, rt, 5 h. (67 %); (ii) Et_3N, bis-enol ether, $CHCl_3$, 60 °C

The dichloro compounds **65** derived from 1,4-oxathiane-3-carboxanilides **64** yield bicyclic β-lactams on treatment with base through an intramolecular nucleophilic substitution facilitated by the neighbouring sulfur atom <99H(50)713>.

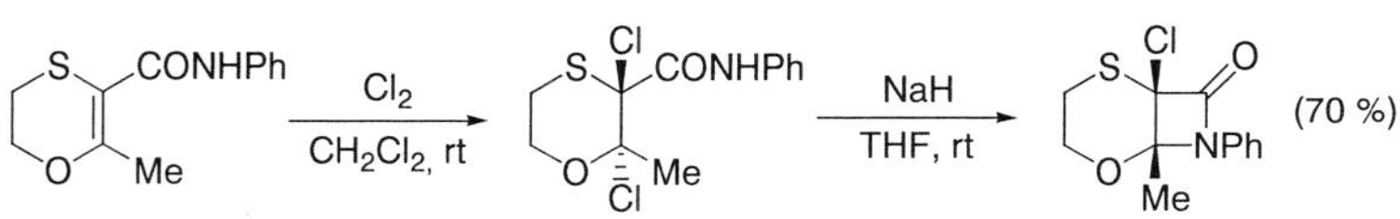

64 **65**

6.4.6 REFERENCES

99AG(E)1435 B. A. Chauder, A. V. Kalinin, N. J. Taylor, V. Snieckus, *Angew. Chem. Int. Ed. Engl.*, **1999**, *38*, 1435.

99AG(E)1604 E. Block, M. Birringer, C. He, *Angew. Chem. Int. Ed. Engl.*, **1999**, *38*, 1604.

99AG(E)2258 A. B. Dounay, R. A. Urbanek, S. F. Sabes, C. J. Forsyth, *Angew. Chem. Int. Ed. Engl.*, **1999**, *38*, 2258.

99AG(E)2398 A. G. Dossetter, T. F. Jamison, E. N. Jacobsen, *Angew. Chem. Int. Ed. Engl.*, **1999**, *38*, 2398.

B-99MI1 The Handbook of Natural Flavonoids, Vol. 1 and 2, eds. J. B. Harborne, H. Baxter, Wiley-VCH, Weinheim, Germany, 1999.

99BCJ73 K. S. Shrestha, K. Honda, M. Asami, S. Inoue, *Bull. Chem. Soc. Jpn.*, **1999**, *72*, 73.

99CC291 W.-C. Zhang, G. S. Viswanathan, C.-J. Li, *J. Chem. Soc. Chem. Commun.*, **1999**, 291.

99CC777 T. Nishinaga, A. Wakamiya, K. Komatsu, *J. Chem. Soc. Chem. Commun.*, **1999**, 777.

99CC793 N. Bushby, C. J. Moody, D. A. Riddick, I. R. Waldron, *J. Chem. Soc. Chem. Commun.*, **1999**, 793.

99CC1001 T.Saito, K. Takekawa, T. Takahashi, *J. Chem. Soc. Chem. Commun.*, **1999**, 1001.

99CC1063 K. Maeda, T. Oishi, H. Oguri, M. Hirama, *J. Chem. Soc. Chem. Commun.*, **1999**, 1063.

99CC1637 C. A. Roeschlaub, N. L. Maidwell, M. R. Rezai, P. G. Sammes, *J. Chem. Soc. Chem. Commun.*, **1999**, 1637.

99CC1673 R. A. Aitken, L. Hill, P. Lightfoot, N. J. Wilson, *J. Chem. Soc. Chem. Commun.*, **1999**, 1673.

99SL717 J. Yang, C.-J. Li, *Synlett.*, **1999**, 717.
99SL1037 K. Fujiwara, K. Saka, D. Takaoka, A. Murai, *Synlett.*, **1999**, 1037.
99SL1480 F. Bois, C. Beney, A.-M. Mariotte, A. Boumendjel, *Synlett.*, **1999**, 1480.
99SL1739 A. Degl'Innocenti, A. Capperucci, P. Scafato, T. Mecca, G. Reginato, A. Mordini, *Synlett.*, **1999**, 1739.
99SL1945 M. A. Leeuwenburgh, C. Kulker, H. S. Overkleeft, G. A. van der Marel, J. H. van Boom, *Synlett.*, **1999**, 1945.
99T359 E. Juaristi, G. Cuevas, *Tetrahedron,* **1999**, *55*, 359.
99T801 C. A. M. Alfonso, M. T. Barros, C. D. Maycock, *Tetrahedron,* **1999**, *55*, 801.
99T3445 S. Rasku, K. Wähälä, J. Koskimies, T. Hase, *Tetrahedron,* **1999**, *55*, 3445.
99T3625 H. O'Dowd, P. Ploypradith, S. Xie, T. A. Shapiro, G. H. Posner, *Tetrahedron,* **1999**, *55*, 3625.
99T4783 I. P. Lokot, F. S. Pashovsky, F. A. Lakhvich, *Tetrahedron,* **1999**, *55*, 4783.
99T4969 M. Kotora, M. Ishikawa, F.-Y. Tsai, T. Takahashi, *Tetrahedron,* **1999**, *55*, 4969.
99T5567 F. Bertha, J. Fetter, M. Kajtar-Peredy, K. Lempert, *Tetrahedron,* **1999**, *55*, 5567.
99T6557 S. Cenini, G. Cravotto, G. B. Giovenzana, G. Palmisano, S. Tollari, *Tetrahedron,* **1999**, *55*, 6557.
99T7555 A. Burgard, H.-J. Lang, U. Gerlach, *Tetrahedron,* **1999**, *55*, 7555.
99T7661 M. A. Brimble, F. A. Fares, *Tetrahedron,* **1999**, *55*, 7661.
99T7847 P. Audin, N. Piveteau, A.-S. Dussert, J. Paris, *Tetrahedron,* **1999**, *55*, 7847.
99T8231 J. S. Clark, J. G. Kettle, *Tetrahedron,* **1999**, *55*, 8231.
99T8253 M. A. Leeuwenburgh, C. Kulker, H. J. Duynstee, H. S. Overkleeft, G. A. van der Marel, J. H. van Boom, *Tetrahedron,* **1999**, *55*, 8253.
99T10295 M. Murakata, H. Tsutsui, N. Takeuchi, O. Hoshino, *Tetrahedron,* **1999**, *55*, 10295.
99T11437 P. H. Dussault, Q. Han, D. G. Sloss, D. J. Symonsbergen, *Tetrahedron,* **1999**, *55*, 11437.
99T12907 R. Hayes, K.-D. Li, P. Leeming, T. W. Wallace, R. C. Williams, *Tetrahedron,* **1999**, *55*, 12907.
99T14719 P. Bovicelli, P. Lupattelli, B. Crescenzi, A. Sanetti, R. Bernini, *Tetrahedron ,* **1999**, *55*, 14719.
99T15011 C. Uncuja, A. Tudose, M. T. Căproiu, M. Plăveji, R. Kakou-Yao, *Tetrahedron,* **1999**, *55*, 15011.
99T15181 A. F. Barrero, E. J. Alvarez-Manzaneda, R. Chahboun, M. Cortés, V. Armstrong, *Tetrahedron,* **1999**, *55*, 15181.
99TL863 I. Hanna, L. Ricard, *Tetrahedron Lett.*, **1999**, *40*, 863.
99TL1083 A. K. Ghosh, R. Kawahama, *Tetrahedron Lett.*, **1999**, *40*, 1083.
99TL1241 J. C. Bussolari, D. C. Rehborn, D. W. Combs, *Tetrahedron Lett.*, **1999**, *40*, 1241.
99TL1257 R. H. Schlessinger, K. W. Gillman, *Tetrahedron Lett.*, **1999**, *40*, 1257.
99TL1747 S. Kamijo, Y. Yamamoto, *Tetrahedron Lett.*, **1999**, *40*, 1747.
99TL1925 K. Tatsuta, T. Tamura, T. Mase, *Tetrahedron Lett.*, **1999**, *40*, 1925.
99TL2145 N. Hori, K. Nagasawa, T. Shimizu, T. Nakata, *Tetrahedron Lett.*, **1999**, *40*, 2145.
99TL2235 F. E. McDonald, P. Vadapally, *Tetrahedron Lett.*, **1999**, *40*, 2235.
99TL2469 A. S. Bhat, J. L. Whetstone, R. W. Brueggemeier, *Tetrahedron Lett.*, **1999**, *40*, 2469.
99TL2811 N. Hori, H. Matsukura, G. Matsuo, T. Nakata, *Tetrahedron Lett.*, **1999**, *40*, 2811.
99TL2817 J. Blanco-Urgoiti, L. Casarrubios, J. Pérez-Castells, *Tetrahedron Lett.*, **1999**, *40*, 2817.
99TL2879 D. A. Evans, J. S. Johnson, C. S. Burgey, K. R. Campos, *Tetrahedron Lett.*, **1999**, *40*, 2879.
99TL3293 M. A. McGuire, S. C. Shilcrat, E. Sorenson, *Tetrahedron Lett.*, **1999**, *40*, 3293.
99TL3523 A. G. Suárez, *Tetrahedron Lett.*, **1999**, *40*, 3523.
99TL3741 T. Iwama, H. Kinoshita, T. Kataoka, *Tetrahedron Lett.*, **1999**, *40*, 3741.
99TL3777 T. Ishikawa, Y. Oku, T. Tanaka, T. Kumamoto, *Tetrahedron Lett.*, **1999**, *40*, 3777.
99TL4319 B. Schmidt, *Tetrahedron Lett.*, **1999**, *40*, 4319.
99TL4421 G. Capozzi, G. Delogu, M. A. Dettori, D. Fabbri, S. Menichetti, C. Nativi, R. Nuti, *Tetrahedron Lett.*, **1999**, *40*, 4421.
99TL4751 A. K. Ghosh, R. Kawahama, *Tetrahedron Lett.*, **1999**, *40*, 4751.
99TL4871 R. G. F. Giles, I. R. Green, C. P. Taylor, *Tetrahedron Lett.*, **1999**, *40*, 4871.
99TL5803 J. P. G. Versleijen, A. M. van Leusen, B. L. Feringa, *Tetrahedron Lett.*, **1999**, *40*, 5803.
99TL6757 S. Bräse, *Tetrahedron Lett.*, **1999**, *40*, 6757.
99TL6761 V. Lokshin, A. Heynderickx, A. Samat, G. Pèpe, R. Guglielmetti, *Tetrahedron Lett.*, **1999**, *40*, 6761.
99TL6903 L.-L. Wei, H. Xiong, C. J. Douglas, R. P. Hsung, *Tetrahedron Lett.*, **1999**, *40*, 6903.
99TL7135 J. C. Anderson, B. P. McDermott, *Tetrahedron Lett.*, **1999**, *40*, 7135.
99TL7455 R. Roggenbuck, P. Eilbracht, *Tetrahedron Lett.*, **1999**, *40*, 7455.
99TL7709 R. Grigg, J. P. Major, F. M. Martin, M. Whittaker, *Tetrahedron Lett.*, **1999**, *40*, 7709.
99TL7911 M. P. Coleman, M. K. Boyd, *Tetrahedron Lett.*, **1999**, *40*, 7911.
99TL7961 P. Lestè-Lasserre, D. N. Harpp, *Tetrahedron Lett.*, **1999**, *40*, 7961.
99TL8019 Y. Mori, H. Furuta, T. Takase, S. Mitsuoka, H. Furukawa, *Tetrahedron Lett.*, **1999**, *40*, 8019.
99TL8383 T. Saito, J. Nishimura, D. Akiba, H. Kusuoku, K. Kobayashi, *Tetrahedron Lett.*, **1999**, *40*, 8383.
99TL8391 C. H. Oh, H. J. Kim, S. H. Wu, H. S. Won, *Tetrahedron Lett.*, **1999**, *40*, 8391.
99TL8543 J. Ma, E. Katz, H. Ziffer, *Tetrahedron Lett.*, **1999**, *40*, 8543.
99TL8859 G. Matsuo, N. Hori, T. Nakata, *Tetrahedron Lett.*, **1999**, *40*, 8859.
99TL9133 P. M. O'Neill, A. Miller, J. F. Bickley, F.Scheinmann, C. H. Ho, G. H. Posner, *Tetrahedron Lett.*, **1999**, *40*, 9133.

Chapter 7

Seven-Membered Rings

David J. LeCount
Formerly of Zeneca Pharmaceuticals UK
1, Vernon Avenue, Congleton, Cheshire, UK
email: DLEC@Compuserve.com

7.1 INTRODUCTION

As in more recent years, the chemistry of seven-membered ring systems has been dominated by the chemistry of oxygen heterocycles in the form of the marine toxins and, to a lesser extent, the antimalarial artemisinin. Indeed, if it were not for the interest in these systems it would have been a sparse year indeed. For this reason the division of this report will be into just three section, nitrogen, oxygen, and other systems.

7.2 SEVEN-MEMBERED SYSTEMS CONTAINING NITROGEN

In spite of what has been stated above, there have been a number of interesting studies on seven-membered systems containing nitrogen. Further studies on the chemistry of *N,N*-dialkylaminoallenes have shown that reaction of the system **1** with dimethyl acetylenedicarboxylate gives the bicyclic system **2** as a single isomer. However, if the *t*-butyl group in **1** is replaced by an allyl group two isomers **3** are formed, when the intermediate 2+2 cycloaddition products were also isolated and characterised <99T1309>.

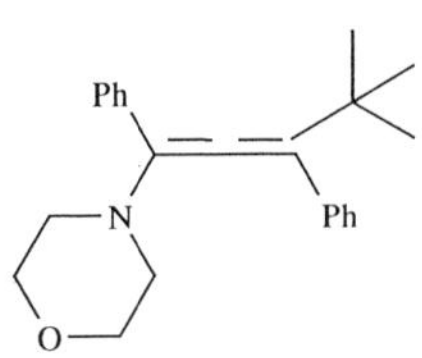

1

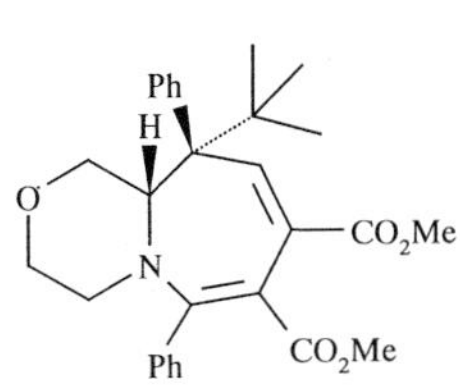

2

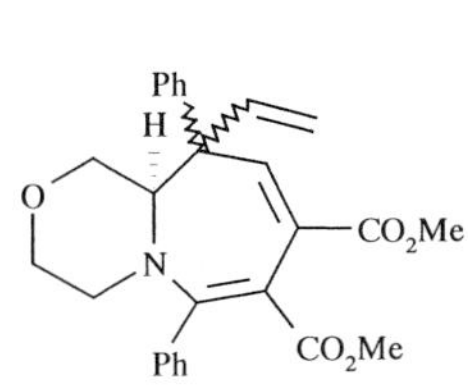

3

It is rare that a year passes without a report on the cyclisation of triene-conjugated nitrile ylides from Sharp and co-workers, and this year is no exception. In this example, cyclisation of the ylides from **4** (R^1, R^3 = Ph; R^2 = H) and **4** (R^1, R^2 = Ph; R^3 = H) at 0 °C yields the very complex isoquinolines **5** (R^1, R^3 = Ph, R^2 = H) and **5** (R^1, R^2 = Ph; R^3 = H) which are intermediates in the formation of the major product, the azabenzo[3,4]barbaralane **6** <99JCS(P1)443>.

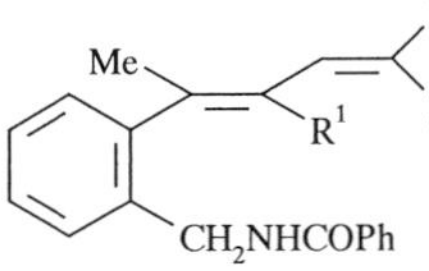

4

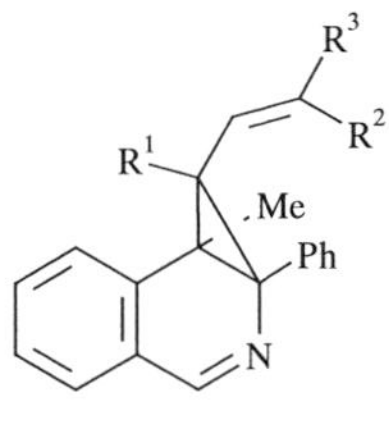

5

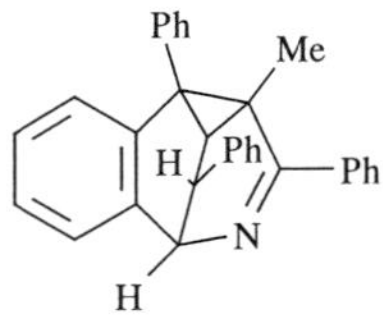

6

Ring-closing metathesis has again proved to be a useful for the preparation of azepine derivatives. Thus the bicyclic derivative **7** is just one example of a series of bicyclic derivatives of varying ring sizes obtained in yields in excess of 80% by this method <99JCS(P1)1695>. A solid state version of the reaction has been reported in the formation of **8** <99S138>. In both cases the ring sizes are not restricted to those illustrated.

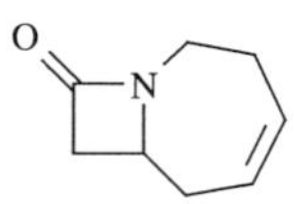

7

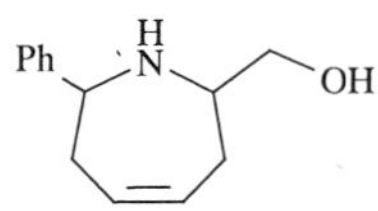

8

Heating the alkyne **9** in the presence of toluene results in the formation of the bicyclic β-lactam **10** in moderate yield (Scheme **1**) <99CC1913>. However, reaction in benzene in the presence of AIBN and triphenyltin hydride results in the formation of **11** (R = $SnPh_3$) as the *Z* isomer in almost quantitative yield. If triphenyltin is replaced by PhSH the yield of **11** (R= SPh) is much reduced and both *Z* and *E* isomers are formed

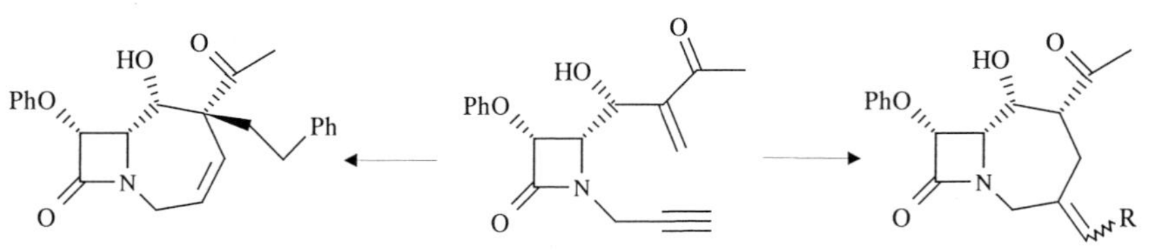

10 **9** **11**

Scheme **1**

Radical cyclisation of the trimethylsilylalkyne **12** affords the macrocycles **13** and **14** (Scheme **2**). These in turn can also be induced to cyclise by treatment with boranedimethylsulphide complex to form **15** and **16**, respectively, which contain the ring systems of the protoberberine alkaloids <99JOC4830>.

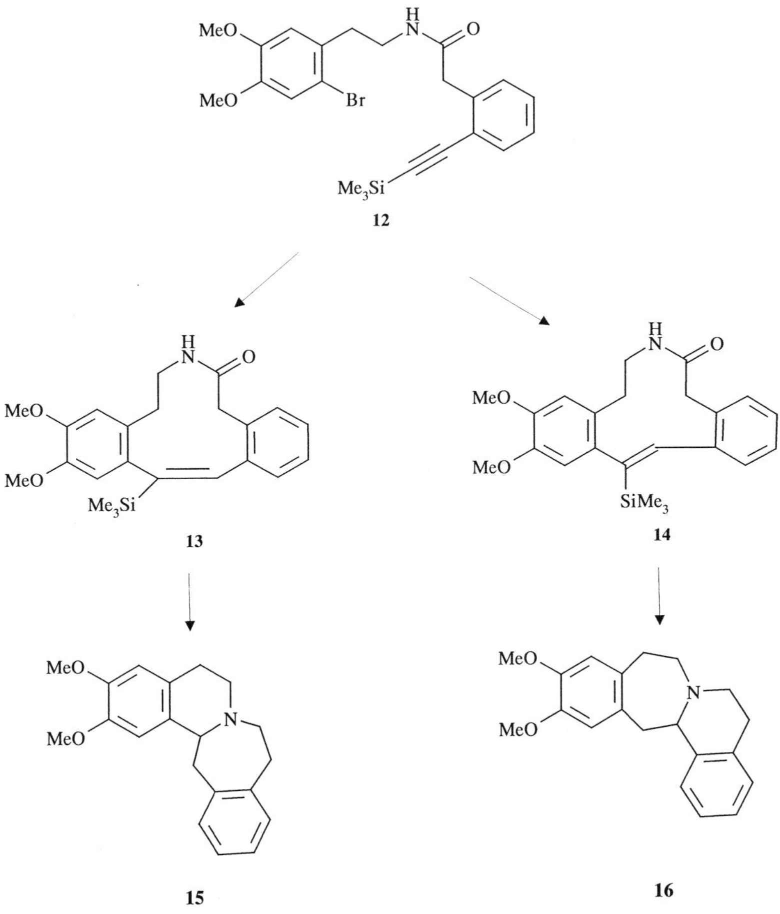

Scheme **2**

Titanium chloride catalysed cyclisation of the methoxylated amide **17** is highly dependent upon the nature of the substituent R <99TL7939>. If R = Me cyclisation to the 6,6 system **18** occurs, but the merest hint of an electron withdrawing character results in a different path for the reaction ending in the formation of the 7,6 system **9** (Scheme **3**).

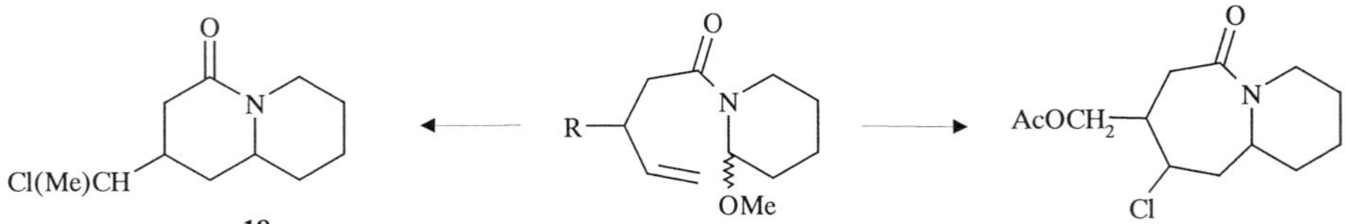

18 17 19

Scheme **3**

An enantioselective synthesis of the methyl esters of (-)-*cis* and (-)-*trans*-clavicipitic esters has been achieved <99T10989>. The key cyclisation step is the acid cyclisation of the amino acid ester **20**. A number of tetracyclic derivatives of the general structure **21** have also been reported <99TL5569>.

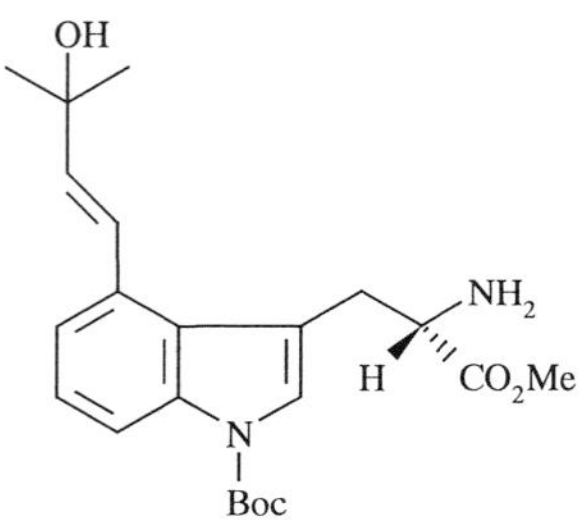

20

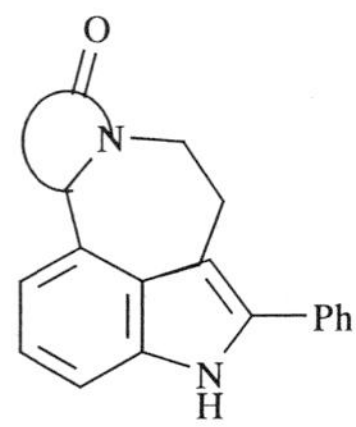

21

Two reports on the preparation of polyhydroxylated [1,3]thiazolo[3,2*a*]azepane derivatives have been published. The one reports the synthesis of the β-turn mimetic **22** from D-glucurono-3,6-lactone and L-cysteine <99TL477>, whilst the other describes, for example the preparation of **23** in 78% yield from 6-*O*-tosyl-2,3-*O* isopropylidene-D-mannofuranose and 2-aminothioethanol <99S839>.

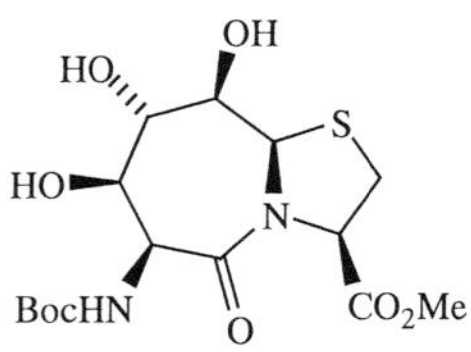

22

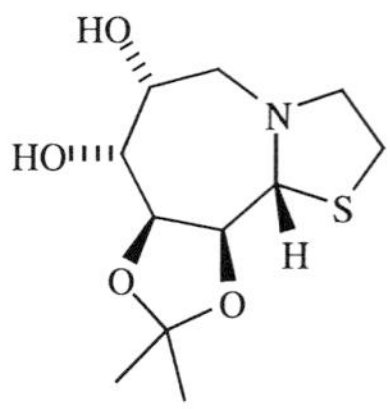

23

Irradiation of the caged oxadiazabicyclo[2.2.3]nonadiene derivative **24** (X = CH) brings about its rearrangement to the isomeric system **25** and the formation of the 1-ethoxycarbonyl-1*H*-azepine (**26**) <99H(51)141>. The latter is also the product of irradiation **25**, but the authors interpret their kinetic results to suggest that the azepine is not derived solely from **25**, but is also formed directly from **24**. Similar studies with **24** (X = N) give 1-ethoxycarbonyl-1*H*-1,2-diazepine (Scheme **4**).

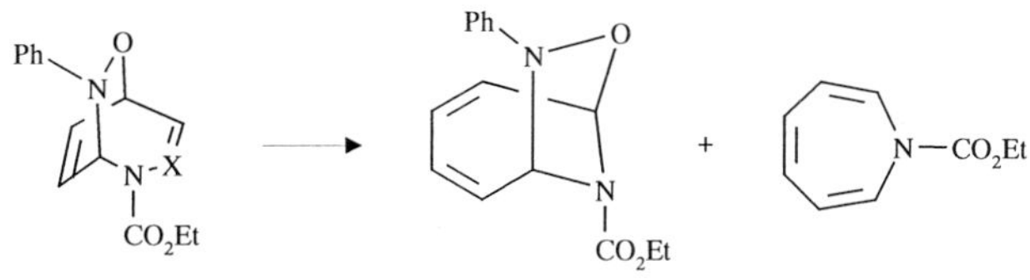

24 **25** **26**

Scheme **4**

Carbonylation of the urea **27** in the presence of Pd(0) and KOAc as base is a useful route to 4-butyl-2-phenyl-2,3,4,5-tetrahydro-1*H*-2,4-benzodiazepine-1,3-dione **(28)** <99TL2623>. Also isolated from this reaction as secondary product is *N*-*n*-butylisoindoline. Pyrolysis of the cyclobutenones **29** afford the diazepines **30** in moderate yield <99JOC707>.

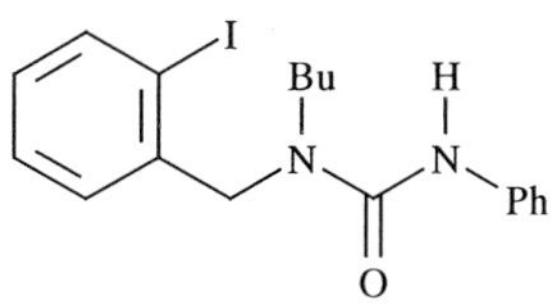

27

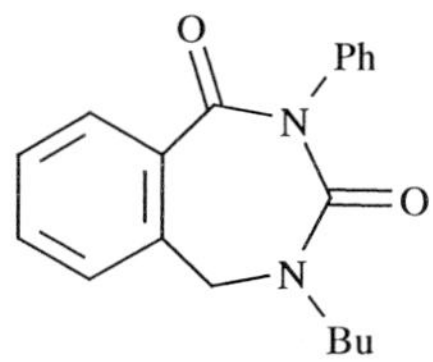

28

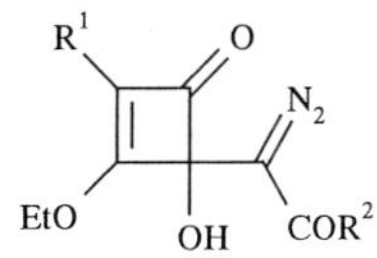

29

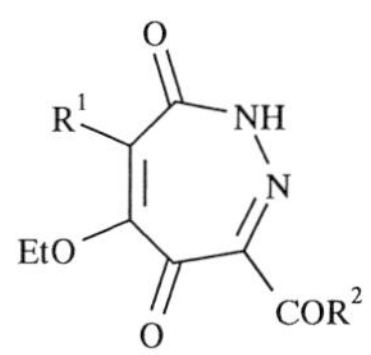

30

When irradiated in an Argon matrix at 10K 2-azido-4,6-dichloro-*s*-triazine yields a triplet nitrene and the corresponding dichlorodidehydrotetrazepine <99CC2113>. This is in complete contrast to the findings with the corresponding 4,6-dimethoxy analogue when no products that may be attributed to the formation of a didehydrotetrazepine have been described.

Photolysis of the amide **31** in methylene chloride at room temperature results in the formation of the tricyclic lactam **32** in both *syn* and *anti* forms <99TL6001>. Treatment of **32** with BF_3 etherate brings about cleavage of the cyclobutane ring in the *syn* isomer only, with the formation of **33**. (Scheme **5**). The corresponding esters undergo similar transformations, if somewhat less efficiently.

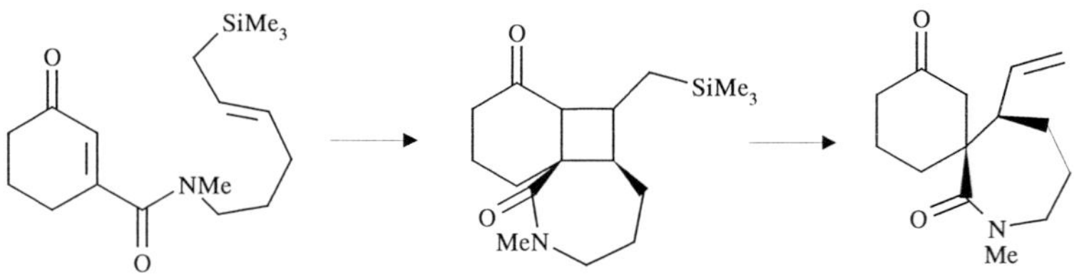

31 **32** **33**

Scheme **5**

7.3 SEVEN-MEMBERED SYSTEMS CONTAINING OXYGEN

There is no doubt that the main emphasis in the year under review has been devoted to the chemistry of seven-membered rings systems containing one oxygen atom. This is to no small extent a consequence of the on-going interest in the synthesis of the marine toxins, particularly the ciguatoxins, which, because of their accumulation in the food chain are increasingly giving rise to public health problems in warm weather climes, notably the Caribbean. So be aware of from where your red snapper comes!

As in previous years, two fundamental approaches to the formation of the ether rings of these compound have been ring-closing metathesis and cobalt carbonyl catalysed cyclisations of alkenes. Typical of the use of RCM are the approaches of Perlmutter and co-workers <99JOC8396> (Ring A) and Hirama and co-workers <99TL5405> (Ring A), <99CC1063> (Rings A and D). The cyclisation of alkyne biscobalthexacarbonyl complexes are illustrated in two reports by Isobe and co-workers <99JOC37>, <99TL191>.

However, these have not been the only approaches to the synthesis of these ring systems. For example, Sasaki et al. were able to use an intramolecular nucleophilic ring opening of an epoxide with sodium dimsylate to form the oxepane ring as illustrated in the conversion of **34** to **35** <99JOC9399>.

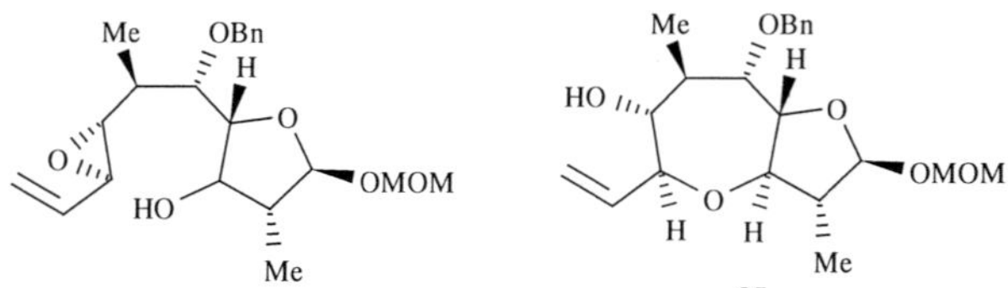

34 **35**

Ketal formation and removal of a methoxy group with $Et_3SiH/BF_3.OEt_2$ has been used by Hirama and co-workers in the formation of ring K (Scheme 6) <99CC2053>. In a further approach Hirama and co-workers have investigated the use of ring expansion reactions for all ring sizes which may be encountered in the synthesis of marine toxins. One example applicable to the seven-membered system involves ozonolysis of the bicycle **36** followed by reduction with triphenylphosphine to afford the diketone **37** <99T7471>.

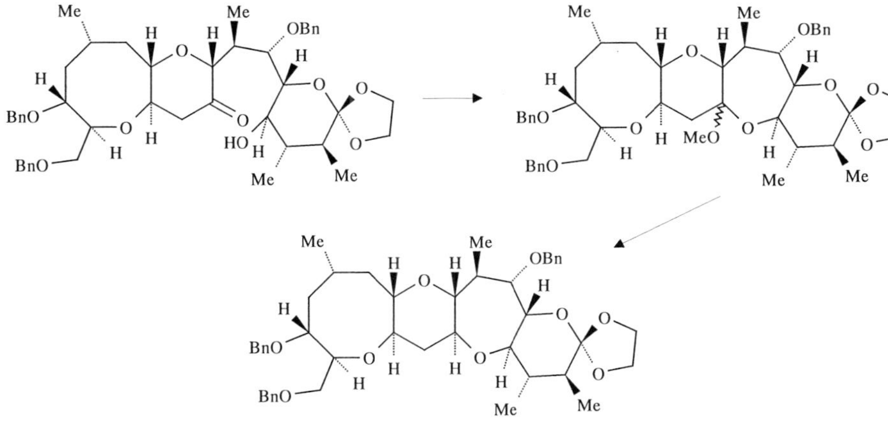

Scheme **6**

The single most extensive work, however, is that reported by Nicelaou et al for their synthesis of brevitoxin A <99CEJ599, 618, 628, 646>. Brevitoxin A differs from the ciguatoxins and, for that matter, from brevitoxin B in that only one ring of the 10 rings (ring D) is an oxepane. Lactonisation formed the critical step in the formation of this ring system.

The other natural product which is perhaps of even greater potential interest is the antimalarial artemisinin. Recent developments on the chemistry and biological activity of artemisinin and its derivatives have been reviewed by Bhattacharya and Sharma <99H(51)1681>. In 1998 Posner and co-workers reported on the synthesis of aryl substituted derivatives **38** <98JMC940> which have in vivo antimalarial activity against chloroquine resistant *P. falciparum* and in vivo activity in mice against *P. berghei*. Enantiomerically pure derivatives of the 4-fluorophenyl analogue have now been prepared to investigate the importance of chirality in the pharmacokinetics and pharmacodynamics of these compounds <99TL9133>. Two 10-(β-hydroxynaphthyl) derivatives, diastereoisomeric at the 9 and 10 positions of the artemisinin have also been prepared as possible probes of biological activity <99JCS(P1)1828>.

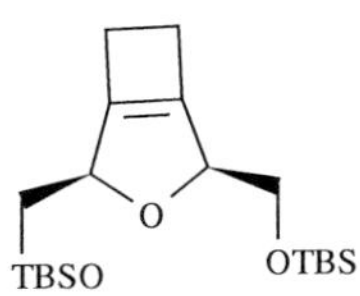

36

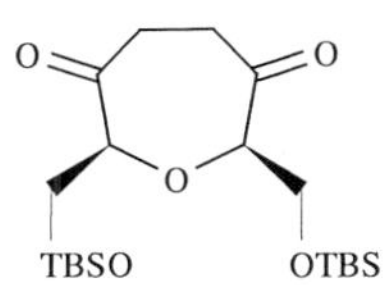

37

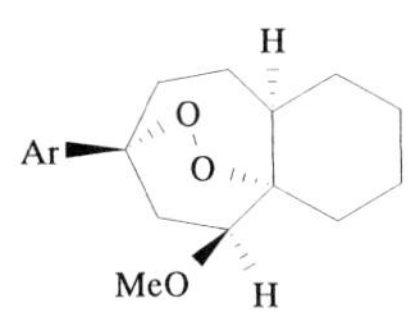

38

Ring-closing metathesis, which has proved to be a popular route to the marine toxins, has found a further application as the key step in the synthesis of the pheromone (-)- and (±)-frontalin <99TL1425>. The precursor in this reaction is a mixture of the *syn*- and *anti*-isomers **39**. Ring closure in the presence of a ruthenium benzylidene catalyst occurs within minutes at room temperature when only the *syn*-isomer cyclises to **40**. The unreacted *anti*-isomer can be re-equilibrated for a further cyclisation.

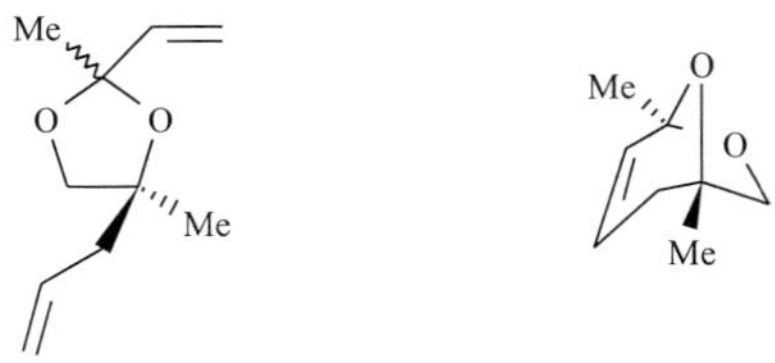

39 **40**

Ring-closing metathesis is also a feature in the formation of a number of oxygenated oxepane surrogates of carbohydrates, as exemplified by the cyclisation of **41** to **42** <99TL8751>. The yields are generally satisfactory, in this case 97%.

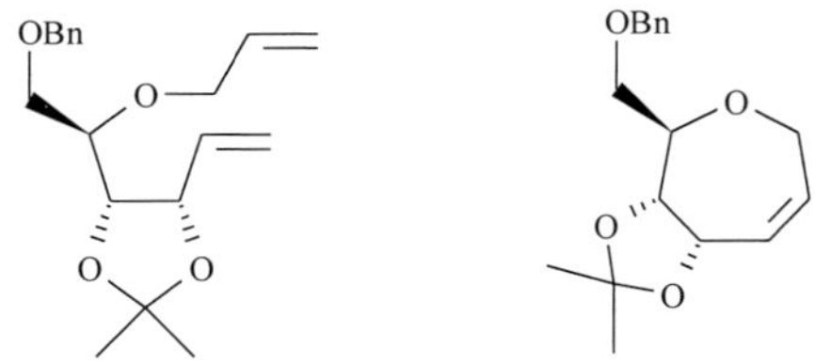

41 **42**

The use of ring-closing metathesis for the preparation of unsaturated heterocyles is now established as a routine method for single cyclisation steps, and the method has been extended to the preparation of bicyclic systems in a single step One example is the cyclisation of the tetraene **43** leading to formation of tetrahydrooxepine **44** in 59% yield <99JOC3354>. The method is equally applicable to the formation of five- and six-membered systems.

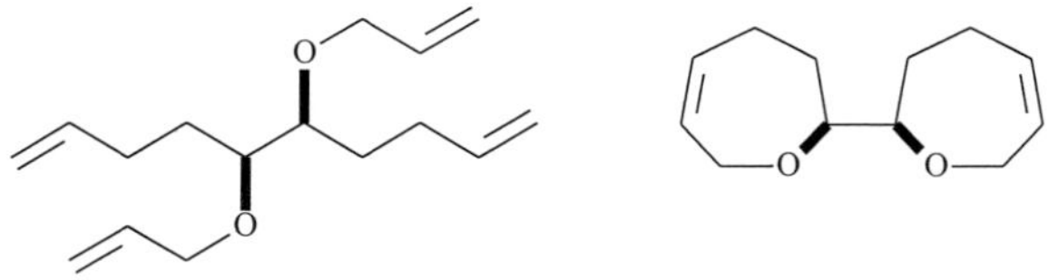

43 **44**

In addition to the examples of diene cyclisations described above there are reports of alkoxyallenes as precursors of five- to seven-membered oxygen heterocycles <99TL1747>. Of interest here is the cyclisation of **45** to **46** in 88% yield in the presence of $Pd(OAc)_2$-dppb complex. The same reagent system has also been used in the regioselective lactonisation of steroids where the aromatic ring of estrone is fused to a seven-membered lactone <99TL1171>.

45 **46**

Rearrangement reactions feature in a number of interesting articles. Thus rearrangement of the tetrahydropyran **47** leads to the oxepane **48** in 90% yield by heating in aqueous acetic acid in the presence of zinc acetate <99TL2145>. Cope rearrangements also feature. Flash vacuum pyrolysis of the sulphone **49** leads to formation of 4,5-dihydrooxepine **51** via the intermediate diene **50** (Scheme 7) <99JCS(P1)605>, and when the diene **52** is heated in dioxane/NEt_3 the fused dihydroderivative **53** is formed <99JOC3806>. The reaction does not proceed to completion as under the reaction conditions used starting material and product are in equilibrium. However, the reaction proceeds to completion when the phenyl group is replaced by an ester function.

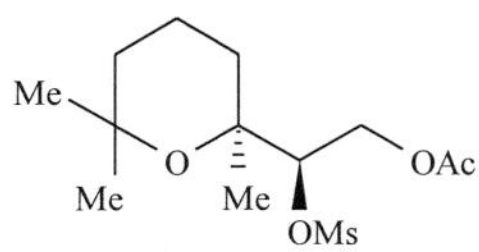

47

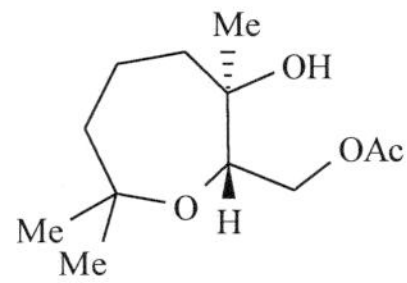

48

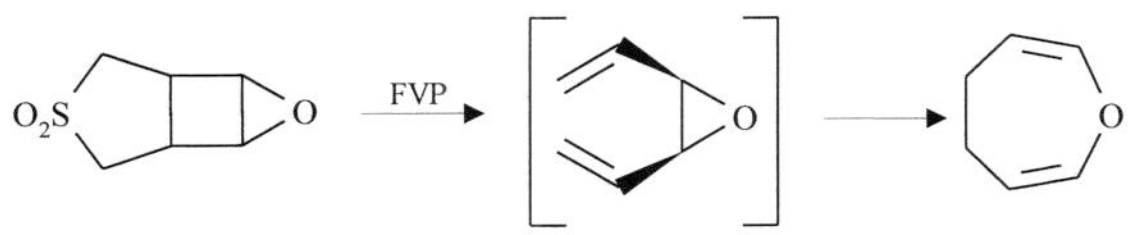

49 **50** **51**

Scheme 7

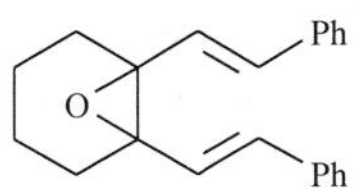

52

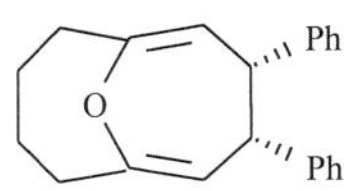

53

Earlier in this review the formation of oxepanes as carbohydrate surrogates was described. Similar derivatives have been reacted with Grignard reagents leading to the formation of alkyl substituted α,ω-hydroxyalkenes <99JOC854>. This does suggest that a combination of these two approaches could lead to a new and general route to polyhydroxy derivatives. The reaction of 2,2,2-triphenyl-1,2l^5-oxaphospholanes with paraformaldehyde in refluxing toluene leads to the formation of 5-methylene-1,3-dioxepanes <99H(50)125>.

7.4 MISCELLANEOUS SEVEN-MEMBERED RING SYSTEMS

The ring-closing metathesis reaction has been applied to the formation of cyclic sulphonamides <99TL4761>. Thus, a ten minute reflux of **54** in methylene chloride under ruthenium catalysis gave **55** in 91% yield. The same methodology has been used in the formation of the disilacycloalkene derivatives **56** in which R = CH_2, O or NPh <99BCJ821>. The reaction appears to be specific to seven and eight-membered rings and also to allyl silanes. Vinyl silanes do not react. This is a interesting observation in the case the 1,3-bis(dimethylvinylsilyl)propane as it has been reported previously that the corresponding butane does undergo cyclisation <98CC699>. Role reversal seems to be the order of the day in the formation of **57**, again by ring-closing metathesis <99TL1429>. In this case the methodology may also be extended to larger rings.

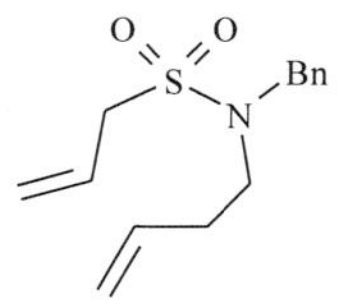

54

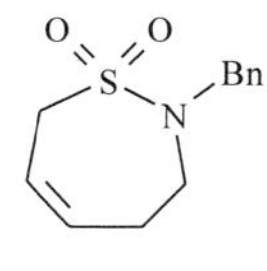

55

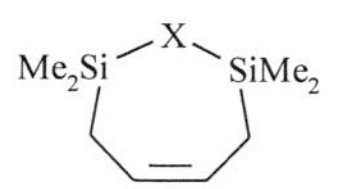

56

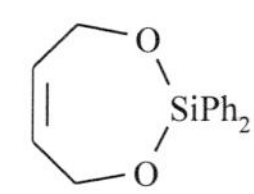

57

Compound **58**, formed by the reaction of triallylborane and bis(dimethylsilyl)ethyne, undergoes slow rearrangement to **59** <99AG(E)124>. A similar reaction path is followed by the product from 1-dimethylsilyl-2-trimethylsilylethyne. Here, two isomers are formed initially and only the isomer which is capable of forming a boron-hydrogen hydrogen bond undergoes the subsequent rearrangement. The ring expansion reaction of chiral diazaphospholidine oxides in which a new seven-members species is obtained has been studies (<99AG(E)1479>. The reaction, illustrated in Scheme **8**, is carried out a -78 °C to room temperature.

Thermal rearrangement of 3,3-dimethyl-3-silathiane S-oxide affords the ring expanded product **60** in what is claimed to be the first example of a thermal sila Pummerer rearrangement of a cyclic organosilicon sulphoxides <99TL185>. In continuation of studies on perisilyl-substituted macrocycles Sekiguchi and co-workers have successfully prepared the tetralithium salt of **61** as a novel 10-centre, 14π electron system stabilised by silicon groups <99CC1981>.

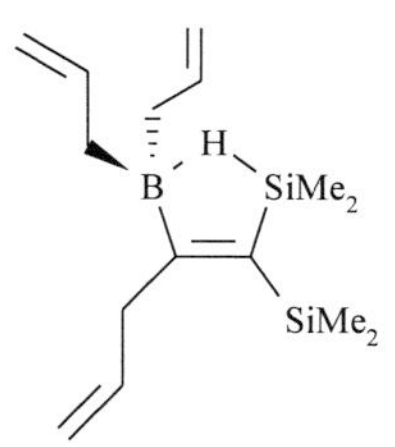

58

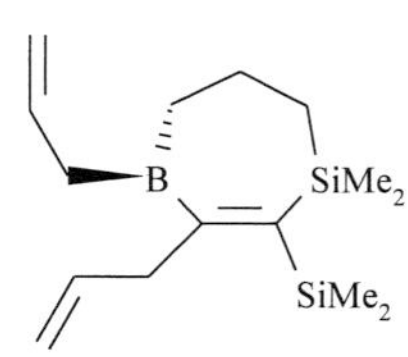

59

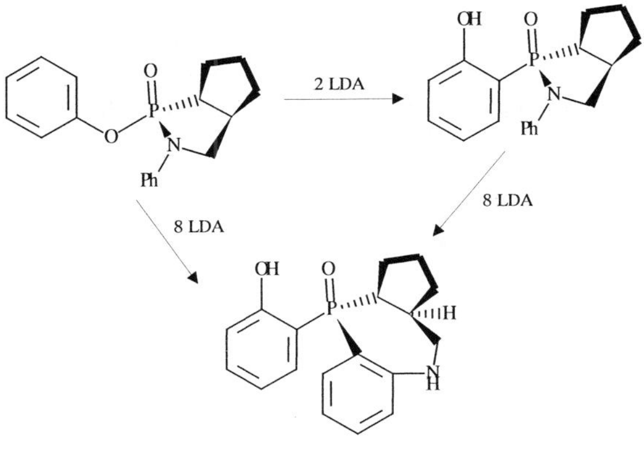

Scheme 8

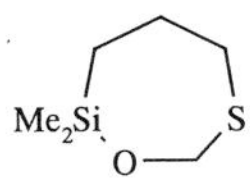

60

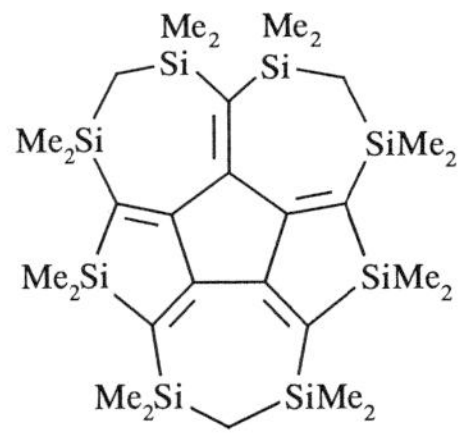

61

7.5 REFERENCES

98CC699 T. Mise, Y. Tagaguchi, T. Umemiya, S. Shimizu, Y. Watatsuki, *Chem. Commun.* **1998**, 699.

98JMC940 G.H. Posner, J.N. Cumming, S.-H. Woo, P. Ploypradith, T.A. Shapiro, *J. Med. Chem.* **1998**, *41*, 940.

99AG(E)123 B. Wrackmeyer, O.L. Tok, Y.N. Bubnov, *Angew. Chem. Int. Ed.* **1999**, *38*, 123.

99AG(E)1479 O. Legrand, J. M. Brunel, G. Buono, *Angew. Chem. Int. Ed.* **1999**, *38*, 1479.

99BCJ821 T. Hoshi, H. Yasuda, T. Sanji, H. Sakurai, *Bull. Chem. Soc. Jpn.* **1999**, *72*, 821.

99CC1063 K. Maeda, T. Oishi, H. Oguri, M. Hirama, *Chem. Commun.* **1999**, 1063.

99CC1913 B. Alcaide, P. Almendros, C. Aragoncillo, *Chem. Commun.* **1999**, 1913.

99CC1981 T. Matsuo, H. Fure, A. Sekiguchi, *Chem. Commun.* **1999**, 1981.

99CC2035 T. Oishi, Y. Nagumo, M. Shoji, J.-Y. Le Brazidec, H. Uehara, M. Hirama, *Chem. Commun.* **1999**, 2035.

99CC2113 G. Bucher, F. Siegler, J.J. Wolff, *Chem. Commun.* **1999**, 2113.

99CEJ599 K.C. Nicolaou, M.E. Bunnage, D.G. McGarry, S. Shi, P.K. Somers, P.A. Wallace, X.-J. Chu, K. A. Agrios, J.L. Gunzer, Z. Yang, *Chem. Eur. J.* **1999**, *5*, 5999.

99CEJ618	K.C. Nicolaou, P.A. Wallace, S. Shi, M.A. Ouellette, M.E. Bunnage, J.L. Gunzer, K.A. Agrios, G.-Q. Shi, P. Gärtner, Z. Yang, *Chem. Eur. J.* **1999**, *5*, 618
99CEJ628	K.C. Nicolaou, G.-Q. Shi, J.L. Gunzer, P. Gärtner, P.A. Wallace, M.A. Ouellette, S. Shi, M.E. Bunnage, K.A. Agrios, C.A. Veale, C.-K. Hwang, J. Hutchinson, C.V.C. Prasad, W.W. Ogilvie, Z. Yang, *Chem. Eur. J.* **1999**, *5*, 628.
99CEJ646	K.C. Nicolaou, J.L. gunzer, G.-Q. Shi, K.A. Agrios, P. Gärtner, Z. Yang, *Chem. Eur. J.* **1999**, *5*, 646.
99H(50)125	K. Okuma, Y. Tanaka, I. Shuzui, K. Shioji, *Heterocycles* **1999**, *50*, 124.
99H(51)141	T. Tomita, H. Ishiguro, K. Saito, *Heterocycles* **1999**, *51*, 141.
99H(51)1681	A.K. Bhattacharya, R.P. Sharma, *Heterocycles* **1999**, *51*, 1681.
99JCS(P1)1695	C.A. Tarling, A.B. Holmes, R. E. Markwell, N. D. Pearson, *J. Chem. Soc., Perkin Trans. 1* **1999**, 1695.
99JCS(P1)1827	D.-Y. Wang, Y. Wu, Y.-L. Wu, Y. Li, F. Shan, *J. Chem. Soc., Perkin Trans. 1* **1999**, 1827.
99JCS(P1)443	J.-P. Strachan, J.T. Sharp, M.J. Crawshaw, *J. Chem. Soc., Perkin Trans 1.* **1999**, 443.
99JCS(P1)605	R.A. Aitken, J.I.G. Cadogen, I. Gosney, C.M. Humphries, L.M. McLaughlin, S.J. Wyse, *J. Chem. Soc., Perkin Trans.1* **1999**, 605.
99JOC37	S. Hosokawa, M. Isobe, *J. Org. Chem.* **1999**, *64*, 37.
99JOC707	M. Ohno, M. Noda, Y. Yamamoto, S. Eguchi, *J. Org. Chem.* **1999**, *64*, 707.
99JOC854	J.A. Adams, N. M. Heron, A.-M. Koss, A.H. Hoveyda, *J. Org. Chem.* **1999**, *64*, 854.
99JOC3354	C. Baylon, M.-P. Heck, C. Mioskowski, *J. Org. Chem.* **1999**, *64*, 3354.
99JOC3806	P. Von Zezschwitz, K. Voigt, A. Lansky, M. Noltemeyer, A. de Meijere, *J. Org. Chem.* **1999**, *64*, 3806.
99JOC4830	G. Rodríguez, L. Castedo, D. Domínguez, C. Saá, *J. Org. Chem.* **1999**, *64*, 4830.
99JOC8396	L. Eriksson, S. Guy, P. Perlmutter, *J. Org. Chem.* **1999**, *64*, 8396.
99JOC9399	M. Sasaki, M. Inoue, K. Takamatsu, K. Tachibana, *J. Org. Chem.* **1999**, *64*, 9399.
99S138	J. Pernerstorfer, M. Schuster, S. Blechert, *Synthesis*, **1999**, 138.
99S839	D. Marek, A. Wadouachi, D. Beaupère, *Synthesis* **1999**, 839.
99T1309	G. Maas, B. Manz, T. Mayer, U. Werz, *Tetrahedron* **1999**, *55*. 1309.
99T7471	T. Oishi, M. Maruyama, M. Shoji, K. Maeda, N. Kumahara, S.-i. Tanaka, M. Hirama, *Tetrahedron*, **1999**, *55*, 7471.
99T10989	H. Shinohara, T. Fukusa, M. Iwao, *Tetrahedron* **1999**, *55*, 10989.
99TL185	S.V. Kirpichenko, E.N. Suslova, A.I. Albanov, B.A. Shainyan, *Tetrahedron Lett.* **1999**, *40*, 185.
99TL477	A. Geyer, D. Bockelmann, K. Weissenbach, H. Fischer, *Tetrahedron Lett.* **1999**, *40*, 477.
99TL1425	M. Scholl, R.H. Grubbs, *Tetrahedron Lett.* **1999**, *40*, 1425.
99TL1429	T.R. Hoye, M. A. Promo, *Tetrahedron Lett.* **1999**, *40*, 1429.
99TL1747	S. Kamijo, Y. Yamamoto, *Tetrahedron Lett.* **1999**, *40*, 1747.
99TL1771	L. Troisi, G. Vasapollo, B. El Ali, G. Mele, S. Florio, V. Capriati, *Tetrahedron Lett.* **1999**, *40*, 1771.
99TL1911	R. Saeeng, M. Isobe, *Tetrahedron Lett.* **1999**, *40*, 1911.
99TL2145	N. Hori, K. Nagasawa, T. Shimizu, T. Nakata, *Tetrahedron Lett.* **1999**, *40*, 2145.
99TL2623	G. Bocelli, M. Catellani, F. Cugini, R. Ferraccioli, *Tetrahedron Lett.* **1999**, *40*, 2623.

99TL4761	P.R. Hanson, D.A. Probst, R.E. Robinson, M. Yau, *Tetrahedron Lett.* **1999**, *40*, 4761.
99TL5405	H. Oguri, S.-y. Sasaki, T. Oishi, M. Hirama, *Tetrahedron Lett.* **1999**, *40*, 5405.
99TL5569	Y.S. Lee, B.J. Min, Y.K. Park, J.Y. Lee, S.J. Lee, H. Park, *Tetrahedron Lett.*, **1999**, *40*, 5569.
99TL6001	S. Faure, S. Piva-Le Blanc, O. Piva, *Tetrahedron Lett.* **1999**, *40*, 6001.
99TL7939	W. Chu, K. D. Moeller, *Tetrahedron Lett.* **1999**, *40*, 7939.
99TL8751	J.C.Y. Wong, P. Lacombe, C.F. Sturino, *Tetrahedron Lett.* **1999**, *40*, 8751.
99TL9133	P.M. O'Neill, A. Miller, J.F. Bickley, F. Scheinmann, C.H. Oh, *Tetrahedron Lett.* **1999**, *40*, 9133.

Chapter 8

Eight-Membered and Larger Rings

George R. Newkome
University of South Florida, Tampa, FL, USA
e-mail: gnewkome@research.usf.edu

8.1 INTRODUCTION

In the first half of the nineties, there has been a continuing trend from synthetic studies of classical "crown ethers" towards polyazamacromolecules and the introduction of multiple heteroatoms, including most recently metal atom centers.

Numerous reviews and perspectives have appeared throughout the year that are of interest to the macroheterocyclic scientist and those delving into supramolecular chemistry at the molecular level, as well as those in supermolecules and crystal engineering: homocalixarenes <99ACR729>, functionalized calix[4]pyrroles<99PAC2401>, rotaxanes as new architectures <99CSR293>, nanoporous and mesoporous materials<99CSR279>, transition metals as switches<99ACR846, 99CCR649>, dinuclear complexes of *bis*-macrocycles <99CCR371>, carbohydrate recognition<99ACIE2979>, biomimetic and supramolecular chemistry <99ACIE2979>, carceplexes and hemicarceplexes<99CR931>, polymerization of pseudo-rotaxanes<99CCR139>, aliphatic polyamine ligands<99CCR243>, cryptand ligands <99CCR347>, luminescent signaling systems <99CCR297>, heterosupramolecular chemistry <99CCR277>, catenanes and molecular knots<99CCR167>, polyammonium macrocyclic receptors<99CCR149>, non-conjugated bichromophoric receptors <99CCR357>, MRI contrast agents<99CCR451>, electrochemistry of coordination compounds<99CCR233>, N_4-donor macrocycles<99CCR37>, redox-active receptor molecules<99CCR3>, interlocking macromolecules<99CR1643>, electrochemical molecular recognition<99JCS(D)1897>, self-assembly of [2]catenanes<99ACR53>, photochemical CO_2 reduction with metal complexes <99CCR373>, NMR of crown ethers<99PNMRS327>, C,N,S macrocyclic complexes <98CCR327>, use in radiometal agents for cancer therapy <99ACR1053>, supramolecular aspects derived from glycoluril, e.g. cucurbituril <99ACR995>, oligomeric porphyrin arrays <99CC1771>, nanometer-sized oligonuclear coordination compounds<99ACIE3463>, metal ion extraction by lariat ethers<99PAC2393>, crown ether polysiloxanes<99JH15>, crown ethers as chiral selectors<99E2605>, oxa-cage compounds<99MI1>, solvent extractions of metals with macrocyclic reagents<97CA63>, paramagnetic complexes for crown ether molecular structural studies<98JSC580>, metal halide-macrocyclic polyethers<99RCR119>, supramolecular systems based on crown ethers and secondary dialkylammonium ions <99ASC1>, template control of supramolecular architectures<99ASC237>, Li^+/14-crown-4 systems<99MI2>, metal ion separations with proton-ionizable lariat ethers<99MI3>, and electroactive polymers containing crown ethers<98CCR1211>.

Because of space limitations, only meso- and macrocycles possessing heteroatoms and/or subheterocyclic rings are reviewed; in general, lactones, lactams, and cyclic imides have been excluded. In view of the delayed availability of some articles appearing in previous years, several have been incorporated.

8.2 CARBON–OXYGEN RINGS

Numerous macrocyclic crown ethers possessing diverse subunits, for example: isobutenyl aromatic rings<99TL6639>, fullerene<99HCA1572>, binaphthyl<99CEJ984, 99EJOC995> biphenyl<99TL6019>, phenolphthalein<99JA3807>, distyrylbenzene-connector<99JA5599>, amidobenzo<99CC1253>, fluorenylidene-diphenyl<99TL5865>, 7-oxanorbornane constructs <99OL199, 99OL203>, a picrylamino-type sidearm attached to the center of a three-carbon bridge<99JOC5341>, *trans*-stilbene<99OL1697>, and polymer backbones (e.g., **1**) <99TL4123, 99M1500, 99JA1466> have been reported.

Chirality also plays an important role in the use of macrocyclic ethers; thus, numerous chiral subunits have been utilized in their construction: e.g. *m*-xylylene moiety as a rigid spacer in intramolecular glycoside bond formation<99JOC6190>. The introduction of a degree of strain into aromatic systems has been shown by the creation of the bent 1,8-dioxa[8](2,7)pyrenophane (**2**) prepared from a suitable metacyclophane-1,9-diene by valence isomerization, followed by dehydrogenation<99CEJ1823>. *H*-bonding hosts (**3**) with a flexible frame possessing four-directed hydroxy moieties have been synthesized in a multistep sequence<99JCS(P1)1885>. The tin chloride catalyzed oxidation of acetone with 30% hydrogen peroxide gave variable yields of the crystalline tetrameric acetone peroxide (**4**) <99JCR(S)288>.

Systematic analysis of stabilities of cyclic dimeric pseudorotaxanes, generated from complementary homoditopic molecules, has been shown<99ACIE143, 99CC789> A new class of water-soluble cyclophanes, pyrenophane, capable of encompassing a neutral cavity has been constructed<99JA1452>.

Calix[4]quinone crown-4-ethers have been formed <99S69> by the reduction of the corresponding lactone. The first example of an 1,2,4-tripodal calix[6]cryptand has been reported by the direct condensation of *p-tert*-butylcalix[6]arene with 1,1,1-*tris*(tosyloxy-ethoxyethoxymethyl)propane<99CL881>. Bridged calix[*n*]arenes with tribenzo crown ethers<99TL691>, dibenzocrown ethers<99JCS(P2)837>, benzo crowns<99CC609>, or simple crown ethers<99TL499>, have been reported and many examples cite enhanced complexation properties. Substituted hexahomotrioxacalix[3]arenes (**5**) have been shown to

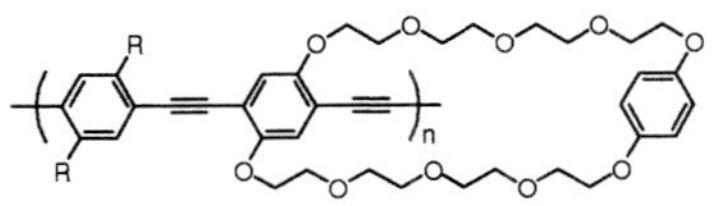

1 R = $CON(C_8H_{17})_2$

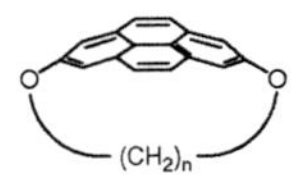

2

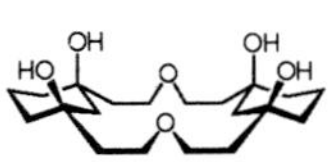

3

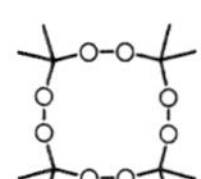

4

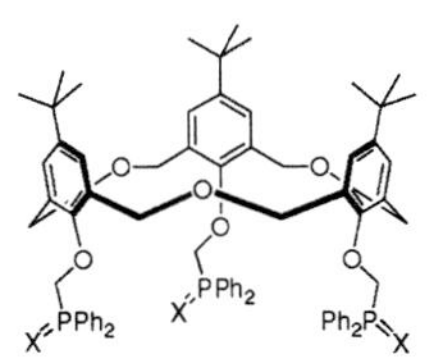

5

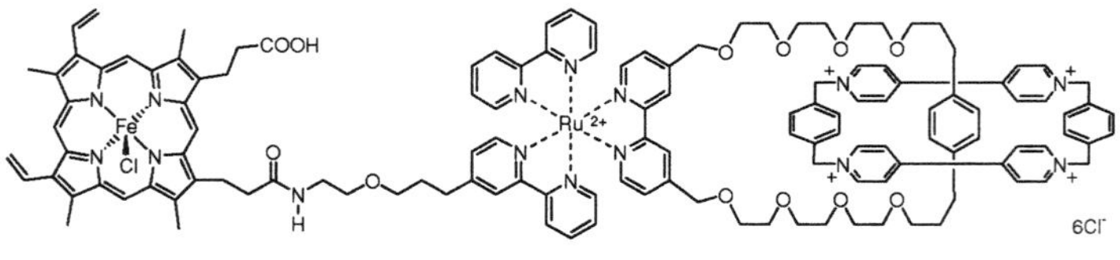

13

sensitizer and covalently linked to the protoheme donor, has been prepared in a stepwise procedure<99CL517>. Preorganized U-shaped calix[4]arene clefts consisting of a cone calix[4]arene and two bipyridines linked with an aliphatic chain have been synthesized and shown to self-assemble by utilizing π-stacking interactions between an hydroquinone and bipyridinium moieties<99JOC3572>. A Sn(IV) porphyrin has been shown to promote dramatic resolution in the NMR spectrum via a combination of dynamic and dispersive effects of a free acid - [2]catenane<99OL1343>. A series of [2]catenanes possessing a 4,4'-azobiphenoxy moiety have been reported<99CEJ860>. An unusual asymmetric [2]catenane has been prepared by means of interlocking an alkyne crown ether with a diimide macrocycle <JCS(P1)1057>. A recognition motif has been utilized to template the formation of a "rotacatenane", which possesses a dumbbell-shaped component threaded through a cavity of one of the two mechanically interlocked macrocyclic components of a [2]catenane <99EJOC1295>. Macrocyclic receptors with two facing π-electron-rich aromatic surfaces binding bipyridinium guests have led to the development of efficient template-directed syntheses of mechanically interlocked molecules; a theoretical evaluation of receptor affinities has appeared<99JA1479>. An efficient synthesis of a [2]catenane with an 87-membered ring with interestingly placed functionality should lead to the formation of poly[2]catenanes <99CEJ1728>.

8.5 CARBON–SULFUR RINGS

The synthesis of the first phosphorylated derivatives of *p-tert*-butylthiocalix[4]arene was accomplished by treatment with PCl_3; a flattened 1,2-alternate conformation of a P(IV) derivative was proven<99TL8461>. Synthesis of *p-tert*-butyltetramercaptotetrathia-calix[4]arene bearing eight sulfur atoms was accomplished in 80% overall yield, based on the preparation of the parent tetrathiacalix[4]arene and subsequent treatment of *N,N*-dimethyl-thiacarbamoyl chloride<99CC2169>. Oxidation of the titanocene dithiolene complex with SO_2Cl_2 afforded 3,4-*bis*(methoxycarbonyl)-1,2-dithiete, the corresponding 1,2,5,6-tetrathiocin, and the related 16-membered cyclic compound **14**<99JOC8489>. Synthetic sequences to two series of dithiametacyclophanes, starting from 2-iodo-1,3-dimethylbenzene and copper phenylacetylide, have been reported<99EJOC791>.

Anthraquinone and the dianion of *bis*(1,3-dithiolyl)diphosphonate reagents gave rise to cyclophane derivatives (**15**)<99CC1835> and other bridged thiacrown ether TTF derivatives<99CC1417>. Treatment of 1,3,5-*tris*(2-mercaptoethyl)benzene with the corresponding bromo analogue gave (66%) the desired mixed π,*n*-donor macrobicycle, described as a *S*-cylindrophane<99ACIE1968>.

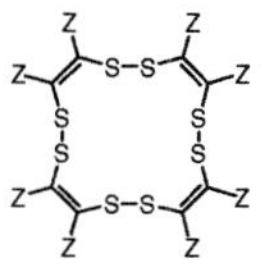

14

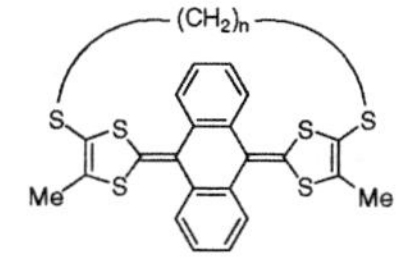

15

8.6 CARBON-SILICON RINGS

Palladium-catalyzed insertion of alkynes into the Si-Si bond gave an expanded series of 4,7,12-trisilabicyclo[8.3.0]trideca-1(13),5,10-triene-2,8-diynes<99O3792>. Cyclic diynes with tetramethyldisilyl groups have been prepared from α,ω-*bis*(chloromethyl)diynes and 1,2-dichloro-1,1,2,2-tetramethyldisilane with lithium in the presence of catalytic quantities of biphenyl<99JCS(P2)2093>. The silapericyclyne, $(Ph_2SiC{\equiv}C)_6$, underwent spontaneous separation into the chair and boat conformations upon crystallization<99CL1235>. A one-pot reaction of 1,3-dibromobenzene and dimethyldichlorosilane with magnesium in THF gave two cyclophanes: the trisila[1.1.1]metacyclophane, possessing the saddle structure, and tetrasila[1.1.1.1]metacyclophane, adopting the 1,2-alternate orientation<99O1465>. The syntheses of a series of related macrocyclic diynes containing one or more dimethylsilyl groups have been reported<99O3615>.

8.7 CARBON–NITROGEN-OXYGEN RINGS

As in the past, the aza-crown ether family is quite diffuse in content due to their ability to create *N*-pendant moieties, which adds to the supramolecular aspects of these materials. In general, a 4-(2-hydroxyethyl)-1,4,7,16,19,22-hexaaza-10,13,25,28-tetraoxacyclotriacontane has been synthesized and its dizinc complex demonstrated ester hydrolysis properties<99IC4115>. Novel *tris*macrocycles e.g., $R[N18N]C_{12}[N18N]C_{12}[N18N]R$, possessing fluorescent residues, R, have been prepared<99JA9043>. The concave pyridine macrocycle **16** has been synthesized (61%) in two simple macrocyclization steps. This 4-functionalized pyridine subunit was easily attached to polymer backbones <99JPC662, 99JPC218> other 4-substituted pyridyl crowns have been reported <99JCS(P2)1739>. Novel protected oxa-azamacrocycles have been prepared by direct alkylation between α,ω-*bis*[(2-mesitylsulfonyl)aminoxy]alkanes and α,ω-*bis*(tosyl)alkanediols in the presence of K_2CO_3 giving a mixture of 1:1- and 2:2-macromolecules<99S1034>. Using 4,7-dichlorophenanthroline with different linker moieties has given rise to a series of novel mono- and *bis*-macrocycles possessing exotopic binding sites<99JPC228, 99SL750>. Macrobicyclic polyethers (**17**) based on the cryptand model and displaying chirality due to a spirane junction have been synthesized<99TL4993> and evaluated for their chiral recognition

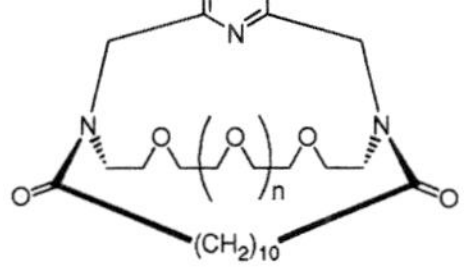

16

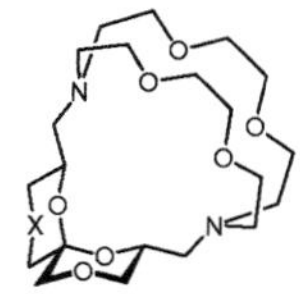

(X = N-CH$_2$-Ph)

17

properties<99TL4997>. Metal-free and nickel phthalocyanines substituted peripherally with four 20-membered diazatetraoxamacrocycles each attached to 15-membered crown ether moieties have been reported<99ACS247>.

The Rh(I)-catalyzed hydroformylation/reductive amination in the presence of α,ω-diamines afforded a simple route to C,N,O-macrocycles<99TL7773>. The Pd(0)-catalyzed cross-coupling of electron deficient aryl- and heteroaryl bromides with aza-crown ethers has been shown<99SL1223>. The *N*-attachment of other functionality has been demonstrated, as shown by several examples: 8-hydroxyquinoline<99JOC8855, 99JOC3162>, dicarboxylic acids<99JCR(S)58>, ferrocenyl units<99O1911>, aminophenolic moieties<99JOC3825>, cavitands<99TL1527>, *bis*-benzyl halide<99IC2064>, 4-acyl-3-methyl-1-phenylpyrazol-5-one<JCS(P1)693>, *bis*[2-(2-pyridyl)ethyl]amine<99IC5755>, and a diverse group of substituents<99JOC8855>. Direct modification of trioxatriaza-18-crown-6 with *N*-phosphonic acid moieties has afforded insight into a viable mechanism for the setting of cement<99JCS(P2)1973>!

The first syntheses of a tetra*kis*(rotaxane) and a *bis*(pretzelane), a new intertwined architecture, have appeared<99S1753>. Different multicomponent molecular systems have been synthesized by means of the three-dimensional template effect, see the elegant works of Professor Sauvage and his coworkers in the area of rotaxanes<99JA4397, 99IC4203>, porphyrin homodimers (**18**)<99CC2419>, and porphyrin-stoppered rotaxanes <99JA3684>.

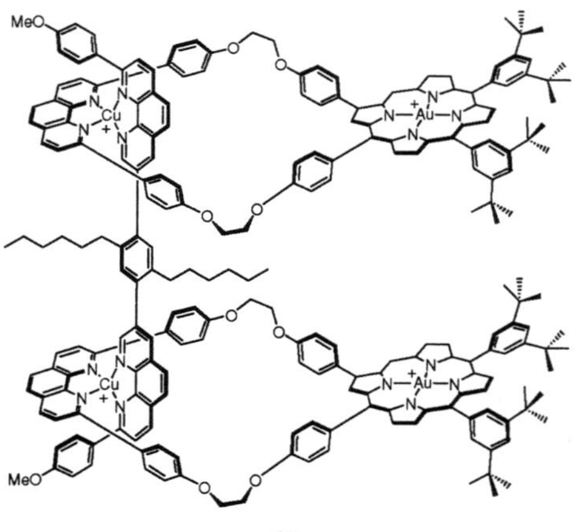

18

8.8 (CARBON-NITROGEN-OXYGEN)^n (n>1) RINGS (CATENANES)

The formation of catenane structures by means of a useful (88-92% yields) ring-closing metathesis procedure with the Grubbs ruthenium catalyst has been detailed<99JOC5463> and applied to the formation of molecular knots containing two tetrahedral or octahedral coordination sites<99JA994>. An efficient synthesis of doubly interlocked [2]catenanes has appeared<99CC615> in which lithium ion is used as the assembly center to generate a double-stranded helical complex; whereas, ruthenium-catalyzed ring-closure metathesis on the helical precursor afforded a 4-crossing [2]catenane in 30% yield. A series of difunctionalized catenate and catenand, possessing macrocycles with 45 atoms, has been copolymerized with a terephthalic acid derivative affording a cyclic oligo[2]catenand and linear poly[2]catenane <99CEJ1841>. New multicomponent complexes containing [2]catenanes and $[Ru(tpy)_2]^{2+}$ moieties within the construct have been reported<99JA5481>.

8.9 CARBON–SULFUR-OXYGEN RINGS

Diethylene glycol monochloride was converted into substituted dithia-9-crown-3-ethers upon treatment with *p*-toluenesulfonyl chloride, followed by 1,2-dimercaptoethane <99SC3939>. Four new C_{4v} tetraoxatetrathiahemicarceplexes have been reported <99TL8905>. A series of homooxacalix[*n*]thiophenes (*n* = 3-7) was prepared from 2,5-thiophenedimethanol in a one-pot procedure by acid-catalyzed dehydration<99TL3749>. A macrocycle with two chiral (*R*)-1,1'-binaphthyl systems bridged by a 2,5-(*p*-tolyl)thiophene spacer has been synthesized and compared with related open tweezer-type molecules <99TL9065>. *Bis*(2-oxy-1,3-propylenedithio)tetrathiafulvalene-containing acyclic polyethers have been prepared and incorporated into pseudorotaxanes and catenanes; charge-transfer interactions between the TTF moiety and the electron-acceptor units of the tetracationic cyclophane have been evaluated<99EJOC985>.

8.10 CARBON–NITROGEN–SULFUR RINGS

The preparation and characterization of *meso*-sapphyrins and rubyrins containing S, O, and/or Se heteroatoms in addition to pyrrole nitrogens have been reported<99JA9053, 99JOC8693>. The 2-thia-5,10,15,20-tetraphenyl-21-carbaporphyrin, possessing an inverted thiophene subunit, was produced by two different procedures<99TL8457>. Cyclic polythiazaalkane derivatives bearing a benzothiazolyl moiety, as a fluorophore, have been synthesized<99JCS(P2)1273>. The air-stable 1,2-*bis*(4-*tert*-butyl-2,6-diformylphenylsulfanyl)ethane has been used in the construction of larger 36- and 40-membered amine-thiophenolate ligands<99EJIC2167>.

8.11 CARBON-OXYGEN-SELENIUM RING

The reaction of diselena-benzocrown ether with Li_2PdCl_4 afforded a novel cationic palladium tetraselena complex<99JCS(D)1039>.

8.12 CARBON-SULFUR-SILICON (PHOSPHORUS) RING

The condensation of *tris*[2-(bromomethyl)phenyl]fluorosilane and 1,3,5-*tris*(mercaptomethyl)benzene gave (0.4%) tribenzo 6-fluoro-6-sila-2,10,19-trithia[$5^{6,14}$][11]metacyclophane (**19**) possessing the *in*-configuration for the fluoro substituent; attempts to prepare the related *in*-phosphane oxide were unsuccessful<99JOC5626>. A novel cyclic tetramer **20** possessing D_4 symmetry was synthesized from a new optically active spirosilane by a Pd-catalyzed cross-coupling procedure<99CL399>. Dimethylsilacalix[4](3,5-diphenylphosphinine-2,6-diyl) was prepared from an equimolar mixture of 2,6-*bis*(phenylethynyldimethylsilyl)phosphinine and *bis*(dimethylsilyl-1,2-azaphosphinine)phosphinine under high-dilution and controlled concentrations<99CEJ2109>.

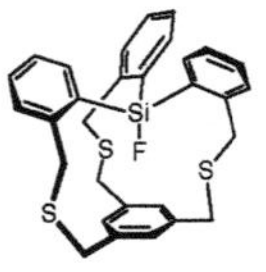

19

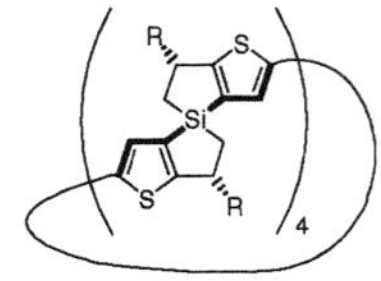

20

8.13 CARBON–NITROGEN–SULFUR-OXYGEN RINGS

The first porphyrin containing furan and thiophene subunits adjacent to each other was accomplished by the 2+2 condensation of dipyrromethane and (*meso*-aryl)furylthienylmethane <99TL8879>.

8.14 CARBON-METAL RINGS

A novel macrocyclic multinuclear acetylide complex was prepared from *o*-diethynylbenzene with equimolar [$PdCl_2(PEt_3)_2$] in the presence of CuCl catalyst in Et_2NH at 25 °C gave **21** in 33% yield<99AGIE174>.

8.15 CARBON-PHOSPHORUS-METAL RINGS

Cationic and neutral rhenium and platinum complexes containing *P*-ligands with the $-(CH_2)CH{=}CH_2$ substituent underwent efficient intramolecular metatheses with $(Cl)_2(Pcy_3)_2Ru({=}CHPh)$ to generate metallomacrocycles<99O955>. The triflate of 1,3,5-*tris*(3-hydroxypropyl)benzene with $LiPPh_2$ gave the corresponding *tris*(phosphane), which

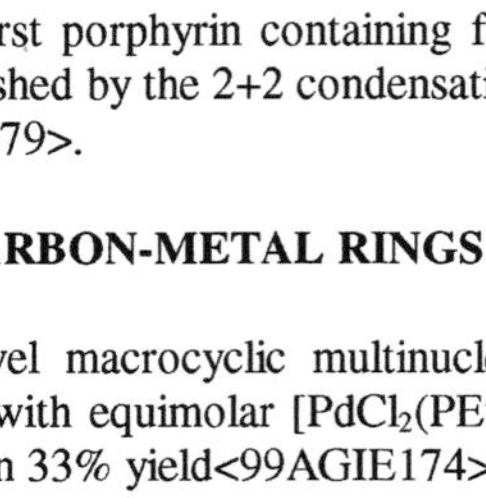

21

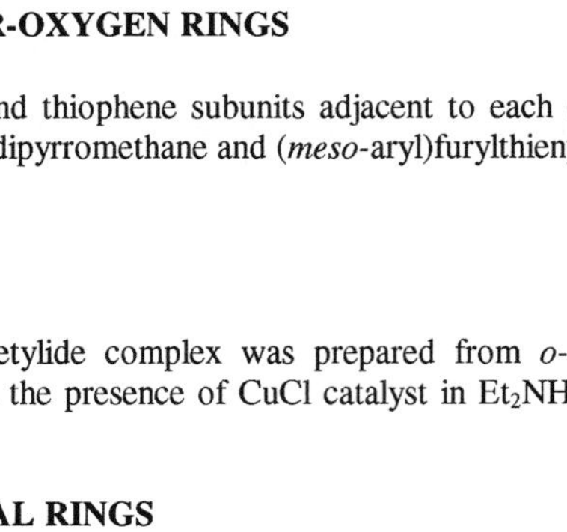

22

under high dilution conditions with $Cl_2Pt(NCPh)_2$ gave the nanoscopic tri- and hexaplatinacyclophanes (e.g., **22**)<99EJIC679>. The synthesis of diphosphino α-cyclodextrin was reported and subsequently treated with transition metals to cap the "torus with precious metals," thus affording macrocyclization<99CC1073>.

8.16 CARBON–NITROGEN-METAL RINGS

The synthesis of $[\{[(Me_3Si)_2N]_4K_2Mg_2(O_2)\}_\infty]$, a peroxo-centered macrocycle linked into infinite chains by intermolecular interactions, represents the first potassium member of a novel class of amide-supported heterobimetallic macrocycle<99CC353>. Molecular squares derived from 4,4'-bipyridine<99IC4149> or extended bipyridines<99IC4181, 99CC449, 99IC921, 99JA557> and metal corners possessing an appropriate 90° binding angle continue to fascinate researchers. Multipyridine ligands with metal connectors offer a route to "dynamic receptor library" and thus further insight into molecular recognition<99JA10239>. The formation of stable dimers of *cis*-azobenzene and -stilbene derivatives within the hydrophobic inner domain of the macromolecular cage **23** has appeared; this process has been termed the "ship-in-a-bottle"<99JA1397>. Other directed *N*-electron pairs, e.g. pyrazolyl <99O3991>, 1,3(4)-$C_6H_4(NHR)_2$<99CC1909>, chiral spacer<99TL531>, angular dipyridinylporphorins<99JA2741>, or dipyridinyl ketone ketal<99O4817>, affording differing bonding angles have been reported in the construction of metallomacrocycles.

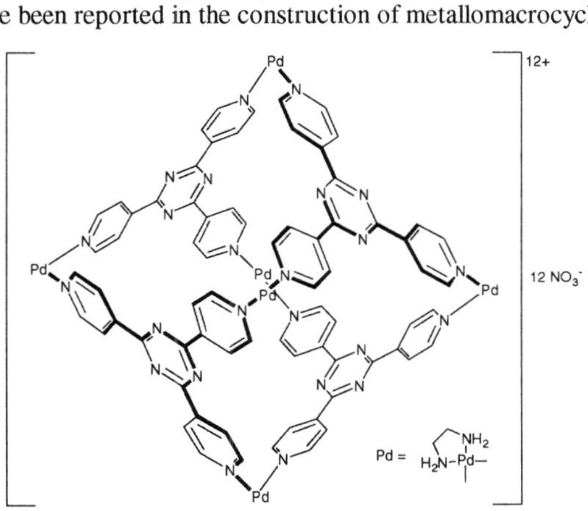

23

The novel chemistry displayed by Professor Sauvage and his coworkers over the years continues to fascinate chemists interested in unique topologies, and coupled with the self-assembly approaches developed by Professor Fujita and his colleagues, make the mixing of template techniques and self-assembly a novel adventure into new catenane strategies. Thus, their reported target **24**, being spontaneously prepared in quantitative yield, is probably the start of many, new and delightful molecular architectures<99JA11014>.

Formation of molecular necklaces [*n*]MN (*n*=4-7) has taken on a new 2+2 approach in which pseudorotaxanes containing two molecular beads, e.g. cucurbituril, are treated with the connective metal centers to afford the desired **25**<99ACIE638>.

<99CC1073> D. Armspach, D. Matt, *Chem. Commun.* **1999**, 1073-1074.
<99CC1253> P. D. Beer, P. K. Hopkins, J. D. McKinney, *Chem. Commun.* **1999**, 1253-1254.
<99CC1417> F. Le Derf, M. Mazari, N. Mercier, E. Levillain, P. Richomme, J. Becher, J. Garín, J. Orduna, A. Gorgues, M. Sallé, *Chem. Commun.* **1999**, 1417-1418.
<99CC1625> Y. Tobe, H. Nakanishi, M. Sonoda, T. Wakabayashi, Y. Achiba, *Chem. Commun.* **1999**, 1625-1626.
<99CC1771> M. G. H. Vicente, L. Jaquinod, K. M. Smith, *Chem. Commun.* **1999**, 1771 1782.
<99CC1835> T. Finn, M. R. Bryce, A. S. Batsanov, J. A. K. Howard, *Chem. Commun.* **1999**, 1835-1836.
<99CC1909> S. Daniele, P. B. Hitchcock, M. F. Lappert, *Chem. Commun.* **1999**, 1909-1910.
<99CC1911> C. B. Dieleman, D. Matt, I. Neda, R. Schmutzler, A. Harriman, R. Yafian, *Chem. Commun.* **1999**, 1911-1912.
<99CC2169> P. Rao, M. W. Hosseini, A. De Cian, J. Fischer, *Chem. Commun.* **1999**, 2169-2170.
<99CC2419> M. Linke, J.-C. Chambron, V. Heitz, J.-P. Sauvage, *Chem. Commun.* **1999**, 2419-2420.
<99CCR3> P. D. Beer, P. A. Gale, G. Z. Chen, *Coord. Chem. Rev.* **1999**, *185-186,* 3-36.
<99CCR37> H. Elias, *Coord. Chem. Rev.* **1999**, *187,* 37-73.
<99CCR139> M. C. T. Fyfe, J. F. Stoddart, *Coord. Chem. Rev.* **1999**, *183,* 139-155.
<99CCR149> F. Pina, A. J. Parola, *Coord. Chem. Rev.* **1999**, *185-186,* 149-165.
<99CCR167> C. Dietrich-Buchecker, G. Rapenne, J.-P. Sauvage, *Coord. Chem. Rev.* **1999**, *185-186,* 167-176.
<99CCR233> M. Venturi, A. Credi, V. Balzani, *Coord. Chem. Rev.* **1999**, *185-186,* 233-256.
<99CCR243> C. Bazzicalupi, A. Bencini, A. Bianchi, C. Giorgi, B. Valtancoli, *Coord. Chem. Rev.* **1999**, *184,* 243-270.
<99CCR277> S. Connolly, S. N. Rao, R. Rizza, N. Zaccheroni, D. Fitzmaurice, *Coord. Chem. Rev.* **1999**, *185-186,* 277-295.
<99CCR297> A. P. de Silva, D. B. Fox, A. J. M. Huxley, N. D. McClenaghan, J. Roiron, *Coord. Chem. Rev.* **1999**, *185-186,* 297-306.
<99CCR347> M. Formica, V. Fusi, M. Micheloni, R. Pontellini, P. Romani, *Coord. Chem. Rev.* **1999**, *184,* 347-363.
<99CCR357> J.-P. Desvergne, H. Bouas-Laurent, E. Perez-Inestrosa, P. Marsau, M. Cotrait, *Coord. Chem. Rev.* **1999**, *185-186,* 357-371.
<99CCR371> T. A. Kaden, *Coord. Chem. Rev.* **1999**, *190-192,* 371-389.
<99CCR373> E. Fujita, *Coord. Chem. Rev.* **1999**, *185-186,* 373-384.
<99CCR451> V. Comblin, D. Gilsoul, M. Hermann, V. Humblet, V. Jacques, M. Mesbahi, C. Sauvage, J. F. Desreux, *Coord. Chem. Rev.* **1999**, *185-186,* 451-470.
<99CCR649> V. Amendola, L. Fabbrizzi, M. Licchelli, C. Mangano, P. Pallavicini, L. Parodi, A. Poggi, *Coord. Chem. Rev.* **1999**, *190-192,* 649-669.
<99CEJ345> W. Boomgaarden, F. Vögtle, M. Nieger, H. Hupfer, *Chem. Eur. J.* **1999**, *5,* 345-355.
<99CEJ860> M. Asakawa, P. R. Ashton, V. Balzani, C. L. Brown, A. Credi, O. A. Matthews, S. P. Newton, F. M. Raymo, A. N. Shipway, N. Spencer, A. Quick, J. F. Stoddart, A. J. P. White, D. J. Williams, *Chem. Eur. J.* **1999**, *5,* 860-875.
<99CEJ984> E. Ishow, A. Credi, V. Balzani, F. Spadola, L. Mandolini, *Chem. Eur. J.* **1999**, *5,* 984-989.
<99CEJ1253> S. Aime, M. Botta, L. Frullano, S. G. Crich, G. B. Giovenzana, R. Pagliarin, G. Palmisano, M. Sisti, *Chem. Eur. J.* **1999**, *5,* 1253-1260.
<99CEJ1728> Ö. Ünsal, A. Godt, *Chem. Eur. J.* **1999**, *5,* 1728-1733.
<99CEJ1823> G. J. Bodwell, J. N. Bridson, T. J. Houghton, J. W. J. Kennedy, M. R. Mannion, *Chem. Eur. J.* **1999**, *5,* 1823-1827.
<99CEJ1841> J.-L. Weidmann, J.-M. Kern, J.-P. Sauvage, D. Muscat, S. Mullins, W. Köhler, C. Rosenauer, H. J. Räder, K. Martin, Y. Geerts, *Chem. Eur. J.* **1999**, *5,* 1841-1851.
<99CEJ2109> N. Avarvari, N. Maigrot, L. Ricard, F. Mathey, P. L. Floch, *Chem. Eur. J.* **1999**, *5,* 2109-2118.
<99CEJ3055> M. R. Suissa, C. Rųmming, J. Dale, *Chem. Eur. J.* **1999**, *5,* 3055-3065.
<99CL399> S. Yamaguchi, H. Ishii, K. Tamao, *Chem. Lett.* **1999**, 399-400.
<99CL517> Y.-Z. Hu, S. Tsukiji, S. Shinkai, I. Hamachi, *Chem. Lett.* **1999**, 517-518.

<99CL881> J.-S. Li, Y.-Y. Chen, Z.-L. Zhong, X.-R. Lu, T. Zhang, J.-L. Yan, *Chem. Lett.* **1999**, 881-882.

<99CL915> N. Watanabe, Y. Furusho, N. Kihara, T. Takata, K. Kinbara, K. Saigo, *Chem. Lett.* **1999**, 915-916.

<99CL1235> M. Unno, T. Saito, H. Matsumoto, *Chem. Lett.* **1999**, 1235-1236.

<99CR931> A. Jaset, J. C. Sherman, *Chem. Rev.* **1999**, *99*, 931-967.

<99CR1643> F. M. Raymo, J. F. Stoddart, *Chem. Rev.* **1999**, *99*, 1643-1663.

<99CSR279> P. J. Langley, J. Hulliger, *Chem. Soc. Rev.* **1999**, *28*, 279-291.

<99CSR293> M.-J. Blanco, M. C. Jiménez, J.-C. Chambron, V. Heitz, M. Linke, J.-P. Sauvage, *Chem. Soc. Rev.* **1999**, *28*, 293-305.

<99E2605> R. Kuhn, *Electrophoresis (Weinheim, Fed. Repub. Ger.)* **1999**, *20*, 2605-2613.

<99EJIC679> E. Lindner, C. Hermann, G. Baum, D. Fenske, *Eur. J. Inorg. Chem* **1999**, 679-685.

<99EJIC2167> B. Kersting, G. Steinfeld, T. Fritz, J. Hausmann, *Eur. J. Inorg. Chem* **1999**, 2167-2172.

<99EJIC2261> M. Ciampolini, M. Formica, V. Fusi, A. Saint-Mauricec, M. Micheloni, N. Nardi, R. Pontellini, F. Pina, P. Romani, A. M. Sabatini, B. Valtancoli, *Eur. J. Inorg. Chem* **1999**, 2261-2268.

<99EJOC791> D. Cao, D. Schollmeyer, H. Meier, *Eur. J. Org. Chem.* **1999**, 791-795.

<99EJOC985> M. Asakawa, P. R. Ashton, V. Balzani, S. E. Boyd, A. Credi, G. Mattersteig, S. Menzer, M. Montalti, F. M. Raymo, C. Ruffilli, J. F. Stoddart, M. Venturi, D. J. Williams, *Eur. J. Org. Chem.* **1999**, 985-994.

<99EJOC995> P. R. Ashton, A. M. Heiss, D. Pasini, F. M. Raymo, A. N. Shipway, J. F. Stoddart, N. Spencer, *Eur. J. Org. Chem.* **1999**, 995-1004.

<99EJOC1295> D. B. Amabilino, P. R. Ashton, J. A. Bravo, F. M. Raymo, J. F. Stoddart, A. J. P. White, D. J. Williams, *Eur. J. Org. Chem.* **1999**, 1295-1302.

<99EJOC2373> H. Scheytza, O. Rademacher, H.-U. Reiβig, *Eur. J. Org. Chem.* **1999**, 2373-2381.

<99EJOC2807> M. B. Nielsen, J. G. Hansen, J. Becher, *Eur. J. Org. Chem.* **1999**, 2807-2815.

<99HCA1572> J.-P. Bourgeois, P. Seiler, M. Fibbioli, E. Pretsch, F. Diederich, L. Echegoyen, *Helv. Chim. Acta* **1999**, *82*, 1572-1595.

<99IC921> A. Mayr, J. Guo, *Inorg. Chem.* **1999**, *38*, 921-928.

<99IC2064> A. Bencini, A. Bianchi, V. Fusi, C. Giorgi, P. Paoletti, J. A. Ramírez, B. Valtancoli, *Inorg. Chem.* **1999**, *38*, 2064-2070.

<99IC4115> C. Bazzicalupi, A. Bencini, E. Berni, A. Bianchi, V. Fedi, V. Fusi, C. Giorgi, P. Paoletti, B. Valtancoli, *Inorg. Chem.* **1999**, *38*, 4115-4122.

<99IC4149> H. A. Cruse, N. E. Leadbeater, *Inorg. Chem.* **1999**, *38*, 4149-4151.

<99IC4181> S.-S. Sun, A. J. Lees, *Inorg. Chem.* **1999**, *38*, 4181-4182.

<99IC4203> P.-L. Vidal, B. Divisia-Blohorn, G. Bidan, J.-M. Kern, J.-P. Sauvage, J.-L. Hazemann, *Inorg. Chem.* **1999**, *38*, 4203-4210.

<99IC5755> R. J. M. K. Gebbink, C. F. Martens, P. J. A. Kenis, R. J. Jansen, H.-F. Nolting, V. A. Solé, M. C. Feiters, K. D. Karlin, R. J. M. Nolte, *Inorg. Chem.* **1999**, *38*, 5755-5768.

<99JA557> S. Bélanger, J. T. Hupp, C. L. Stern, R. V. Slone, D. F. Watson, T. G. Carrell, *J. Am. Chem. Soc.* **1999**, *121*, 557-563.

<99JA994> G. Rapenne, C. Dietrich-Buchecker, J.-P. Sauvage, *J. Am. Chem. Soc.* **1999**, *121*, 994-1001.

<99JA1397> T. Kusukawa, M. Fujita, *J. Am. Chem. Soc.* **1999**, *121*, 1397-1398.

<99JA1452> M. Inouye, K. Fujimoto, M. Furusyo, H. Nakazumi, *J. Am. Chem. Soc.* **1999**, *121*, 1452-1458.

<99JA1466> I. A. Levitsky, J. Kim, T. M. Swager, *J. Am. Chem. Soc.* **1999**, *121*, 1466-1472.

<99JA1479> K. N. Houk, S. Menzer, S. P. Newton, F. M. Raymo, J. F. Stoddart, D. J. Williams, *J. Am. Chem. Soc.* **1999**, *121*, 1479-1487.

<99JA2741> J. Fan, J. A. Whiteford, B. Olenyuk, M. D. Levin, P. J. Stang, E. B. Fleischer, *J. Am. Chem. Soc.* **1999**, *121*, 2741-2752.

<99JA3684> N. Solladié, J.-C. Chambron, J.-P. Sauvage, *J. Am. Chem. Soc.* **1999**, *121*, 3684-3692.

<99JA3807> K. Fuji, K. Tsubaki, K. Tanaka, N. Hayashi, T. Otsubo, T. Kinoshita, *J. Am. Chem. Soc.* **1999**, *121*, 3807-3808.

<99JA4397> N. Armaroli, V. Balzani, J.-P. Collin, P. Gaviña, J.-P. Sauvage, B. Ventura, *J. Am. Chem. Soc.* **1999**, *121*, 4397-4408.

<99JA5481> D. J. Cárdenas, J.-P. Collin, P. Gaviña, J.-P. Sauvage, A. De Cian, J. Fischer, N. Armaroli, L. Flamigni, V. Vicinelli, V. Balzani, *J. Am. Chem. Soc.* **1999**, *121*, 5481-5488.

<99JA5599> W.-S. Xia, R. H. Schmehl, C.-J. Li, *J. Am. Chem. Soc.* **1999**, *121*, 5599-5600.

<99JA9043> E. Abel, G. E. M. Maguire, O. Murillo, I. Suzuki, S. L. De Wall, G. W. Gokel, *J. Am. Chem. Soc.* **1999**, *121*, 9043-9052.

<99JA9053> S. J. Narayanan, B. Sridevi, T. K. Chandrashekar, A. Vij, R. Roy, *J. Am. Chem. Soc.* **1999**, *121*, 9053-9068.

<99JA10239> S. Hiraoka, M. Fujita, *J. Am. Chem. Soc.* **1999**, *121*, 10239-10240.

<99JA11014> F. Ibukuro, M. Fujita, K. Yamaguchi, J.-P. Sauvage, *J. Am. Chem. Soc.* **1999**, *121*, 11014-11015.

<99JCR(S)58> A. L. Bate, T. P. Carter, J. W. Wheeler, *J. Chem. Res., Synop.* **1999**, 58-59.

<99JCR(S)288> H. Jiang, G. Chu, H. Gong, Q. Qiao, *J. Chem. Res., Synop.* **1999**, 288-289.

<99JCS(D)1039> C. Bornet, R. Amardeil, P. Meunier, J. C. Daran, *J. Chem. Soc., Dalton Trans.* **1999**, 1039-1040.

<99JCS(D)1897> P. D. Beer, P. A. Gale, G. Z. Chen, *J. Chem. Soc., Dalton Trans.* **1999**, 1897-1909.

<99JCS(P1)693> S. Yamazaki, M. Hanada, Y. Yanase, C. Fukumori, K. Ogura, T. Saeki, S. Umetani, *J. Chem. Soc., Perkin Trans. 1* **1999**, 693-696.

<99JCS(P1)1057> D. G. Hamilton, L. Prodi, N. Feeder, J. K. M. Sanders, *J. Chem. Soc., Perkin Trans. 1* **1999**, 1057-1065.

<99JCS(P1)1885> H. Takagi, T. Hayashi, T. Mizutani, H. Masuda, H. Ogoshi, *J. Chem. Soc., Perkin Trans. 1* **1999**, 1885-1892.

<99JCS(P2)837> J. S. Kim, J. H. Pang, I. Y. Yu, W. K. Lee, I. H. Suh, J. K. Kim, M. H. Cho, E. T. Kim, D. Y. Ra, *J. Chem. Soc., Perkin Trans. 2* **1999**, 837-846.

<99JCS(P2)1273> J. Ishikawa, H. Sakamoto, H. Wada, *J. Chem. Soc., Perkin Trans. 2* **1999**, 1273-1279.

<99JCS(P2)1739> A. Tahri, E. Cielen, K. J. V. Aken, G. J. Hoornaert, F. C. De Schryver, N. Boens, *J. Chem. Soc., Perkin Trans. 2* **1999**, 1739-1748.

<99JCS(P2)1973> J. L. W. Griffin, P. V. Coveney, A. Whiting, R. Davey, *J. Chem. Soc., Perkin Trans. 2* **1999**, 1973.

<99JCS(P2)2093> G. Haberhauer, R. Gleiter, H. Irngartinger, T. Oeser, F. Rominger, *J. Chem. Soc., Perkin Trans. 2* **1999**, 2093-2097.

<99JCS(P2)2415> L. J. Govenlock, J. A. K. Howard, J. M. Moloney, D. Parker, R. D. Peacock, G. Siligardi, *J. Chem. Soc., Perkin Trans. 2* **1999**, 2415-2418.

<99JH15> S. Liu, Y. Bi, L. Xu, *Jingxi Huagong* **1999**, *16*, 15-20.

<99JOC1335> E. Bardazzi, M. Ciampolini, V. Fusi, M. Micheloni, N. Nardi, R. Pontellini, P. Romani, *J. Org. Chem.* **1999**, *64*, 1335-1337.

<99JOC3162> Z. Yang, J. S. Bradshaw, X. X. Zhang, P. B. Savage, K. E. Krakowiak, N. K. Dalley, N. Su, R. T. Bronson, R. M. Izatt, *J. Org. Chem.* **1999**, *64*, 3162-3170.

<99JOC3572> Z.-T. Li, G.-Z. Ji, C.-X. Zhao, S.-D. Yuan, H. Ding, C. Huang, A.-L. Du, M. Wei, *J. Org. Chem.* **1999**, *64*, 3572-3584.

<99JOC3825> N. Su, J. S. Bradshaw, X. X. Zhang, P. B. Savage, K. E. Krakowiak, R. M. Izatt, *J. Org. Chem.* **1999**, *64*, 3825-3829.

<99JOC5341> R. A. Bartsch, H.-S. Hwang, V. S. Talanov, G. G. Talanova, D. W. Purkiss, R. D. Rogers, *J. Org. Chem.* **1999**, *64*, 5341-5349.

<99JOC5463> M. Weck, B. Mohr, J.-P. Sauvage, R. H. Grubbs, *J. Org. Chem.* **1999**, *64*, 5463-5471.

<99JOC5626> S. Dell, D. M. Ho, R. A. Pascel, Jr., *J. Org. Chem.* **1999**, *64*, 5626-5633.

<99JOC6135> V. J. Arán, M. Kumar, J. Molina, L. Lamarque, P. Navarro, E. García-España, J. A. Ramírez, S. V. Luis, B. Escuder, *J. Org. Chem.* **1999**, *64*, 6135-6146.

<99JOC6190> M. Müller, U. Huchel, A. Geyer, R. R. Schmidt, *J. Org. Chem.* **1999**, *64*, 6190-6201.

<99JOC8489> T. Shimizu, H. Murakami, N. Kamigata, *J. Org. Chem.* **1999**, *64*, 8489-8494.

<99JOC8693> A. Srinivasan, V. G. Anand, S. J. Narayanan, S. K. Pushpan, M. R. Kumar, T. K. Chandrashekar, K.-i. Sugiura, Y. Sakata, *J. Org. Chem.* **1999**, *64*, 8693-8697.

<99JOC8855> N. Su, J. S. Bradshaw, X. X. Zhang, H. Song, P. B. Savage, G. Xue, K. E. Krakowiak, R. M. Izatt, *J. Org. Chem.* **1999**, *64*, 8855-8861.

<99JPC218> B. König, T. Fricke, U. Lüning, M. Hagen, M. Müller, M. Bahadir, *J. Prakt. Chem.* **1999**, *314*, 218-221.

<99JPC228> M. Schmittel, C. Michel, A. Ganz, M. Herderich, *J. Prakt. Chem.* **1999**, *341,* 228-236.

<99JPC662> U. Lüning, W. Hacker, *J. Prakt. Chem.* **1999**, *341,* 662-667.

<99M1500> J. Kim, S. K. McHugh, T. M. Swager, *Macromolecules* **1999**, *32,* 1500-1507.

<99MI1> H.-J. Wu, *Adv. Strained Interesting Org. Mol.* **1999**, 167-197.

<99MI2> R. A. Sachleben, B. A. Moyer, *ACS Symp. Ser.* **1999**, *716,* 114-132.

<99MI3> R. A. Bartsch, *ACS Symp. Ser.* **1999**, *716,* 146-155.

<99O955> J. M. Martín-Alvarez, F. Hampel, A. M. Arif, J. A. Gladysz, *Organometallics* **1999**, *18,* 955-957.

<99O1465> M. Yoshida, M. Goto, F. Nakanishi, *Organometallics* **1999**, *18,* 1465-1470.

<99O1887> K. K. Hii, M. Thornton-Pett, A. Jutand, R. P. Tooze, *Organometallics* **1999**, *18,* 1887-1896.

<99O1911> T.-Y. Dong, C.-K. Chang, C.-H. Cheng, K.-J. Lin, *Organometallics* **1999**, *18,* 1911-1922.

<99O3615> R. Gleiter, G. Haberhauer, H. Irngartinger, T. Oeser, F. Rominger, *Organometallics* **1999**, *18,* 3615-3622.

<99O3792> S. Tsutsui, E. Toyoda, T. Hamaguchi, J. Ohshita, F. Kanetani, A. Kunai, A. Naka, M. Ishikawa, *Organometallics* **1999**, *18,* 3792-3795.

<99O3991> S.-W. Lai, M. C. W. Chan, K.-K. Cheung, S.-M. Peng, C.-M. Che, *Organometallics* **1999**, *18,* 3991-3997.

<99O4817> M. Schmitz, S. Leininger, J. Fan, A. M. Arif, P. J. Stang, *Organometallics* **1999**, *18,* 4817-4824.

<99O4856> J. R. Farrell, A. H. Eisenberg, C. A. Mirkin, I. A. Guzei, L. M. Liable-Sands, C. D. Incarvito, A. L. Rheingold, C. L. Stern, *Organometallics* **1999**, *18,* 4856-4868.

<99OL199> R. N. Warrener, D. Margetic, A. S. Amarasekara, D. N. Butler, I. B. Mahadevan, R. A. Russell, *Org. Lett.* **1999**, *1,* 199-202.

<99OL203> R. N. Warrener, D. Margetic, A. S. Amarasekara, R. A. Russell, *Org. Lett.* **1999**, *1,* 203-206.

<99OL1035> E. Alcalde, S. Ramos, L. Pérez-García, *Org. Lett.* **1999**, *1,* 1035-1038.

<99OL1343> Y. Tong, D. G. Hamilton, J.-C. Meillon, J. K. M. Sanders, *Org. Lett.* **1999**, *1,* 1343-1346.

<99OL1697> S. P. Gromov, E. N. Ushakov, A. I. Vedernikov, N. A. Lobova, M. V. Alfimov, Y. A. Strelenko, J. K. Whitesell, M. A. Fox, *Org. Lett.* **1999**, *1,* 1697-1699.

<99PAC2393> R. A. Bartsch, S. N. Ivy, J. Lu, V. J. Huber, V. S. Talanov, W. Walkowiak, C. Park, B. Amiri-Eliasi, *Pure Appl. Chem.* **1999**, *70,* 2393-2400.

<99PAC2401> J. L. Sessler, P. Anzenbacher, Jr., K. Jursíková, H. Miyaji, J. W. Genge, N. A. Tvermoes, W. E. Allen, J. A. Shriver, P. A. Gale, V. Král, *Pure Appl. Chem.* **1999**, *70,* 2401-2408.

<99PNMRS327> G. W. Buchanan, *Prog. Nucl. Magn. Reson. Spectrosc.* **1999**, *34,* 327-377.

<99RCR119> V. K. Bel'sky, B. M. Bulychev, *Russ. Chem. Rev.* **1999**, *68,* 119-135.

<99S69> C.-F. Chen, Q.-Y. Zheng, Z.-T. Huang, *Synthesis* **1999**, 69-71.

<99S295> H. Schwierz, F. Vögtle, *Synthesis* **1999**, 295-305.

<99S849> C. Kauffmann, W. M. Müller, F. Vögtle, S. Weinman, S. Abramson, B. Fuchs, *Synthesis* **1999**, 849-853.

<99S1034> V. Kuksa, C. Marshall, S. Wardell, P. K. T. Lin, *Synthesis* **1999**, 1034-1038.

<99S1753> A. Mohry, H. Schwierz, F. Vögtle, *Synthesis* **1999**, 1753-1758.

<99SC2817> Z. Kovacs, E. A. Archer, M. K. Russell, A. D. Sherry, *Syn. Commun.* **1999**, *29,* 2817-2822.

<99SC3939> M. R. Younes, M. M. Chaabouni, *Syn. Commun.* **1999**, *29,* 3939-3948.

<99SL750> M. Schmittel, H. Ammon, *Synlett* **1999**, 750-752.

<99SL1223> B. Witulski, *Synlett* **1999**, 1222-1226.

<99TL499> Z. Asfari, P. Thuéry, M. Nierlich, J. Vicens, *Tetrahedron Lett.* **1999**, *40,* 499-502.

<99TL531> G. Ma, Y. S. Jung, D. S. Chung, J.-I. Hong, *Tetrahedron Lett.* **1999**, *40,* 531-534.

<99TL691> Z. Asfari, V. Lamare, J.-F. Dozol, J. Vicens, *Tetrahedron Lett.* **1999**, *40,* 691-694.

<99TL1527> F. Hamada, S. Ito, M. Narita, N. Nasirozawa, *Tetrahedron Lett.* **1999**, *40,* 1527-1530.

<99TL3749> N. Komatsu, A. Taniguchi, H. Suzuki, *Tetrahedron Lett.* **1999**, *40,* 3749-3752.

<99TL4123> B. P. S. Chauhan, P. Boudjouk, *Tetrahedron Lett.* **1999**, *40,* 4123-4126.

<99TL4993> C. Garcia, Y. Pointud, G. Jeminet, D. Dugat, *Tetrahedron Lett.* **1999**, *40,* 4993-4996.

<99TL4997> C. Garcia, J. Guyot, G. Jeminet, E. Leize-Wagner, H. Nierengarten, A. V. Dorsselaer, *Tetrahedron Lett.* **1999**, *40,* 4997-5000.

<99TL5865> A. Nehlig, G. Kaufmann, Z. Asfari, J. Vicens, *Tetrahedron Lett.* **1999**, *40,* 5865-5868.

<99TL5901> A. Lorente, M. Fernández-Saiz, F. Herraiz, J.-M. Lehn, J.-P. Vigneron, *Tetrahedron Lett.* **1999**, *40,* 5901-5904.

<99TL6019> Y. Kubo, Y. Murai, J.-i. Yamanaka, S. Tokita, Y. Ishimaru, *Tetrahedron Lett.* **1999** , *40,* 6019-6023.

<99TL6431> H. Takemura, Y. Kozima, T. Inazu, *Tetrahedron Lett.* **1999**, *40,* 6431-6434.

<99TL6639> H. Houjou, K. Hiratani, Y. Nagawa, M. Goto, *Tetrahedron Lett.* **1999**, *40,* 6639-6642.

<99TL7641> F. Wang, A. W. Schwabacher, *Tetrahedron Lett.* **1999**, *40,* 7641-7644.

<99TL7687> B. M. Kim, S. M. So, *Tetrahedron Lett.* **1999**, *40,* 7687-7690.

<99TL7773> C. L. Kranemann, B. Costisella, P. Eilbracht, *Tetrahedron Lett.* **1999**, *40,* 7773-7776.

<99TL8457> N. Sprutta, L. Latos-Grażyński, *Tetrahedron Lett.* **1999**, *40,* 8457-8460.

<99TL8461> I. S. Antipin, I. I. Stoikov, A. T. Gubaidullin, I. A. Litvinov, D. Weber, W. D. Habicher, A. I. Konovalov, *Tetrahedron Lett.* **1999**, *40,* 8461-8464.

<99TL8879> C.-H. Lee, W.-S. Cho, *Tetrahedron Lett.* **1999**, *40,* 8879-8882.

<99TL8905> K. Paek, H. Ihm, S. Yun, H. C. Lee, *Tetrahedron Lett.* **1999**, *40,* 8905-8909.

<99TL9065> K. Ogura, H. Matsui, S. Matsumoto, M. Akazome, N. Harada, *Tetrahedron Lett.* **1999**, *40,* 9065-9068.

INDEX